9783642459023

W0082811

ENCYCLOPEDIA OF PHYSICS

EDITED BY

S. FLÜGGE

VOLUME XXXVIII/1

EXTERNAL PROPERTIES
OF ATOMIC NUCLEI

WITH 160 FIGURES

SPRINGER-VERLAG

BERLIN · GÖTTINGEN · HEIDELBERG

1958

HANDBUCH DER PHYSIK

HERAUSGEGEBEN VON

S. FLÜGGE

BAND XXXVIII/1

ÄUSSERE EIGENSCHAFTEN
DER ATOMKERNE

MIT 160 FIGUREN

SPRINGER-VERLAG

BERLIN · GÖTTINGEN · HEIDELBERG

1958

Druck der Universitätsdruckerei H. Stürtz AG., Würzburg

Atomic Masses of Nuclides.

By

A. H. WAPSTRA.

With 12 Figures.

A. Reasons and methods for measuring masses.

1. Introduction. In 1815 PROUT remarked that many atomic weights, expressed in the hydrogen atomic weight as a unit, were nearly whole numbers. He therefore suggested that the atoms of all elements were tightly bound combinations of hydrogen atoms. When slightly later it was discovered that some atomic weights were definitely non-integral numbers, one had either to abandon this hypothesis or to assume that atoms of one element could have different masses. The first alternative was chosen. In 1906, however, BOLTWOOD[1] discovered ionium, which proved to be chemically inseparable from thorium, whereas the radioactive properties and the atomic weights of both substances were different. Within a short time many more examples of such a behaviour became known, and in 1910 SODDY[2] concluded that atoms of one element could have different masses; he suggested the name isotopes for bodies with the same chemical properties but different atomic weights. This discovery revived PROUT's hypothesis.

SODDY[2] also suggested that the occurrence of isotopy should not be limited to heavy elements. This suggestion was confirmed in 1912 when THOMPSON[3] discovered a second isotope of neon with mass 22 besides the most abundant one with mass 20.

It was known that the ratio of the atomic weights of hydrogen and oxygen was 0,8% higher than $\frac{1}{16}$. This deviation is not due to isotopic impurity in these elements, as STERN and VOLMER[4] concluded in 1919 from their failure to change the atomic weights by fractionated diffusion. Moreover ASTON[5] proved in 1920 with his mass spectrograph that the mass ratio of mass separated hydrogen and oxygen-atoms was $(1.008 \pm 0.001):16$. Therefore, if an oxygen atom is thought to consist of hydrogen atoms, some mass has disappeared in its formation. This can be understood on the basis of EINSTEIN's[6] theoretical argument that "if a body parts with an energy $E(\ldots)$, then its mass is reduced by a value E/c^2 where c is the speed of light"[7].

The atomic hypothesis of RUTHERFORD[8] has given us a model in which an atom consists of a nucleus surrounded by a cloud of electrons; after the discovery

---

[1] B. B. BOLTWOOD: Amer. J. Sci. **24**, 370 (1907).
[2] F. SODDY: Ann. Rep. Chem. Soc. **7**, 285 (1910).
[3] J. J. THOMPSON: Proc. Cambridge Phil. Soc. **17**, 201 (1912).
[4] O. STERN and M. VOLMER: Ann. Phys., Lpz. **59**, 225 (1919).
[5] F. W. ASTON: Phil. Mag. **39**, 611 (1920).
[6] A. EINSTEIN: Ann. Phys., Lpz. **18**, 639 (1905).
[7] To the writers best knowledge the explanation of the mass defect in terms of the EINSTEIN equation has first been given by W. D. HARKINS and E. D. WILSON: Phil. Mag. **30**, 723 (1915).
[8] E. RUTHERFORD: Phil. Mag. **21**, 669 (1911).

1

of the neutron by CHADWICK the nucleus is considered to be built up of protons (hydrogen nuclei) and neutrons (neutral particles with nearly the same mass)[1]. Nevertheless the explanation of the difference in mass between the atom and the constituent particles as a binding energy is still considered to be true[2].

The binding energies of the nuclei being important constants, it is highly desirable to determine the atomic masses with high precision. In the following sections of this Chap. A we will describe the ways in which this problem can be solved experimentally. In Chap. B a list is given of masses and binding energies with a discussion of its trustworthiness. In Chap. C we shall discuss some conclusions that can be drawn from the trends of these binding energies and of some related data.

2. Definitions and constants. Atomic nuclei are thought to consist of a certain number of *protons* Z, a number of *neutrons* N and a *binding energy* E. Neutrons and protons together are called *nucleons*. The *nuclear charge*, resp. *mass* are determined by Z, resp. $A = Z + N$; therefore these quantities are called the *charge number*, resp. *mass number*; $I = N - Z$ is called the *neutron excess*. Assemblies consisting of atoms with the same nuclear charge and mass will be called *nuclides*[3]. Nuclides with the same nuclear charge but different masses will be called *isotopes*[4]; in the same way nuclides with the same number of neutrons, mass number or neutron excess are called *isotones*, *isobars* and *isodiaspheres* respectively. *Isomers* are states of one nuclide with a measurable lifetime (say $> 10^{-10}$ sec); (sometimes only excited states are called isomers in contrast to the ground state).

Assemblies of isotopes are called *elements*; elements are indicated by chemical symbols which stand as abbreviations for their names (e.g., O for Oxygen). Isotopes are indicated by the chemical symbol with the mass number attached to it (e.g. O^{16})[5]; sometimes the numbers of neutrons and (or) protons are also attached: $_Z El_N^A$.

Nuclei with an odd number of protons and an even number of neutrons will be called *odd-even nuclei*, sometimes shortened o-e nuclei; in the same way e-e, e-o and o-o nuclei are defined.

Nuclear energies will be measured in *electron-volts*, 1 eV being the energy gained by an electron in traversing a potential difference of 1 volt. Our *mass unit* (1 MU) will be $\frac{1}{16}$ of the mass of one neutral O^{16} atom; by this definition all atomic masses of nuclides M become nearly equal to the mass numbers. This "physical scale" mass unit is different from that of the chemical mass scale in which $\frac{1}{16}$ of the average mass of the oxygen isotopes in the natural isotopic mixture is taken as a unit. According the to most recent measurements[6] natural oxygen contains $99.758_7\%$ O^{16}, $0.037_4\%$ O^{17} and $0.203_9\%$ O^{18}, so that the ratio of the chemical and physical mass units is

$$1 \text{ MU (chem)} = 1.0002783 \pm 5 \text{ MU (phys)}.$$

---

[1] J. CHADWICK: Nature, Lond. **129**, 312 (1932). — W. HEISENBERG: Z. Physik **77**, 1 (1932).

[2] It is even so that a comparison of nuclear masses computed from energy measurements using EINSTEIN's relation and those measured directly constitutes the most sensitive test for the validity of EINSTEIN's hypothesis [4].

[3] T. P. KOHMAN: Amer. J. Phys. **15**, 356 (1947).

[4] This word is often used synonymously with "nuclides", but it appears to be better to avoid this confusion.

[5] In most European journals the notation ^{16}O is used.

[6] A. O. NIER: Phys. Rev. **77**, 789 (1950).

The connection between energy and mass units is[1]:

$$1\ MU = 931\,162 \pm 24\ keV,$$

$$1\ MeV = 1073.927 \pm 26\ \mu MU,$$

$$1\ Joule = (6.24192 \pm 27) \cdot 10^{12}\ MeV = (6.70337 \pm 29) \cdot 10^9\ MU.$$

The difference between the atomic mass of a nuclide in MU and the mass number $M - A$ will be called its *mass excess*, the difference between its mass and the mass of the constituent particles $NM_n + ZM_H - M$ the *binding energy*, whether expressed in mass units or in MeV[2].

3. Mass spectrometers[3]. The path of a particle with a charge e and a mass m, moving initially with a velocity v in a direction perpendicular to an electric or magnetic field E or B is a parabola with a latus rectum eE/mv^2 or a circle with a radius mv/eB respectively. These properties, separately or taken together, can be used to measure the mass m. Instruments based on this principle are called *mass spectrometers* (or mass spectrographs if the spectra are recorded photographically).

The time it takes a particle to complete a revolution in a magnetic field is $2\pi m/eB$, which is also a measure for the mass. GOUDSMIT[4] proposed an instrument measuring this time of flight and called it a chronotron. In a modification of this instrument, the mass synchrometer, the rotation frequency (better: a high harmonic of this frequency) is measured; in this modification the instrument has shown to be capable of very high precision[5].

Mass spectrometers are used to measure mass doublets, that is differences in m/e between ions having the same ratio of mass number and charge; commonly the difference of two well-known ions with different values of A/e is taken as a standard. From a combination of such doublets the masses can be computed. The doublet distances being of the order of 0.1 % of the mass numbers, the relative errors in the computed masses are about 1000 times less than those in the measured doublets.

Mass spectrometers and their measuring techniques are described elsewhere in this Encyclopedia[3]; a recent table of mass doublets is given by DUCKWORTH *et al.*[6].

4. Microwave mass measurements. The rotation and vibration frequencies of molecules are functions of the masses of the constituent atoms, and therefore in principle mass data can be derived from the study of molecular spectra. A technique similar to the mass doublet measurement mentioned above is not available in this case; therefore only the recent advances in measuring molecular spectra with microwave methods made it possible to obtain useful results. In the case of certain diatomic molecules (Cl_2, Br_2, KCl) the ratio of two isotopic

---

[1] J. W. M. DuMOND and E. R. COHEN: Rev. Mod. Phys. **25**, 691 (1953). — The error limits are written in an abridged form giving them in units of the last digit, e.g. 1073.927 ± 26 means: 1073.927 ± 0.026, etc.

[2] This choice has been made in order to avoid the use of the ambiguous term "mass defect" which was originally intended to indicate the binding energy expressed in mass units, but later also used for $A - M$ and even for $M - A$. We also prefer to avoid the term *packing fraction* for $(A-M)/A$ which term suggests a physical meaning not present.

[3] For detailed information see H. EWALD's contribution to Vol. XXXIII of this Encyclopedia.

[4] S. A. GOUDSMIT: Phys. Rev. **74**, 622 (1948).

[5] L. G. SMITH and C. C. DAMM: Phys. Rev. **90**, 324 (1953).

[6] H. E. DUCKWORTH, B. J. HOGG and E. M. PENNINGTON: Rev. Mod. Phys. **26**, 463 (1954).

1*

masses could be measured, and for some types of polyatomic molecules (OCS, GeH_3Cl) some ratios of isotopic mass differences. For a more complete description of methods and results we refer to reviews by GESCHWIND et al.[1].

5. Nuclear reaction energies. In bombardments of nuclei with some kinds of particles other nuclides may be formed under the emission of secondary particles. From the energy balance in such nuclear reactions the difference in binding energy of the initial and the final nuclides can be computed.

This energy balance is almost always derived from the energies of the incident particles and those of the secondary particles emitted under a certain angle with the incident ones, using the principles of conservation of energy and momentum. The energy of charged incident particles is mostly determined by accelerating them through a well-known voltage difference; that of γ quanta by producing them as bremsstrahlung of electrons with a well-known energy. Neutrons are either used as thermal neutrons (energy practically zero) or as a result of another nuclear reaction under specified conditions allowing a computation of their energy. Energies of charged secondary particles are measured with magnetic or electric spectrographs, proportional counters or scintillation spectrometers[2] or by determining the range of the particles in certain absorbing substances (e.g. photographic emulsions), the first technique being the most accurate. The energy of secondary γ quanta is best determined by using electron pair spectrographs; scintillation spectrometers can also be used. Secondary neutrons are mostly measured by determining the energy of protons set in motion by collision with the neutrons.

A difficulty in determining nuclear binding energies from nuclear reaction energy measurements is, that it may be difficult to determine whether a certain secondary particle group belongs to the ground state transition or, instead, leads to an excited state of the final nucleus. This uncertainty can, however, be reduced largely by studying more nuclear reactions and radioactive decay processes to the same final nucleus and comparing the nuclear level structure as found from these different data. In addition the new insight in nuclear structure gained from the nuclear shell model[3] and some empirical regularities in level structures make it sometimes possible to reach a decision.

Details about nuclear reaction measurements can be found in Vol. XL of this Encyclopedia[4]; a recent compilation of reaction energies is given by VAN PATTER and WHALING[5].

6. Beta decay. Many nuclides decay by emission of negative or positive electrons or by nuclear capture of atomic electrons (changing nuclear neutrons in protons or reverse) to neighbouring isobars; the energy set free in these transitions is connected with the difference in binding energy of the initial and final nucleus.

In every electron emission process a neutrino is emitted together with the electron; the electrons are therefore emitted with a certain, theoretically mostly well-known energy distribution with a maximum energy equal to the transition energy[6] (the mass of the neutrino being zero within narrow limits). This transition energy can therefore best be determined from an analysis of the energy

---

[1] S. GESCHWIND, G. R. GUNTHER-MOHR and C. H. TOWNES: Rev. Mod. Phys. **26**, 444 (1954). — S. GESCHWIND: This volume next chapter.

[2] This Encyclopedia, Vol. XLV.

[3] Article by LANE and ELLIOTT: This Encyclopedia, Vol. XXXIX.

[4] Cf. especially article by W. E. BURCHAM.

[5] D. M. VAN PATTER and W. WHALING: Rev. Mod. Phys. **26**, 402 (1954).

[6] E. GREULING and L. W. NORDHEIM: This Encyclopedia, Vol. XLII.

distribution of the electrons, measured as in the case of secondary particles in nuclear reactions[1] (Sect. 5). Another, less precise method is the analysis of an absorption curve in certain materials.

The γ rays accompanying most beta decays can be investigated with the same methods after (internal or external) conversion in electrons, or directly with a γ scintillation counter[2]. With help of intensity considerations, energy sum rules and especially of coincidence counting methods[2] the relation between β and γ transitions can be established and the total transition energy determined.

In most cases of electron capture the decay energy is carried away by a neutrino. In a small fraction of the number of decays a second order process occurs in which the decay energy is shared between the neutrino and a γ quantum; therefore a weak continuous γ spectrum is emitted with a theoretically known energy distribution[3]. Analysis of the experimentally found γ spectrum may yield the maximum energy of this spectrum, which is again equal to the transition energy. In practice this method is only usable if no or only very few γ transitions follow the electron capture process.

The ratio of capture of K shell electrons to that of electrons from higher shells varies with the transition energy; this fact has also been used to estimate electron capture decay energies.

The transition energy in β^- processes is equal to the difference in atomic mass of initial and final nuclide:

$$Q_{\beta^-} = M_i - M_f \tag{6.1}$$

because, though one electron has been emitted, another has to be added to the atom in order to make it neutral again. The same relation applies to electron capture. In the case of positron emission, however, the atom looses a negative electron in addition to the positive one, and therefore

$$Q_{\beta^+} = M_i - M_f - 2m_e \tag{6.2}$$

m_e being the electron mass. For this reason often not the maximum energy of the positons Q_{β^+}, but $Q_{\beta^+} + 2m_e$ is considered to be the transition energy.

A table of β transition energies has recently been published by KING[4].

7. Alpha decay. Most nuclides with $Z > 82$ and some with lower nuclear charge decay spontaneously by emission of α particles (He^4 nuclei) into isodiaspheres with a mass four units lower. The difference in binding energy of the initial and the final nuclide can be computed from the α decay energy by the relation

$$E_i - E_f = E_\alpha - Q_\alpha \tag{7.1}$$

E_α being the binding energy of the He^4 atom and Q_α the kinetic energy of the α particle augmented by the kinetic energy of the recoiling daughter nucleus.

The α particle energy can be measured in the same way as that of charged secondary particles in nuclear reactions (Sect. 5); the relation between α transitions to, and γ transitions in the final nuclide is established in the same way as in the case of β transitions (Sect. 6). In the special case of α decay between even-even nuclides the ground state transition is always the most intensive one, and it is accompanied by a transition to the first excited state of the final nuclide, the position of which can be estimated fairly accurately from the systematics of these first excited states; it can therefore be identified

[1] T. R. GERHOLM: This Encyclopedia, Vol. XXXIII.
[2] This Encyclopedia, Vol. XLV, and the article by KOFOED-HANSEN in Vol. XLII.
[3] E. GREULING and L. W. NORDHEIM: This Encyclopedia, Vol. XLII.
[4] R. W. KING: Rev. Mod. Phys. 26, 327 (1954).

with certainty. In cases where the resolution was insufficient it is necessary to make a correction for the presence of this second component[1].

A table of α decay energies was recently compiled by ASARO and PERLMAN[2].

B. Atomic masses of nuclides.

8. Historical introduction. The first collection of data on nuclear masses allowing significant deductions about the trends of the nuclear binding energies was ASTON's[3] first packing fraction curve, in 1927. These results, and also the improved packing fraction curve of DEMPSTER[4] were obtained by mass spectroscopy. The first author comparing mass spectroscopic values with nuclear reaction energy measurements was BAINBRIDGE[5], in 1933; he found EINSTEIN's relation $E = mc^2$ to be correct within 3% by comparing the reaction energy of the reaction $Li^7(p, \alpha) He^4$ with his mass values for H^1, He^4 and Li^7. Shortly thereafter several authors published lists of masses of light nuclides, using reaction data in combination with mass measurements[6]. Later more elaborate tables were given by FLÜGGE and MATTAUCH, BETHE, and WAPSTRA[7].

After 1947 very notable advances were made in the techniques of both nuclear reaction energy measurements and mass spectroscopy. This afforded renewed consideration of the consistency of these measurements. It appeared that the reaction energy measurements agreed well among themselves, but not with the new mass spectrographic data measured by different groups, which also disagreed with one another. Therefore new computations were made, using reaction energies only[8].

At present, combination of the nuclear reaction data for light nuclides ($A < 25$) with the most precise mass measurements yields a mass-energy conversion factor deviating by a factor 1.0006 ± 0.0001 from the conversion factor derived from the equation $E = mc^2$ and the recent values for the atomic constants [4], if the calibrations in all nuclear reaction energy measurements are reduced to BRIGGS'[9] measurement of the α ray energy of RaC'. It must, however, be noted that recently BRIGGS himself was forced to revise his older α ray energies of a number of natural radioactive substances by 0.05%. Assuming this correction to apply also to RaC', it is found that the EINSTEIN relation should be correct with a precision of about 2 parts in 10000.

9. List of masses and binding energies. In Table 1 a list is given of experimentally determined atomic masses and nuclear binding energies. This table is essentially taken from the computations of WAPSTRA and HUIZENGA [4], [5], [6]. A few values were corrected and some new values added because some new data became available; these new data are enumerated in Table 2.

---

[1] A. H. WAPSTRA: Thesis Amsterdam 1953.

[2] F. ASARO and I. PERLMAN: Rev. Mod. Phys. **26**, 456 (1954).

[3] F. W. ASTON: Proc. Roy. Soc. Lond. **115**, 487 (1927).

[4] A. J. DEMPSTER: Phys. Rev. **53**, 869 (1938).

[5] K. T. BAINBRIDGE: Phys. Rev. **44**, 123 (1933).

[6] M. L. E. OLIPHANT: Nature, Lond. **137**, 396 (1936). — H. A. BETHE: Phys. Rev. **47**, 633 (1935). — M. S. LIVINGSTON and H. A. BETHE: Rev. Mod. Phys. **9**, 245 (1937). — T. W. BONNER and W. M. BRUBAKER: Phys. Rev. **50**, 308 (1936).

[7] S. W. FLÜGGE and J. MATTAUCH: Phys. Z. **43**, 1 (1942). — Nuclear Physics Tables. Berlin 1942. — Phys. Z. **44**, 161 (1943). — H. A. BETHE: Elementary Nuclear Theory. New York 1947. — A. H. WAPSTRA, in L. ROSENFELD: Nuclear Forces. Amsterdam 1948.

[8] J. MATTAUCH and A. FLAMMERSFELD: Isotopic Report 1949 (Z. Naturforsch., special issue). — C. W. LI, W. WHALING, W. A. FOWLER and E. C. LAURITSEN: Phys. Rev. **83**, 512 (1951). — C. W. LI: Phys. Rev. **88**, 1038 (1952). — M. O. STERN: Rev. Mod. Phys. **21**, 316 (1949). — A. H. WAPSTRA: Physica, Haag **17**, 628 (1951). — Phys. Rev. **84**, 837 (1951). — Thesis Amsterdam 1953.

[9] G. H. BRIGGS: Rev. Mod. Phys. **26**, 1 (1955).

Table 1. *Nuclear binding energies and atomic masses computed from experimental data.*

	Binding energy keV	Mass excesses keV	Mass excesses µMU		Binding energy keV	Mass excesses keV	µMU
n		8367_5	$8986_1\pm$ 1_5	Na^{25}	$202642\pm$ 160	-2068	$-2219\pm$ 180
H^1		7584_5	$8145_2\pm$ 1_5	Mg^{25}	$205609\pm$ 40	-5818	$-6248\pm$ 15
H^2	$2226_4\pm$ 1_8	13725_6	$14740_3\pm$ 2_8	Al^{25}	$200580\pm$ 65	-1572	$-1688\pm$ 60
H^3	$8485\pm$ 4	15835	$17005\pm$ 5	Mg^{26}	$216728\pm$ 43	-8569	$-9202\pm$ 23
He^3	$7720\pm$ 4	15817	$16986\pm$ 5	Al^{26}	$211920\pm$ 45	-4544	$-4880\pm$ 23
He^4	$28296_5\pm$ 4_8	3607_3	$3873_9\pm$ 2_6	Mg^{27}	$223167\pm$ 45	-6641	$-7132\pm$ 23
He^5	$27340\pm$ 30	12932	$13888\pm$ 32	Al^{27}	$224979\pm$ 44	-9236	$-9919\pm$ 16
Li^5	$26500\pm$ 200	12988	$13948\pm$ 220	Si^{27}	$219369\pm$ 45	-4409	$-4735\pm$ 19
He^6	$29242\pm$ 29	19398	$20831\pm$ 31	Mg^{28}	$231678\pm$ 50	-6784	$-7286\pm$ 32
Li^6	$31994\pm$ 6	15862	$17034\pm$ 5	Al^{28}	$232705\pm$ 44	-8594	$-9229\pm$ 16
Li^7	$39247\pm$ 7	16977	$18232\pm$ 6	Si^{28}	$236574\pm$ 45	-13246	$-14225\pm$ 19
Be^7	$37601\pm$ 7	17840	$19159\pm$ 5	P^{28}	$222000\pm$ 300	545	$585\pm$ 300
Li^8	$41281\pm$ 8	23310	$25033\pm$ 5	Al^{29}	$241860\pm$ 100	-9382	$-10075\pm$ 110
Be^8	$56499\pm$ 10	7309	$7849\pm$ 5	Si^{29}	$245048\pm$ 45	-13353	$-14340\pm$ 20
B^8	$38100\pm$ 400	24925	$26768\pm$ 430	P^{29}	$239298\pm$ 46	-8386	$-9006\pm$ 23
Be^9	$58165\pm$ 11	14010	$15046\pm$ 6	Si^{30}	$255658\pm$ 46	-15595	$-16748\pm$ 19
B^9	$56312\pm$ 11	15081	$16195\pm$ 6	P^{30}	$250560\pm$ 65	-11280	$-12115\pm$ 48
Be^{10}	$64976\pm$ 12	15566	$16716\pm$ 7	Si^{31}	$262255\pm$ 47	-13825	$-14847\pm$ 23
B^{10}	$64750\pm$ 12	15010	$16119\pm$ 6	P^{31}	$262955\pm$ 48	-15308	$-16439\pm$ 24
C^{10}	$60130\pm$ 100	18847	$20240\pm$ 110	S^{31}	$256730\pm$ 85	-9866	$-10595\pm$ 80
B^{11}	$76213\pm$ 15	11914	$12795\pm$ 6	Si^{32}	$271571\pm$ 75	-14773	$-15866\pm$ 60
C^{11}	$73450\pm$ 15	13895	$14922\pm$ 7	P^{32}	$270888\pm$ 49	-14873	$-15972\pm$ 26
B^{12}	$79577\pm$ 16	16917	$18168\pm$ 6	S^{32}	$271811\pm$ 49	-16579	$-17804\pm$ 26
C^{12}	$92170\pm$ 19	3541	$3803\pm$ 5	Cl^{32}	$258000\pm$ 400	-3551	$-3814\pm$ 430
N^{12}	$73720\pm$ 100	21209	$22776\pm$ 110	P^{33}	$280997\pm$ 51	-16615	$-17844\pm$ 30
C^{13}	$97116\pm$ 19	6963	$7478\pm$ 5	S^{33}	$280463\pm$ 51	-16864	$-18111\pm$ 30
N^{13}	$94112\pm$ 19	9185	$9864\pm$ 5	Cl^{33}	$274228\pm$ 140	-11412	$-12256\pm$ 140
C^{14}	$105289\pm$ 21	7157	$7687\pm$ 3	P^{34}	$287537\pm$ 230	-14787	$-15880\pm$ 230
N^{14}	$104661\pm$ 21	7002	$7520\pm$ 3	S^{34}	$291857\pm$ 60	-19890	$-21360\pm$ 50
O^{14}	$98711\pm$ 26	12169	$13069\pm$ 15	Cl^{34}	$285552\pm$ 70	-14368	$-15430\pm$ 60
C^{15}	$107627\pm$ 55	13187	$14162\pm$ 50	S^{35}	$298879\pm$ 55	-18545	$-19915\pm$ 35
N^{15}	$115503\pm$ 23	4528	$4862\pm$ 5	Cl^{35}	$298263\pm$ 55	-18712	$-20095\pm$ 35
O^{15}	$112015\pm$ 23	7233	$7767\pm$ 6	A^{35}	$292070\pm$ 80	-13302	$-14285\pm$ 65
N^{16}	$117997\pm$ 30	10402	$11171\pm$ 13	S^{36}	$308778\pm$ 120	-20076	$-21560\pm$ 120
O^{16}	$127615\pm$ 24	0	0	Cl^{36}	$306833\pm$ 60	-18914	$-20312\pm$ 40
N^{17}	$123743\pm$ 200	13022	$13984\pm$ 210	A^{36}	$306764\pm$ 60	-19628	$-21079\pm$ 40
O^{17}	$131760\pm$ 25	4222	$4534\pm$ 5	S^{37}	$313784\pm$ 320	-16714	$-17950\pm$ 330
F^{17}	$128210\pm$ 25	6989	$7506\pm$ 4	Cl^{37}	$317200\pm$ 65	-20914	$-22460\pm$ 45
O^{18}	$139830\pm$ 28	4521	$4855\pm$ 8	A^{37}	$315601\pm$ 65	-20098	$-21584\pm$ 45
F^{18}	$137380\pm$ 28	6188	$6646\pm$ 9	K^{37}	$308688\pm$ 300	-13967	$-15000\pm$ 300
Ne^{18}	$132370\pm$ 200	10415	$11185\pm$ 220	Cl^{38}	$323310\pm$ 75	-18656	$-20035\pm$ 60
O^{19}	$143788\pm$ 28	8931	$9591\pm$ 13	A^{38}	$327346\pm$ 70	-23475	$-25210\pm$ 50
F^{19}	$147793\pm$ 31	4142	$4448\pm$ 7	K^{38}	$320687\pm$ 80	-17599	$-18900\pm$ 60
Ne^{19}	$143754\pm$ 32	7398	$7945\pm$ 9	Cl^{39}	$331813\pm$ 100	-18791	$-20180\pm$ 80
F^{20}	$154398\pm$ 34	5904	$6340\pm$ 11	A^{39}	$333987\pm$ 90	-21748	$-23356\pm$ 70
Ne^{20}	$160666\pm$ 33	-1146	$-1231\pm$ 9	K^{39}	$333769\pm$ 90	-22313	$-23963\pm$ 70
Na^{20}	$144550\pm$ 200	14187	$15236\pm$ 220	Ca^{39}	$326130\pm$ 400	-15457	$-16600\pm$ 400
F^{21}	$162300\pm$ 100	6370	$6840\pm$ 110	A^{40}	$343838\pm$ 70	-23232	$-24950\pm$ 50
Ne^{21}	$167422\pm$ 34	465	$499\pm$ 10	K^{40}	$341563\pm$ 80	-21740	$-23347\pm$ 60
Na^{21}	$163117\pm$ 45	3987	$4281\pm$ 35	Ca^{40}	$342105\pm$ 80	-23065	$-24770\pm$ 60
Ne^{22}	$177788\pm$ 36	-1533	$-1646\pm$ 12	Sc^{40}	$327336\pm$ 500	-9079	$-9750\pm$ 500
Na^{22}	$174165\pm$ 36	1307	$1404\pm$ 14	A^{41}	$349897\pm$ 80	-20923	$-22470\pm$ 60
Ne^{23}	$182980\pm$ 37	1643	$1764\pm$ 12	K^{41}	$351693\pm$ 80	-23503	$-25240\pm$ 60
Na^{23}	$186583\pm$ 38	-2744	$-2947\pm$ 11	Ca^{41}	$350472\pm$ 80	-23065	$-24770\pm$ 60
Mg^{23}	$181703\pm$ 39	1353	$1453\pm$ 14	Sc^{41}	$343730\pm$ 90	-17105	$-18370\pm$ 70
Ne^{24}	$191900\pm$ 110	1090	$1170\pm$ 110	K^{42}	$359065\pm$ 90	-22506	$-24170\pm$ 70
Na^{24}	$193543\pm$ 39	-1336	$-1435\pm$ 14	Ca^{42}	$361950\pm$ 80	-26175	$-28110\pm$ 60
Mg^{24}	$198277\pm$ 39	-6853	$-7360\pm$ 15	K^{43}	$368828\pm$ 100	-23930	$-25670\pm$ 80
Al^{24}	$183480\pm$ 300	7161	$7691\pm$ 320	Ca^{43}	$369889\pm$ 100	-25747	$-27650\pm$ 80

Table 1. (Continued.)

	Binding energy	Mass excesses			Binding energy	Mass excesses	
	keV	keV	μMU		keV	keV	μMU
Sc^{43}	366890 ± 100	-23530	-25270 ± 80	Ni^{59}	515437 ± 250	-43678	-46907 ± 250
K^{44}	375734 ± 230	-22441	-24100 ± 230	Cu^{59}	510234 ± 500	-39258	-42160 ± 530
Ca^{44}	381060 ± 100	-28550	-30660 ± 80	Co^{60}	524797 ± 250	-43888	-47132 ± 250
Sc^{44}	376627 ± 100	-24900	-26740 ± 80	Ni^{60}	526833 ± 250	-46707	-50160 ± 250
Ca^{45}	388487 ± 100	-27609	-29650 ± 80	$Cu^{6)}$	519774 ± 250	-40431	-43420 ± 250
Sc^{45}	387959 ± 100	-27865	-29925 ± 80	Co^{61}	534633 ± 250	-45357	-48710 ± 250
Ti^{45}	385132 ± 100	-25820	-27730 ± 80	Ni^{61}	535340 ± 250	-46847	-50310 ± 250
Sc^{46}	396872 ± 100	-28410	-30510 ± 80	Cu^{61}	532328 ± 250	-44618	-47916 ± 250
Ti^{46}	398450 ± 100	-30771	-33046 ± 80	Zn^{61}	525674 ± 350	-38746	-41610 ± 360
V^{46}	390268 ± 400	-23372	-25100 ± 400	Co^{62}	541464 ± 400	-43820	-47060 ± 400
Ca^{47}	406256 ± 110	-28643	-30760 ± 100	Ni^{62}	545682 ± 250	-48821	-52430 ± 250
Sc^{47}	407414 ± 110	-30584	-32845 ± 95	Cu^{62}	540967 ± 250	-44889	-48207 ± 250
Ti^{47}	407240 ± 100	-31194	-33500 ± 80	Zn^{62}	538487 ± 250	-43192	-46385 ± 250
V^{47}	403545 ± 100	-28282	-30373 ± 80	Ni^{63}	552245 ± 210	-47016	-50492 ± 200
Ca^{48}	416057 ± 200	-30077	-32300 ± 200	Cu^{63}	551525 ± 210	-47080	-50560 ± 200
Sc^{48}	415544 ± 110	-30347	-32590 ± 90	Zn^{63}	547400 ± 210	-43738	-46971 ± 200
Ti^{48}	418755 ± 100	-34341	-36880 ± 80	Ni^{64}	561895 ± 210	-48299	-51870 ± 200
V^{48}	413938 ± 100	-30307	-32547 ± 80	Cu^{64}	559434 ± 210	-46621	-50066 ± 200
Cr^{48}	411714 ± 210	-28866	-31000 ± 210	Zn^{64}	559222 ± 210	-47192	-50680 ± 200
Ca^{49}	421202 ± 200	-26855	-28840 ± 200	Ga^{64}	551194 ± 600	-39947	-42900 ± 600
Sc^{49}	425652 ± 120	-32088	-34460 ± 100	Ni^{65}	567916 ± 210	-45953	-49350 ± 200
Ti^{49}	426871 ± 100	-34090	-36610 ± 80	Cu^{65}	569229 ± 210	-48048	-51600 ± 200
V^{49}	425480 ± 100	-33482	-35957 ± 80	Zn^{65}	567098 ± 210	-46701	-50153 ± 200
Cr^{49}	422136 ± 100	-30920	-33206 ± 80	Ga^{65}	563193 ± 230	-43578	-46800 ± 220
Ti^{50}	437855 ± 100	-36706	-39420 ± 80	Ni^{66}	576816 ± 200	-46484	-49920 ± 220
V^{50}	434707 ± 190	-34341	-36880 ± 180	Cu^{66}	576330 ± 210	-46782	-50240 ± 200
Cr^{50}	435300 ± 160	-35719	-38360 ± 150	Zn^{66}	578172 ± 210	-49407	-53060 ± 200
Mn^{50}	426735 ± 500	-27935	-30000 ± 500	Ga^{66}	572222 ± 210	-44240	-47510 ± 200
Ti^{51}	444286 ± 130	-34770	-37340 ± 120	Cu^{67}	585299 ± 210	-47383	-50886 ± 200
V^{51}	445943 ± 130	-37210	-39960 ± 120	Zn^{67}	585087 ± 210	-47955	-51500 ± 200
Cr^{51}	444411 ± 130	-36461	-39156 ± 120	Ga^{67}	583302 ± 210	-46953	-50424 ± 200
Mn^{51}	440410 ± 150	-33243	-35700 ± 140	Ge^{67}	578093 ± 400	-42526	-45670 ± 400
V^{52}	453245 ± 130	-36144	-38816 ± 120	Zn^{68}	595299 ± 210	-49799	-53480 ± 200
Cr^{52}	456367 ± 140	-40049	-43010 ± 130	Ga^{68}	591610 ± 210	-46893	-50360 ± 200
Mn^{52}	450854 ± 150	-35319	-37930 ± 140	Zn^{69}	601792 ± 210	-47924	-51467 ± 200
Fe^{52}	448078 ± 160	-33326	-35790 ± 150	Ga^{69}	601905 ± 210	-48821	-52430 ± 200
Cr^{53}	464297 ± 140	-39612	-42540 ± 130	Ge^{69}	598888 ± 210	-46586	-50030 ± 200
Mn^{53}	462918 ± 140	-39016	-41900 ± 130	Zn^{70}	611186 ± 210	-48951	-52570 ± 200
Fe^{53}	458271 ± 200	-35151	-37750 ± 200	Ga^{70}	609742 ± 210	-48290	-51860 ± 200
Cr^{54}	474006 ± 140	-40953	-43980 ± 130	Ge^{70}	610607 ± 210	-49938	-53630 ± 200
Mn^{54}	472022 ± 200	-39752	-42690 ± 200	Zn^{71}	616696 ± 300	-46093	-49500 ± 300
Fe^{54}	471862 ± 160	-40375	-43360 ± 150	Ga^{71}	618826 ± 210	-49007	-52630 ± 200
Co^{54}	462177 ± 700	-31473	-33800 ± 700	Ge^{71}	617798 ± 210	-48761	-52366 ± 200
Cr^{55}	480129 ± 180	-38708	-41570 ± 180	As^{71}	614994 ± 240	-46740	-50195 ± 230
Mn^{55}	482167 ± 160	-41530	-44600 ± 150	Ga^{72}	625769 ± 250	-47582	-51100 ± 250
Fe^{55}	481155 ± 160	-41301	-44354 ± 150	Ge^{72}	628990 ± 250	-51586	-55400 ± 250
Co^{55}	476914 ± 160	-37842	-40640 ± 150	As^{72}	623850 ± 250	-47229	-50720 ± 250
Mn^{56}	489427 ± 170	-40422	-43410 ± 160	Ga^{73}	635022 ± 170	-48467	-52050 ± 150
Fe^{56}	492322 ± 160	-44100	-47360 ± 150	Ge^{73}	635635 ± 130	-49864	-53550 ± 100
Co^{56}	486911 ± 160	-39472	-42390 ± 150	As^{73}	634480 ± 130	-49491	-53150 ± 110
Fe^{57}	499763 ± 160	-43374	-46580 ± 150	Se^{73}	630922 ± 140	-46716	-50170 ± 110
Co^{57}	498659 ± 330	-42852	-46020 ± 330	Ge^{74}	645735 ± 130	-51596	-55410 ± 100
Ni^{57}	494635 ± 330	-39612	-42540 ± 330	As^{74}	642386 ± 130	-49030	-52655 ± 100
Fe^{58}	510146 ± 170	-45189	-48530 ± 160	Se^{74}	642958 ± 130	-50385	-54110 ± 100
Co^{58}	507063 ± 170	-42889	-46060 ± 160	Ge^{75}	652212 ± 130	-49705	-53380 ± 100
Ni^{58}	506439 ± 250	-43048	-46230 ± 250	As^{75}	652564 ± 130	-50841	-54600 ± 100
Cu^{58}	496130 ± 900	-33522	-36000 ± 900	Se^{75}	650913 ± 130	-49973	-53667 ± 100
Fe^{59}	516516 ± 250	-43191	-46385 ± 250	Br^{75}	647404 ± 130	-47247	-50740 ± 100
Co^{59}	517293 ± 250	-44752	-48060 ± 250	Ge^{76}	661781 ± 170	-50907	-54670 ± 150

Table 1. (Continued.)

	Binding energy	Mass excesses			Binding energy	Mass excesses	
	keV	keV	μMU		keV	keV	μMU
As^{76}	659880 ± 130	-49789	-53470 ± 100	Y^{90}	782877 ± 230	-60339	-64800 ± 200
Se^{76}	662068 ± 130	-52760	-56660 ± 100	Zr^{90}	784292 ± 230	-62537	-67160 ± 200
Br^{76}	656694 ± 150	-48169	-51730 ± 130	Nb^{90}	779635 ± 240	-58663	-63000 ± 210
Ge^{77}	667644 ± 160	-48403	-51980 ± 140	Mo^{90}	776310 ± 240	-56121	-60270 ± 230
As^{77}	669584 ± 130	-51125	-54905 ± 100	Sr^{91}	788800 ± 220	-57111	-61333 ± 200
Se^{77}	669485 ± 130	-51810	-55640 ± 100	Y^{91}	790682 ± 220	-59776	-64195 ± 200
Br^{77}	667345 ± 130	-50452	-54182 ± 100	Zr^{91}	791449 ± 220	-61326	-65860 ± 200
Kr^{77}	663683 ± 130	-47573	-51090 ± 100	Nb^{91}	789279 ± 320	-59939	-64370 ± 320
As^{78}	676652 ± 160	-49826	-53510 ± 150	Mo^{91}	784901 ± 340	-56345	-60510 ± 340
Se^{78}	679967 ± 130	-53924	-57910 ± 100	Sr^{92}	796252 ± 260	-56196	-60350 ± 240
Br^{78}	675682 ± 200	-50422	-54150 ± 180	Y^{92}	797396 ± 250	-58123	-62420 ± 240
Kr^{78}	675793 ± 210	-51316	-55110 ± 200	Zr^{92}	800114 ± 230	-61624	-66180 ± 210
As^{79}	685430 ± 200	-50236	-53950 ± 180	Nb^{92}	797665 ± 310	-59958	-64390 ± 300
Se^{79}	686947 ± 180	-52536	-56420 ± 150	Mo^{92}	797254 ± 300	-60330	-64790 ± 290
Br^{79}	686322 ± 180	-52694	-56590 ± 150	Tc^{92}	790074 ± 750	-53933	-57920 ± 750
Kr^{79}	683919 ± 180	-51074	-54850 ± 150	Y^{93}	804758 ± 330	-57118	-61340 ± 330
Se^{80}	696804 ± 220	-54026	-58020 ± 200	Zr^{93}	807079 ± 260	-60221	-64673 ± 240
Br^{80}	694131 ± 220	-52136	-55990 ± 200	Nb^{93}	806358 ± 260	-60284	-64740 ± 240
Kr^{80}	695350 ± 220	-54138	-58140 ± 200	Mo^{93}	805137 ± 300	-59846	-64270 ± 290
Se^{81}	703682 ± 230	-52536	-56420 ± 210	Tc^{93}	801216 ± 300	-56708	-60900 ± 290
Br^{81}	704295 ± 220	-53933	-57920 ± 200	Y^{94}	810388 ± 480	-54380	-58400 ± 480
Kr^{81}	703354 ± 280	-53775	-57750 ± 270	Zr^{94}	815006 ± 360	-59781	-64200 ± 350
Rb^{81}	700374 ± 290	-51577	-55390 ± 280	Nb^{94}	813552 ± 270	-59110	-63480 ± 250
Se^{82}	712952 ± 180	-53439	-57390 ± 150	Mo^{94}	814808 ± 300	-61149	-65670 ± 270
Br^{82}	712104 ± 230	-53374	-57320 ± 210	Tc^{94}	809695 ± 400	-56820	-61020 ± 400
Kr^{82}	714375 ± 220	-56428	-60600 ± 200	Zr^{95}	821413 ± 400	-57820	-62094 ± 400
Rb^{82}	709728 ± 500	-52564	-56450 ± 500	Nb^{95}	821753 ± 400	-58943	-63300 ± 400
Br^{83}	721671 ± 220	-54574	-58608 ± 200	Mo^{95}	821901 ± 400	-59874	-64300 ± 400
Kr^{83}	721868 ± 220	-55554	-59660 ± 200	Tc^{95}	819486 ± 400	-58243	-62549 ± 400
Br^{84}	728462 ± 220	-52903	-56915 ± 200	Ru^{95}	816609 ± 450	-56149	-60300 ± 450
Kr^{84}	732358 ± 220	-57676	-61940 ± 200	Zr^{96}	829198 ± 500	-57238	-61470 ± 500
Rb^{84}	728930 ± 220	-55031	-59100 ± 200	Nb^{96}	828666 ± 430	-57489	-61739 ± 430
Sr^{84}	729116 ± 270	-56000	-60140 ± 250	Mo^{96}	831013 ± 430	-60619	-65100 ± 430
Br^{85}	737272 ± 250	-53439	-57390 ± 230	Tc^{96}	827129 ± 600	-57518	-61770 ± 600
Kr^{85}	739292 ± 220	-56242	-60400 ± 200	Ru^{96}	826634 ± 500	-57806	-62080 ± 500
Rb^{85}	739179 ± 220	-56913	-61120 ± 200	Zr^{97}	834875 ± 500	-54545	-58580 ± 500
Sr^{85}	737353 ± 500	-55870	-60000 ± 500	Nb^{97}	836764 ± 500	-57220	-61450 ± 500
Kr^{86}	748953 ± 220	-57536	-61790 ± 200	Mo^{97}	837918 ± 500	-59257	-63530 ± 500
Rb^{86}	747881 ± 220	-57247	-61479 ± 200	Mo^{98}	846202 ± 500	-59073	-63440 ± 500
Sr^{86}	748868 ± 220	-59017	-63380 ± 200	Ru^{98}	844105 ± 650	-58542	-62870 ± 650
Y^{86}	743886 ± 240	-54818	-58870 ± 230	Mo^{99}	851369 ± 900	-55873	-60003 ± 900
Br^{87}	747247 ± 600	-46679	-50130 ± 600	Tc^{99}	851966 ± 900	-57252	-61485 ± 900
Kr^{87}	754462 ± 210	-54678	-58720 ± 180	Ru^{99}	851477 ± 900	-57546	-61800 ± 900
Rb^{87}	757786 ± 190	-58784	-63130 ± 150	Mo^{100}	861335 ± 500	-57471	-61720 ± 500
Sr^{87}	757277 ± 190	-59058	-63424 ± 150	Ru^{102}	878246 ± 500	-59213	-63590 ± 500
Y^{87}	754804 ± 280	-57369	-61610 ± 270	Rh^{102}	875284 ± 500	-57034	-61250 ± 500
Zr^{87}	750511 ± 280	-53858	-57840 ± 270	Pd^{102}	875637 ± 500	-58170	-62470 ± 500
Kr^{88}	761722 ± 250	-53570	-57530 ± 240	Ru^{103}	884477 ± 500	-57077	-61296 ± 500
Rb^{88}	763891 ± 230	-56522	-60700 ± 210	Rh^{103}	884443 ± 500	-57825	-62100 ± 500
Sr^{88}	768265 ± 220	-61689	-66250 ± 200	Pd^{103}	883100 ± 500	-57266	-61500 ± 500
Y^{88}	763749 ± 230	-57946	-62230 ± 200	Ru^{104}	893686 ± 700	-57918	-62200 ± 700
Kr^{89}	767929 ± 360	-51409	-55210 ± 360	Rh^{104}	891227 ± 500	-56242	-60400 ± 500
Rb^{89}	771150 ± 350	-55413	-59510 ± 350	Pd^{104}	892986 ± 500	-58784	-63130 ± 500
Sr^{89}	774869 ± 230	-59916	-64345 ± 200	Ag^{104}	888479 ± 510	-55060	-59130 ± 510
Y^{89}	775553 ± 230	-61382	-65920 ± 200	Cd^{104}	885693 ± 520	-53058	-56980 ± 520
Zr^{89}	771929 ± 230	-58541	-62869 ± 200	Ru^{105}	899218 ± 500	-55083	-59155 ± 500
Nb^{89}	767274 ± 250	-54669	-58710 ± 250	Rh^{105}	900440 ± 500	-57088	-61308 ± 500
Rb^{90}	778205 ± 350	-54101	-58100 ± 350	Pd^{105}	900227 ± 500	-57658	-61920 ± 500
Sr^{90}	783127 ± 230	-59806	-64227 ± 200	Ag^{105}	897190 ± 750	-55404	-59500 ± 750

Table 1. (Continued.)

	Binding energy	Mass excesses			Binding energy	Mass excesses	
	keV	keV	µMU		keV	keV	µMU
Cd^{105}	893409 ± 800	-52406	-56280 ± 800	Sn^{122}	1035599 ± 500	-53914	-57900 ± 500
Ru^{106}	908157 ± 500	-55654	-59768 ± 500	Sb^{122}	1033345 ± 500	-52443	-56320 ± 500
Rh^{106}	907413 ± 500	-55693	-59810 ± 500	Te^{122}	1034536 ± 500	-54417	-58440 ± 500
Pd^{106}	910159 ± 500	-59222	-63600 ± 500	I^{122}	1029609 ± 510	-50273	-53990 ± 510
Ag^{106}	906405 ± 500	-56251	-60410 ± 500	Sn^{123}	1041620 ± 500	-51568	-55380 ± 500
Cd^{106}	905688 ± 500	-56317	-60480 ± 500	Sb^{123}	1042253 ± 500	-52983	-56900 ± 500
Rh^{107}	915733 ± 410	-55646	-59760 ± 410	Te^{123}	1041284 ± 500	-52797	-56700 ± 900
Pd^{107}	916154 ± 400	-56849	-61052 ± 400	Sn^{124}	1050062 ± 500	-51642	-55460 ± 500
Ag^{107}	915406 ± 400	-56885	-61090 ± 400	Sb^{124}	1048632 ± 500	-50995	-54765 ± 500
Cd^{107}	913188 ± 400	-55450	-59549 ± 400	Te^{124}	1050764 ± 500	-53910	-57890 ± 500
In^{107}	909073 ± 450	-52117	-55970 ± 450	I^{124}	1046754 ± 500	-50683	-54430 ± 500
Pd^{108}	925590 ± 400	-57918	-62200 ± 400	Xe^{124}	1046138 ± 500	-50850	-54610 ± 500
Ag^{108}	922675 ± 400	-55786	-59910 ± 400	Sn^{125}	1055748 ± 500	-48960	-52580 ± 500
Cd^{108}	923652 ± 400	-55546	-61800 ± 400	Sb^{125}	1057312 ± 500	-51307	-55100 ± 500
Pd^{109}	931434 ± 400	-55395	-59490 ± 400	Te^{125}	1057292 ± 500	-52071	-55920 ± 500
Ag^{109}	931741 ± 400	-56484	-60660 ± 400	I^{125}	1056360 ± 510	-51922	-55760 ± 510
Cd^{109}	930799 ± 410	-56326	-60490 ± 410	Te^{126}	1066107 ± 500	-52518	-56400 ± 500
In^{109}	928247 ± 420	-54557	-58590 ± 420	I^{126}	1063228 ± 500	-50422	-54150 ± 500
Pd^{110}	940649 ± 500	-56242	-60400 ± 500	Xe^{126}	1063702 ± 500	-51679	-55500 ± 500
Ag^{110}	938199 ± 400	-54575	-58610 ± 400	Cs^{126}	1058115 ± 650	-46875	-50340 ± 650
Cd^{110}	940294 ± 400	-57453	-61700 ± 400	Te^{127}	1072556 ± 520	-50599	-54340 ± 520
In^{110}	935581 ± 400	-53523	-57480 ± 400	I^{127}	1072574 ± 500	-51400	-55200 ± 500
Pd^{111}	945954 ± 530	-53179	-57110 ± 530	Xe^{127}	1070785 ± 900	-50394	-54120 ± 900
Ag^{111}	947321 ± 520	-55330	-59420 ± 520	Cs^{127}	1067879 ± 900	-48271	-51840 ± 900
Cd^{111}	947590 ± 500	-56382	-60550 ± 500	Te^{128}	1080513 ± 500	-50190	-53900 ± 500
In^{111}	945922 ± 750	-55497	-59600 ± 750	I^{128}	1079200 ± 500	-49659	-53330 ± 500
Sn^{111}	942616 ± 750	-52974	-56890 ± 750	Xe^{128}	1080437 ± 500	-51679	-55500 ± 500
Pd^{112}	954432 ± 510	-53290	-57230 ± 510	Cs^{128}	1075557 ± 650	-47582	-51100 ± 650
Ag^{112}	953947 ± 510	-53588	-57550 ± 510	Te^{129}	1087503 ± 520	-48812	-52420 ± 520
Cd^{112}	957094 ± 500	-57518	-61770 ± 500	I^{129}	1088422 ± 500	-50514	-54248 ± 500
In^{112}	953769 ± 510	-54976	-59040 ± 510	Xe^{129}	1087827 ± 500	-50702	-54450 ± 500
Sn^{112}	953641 ± 510	-55631	-59744 ± 510	Cs^{129}	1085936 ± 750	-49594	-53260 ± 750
Ag^{113}	962473 ± 510	-53747	-57720 ± 510	Ba^{129}	1082555 ± 800	-46996	-50470 ± 800
Cd^{113}	963552 ± 500	-55609	-59720 ± 500	Te^{130}	1095665 ± 500	-48607	-52200 ± 500
In^{113}	962918 ± 500	-55758	-59880 ± 500	I^{130}	1094873 ± 500	-48597	-52190 ± 500
Cd^{114}	972600 ± 500	-56289	-60450 ± 500	Xe^{130}	1097061 ± 500	-51568	-55380 ± 500
In^{114}	969908 ± 600	-54380	-58400 ± 600	Cs^{130}	1093289 ± 500	-48579	-52170 ± 500
Sn^{114}	971110 ± 600	-56365	-60532 ± 600	Ba^{130}	1092948 ± 500	-49021	-52645 ± 500
Ag^{115}	976415 ± 550	-50953	-54720 ± 550	Te^{131}	1101883 ± 500	-46456	-49890 ± 520
Cd^{115}	978631 ± 500	-53952	-57940 ± 500	I^{131}	1103381 ± 500	-48737	-52340 ± 500
In^{115}	979300 ± 500	-55404	-59500 ± 500	Xe^{131}	1103566 ± 500	-49705	-53380 ± 500
Sn^{115}	979020 ± 500	-55907	-60040 ± 500	Cs^{131}	1102432 ± 500	-49355	-53004 ± 500
Cd^{116}	987093 ± 500	-54147	-58150 ± 500	Te^{132}	1110110 ± 510	-46316	-49740 ± 510
In^{116}	985777 ± 500	-53514	-57470 ± 500	I^{132}	1109737 ± 510	-46726	-50180 ± 510
Sn^{116}	988309 ± 500	-56829	-61030 ± 500	Xe^{132}	1112515 ± 500	-50283	-54000 ± 500
Sb^{116}	982805 ± 510	-52108	-55960 ± 510	Cs^{132}	1109931 ± 650	-48486	-52070 ± 650
Cd^{117}	992814 ± 600	-51400	-55200 ± 600	Te^{133}	1116066 ± 720	-43904	-47150 ± 720
In^{117}	994850 ± 500	-54219	-58210 ± 500	I^{133}	1118282 ± 710	-46903	-50370 ± 710
Sn^{117}	995531 ± 500	-55687	-59800 ± 500	Xe^{133}	1119333 ± 700	-48737	-52340 ± 700
Sb^{117}	992961 ± 500	-53896	-57880 ± 500	Cs^{133}	1118977 ± 700	-49165	-52800 ± 700
Te^{117}	988658 ± 510	-50376	-54100 ± 510	I^{134}	1125122 ± 650	-45376	-48730 ± 650
Sn^{118}	1004737 ± 500	-56522	-60700 ± 500	Xe^{134}	1127738 ± 500	-48774	-52380 ± 500
Sb^{118}	999940 ± 550	-52508	-56390 ± 550	Cs^{134}	1125707 ± 800	-47527	-51040 ± 800
Sn^{119}	1011521 ± 500	-54939	-59000 ± 500	Ba^{134}	1126975 ± 800	-49578	-53243 ± 800
Sn^{120}	1020764 ± 500	-55814	-59940 ± 500	La^{134}	1122492 ± 820	-45878	-49270 ± 820
Sb^{120}	1017262 ± 500	-53095	-57020 ± 500	I^{136}	1136595 ± 540	-40114	-43080 ± 540
Te^{120}	1016907 ± 500	-53523	-57480 ± 500	Xe^{136}	1142210 ± 500	-46512	-49950 ± 500
Sn^{121}	1026943 ± 500	-53625	-57589 ± 500	Xe^{137}	1145782 ± 1400	-41716	-44800 ± 1400
Sb^{121}	1026542 ± 500	-54007	-58000 ± 500	Cs^{137}	1149004 ± 1010	-45721	-49101 ± 1010

Table 1. (Continued.)

	Binding energy	Mass excesses			Binding energy	Mass excesses	
	keV	keV	µMU		keV	keV	µMU
Ba^{137}	1149403 ± 1010	-46903	-50370 ± 1010	Gd^{157}	1290403 ± 1000	-26817	-28800 ± 1000
Cs^{138}	1154577 ± 1000	-42927	-46100 ± 1000	Gd^{158}	1296703 ± 1000	-24750	-26580 ± 1000
Ba^{138}	1158636 ± 1000	-47769	-51300 ± 1000	Gd^{160}	1309406 ± 1000	-20718	-22250 ± 1000
La^{138}	1156549 ± 1020	-46465	-49900 ± 1020	Tb^{160}	1308619 ± 2000	-20714	-22245 ± 2000
Ce^{138}	1156762 ± 1040	-47461	-50970 ± 1040	Dy^{160}	1309656 ± 2000	-22534	-24200 ± 2000
Pr^{138}	1152478 ± 1060	-43960	-47210 ± 1060	Dy^{162}	1325311 ± 2000	-21454	-23040 ± 2000
Ba^{139}	1163875 ± 1000	-44640	-47940 ± 1000	Dy^{164}	1338824 ± 2000	-18232	-19580 ± 2000
La^{139}	1165476 ± 1000	-47023	-50500 ± 1000	Ho^{164}	1335853 ± 1000	-16044	-17230 ± 1000
Ce^{139}	1164571 ± 1020	-46903	-50370 ± 1020	Er^{164}	1336048 ± 1000	-17022	-18280 ± 1000
Ba^{140}	1170091 ± 1050	-42489	-45630 ± 1050	Dy^{165}	1345311 ± 2000	-16351	-17560 ± 2000
La^{140}	1170612 ± 1000	-43793	-47030 ± 1000	Ho^{165}	1345775 ± 2000	-17599	-18900 ± 2000
Ce^{140}	1173618 ± 1000	-47582	-51100 ± 1000	Er^{168}	1367469 ± 1000	-14973	-16080 ± 1000
Pr^{140}	1169586 ± 1000	-44333	-47610 ± 1000	Er^{170}	1379139 ± 2000	-9908	-10640 ± 2000
Nd^{140}	1168700 ± 1010	-44230	-47500 ± 1010	Yb^{172}	1399448 ± 3000	-15048	-16160 ± 3000
La^{141}	1177713 ± 1000	-42526	-45670 ± 1000	Yb^{174}	1419060 ± 3000	-17925	-19250 ± 3000
Ce^{141}	1179356 ± 1000	-44953	-48276 ± 1000	Lu^{176}	1419536 ± 2000	-2449	-2630 ± 2000
Pr^{141}	1179155 ± 1000	-45534	-48900 ± 1000	Hf^{176}	1419749 ± 2000	-3445	-3700 ± 2000
Nd^{141}	1176621 ± 1000	-43783	-47020 ± 1000	Hf^{178}	1433262 ± 2000	-223	-240 ± 2000
Ce^{142}	1186498 ± 1000	-43727	-46960 ± 1000	Hf^{180}	1447912 ± 2000	1862	2000 ± 2000
Pr^{142}	1184993 ± 1000	-43005	-46184 ± 1000	Ta^{180}	1446942 ± 2000	2049	2200 ± 2000
Nd^{142}	1186376 ± 1000	-45171	-48510 ± 1000	W^{180}	1446858 ± 2000	1350	1450 ± 2000
Ce^{143}	1191578 ± 1000	-40439	-43429 ± 1000	Hf^{181}	1454416 ± 2000	3725	4000 ± 2000
Pr^{143}	1192240 ± 1000	-41884	-44980 ± 1000	Ta^{181}	1454658 ± 2000	2700	2900 ± 2000
Nd^{143}	1192386 ± 1000	-42813	-45978 ± 1000	W^{181}	1453783 ± 2000	2793	3000 ± 2000
Ce^{144}	1198200 ± 1000	-38694	-41555 ± 1000	Ta^{182}	1460726 ± 2000	5000	5370 ± 2000
Pr^{144}	1197720 ± 1000	-38997	-41880 ± 1000	W^{182}	1461675 ± 2000	3268	3510 ± 2000
Nd^{144}	1199907 ± 1000	-41967	-45070 ± 1000	W^{183}	1468375 ± 2000	4935	5300 ± 2000
Sm^{144}	1197345 ± 1600	-40971	-44000 ± 1600	Ta^{183}	1468091 ± 2000	6003	6447 ± 2000
Ce^{146}	1209483 ± 1030	-33242	-35700 ± 1030	W^{184}	1475812 ± 2000	5866	6300 ± 2000
Pr^{146}	1209715 ± 1010	-34257	-36790 ± 1010	W^{186}	1488729 ± 2300	9684	10400 ± 2000
Nd^{146}	1213132 ± 1000	-38457	-41300 ± 1000	Re^{186}	1487669 ± 2300	9961	10697 ± 2300
Pm^{146}	1212116 ± 1010	-38224	-41050 ± 1010	Os^{186}	1487954 ± 2300	8893	9550 ± 2300
Sm^{146}	1212032 ± 1000	-38923	-41800 ± 1000	W^{187}	1495179 ± 2200	11602	12460 ± 2200
Nd^{147}	1218864 ± 1000	-35822	-38470 ± 1000	Re^{187}	1495707 ± 2200	10291	11052 ± 2200
Pm^{147}	1218992 ± 1000	-36732	-39448 ± 1000	Os^{187}	1494930 ± 2200	10285	11045 ± 2200
Sm^{147}	1218434 ± 1000	-36958	-39690 ± 1000	Re^{188}	1499159 ± 2000	15206	16330 ± 2000
Nd^{148}	1224932 ± 1000	-33522	-36000 ± 1000	Os^{188}	1500453 ± 2000	13129	14100 ± 2000
Pm^{148}	1223963 ± 1050	-33336	-35800 ± 1050	Os^{189}	1505077 ± 2000	16873	18120 ± 2000
Sm^{148}	1225880 ± 1000	-36036	-38700 ± 1000	Os^{190}	1514115 ± 2000	16202	17400 ± 2000
Gd^{148}	1222386 ± 1600	-34108	-36630 ± 1600	Os^{191}	1518664 ± 2100	20020	21500 ± 2100
Nd^{149}	1230785 ± 1000	-31008	-33300 ± 1000	Ir^{191}	1518124 ± 2100	19878	21240 ± 2100
Pm^{149}	1231678 ± 1000	-32684	-35100 ± 1000	Os^{192}	1526101 ± 3000	20951	22500 ± 3000
Sm^{149}	1232236 ± 1000	-34025	-36540 ± 1000	Ir^{192}	1523269 ± 2050	23000	24700 ± 2050
Nd^{150}	1238035 ± 1000	-29890	-32100 ± 1000	Pt^{192}	1523976 ± 2050	21510	23100 ± 2050
Pm^{150}	1236135 ± 1010	-28773	-30900 ± 1010	Os^{193}	1530865 ± 2000	24555	26370 ± 2000
Sm^{150}	1240660 ± 1000	-34081	-36600 ± 1000	Ir^{193}	1531171 ± 2000	23465	25200 ± 2000
Eu^{150}	1237269 ± 1000	-31473	-33800 ± 1000	Pt^{193}	1530295 ± 1550	23558	25300 ± 1550
Gd^{150}	1237557 ± 1000	-32544	-34950 ± 1000	Ir^{194}	1538421 ± 1510	24583	26400 ± 1510
Sm^{152}	1253763 ± 1000	-30449	-32700 ± 1000	Pt^{194}	1539873 ± 1500	22348	24000 ± 1500
Sm^{154}	1267500 ± 1000	-27451	-29480 ± 1000	Au^{194}	1536520 ± 1510	24918	26760 ± 1510
Eu^{154}	1264482 ± 1010	-25216	-27080 ± 1010	Ir^{195}	1544685 ± 1510	26687	28660 ± 1510
Gd^{154}	1266679 ± 1000	-28196	-30280 ± 1000	Pt^{195}	1546005 ± 1500	24583	26400 ± 1500
Sm^{155}	1273083 ± 1050	-24666	-26490 ± 1050	Pt^{196}	1553925 ± 1500	25030	26880 ± 1500
Eu^{155}	1274501 ± 1060	-26868	-28854 ± 1060	Au^{196}	1552007 ± 3010	26166	28100 ± 3010
Gd^{155}	1273966 ± 1060	-27115	-29120 ± 1060	Hg^{196}	1551923 ± 3010	25467	27350 ± 3010
Sm^{156}	1280016 ± 1020	-23232	-24950 ± 1020	Pt^{197}	1560059 ± 3000	27264	29280 ± 3000
Eu^{156}	1280137 ± 1010	-24136	-25920 ± 1010	Au^{197}	1560030 ± 3000	26510	28470 ± 3000
Gd^{156}	1281756 ± 1000	-26538	-28500 ± 1000	Ir^{198}	1565094 ± 2010	31380	33700 ± 2010
Eu^{157}	1289482 ± 1010	-25113	-26970 ± 1010	Pt^{198}	1568688 ± 2000	27004	29000 ± 2000

Table 1. (Continued.)

	Binding energy keV	Mass excesses keV	Mass excesses µMU		Binding energy keV	Mass excesses keV	Mass excesses µMU
Au^{198}	$1\,566\,526 \pm 3000$	28382	30480 ± 3000	Em^{217}	$1\,681\,530 \pm 70$	66880	71824 ± 80
Hg^{198}	$1\,567\,121 \pm 3000$	27004	29000 ± 3000	Fr^{217}	$1\,679\,923 \pm 160$	67704	72709 ± 170
Au^{199}	$1\,574\,379 \pm 3050$	28897	31033 ± 3050	Po^{218}	$1\,686\,624 \pm 70$	71719	77021 ± 80
Hg^{199}	$1\,574\,046 \pm 3050$	28447	30550 ± 3050	At^{218}	$1\,686\,191 \pm 220$	71369	76645 ± 240
Au^{200}	$1\,579\,630 \pm 3030$	32013	34380 ± 3030	Em^{218}	$1\,688\,068 \pm 40$	68709	73788 ± 40
Hg^{200}	$1\,581\,147 \pm 3010$	29713	31910 ± 3010	Fr^{218}	$1\,685\,486 \pm 90$	70508	75720 ± 100
Au^{201}	$1\,586\,861 \pm 3010$	33149	35600 ± 3010	Em^{219}	$1\,692\,481 \pm 100$	72664	78036 ± 110
Hg^{201}	$1\,587\,568 \pm 3000$	31660	34000 ± 3000	Fr^{219}	$1\,691\,955 \pm 70$	72407	77760 ± 80
Tl^{202}	$1\,592\,904 \pm 420$	33908	36415 ± 450	Ra^{219}	$1\,690\,463 \pm 160$	73116	78521 ± 170
Pb^{202}	$1\,592\,041 \pm 430$	33988	36501 ± 460	Em^{220}	$1\,698\,823 \pm 40$	74689	80211 ± 40
Hg^{203}	$1\,602\,000 \pm 370$	33963	36474 ± 400	Fr^{220}	$1\,697\,173 \pm 60$	75556	81142 ± 60
Tl^{203}	$1\,601\,704 \pm 370$	33476	35951 ± 400	Ra^{220}	$1\,697\,660 \pm 50$	74286	79778 ± 50
Hg^{204}	$1\,609\,576 \pm 210$	34754	37323 ± 230	Fr^{221}	$1\,703\,546 \pm 70$	77551	83284 ± 80
Tl^{204}	$1\,608\,463 \pm 120$	35084	37678 ± 130	Ra^{221}	$1\,702\,993 \pm 80$	77321	83037 ± 90
Pb^{204}	$1\,608\,442 \pm 120$	34322	36859 ± 130	Ac^{221}	$1\,700\,476 \pm 210$	79055	84899 ± 230
Hg^{205}	$1\,615\,117 \pm 150$	37581	40359 ± 160	Em^{222}	$1\,709\,330 \pm 80$	80917	86899 ± 80
Tl^{205}	$1\,616\,084 \pm 120$	35831	38480 ± 130	Ra^{222}	$1\,709\,687 \pm 50$	78994	84834 ± 50
Pb^{205}	$1\,615\,246 \pm 50$	35886	38539 ± 50	Ac^{222}	$1\,706\,689 \pm 110$	81209	87213 ± 120
Tl^{206}	$1\,622\,619 \pm 20$	37663	40447 ± 20	Fr^{223}	$1\,714\,403 \pm 120$	83429	89597 ± 130
Pb^{206}	$1\,623\,346 \pm 10$	36153	38826 ± 10	Ra^{223}	$1\,714\,807 \pm 110$	82242	88322 ± 120
Bi^{206}	$1\,618\,963 \pm 200$	39753	42692 ± 210	Ac^{223}	$1\,713\,488 \pm 80$	82778	88898 ± 90
Tl^{207}	$1\,629\,413 \pm 40$	39237	42138 ± 40	Th^{223}	$1\,711\,066 \pm 190$	84417	90658 ± 200
Pb^{207}	$1\,630\,080 \pm 10$	37787	40580 ± 10	Ra^{224}	$1\,721\,332 \pm 40$	84084	90300 ± 40
Bi^{207}	$1\,626\,897 \pm 40$	40187	43160 ± 40	Ac^{224}	$1\,719\,186 \pm 80$	85447	91764 ± 90
Tl^{208}	$1\,633\,247 \pm 20$	43770	47006 ± 20	Th^{224}	$1\,718\,693 \pm 60$	85157	91452 ± 60
Pb^{208}	$1\,637\,460 \pm -$	38774	$41640 \pm -$	Ra^{225}	$1\,726\,362 \pm 90$	87422	93885 ± 100
Bi^{208}	$1\,633\,747 \pm 70$	41704	44787 ± 70	Ac^{225}	$1\,725\,939 \pm 80$	87062	93498 ± 90
Po^{208}	$1\,631\,527 \pm 110$	43141	46330 ± 120	Th^{225}	$1\,724\,596 \pm 90$	87622	94143 ± 100
Tl^{209}	$1\,638\,197 \pm 60$	47188	50676 ± 60	Ra^{226}	$1\,732\,760 \pm 80$	89391	95999 ± 90
Pb^{209}	$1\,641\,330 \pm 50$	43272	46471 ± 50	Ac^{226}	$1\,731\,147 \pm 320$	90221	96891 ± 340
Bi^{209}	$1\,641\,177 \pm 50$	42642	45794 ± 50	Th^{226}	$1\,731\,530 \pm 60$	89055	95639 ± 60
Po^{209}	$1\,638\,567 \pm 80$	44469	47756 ± 90	Pa^{226}	$1\,728\,052 \pm 130$	91750	98533 ± 140
Tl^{210}	$1\,641\,965 \pm 30$	51787	55615 ± 30	Ra^{227}	$1\,737\,138 \pm 120$	93381	100177 ± 130
Pb^{210}	$1\,646\,569 \pm 20$	46400	49830 ± 20	Ac^{227}	$1\,737\,665 \pm 120$	92071	98878 ± 130
Bi^{210}	$1\,645\,850 \pm 20$	46336	49761 ± 20	Th^{227}	$1\,736\,962 \pm 110$	91991	98792 ± 120
Po^{210}	$1\,646\,237 \pm 20$	45166	48505 ± 20	Pa^{227}	$1\,735\,201 \pm 80$	92969	99842 ± 90
At^{210}	$1\,641\,630 \pm 210$	48990	52611 ± 230	U^{227}	$1\,732\,439 \pm 230$	94948	101967 ± 250
Pb^{211}	$1\,650\,352 \pm 90$	50985	54754 ± 100	Ra^{228}	$1\,743\,391 \pm 80$	95495	102555 ± 90
Bi^{211}	$1\,650\,959 \pm 50$	49595	53261 ± 50	Ac^{228}	$1\,742\,648 \pm 60$	95455	102512 ± 60
Po^{211}	$1\,650\,787 \pm 20$	48984	52605 ± 20	Th^{228}	$1\,744\,105 \pm 50$	93215	100106 ± 50
At^{211}	$1\,649\,215 \pm 50$	49773	53453 ± 50	Pa^{228}	$1\,741\,279 \pm 80$	95258	102300 ± 90
Pb^{212}	$1\,655\,538 \pm 30$	54166	58170 ± 30	U^{228}	$1\,740\,196 \pm 70$	95558	102622 ± 80
Bi^{212}	$1\,655\,337 \pm 30$	53584	57545 ± 30	Th^{229}	$1\,749\,545 \pm 110$	96143	103251 ± 120
Po^{212}	$1\,656\,804 \pm 10$	51334	55129 ± 10	Pa^{229}	$1\,748\,442 \pm 100$	96463	103594 ± 110
Em^{212}	$1\,653\,438 \pm 120$	53134	57062 ± 120	U^{229}	$1\,746\,359 \pm 100$	97763	104990 ± 110
Bi^{213}	$1\,660\,520 \pm 60$	56769	60966 ± 60	Th^{230}	$1\,756\,288 \pm 90$	97767	104995 ± 100
Po^{213}	$1\,661\,127 \pm 60$	55379	59473 ± 60	U^{230}	$1\,753\,831 \pm 60$	98658	105951 ± 60
At^{213}	$1\,660\,090 \pm 120$	55633	59746 ± 130	Th^{231}	$1\,761\,286 \pm 120$	101137	108614 ± 130
Pb^{214}	$1\,664\,441 \pm 70$	61998	66581 ± 80	Pa^{231}	$1\,760\,827 \pm 120$	100813	108266 ± 130
Bi^{214}	$1\,664\,648 \pm 40$	61008	65518 ± 40	U^{231}	$1\,759\,705 \pm 130$	101152	108630 ± 140
Po^{214}	$1\,667\,035 \pm 30$	57838	62114 ± 30	Np^{231}	$1\,757\,106 \pm 110$	102968	110580 ± 120
At^{214}	$1\,665\,193 \pm 60$	58897	63251 ± 60	Th^{232}	$1\,767\,620 \pm 60$	103170	110797 ± 60
Po^{215}	$1\,671\,122 \pm 90$	62119	66711 ± 100	Pa^{232}	$1\,766\,480 \pm 60$	103527	111180 ± 60
At^{215}	$1\,671\,102 \pm 60$	61356	65892 ± 60	U^{232}	$1\,766\,987 \pm 50$	102237	109795 ± 50
Em^{215}	$1\,670\,320 \pm 110$	61355	65891 ± 120	Pu^{232}	$1\,761\,789 \pm 100$	105869	113696 ± 110
Po^{216}	$1\,676\,928 \pm 30$	64680	69462 ± 30	Th^{233}	$1\,772\,707 \pm 120$	106451	114321 ± 130
At^{216}	$1\,675\,690 \pm 50$	65135	69950 ± 50	Pa^{233}	$1\,773\,154 \pm 120$	105221	113000 ± 130
Em^{216}	$1\,676\,937 \pm 40$	63105	67770 ± 40	U^{233}	$1\,772\,931 \pm 110$	104661	112398 ± 120
At^{217}	$1\,681\,673 \pm 70$	67520	72512 ± 80	Np^{233}	$1\,771\,105 \pm 120$	105704	113518 ± 130

Table 1. (Continued.)

	Binding energy	Mass excesses			Binding energy	Mass excesses	
	keV	keV	µMU		keV	keV	µMU
Th^{234}	$1\,778\,786\pm\ 100$	$108\,739$	$116\,778\pm\ 110$	Am^{239}	$1\,806\,413\pm\ 150$	$119\,035$	$127\,835\pm\ 160$
Pa^{234}	$1\,778\,199\pm\ 340$	$108\,543$	$116\,567\pm\ 370$	U^{240}	$1\,813\,430\pm\ 100$	$122\,734$	$131\,807\pm\ 110$
U^{234}	$1\,779\,736\pm\ \ 90$	$106\,223$	$114\,076\pm\ 100$	Np^{240}	$1\,813\,007\pm\ \ 90$	$122\,374$	$131\,421\pm\ 100$
Pu^{234}	$1\,775\,824\pm\ \ 90$	$108\,569$	$116\,595\pm\ 100$	Pu^{240}	$1\,814\,380\pm$	$120\,218$	$129\,105\pm\ 100$
Pa^{235}	$1\,784\,302\pm\ 260$	$110\,808$	$119\,000\pm\ 280$	Cm^{240}	$1\,811\,363\pm\ 100$	$121\,669$	$130\,664\pm\ 110$
U^{235}	$1\,784\,919\pm\ 130$	$109\,408$	$117\,496\pm\ 140$	Pu^{241}	$1\,819\,915\pm\ 140$	$123\,051$	$132\,148\pm\ 150$
Np^{235}	$1\,783\,970\pm\ 130$	$109\,574$	$117\,674\pm\ 140$	Am^{241}	$1\,819\,152\pm\ 140$	$123\,031$	$132\,126\pm\ 150$
Pu^{235}	$1\,782\,048\pm\ 150$	$110\,713$	$118\,898\pm\ 160$	Pu^{242}	$1\,826\,139\pm\ 110$	$125\,194$	$134\,449\pm\ 120$
U^{236}	$1\,791\,336\pm\ \ 90$	$111\,358$	$119\,590\pm\ 100$	Am^{242}	$1\,824\,669\pm\ 100$	$125\,881$	$135\,187\pm\ 110$
Np^{236}	$1\,789\,703\pm\ \ 70$	$112\,208$	$120\,503\pm\ \ 80$	Cm^{242}	$1\,824\,520\pm\ 100$	$125\,247$	$134\,506\pm\ 110$
Pu^{236}	$1\,789\,430\pm\ \ 70$	$111\,698$	$119\,955\pm\ \ 80$	Pu^{243}	$1\,831\,120\pm\ 150$	$128\,581$	$138\,087\pm\ 160$
U^{237}	$1\,796\,760\pm\ 140$	$114\,302$	$122\,752\pm\ 150$	Am^{243}	$1\,830\,903\pm\ 140$	$128\,015$	$137\,479\pm\ 150$
Np^{237}	$1\,796\,487\pm\ 140$	$113\,792$	$122\,204\pm\ 150$	Cm^{243}	$1\,830\,116\pm\ 140$	$128\,019$	$137\,483\pm\ 150$
Am^{237}	$1\,793\,288\pm\ 140$	$115\,425$	$123\,958\pm\ 150$	Bk^{243}	$1\,827\,876\pm\ 170$	$129\,476$	$139\,048\pm\ 180$
U^{238}	$1\,802\,826\pm\ 100$	$116\,603$	$125\,223\pm\ 110$	Am^{244}	$1\,836\,061\pm\ 160$	$131\,224$	$140\,925\pm\ 170$
Np^{238}	$1\,801\,919\pm\ 100$	$116\,727$	$125\,356\pm\ 110$	Cm^{244}	$1\,836\,778\pm\ 100$	$129\,724$	$139\,314\pm\ 110$
Pu^{238}	$1\,802\,440\pm\ 100$	$115\,423$	$123\,956\pm\ 110$	Cf^{244}	$1\,832\,386\pm\ 120$	$132\,550$	$142\,349\pm\ 130$
Cm^{238}	$1\,797\,507\pm\ 130$	$118\,790$	$127\,572\pm\ 140$	Bk^{245}	$1\,841\,005\pm\ 160$	$133\,082$	$142\,920\pm\ 170$
U^{239}	$1\,807\,528\pm\ 140$	$120\,269$	$129\,160\pm\ 150$	Cf^{246}	$1\,845\,948\pm\ 110$	$135\,723$	$145\,757\pm\ 120$
Np^{239}	$1\,808\,040\pm\ 140$	$118\,974$	$127\,769\pm\ 150$	99^{247}	$1\,848\,749\pm\ 210$	$140\,507$	$150\,894\pm\ 230$
Pu^{239}	$1\,807\,974\pm\ 140$	$118\,257$	$126\,999\pm\ 150$	Cf^{248}	$1\,858\,711\pm\ 120$	$139\,695$	$150\,022\pm\ 130$

Table 2. *New nuclear data used in correcting Table 1.*

Nuclide	Reaction	Reaction energy (keV)	Reference	Nuclide	Reaction	Reaction energy (keV)	Reference
F^{21}	$F^{19}\ (t,\ p)$	—	1	Mo^{90}	β^+	$2542(\pm\ \ 50)$	10
Ne^{24}	β^-	$2420(\pm\ 100)$	2	In^{117}	β^-	$1465\ \pm\ \ 10$	11
Ca^{47}	β^-	$1940\ \pm\ \ 20$	3	Sb^{117}	β^+	$1790\ \pm\ \ 30$	11
Sc^{47}	β^-	$610\ \pm\ \ \ 5$	3	Te^{117}	β^+	$3782\ \pm\ 100$	12
Cr^{48}	K	$1450\ \pm\ 200$	4	Sb^{118}	β^+	$4014\ \pm\ \ 50$	12
Cu^{59}	β^+	$4420\ \pm\ 500$	5	Te^{131}	β^-	$2280\ \pm\ \ 50$	13
Zn^{61}	β^+	$5880\ \pm\ 250$	6 ,5	Ce^{143}	β^-	$1446\ \pm\ \ \ 5$	14
Zn^{71}	β^-	$2910\ \pm\ 200$	7	W^{181}	K	$92\ \pm\ \ 10$	15
As^{71}	β^+	$2012\ \pm\ \ 10$	8	Ta^{183}	β^-	$1068(\pm\ \ 20)$	16
Nb^{89}	β^+	$3870\ \pm\ 150$	9	Au^{194}	β^+	$2572(\pm\ \ 50)$	17
Nb^{90}	β^+	$3870(\pm\ 100)$	10				

1 BIGHAM *et al.*: Bull. Amer. Phys. Soc. **30**, No. 3, K 5 (1955).
2 DROPESKY and SCHARDT: Bull. Amer. Phys. Soc. **30**, No. 5, F 3 (1955).
3 LIDOFSKY *et al.*: Bull. Amer. Phys. Soc. **30**, No. 3, V 2 (1955).
4 VAN LIESHOUT *et al.*: Private communication.
5 LINDNER *et al.*: Private communication.
6 CUNNING: Bull. Amer. Phys. Soc. **30**, No. 4, L 2 (1955).
7 LE BLANC *et al.*: Phys. Rev. **97**, 750 (1955).
8 GRAVES and MITCHELL: Phys. Rev. **97**, 1033 (1955).
9 MATHUR *et al.*: Phys. Rev. **97**, 117 (1955).
10 MATHUR and HYDE: Phys. Rev. **98**, 79 (1955).
11 McGINNIS: Phys. Rev. **97**, 93 (1955).
12 FINK: Thesis Rochester 1955.
13 HEBB: Phys. Rev. **97**, 987 (1955).
14 MARTIN *et al.*: Bull. Amer. Phys. Soc. **30**, No. 3, YA 1 (1955).
15 BISI *et al.*: Nuovo Cim. **1**, 291, 651 (1955).
16 MURRAY *et al.*: Phys. Rev. **97**, 1007 (1955).
17 THIEME and BLEULER: Bull. Amer. Phys. Soc. **30**, No. 4, L 6 (1955).

For mass numbers below 33 binding energies were computed from nuclear reaction and decay energy data only [4], corrected for the slight change in calibration mentioned at the end of Sect. 8. The agreement with the most modern mass spectroscopical results for $A < 17$ is not unsatisfactory; between mass numbers 16 and 33 the difference with mass spectroscopic data is of the order of two times the errors assigned to the latter values. In the mass region $33 < A < 202$ [5] the mass excesses were computed from mass spectroscopic results combined with energy measurements. The agreement between the two sets of data was not very good, the differences being of the order of three times the errors assigned to the mass spectroscopic values. The disagreement is probably due to uncertainties in the mass spectroscopic results, according to the discussion of a few cases where the energy measurements were very reliable [5].

For mass numbers above 201 [6] the existence of extended families of α- and β-active nuclides decaying into one another together with a few dependable reaction energy measurements allows a pretty accurate computation of relative binding energies; the absolute values were determined by comparing the values obtained from these energy measurements with a few mass measurements which in this case turned out to agree pretty well with the energy measurements. The errors in masses and binding energies given in this part of the table do *not* include the error in the mass spectroscopic absolute calibration, which is estimated to be about 1500 keV.

It seems at first sight that the values given in Table 1 are often given in more decimals than is warranted by the assigned errors. This is due in some cases to the fact that differences in masses or binding energies are known much more accurately than the masses or binding energies themselves, in other cases to our wish to preserve the correct relation between the values given in the three columns.

C. Some trends in atomic masses of nuclides.

10. Main trends in nuclear binding energies. Fig. 1 shows the binding energy per nucleon E/A for the most β stable nuclides plotted versus the mass number. The first thing apparent from this graph is that (for $A > 11$) the binding energy per nucleon is remarkably constant: within 10% equal to 8.1 MeV. It reaches a maximum of 8.8 MeV in the neighbourhood of $A = 60$ and decreases rather continuously to both higher and lower mass numbers. This overall behaviour will be explained in Sect. 15.

Fig. 1 shows, in addition, that the even-even nuclides have consistently higher binding energies than those of odd mass with neighbouring charges and masses. We have only plotted few data for odd-odd nuclides, since only four β stable o-o nuclides exist (insert in Fig. 1), but it may be remarked already here that the binding energies of o-o nuclides are about as much lower than those of odd mass ones as the binding energies of e-e nuclides are higher (compare, however, Sect. 19).

Fig. 2 illustrates the behaviour of the binding energies of isobars as a function of the neutron excess. For odd mass numbers the masses or binding energies lie near one smooth curve which can often be approximated rather well by a parabola, as shown in the left hand part of Fig. 2. There is a systematic difference between the binding energies of e-e nuclides and o-o ones; the right hand part of Fig. 2 shows that it is often possible to approximate the binding energies of even mass nuclides by two parallel parabolae:

$$E = E_0(A) - \tfrac{1}{2} b(A)\{I - I_0(A)\}^2 + \delta,$$
$$\delta = + \delta_e(A) \quad \text{for even } Z, \quad \delta = - \delta_e(A) \quad \text{for odd } Z. \qquad (10.1)$$

There is also a minor but systematic difference between e-o and o-e nuclides, the first ones being in the average slightly more stable (see below). We will therefore apply Eq. (10.1) to odd mass nuclides too, introducing a new function δ_0 of A.

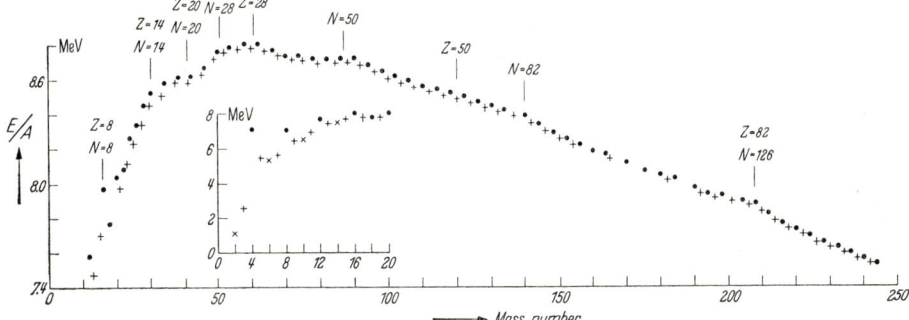

Fig. 1. Binding energy per nucleon E/A of the most β stable isobars as a function of mass number. Points refer to even even nuclides, crosses to such of odd mass; each symbol represents an average value for a few neighbouring nuclides of the type indicated. The positions of magic numbers are indicated. Insert: detailed picture for the lowest masses; in this part the four known β stable odd-odd nuclides are indicated by oblique crosses.

Later sections will show that in a good approximation the parameter functions E_0, b and I_0 are the same for all four types of nuclides.

Let us call δ the spin term, b the parabolic constant. I_0 is the neutron excess for which the binding energy is a maximum for isobars. This is not the same as

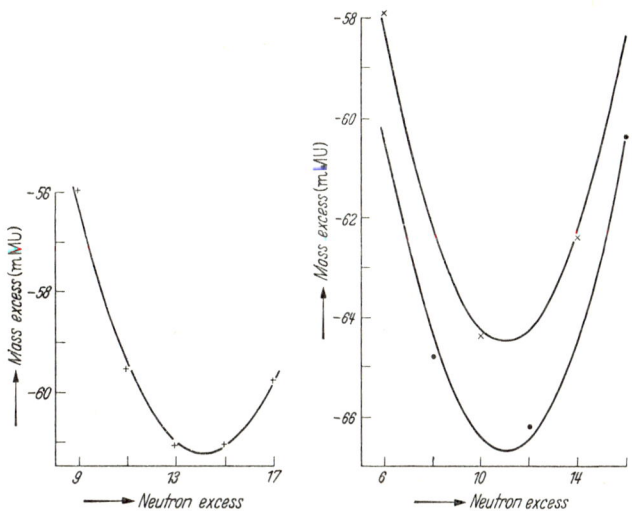

Fig. 2. Mass excesses of isobaric nuclides as a function of the neutron excess. Left hand part: odd mass ($A = 107$) right hand part: even mass ($A = 92$).

the neutron excess I_s for the most β stable isobar. The β decay energy being the difference in mass of initial and final nuclide (see Sect. 6), the most β stable isobar is that for which the atomic mass is a minimum[1].

The relation between nuclear binding energy and atomic mass is

$$M = \tfrac{1}{2} A (n + H) + \tfrac{1}{2} I (n - H) - E \tag{10.2}$$

[1] Here and the following we shall consider M, E etc. to be continuous functions of I and A; in most cases therefore I_s is no integer.

(n and H indicate the masses of the neutron and of the H^1 atom; we neglect the influence of electron binding energies which are at most about 0.03% of the nuclear binding energies). Using Eq. (10.1) we find[1]

$$M = M_0 + \tfrac{1}{2} b (I - I_s)^2 - \delta, \tag{10.3}$$

$$I_s = I_0 - (n - H)/2b. \tag{10.4}$$

Eq. (10.1) defines four surfaces in a space with coordinates E, I and A (or E, Z and N); we shall call them the *binding energy surfaces*. Often not I but Z is used as a parameter in Eqs. (10.1) to (10.3); the alterations due to this change in coordinates are evident. Due to the slow variation of $(M_0 - A)$ with A (cf. Fig. 1) the isotopic cross sections (Z constant) of the energy surfaces are also nearly parabolae. This point has beautifully been demonstrated by Collins, Johnson and Nier[2].

In some places ($A \approx 60$, 90, 140, 208) the curve in Fig. 1 shows humps in the direction of higher binding energies. This phenomenon is connected with the fact that nuclides with numbers of neutrons or protons in the neighbourhood[3] of the so-called "magic numbers" (8, 14, 20, 28, 50, 82, 126) have higher binding energies than normal by amounts following a general shape as depicted in Fig. 3; we shall for brevity call them the magic peaks[4]. As a result the functions E_0, b, I_0 and I_m defined

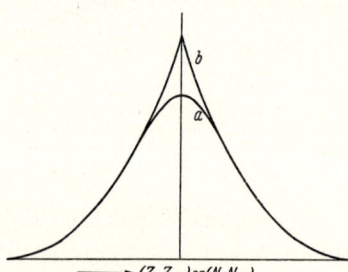

Fig. 3. Possible shapes of the increase in nuclear binding energies in the neighbourhood of a magic number Z_m or N_m.

in Eqs. (10.1) to (10.3) fluctuate rather irregularly; let us therefore distinguish between the quantities E_0' etc. obtained from experimental data and those E_0 etc. defining an energy surface abstracting from magic number effects. Fig. 1 then gives essentially information about E_0' or M_0'. According to Green and Engler [9] M_0 can be well represented by

$$M_0 = A - 0.064 + 10^{-5}(A - 100)^2 \, \text{MU}. \tag{10.5}$$

Another equation for M_0 can be derived from semi-theoretical considerations [Sect. 15, Eq. (15.12)].

In the following sections we shall derive information about the other parameter functions in Eqs. (10.1) to (10.3) from experimental data, following (and extending sometimes) the work of various authors [14] to [22]. We shall also obtain information about the shape and dimensions of the magic peaks; among others about the controversial question whether these peaks are rounded at the top [8] (a in Fig. 3) or peaked instead [17] (b in Fig. 3).

11. Various systematics. If the binding energies lie on smooth binding energy surfaces, the same should be true for the binding energies of the last added protons and neutrons (briefly: proton and neutron binding energies and designated B_Z and B_N) and β and α decay energies, which are connected with differentials of

---

[1] A nice graphical representation of nuclear masses along the lines of this formula was given by H. Ewald: Z. Naturforsch. **8a**, 116 (1953).

[2] T. L. Collins, W. H. Johnson jr. and A. O. Nier: Phys. Rev. **94**, 398 (1954).

[3] A. H. W. Aten jr.: Science, Lancaster, Pa. **110**, 260 (1949) was the first who pointed to the fact that the influence of magic numbers extended some distance from these numbers.

[4] The fact that the effect of magic numbers can largely be accounted for by the adding of magic peaks will be demonstrated below, especially in Sect. 16β (Fig. 10).

the binding energy in the direction of the coordinates Z, N, I and A respectively:

$$\left.\begin{aligned} B_Z(Z, N) &= E(Z, N) - E(Z-1, N), \\ B_N(Z, N) &= E(Z, N) - E(Z, N-1), \\ B_A(A, I) &= E(A, I) - E(A-2, I), \\ B_I(A, I) &= E(A, I) - E(A, I-2), \end{aligned}\right\} \tag{11.1}$$

$$E_\alpha = \alpha - B_A(A) - B_A(A-2), \tag{11.2}$$

$$E_\beta = (n - H) - B_I \tag{11.3}$$

(α is the binding energy of the He⁴ nuclide, $n-H$ is defined as in Sect. 10; E_α and E_β are the α and β decay energies of the nuclide considered, including the energy of the recoiling nucleus which is important in the case of α decay). The quantities B are similar to the partial derivates $\frac{\partial E}{\partial Z}$, $\frac{\partial E}{\partial N}$, $\frac{\partial E}{\partial A}$ and $\frac{\partial E}{\partial I}$ respectively; between them exist relations[1] similar to $2\frac{\partial E}{\partial I} = \frac{\partial E}{\partial N} - \frac{\partial E}{\partial Z}$, and $\frac{\partial^2 E}{\partial Z \partial N} = \frac{\partial^2 E}{\partial N \partial Z}$, the last one reading

$$\Delta(Z, N) = B_Z(Z, N) - B_Z(Z, N-1) = B_N(Z, N) - B_N(Z-1, N). \tag{11.4}$$

In view of the relations (11.1) to (11.3) it is clear that the systematics of experimental proton and neutron binding energies [10], [16], [29] and α [26], [27], [28] and β decay energies [14] to [22] can be used to derive information about nuclear binding energies. Some of them are advantageous because they accentuate the influence of some of the parameter functions defined in Sect. 10 (e.g. the spin term in β systematics, see Sect. 13). Others are useful because they just eliminate the influence of some of these quantities (e.g. the spin term in α systematics). They can, however, only give information about the slope of the magic peaks; for investigating the heights and the widths of these peaks the binding energies themselves should be considered (see Sect. 16β).

Sometimes even more complicated derivations of the nuclear energy surfaces are used for special purposes. So, e.g. FEATHER [16] defines a function

$$_D\Delta_p' = \Delta(A, I+2) - \Delta(A, I) \tag{11.5}$$

[Δ defined in Eq. (11.4)] which he considers to be especially suited for detecting nuclear shell effects, because this function is very nearly zero if such effects are absent.

12. Beta stability systematics. α) *The spin terms.* Some information about the energy surfaces can already be obtained from the statistics of the β stable nuclides. Fig. 4 shows all nuclides (with $A > 20$) for which the β stability is reasonably certain in a plot of neutron excess versus mass number. This diagram is essentially identical with one published earlier [8] except for some recent extensions in the high mass region in which we mainly followed the work of HUIZENGA [6]. Fig. 4 shows the following regularities:

1. For each odd mass number, there exists only one stable isobar[2].

For mass numbers 87, 113, 123 and 187 two isobars occur in nature; but from these Rb⁸⁷ is demonstrated to be β active, Re¹⁸⁷ is found to yield Os¹⁸⁷ in geologic times (though its decay data are still controversial); Cd¹¹³ is known to be β unstable energetically from the decay data of an isomer of this nuclide, and Te¹²³ is suspected to be β unstable too [21].

---

[1] N. FEATHER: Phil. Mag. **43**, 133 (1952).
[2] J. MATTAUCH: Z. Physik **91**, 361 (1934).

2. *There exist often two, sometimes even three stable even mass isobars; all of them are even-even nuclides*[1].

Real exceptions are H^2, Li^6, B^{10} and N^{14}; of the four other odd-odd isotopes shown to occur in nature, La^{138} and Lu^{176} are known to be β active, V^{50} is energetically β unstable,

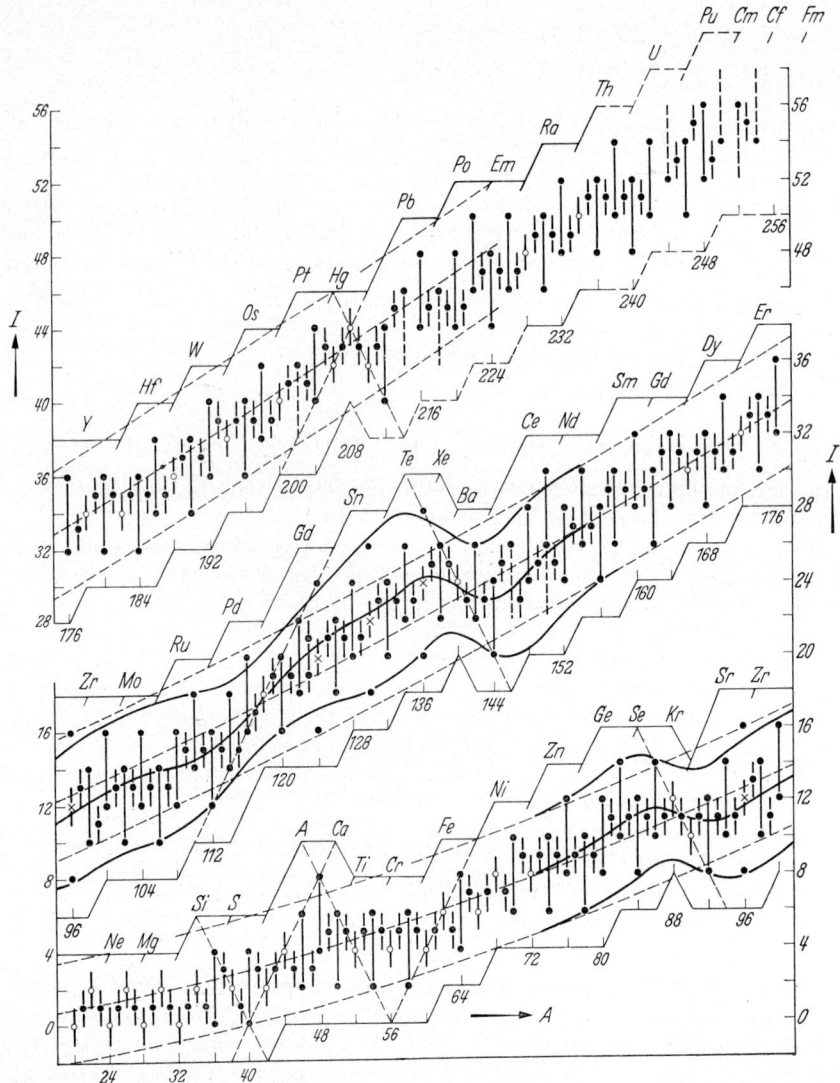

Fig. 4. Diagram of β stable nuclides. Odd mass nuclides are represented by dots on vertical lines; e-e nuclides are plotted: as circles on vertical lines if only one stable isobar exists; as dots joined by vertical lines if a doublet exists; and as crosses on vertical lines for the central one of an isobaric triplet. The nearest β unstable e-e isotopes are joined by two broken lines. The positions of magic numbers 20, 28, 50, 82 and 126 are indicated by dashed lines. The meaning of the smooth curves is explained in Sect. 12 β.

while the same probably is also true for the recently discovered Ta^{180} [5]. Triplets of stable isobars occur for mass numbers 96, 124, 130, 136, and perhaps 150. (Gd^{150} is α active but may be β stable); all triplets would probably not exist but for the influence of magic numbers.

These facts indicate that in the average the binding energies of e-e nuclides are larger than those of o-o nuclides, but that the difference between e-o and o-e

[1] W. D. HARKINS: Proc. Nat. Acad. Sci. Wash. **19**, 307 (1933).

nuclides should be much smaller. Numerical information about these differences can be gained from the statistics of β stable nuclides[1] (Tables 3 and 4). In their interpretation the following considerations are of some help.

The limits of the neutron excesses between which nuclides with even proton number are β stable can be computed from Eq. (10.3) and from the condition $M_e(I_b) = M_c(I_b \pm 2)$ (meaning that β decay occurs when it is energetically possible). We then find:

$$I_b = I_s \pm (1 + \delta/b). \quad (12.1)$$

Therefore the probability for the occurrence of even mass doublets is

$$W_d = \tfrac{1}{2}(-1 + \delta_e/b), \quad (12.2)$$

and the probability for the occurrence of stable e-o nuclides

$$W_e = \tfrac{1}{2}(1 + \delta_o/b). \quad (12.3)$$

With help of these two formulae the spin term can be expressed approximately in terms of the parabolic constant (last columns in Tables 3 and 4). The last constant is approximately the same for even and

Table 3. *Statistics of β stable odd mass nuclides.*
Column 1: mass region; 2 and 3: number of β stable nuclides of each kind; 4: probability for e-o nuclides being stable as derived from columns 2 and 3; 5: spin term in terms of the parabolic constants.

A	N_{eo}	N_{oe}	W_e [2]	δ_o/b
<100	24	26	0.48 ± 0.07	-0.04 ± 0.14
$100-206$	30	23	0.56 ± 0.06	$+0.15 \pm 0.13$
>206	12	7	0.62 ± 0.10	$+0.24 \pm 0.20$

Table 4. *Statistics of β stable even-even isobaric pairs.*
Column 1: mass region; 2: number of single stable isobars; 3: number of isobaric doublets (for reasons explained in the text the 4 triplets have been counted as doublets); 4: probability for isobaric pairs; 5: spin term in terms of the parabolic constant.

A	N_s	N_d	W_d [2]	δ_e/b
<35	17	0	0.05 ± 0.05	$<1.1 \pm 1.1$ [3]
$35-95$	12	18	0.59 ± 0.09	2.18 ± 0.18
$95-205$	10	45	0.81 ± 0.05	2.62 ± 0.10
>205	5	16	0.74 ± 0.09	2.48 ± 0.18

odd neighbouring masses (see Sect. 13 α). These tables therefore clearly demonstrate the difference in magnitude of the spin terms for even and odd masses.

$\beta)$ *The line of maximum β stability.* In addition to rules 1 and 2 in Sect. 12α, Fig. 4 demonstrates that

3. *If isotopes (isotones) with masses $A+2$ and $A-2$ are stable, that with mass A is stable, too.*

The one apparent exception to this rule, Sm[146], probably does not occur in nature because of its α instability.

4. *If a nuclide with mass $2n+1$ and an even number of neutrons (protons) is stable, its isotones (isotopes) with masses $2n$ and $2n+2$ are stable, too.*

5. *The β stable nuclides concentrate along a smooth curve $[I_s'(A)$ according to the definition in Sect. 10]. For low mass numbers I_s' tends to zero, or higher masses I_s' increases steadily.*

Fig. 4 can be used to obtain an estimate for I_s' as a function of A. In order to visualize this curve distinctly the vertical lines through the points in Fig. 4 are plotted in such a way that the line of maximum β stability should cross all of them (assuming the isobaric cross sections to be symmetric and δ_o to be small).

---

[1] T. P. KOHMAN: Phys. Rev. **73**, 16 (1948).

[2] If in N events a condition occurs n times, the chance to occur is

$$\frac{1}{N+2}\left\{(n+1)\pm \sqrt{(n+1)(N-n+1)/(N+3)}\right\}.$$

[3] Together with the fact that in the region $15 < A < 35$ no β stable odd-odd nuclides exist we find in this region $\delta_e/b = 1.0 \pm 0.3$.

The lines representing the limits of β stability for even mass isobars should pass between the broken lines in Fig. 4 and the outer points. The thin dotted lines in Fig. 4 represent estimates for I_s and I_b obtained from a semi-empirical binding energy formula [8] [compare Eq. (15.9)]. A somewhat simpler expression for I_s was proposed by GREEN and ENGLER [9]

$$I_s = \frac{0.4\,A^2}{A+200}. \tag{12.4}$$

Fig. 4 demonstrates that the points lie somewhat high in the regions between magic numbers $Z=28$ and $N=50$ and between $Z=50$ and $N=82$, and somewhat low between $Z=50$ and $N=50$. It has been shown [8] that these deviations can be explained numerically by postulating magic peaks of the shape $1/(1+x^2)$ (x proportional to $Z-Z_m$ or $M-M_m$) with heights of about 6 MeV decreasing to half this value for $(Z-Z_m)$ or $(N-N_m)$ equal to about $3\frac{1}{2}$. The same procedure together with an estimate for δ_e/b could also explain the experimental course of the limits of β stability [8]; the values found in this way for both I_s and I_b are indicated in Fig. 4 by smooth lines.

These computations were thus made with magic peaks with rounded tops; they could, however, have been made equally successful with peaks with a sharp top (b in Fig. 3). In this case the I_s' curve would show discontinuities at the magic numbers, which e.g. CORYELL [17] considers to agree better with empirical data. GREEN and EDWARDS [10] postulate correction terms to the mass formula of the form

$$-\{N-\tfrac{1}{2}(N_{i+1}+N_i)\}^2/(N_{i+1}-N_i)-\{Z-\tfrac{1}{2}(Z_{j+1}+Z_j)\}^2/(Z_{j+1}-Z_j)+K_{ij} \tag{12.5}$$

(to be valid in the region between magic numbers N_i-N_{i+1} and Z_j-Z_{j+1}) which could equally well explain the experimental course of I_s' (see Fig. 4 in [10]). It should, however, be considered as a disadvantage that this correction term, at least together with the constants proposed by GREEN and EDWARDS, introduces at magic numbers discontinuities of some MeV in the binding energies (see Fig. 5 in [10]), in serious disagreement with the experimental data (compare our Fig. 10).

The discussed magic number effects on I_s' explain the fact that no β stable odd mass isobars exist for proton numbers 18 (A) and 58 (Ce). For proton numbers 43, 61, 85 (?) and 97 no β stable isotopes occur at all; this phenomenon is partly connected with other effects caused by the existence of magic numbers (cf. Sect. 13 β).

Earlier empirical determinations of the function I_s' were made by various authors [14] to [22]; most of them computed narrower limits on I_s' by the procedure outlined in Sect. 13 α. It should be remarked that CORYELL [17] and BOUCHEZ, ROBERT and TOBAILEM [18] find that it is possible to represent $I_s'(A)$, between the breaks caused by the magic numbers, by straight lines.

13. Beta decay energy systematics. $\alpha)$ *The parabolic constant.* If masses of four isobars are known, it is possible to construct through them two parallel parabolae [Eq. (10.3)] or, what means the same, to compute the quantities b, I_s and δ at this mass number. The easiest way for doing so [22] is to plot the β decay energies between the isobars:

$$E_\beta = 2b(I-I_s-1)-2\delta. \tag{13.1}$$

This formula can also be applied to K capture or positron decay if their decay energies are formally considered as negative β^- decay energies; the sign before the spin term δ is then determined by the odd-even characteristics of the "initial" nuclide of the β^- decay. The parabolic constant b is then already determined by two isobaric β^- decay energies if the "initial" nuclides have the same odd-even characteristics; if a third decay energy of the other kind is known, I_s and δ are also determined. If more than four isobaric masses (or three β decay energies) are known, one can often represent them reasonably well by two parallel lines in a plot of E_β versus I and thus give compromise values for b'.

The values b' obtained in this way (essentially from the data used in computing Table 1) are shown in Fig. 5. We have plotted the product of parabolic constant and mass number $b'A$, versus A in a scale in which $A^{\frac{2}{3}}$ is linear: according to Eq. (15.8) the points should then be expected to lie on a straight line if magic number effects are neglected. For $A < 35$ only three isobaric masses were known; the points in this region were computed assuming δ_0 to be small

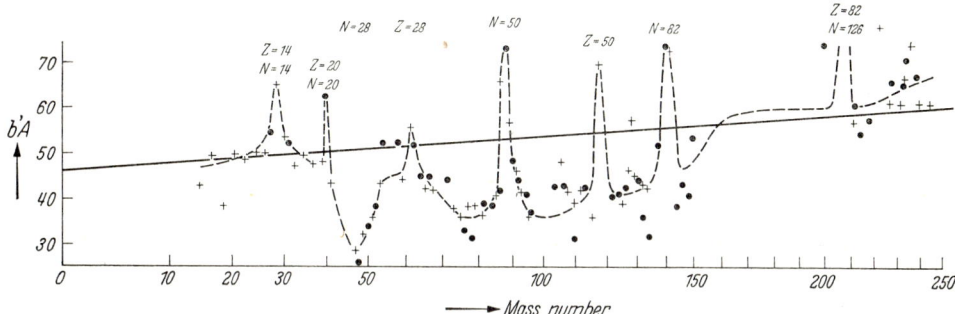

Fig. 5. The parabolic constant b' as a function of the mass number, obtained from β systematics. The positions where magic numbers should show their influence have been indicated. Dots mark values obtained from even mass, crosses those from odd mass nuclides. The reason for the choice of the scales is explained in the text.

(compare Table 3). For mass numbers between 150 and 200 no sufficient data were known too; in some cases, however, limiting values for β decay energies[1] could be used to determine lower limits for $b'A$. These limits appeared to be rather high (up to $b'A \approx 55$ MeV), this fact guided our hand in drawing the dashed curve in this mass region.

Fig. 5 shows that there is no systematic difference between the values b' for odd and for even nuclides. The influence of magic numbers is striking. It will be clear that b' will be high if computed from experimental points embracing the magic number. If the magic peak has a sharp top (Fig. 3b) all other values for b' will be low, even if the binding energy for the magic number is included as a border point. If the magic peak has a rounded top (Fig. 3a) the values b' for sets of points near the magic number may be high too. The experimental evidence in Fig. 5 indicates a pointed shape, as already remarked earlier by CORYELL [17] on other grounds.

Earlier empirical determinations of b' were made by several authors [9], [14], [17], [18], [22]. Of these only BOUCHEZ, ROBERT and TOBAILEM [18] reported some influence of the magic numbers on b'. They limit themselves, however, to values derived from experimental points on one side of the magic number; in doing so they probably find values b' which are systematically low if compared with the values b (in which the magic number effects are corrected for) as explained above.

β) *The spin term.* Values for the spin term can be derived from the analysis demonstrated in Sect. 13α. Information about δ can, however, be obtained more

---

[1] R. W. KING: Rev. Mod. Phys. **26**, 327 (1954).

easily by plotting the β decay energies of isodiapheres $[19]$ as a function of (e.g.) the mass number. We can join the points in this plot by pretty smooth lines which are, according to Eq. (13.1), a distance $4\delta'$ apart.

Fig. 6 shows the values of the spin term for even mass isotopes δ'_e obtained in this way, and plotted logarithmically as a function of $\log A$. It may be added that the values obtained by the analysis in Sect. 13α agree with Fig. 6. The data can be represented approximately by the formula

$$\delta_e = 11.2\,A^{-\frac{1}{2}}\,\text{MeV} = 12.04\,A^{-\frac{1}{2}}\,\text{mMU} \tag{13.2}$$

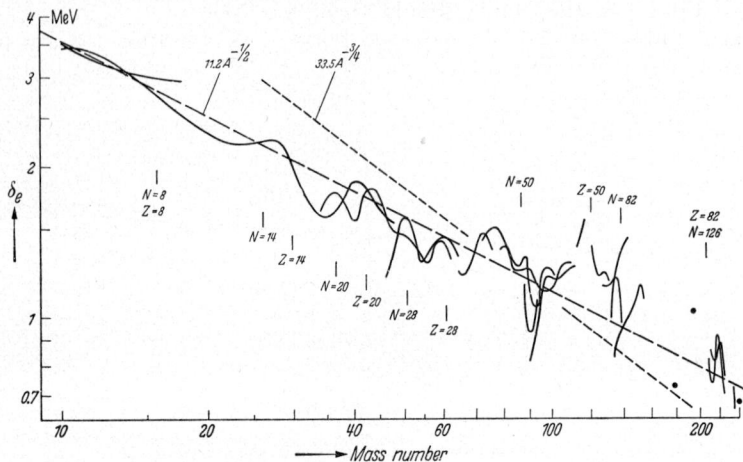

Fig. 6. The even mass spin term, obtained from isodiaspheric β^- decay energies. Each separate curve has been computed from cases in which the initial nuclides have the same neutron excess. The positions of magic numbers have been indicated.

with an average accuracy of about 10% or about 100 keV (though some deviations up to four times these amounts occur). This formula is quite close to the estimate $\delta = 12\,A^{-\frac{1}{2}}\,\text{mMU}$ of Green and Edwards $[10]$ derived from an analysis of neutron binding energies. For comparison Fig. 6 shows also the curve

$$\delta_e = 33.5\,A^{-\frac{3}{4}}\,\text{MeV} \tag{13.3}$$

due to Fermi[1], which appears to be less accurate.

Bouchez, Robert and Tobailem $[18]$ report that magic numbers are connected with minima in δ'_e. This effect is not very apparent in Fig. 6, which shows at most a slight downward trend in δ_e after passing some magic numbers.

The existence of an odd mass spin term was first demonstrated by Glueckauf $[26]$ from α decay energy systematics. It was postulated by Kohman $[7]$ in order to explain the β stability statistics and by Kowarski[2] to explain the fact that for some charge numbers (43, 61) no stable isotopes occur in nature. The last explanation was specified by Suess and Jensen $[19]$ who postulated that δ_o should be positive after passing a magic neutron number, and negative after passing a magic proton number (except 20; this rule follows from the nuclear shell model interpretation of magic numbers: the pairing energy from both protons and neutrons should be lowered after passing a shell closure $[3]$). This effect was demonstrated empirically by Suess and Jensen $[19]$ and by Bouchez, Robert and Tobailem $[18]$. Fig. 7 shows information on this subject (obtained

---

[1] E. Fermi: Nuclear Physics. Chicago: University of Chicago Press 1949.
[2] L. Kowarski: Phys. Rev. 78, 477 (1950).

in the same way as in Fig. 6 for even masses) which corroborates this conclusion, as well as the more general one (cf. Sect. 12α) that δ_o tends to be positive (in the average about 60 keV). Not too much weight should be given to the absolute values plotted in Fig. 7. Drawing smooth lines through the experimental β decay energies is a somewhat arbitrary procedure; moreover the largest maxima in the curves are caused by the deviating decay energies of in each case essentially one nuclide (Al^{29}, Ca^{45} and Zr^{95}; compare also Sect. 18).

A more elaborate investigation along the lines of the nuclear shell model was published by GREEN and EDWARDS ([10]; compare also Sect. 19).

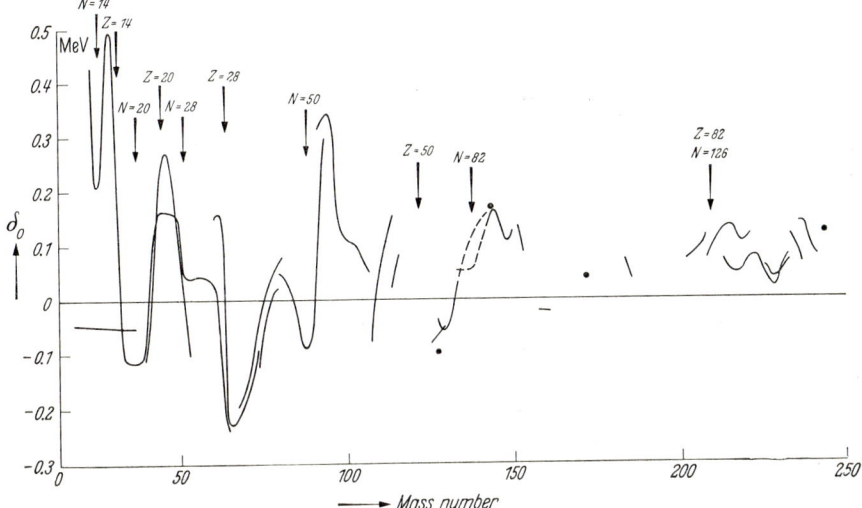

Fig. 7. The odd mass spin term derived from isodiaspheric β decay energies.

14. Nuclear radii[1]. The part of the nuclear binding energies due to the COULOMB interaction between the proton charges depends on the distribution of these charges in the nuclei, and especially on the nuclear dimensions. Information about this subject indicates that the nuclear volume is in a good approximation proportional to the number of nucleons, and therefore, if the nuclei are assumed to be nearly spherical, their radii are described by a function

$$r = r_0 A^{\frac{1}{3}}. \tag{14.1}$$

For the radius constant the following values were found[2].

1. From nucleon scattering experiments $r_0 = 1.4 \times 10^{-13}$ cm.
2. From α decay energies and lifetimes $r_0 = 1.35 \times 10^{-13}$ cm.
3. From analysis of nuclear reaction yields $r_0 = 1.5 \times 10^{-13}$ cm.
4. From β decay energies between mirror nuclides $r_0 = 1.43 \times 10^{-13}$ cm.
5. From the COULOMB term in the BETHE-WEIZSÄCKER binding energy formula $r_0 = 1.42 \times 10^{-13}$ cm. [13].

Recently some new methods for measuring nuclear sizes gave rather smaller results.

6. Electron scattering experiments[3] gave $r_0 = 1.1 \times 10^{-13}$ cm. In addition they gave the interesting result that the charge distribution in the nuclei should be nearly uniform.

7. An analysis of the energies of X rays emitted by μ^- mesons in transitions between different orbits in μ mesonic atoms[4] yielded $r_0 = 1.19 \times 10^{-13}$ cm.

---

[1] Cf. the contribution of D. L. HILL in Vol. XXXIX of this Encyclopedia for details.

[2] J. M. BLATT and V. F. WEISKOPF: Theoretical Nuclear Physics, p. 14ff. New York: Wiley & Sons 1953.

[3] F. BITTER and H. FESHBACH: Phys. Rev. **92**, 837 (1953). — R. W. PIDD, C. L. HAMMER and E. C. RAKA: Phys. Rev. **92**, 436 (1953). — R. HOFSTADTER, H. R. FECHTER and J. A. McINTYRE: Phys. Rev. **92**, 978 (1953). — R. HOFSTADTER, B. HAHN, A. W. KNUDSEN and J. A. McINTYRE: Phys. Rev. **95**, 512 (1953).

[4] V. L. FITCH and J. RAINWATER: Phys. Rev. **92**, 789 (1953).

A difference between the last measurements and methods 1, 2 and 3 is that the first measure the distance at which an incoming particle falls under the influence of nuclear *forces*, whereas methods 6 and 7 (and also 4 and 5) determine an (average) radius of the actual *charge* distributions. The smaller radii obtained by the latter methods are therefore sometimes considered to be an indication that the protons are somewhat concentrated towards the centre of the nucleus. A semi-theoretical explanation for such an effect has been given by Johnson and Teller[1] and investigated somewhat more completely by Mittelstaedt[2]; Hess and Moyer[3] have found experimental evidence that near the surface of heavy nuclei considerably more neutrons than protons are present.

On the other hand it should be remarked that new analyses of the types 1 to 4 mentioned above gave also much smaller radii than found before[4,5,6,7]. It seems therefore not impossible that all data could be reconciled with a value r_0 of nearly 1.2×10^{-13} cm (if the finite range of nuclear forces is taken into account together with the fact that the nuclear density will fall off to zero gradually near the "surface" of a nucleus). The bearing of the experimental course of nuclear masses on this problem will be discussed in Sect. 16.

15. Semi-empirical binding energy formula. In Sect. 10 we have seen that the nuclear binding energy is approximately proportional to the number of nucleons; Sect. 14 showed the nuclear volume also to be proportional to A. Combining these facts we are led to suppose that nuclei behave more or less like drops of a liquid. We can use this nuclear drop model to derive a formula for the binding energies [12]. The first term will be the volume energy:

$$E_v = a_v A \qquad (15.1)$$

which will, however, have to be corrected for surface effects giving a decrease of the binding energy proportional to the surface

$$E_s = - a_s A^{\frac{2}{3}}. \qquad (15.2)$$

The charge of the protons gives rise to a Coulomb interaction energy

$$E_c = - \frac{3}{5} \frac{(Ze)^2}{r} = - a_c (A-I)^2 A^{-\frac{1}{3}}, \qquad (15.3)$$

$$a_c = \frac{3}{20} \frac{e^2}{r_0} = 0.216/r_0 \text{ MeV} \qquad (15.3')$$

(if r_0 is expressed in 10^{-13} cm). The factor $\frac{3}{5}$ is only correct if the charge is distributed homogeneously through the nuclear volume, as indicated by electron scattering experiments (see Sect. 14): if the charge were concentrated on the surface it should be changed into $\frac{1}{2}$. For low charge numbers Z^2 in Eq. (15.3) should better be replaced by $Z(Z-1)$; we shall not apply this small correction.

If no further terms were added only polyneutrons should be stable $(I = A)$, whereas Sect. 12 showed that I tends to zero, at least for low mass numbers. This fact can be explained by the existence of a so-called symmetry term

$$E_I = - a_I I^2 A^{-1}. \qquad (15.4)$$

Below we give an explanation of the form of this term, following a discussion by Fermi[8].

---

[1] M. H. Johnson and E. Teller: Phys. Rev. **93**, 357 (1954).

[2] P. Mittelstaedt: Z. Naturforsch. **10**a, 379 (1955).

[3] W. N. Hess and B. J. Moyer: Phys. Rev. **96**, 859 (1954).

[4] T. Coor, D. A. Hill, W. F. Hornyak, L. W. Smith and G. Snow: Phys. Rev. **98**, 1369 (1955). — R. W. Williams: Phys. Rev. **98**, 1387 (1955).

[5] H. A. Tolhoek and P. J. Brussaard: Physica, Haag **21**, 449 (1955).

[6] J. R. Holt and T. N. Marsham: Proc. Phys. Soc. Lond. A **66**, 1032 (1955).

[7] D. C. Peaslee: Phys. Rev. **95**, 716 (1954). — B. C. Carlson and I. Talmi: Phys. Rev. **96**, 436 (1954).

[8] E. Fermi: Nuclear Physics. Chicago: University of Chicago Press 1949.

We consider the nucleus as a volume $\frac{4}{3}\pi r_0^3 A$ containing a cold FERMI gas of N neutrons and Z protons. The number of proton states with a momentum less than p is

$$n = 2 \cdot \tfrac{4}{3}\pi p^3 \cdot \tfrac{4}{3}\pi r_0^3 A/h^3$$

and therefore the maximum proton energy present is

$$E_0 = \frac{1}{8}\left(\frac{3}{2\pi}\right)^{\frac{1}{3}}\frac{h^2}{M r_0^2}\left(\frac{Z}{A}\right)^{\frac{2}{3}}.$$

The total proton energy is then

$$E_p = \tfrac{3}{5} Z E_0.$$

Add a similar term for the neutrons, substitute $\frac{1}{2}(1+I/A)$ for N/A and $\frac{1}{2}(1-I/A)$ for Z/A and expand in a power series of I/A. Disregarding the 0.1% difference in mass between proton and neutron (and assuming equal radii for the volumes occupied by protons and neutrons) we obtain

$$E = a_I A \left\{\tfrac{9}{5} + (I/A)^2 + \tfrac{1}{27}(I/A)^4 + \cdots\right\} \qquad (15.4'')$$

with

$$a_I = \frac{1}{96}\left(\frac{3}{\pi}\right)^{\frac{1}{3}}\frac{h^2}{M r_0^2} = 16.0/r_0^2\ \text{MeV} \qquad (15.4')$$

(if again r_0 is expressed in 10^{-13} cm). The first term in Eq. (15.4'') can be considered to be contained in the volume energy [Eq. (15.1)]; the third term is smaller than 0.3% of the second one for all known nuclei and can therefore be neglected together with the following terms. The remaining term has the shape given in Eq. (15.4).

If we consider the possibility of a difference in the volume occupied by protons and neutrons (see Sect. 14), the odd power terms in the power series expansion do not vanish, giving rise to an additional term

$$- 3 a_I \frac{\Delta r}{r} I. \qquad (15.4''')$$

$\Delta r/r$ indicating the fractional difference in radius of the volumes occupied by neutrons and protons. We shall consider the influence of this term in Sect. 16β.

The complete binding energy equation then becomes

$$E = a_v A - a_s A^{\frac{2}{3}} - a_c (A - I)^2 A^{-\frac{1}{3}} - a_I I^2 A^{-1} + \delta. \qquad (15.5)$$

The maximum binding energy for isobars $(\partial E/\partial I = 0)$ occurs for a neutron excess

$$I_0 = A \left/ \left(1 + \frac{a_I}{a_c} A^{-\frac{2}{3}}\right)\right. . \qquad (15.6)$$

Inserting this quantity in Eq. (15.5) we obtain

$$E = a_v A - a_s A^{\frac{2}{3}} - a_I I_0 - \tfrac{1}{2} b (I - I_0)^2 + \delta \qquad (15.7)$$

and

$$b = \frac{2}{A}(a_c A^{\frac{2}{3}} + a_I) = 2 a_c A^{\frac{2}{3}}/I_0. \qquad (15.8)$$

Using Eq. (10.3) we can convert (15.7) into a formula for nuclear masses. The neutron excess for maximum β stability then becomes

$$I_s = I_0 \{1 - (n - H)/4 a_c A^{\frac{2}{3}}\}. \qquad (15.9)$$

Often not the neutron excess but the charge number for maximum β stability $Z_s = (A - I_s)/2$ is given:

$$Z_s = \tfrac{1}{2} A \{a_I + \tfrac{1}{4}(n - H)\}/(a_I + a_c A^{\frac{2}{3}}). \qquad (15.10)$$

The following relation exists between these quantities:

$$I_s/Z_s = 2(-C + a_c/a_I \cdot A^{\frac{2}{3}})/(1 + C); \quad C = (n - H)/4 a_I. \qquad (15.11)$$

With help of the parameter Z_s the mass equation can be written

$$M = (M_n - a_v + a_I) A + a_s A^{\frac{2}{3}} - 2\{a_I + \tfrac{1}{4}(n - H)\} Z_s \{1 - (1 - Z/Z_s)^2\} - \delta. \qquad (15.12)$$

In this form the formula was tabulated by METROPOLIS and REITWIESNER [23]; another similar tabulation was made by MARTIN [24].

The above formulae (15.5) to (15.12) were derived in a somewhat different form by various authors from the statistical model of the nuclei. The constants computed from this model, however, did not agree with the empirical data. WEIZSÄCKER[1] therefore suggested a semi-empirical approach: The binding energy is assumed to have a form as indicated by the model, but the constants are left arbitrary and are determined from nuclear data. Later the formula was put in a somewhat simplified form, equivalent to Eq. (15.5), by BETHE and BACHER [12, Eq. (182)]; for these reasons the formula is mostly quoted as the BETHE-WEIZSÄCKER formula. The δ term, already considered by HEISENBERG[2] and BETHE and BACHER (§ 10 of [12]) in another context, was definitely added to this formula by BOHR and WHEELER[3]. Sometimes a term accounting for the compressibility of the nuclear "fluid" is added [15]. A somewhat different formula also used in comparisons with experimental data[4] was proposed by WIGNER [25].

16. Adjustment of constants in the BETHE-WEIZSÄCKER formula. α) *The line of maximum beta stability.* It is customary to adjust the constants in the BETHE-WEIZSÄCKER formula in such a way that (α) it fits the line of maximum β stability, and (β) it represents well the course of the nuclear binding energies along this line. The easiest way to use condition (α) is obtained from Eq. (15.11). In Fig. 8 we have plotted the β stable odd mass nuclides in a diagram of I/Z versus $A^{\frac{2}{3}}$. In this plot Eq. (15.11) represents a straight line with an inclination $2a_c/a_I(1+C)$ through a point $y=-C$ of the Y axis. The quantity C is small (~ 0.008, see below); and uncertainty in C therefore introduces only a small deviation in the value for a_c/a_I obtained in this way. We find

$$a_I/a_c = 130. \tag{16.1}$$

The line of maximum stability obtained in this way runs progressively high in the region of the highest mass numbers as compared with Eq. (15.11) ($A > 210$, see Fig. 8), which is the reason that mostly a larger value is given for a_I/a_c (see Table 5). As it appears that anyhow the binding energies in this region are not too well represented by the BETHE-WEIZSÄCKER formula (see Sect. 16β) we choose this value, which makes it possible to adjust better both the empirical lines of maximum stability and of the stability limits for lower mass numbers after adding magic number correction terms (see Fig. 4).

The ratio of the constants in the symmetry and the COULOMB term derived theoretically in the preceding section [Eqs. (15.3') and (15.4') is

$$a_I/a_c = 74/r_0 \quad (r_0 \text{ in } 10^{-13} \text{ cm}). \tag{16.2}$$

Comparing this result with Eq. (16.1) would yield $r_0 = 0.57 \cdot 10^{-13}$ cm, which is only half the value found experimentally (see Sect. 14). It will appear below that the COULOMB term is in a reasonable agreement with the radii of the charge distributions found experimentally. The result above indicates, therefore, that the empirical magnitude of the symmetry term is about twice the theoretical amount. According to e.g. WILDERMUTH[5] the remaining part should be due to the potential energy of the nuclear forces.

[1] C. F. v. WEIZSÄCKER: Z. Physik 96, 431 (1935).
[2] W. HEISENBERG: Z. Physik 78, 156 (1932).
[3] N. BOHR and J. A. WHEELER: Phys. Rev. 56, 426 (1939).
[4] T. L. COLLINS, A. O. NIER and W. H. JOHNSON jr.: Phys. Rev. 86, 408 (1952).
[5] K. WILDERMUTH: Z. Naturforsch. 9a, 1047 (1954).

Another interesting problem is that of the influence of a possible difference in the distribution of protons and neutrons in the nucleus, which would give the additional term (15.4‴) in the BETHE-WEIZSÄCKER formula. Then the constant C in the Eq. (15.11) for the line of maximum β stability should be replaced by

$$ C' = C - \frac{1}{2} \cdot \frac{3}{2} \cdot \frac{\Delta r}{r}. \tag{16.3} $$

The factor $\frac{1}{2}$ in this equation accounts in an approximate way for the fact that only half the empirical symmetry term is due to the kinetic energies considered in Sect. 15. An analysis of Fig. 8 shows that, Eq. (16.3) being valid, the difference in radius for proton and neutron distributions should in the average be smaller than 3%. Even taking into account that the influence of the term (15.4‴) may be partly compensated by a similar term in the potential

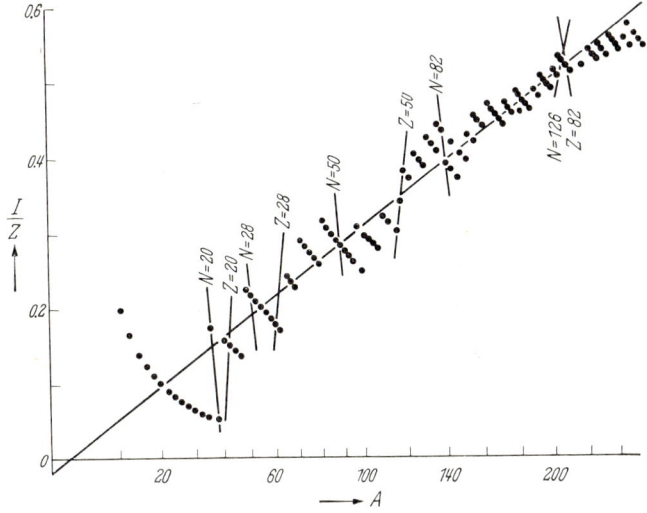

Fig. 8. Adjustment of the line of maximum β stability. This diagram shows the β stable odd mass nuclides in a plot of I/Z versus A in a scale in which $A^{\frac{2}{3}}$ is linear. Points for successive (odd) values of the neutron excess lie on curves rather similar to hyperbolae. The positions of magic numbers have been indicated.

energies, this result makes it quite improbable that there would be a real difference of 0.2×10^{-13} cm between the radius constants for the proton and neutron distributions, as was sometimes supposed (see Sect. 14).

β) *The binding energies.* In applying condition (β) (see Sect. 16α) it is a problem whether the constants in the BETHE-WEIZSÄCKER formula should be adjusted in such a way that (a) it represents the average course of the binding energies, or (b) it fits to the valleys between the magic peaks.

In our opinion (b) should be preferred, because 1° it offers the practical advantage that then the total course of the binding energies can be represented by formula (15.5) with the addition of some estimate for the magic peaks (compare Sect. 20), and 2° from the theoretical point of view it should be expected according to its derivation from the nuclear drop model that Eq. (15.5) would be more nearly valid far from magic numbers, than near magic numbers where the nuclear drop model is superseded by the independent particle model [3].

We therefore constructed a set of parameters according to condition (a) above, essentially by a trial and error procedure. The resulting binding energy formula becomes:

$$ \left. \begin{aligned} E &= 15.835\,A - 18.33\,A^{\frac{2}{3}} - 0.1785\,(A - I)^2\,A^{-\frac{1}{3}} - 23.20\,I^2\,A^{-1} + \delta \text{ MeV} \\ \delta &= \pm\, 11.2\,A^{-\frac{1}{2}}\,\text{Mev}, + \text{ for e-e nuclide}, - \text{ for o-o nuclides}; \end{aligned} \right\} \tag{16.4} $$

we here neglect δ_0.

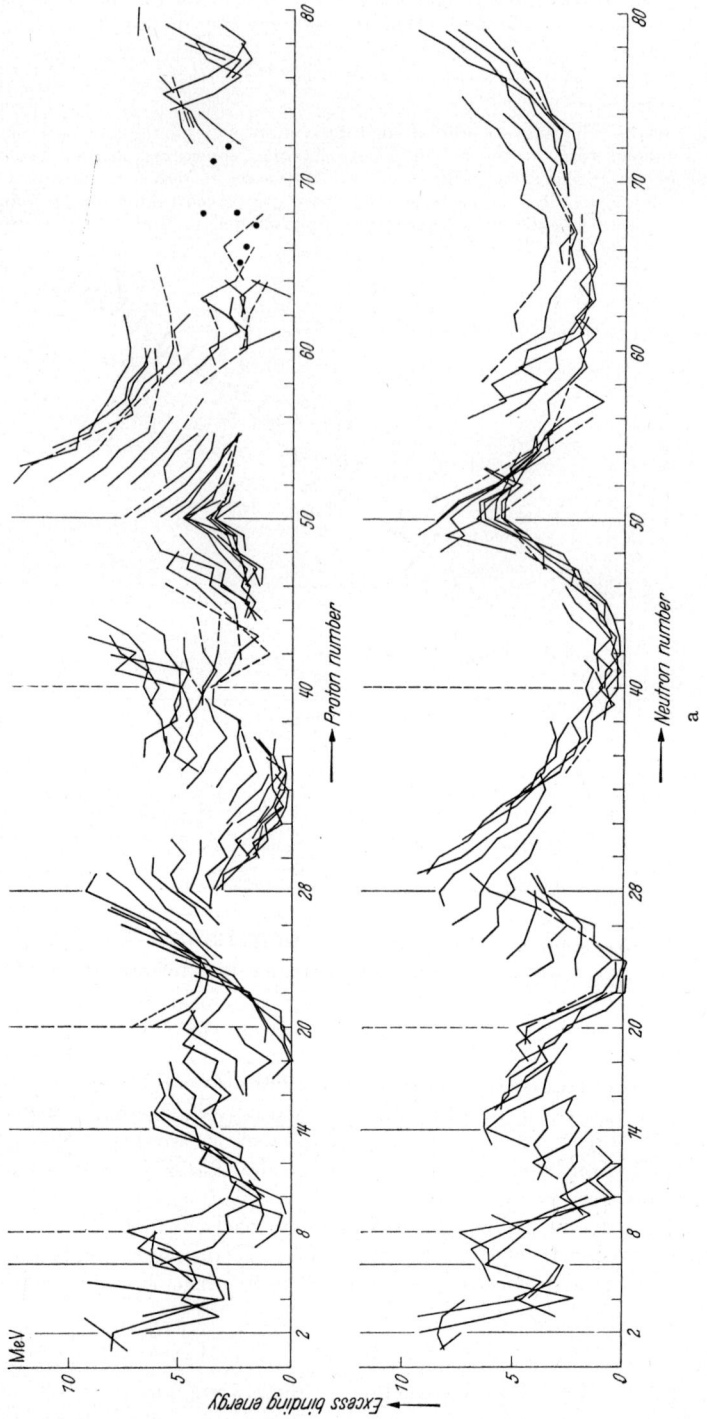

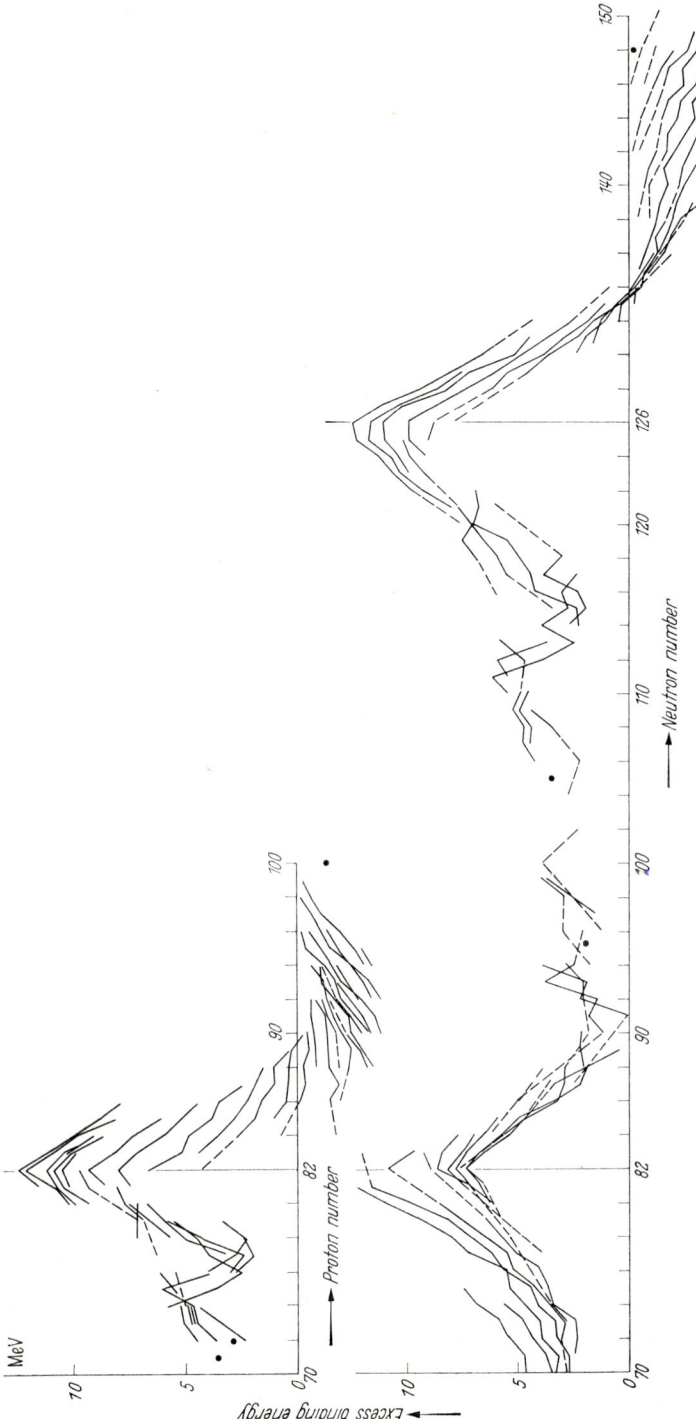

Fig. 9 a and b. Excesses of experimental binding energies over the BETHE-WEISZÄCKER formula (16.4). The upper parts show this quantity for isotones as a function of Z, the lower parts as a function of N for isotopes. If two points are joined by dotted lines, the intermediate point is unknown. The magic number series $\frac{1}{2}n(n^2-1)$ (L-S coupling) is indicated by dotted lines, the series $\frac{1}{2}n(n^2+5)$ (j-j coupling) by full lines.

Fig. 9 shows the differences between the experimental values for the binding energies (Table 1) and those computed from Eq. (16.4). In the upper parts this quantity is plotted as a function of Z; points belonging to isotones are joined. In this way the influences of magic proton and neutron numbers can be separated: the first determines the direction, the second the height of the lines. This makes it possible to obtain an estimate of the magic proton peaks by drawing a curve as well as possible parallel to all line fragments; the results will be discussed in Sect. 20. In the lower parts of Fig. 9 the roles of Z and N are interchanged.

For the purpose of comparisons like these it should be possible to use a simpler function in order to get rid of the main course of the binding energies. This point was stressed by GREEN and ENGLER [9] who propose the function:

$$M = \{A - 0.064 + 10^{-5}(A - 100)^2\} + \frac{0.025}{A} \{I - 0.4\,A^2/(A + 200)\}^2 \text{ MU}. \quad (16.5)$$

which indeed fits the main course of the binding energies quite well. One of their reasons to prefer this formula was, that the constants in the BETHE-WEIZSÄCKER formula (used as a semiempirical formula) cannot be said to represent real physical quantities. Nevertheless, GREEN and ENGLER have also given a readjustment of the constants in the BETHE-WEIZSÄCKER formula, and GREEN [11] has used his value for the COULOMB constant in order to derive a new value for the nuclear radius parameter r_0

$$r_0 = 1.23 \cdot 10^{-13} \text{ cm}$$

fitting well with the newest direct measurements (Sect. 14). It is interesting to note that this value, obtained by an adjustment using criterion (a) defined above, is so near to our value derived with condition (b)

$$r_0 = 1.21 \cdot 10^{-13} \text{ cm}.$$

Fig. 9 shows, that the binding energies are somewhat high in the region around $N = 100$. This effect was discovered by HOGG and DUCKWORTH[1] and discussed by them in terms of a mid-shell extra stability. The deviations are, however, of the order of the errors that we felt obliged to assign to the binding energies in this region on the basis of the consistency between different experimental data (see Table 1; [5]). Moreover the sudden drop in the binding energy excesses between mass numbers 187 and 188 ($N \approx 112$, see Fig. 9) suggest that this apparent extra-stability could perhaps be due to an accumulation of experimental errors. The progressive deviation for the highest mass numbers ($Z > 85$, $N > 135$) should, however, almost certainly be ascribed to a real failure of the BETHE-WEIZSÄCKER formula in this region.

Apart from the deviations discussed above and magic number effects, the BETHE-WEIZSÄCKER formula allows a rather nice fit to the line of maximum β stability and the course of the binding energies along this line. Binding energy differences, especially those of isobars, are represented worse: the parabolic constant (b) varies in a more complicated way (compare Fig. 5) than is predicted by this formula [Eq. (15.8)]. It is therefore not excluded that the BETHE-WEIZSÄCKER formula might be replaced by a better one.

Table 5 (taken from [9]) summarizes older values for the constants in the BETHE-WEIZSÄCKER formula. It can be seen that most earlier values are rather lower than our values (mentioned here in mass units). This difference is not only

---

1 B. J. HOGG and H. E. DUCKWORTH: Canad. J. Phys. **32**, 65 (1954).

caused by our using a new criterion (b) in adjusting the mass equation: the values given by GREEN and FOWLER using criterion (a) are also much higher than older values. This difference might be due to the fact that most earlier workers started to compute the COULOMB constant from the nuclear radius parameter, for which parameter they then used the old large value.

Table 5. *Constants in the* BETHE-WEIZSÄCKER *formula (in mMU) (taken from [9]).*

		a_v	a_s	a_I	a_c	a_I/a_c
BETHE . . .	(1936)	14.885	14.176	20.943	0.1558	134.4
FLÜGGE . .	(1942)	15.74	16.5	22.06	0.1618	136.3
FERMI[1] . .	(1945)	15.04	14.00	20.75	0.1568	132.5
FEENBERG .	(1947)	15.035	14.069	19.439	0.1568	124.0
PRYCE . . .	(1950)	15.089	15.035	21.050	0.1638	128.5
FOWLER . .	(1952)	16.432	17.989	24.218	0.1853	130.7
GREEN . . .	(1953)	16.918	19.120	25.445	0.1907	133.4
WAPSTRA . .	(1955)	17.006	19.685	24.915	0.1917	130.0

17. α-decay energy systematics. Regularities in the course of the α-decay energies were first noticed by HEISENBERG[2]; later on remarks about this subject were made by GLUECKAUF [26], PRYCE [27], PERLMAN, GHIORSO and SEABORG [28] and HAXEL, JENSEN and SUESS [20]. α-decay occurs between nuclides with the same even-odd characteristics and therefore is negligibly influenced by the spin terms. Application of Eq. (10.1) together with the empirical course of the parameter functions in this equation, but abstracting from magic number corrections, yields the following regularities:

(a) α-decay energies of isotopes increase with the mass (or neutron) number; those of isotones decrease with increasing A or Z.

(b) α-decay energies of nuclides near the line of maximum β stability increase with the mass number.

The last rule is the main reason that until rather recently α-decay was only found in nuclides with high mass numbers ($A > 210$).

Fig. 10a and b show the experimental α-decay energies of isotopes and isotones as a function of the number of neutrons and protons respectively. It is seen that rule (a) holds except very near the magic numbers $Z = 82$ and $N = 126$; rule (b), however, fits only rather far away from these magic numbers. This fact, together with rule (a) above, explains the sudden appearance of α-activity in nuclides with $Z > 82$.

In trying to explain the trends in the α-decay energies in terms of magic peaks, we have to consider again whether these peaks have a rounded or a pointed top (a or b in Fig. 3). In the last case the α-decay energies would be increased mostly for nuclides decaying to a magic number nuclide ($Z = 84$ or $N = 128$); if the top were rounded the increase could even be larger for higher values of Z or N. Similar arguments apply to the decreases in α-decay energy below the magic numbers. The evidence from Fig. 10a is not clear cut: $N > 127$ indicates a pointed shape, whereas $N < 127$ points to a more rounded peak (the last fact in agreement with the conclusion of PRYCE [27] drawn from the course of neutron binding energies in this region). Fig. 10b, too, does not give a quite unambiguous answer, though it could perhaps be explained best by a pointed top.

---

[1] Essentially this set of parameters was used in the tabulation of METROPOLIS and REITWIESNER [23] and MARTIN [24].

[2] W. HEISENBERG: Rapport VIIme Congrès Solvay 1934.

Another reversal in the normal trends of α-decay energies occurs in the region of the californium isotopes ($Z = 98$). The experimental data (see Fig. 10b) point into the direction of a slightly magic character for $N = 152^1$.

Due to the influence of magic number $N = 82$ α-decay occurs also in some neutron deficient nuclides with mass numbers around 150.

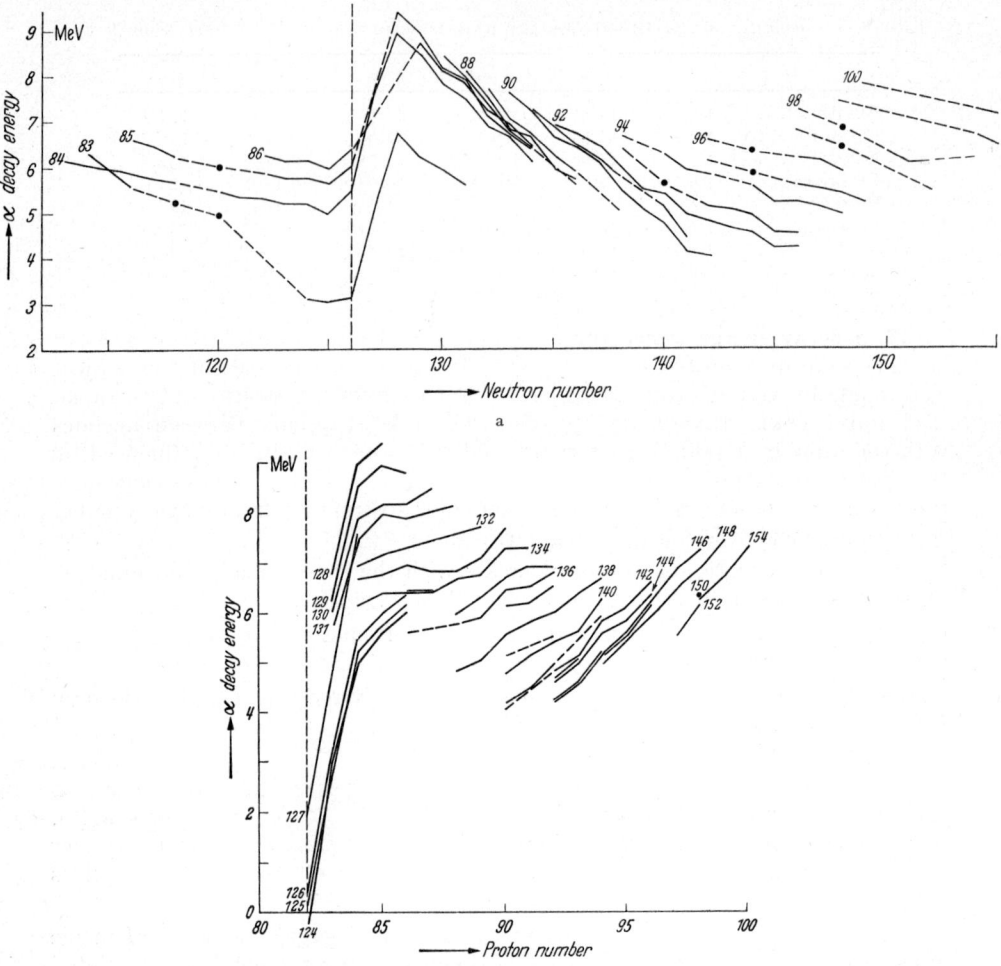

Fig. 10a and b. (a) α-decay energies of isotopes as a function of the number of neutrons. If two points are joined by a broken line, the α-decay energy of the intermediate isotope is unknown. (b) α-decay energies of isotones as a function of Z.

18. β-decay energy systematics. Results of β-decay energy systematics concerning the parameter functions in Eqs. (10.1) to (10.3) have already been described in Sect. 13. Suess and Jensen [19] were the first authors considering in detail the influence of magic number effects. Later important contributions to this subject were made by Coryell [17], Bouchez, Robert and Tobailem [18] and Way and Wood [21].

Fig. 11 shows a fragment of a plot of β-decay energies as suggested by SUESS and JENSEN [19]. The nature of the deviations at magic numbers ($N = 50$ in Fig. 11) depends again on the shape of the magic peak, in the same way as discussed in the preceding section. CORYELL [17] found that β-systematics indicate a sharp peak; Fig. 11 points in the same direction (as did the discussion in Sect. 13α).

The SUESS-JENSEN diagram Fig. 11 shows nicely that after passing the magic number $N = 50$ the decay energy of e-o nuclides is depressed as compared to that of o-e nuclides (compare Sect. 13β).

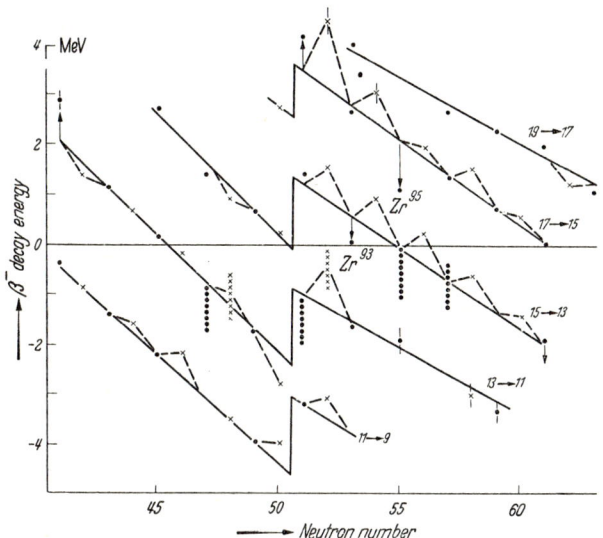

Fig. 11. β^- decay energies of odd mass isodiaspheres as a function of the number of neutrons in the "initial" nuclides; K and β^+ decays are represented as β^- decays with negative decay energies. Points indicate even-odd nuclides, crosses odd-even ones; 1 MeV high columns of points or crosses represent estimates of electron capture decay energies. The neutron excesses of "initial" and "final" nuclides are indicated.

Some points in Fig. 11 deviate rather much from the lines drawn through the points for isodiaspheres. In some cases this could be due to experimental errors; in other cases (e.g. Zr^{93} and Zr^{95} in Fig. 11) this is certainly not true. It is difficult to decide whether this particular deviation is due to a large fluctuation in the spin term (as assumed in drawing Fig. 7) or to a slightly magic character of the proton number 40 in these two isotopes (40 should be magic according to the L-S coupling model).

19. Nucleon binding energies. Remarks on the influence of magic numbers on nucleon binding energies were made by WAY[1], HARVEY [29], JENSEN, HAXEL and SUESS [20], FEATHER [16] and GREEN and EDWARDS [10]. Mostly neutron binding energies are considered, because they can be measured directly whereas proton binding energies can in most cases only be obtained by combining measured neutron binding energies with β-decay energies.

Fig. 12 shows the differences between the experimental neutron binding energies (obtained from Table 1) and the values computed from the BETHE-WEIZSÄCKER formula, Eq. (16.4), as a function of the number of neutrons in the nucleus. There is a considerable scatter in the position of these points in Fig. 12; this may partly be due to the fact that the BETHE-WEIZSÄCKER formula

---

[1] K. WAY: Phys. Rev. **75**, 1448 (1949).

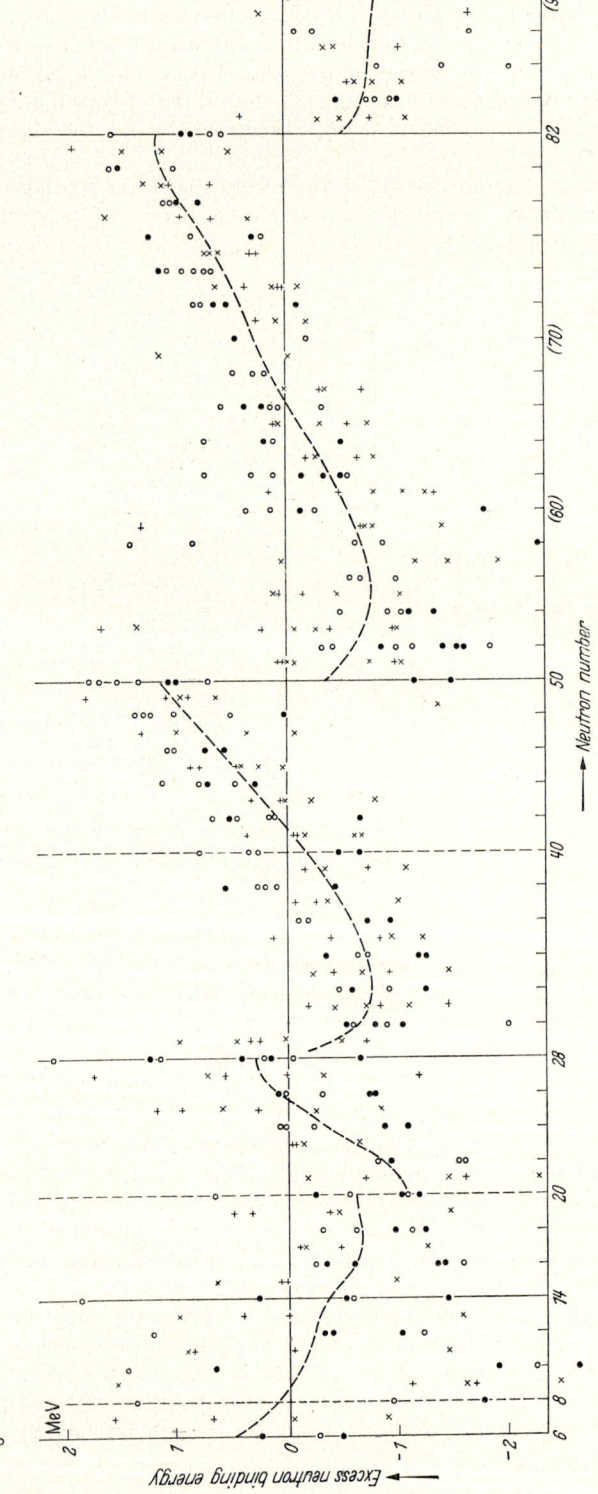

Fig. 12. Neutron binding energy excesses. This plot shows the differences between the experimental neutron binding energies (Table 1) and those computed from the semi-empirical formula (16.4). Open circles indicate e-e nuclides, points o-e ones; e-o and o-o nuclides are indicated by oblique and standing crosses respectively. The curve represents the average course of the neutron binding energies.

overestimates the parabolic constant, though rather probably the largest deviations are due to experimental errors. The scatter is especially large in the low mass region; it is, however, not to be expected that any statistical description could reasonably describe nuclei containing only a few nucleons. Nevertheless the influence of magic numbers 28, 50 and 82 can be seen clearly in Fig. 12.

A plot like Fig. 12 can be used to derive more information about the relative position of the four binding energy surfaces. Until now, we assumed silently that the e-e and o-o binding energies were situated symmetrically about the e-o and o-e surfaces. We will now drop this assumption and write (following GREEN and EDWARDS $[10]$):

$$\left.\begin{aligned} \text{o-o nuclides: } E &= f(A, I), \\ \text{o-e nuclides: } E &= f(A, I) + \nu, \\ \text{e-o nuclides: } E &= f(A, I) + \pi, \\ \text{e-e nuclides: } E &= f(A, I) + \tau. \end{aligned}\right\} \tag{19.1}$$

In this notation π and $(\tau - \nu)$ should be considered as the pairing energies of two protons in nuclei containing an odd or an even number of neutrons respectively; ν and $(\tau - \pi)$ are neutron pairing energies. The binding energies of the last neutrons in these different kinds of nuclei would be

$$\left.\begin{aligned} \text{e-e and e-o nuclides: } B_n &= g(A, I) \pm (\tau - \pi), \\ \text{o-e and o-o nuclides: } B_n &= g(A, I) \pm \nu. \end{aligned}\right\} \tag{19.2}$$

By drawing a line through the points for e-e nuclides in Fig. 12 one should therefore obtain a line $2(\tau - \pi)$ higher than the line through the o-o nuclides (corrected for the δ_e term already used in drawing Fig. 12). In a similar way the lines through the o-e and e-o nuclides would yield ν. This analysis would only be permitted if the consistency condition

$$(e\ e) - (o\ e) = (e\ o) - (o\ o) \quad (= (\tau - \pi) - \nu) \tag{19.3}$$

were found to apply [the symbols (e e) standing for the curves through the points in Fig. 12 for even-even nuclides, etc.].

A complete analysis along these lines has not yet been published. We have attempted to draw such curves in Fig. 12. The result was that the condition (19.3) applies pretty well. After passing magic numbers 28 and 50 ν appears to have a minimum, after magic number 20 a small maximum, in agreement with the current explanation of the behaviour of δ_o after shell crossings (cf. Sect. 13). The quantity defined in Eq. (19.3) (the difference in neutron pairing energy in nuclides with an even and an odd number of protons) appears to be slightly positive in the average [1] (of the order of a few hundred keV), meaning that the average of the e-e and o-o energy surfaces lies slightly higher than that of the odd mass nuclides.

20. Final comments about magic number effects. It is shown in the preceding sections that the main effect of the magic numbers is an excess of the nuclear binding energies in the shape of a peak approximately symmetrical about the magic number. Fig. 9 suggests that it may be possible to describe most binding energies with an accuracy of a few MeV by the BETHE-WEIZSÄCKER formula if some

---

[1] GREEN and EDWARDS $[10]$ report $\tau - \pi - \nu$ to be about 1 MeV for nuclides with mass numbers around 230; this result depends, however, rather critically on the relative adjustment of the binding energies of the different α radioactive series.

additional terms are added to account for magic number effects. Assuming Gaussian shapes for these terms, they are found to have a nearly uniform half-width so that they can be represented by the formula

$$E_m = A_m \exp - \left(\frac{Z - Z_m}{4}\right)^2$$

in which the height is found to be

4, 3 and 5 MeV for $Z = 28$, 50 and 82

5, 6, 7 and 8 MeV for $N = 28$, 50, 82 and 126 respectively.

These values represent only preliminary estimates. It should be kept in mind that according to most evidence (Sect. 13α, 17 and 18) the magic peaks should be more pointed than GAUSSIAN shapes. A correction term showing a pointed character proposed by GREEN and EDWARDS [Eq. (12.5)] has the disadvantage that it shows a strong discontinuity at the tops, which does not seem to agree with Fig. 9. An exponential correction term was proposed by STERN[1], correcting for the upper side of the magic peaks at $Z = 82$ and $N = 126$ as well as for the general deviation in this region; it does not seem possible to extend this suggestion into a reasonable magic number correction term. The conclusion must be that a completely satisfactory correction term has not yet been proposed. Perhaps this should wait until a corrected BETHE-WEIZSÄCKER formula will be available (cf. Sect. 16β). A better determination of magic number effects on the spin terms (cf. Sect. 13β and 19) could perhaps also be of some help.

General references.

[1] FEATHER, N.: Nuclear Stability Rules. Cambridge: The University Press 1952.
[2] HUNTLEY, H. E.: Nuclear Species. London: McMillan 1954.
[3] MAYER, M. G., and J. H. D. JENSEN: Elementary Theory of Nuclear Shell structure. New York: Wiley & Sons.
[4] WAPSTRA, A. H.: Physica, Haag 21, 367 (1955).
[5] WAPSTRA, A. H.: Physica, Haag 21, 385 (1955).
[6] HUIZENGA, J. R.: Physica, Haag 21, 410 (1955). — The above three papers together represent a complete recomputation of binding energies from experimental data.
[7] KOHMAN, T. P.: Phys. Rev. 85, 530 (1952). — A discussion of β stability as connected with the spin terms.
[8] WAPSTRA, A. H.: Physica, Haag 18, 83 (1952). — An attempt to explain the course of β stability by adding magic number terms to the BETHE-WEIZSÄCKER formula.
[9] GREEN, A. E. S., and N. A. ENGLER: Phys. Rev. 91, 40 (1953).
[10] GREEN, A. E. S., and D. F. EDWARDS: Phys. Rev. 91, 46 (1953).
[11] GREEN, A. E. S.: Phys. Rev. 95, 1006 (1954). — These three papers consider various aspects of the problem of total binding energies and neutron binding energies.
[12] BETHE, H. A., and R. F. BACHER: Rev. Mod. Phys. 8, 82 (1936). — This review of nuclear physics contains the first derivation of the BETHE-WEIZSÄCKER formula in the form in which it is mostly used.
[13] MATTAUCH, J., and S. FLÜGGE: Kernphysikalische Tabellen. Berlin: Springer 1942 (Nuclear Physics Tables. New York: Interscience Publ. Inc. 1946). — A table of nuclear properties with an introduction into the state of nuclear physics in 1942.
[14] JOLIOT-CURIE, I.: J. Phys. Radium 6, 209 (1945). — An earlier empirical evaluation of the parameter functions in the binding energy equation.
[15] FEENBERG, E.: Rev. Mod. Phys. 19, 239 (1947). — A discussion of the BETHE-WEIZSÄCKER formula in the light of empirics.
[16] FEATHER, N.: Proc. Roy. Soc. Edinburgh 63, 242 (1952). — Phil. Mag. Suppl. 2, 141 (1953). — Papers about various aspects of β decay and neutron binding energy systematics.
[17] CORYELL, C. D.: Ann. Rev. Nucl. Sci. 2, 305 (1953).

---

1 O. STERN: Rev. Mod. Phys. 21, 316 (1949).

[18] BOUCHEZ, R., J. ROBERT and J. TOBAILEM: J. Phys. Radium 14, 281 (1953). — Recent discussions of β systematics with empirical evaluations of the functions in the binding energy equation with a special view to computing β decay energies.

[19] SUESS, H. E., and J. H. D. JENSEN: Phys. Rev. 81, 1071 (1951). — Ark. Fys. 3, 577 (1951). — Papers about β systematics, giving the first detailed explanation of the absence of β stable isotopes of elements 43 and 61.

[20] HAXEL, O., J. H. D. JENSEN and H. E. SUESS: Ergebn. exakt. Naturw. 26, 244 (1952). — Discusses (among other things) magic number influences on α decay, β decay and neutron binding energy systematics.

[21] WAY, K., and M. WOOD: Phys. Rev. 94, 119 (1954). — A recent paper on β decay systematics.

[22] KUMAR, K., and M. A. PRESTON: Canad. J. Phys. 32, 298 (1955). — Gives an empirical derivation of the functions in the binding energy equation together with a discussion upon the validity of the BETHE-WEIZSÄCKER formula.

[23] METROPOLIS, N., and G. REITWIESNER: Table of Atomic Masses, Argonne National Laboratory.

[24] MARTIN, C. N.: Tables Numériques de Physique Nucléaire. Paris: Gauthier-Villars 1954. — Tables giving nuclear masses etc. computed from the BETHE-WEIZSÄCKER formula; both use, however, older values for the constants.

[25] WIGNER, E.: Phys. Rev. 51, 106, 947 (1937). — Derives a binding energy formula used sometimes instead of the BETHE-WEIZSÄCKER formula.

[26] GLUECKAUF, E.: Proc. Phys. Soc. Lond. 61, 25 (1948). — Discusses α decay energy systematics; describes the discovery of the odd-mass spin term.

[27] PRYCE, M. H. L.: Proc. Phys. Soc. Lond. A 63, 692 (1950). — A discussion of the bearing of α systematics on the magic number problem.

[28] PERLMAN, I., A. GHIORSO and G. T. SEABORG: Phys. Rev. 77, 26 (1950). — A standard paper on α desintegration systematics; for a more recent discussion see I. PERLMAN and F. ASARO: Ann. Rev. Nucl. Sci. 4, 157 (1954).

[29] HARVEY, J. A.: Phys. Rev. 81, 353 (1951). — One of the first papers on neutron binding energy systematics.

Determination of Atomic Masses by Microwave Methods.

By

S. GESCHWIND.

With 3 Figures.

1. Introduction. The development of microwave techniques in the last decade has resulted in another tool for the accurate determination of atomic masses. The characteristic rotational frequencies of molecules lie in the microwave region (electromagnetic spectrum from about 1 to 300 mm in wavelength) and comprise the bulk of the microwave spectra of gases. By measuring the frequency shift which occurs in the pure rotational absorption spectrum of a molecule when another isotope is substituted for one of the atoms, accurate information about the relative masses of the isotopes can be obtained.

Prior to the advent of microwave spectroscopy, the rotational frequencies of molecules were mainly studied as a fine structure superimposed upon their vibrational spectra which lie in the infrared region [5]. In 1932, HARDY, BARKER and DENNISON[1] determined the mass of deuterium to 0.1 mMU by comparing the spacings of the rotational lines in the DCl vibrational spectrum with those of HCl. Since the separation of the rotational lines of a diatomic molecule is inversely proportional to its reduced mass, which for DCl is about twice that of HCl, the isotope shift in this case was quite large. For heavier atoms, however, even the most favorable cases give isotopic shifts which are considerably less pronounced. In ICl, for example, the replacement of Cl^{35} by Cl^{37} shifts the rotational frequency by only 5%. With the limited resolution of infrared techniques, a measurement of this shift sufficiently accurate for a mass determination is very difficult if not impossible.

However, since pure rotational absorption spectra (many of which lie in the region near 1 cm^{-1}) can now be directly observed in the microwave region where high resolution resulting from very monochromatic oscillator sources is coupled with accuracy of frequency measurement, accurate mass determinations by isotope shifts become practical. For example, the aforementioned shift of 5% in ICl corresponds to a displacement[2] of 1000 Mc of the ICl^{35} line from ICl^{37} in the region of 24000 Mc (0.8 cm^{-1}). Frequency separations and absolute frequencies in this region can be measured to an accuracy of at least 0.01 Mc. This mass change of two units in Cl can therefore be determined to 0.01 Mc/1000 Mc or 1 part in 10^5, or the Cl^{37} mass determined to 0.02 thousandths of a mass unit (0.02 mMU) relative to Cl^{35}.

The most general variety of molecule that has been used in the determination of atomic masses from microwave spectra has been the symmetric top whose energy levels are given by

$$W_r = h\, B\, J(J+1) + h\,(A-B)\, K^2. \tag{1.1}$$

---

[1] HARDY, BARKER and DENNISON: Phys. Rev. **42**, 279 (1932).
[2] Frequencies are measured in Mc, i.e. in megacycles per second. 10^{-3} Mc $= 1$ kc.

Here $B = h/8\pi^2 I_b$ and $A = h/8\pi^2 I_a$ where I_a and I_b are the moments of inertia parallel and perpendicular respectively to the symmetry axis. $\frac{h}{2\pi}\sqrt{J(J+1)}$ is the total angular momentum of the molecule and $K\frac{h}{2\pi}$ its projection on the symmetry axis. If the molecule possesses a permanent electric dipole moment it can absorb a quantum of microwave energy $h\nu$ and make a transition to a higher rotational state governed by the selection rules $\Delta J = +1, \Delta K = 0$. The pure rotational absorption frequency is then given by

$$\nu = 2BJ \qquad (1.2)$$

where J is the angular momentum quantum number for the upper state of the transition. If an isotopic substitution is made for an atom lying along the symmetry axis of the molecule, thus changing its moment of inertia and hence its B value, the rotational absorption frequency will shift. The program of mass determination from microwave spectra is to correlate this frequency shift with the change in mass of the substituted isotope. These shifts can range from a few Mc to several thousands of Mc for a change of one mass unit, being larger generally for the lighter atoms. Typical isotopic shifts in the middle mass region ($A = 60$ to 130) are 100 Mc per mass unit. As mentioned above, frequency shifts in the 24000 Mc region can be measured to at least an accuracy of 0.01 Mc, so that this change of one mass unit can easily be determined to 1 part in 10^4 or to 0.1 mMU. Current improvements in techniques allow an increase in accuracy of frequency measurement to 0.002 Mc, so that for these typical shifts relative masses can now be determined to an accuracy of 0.02 mMU. This accuracy compares favorably with recent mass spectrographic results in this same middle mass region[1].

The electric quadrupole moments, and to a much smaller extent the magnetic moments of many nuclei, interact with the electric and magnetic fields in a molecule so as to cause a splitting of the rotational energy levels into a number corresponding to the quantum mechanically allowed orientations of the nuclear spin $\boldsymbol{I}$ with respect to the molecular rotation $\boldsymbol{J}$. Calling this interaction energy $W_i(I, J, K, F)$, where $F = |\boldsymbol{I}+\boldsymbol{J}|$, the split levels are[2]

$$W = W_r + W_i (I, J, K, F) \qquad (1.3)$$

giving rise to several closely spaced frequencies given by

$$\nu = 2BJ + \left[\frac{W_i(I, J, K, F') - W_i(I, J - 1, K, F)}{h} \right] \qquad (1.4)$$

where $F' - F = 0, \pm 1$.

Generally $W_i \ll W_r$ and does not depend on B, but in some cases W_i is sufficiently large to perturb the rotation and hence depends upon B. However, even in these cases the theory of this interaction has been worked out precisely enough[3] so that an examination of the spectra allows an accurate extrapolation to the rotational frequency as it would be in the absence of hyperfine splitting. In all that follows it will be assumed that no hyperfine structure occurs, or that it has been corrected for, so that only the rotational energy of the molecule need be considered.

[1] COLLINS, NIER and JOHNSON: Phys. Rev. **94**, 398 (1954).
[2] J. BARDEEN and C. H. TOWNES: Phys. Rev. **73**, 97 (1948).
[3] J. BARDEEN and C. H. TOWNES: Phys. Rev. **73**, 627 (1948).

2. Theory of mass determination from rotational spectra of diatomic molecules.
The correlation between the shift in the rotational frequency of a molecule with
the isotopic change of mass of one of the atoms is most easily made in the case
of a diatomic molecule. The pure rotational absorption frequencies of a diatomic
polar gas are given to a sufficient approximation by

$$\nu = 2J[B_e - \alpha(v + \tfrac{1}{2})] \tag{2.1}$$

where

$$B_e = h/(8\pi^2 I_e),$$

and I_e is the equilibrium moment of inertia assuming the nuclei are stationary
at their equilibrium separation r_e. $I_e = \mu r_e^2$, with μ the reduced mass of the
molecule. α gives a rotation-vibration correction representing the change in
effective value of B due to vibration of the molecule. v is the vibrational quantum
number.

By measurement of the frequency separation between two absorption lines
having the same rotational transition but different vibrational states (for example
the ground and first excited vibrational states), α can be determined and B_e
evaluated. If the excited vibrational state of one of the isotopic species is insufficiently populated so that the transition is too weak to be observed, α for
this species can be calculated if it is known for the more abundant species since
$\alpha \sim \mu^{-\frac{3}{2}}$. With α calculated in this way B_e can be determined for this isotopic
species by measurement of its rotational frequency in the ground vibrational
state.

The equilibrium internuclear distance r_e is determined to a very high approximation only by the electronic structure of the molecule so that it is the same
for different isotopic species. Thus by measuring the B_e's corresponding to two
different isotopic forms of the molecule, say with masses M and $m^{(0)}$ and M
and $m^{(1)}$, respectively, the ratio of the reduced masses can be expressed directly
in terms of the B_e's:

$$\frac{B_e^{(0)}}{B_e^{(1)}} = \frac{\mu^{(1)}}{\mu^{(0)}} = \frac{m^{(1)}(M + m^{(0)})}{m^{(0)}(M + m^{(1)})} \tag{2.2}$$

or

$$\frac{m^{(1)}}{m^{(0)}} = \frac{\dfrac{M}{m^{(0)}}\dfrac{B_e^{(0)}}{B_e^{(1)}}}{1 + \dfrac{M}{m^{(0)}} - \dfrac{B_e^{(0)}}{B_e^{(1)}}}. \tag{2.3}$$

In expressing the reduced masses as in Eq. (2.2), it is implied that the mass
of the entire neutral atom, i.e. nucleus plus electrons, is concentrated at a point.
It will be shown below that this assumption introduces an even smaller error
in the mass ratio than one might first expect. By inserting the measured ratios
of the B_e's in Eq. (2.3), the mass ratio $m^{(1)}/m^{(0)}$ can be very accurately determined.

$M/m^{(0)}$ need not be known to especially high accuracy in order to make its
contribution to the uncertainty in $m^{(1)}/m^{(0)}$ negligible. If the uncertainty in
$M/m^{(0)}$ is Δ, then the corresponding uncertainty in $m^{(1)}/m^{(0)}$ is

$$\approx \frac{m^{(1)}\Delta}{M}\left[\frac{m^{(0)} - m^{(1)}}{m^{(0)} + M}\right],$$

assuming $m^{(0)} - m^{(1)}$ is small. For example, in ICl, an uncertainty of 1 mMU
in both M and $m^{(0)}$ produces an error in the Cl^{35}/Cl^{37} ratio of only 1 part in 10^7.
Note that no information about internuclear distances or atomic constants such

as h are needed. With this method the mass ratio Cl^{35}/Cl^{37} has been determined to a few parts in a million for several diatomic molecules. These results will be listed and discussed in Sect. 6.

3. Approximations contributing to error in mass ratio determinations from spectra of diatomic molecules. There are several possible sources of error in determining mass ratios from $(2.3)^1$ (other than simply inaccurate measurements in B_e). They are associated mainly with inaccuracies in the approximations used to obtain this equation. As the discussion which follows will indicate, they generally offer no serious limitation to the accuracy of mass measurements as they can usually be properly corrected for.

α) *Anharmonicity of the potential function.* Eq. (2.1) is a first approximation for the effective rotational constant of a diatomic molecule, although usually a very satisfactory one. The more complete Bohr theory expansion of the energy levels of a rotating vibrator appearing in band spectra analysis is given e.g. by [5]:

$$\left.\begin{aligned} W_{v,J}/h = {} & \omega_e\left(v+\tfrac{1}{2}\right) - \omega_e\,x_e\left(v+\tfrac{1}{2}\right)^2 + \omega_e\,y_e\left(v+\tfrac{1}{2}\right)^3 + \cdots \\ & + \left[B_e - \alpha_e\left(v+\tfrac{1}{2}\right) + \gamma_e\left(v+\tfrac{1}{2}\right)^2 + \cdots\right] J(J+1) \\ & - \left[D_e + \beta_e\left(v+\tfrac{1}{2}\right) + \right] J^2(J+1)^2 + \cdots \end{aligned}\right\} \tag{3.1}$$

$D_e = 4B_e^3/\omega_e^2$, and is the term which takes into account the change in moment of inertia due to the centrifugal distortion of the rotating molecule, and ω_e is the vibrational frequency. However, even this expansion is based upon a restricted potential function which does not take into complete account the anharmonicity of the potential, i.e. not all the coefficients are independent.

Dunham^2 has calculated the energy levels of a vibrating rotator by a W.K.B. method for any potential which can be expanded as a power series in $(r - r_e)$ in the neighborhood of the potential minimum of the form

$$V = \sum_{i=2}^{\infty} a_i \left(\frac{r - r_e}{r_e}\right)^i .$$

He shows that in this case the energy levels can be expressed by

$$W_{v,J}/h = \sum_{l,j} Y_{lj}\left(v+\tfrac{1}{2}\right)^l J^j (J+1)^j \tag{3.2}$$

where l and j are summation indices and the Y_{lj}'s coefficients which depend on molecular constants. If the ratio B_e/ω_e is small so that terms of order $(B_e/\omega_e)^2$ can be neglected, the following identification can be made between the Y_{lj}'s and the ordinary band spectrum constants.

$$Y_{01} \sim B_e, \qquad Y_{11} \sim -\alpha_e, \qquad Y_{10} \sim \omega_e,$$
$$Y_{02} \sim -D_e, \qquad Y_{20} \sim -\omega_e\,x_e, \qquad Y_{21} \sim \gamma_e.$$

Actually

$$Y_{01} = B_e\left[1 + \frac{(B_e)^2}{(\omega_e)^2}\beta_{01}\right] \tag{3.3}$$

where

$$\beta_{01} = \frac{Y_{10}^2\,Y_{21}}{4Y_{01}^3} + \frac{16a_1 Y_{20}}{3Y_{01}} - 8a_1 - 6a_1^2 + 4a_1^3$$

^{1} J. H. van Vleck: J. Chem. Phys. **4**, 327 (1936).
^{2} J. L. Dunham: Phys. Rev. **41**, 721 (1932).

and

$$a_1 = \frac{Y_{11} Y_{10}}{6 Y_{01}^2} - 1.$$

The ratio of reduced masses for two isotopic species is then more correctly given by

$$\frac{\mu^{(1)}}{\mu^{(0)}} = \frac{B_e^{(0)}}{B_e^{(1)}} = \frac{Y_{01}^{(0)}}{Y_{01}^{(1)}} \left[1 + \beta_{01} \left(\frac{B_e}{\omega_e}\right)^2 \left(\frac{\mu^{(1)}}{\mu^{(0)}} - 1\right)\right]^{-1} \tag{3.4}$$

Neglect of terms of order $(B_e/\omega_e)^2$ in ICl and KCl[1] gives an error of 5×10^{-7} in the Cl^{35}/Cl^{37} mass ratio and in RbCl an error of 1×10^{-7} in the Rb^{85}/Rb^{87} mass ratio[2]. In these cases these errors are smaller than the experimental errors which were several parts in 10^6 (see Table 1). In lighter molecules however, this correction may be larger and should be taken into account. When this correction is important and a sufficient number of the Y's can not be obtained from the microwave spectrum, the constants ω_e, γ_e, and $\omega_e x_e$ can then usually be obtained with sufficient accuracy from optical band spectra to determine β_{01} for substitution into Eq. (3.4).

β) *Contribution of electrons to moment of inertia.* In calculating the moment of inertia of the molecule, it was assumed that the mass of an entire neutral atom, nucleus plus electrons is concentrated at a point. However, the electrons obviously have an extension in space about the nuclei, so that the contribution to the molecular moment of inertia due to the spatial distribution and motion of the electrons must be considered.

If one imagines the electrons to be more or less spherically distributed about their nuclei and rotating rigidly with the molecule, one might expect the moment of inertia to be greater than that given by the point mass assumption by approximately an amount equal to the moment of inertia of the electrons about their respective nuclei. This contribution to the moment of inertia would be large but is fortunately really not present because the orientation in space of a completely spherical shell of electrons remain fixed as the molecule rotates[3]. The slipping of closed shell electrons as the molecule rotates can be compared to the motion of a chair on a ferris wheel.

Valence electrons, on the other hand, are not spherically distributed and so do not slip and give an extra contribution, above the point mass assumption, to the moment of inertia of $n m \overline{r_e^2}$. Here n is the number of valence electrons, m the electron mass and $\overline{r_e^2}$ the average, taken over all the valence electrons, of the square of their distance from the nucleus with which they are associated. If μ_{red} is the reduced mass of the molecule and r the internuclear distance, then the fractional correction to the moment of inertia I_{atom} calculated using atomic masses would be

$$\frac{\Delta I}{I_{\text{atom}}} = \frac{n \, m \, \overline{r_e^2}}{\mu_{\text{red}} \, r^2}. \tag{3.5}$$

This correction is generally negligible for heavier atoms but becomes important in lighter ones. For example, if one assumes $n = 1$ and $r_e \approx r$ then the uncertainty in electron behavior would introduce an error of 8×10^{-7} in the Cl^{35}/Cl^{37} mass ratio in ICl. On the other hand, under these same assumptions in lighter molecules like CO, it would introduce an error of 1.0×10^{-5} in the C^{12}/C^{13} ratio, which is two orders of magnitude larger than the experimental error.

[1] Lee, Fabricand and Carlson: Phys. Rev. **91**, 1395 (1953).
[2] J. W. Trischka and R. Braunsteen: Phys. Rev. **96**, 968 (1954).
[3] G. C. Wick: Phys. Rev. **73**, 51 (1948).

However, there is a direct connection between the contribution of the electrons to the moment of inertia of a $^1\Sigma$ molecule (ground state with zero electronic angular momentum) and their contribution to the molecular magnetic moment, such that a measurement of the magnetic moment of the molecule yields all the necessary information to very accurately correct for the electronic contribution to the moment of inertia. This relation is based essentially on nothing more than the gyromagnetic ratio of rotating charges. Proceeding with the above model, the closed shell electrons which slip as the molecule rotates will cancel an equal amount of nuclear charge as far as the production of magnetic moment is concerned. The resultant magnetic moment of the rotating molecule will be due to the nuclear charge not cancelled by slipping electrons plus an equal amount of valence electrons rotating with the molecule and is ,

$$\mu_J = \frac{-n\,e}{2c}\,\omega\,\overline{r_e^2} = \frac{-n\,e}{2c}\,\frac{J\hbar}{\mu_{\mathrm{red}}}\,\frac{\overline{r_e^2}}{r^2} = \frac{-n\,\mu_n J}{\mu_{\mathrm{red}}}\,\frac{\overline{r_e^2}}{r^2}\,M_p \qquad (3.6)$$

where μ_n is the nuclear magneton, $J\hbar$ the angular momentum of the molecule, M_p the mass of the proton and the other symbols as in (3.5). Note that with this picture the sign of the magnetic moment will generally be that given by rotating negative charges.

Combining (3.5) and (3.6) we have

$$\frac{\Delta I}{I_{\mathrm{atom}}} = \frac{-m}{M_p}\,\frac{\mu_J}{\mu_n}\,\frac{1}{J}\,. \qquad (3.7)$$

This basic expression giving the correction to the moment of inertia of a diatomic molecule as calculated with point atomic masses in terms of the magnetic moment of the molecule has been derived more rigorously by several authors[1, 2, 3].

A simple second order (in energy) perturbation calculation shows that the contribution of the electrons to the rotational energy of a linear molecule is given by

$$W_r^e = \frac{\hbar^2 J(J+1)}{I_N^2}\sum_n \frac{\hbar^2|(0|L_x|n)|^2}{W_0 - W_n} \qquad (3.8)$$

where W_0 is the energy of the unperturbed electronic state and W_n the energy of the n-th excited state. L_x is one component of the orbital electronic angular momentum (in units of $\hbar$) perpendicular to the molecular axis, and I_n the moment of inertia of the nuclear frame. The total rotational energy of nuclei plus electrons is then

$$W_r = \frac{\hbar^2 J(J+1)}{2 I_N} + \frac{\hbar^2 J(J+1)}{I_N^2}\sum_n \frac{\hbar^2|(0|L_x|n)|^2}{W_0 - W_n} \qquad (3.9)$$

and the effective moment of inertia follows from[4]

$$\frac{1}{I_{\mathrm{eff}}} = \frac{1}{I_N} + \frac{2\hbar^2}{I_N^2}\sum \frac{|(0|L_x|n)|^2}{W_0 - W_n}\,. \qquad (3.10)$$

---

[1] H. R. JOHNSON and M. W. P. STRANDBERG: J. Chem. Phys. **20**, 687 (1952).

[2] M. W. P. STRANDBERG [2], Chap. VI and VII.

[3] TOWNES and SCHAWLOW [1], Chap. VIII.

[4] The slippage of closed shell electrons is illustrated by evaluation of $\dfrac{|(0|L_x|n)|^2}{W_0 - W_n}$ for electrons spherically distributed about nuclei, [1], p. 213, which shows that this second term adds to $1/I_N$ as if the electrons were indeed concentrated at these nuclei.

As expected, the effect of the electrons is to increase I_{eff} since the second term is always negative.

The same calculation including an external magnetic field gives for the rotationally induced electronic contribution to the magnetic moment of the molecule

$$\mu^e = \frac{2J\hbar^2}{I_N}\mu_0 \sum_n \frac{|(0|L_x|n)|^2}{W_0 - W_n}. \tag{3.11}$$

The magnetic moment of the rotating nuclear frame is given by

$$\mu^N = \frac{e}{2c}\sum_k Z_k r_k^2 \omega \tag{3.12}$$

where ω is the angular velocity of the nuclear frame, Z_k is the number of protons in the k-th nucleus, r_k is the distance of the nucleus to the center of gravity of the molecule, and the summation is over all the nuclei. Multiplying numerator and denominator of (3.12) by the mass of the electron and substituting $\omega \approx \hbar J/I_N$ we have

$$\mu^N = +\frac{e}{2mc}\sum_k Z_k m r_k^2 \frac{\hbar J}{I_N} = +\mu_0 \frac{J(I^e)_{\text{conc}}}{I_N} \tag{3.13}$$

where $(I^e)_{\text{conc.}}$ is the moment of inertia the electrons would have if they were concentrated at their nuclei so as to make them electrically neutral. μ_0 is the Bohr magneton. The total molecular magnetic moment is therefore given by

$$\mu_J = \mu^e + \mu^N = \frac{2J\hbar^2}{I_N}\mu_0 \sum_n \frac{|(0|L_x|n)|^2}{W_0 - W_n} + \mu_0 \frac{J(I^e)_{\text{conc}}}{I_N}. \tag{3.14}$$

Eliminating the matrix elements of L_x between (3.10) and (3.14) and using the fact that $(I^e)_{\text{conc}} \ll I_N$,

$$\left(\frac{1}{I}\right)_{\text{eff}} = \frac{1}{I_N + (I^e)_{\text{conc}}} + \frac{\mu_J}{\mu_0 I J}. \tag{3.15}$$

where $I_N + (I^e)_{\text{conc}} = I_{\text{atom}}$, i.e. the moment of inertia calculated with point atomic masses. If (3.15) is rewritten using μ_n, the nuclear magneton instead of μ_0 a result essentially identical to (3.7) is obtained

$$\left(\frac{1}{I}\right)_{\text{eff}} = \left(\frac{1}{I}\right)_{\text{atom}} + \frac{m}{M_p}\frac{\mu_J}{\mu_n I J}. \tag{3.16}$$

It is interesting to note that (3.16) even applies to an asymmetric rotor where it is to be regarded as a component equation, i.e. $1/I$ would be the reciprocal moment of inertia in the direction of one of the principal axes and μ the component of the molecular magnetic moment along this axis (cf. [2], Chap. VII). This correction is however, relatively less important in mass determinations in molecules other than diatomic where it is masked by errors due to zero point vibrations.

The first application of this electronic correction in terms of the molecular magnetic moment has been made by Rosenblum and Nethercot in the determination of C^{12}/C^{13} and O^{18}/O^{16} mass ratios from the $J=0\rightarrow1$ transition in CO at 114000 Mc. They measured the Zeeman splitting of this line from which they found $\mu_J = -0.269\,\mu_n$. Insertion of this value of μ_J into (3.16) gives a fractional correction to the B value of 1 part in 6800 and a correction of 12.4 and 18.6 parts per million in the C^{13}/C^{12} and O^{18}/O^{16} mass ratios, respectively. When this correction is applied with the "wobble stretching correction"[1] their result is found to be in excellent agreement with the results from (d, p) nuclear reaction data (see Table 1 and discussion in Sect. 6 under carbon and sulphur).

4. Polyatomic linear and symmetric top molecules. α) *Mass-difference ratios.* The one structural parameter of a diatomic molecule, i.e. its internuclear distance, does not appear explicitly in the expression for isotopic mass ratio (2.3). One may regard it as having been determined and eliminated from (2.3) by measurement of the B_e value of the first isotopic species. Similarly consider a polyatomic molecule with n structural parameters, i.e. internuclear distances and bond angles, and suppose the mass ratio $m_j^{(1)}/m_j^{(0)}$ of two isotopes of an atom in the position labeled j is to be determined. In principle, the B_e's corresponding to n isotopic forms of the molecule all with atom $m_j^{(0)}$ would determine the structural parameters so that a further measurement of B_e for an isotopic species with $m_j^{(1)}$ would determine the ratio $m_j^{(1)}/m_j^{(0)}$ in terms of the B_e's and other masses. However, in a molecule more complex than a diatomic molecule, zero point vibrations prevent the determination of the equilibrium B values.

As an obvious extension of vibrational effects in a diatomic molecule, the effective B value, B_0, of polyatomic linear molecules and symmetric tops may be written [6] as

$$B_0 = B_e - \sum_i \alpha_i \left(v_i + \frac{d_i}{2} \right) \tag{4.1}$$

where α_i, v_i and d_i are the rotation-vibration constant, the vibrational quantum number and the degree of degeneracy respectively of the i-th vibrational mode.

To determine B_e the α's corresponding to all the vibrational modes of the molecule would have to be determined by observing the same rotational transition in the ground and first excited state of each mode of vibration. However, this involves severe experimental difficulties since some of these vibrational frequencies are sufficiently high so that only a small fraction of the molecules is in the excited vibrational state at room temperatures. The corresponding absorption line is then too weak to be detected by present day microwave spectrographs. To date, the α's of a polyatomic molecule have been measured for only one isotopic species of OCS whose vibrational modes are shown in Fig. 1[2]. In addition, often more than one excited state of a given mode of vibration is needed in order to correct for FERMI resonance[3]. Moreover, in a polyatomic molecule the α's depend in a rather complex way upon anharmonic potential constants and mass so that

---

[1] ROSENBLUM, NETHERCOT and TOWNES (to be published in Phys. Rev.) have shown that the precession of electronic angular momentum about the internuclear axis in $^1\Sigma$ molecules, which gives rise to a "wobble" of the internuclear axis, causes a centrifugal stretching which changes the effective moment of inertia of the molecule and hence the rotational energy, in analogous fashion to the more familiar centrifugal distortion associated with rotation of the nuclei. This energy is given by

$$\frac{\hbar^2}{4} J(J+1) \sum_n \left(0 \left| \frac{1}{I_N} \right| n \right) \left(n \left| \frac{L^2}{I_N} \right| 0 \right) \Big/ (W_0 - W_n)$$

where the symbols are the same as above and the summation is over all vibrational states in all $^1\Sigma$ states, except the ground vibrational state in the ground $^1\Sigma$-state. By comparing the value of C^{14}/C^{12} obtained from the microwave CO spectrum with the nuclear reaction value which is assumed to be correct, this correction for "wobble stretching" in CO is evaluated and amounts to 1.6×10^{-6} and 2.4×10^{-6} in the C^{12}/C^{13} and O^{18}/O^{16} mass ratios respectively.

[2] M. PETER and M. W. P. STRANDBERG: Phys. Rev. 95, 622 (1954).

[3] W. LOW and C. H. TOWNES: Phys. Rev. 79, 224 (1950).

even knowing their value for one isotopic species, one cannot predict from theory their values for another isotopic form.

Nevertheless, while zero-point vibrations prevent one from determining the B_e's and hence mass ratios in polyatomic molecules, accurate mass information can still be obtained in terms of *mass difference ratios*. Moreover in a symmetric top, the mass difference ratio of three isotopes can be determined by the measurement of only three rotational frequencies regardless of the number of structural parameters of the molecule.

The moment of inertia through the center of mass of a molecule and along a direction fixed in the molecule and defined as the Z axis is

$$I = \sum_i m_i (x_i^2 + y_i^2) \tag{4.2}$$

where m_i is the mass of the i-th atom and $(x_i^2 + y_i^2)$ is the square of its distance from the z-axis. If the j-th atom with mass $m_j^{(0)}$ is replaced by another isotope of mass $m_j^{(1)}$, the x-y component of the shift of the center of mass of the molecule will be

$$\delta = \frac{m_j^{(1)} - m_j^{(0)}}{M^{(1)}} (x_j^2 + y_j^2)^{\frac{1}{2}} \tag{4.3}$$

where $M^{(1)}$ is the total mass of the molecule with mass $m_j^{(1)}$. Calling $I^{(1)}$ the new moment of inertia through the new center of mass about an axis parallel to the z-axis and $I^{(0)}$ the former moment of inertia, then by the parallel axis theorem

$$I^{(1)} + M^{(1)} \delta^2 = I^{(0)} + (m_j^{(1)} - m_j^{(0)}) (x_j^2 + y_j^2). \tag{4.4}$$

Substituting (4.3) into (4.4) gives

$$I^{(1)} - I^{(0)} = \frac{(m_j^{(1)} - m_j^{(0)}) M^{(0)} (x_j^2 + y_j^2)}{M^{(1)}} \tag{4.5}$$

where $M^{(0)}$ is the total mass of the molecule with atom $m_j^{(0)}$. An equation similar to (4.5) applies to isotope $m_j^{(2)}$ so that

$$\frac{m^{(1)} - m^{(0)}}{m^{(2)} - m^{(0)}} = \frac{M^{(1)}}{M^{(2)}} \frac{I^{(1)}}{I^{(2)}} \frac{\frac{1}{I^{(0)}} - \frac{1}{I^{(1)}}}{\frac{1}{I^{(0)}} - \frac{1}{I^{(2)}}}. \tag{4.6}$$

This type of expression was given by Strandberg, Wentink and Hill[1] for an isotopic substitution of an end atom in a linear XYZ type molecule. However, it is quite general and might even be applied to asymmetric rotors. No mass measurements have been made on asymmetric rotors as yet because of the complexity of their spectra.

When (4.6) is applied to linear molecules and symmetric tops with isotopic substitutions on the symmetry axis only and if the z-axis is taken perpendicular to the axis of the molecule or the symmetry axis, it may be written as

$$\frac{m^{(1)} - m^{(0)}}{m^{(2)} - m^{(0)}} = \frac{M^{(1)} B_e^{(2)} (B_e^{(0)} - B_e^{(1)})}{M^{(2)} B_e^{(1)} (B_e^{(0)} - B_e^{(2)})}. \tag{4.7}$$

The B_e's appearing in (4.7) are tne values of the rotational constants for equilibrium internuclear distances while the measured rotational frequencies give the effective rotational constant B_0 which includes the effects of zeropoint vibrations. Since the B_e's are so difficult to determine in polyatomic molecules, the effective rotational constants, which are proportional to the measured frequencies, are used instead in (4.7), the error thereby being small as will be indicated

---

[1] Strandberg, Wentink and Hill: Phys. Rev. **75**, 827 (1949).

below. A similar replacement of B_e by B_0 in a mass ratio determination would lead to an order of magnitude larger error.

Thus if three rotational frequencies of a molecule corresponding to three isotopic substitutions of the same atom are measured and if the masses of two of the isotopes are assumed, the mass of the third isotope can be determined with very good accuracy. Note that no information about the structural parameters of the molecule is required. About all that need be known concerning the molecule is the masses of the other atoms, and even this is not required to high precision.

β) *Error in mass difference ratio arising from neglect of zero-point vibrations.* The rotational constants B_0 that are inserted in (4.7) are those for the ground vibrational state and for a particular isotopic species are given by (4.1) with $v_i = 0$. Henceforth the sum, $\sum_i \alpha_i \frac{d_i}{2}$ will be abbreviated as α.

If α and B had the same functional dependence upon mass, then no error at all would be introduced in the mass difference ratio by using B_0 instead of B_e. However, even though α and B do not have the same functional dependence upon mass, to a first approximation they both vary linearly with small fractional changes of mass and since $\alpha \ll B$, the error introduced by neglecting α is small. This error can be estimated by expanding α and B about their values when $m = m^{(0)}$ in powers of the change in isotopic mass Δm:

$$\alpha = \alpha^{(0)} + \alpha' \Delta m + \tfrac{1}{2}\alpha''(\Delta m)^2 + \cdots, \tag{4.8}$$[1]

$$B = B_e^{(0)} + B' \Delta m + \tfrac{1}{2} B''(\Delta m)^2 + \cdots. \tag{4.9}$$

With this expansion

$$\frac{B_0^{(1)} - B_0^{(0)}}{B_0^{(2)} - B_0^{(0)}} \approx \left[1 - \frac{1}{4}\frac{\alpha'}{B'}\left(\frac{B''}{B'} - \frac{\alpha''}{\alpha'}\right)(m^{(2)} - m^{(1)})\right]\left[\frac{B_e^{(1)} - B_e^{(0)}}{B_e^{(2)} - B_e^{(0)}}\right]. \tag{4.10}$$

Similarly, expanding $B_e^{(1)}$ and $\alpha^{(1)}$ and $\alpha^{(2)}$ respectively

$$\frac{B_0^{(2)}}{B_0^{(1)}} \approx \left[1 + \frac{1}{2}\frac{\alpha}{B}\left(\frac{B'}{B} - \frac{\alpha'}{\alpha}\right)(m^{(2)} - m^{(1)})\right]\frac{B_e^{(2)}}{B_e^{(1)}}. \tag{4.11}$$

The experimental value of the mass difference ratio obtained by using B_0 in (4.7) is then related to the true value which would result from using B_e by

$$\left(\frac{m^{(1)} - m^{(0)}}{m^{(2)} - m^{(0)}}\right)_{exp} = \left(\frac{m^{(1)} - m^{(0)}}{m^{(2)} - m^{(0)}}\right)_{true}\left\{1 + \left[-\frac{1}{4}\frac{\alpha'}{B'}\left(\frac{B''}{B'} - \frac{\alpha''}{\alpha'}\right) + \right.\right.$$
$$\left. + \frac{1}{2}\frac{\alpha}{B}\left(\frac{B'}{B} - \frac{\alpha'}{\alpha}\right)\right][m^{(2)} - m^{(1)}] + \tag{4.12}$$
$$\left. + \text{terms in higher powers of mass differences}\right\}.$$

B' and B'' are easily determined experimentally, but an exact evaluation of α' and α'' is very difficult. In fact if they could be determined they could be substituted in (4.12) so that the error in mass difference ratio due to neglect of zero point vibrations could be eliminated. However, even an approximate knowledge of α' and α'' allows a reasonable estimate of this error. B and α are positive in all known cases, B' and α' are negative and B'' and α'' positive. In

---

[1] Here and in Eqs. (4.10) to (4.12) B'_e and B''_e should be B'_0 and B''_0. However as they appear as corrections and the differences between them and B'_0 and B''_0 are small, one may use B'_0 and B''_0 in (4.12).

this respect α and B are similar if not identical functions of m, so that the two terms in each bracket

$$\left(\frac{B''}{B'} - \frac{\alpha''}{\alpha}\right) \quad \text{and} \quad \left(\frac{B'}{B} - \frac{\alpha'}{\alpha}\right)$$

will partially cancel. In addition, the two error terms in (4.12) are of opposite sign and will partially cancel.

OCS, whose vibrational modes are shown in Fig. 1, is the only polyatomic molecule for which sufficient experimental evidence on the α's is available to allow an estimate of these error terms. A detailed estimate[1] of the error terms in (4.12) for this molecule has been made and indicates an error of less than 1 part in 15000 in the $(S^{33} - S^{32})/(S^{34} - S^{32})$ mass ratio. This corresponds to an error of 0.03 mMU in the determination of the S^{33} mass assuming S^{32} and S^{34} as known. A similar numerical estimate for more complex molecules such as GeH_3Cl and SiH_3Cl is almost impossible. One can only surmise that in most cases the error would be of the same order of magnitude.

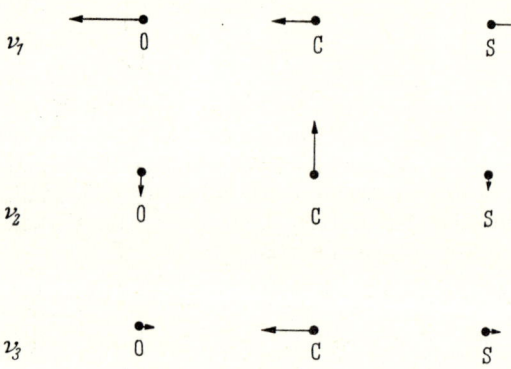

Fig. 1. Vibrational modes of the linear molecule OCS.

It is possible to find exceptional cases where the errors given by (4.12) are important. These will occur where the isotopes being compared are located near the center of gravity of the molecule such as the central nitrogen in NNO. In this case B' would be small, but α' and α'' may not be so that the error terms of (4.12) would be serious.

The best evidence for the smallness of error due to neglect of zero-point vibrations in determining mass difference ratios in polyatomic molecules is the very good agreement between the microwave results and those obtained from mass spectrographs and nuclear reactions as shown in Table 2. The case of O^{16}, O^{17} and O^{18} measured in OCS affords a particularly good comparison because the oxygen atom is light and its vibrational motion large so that one might expect an error due to zero-point vibration to show up in this case.

γ) *Error in mass difference ratio caused by uncertainty in assumed reference masses and remaining masses in the molecule.* If $M^{(1)}/M^{(2)}$ is written as $(M + m^{(1)})/(M + m^{(2)})$ then it can readily be shown that the error introduced in the mass difference ratio (4.7) by an uncertainty ΔM in M is equal to

$$\frac{\Delta M (m^{(2)} - m^{(1)})}{(M + m^{(2)})^2}.$$

Since ΔM is invariably known to a few milli-mass units, the resultant error in mass difference ratio is usually of the order of 1 part in 10^6 and is negligible.

An uncertainty δ in the difference between the two reference masses, $m^{(2)} - m^{(1)}$, on the other hand, produces an error equal to

$$\frac{\delta}{M + m^{(2)}}$$

[1] Geschwind, Gunther-Mohr and Townes: Rev. Mod. Phys. 26, 444 (1954).

in the mass difference ratio. The maximum uncertainty in δ is generally of the order of a few milli-mass units so that the resulting error will be of the order of one part in 30000, but is invariably considerably less and usually negligible.

δ) *Odd-even mass difference.* Information on the odd-even mass difference of isotopes can be obtained which is not subject to the small errors of zero-point vibrations. This and the general application of the method are illustrated in the microwave determination of the masses of the selenium isotopes. The frequencies of the $J = 1 \rightarrow 2$ rotational transition in OCSe corresponding to the six stable selenium isotopes were measured to an accuracy of 0.15 Mc[1]. In addition subsequent measurements of the rotational lines in the same microwave spectrum of OCSe corresponding to radioactive Se^{75} and Se^{79} were made[2,3]. Since the measurements [when substituted into (4.7)] furnish information only on mass difference ratios, the masses of two of the isotopes must be assumed in order to determine the others. Using the mass spectrographic values[4] of Se^{76} and Se^{80}, the remaining Se masses were found and are plotted in Fig. 2 in terms of mass defect vs. mass number.

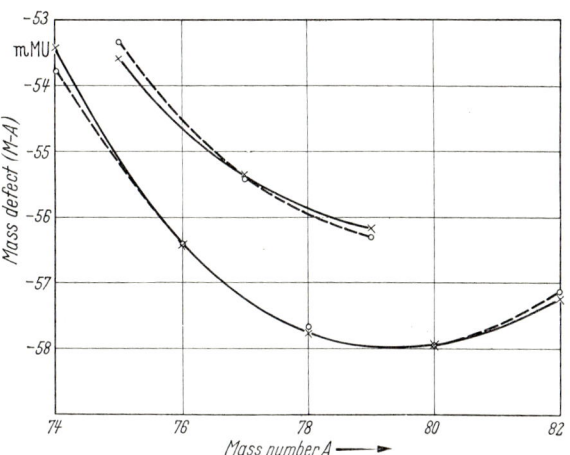

Fig. 2. Mass defects of the Se isotopes, showing odd-even mass differences. Crosses: microwave values. Dots: mass spectroscopic and nuclear reaction values.

The curves for both the even and odd masses illustrate the decrease in binding energy of successively added pairs of neutrons. In addition, the curve for the odd isotopes is seen to be above that for the even isotopes. The distance that the odd isotopes lie above the curve for the even isotopes (odd-even mass difference) is essentially unaffected by errors in the assumed masses of Se^{76} and Se^{80} or by uncertainties due to zero-point vibrations. The odd-even mass difference can therefore be determined with an accuracy limited only by the experimental errors in frequency measurement.

5. Experimental techniques for measuring microwave absorption lines. Since there exists an extensive literature describing the experimental techniques of microwave spectroscopy[5] [1] to [4] only brief mention will be made here of the three different catagories of spectrometers with a view to emphasizing the relative resolution and accuracy of each.

1. The general outline of the most widely used type with which the large majority of mass measurements have been made is the waveguide gas absorption cell shown in Fig. 3a. The molecule being studied is contained in gaseous form at a pressure of 10^{-2} to 10^{-3} mm Hg in a hollow waveguide. Microwave radiation

¹ GESCHWIND, MINDEN and TOWNES: Phys. Rev. **78**, 174 (1950).
² L. C. AAMODT and P. C. FLETCHER: Phys. Rev. **98**, 1224 (1955).
³ HARDY, SILVEY, TOWNES, BURKE, STRANDBERG, PARKER and COHEN: Phys. Rev. **92**, 1532 (1953).
⁴ COLLINS, NIER and JOHNSON: Phys. Rev. **94**, 398 (1954).
⁵ For details cf. the article of W. GORDY in Vol. XXVIII of this Encyclopedia.

incident on the cell is absorbed at one of the characteristic rotational frequencies of the molecule. This absorption decreases the power reaching the silicon detector at this frequency. The highest resolution instruments of this type[1] have absorption line widths (defined as the frequency separation between points of half-maximum intensity of absorption) limited mainly by Doppler broadening and collisions of the molecules with the walls. At room temperature for such typical molecules as OCS and SiD_3F linewidths are about 60 to 80 kc, so that absorption frequencies can be conveniently measured to an accuracy of 2 kc in the 24000 Mc region (1 part in 10^7).

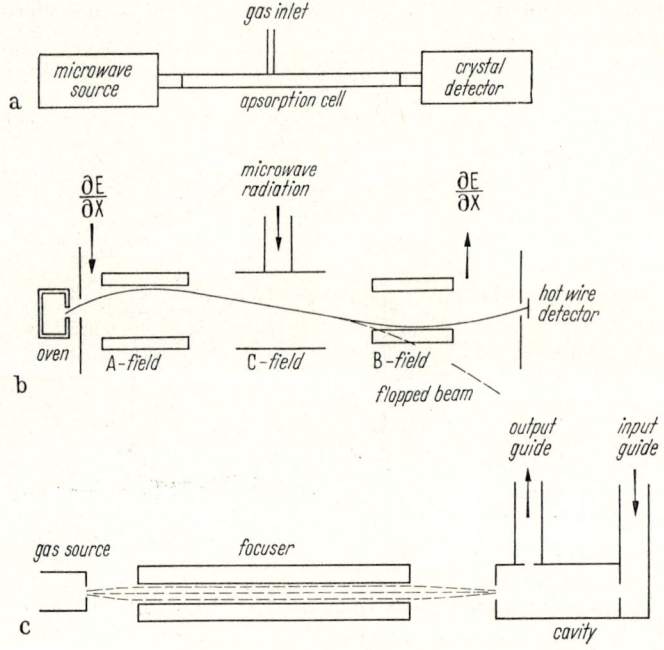

Fig. 3a—c. Three types of microwave spectrometers (a) microwave gas absorption cell, (b) molecular beam, (c) molecular beam microwave emission spectrometer.

2. Several mass determinations have been made with a molecular beam electric resonance apparatus shown schematically in Fig. 3b[2]. In this method a beam of molecules successively traverses regions of oppositely directed inhomogeneous electric fields separated by an intermediate region (the C-field) which for an examination of rotational transitions is usually field free. The deflection of the molecule in an inhomogeneous electric field is proportional to the average dipole moment of the molecule which is a function among other things of the particular rotational state, J, of the molecule. Thus by a suitable arrangement of stops, one particular rotational state is refocused at the detector. Microwave radiation of the proper frequency, introduced in the C-region, causes the molecule to make a transition from the rotational state J to $J+1$. The resultant change in J changes the magnitude of the deflection in the second inhomogeneous field

[1] S. Geschwind: N. Y. Acad. Sci. 55, 751 (1952). — R. L. White: J. Chem. Phys. 23, 249 (1955).
[2] Lee, Fabricand, Carlson and Rabi: Phys. Rev. 91, 1395 (1953). — J. W. Trischka and R. Braunstein: Phys. Rev. 96, 968 (1954). — For molecular beam techniques cf. P. Kusch and V. W. Hughes, Vol. XXXVII of this Encyclopedia.

so that it is not refocused at the detector causing a decrease in beam intensity. Beam intensity is observed as a function of frequency. Since the beam travels in essentially one direction relative to the radiation field no DOPPLER broadening is produced. The line width, $\Delta \nu$, is given by $\Delta \nu \cdot \tau \sim 1$ where τ is the length of time the molecule spends in the radiation field. Line widths from 10 to 20 kc are typical. Resolution can be improved by increasing the length of the interaction region or by splitting this region into two widely separated parts[1]. Line widths as narrow as 1 kc or less appear practical with the possibility of very high accuracy of measurement of frequency, although no mass determinations have been made by this split field technique as yet.

3. The third general method[2] is a combination of the above mentioned methods in that a molecular beam is used but a change in radiation power is measured. Referring to Fig. 3c, a beam of molecules passes through a set of focusing electrodes which focuses those molecules in a particular energy state (above the ground state) into the microwave cavity tuned to a frequency corresponding to an allowable transition to a lower state. The radiation field in the cavity induces transitions to the lower state, so that the stimulated emission of the molecules enhances the output radiation. As in the molecular beam spectroscope, the molecules move in essentially one direction with respect to the radiation field so that DOPPLER broadening is reduced. Again the line width $\Delta \nu$ is approximately $1/\tau$, where τ is the time required for the molecule to traverse the cavity. A line width of 6 to 8 kc was obtained for NH_3 at room temperature. While no mass measurements have been made as yet on this type of instrument, this method holds promise of enabling one to determine transition frequencies to a few parts in 10^8 in the microwave region.

6. Experimental results. Mass information has been obtained from microwave spectra for all the stable isotopes of Li, C, O, Si, S, Cl, K, Br, Ge, Se, Rb and Te, with the exception of Ge^{73} and Te^{120}. To this list must be added the small but increasing number of radioactive isotopes H^3, S^{35}, Cl^{36}, Se^{75} and Se^{79}. This information is compiled in Tables 1 and 2, with mass ratios obtained from diatomic molecules gathered in Table 1, and mass difference ratios obtained from polyatomic molecules listed in Table 2.

The microwave results are compared with the mass spectrometer and nuclear reaction results where available, indicating excellent agreement in most cases. In quoting the different results, the values and assigned errors given in the original work are used. No effort is made to average the different results or their errors as is often done in tabulations of masses. It is felt that in this way one is afforded the fairest comparison of the relative precision of the different techniques of mass measurement.

The tables are supplemented by a discussion of the results, which follows, listed in order of increasing atomic number.

H. — HBr and HI have considerably smaller moments of inertia than most of the other molecules listed. The electronic contribution to the moment of inertia should, therefore, be fairly pronounced in these molecules. Unfortunately the magnetic moment of this molecule has not been measured as yet so as to allow a correction for this effect (Sect. 3). A large uncertainty in the H^3/H^2 mass

[1] N. F. RAMSEY: Phys. Rev. **76**, 996 (1949).
[2] GORDON, ZEIGER and TOWNES: Phys. Rev. **95**, 282 (1954). — Phys. Rev. **99**, 1264 (1955). — H. R. JOHNSON and M. W. P. STRANDBERG: Phys. Rev. **85**, 503 (1952). — M. W. P. STRANDBERG and H. DREICER: Phys. Rev. **94**, 1393 (1954).

Table 1. *Mass ratios from the spectra of diatomic molecules.*

Mass ratio	Molecule in which measured	Microwave spectra	Ref.	Mass spectrograph	Ref.	Nuclear reactions	Ref.
H^3/H^2	HBr	1.497469 ± 7	a)[1]			1.497466 ± 7	c)
		1.497470 ± 8	b)[1]			1.497469 ± 4	d)
	HI	1.497467 ± 12	a)[1]				
		1.497448 ± 60	a)				
Li^6/Li^7	LiI	0.8573423 ± 20	e)			0.8573437 ± 6	f)
						0.8573425 ± 30	c)
C^{12}/C^{13}	CO	0.9228388 ± 3	g)	0.9228359 ± 11	i)	0.9228387 ± 3	k)
	CS	0.9228447 ± 20	h)[1]	0.9228347 ± 14	j)	0.9228389 ± 4	l)
				0.9228386 ± 1 [2]		0.9228384 ± 9	m)
O^{18}/O^{16}	CO	1.1253047 ± 3	g)	1.1253047 ± 8	n)	1.1253049 ± 7	o),p)
				1.1253052 ± 13	i)		
S^{32}/S^{33}	CS	0.9696909 ± 32	h)[1]	0.9696905 ± 10	i)	0.9696896 ± 12	d)
		0.9696884 ± 32	q)	0.9696838 ± 14	j)		
S^{32}/S^{34}	CS	0.9412462 ± 23	h)[1]	0.9412446 ± 6	i)		
		0.9412435 ± 25	q)	0.9412428 ± 14	n)		
				0.9412414 ± 15	j)		
Cl^{35}/Cl^{37}	ICl	0.9459801 ± 50	r)	0.9459777 ± 20	j)	0.9459762 ± 10	u)
	FCl	0.9459775 ± 40	s)	0.9459779 ± 9	i)		
	CsCl	0.9459781 ± 30	e)				
	$K^{39}Cl$	0.9459803 ± 15	t)				
K^{39}/K^{41}	KI	0.9512250 ± 70	e)	0.9512198 ± 25	v)	0.9512204 ± 10	w)
	KCl^{35}	0.9512189 ± 15	t)	0.9512180 ± 21	z)		
Br^{79}/Br^{81}	RbBr	0.9752999 ± 65	e)				
	CsBr	0.9753068 ± 45	e)	0.9753075 ± 10	y)		
	$K^{39}Br$	0.9753088 ± 20	x)				
Rb^{85}/Rb^{87}	RbI	0.9770177 ± 45	e)	0.9770191 ± 20			
	$RbBr^{79}$	0.9770146 ± 55	e)				
	$RbCl^{35}$	0.9770163 ± 45	z)				

References for Table 1.

a) B. Rosenblum and A. H. Nethercot jr.: Phys. Rev. **97**, 84 (1955).
b) Burrus, Gordy, Benjamin and Livingston: Phys. Rev. **97**, 1661 (1955).
c) Li, Whaling, Fowler and Lauritsen: Phys. Rev. **83**, 512 (1951).
d) Strait, van Patter, Buechner and Sperduto: Phys. Rev. **81**, 747 (1951).
e) Honig, Mandel, Stitch and Townes: Phys. Rev. **96**, 629 (1954).
f) Collins, McKenzie and Ramm: Proc. Roy. Soc. Lond., Ser. A **216**, 242 (1953).
g) B. Rosenblum and A. N. Nethercot jr.: Private communication.
h) R. C. Mockler and G. R. Bird: Phys. Rev. **98**, 1937 (1955).
i) K. Ogata and H. Matsuda: Phys. Rev. **89**, 27 (1953).
j) Collins, Johnson and Nier: Phys. Rev. **84**, 717 (1951).
k) Elliot and Livesey: Proc. Roy. Soc. Lond. **224**, 129 (1954).
l) K. F. Familiaro and G. C. Philips: Phys. Rev. **91**, 1195 (1953).
m) Sperduto, Buechner, Beckelman and C. P. Brune: Phys. Rev. **96**, 1316 (1954).
n) H. Ewald: Z. Naturforsch. 6a, 293 (1951).
o) Strait, van Patter, Beuchner and Sperduto: Phys. Rev. **86**, 951 (1952).
p) K. Ahnlund: Phys. Rev. **96**, 999 (1954).
q) See discussion of sulfur.
r) Townes, Merritt and Wright: Phys. Rev. **73**, 1334 (1948).
s) Gilbert, Roberts and Griswold: Phys. Rev. **76**, 1723 (1949).

[1] These values are questionable as they are uncorrected for electron effects. See discussion of hydrogen, carbon and sulfur in text.
[2] Added in Proof (Jan. 1957): Scolman, Quisenberry and Nier: Phys. Rev. **102**, 1076 (1956).

t) LEE, FABRICAND, CARLSON and RABI: Phys. Rev. **91**, 1395 (1953).
u) See discussion of chlorine in text.
v) A. HENGLEIN: Z. Naturforsch. **6**a, 743 (1951).
w) See discussion of potassium.
x) FABRICAND, CARLSON, LEE and RABI: Phys. Rev. **91**, 1403 (1953).
x) COLLINS, NIER and JOHNSON: Phys. Rev. **94**, 398 (1954).
z) J. W. TRISCHKA and R. BRAUNSTEIN: Phys. Rev. **96**, 968 (1954).

ratio is thereby introduced from this source. ROSENBLUM and NETHERCOTT[1] have estimated this error as 50 parts per million.

In addition, again because these molecules are light, higher order terms in the potential must be considered. α and γ appearing in (3.1) could not be obtained from the microwave spectra and so were taken from infrared data. Uncertainty in these quantities contribute an error of 20×10^{-6} to the result, again considerably larger than the experimental error associated solely with the frequency measurement.

Although the microwave values of H^3/H^2 listed in Table 1 and the associated errors are considerably different, the actual measurements of ROSENBLUM and NETHERCOTT on the one hand and BURRUS et al.[2] on the other are in very good agreement. The apparent difference arises from the fact that BURRUS et al. used an older and probably less correct value of α and in addition neglected the effects of the electronic contribution to the moment of inertia. The very good agreement of BURRUS et al.'s value with the nuclear reaction value probably results from a fortuitous cancellation of these two corrections. A determination of the magnetic moment of these molecules would help resolve this uncertainty.

Li.—The lithium atom is light so that the electronic correction would here also be significant in the Li^6/Li^7 mass ratio. While the magnetic moment of LiI has not been measured, a good approximation to the necessary correction is obtained if the molecule is regarded as composed of two ions rather than two atoms. Actually the Li—I bond was taken to have 85% ionic character[3]. This ionic character can be estimated to an accuracy of about 10% from the quadrupole coupling in the molecule[4]. Although this approach is not exact, in that it does not account completely for polarization of the electron core and its consequent departure from spherical symmetry, it should still be fairly good.

C.—CO and CS are again light molecules where electronic effects are important. The magnetic moment of CO was measured (Sect. 3) and the electronic correction made for the C^{12}/C^{13} mass ratio, amounting to 12.4 parts per million (100 times greater than the experimental error). This correction plus the "wobble stretching" correction of 1.6 parts per million yields a microwave value of C^{12}/C^{13} in excellent agreement with nuclear reaction results. Half of the assigned error in the microwave result is due to the uncertainty in the nuclear reaction C^{14}/C^{12} ratio which is used to calibrate for the "wobble stretching".

The C^{12}/C^{13} ratio determined in the molecule CS, however, was uncorrected for the electronic modification of the moment of inertia and is, therefore, considerably different and in error.

O.—The O^{18}/O^{16} mass ratio determined from the microwave spectrum of CO by ROSENBLUM and NETHERCOT[5] was corrected for electron effects which amounted

[1] B. ROSENBLUM and A. H. NETHERCOT jr.: Phys. Rev. **97**, 84 (1955).
[2] BURRUS, GORDY, BENJAMIN and LIVINGSTON: Phys. Rev. **97**, 1661 (1955).
[3] HONIG, MANDEL, STITCH and TOWNES: Phys. Rev. **96**, 629 (1954).
[4] B. P. DAILEY and C. H. TOWNES: J. Chem. Phys. **23**, No. 1, 118, (1955).
[5] See end of Sect. 3.

Table 2. *Mass difference ratios.*

Atom	Molecule in which measured	Mass difference ratio	Microwave value	Ref.	Other values	Ref.
O	$OC^{12}S^{32}$	$\dfrac{O^{17}-O^{16}}{O^{18}-O^{16}}$	$0.501042 \pm 8^*$	a)	0.501032 ± 9 0.501045 ± 3	b) c)
Si	SiH_3Cl	$\dfrac{Si^{30}-Si^{29}}{Si^{30}-Si^{28}}$	0.49941 ± 5	d)	0.499424 ± 15 0.49943 ± 3 0.49934 ± 20	e) f) g)
	SiD_3F	$\dfrac{Si^{30}-Si^{29}}{Si^{30}-Si^{28}}$	0.49934 ± 3	h)		
S	$O^{16}C^{12}S$	$\dfrac{S^{33}-S^{32}}{S^{34}-S^{32}}$	0.500714 ± 30	d)	0.500727 ± 20 0.500820 ± 30 0.500759 ± 37	f) i) j)
		$\dfrac{S^{35}-S^{32}}{S^{35}-S^{34}}$	2.993825 ± 85	k)	2.993375 ± 75	j)
		$\dfrac{S^{36}-S^{32}}{S^{36}-S^{34}}$	1.998320 ± 30	l)		
Cl	CH_3Cl	$\dfrac{Cl^{36}-Cl^{35}}{Cl^{37}-Cl^{36}}$	1.00168 ± 40	m)	1.001790 ± 66	n)
	$ClCN$	$\dfrac{Cl^{36}-Cl^{35}}{Cl^{37}-Cl^{36}}$	1.002260 ± 800	o)		
Ge	GeH_3Cl^{35}	$\dfrac{Ge^{76}-Ge^{74}}{Ge^{76}-Ge^{72}}$	0.500127 ± 35	d)	0.500111 ± 20	p)
		$\dfrac{Ge^{72}-Ge^{70}}{Ge^{74}-Ge^{70}}$	0.499852 ± 35	d)	0.499776 ± 20	p)
Se	$O^{16}C^{12}Se$	$\dfrac{Se^{74}-Se^{76}}{Se^{74}-Se^{80}}$	0.333125 ± 18	q)	0.333082 ± 25	p)
		$\dfrac{Se^{77}-Se^{76}}{Se^{80}-Se^{77}}$	0.333947 ± 23	q)	0.333943 ± 35	p)
		$\dfrac{Se^{78}-Se^{76}}{Se^{80}-Se^{78}}$	0.999510 ± 50	q)	0.999380 ± 100	p)
		$\dfrac{Se^{82}-Se^{76}}{Se^{82}-Se^{80}}$	2.995441 ± 110	q)	2.998541 ± 100	p)
		$\dfrac{Se^{75}-Se^{76}}{Se^{75}-Se^{80}}$	0.199566 ± 30	r)	0.199601 ± 100	s)
		$\dfrac{Se^{79}-Se^{76}}{Se^{80}-Se^{79}}$	3.00557 ± 21	t)	3.00525 ± 18	s)
Te	$S^{32}C^{12}Te$	$\dfrac{Te^{128}-Te^{122}}{Te^{124}-Te^{122}}$	3.00127 ± 60	u)	3.00100 ± 25 3.00300 ± 90	v) w)
		$\dfrac{Te^{128}-Te^{123}}{Te^{124}-Te^{123}}$	3.00826 ± 100	u)	3.00731 ± 160	v)
		$\dfrac{Te^{128}-Te^{125}}{Te^{125}-Te^{124}}$	2.99509 ± 60	u)	2.99645 ± 70	v)
		$\dfrac{Te^{128}-Te^{126}}{Te^{126}-Te^{124}}$	1.00030 ± 14	u)	1.00043 ± 10	v)
		$\dfrac{Te^{128}-Te^{130}}{Te^{124}-Te^{130}}$	0.33346 ± 3.5	u)	0.33337 ± 2.2 0.33351 ± 11	v) w)

References for Table 2.

a) GESCHWIND, GUNTHER-MOHR and TOWNES: Rev. Mod. Phys. **26**, 444 (1955).

b) STRAIT, VAN PATTER, BEUCHNER and SPERDUTO: Phys. Rev. **86**, 951 (1952). — K. AHNLUND: Phys. Rev. **96**, 999 (1954).

c) H. EWALD: Z. Naturforsch. **6a**, 293 (1951).

d) S. GESCHWIND and G. R. GUNTHER-MOHR: Phys. Rev. **81**, 882 (1951).

e) STRAIT, VAN PATTER, BUECHNER and SPERDUTO: Phys. Rev. **81**, 747 (1951). — VAN PATTER, SPERDUTO, ENDT, BUECHNER and ENGE: Phys. Rev. **85**, 142 (1952).

f) K. OGATA and H. MATSUDA: Phys. Rev. **89**, 27 (1953).

g) DUCKWORTH, PRESTON and WOODCOCK: Phys. Rev. **79**, 402 (1950); **99**, 188 (1950).

h) R. L. WHITE and C. H. TOWNES: Phys. Rev. **92**, 1256 (1953).

i) COLLINS, JOHNSON and NIER: Phys. Rev. **84**, 717 (1951).

j) See discussion of sulfur in part D.

k) KOSKI, WENTINK and COHEN: Phys. Rev. **81**, 948 (1951).

l) W. A. HARDY: Private communication.

m) L. C. AAMODT and P. C. FLETCHER: Phys. Rev. **98**, 1317 (1955).

n) See discussion of chlorine.

o) Low and TOWNES quote R. G. SCHULMAN: Phys. Rev. **80**, 608 (1950).

p) COLLINS, NIER and JOHNSON: Phys. Rev. **94**, 398 (1954).

q) GESCHWIND, MINDEN and TOWNES: Phys. Rev. **78**, 174 (1950).

r) L. C. AAMODT and P. C. FLETCHER: Phys. Rev. **98**, 1224 (1955).

s) See discussion of selenium in part D.

t) HARDY, SILVEY, TOWNES, BURKE, STRANDBERG, PARKER and COHEN: Phys. Rev. **95**, 385 (1954).

u) W. G. HARDY and G. SILVEY: Phys. Rev. **95**, 385 (1954).

v) R. HALSTED: Phys. Rev. **88**, 666 (1952).

w) B. G. HOGG and H. E. DUCKWORTH: Canad. J. Phys. **30**, 628 (1952).

to 18.6 parts per million for this mass ratio. This microwave value of the mass ratio as seen from Table 1 is in good agreement with the other values.

The microwave oxygen mass difference ratio given in Table 2 is in excellent agreement with the nuclear reaction value obtained from the $O^{16}(d, p) O^{17}$ and $O^{17}(d, p) O^{18}$ reactions.

In OCS, as a result of zero-point vibrations, the O atom undergoes relatively large displacements from its equilibrium position. One might, therefore, expect zero point vibrations to show up in this case. The very good agreement between the microwave value and the nuclear reaction result in this case is further evidence for the smallness of the error due to neglect of zero-point vibrations in microwave mass difference ratio determinations.

Si.—The microwave measurement of the Si mass difference ratio determined in SiH_3Cl is in better agreement with the best values from other sources than that from SiD_3F. This may be due to a small systematic error arising from the use of STARK modulation in the latter measurement to increase sensitivity ([*1*], p. 418). STARK modulation was not used in the SiH_3Cl study and hence the mass determination in this case was not subject to this type of error.

S.—In reviewing the microwave work[1] on the molecule CS, different mass ratios than those given in the original work were obtained from the stated B_e values. The recalculated mass ratios[2] are $S^{32}/S^{33} = 0.969\,691\,7 \pm 32$, $S^{32}/S^{34} = 0.941\,250\,6 \pm 23$ and $C^{12}/C^{13} = 0.922\,846\,7 \pm 20$. One can apply here the electronic correction in reverse by using the very accurately known C^{12}/C^{13} ratio (from nuclear reaction and microwave CO data) to calculate the magnetic moment of CS from the difference between this assumed mass ratio and the recalculated

---

[1] R. C. MOCKLER and G. R. BIRD: Phys. Rev. **98**, 1837 (1955).

[2] Error limits are given in units of the last digit.

value from CS given immediately above [see Eq. (2.16)]. Using an average value of $C^{12}/C^{13} = 0.9228387$ gives a value of $\mu_J = -0.195\,\mu_n$ for CS. With this derived value of μ_J, the sulfur mass ratios can be corrected using Eq. (3.16). The corrected values are $S^{32}/S^{33} = 0.9696884 \pm 32$ and $S^{32}/S^{34} = 0.9412435 \pm 25$.

The S^{35} mass can be rather accurately determined relative to Cl^{35} from the end-point[1] of the β decay of S^{35}. If one uses OGATA's value[2] for Cl^{35}, one obtains 34.980244 ± 22 for the S^{35} mass. From this value of S^{35} and OGATA's values for S^{32} and S^{34}, one obtains $(S^{35} - S^{32})/(S^{35} - S^{34}) = 2.2993360 \pm 75$. The microwave value is 2.993806 ± 85, corresponding to a disagreement by 0.4 mMU in the S^{35} mass. This discrepancy would require an error in the microwave frequency measurement of 10 kc, which is enormously greater than the stated error. Unfortunately, the microwave measurement cannot easily be redone, since radioactive S^{35} in the form OCS is necessary.

No mass measurement other than the microwave one exists for S^{36}. The frequency of the $J = 1 \rightarrow 2$ transition in OCS is 23198.76 ± 0.02 Mc[3].

Cl.—The internal consistency of the microwave values of the Cl^{35}/Cl^{37} mass ratio determined in four different molecules where corrections due to electronic structure should differ demonstrates that these effects are not important to the accuracy quoted.

The nuclear reaction value for Cl^{35}/Cl^{37} is taken from the value for the chain

$$Cl^{35}\,(n,\gamma)\,Cl^{36}\,(\beta^-)\,A^{36}\,(d,p)\,A^{37}\,(K)\,Cl^{37}$$

given by WAPSTRA [8].

Cl^{36} is very accurately connected by the end point of its β-decay spectrum to A^{36}. By use of COLLINS et al.'s[4] value for A^{36} and FELDMAN and WU's[5] value for the β-decay end point one obtains $Cl^{36} = 35.979764 \pm 30$. This value taken with OGATA's Cl^{35} and Cl^{37} gives the result quoted in Table 2, i.e. $(Cl^{36} - Cl^{35})/(Cl^{37} - Cl^{36}) = 1.001790 \pm 66$.

K.—The most accurate microwave determination of K^{39}/K^{41} mass ratio was made using the molecular beam technique[6]. It is in very good agreement with the two best mass spectroscopic results and just about equal to their average. The nuclear reaction value is that given by WAPSTRA [8] for the chain

$$K^{39}\,(n,\gamma)\,K^{40}\,(\beta^-)\,Ca^{40}\,(d,p)\,Ca^{41}\,(n,p)\,K^{41}.$$

Ge.—The discrepancy of 0.6 mMU between the microwave and mass spectroscopic value of Ge^{70} is an interesting one. Ge^{72} has 40 neutrons for which the closing of a subshell is expected. The value of Ge^{70} given by COLLINS et al.[7] seems to indicate such a break in the mass defect curve for Ge at 40 neutrons of approximately 600 kev, while the microwave value indicates a maximum break of 100 kev.

The rotational transition of GeH_3Cl for Ge^{73} is split by nuclear quadrupole hyperfine interaction into components which were too weak to be measured with the spectrometer used. No microwave data are, therefore, available for Ge^{73}.

[1] P. M. ENDT and J. C. KLUGER: Rev. Mod. Phys. **26**, 95 (1954).
[2] K. OGATA and H. MATSUDA: Phys. Rev. **89**, 27 (1953).
[3] W. A. HARDY: Private communication.
[4] COLLINS, JOHNSON and NIER: Phys. Rev. **84**, 717 (1951).
[5] L. FELDMAN and C. S. WU: Phys. RIv. **87**, 1091 (1952).
[6] LEE, FABRICAND, CARLSON and RABI: Phys. Rev. **91**, 1395 (1953).
[7] COLLINS, JOHNSON and NIER: Phys. Rev. **84**, 717 (1951).

Se.—The assigned error in the microwave Se mass difference ratio given in reference q of Table 2, contains errors due to zero-point vibration as well as the experimental error in frequency measurement. Since the errors allowed for zero-point vibrations represent an absolute upper limit and may be too large, only the experimental errors are listed in Table 2. From a plot of the mass defect of the Se isotopes one obtains the following microwave determined odd-even mass differences

$$Se^{75} - 1.5 \pm 0.2 \text{ mMU},$$

$$Se^{77} - 2.0 \pm 0.12 \text{ mMU},$$

$$Se^{79} - 1.8 \pm 0.05 \text{ mMU}.$$

Br.—There is very good agreement between the microwave values and the mass spectroscopic value.

Rb.—Attention is called to the excellent internal consistency between the microwave results obtained in three different molecules. The microwave values are in turn in agreement with the mass spectroscopic result which is just slightly higher.

Te.—With the exception of $(Te^{128} - Te^{130})/(Te^{124} - Te^{130})$ the microwave results are in good agreement with the mass spectroscopic values. If the masses of Te^{124} and Te^{128} are assumed known, there is a discrepancy of 1 mMU in the Te^{130} mass.

7. Conclusion. The precision of mass determination by microwave methods, in general, compares favorably with the better known methods. Although the precision of mass spectrographic and nuclear reaction results are often very high, deviations between the various mass spectrographic and nuclear reaction values are often considerably greater than the quoted errors, indicating the presence of troublesome systematic errors. The continuing development of microwave methods now allows considerable improvement in the microwave results with greater emphasis being placed on the corrections for the electronic structure of the molecule in the case of light atoms.

Acknowledgement.

This work is a modification and revision of an earlier report on this subject written by the author with Dr. R. G. GUNTHER-MOHR and Professor C. H. TOWNES which appeared in the Reviews of Modern Physics **26**, 444 (1955).

Bibliography.

Microwave spectroscopy.

[1] TOWNES, C. H., and A. SCHAWLOW: Microwave Spectroscopy. New York: McGraw Hill 1955. — This excellent book is the most complete and detailed treatment to be found of both the theoretical and experimental aspects of the microwave spectroscopy of gases. Parts of chapters 1, 2, 3 and 8 contain accounts and information dealing with microwave mass determinations.

[2] STRANDBERG, M. W. P.: Microwave Spectroscopy. London: Methuen & Co. Ltd.; New York: John Wiley & Sons, Inc. 1954. — A compact treatment emphasizing quantum mechanical matrix methods in deriving the energy levels of a rotating molecule. The electronic contribution to the moment of inertia of molecules is discussed in Chap. VI and VII.

[3] Gordy, W., W. V. Smith and R. F. Trambarulo: Microwave Spectroscopy. New York: John Wiley & Sons, Inc. — A general survey including accounts of microwave spectra of solids such as paramagnetic and ferromagnetic resonance.
[4] "Symposium on Microwave Spectroscopy". Ann. N.Y. Acad. Sci. 55, 751 (1952). — Collection of papers on various topics in microwave spectroscopy presented at the Conference.

Molecular spectra.

[5] Herzberg, G.: Spectra of Diatomic Molecules. New York: Van Nostrand Co. Inc. 1950.
[6] Herzberg, G.: Infrared and Raman Spectra. New York: Van Nostrand Co. Inc. 1945. — Polyatomic molecules.

Nuclear mass compilations.

[7] "Mass Differences — A compilation of experimental atomic mass differences from β-decay, reaction energies, microwave data, alpha decay and mass doublets". Rev. Mod. Phys. 26, 327 (1954).
[8] Wapstra, A. H.: Isotopic Masses, Physica, Haag 21, 367 (1955). — Part I contains a computation of isotopic masses for $A < 34$ obtained from nuclear reaction data. Part II contains a computation of masses for $A = 33$ to $A = 202$ using data from all the techniques of mass determination. (See also Wapstra's preceding article in this volume.)

Determination of Nuclear Spins and Magnetic Moments by Spectroscopic Methods.

By

F. M. KELLY.

With 11 Figures.

Introduction.

Many of the lines which occur in atomic spectra consist of a number of closely spaced components. Such lines are said to shown *hyperfine structure*. There are two distinct types of hyperfine structure. The first, called *isotope shift*, is attributed to small displacements of the energy levels of the isotopes relative to one another, while the second, called *magnetic hyperfine structure,* is due to an interaction between the magnetic moment of the nucleus and the electrons. This interaction splits the energy levels of a single isotope into a number of states. Frequently combinations of both types are observed in a single spectral line.

Two causes of isotope shift are recognized, the *mass effect* [6] and the *field or volume effect* [16]. For hydrogen-like atoms the mass effect is completely accounted for by a simple reduced mass calculation. However, if the atom has more than one electron the specific mass effect, which arises from a coupling between the electrons, must also be taken into account. The mass effects decrease with increasing atomic number and in the heavy elements the observed shifts can be accounted for by field effect alone. The field effect arises from the fact that the field experienced by the electrons departs from a COULOMB field within the nuclear volume. This departure from a COULOMB field also affects the coupling between the electrons and the nucleus so that an allowance must be made for it when the nuclear magnetic moments are calculated from spectroscopic data.

Magnetic hyperfine structure has many features which are similar to fine structure which occurs in atomic spectra. It is observed most easily in the spectra of elements, for example bismuth, where there is only one isotope. For a given element the largest structures are found in levels arising from electron configurations which contain an unpaired s electron. Its explanation was first suggested by PAULI[1] when he postulated a spin angular momentum and an associated magnetic moment for the nucleus. The interaction between the nuclear magnetic moment and the valence electrons causes the energy levels of the atom to split into a number of hyperfine structure states, which in turn produce the hyperfine structure observed in the spectral lines.

Part A of this article discusses the determination of nuclear spins and the methods for estimating nuclear magnetic moments from observed hyperfine structure patterns. Part B contains a brief discussion of the application of the atomic beam magnetic resonance method for more accurate determination of

---

[1] W. PAULI: Naturwiss. **12**, 741 (1924).

atomic hyperfine structures. The existence of a nuclear spin affects the relative intensities of the rotational lines in the spectra of homonuclear diatomic molecules. This method of determination of nuclear spin is discussed in Part. C.

A. Atomic spectra.

a) Interaction between the nucleus and the electrons.

1. Electrostatic interaction. The electrostatic interaction between the nuclear charge and the electrons of an isolated atom in a given energy state may be written in the form;

$$V_{\mathrm{el}} = -e^2 \iint \frac{\varrho_e(\mathbf{r}_e)\,\varrho_n(\mathbf{r}_n)}{r}\,d\tau_e\,d\tau_n \qquad (1.1)$$

where $-e\varrho(\mathbf{r}_e)$ is the charge density of the electrons at the position $\mathbf{r}_e$ and $e\varrho(\mathbf{r}_n)$ is the charge density of the nucleus at position $\mathbf{r}_n$ when the centre of the nucleus is taken as the origin and r is the magnitude of the distance between the two volume elements $d\tau_e$ and $d\tau_n$ as shown in Fig. 1. If the integration of the electronic charge is limited to the region outside the nucleus the function $1/r$ can be expanded as a power series in terms of the LEGENDRE polynomials, P_i, as follows:

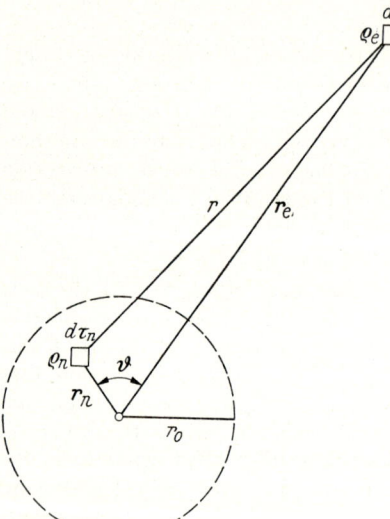

$$\frac{1}{r} = \frac{1}{r_e} + \frac{r_n}{r_e^2}\,P_1 + \frac{r_n^2}{r_e^3}\,P_2 + \cdots. \qquad (1.2)$$

P_1 and P_2 are equal to $\cos\vartheta$ and $\frac{1}{2}(3\cos^2\vartheta - 1)$ respectively. The first term in this series represents the energy of interaction between the nucleus and the electrons when the nuclear radius is taken as zero. It determines the mean basic energy level of the atom in a given state and so is not of direct interest here. The second term represents the interaction with a nuclear electric dipole moment. Since the nuclear electric dipole moment vanishes the inter-

Fig. 1. The electrostatic interaction between the electrons and the atomic nucleus.

action term is zero. The third term in the expansion is the quadrupole term and this interaction is the subject of another article in this volume by C.H. TOWNES [19]. The electric quadrupole interaction, however, acts as a perturbation on the magnetic dipole interaction and in certain circumstances must be taken into account in the determination of spins and nuclear magnetic moments.

From general considerations, which are outside the scope of this article, it can be shown [2] that if the nucleus has a spin I the highest order of multipole which can occur is given by 2^{2I}. Also the multipoles alternate between the electric and magnetic types. The pole of the lowest order is the electric monopole which is the only pole which appears when $I = 0$. For larger values of I there is always an electric monopole, but for reasons of symmetry the electric dipole moment vanishes. An electric quadrupole can be observed when $I \geq 1$. The lowest magnetic pole is the dipole. It can be observed if $I \geq \frac{1}{2}$. The next highest order of magnetic pole is the octupole which requires $I \geq \frac{3}{2}$.

2. Magnetic interaction. The magnetic field produced by the electrons at the position of the nucleus arises from the orbital motion of the electrons and from the magnetic moments of the electrons. Since the electrons precess about the direction of the total angular momentum, J, of the electrons, the total magnetic field at the nucleus is parallel to J. Nuclear magnetic moments are due to the orbital motion of the positive charge within the nucleus and to the magnetic moments of the protons and neutrons. Thus, as in the atomic case, the nuclear magnetic moment of the nucleus is parallel to the total angular momentum of the nucleus. The angular momentum of the nucleus can by represented by a dimensionless vector I which gives the angular momentum in units $\hbar$. The spin, I, of the nucleus is the maximum possible component of I in a fixed direction.

Since the nuclear magnetic moment is parallel to I it may be written in the form

$$\mu_I = g_I \mu_N I \tag{2.1}$$

where g_I is the nuclear g factor, and μ_N is the nuclear magneton. The nuclear magneton is defined by

$$\mu_N = \frac{e\,\hbar}{2\,M\,c} \tag{2.2}$$

where M is the mass of the proton, e the electronic charge, c the speed of light and $\hbar$ PLANCK's constant divided by 2π. The ratio of the mass of the proton to the mass of the electron is 1836 so that (2.2) may be written as

$$\mu_N = \frac{e\,\hbar}{2\,m\,c}\,\frac{1}{1836} = \frac{\mu_0}{1836} \tag{2.3}$$

where μ_0 is the BOHR magneton. Nuclear magnetic moments are usually given in nuclear magnetons and the magnitude of the moment in this unit is

$$\mu_I = g_I I. \tag{2.4}$$

The energy of interaction between the nuclear magnetic moment and the magnetic field produced by the electrons depends on the orientation of the nuclear magnetic moment in the field. The orientation of the nuclear magnetic moment is limited by the rules of quantum mechanics, so the energy levels of the atom are split into hyperfine structure states which produce the magnetic hyperfine structure observed in spectral lines.

The magnetic interaction between the nucleus and the electrons may be given in terms of a vector potential A and a current density σ. The resulting interaction is,

$$V_{\mathrm{mag}} = -\frac{1}{c}\int A \cdot \sigma\,d\tau. \tag{2.5}$$

The nuclear vector potential may be expanded in a power series similar to (1.2), but in the magnetic case the first term which appears is that due to a dipole. The next highest magnetic multipole is the octupole and the high accuracy of observation which can be obtained from atomic beam experiments has permitted the observation of this effect. The octupole term appears as a perturbation of the dipole interaction in atomic spectra.

The vector potential of the nuclear magnetic dipole moment, μ_I, is

$$A = \frac{\mu_I \times r}{r^3}. \tag{2.6}$$

On substitution in (2.5) the magnetic interaction becomes

$$V_{mag} = -\frac{1}{c}\int \frac{\boldsymbol{\mu}_I \times \boldsymbol{r}}{r^3}\cdot \boldsymbol{\sigma}\, d\tau \left.\begin{matrix}\\ \\ \\ \end{matrix}\right\}$$
$$= -\frac{1}{c}\int \boldsymbol{\mu}_I\cdot\left(\frac{\boldsymbol{r}\times\boldsymbol{\sigma}}{r^3}\right)d\tau. \qquad (2.7)$$

In this equation we may write

$$\frac{1}{c}\int \frac{\boldsymbol{r}\times\boldsymbol{\sigma}}{r^3}\, d\tau = \boldsymbol{H}_J \qquad (2.8)$$

where $\boldsymbol{H}_J$ is the magnetic field produced by the electrons at the nucleus. Then, the instantaneous interaction between the electrons and the nuclear magnetic moment is

$$V_{mag} = -\boldsymbol{\mu}_I\cdot\boldsymbol{H}_J. \qquad (2.9)$$

Since the nuclear magnetic moment is parallel to $\boldsymbol{I}$, and $\boldsymbol{H}_J$ is parallel to the total angular momentum, $\boldsymbol{J}$, of the electrons the interaction is

$$V_{mag} = -\frac{\mu_I H_J}{I J}\boldsymbol{I}\cdot\boldsymbol{J}. \qquad (2.10)$$

The interaction must be averaged over the motion of the electron. When this is done the average value of the energy can be written as

$$W = A_J\boldsymbol{I}\cdot\boldsymbol{J} \qquad (2.11)$$

where A_J is a constant known as the interval factor and has the magnitude

$$A_J = -\frac{\mu_I H_J}{I J} = -g_I\mu_N\frac{\boldsymbol{H}_J\cdot\boldsymbol{J}}{\boldsymbol{J}\cdot\boldsymbol{J}}. \qquad (2.12)$$

The two angular momentum vectors $\boldsymbol{I}$ and $\boldsymbol{J}$, which give the total angular momentum of the nucleus and the electrons respectively, couple together to form another vector $\boldsymbol{F}$, the total angular momentum of the atom. From the quantum theoretical rules for vector addition [3], the quantum number F for the vector $\boldsymbol{F}$ may take the following values:

$$F = I+J, \quad I+J-1, \ldots, |I-J|. \qquad (2.13)$$

There are $2I+1$ or $2J+1$ values of F depending on whether $J\geq I$ or $I\geq J$.

Since the total angular momentum $\boldsymbol{F}$ is the vector sum of $\boldsymbol{I}$ and $\boldsymbol{J}$, the factor $\boldsymbol{I}\cdot\boldsymbol{J}$ in (2.11) can be evaluated from

$$\boldsymbol{F}^2 = (\boldsymbol{I}+\boldsymbol{J})^2 = \boldsymbol{I}^2 + \boldsymbol{J}^2 + 2\boldsymbol{I}\cdot\boldsymbol{J} \qquad (2.14)$$

so that

$$\begin{aligned}\boldsymbol{I}\cdot\boldsymbol{J} &= \tfrac{1}{2}(\boldsymbol{F}^2 - \boldsymbol{I}^2 - \boldsymbol{J}^2)\\ &= \tfrac{1}{2}\{F(F+1) - I(I+1) - J(J+1)\}\\ &= \tfrac{1}{2}C.\end{aligned} \left.\begin{matrix}\\ \\ \\ \end{matrix}\right\} \qquad (2.15)$$

The energy of interaction between the nuclear magnetic moment and the magnetic field produced by the electrons then is

$$W = \tfrac{1}{2}A_J C. \qquad (2.16)$$

The atomic energy level, with total angular momentum $\boldsymbol{J}$, is split into a number of states which can be designated by the values of F. Eq. (2.16) is a mathematical expression of the *interval rule,* which also applies (with suitable changes in symbols) to atomic fine structure. Two adjacent hyperfine structure states have F values which differ by unity. Application of (2.16) shows that the energy difference between any two adjacent states is equal to $A_J F$ where F is the greater of the two F values belonging to the two adjacent states.

If the angular momenta of a number of electrons add up in such a manner that the resultant $\mathbf{J}$ is zero, the average value of the magnetic field produced at the nucleus is also zero. For closed subgroups of electrons $\mathbf{J}$ is always zero so that in calculating the magnetic field at the nucleus only those electrons outside closed subgroups need to be considered.

The magnetic field produced by the valence electrons at the nucleus arises from two factors, the orbital motion of the electrons and the field of the magnetic dipole moments of the electrons. Since an electron carries a negative charge the magnetic field due to the orbital motion is antiparallel to the orbital angular momentum. The magnetic moment of the electron is antiparallel to the electron spin, so that the field due to the magnetic moment of the electron will increase the field due to the orbital motion if the electron spin is antiparallel to the orbital angular momentum. If the spin of the electron is parallel to the orbital angular momentum, the field at the nucleus is decreased. In either case, however, the resultant magnetic field at the nucleus is antiparallel to $\mathbf{J}$ since the field due to the orbital motion is larger than that due to the electron magnetic moment. For an s electron the orbital angular momentum is zero, but the magnetic field at the nucleus is still antiparallel to $\mathbf{J}$. In some configurations with more than one electron, the magnetic field is reversed in direction, but these are special cases and need not be considered here (see KOPFERMANN [9], p. 126).

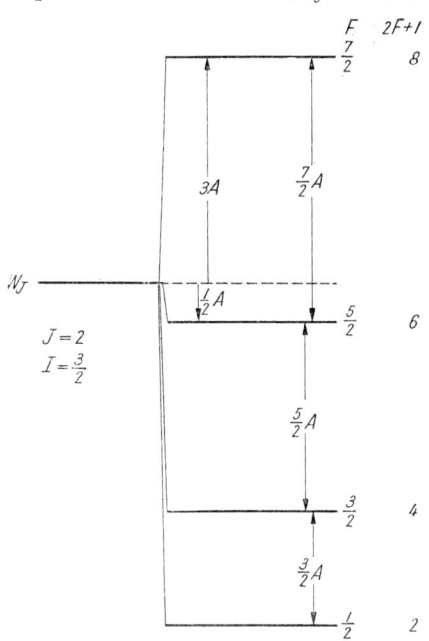

Fig. 2. Hyperfine structure of the atomic energy level W_J with the assumption that the nuclear moment is positive and with $J = 2$ and $I = \frac{3}{2}$. The separation of the states follows the interval rule. If the nuclear moment is negative the order of the states is inverted.

In Eq. (2.1) the nuclear magnetic moment is taken to be parallel to the total angular momentum of the nucleus. In this case the nuclear magnetic moment is said to be positive. If the nuclear magnetic moment is antiparallel to the angular momentum of the nucleus, the magnetic moment is negative. In the normal case with the magnetic field of the electrons antiparallel to their total angular momentum and with a positive nuclear moment the hyperfine state with the highest value of F lies highest in the energy level diagram. The mean energy of the hyperfine states is the electronic energy level of the atom when the effect of the nuclear moment is neglected. The statistical weight of the state F is $2F + 1$. Fig. 2 shows the hyperfine structure states with the assumptiou that $I = \frac{5}{2}$ and $J = 2$. If the nuclear magnetic moment is negative the order of the hyperfine structure states is inverted.

3. The interval factor for hydrogen-like atoms. The measurement of the separation of two hyperfine structure states determines the interval factor, provided the values of F are known for the two states. Eq. (2.12) shows that the interval factor is proportional to the nuclear g factor and that the constant of proportionality contains the magnetic field produced by the electrons at the position of the nucleus. Consequently, a determination of the nuclear magnetic moment from the observed hyperfine structure requires a calculation of this magnetic field.

The magnetic field, H_j, arises from both the orbital motion of the electrons and from their magnetic moments. An exact calculation of the magnetic field can be made in the case of hydrogen-like atoms. For elements with more than one electron, however, an approximate model must be assumed.

For a hydrogenic atom the magnetic field may be determined from a classical calculation using the vector model. The calculation is analogous to the one used to evaluate the electron spin-orbital interaction for the fine structure in atomic spectra [15] (see also KOPFERMANN [9], p. 104 and RAMSEY [14], p. 24). The single electron[1] moves the electrical field of the nucleus with orbital angular momentum l defined by

$$l\hbar = m\,\boldsymbol{r} \times \boldsymbol{v} \qquad (3.1)$$

where $\boldsymbol{r}$ is the radius vector of the electron with the origin at the nucleus and $\boldsymbol{v}$ its velocity. The moving electron produces a magnetic field

$$\boldsymbol{H}_l = -\frac{e}{c}\frac{\boldsymbol{r} \times \boldsymbol{v}}{r^3} \qquad (3.2)$$

at the position of the nucleus. The magnetic field is then proportional to the orbital angular momentum of the electron and

$$\left.\begin{aligned}\boldsymbol{H}_l &= -\frac{e\hbar}{mc}\frac{1}{r^3}\,\boldsymbol{l}\\[2mm]&= -2\mu_0\frac{1}{r^3}\,\boldsymbol{l},\end{aligned}\right\} \qquad (3.3)$$

where μ_0 is the BOHR magneton.

The magnetic dipole moment of the electron is

$$\mu = -2\mu_0\,\boldsymbol{s}, \qquad (3.4)$$

where $\boldsymbol{s}$ is the vector representing the spin angular momentum of the electron The magnetic field due to a dipole is

$$\boldsymbol{H}_s = -\frac{1}{r^3}\left[\mu - \frac{3(\mu \cdot \boldsymbol{r})\,\boldsymbol{r}}{r^2}\right] = \frac{2\mu_0}{r^3}\left[\boldsymbol{s} - \frac{3(\boldsymbol{s} \cdot \boldsymbol{r})}{r^2}\,\boldsymbol{r}\right]. \qquad (3.5)$$

The total magnetic field at the nucleus is the sum of these two vectors so that

$$\boldsymbol{H}_j = \boldsymbol{H}_l + \boldsymbol{H}_s \qquad (3.6)$$

and

$$\boldsymbol{H}_j \cdot \boldsymbol{j} = (\boldsymbol{H}_l + \boldsymbol{H}_s) \cdot (\boldsymbol{l} + \boldsymbol{s}) = -\frac{2\mu_0}{r^3}\left[\boldsymbol{l} - \boldsymbol{s} + \frac{3(\boldsymbol{s} \cdot \boldsymbol{r})\,\boldsymbol{r}}{r^2}\right] \cdot (\boldsymbol{l} + \boldsymbol{s}). \qquad (3.7)$$

In the scalar product the term which contains $\boldsymbol{l} \cdot \boldsymbol{r}$ vanishes since these two vectors are perpendicular to one another.

Then

$$\boldsymbol{H}_j \cdot \boldsymbol{j} = -\frac{2\mu_0}{r^3}\left[l^2 - s^2 + 3\left(\boldsymbol{s} \cdot \frac{\boldsymbol{r}}{r}\right)^2\right]. \qquad (3.8)$$

---

[1] For a single electron lower case letters are used for the various angular momenta and the corresponding quantum numbers. Upper case letters imply that two or more electrons are being considerd.

This quantity must be averaged over the motion of the electron so that the average value of $\frac{1}{r^3}$ must be used in Eq. (3.8). The factor $\left(\mathbf{s} \cdot \dfrac{\mathbf{r}}{r}\right)$ is the component of the electron spin in the direction of $\mathbf{r}/r$ and since $\mathbf{s} \cdot \mathbf{r}$ commutes with the operators for $\mathbf{j}^2$ and $\mathbf{s}^2$ its value is $\pm\frac{1}{2}$. Thus, the proper values of $\mathbf{s}^2$ and $3\left(\mathbf{s} \cdot \dfrac{\mathbf{r}}{r}\right)^2$ are both $\frac{3}{4}$ and the two factors cancel. Then the average value of $\mathbf{H}_j \cdot \mathbf{j}$ is

$$\mathbf{H}_j \cdot \mathbf{j} = -2\mu_0 \overline{\left(\frac{1}{r^3}\right)} l(l+1), \tag{3.9}$$

where $\overline{(1/r^3)}$ is the average value of $1/r^3$.

The average values of various powers of r have been calculated for hydrogen-like atoms (CONDON and SHORTLEY [3], p. 117) and we find that

$$\overline{\left(\frac{1}{r^3}\right)} = \frac{Z^3}{a_H^3\, n^3\, l\,(l+\frac{1}{2})\,(l+1)}, \tag{3.10}$$

where a_H is the radius of the first BOHR orbit in hydrogen and n is the principal quantum number. Substitution in (2.12) gives the value of a_j

$$a_j = g_I\,\mu_N\, \overline{\left(\frac{1}{r^3}\right)}\, \frac{2\mu_0\, l(l+1)}{j(j+1)}. \tag{3.11}$$

The RYDBERG constant $R = \dfrac{2\pi^2\, m\, e^4}{h^3\, c}$ cm^{-1}, the radius of the first BOHR orbit in hydrogen $a_H = \dfrac{h^2}{4\pi^2\, m\, e^2}$, and the fine structure constant $\alpha = \dfrac{2\pi\, e^2}{h\, c}$, so that

$$h\, c\, a_H^3\, R\, \alpha^2 = \frac{1}{2}\, \frac{e^2\, h^2}{8\pi^2\, m^2\, c^2} = 2\mu_0^2. \tag{3.12}$$

Then the interval factor becomes

$$a_j = g_I\, \frac{R\,\alpha^2\, h\, c\, Z^3}{1836\, n^3\, (l+\frac{1}{2})\, j(j+1)}. \tag{3.13}$$

Eq. (3.10) is indeterminate for an s electron where $l=0$. This means that the interval factor a_j is not rigorously determined for an s electron. However, if $l=0$ and $j=\frac{1}{2}$ are substituted into (3.13) the interval factor becomes

$$a_s = g_I\, \frac{8}{3}\, \frac{R\,\alpha^2\, h\, c\, Z^3}{1836\, n^3}, \tag{3.14}$$

which is the result which can be obtained from a strict quantum mechanical calculation.

A quantum mechanical derivation of the interval factor using the DIRAC equation has been given by FERMI[1]. Reviews of the calculation may be found in BETHE and BACHER [1] and KOPFERMANN [9]. For an s electron the interval factor is

$$a_s = \frac{8\pi}{3}\, 2\mu_0^2\, g_I\, |\psi_{n,0}(0)|^2 \tag{3.15}$$

where $|\psi_{n,0}(0)|^2$ is the square of the value of the eigenfunction of the electron at the centre of the nucleus. The value of the square of the eigenfunction is

$$|\psi_{n,0}(0)|^2 = \frac{1}{\pi\, a_H^3} \cdot \frac{Z^3}{n^3}. \tag{3.16}$$

---

[1] E. FERMI: Z. Physik **60**, 320 (1930).

Substitution of (3.16), and the use of (3.12) in Eq. (3.15) leads immediately to the result given in (3.14) for the interval factor of an s electron. The quantum mechanical treatment for a non-s electron gives Eq. (3.11), which involves the average value of $1/r^3$, so that (3.13) follows directly.

4. The interval factor for atoms with more than one electron. For alkali and alkali-like atoms the single valence electron moves in a field determined by the nuclear charge and the charges of the closed shells of electrons. The electron is considered to move in a region outside the closed subgroups of electrons where the effective nuclear charge is $Z_0 e$ and in an inner region where the closed subgroups of electrons are penetrated and where the effective nuclear charge is $Z_i e$. For alkalis $Z_0 = 1$; for singly ionized alkaline earths $Z_0 = 2$, etc. A classical treatment [13] shows that the electron moves in two almost complete elliptic orbits, one in the inner region and the other in the outer region. The average value of $1/r^3$ must now be determined over a complete cycle of the motion and it can be shown that in Eq. (3.10) Z^3 is replaced by $Z_i Z_0^2$ and n^3 by n^{*3}, where n^* is the effective principle quantum number defined by

$$T = \frac{R Z_0^2}{n^{*2}} = \frac{R Z_0^2}{(n - \sigma)^2} \tag{4.1}$$

where T is the term value of the energy level of the atom in cm^{-1}, R is the Rydberg constant and σ is the quantum defect. The interval factor now becomes

$$a_j = g_I \frac{R \alpha^2 h c Z_i Z_0^2}{1836 n^{*3} (l + \frac{1}{2}) j (j + 1)}. \tag{4.2}$$

Eq. (4.2) may be modified with the use of the expression for the fine structure doublet splitting [13]

$$\Delta \nu = \frac{R \alpha^2 h c Z_i^2 Z_0^2}{n^{*3} l (l + 1)} \tag{4.3}$$

so that

$$a_j = g_I \frac{\Delta \nu \, l (l + 1)}{Z_i 1836 (l + \frac{1}{2}) j (j + 1)}. \tag{4.4}$$

A more exact relativistic treatment of the problem[1] shows that the interval factor is

$$a_j = g_I \frac{\Delta \nu \, l (l + 1)}{Z_i 1836 (l + \frac{1}{2}) j (j + 1)} \frac{\varkappa (j, Z_i)}{\lambda (l, Z_i)}$$

where

$$\varkappa (j, Z_i) = \frac{4 j (j + \frac{1}{2}) (j + 1)}{(4 \varrho^2 - 1) \varrho},$$

$$\varrho = \sqrt{(j + \tfrac{1}{2})^2 - (\alpha Z_i)^2},$$

$$\lambda (l, Z_i) = \left[\frac{2 l (l + 1)}{\alpha^2 Z_i^2} \right] \left\{ \sqrt{(l + 1)^2 - (\alpha Z_i)^2} - 1 - \sqrt{l^2 - \alpha^2 Z_i^2} \right\}. \tag{4.5}$$

For an s electron we may take

$$|\psi_{n,0}(0)|^2 = \frac{1}{\pi a_H^3} \cdot \frac{Z_i Z_0^2}{n^{*3}}. \tag{4.6}$$

This value of the wave function must be corrected by the factor $\varkappa (j, Z_i)$ and, according to Fermi and Segrè[2], must be multiplied by the additional factor

---

[1] G. Breit: Phys. Rev. **35**, 1447 (1930). — G. Racah: Z. Physik **76**, 431 (1931).
[2] E. Fermi and E. Segrè: Z. Physik **82**, 729 (1933). — Rend. Accad. Ital., Sci. fisiche, mat. e natur., Memorie **4**, 131 (1933).

$\left(1 - \dfrac{d\sigma}{dn}\right)$ where σ is the quantum defect defined by Eq. (4.1). Thus, the interval factor for an s electron becomes

$$a_s = g_I \frac{8R\alpha^2 hc Z_i Z_2^0}{3n^{*3} 1836}\left(1 - \frac{d\sigma}{dn}\right)\varkappa(j, Z_i).\tag{4.7}$$

For an s electron Z_i may be taken equal to the atomic number Z.

Spectroscopic measurements are usually made in units of wave-numbers ccm^{-1}) in which case the right-hand side of (4.7) should be divided by hc. In Eq. (4.5) a_i will be in the units of $\Delta\nu$.

If the atom has more than one electron outside the closed subgroups, the interaction of each electron with the nucleus must be considered. The vector model of the atom with the principle of energy sums enables one to obtain explicit formulae for both LS and jj coupling between the valence electrons. The simplest case is the configuration of two non-equivalent s electrons which give rise to a $^3S_1$ level. The interval factor can be shown to be

$$A(^3S_1) = \tfrac{1}{2}(a_{s_1} + a_{s_2}).\tag{4.8}$$

Formulae for other configurations are given by KOPFERMANN [9]. If the coupling between the electrons is intermediate a quantum mechanical calculation is required. Formulae for various configurations have been derived by BREIT and WILLS[1].

In some cases, particularly in heavy elements, the assumption, that the energy of interaction between the nuclear magnetic moment and the electrons is small compared to the magnetic spin-orbital interaction of the electrons, is not valid. In this case a perturbation arises due to the proximity of the energy levels of the same configuration. The interval rule is violated and the intensities of the various components are altered. The energy levels must be specified by the total angular momentum F, and J is no longer a good quantum number. Calculations which take this perturbation into account have been given by GOUDSMIT and BACHER[2].

The theory has been applied to the hyperfine structure of the $6s\,6d$ configuration in HgI and the $6s\,7p$ configuration in Bi IV.

b) Determination of nuclear spin.

5. From the number of hyperfine structure components. The number of hyperfine states of a given energy level is the smaller of $2I+1$ and $2J+1$ [Eq. (2.13)]. Then, if there is a transition involving a level which has $J > I$ and which has an observable hyperfine structure, the nuclear spin can be determined by simply counting the number of components in the transition.

For example, in studies[3] of the spectra of Am^{241} and Am^{243} with a large grating spectrograph it has been found that a large number of levels have a maximum of six hyperfine components. Many of the levels in this spectrum may be expected to have large values of J, so that the nuclear spin is $\tfrac{5}{2}$ for both isotopes. An extensive study of many lines in the spectrum of Pb showed[4] that the levels of Pb^{207} were split into at most two hyperfine structure states. The nuclear spin of Pb^{207} is then equal to $\tfrac{1}{2}$.

[1] G. BREIT and L. A. WILLS: Phys. Rev. **44**, 470 (1933).
[2] S. GOUDSMIT and R. F. BACHER: Phys. Rev. **43**, 894 (1933).
[3] M. FRED and F. S. TOMPKINS: Phys. Rev. **89**, 318 (1953). — J. G. CONWAY and R. D. McLAUGHLIN: Phys. Rev. **94**, 498 (1954).
[4] H. KOPFFRMANN: Z. Physik **75**, 36 (1932).

This method is applied, in most cases, to complicated spectra such as the rare earths, where many levels have large J values. Where possible, transitions between one level with large hyperfine structure splittings and another with negligible splitting should be observed. Care must be taken to see that unresolved structure does not mask one or more of the weaker components in the pattern.

6. From the interval rule. The interval rule states that the energy difference between two adjacent hyperfine states is equal to $A_J F$ where A_J is the interval factor and F is the greater of the two F values belonging to the two adjacent states. Then, if the relative separations of three hyperfine structure states can be measured, the ratio of the observed separations gives the ratio of two values of F, and I can be determined, provided J is known. In some special cases it is possible to determine the centroid of the hyperfine structure and the interval rule can be applied to the measured separations of two states from the centroid. Deviations from the interval rule may arise from quadrupole terms, or from perturbations from neighbouring levels.

In two-electron spectra a convenient transition is the one involving the lowest ${}^3S_1$ level, arising from a $n_1s\, n_2s$ configuration, and the ${}^3P_0$ level of an $ns\, np$ configuration. The ${}^3P_0$ level has no hyperfine structure since $J=0$. Then, provided $I>\frac{1}{2}$, the transition will have three hyperfine components corresponding to the structure of the ${}^3S_1$ level. There will be no deviation from the interval rule arising from the electric quadrupole, effect, since these terms vanish for this configuration.

In a study of hyperfine structure in Al II, transitions involving ${}^3S_1$ and ${}^3P_0$ levels have been observed[1]. The $3s\, 4s\ {}^3S_1$ was found to have three states. The ratio of the separation of the states is $\frac{321}{227}=1.41$. For $I=\frac{5}{2}$ the theoretical value of this ratio is $\frac{7}{2}:\frac{5}{2}=1.40$. The $3s\, 5s\ {}^3S_1$ level also has three states and the ratio of the observed separations is $\frac{305}{211}=1.44$ again confirming the value $I=\frac{5}{2}$ for the nuclear spin of Al^{27}.

For non-S states with $J>\frac{1}{2}$, deviations from the interval rule are used to calculate the values of nuclear quadrupole moments [19].

7. From the intensities of hyperfine structure components. Direct observation of the number of hyperfine components for the determination of spin is restricted to levels which have $J>I$. The interval rule cannot be applied to levels with $J=\frac{1}{2}$ or to nuclei with $I=\frac{1}{2}$ since in both cases there are at the most two hyperfine states. A method which can be applied to any transition is to measure the relative intensities of the hyperfine components.

The relative intensities of the components in a given transition can be derived in the same manner as the relative intensities of all the lines belonging to a given multiplet. The selection rule for F is

$$\Delta F = \pm 1, 0 \tag{7.1}$$

but with $F=0 \to F=0$ forbidden. The relative intensities of the permitted transitions have been tabulated (e.g. CONDON and SHORTLEY [3], p. 241).

In many cases the intensity ratios can be determined from an application of the sum rule which states that the sum of the strengths of the components having a given initial state is proportional to the quantum statistical weight $(2F+1)$ of that initial state, and the sum of the strengths of the components having a given final state is proportional to the statistical weight of that final

---

[1] M. HEYDEN and R. RITSCHL: Z. Physik **108**, 739 (1938).

state. In the simplest cases, only one of the two levels of the transition has observable hyperfine structure, and the intensities of the components are proportional to the statistical weight of the state from which the component arises. A measurement of the intensity ratios of the components enables one to determine the nuclear spin, provided the value of J for the level is known.

For example, in alkali-like atoms the level with the largest hyperfine structure is the ground $^2S_{\frac{1}{2}}$ level. Since $J=\frac{1}{2}$ there are only two hyperfine structure states regardless of the value of I. A comparison of Eqs. (4.2) and (4.7) shows that the interval factor for a p electron is much smaller than that for an s electron. Thus, in the first approximation the structure of the $^2P_{\frac{1}{2},\frac{3}{2}}$ levels may be neglected in comparison to the $^2S_{\frac{1}{2}}$ level. The resonance lines of the alkali atoms can then be expected to have two components with an intensity ratio equal to the ratio of the statistical weights of the two states of the $^2S_{\frac{1}{2}}$ level. The two values of F are $I+\frac{1}{2}$ and $I-\frac{1}{2}$ so that the ratio of the statistical weights is $(I+1):I$. For $I=\frac{1}{2}$ this ratio is $3:1$ which can be easily distinguished from the next possible value of the ratio, $2:1$, which occurs for $I=1$. For larger values of I the intensity ratio becomes more closely equal to unity and it becomes correspondingly more difficult to distinguish between different values of I.

Deviations from predicted intensity ratios may arise from perturbations due to neighbouring levels. These discrepancies can be explained by more complete calculations. However, in the experimental work it is important to make certain that the intensity ratios are not invalidated by self-absorption in the source. The observed intensities are very sensitive to self-absorption and are always made more nearly equal. For example, in a study of the transition $4s\,4p\,^3P_0$ to $4s\,5s\,^3S_1$ in the spectrum of calcium, enriched with Ca^{43}, it was found[1] that the intensity ratio of the two components due to Ca^{43} with the highest and lowest frequency was reduced from a value of 1.67 to be expected for $I=\frac{7}{2}$ to an observed value of 1.52. The effect of self-absorption on the intensity ratio of the hyperfine components can be estimated from a measurement of the departure from the theoretical value of the intensity ratio between the line in which the hyperfine structure is observed and another line belonging to the same multiplet.

8. **From the PASCHEN-BACK effect of the hyperfine structure.** When an atom containing a nucleus with non-zero nuclear spin is placed in a very weak magnetic field the total angular momentum, $\boldsymbol{F}$, orients itself in the magnetic field according to the magnetic quantum number, m_F, in complete analogy to the ZEEMAN effect of atomic energy levels. As the magnetic field is increased the coupling between $\boldsymbol{I}$ and $\boldsymbol{J}$ is broken down so that F is no longer a good quantum number. $\boldsymbol{I}$ and $\boldsymbol{J}$ orient themselves independently in the applied field, and the system is described by the magnetic quantum numbers m_I and m_J, following the rules of the PASCHEN-BACK effect. Since the coupling between $\boldsymbol{I}$ and $\boldsymbol{J}$ is of the order of 2000 times smaller than the coupling between $\boldsymbol{L}$ and $\boldsymbol{S}$ the PASCHEN-BACK effect of the hyperfine structure occurs at values of the magnetic field which are easily obtainable. Each ZEEMAN level, specified by a value of m_J, is then split into $2I+1$ hyperfine states. The transitions are limited by the selection rules $\varDelta m_I=0$, and $\varDelta m_J=\pm1,0$. If this hyperfine structure can be resolved the value of I can be determined by counting the number of hyperfine components.

JACKSON and KUHN[2] studied the PASCHEN-BACK effect of the resonance line of NaI, $\lambda=5890$ ($3s\,^2S_{\frac{1}{2}}$ to $3p\,^2P_{\frac{3}{2}}$), utilizing the absorption of the radiation

---

[1] F. M. KELLY, H. KUHN and ANNE PERY: Proc. Phys. Soc. Lond. A **67**, 450 (1954).
[2] D. A. JACKSON and H. KUHN: Proc. Roy. Soc. Lond., Ser. A **167**, 205 (1938).

in an atomic beam. The π components ($\Delta m_J = 0$) only were observed. The structure of the upper, $^2P_{\frac{3}{2}}$, level is small and can be neglected. There should be two groups of $2I + 1$ components. Two groups of four components were observed showing that the spin of Na^{23} is $\frac{3}{2}$. The magnitudes of the separations were found to be in good agreement with theoretical predictions. The results are illustrated in Fig. 3.

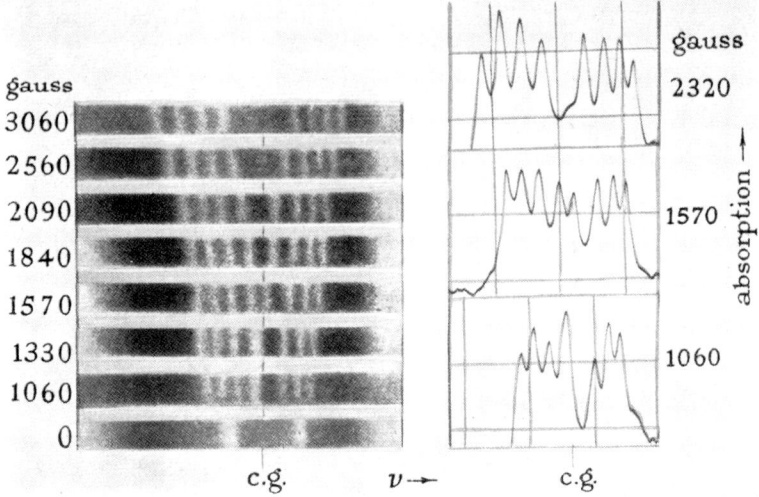

Fig. 3. The PASCHEN-BACK effect of the hyperfine structure of the NaI resonance line $3s\ ^2S_{\frac{1}{2}} - 3p\ ^2P_{\frac{3}{2}}$ (5890 Å).

c) Determination of the nuclear magnetic moment.

9. The sign of the nuclear magnetic moment. The nuclear magnetic moment is taken to be positive if the magnetic moment and the total angular momentum of the nucleus are parallel to one another. If the magnetic moment is antiparallel to the angular momentum, the magnetic moment of the nucleus is negative. In all but a few special cases (see Sect. 2) the magnetic field produced by the electrons at the nucleus is antiparallel to $\mathbf{J}$. Then, in the normal case with the magnetic field of the electron antiparallel to $\mathbf{J}$, and, with a positive nuclear magnetic moment the hyperfine structure state with the largest value of F ($\mathbf{I}$ and $\mathbf{J}$ parallel) lies highest in the energy level diagram. If the nuclear magnetic moment is negative the order of the hyperfine states is inverted and the largest value of F is the lowest in the energy level diagram. In a transition the strongest intensities of the hyperfine components are associated with the largest values of F so that the sign of the nuclear magnetic moment can usually be determined by visual inspection of the hyperfine components.

In MgI the $3s\ 4s\ ^3S_1$ level lies above the $3s\ 3p\ ^3P_0$ level as illustrated in Fig. 4a. The hyperfine structure arises from the $^3S_1$ level. Fig. 4 has been constructed with the assumption that the nuclear moment of Mg^{25} is negative and that $I = \frac{5}{2}$. Thus, the component with the highest frequency in Fig. 4a is the least intense. In Fig. 4b the $^3P_0$ level is assumed to lie above the $^3S_1$ level, and in this case the component with the highest frequency is the strongest. For a positive nuclear moment the intensity would be in the reverse order compared to that given above.

10. Calculation of the nuclear magnetic moment. The largest values of the hyperfine structure occur in levels involving a single unpaired s electron with the smallest possible principal quantum number n. The interval factor for a level due to a $p_{\frac{1}{2}}$ electron is larger than that due to a $p_{\frac{3}{2}}$ electron, not only due to the factor involving j explicitly in Eq. (4.5), but also due to the larger size of the relativistic factor $\varkappa(j, Z_i)$. For a d electron the interval factor is again reduced and the splitting of a level arising from a single d electron is usually beyond the limit of observation by optical means.

In alkali spectra the resonance lines arise from the transition between the ground $^2S_{\frac{1}{2}}$ level and the first excited ^2P levels. In the first approximation the structure of the ^2P levels may be neglected and each line has two components arising from the $^2S_{\frac{1}{2}}$ level in which F has the two values, $I + \frac{1}{2}$ and $I - \frac{1}{2}$. I can

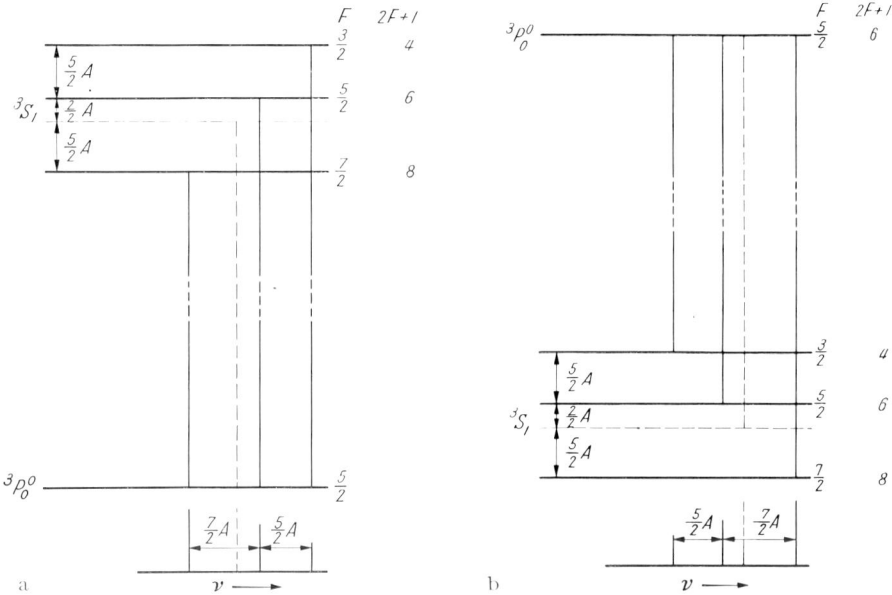

Fig. 4 a and b. Hyperfine structure in a transition between a $^3S_1$ level and a $^3P_0^0$-level. The nuclear magnetic moment has been assumed to be negative in each case. In (a) the $^3P_0$ level lies lowest in the energy level diagram and for a negative magnetic moment the weakest component has the highest frequency. In (b) the $^3P_0$ level is above the $^3S_1$ and with a negative magnetic moment the strongest component has the highest frequency. The intensities are proportional to the heights of the lines representing the components.

be determined from the ratio of the intensities of the two components. The total observed separation, with a small correction[1] for the ^2P structure, is equal to $a_s(I + \frac{1}{2})$ so that the value of a_s can be evaluated and used in Eq. (4.7) to calculate g_I, the nuclear g factor.

For an s electron, Z_i in Eq. (4.7) may be taken equal to Z, the atomic number. For alkalis, $Z_0 = 1$, while for singly ionized alkaline earths $Z_0 = 2$ and so forth. n^* is calculated from the term value [Eq. (4.1)]. $\varkappa(j, Z_i)$ can be calculated from its defining Eq. (4.5) or obtained from tabulated values[2].

Following CRAWFORD and SCHAWLOW[3], unperturbed energy levels of an atom satisfy a RYDBERG-RITZ series formula

$$T = \frac{R Z_0^2}{(n - \alpha - \beta T)^2} \tag{10.1}$$

[1] D. A. JACKSON: Proc. Roy. Soc. Lond., Ser. A **147**, 500 (1934).
[2] S. GOUDSMIT: Phys. Rev. **43**, 636 (1933).
[3] M. F. CRAWFORD and A. L. SCHAWLOW: Phys. Rev. **76**, 1310 (1949).

so that
$$\sigma = \alpha + \beta T \tag{10.2}$$
and
$$\frac{d\sigma}{dn} = \frac{\beta}{\beta - n^*/2\,T} . \tag{10.3}$$

The values of the quantum defect, σ, are calculated from the term values. A plot of σ versus T gives a straight line provided the energy levels are not perturbed and the slope of the line determines β. The FERMI-SEGRÈ factor, $\left(1 - \dfrac{d\sigma}{dn}\right)$, can then be evaluated with the use of (10.3).

A check on the accuracy of the FERMI-SEGRÈ factor can be made from a direct comparison of the values of the nuclear magnetic moment determined from spectroscopic data and the nuclear moment determined by the method of nuclear induction, or, by a comparison of two different s levels which have widely different $\left(1 - \dfrac{d\sigma}{dn}\right)$ factors. The latter method can be applied to the $6s$ and $7s$ levels in Au^{197}. The results [1] are summarized in Table 1. The good agreement of the values from the two levels with widely different FERMI-SEGRÈ factors justifies the use of this factor when evaluated by Eq. (10.3).

Table 1.

Level	Interval factor	$1 - \dfrac{d\sigma}{dn}$	μ_o
$6s$	0.102	1.424	0.132
$7s$	0.0105	1.038	0.133

Hyperfine structures of levels arising from a single p electron have been observed. For those atoms which have a p electron in the ground state reliable data can be obtained. For alkali and alkali-like atoms the p levels are excited states and the possibility of discrepancies due to interconfiguration perturbations must be considered. Eq. (4.5) may be used to estimate the nuclear g factor. In this equation $\Delta \nu$ is the fine-structure separation of the two j levels, but a value of Z_i must be estimated. For the fine structure separation $Z_i = Z - 4$ has been found to work very well. However, in hyperfine structure the average of a different power of Z is involved and this assumption is not as satisfactory. If the value of g_I is known from other experiments the observed hyperfine structure interval factor for a p electron can be used to evaluate $\overline{(1/r^3)}$ which quantity, is needed for the calculation of nuclear quadrupole moments [2].

Nuclear moments may be evaluated from the observed hyperfine structure splittings of levels arising from configurations involving two or more electrons. The largest structure are always found in configurations which have at least one unpaired s electron. The interval factor for the level involves the interval factors for the individual electrons. Uncertainties arise in the numerical approximations which must be made to allow for the screening of one of the valence electrons by the others, so that the results are usually not as reliable as those obtained from single electron configurations. An example of the method for dealing with the calculation of the magnetic moment from structures in two electron spectra is given in the calculation [3] of the moment of P^{31}.

11. The effect of finite nuclear size. For the light elements fairly good agreement has been found between the values of the nuclear magnetic moments measured by spectroscopic methods and those determined by the more accurate

[1] F. M. KELLY: Proc. Phys. Soc. Lond. A **65**, 250 (1952). See also G. WESSEL and H. LEW: Phys. Rev. **92**, 641 (1953).

[2] See for example H. LEW and G. WESSEL: Phys. Rev. **90**, 1 (1953).

[3] M. F. CRAWFORD and J. LEVINSON: Canad. J. Res. A **27**, 156 (1949).

nuclear induction method. However, for the heavy elements there is a marked disagreement between the spectroscopic results, calculated from Eqs. (4.5) and (4.7) and the induction values. This discrepancy may be largely attributed to the assumption that the nucleus is a point charge. In reality, the nucleus occupies a finite, though small, volume and the wave functions for the electrons in the region within the nucleus should be calculated using the electrical potential which corresponds to the actual charge distribution of the nucleus rather than with the usual COULOMB potential.

ROSENTHAL and BREIT [16] calculated the effect of the finite size of the nucleus on the interaction between the electron and the nuclear magnetic moment, but had no experimental data to test the theory. The work was extended by CRAWFORD and SCHAWLOW [16] to include cases where the nuclear charge is concentrated on the surface of the nucleus and where the nuclear charge is uniformly distributed throughout the volume of the nucleus with the assumption that the non-electrical forces between electrons and nucleons are relatively small.

The coupling of the electron and the nuclear magnetic moment is proportional to the integral

$$I = \int_0^\infty \varphi_1 \varphi_2 \, y^{-2} \, dy \qquad (11.1)$$

where φ_1 and φ_2 are the DARWIN-GORDON radial functions for the valence electron, and $y = 2Zr/a_H$,

Table 2.

Z	δ	Z	δ
20	0.003	60	0.033
30	0.006	70	0.058
40	0.011	80	0.101
50	0.019	90	0.180

with r measured from the centre of the nucleus. The numerical value of $\varphi_1\varphi_2$ may be expected to be smaller for a finite nucleus than for a point nucleus since the binding of the electron to the finite nucleus is smaller than the binding to the point nucleus. From the form of (11.1) it is evident that the integral is weighted most heavily in the region where the radial functions depart most from their values in a purely COULOMB field.

The wave functions φ_2 and φ_2 can be evaluated and a correction factor, δ, may be determined such that

$$\mu = \frac{\mu_0}{1 - \delta} \qquad (11.2)$$

where μ_0 is the value of the magnetic moment calculated with Eqs. (4.5) or (4.7) and μ is the value of the moment when the finite size is taken into account. The values of δ calculated for a uniform distribution of nuclear charge and for an s electron using a formula derived by A. BOHR[1] are given in Table 2. In this calculation the nuclear radius was taken to be $1.2\times10^{-13} A^{\frac{1}{3}}$ in accordance with the results from elastic scattering of high energy electrons by nuclei[2]. The effect is negligible for an electron with $j > \frac{1}{2}$.

12. The effect of the volume distribution of nuclear magnetic moment. An anomaly between the ratio of the nuclear magnetic moments of Rb[85] and Rb[87] determined with precision by the magnetic resonance method, and the ratio of accurate values of the hyperfine structure splittings determined by an atomic beam method led to the suggestion[3] that the nuclear magnetic moment is distributed throughout the volume of the nucleus, rather than being concentrated at a point. A calculation[4] of this effect has been carried out. This anomaly

---

[1] A. BOHR: Lecture notes, Columbia University 1951.
[2] K. W. FORD and D. HILL: Ann. Rev. Nucl. Sci. **5**, 25 (1955). Cf. the article by D. HILL in Vol. XXXIX of this Encyclopedia
[3] F. BITTER: Phys. Rev. **77**, 150 (1949).
[4] A. BOHR and V. F. WEISSKOPF: Phys. Rev. **77**, 94 (1950). — A. BOHR: Phys. Rev. **81**, 331 (1951).

cannot be ascribed to the volume distribution of the nuclear charge since an explanation on this basis would require a very large difference in the radii of the two odd isotopes of Rb.

A correction to be applied to the values of the nuclear moment derived from Eqs. (4.5) and (4.7) can be calculated from the formula given by BOHR and WEISS-KOPF and under the assumptions which are discussed by these authors. Then the final value of the nuclear moment is given by

$$\mu = \frac{\mu_0}{(1-\delta)(1-\varepsilon)} \tag{12.1}$$

where ε is the BOHR-WEISSKOPF factor which accounts for the volume distribution of magnetic moment in the nucleus.

13. Comparison between the spectroscopic and nuclear induction values of the magnetic moment. Comparisons of the values of the nuclear magnetic moments calculated from Eqs. (4.5) and (4.7) and the more accurate nuclear induction values have been made for several heavy elements[1]. When the correction for the finite size of the nucleus and the volume distribution of the nuclear magnetic moment are taken into account the discrepancy of about 20% between the spectroscopic and nuclear induction value of the moments for the heavy elements is much reduced. The calculation for Tl, Pb and Bi given in the literature were made with the nuclear radius equal to $1.5 \times 10^{-13} A^{\frac{1}{3}}$. The results given in Table 2 are computed with the more recently accepted value, $1.2 \times 10^{-13} A^{\frac{1}{3}}$. This change has decreased the finite size correction so that the good agreement between the two values of the nuclear moment for the heavy elements no longer exists and the spectroscopic values are about 2% too small. Selected examples[2] of comparisons of the two values are listed in Table 3.

Table 3.

Nucleus	Electron	μ_0	μ Spect.	μ Induction	Induction / Spect.
Na^{23}	$3s$	1.99	1.99	2.217	1.11
K^{39}	$4s$	0.365	0.365	0.391	1.07
Ca^{43}	$4s$	-1.27	-1.27	-1.3153	1.04
Ag^{107}	$5s$	-0.109	-0.111	-0.1130	1.02
Ag^{109}	$5s$	-0.127	-0.129	-0.12995	1.01
Ba^{135}	$6s, 7s$	0.788	0.820	0.8346	1.02
Ba^{137}	$6s, 7s$	0.881	0.918	0.9351	1.02
La^{139}	$6s$	2.65	2.73	2.776	1.02
Tl^{205}	$6s, 7s$ $8s, 9s$	1.41	1.59	1.627	1.02
Pb^{207}	$6s$	0.49(7)	0.57	0.5894	1.03
Bi^{209}	$6s, 7s$ $8s$	3.46	4.01	4.080	1.02

For the light elements the finite size correction is negligible. However, the discrepancy between the two values of the moments tends to be somewhat

---

[1] Tl^{205}. M. F. CRAWFORD and A. L. SCHAWLOW: Phys. Rev. **76**, 1310 (1949). — Pb^{207}. A. L. SCHAWLOW, J. N. P. HUME and M. F. CRAWFORD: Phys. Rev. **76**, 1876 (1949). — Bi^{209}. F. M. KELLY, R. RICHMOND and M. F. CRAWFORD: Phys. Rev. **80**, 295 (1950).
[2] A list of experimental values of hyperfine structure splittings determined up to 1952 given by P. BRIX and H. KOPFERMANN in LANDOLT-BÖRNSTEIN, Zahlenwerte und Funktionen. 6. Aufl., Bd. I, Teil 5. Berlin: Springer 1952.

larger. The reason for the disagreement may lie in some factor, such as a quantum electrodynamic correction, which has been neglected, or, the model on which the GOUDSMIT-FERMI-SEGRÈ formula is based may not be adequate.

d) Experimental methods.

14. Sources. A major problem in the spectroscopic investigations of hyperfine structure is the production of spectral lines with sufficiently small half-widths. The various causes of line broadening are discussed by TOLANSKY [18]. In sources which are used in hyperfine structure studies the most important effect is the DOPPLER broadening brought about by the random motion of the emitting atoms. A calculation[1] shows that the half-width of a spectral line emitted by the atoms of a vapour with molecular weight M and at absolute temperature T is

$$\Delta \nu = 2 (2R \log 2)^{\frac{1}{2}} (\nu/c) (T/M)^{\frac{1}{2}} \text{ cm}^{-1} \tag{14.1}$$

where R is the universal gas constant, c the speed of light and ν the wave-number of the line in question. Then, if there is a choice of more than one transition which will exhibit the hyperfine structure it is advantageous to choose the line with the smaller wave-number. The only parameter which the experimenter can control is the temperature of the vapour which emits the radiation.

A commonly used source where the temperature may be reduced by the use of a refrigerant is a modified SCHÜLER hollow-cathode. The material under study is placed in a cylindrical hole in a metal block which forms the cathode. An electrical discharge, carried by an inert gas, concentrates in the hollow cathode and sputters the material under study into the gas so that the desired spectrum can be observed. Suitable hollow-cathodes can be simply made from a glass-to-metal seal[2]. The cathode is mounted within the metal part of the seal which in turn is immersed in the refrigerant. This type of source is commonly cooled with liquid nitrogen but liquid hydrogen has been used successfully. It is important to keep the power input to the source as low as possible. The hollow-cathode source is employed for investigations with small samples of separated or enriched rare isotopes.

In order to reduce the half-width of a line produced by a source operated at room temperature by a factor of ten it would be necessary to run the source at $3°$ K. Since liquid helium has a low latent heat of vapourization a method of direct refrigeration can be used only with very low current densities[3]. However, effective temperatures of this order are easily obtained with an atomic beam source where the atoms are collimated to form a narrow beam. If the source is used in emission, the atomic beam is excited by electrons moving at right angles to the beam and the radiation is observed in a direction perpendicular to both the beam and the electron stream. Several such sources have been described in the literature [12]. A considerable simplification of the apparatus is achieved if the atomic beam is used to absorb radiation from an auxiliary source [7]. However, the experiments are then restricted to resonance radiation. An atomic beam source can produce spectral lines with half-widths approaching the natural half-widths of the radiation.

[1] R. MINKOWSKY and H. BRUCK: Z. Physik **95**, 274 (1935).
[2] H. KUHN and K. WOODGATE: Proc. Phys. Soc. Lond. A **64**, 1090 (1951). See also O. H. ARROE and J. E. MACK: J. Opt. Soc. Amer. **40**, 387 (1950).
[3] J. BROCHARD, R. CHABBAL, H. CHANTREL and P. JACQUINOT: C. R. Acad. Sci., Paris **241**, 935 (1955).

15. Optical instruments. If the hyperfine structures to be resolved are fairly large, a grating spectrograph can be employed with advantage. For small hyperfine structures the high resolving power is usually obtained with a FABRY-PÉROT interferometer using fixed spacers [18]. The separation between two consecutive orders of the same wave-length, or the usable spectral range, measured in wavenumbers, is $1/2t$ where t is the distance between the two interferometer plates The fraction of the interorder separation which can be resolved depends on the reflection coefficient of the metallic films on the interferometer plates and on the optical perfection of the plates themselves. In practice under optimum conditions about $\frac{1}{50}$ of an order can be resolved when silver films are used in the red or infra-red spectral regions. The highest obtainable reflection coefficients

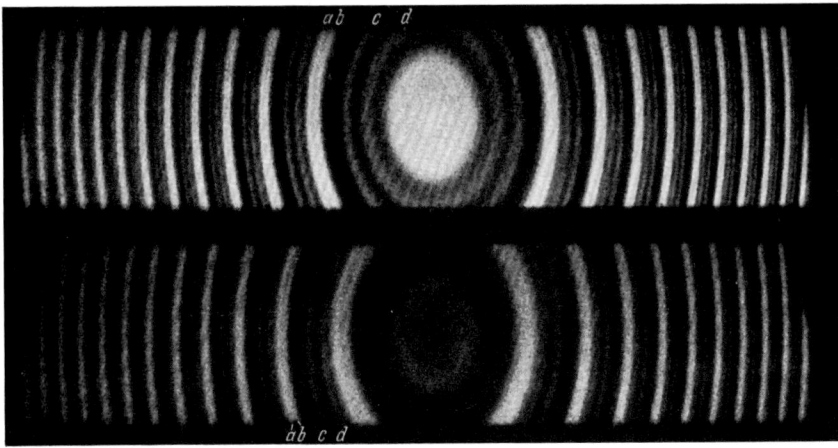

Fig. 5. Hyperfine structure in the resonance lines of AgI observed in the radiation from an atomic beam source and with the high resolution supplied by a FABRY-PÉROT interferometer using a 5 cm etalon.

decrease with decreasing wave-lengths. For the ultra-violet region of the spectrum evaporated aluminium films are in general use, but the resolution is never as good as that obtained with silver films in the red region of the spectrum [10]. The maximum resolving power of the FABRY-PÉROT etalon is closely matched by the optimum line width obtainable from an atomic beam source. Dielectric films are sometimes used to obtain high reflecting coatings [10].

The interferometer must be used in conjunction with a spectrograph which serves only to isolate the line or lines to be examined from radiation of other wavelengths.

An example of the interference patterns which are obtained is given in Fig. 5 which shows the hyperfine structure in the resonance lines of AgI. The upper (more intense) pattern shows the structure of the transition $5s^2S_{\frac{1}{2}}$—$5p^2P_{\frac{3}{2}}$ (3281 Å) and the lower one shows the structure of $5s^2S_{\frac{1}{2}}$—$5p^2P_{\frac{1}{2}}$ (3383 Å). There are two odd isotopes of silver. Four components marked a, b, c and d are observed. Components a and c can be assigned to Ag^{107}, while components b and c arise from Ag^{109} which is 5% less abundant. The intensity ratio a/c and b/d, with a correction for self-absorption, are equal to $3:1$ so that the nuclear spin is $\frac{1}{2}$ and the nuclear magnetic moment is negative. The magnitude of the moment can be calculated using Eq. (4.7). An atomic beam source was used to obtain this photograph. The FABRY-PÉROT etalon had a 5 cm spacer. The pattern repeats itself in various orders in the manner characteristic of this interferometer.

B. Atomic beams.

16. General. An atomic beam light source (Sect. 14) can be used with great advantage to produce radiation with a small half-width by observing in a direction at right angles to the direction of motion of the atoms in the beam. In comparison with the radiation from a gas, where the atoms move with random velocities, much smaller hyperfine structures can be observed. Radiative transitions between hyperfine structure states belonging to the same J value are permitted and a large increase in accuracy can be expected if these are directly observed. In the atomic beam method these transitions are observed by detecting the change in the beam intensity after a transition between two hyperfine states has taken place. This is in contrast to all other spectroscopic methods which detect the radiation, rather than detecting the atoms which have radiated.

The lifetime of the states which are involved in the transitions in the atomic beam apparatus must be sufficiently long to allow the atoms to travel through the deflecting fields and through the region where the energy corresponding to the transition between two hyperfine states is either absorbed or radiated. This time of flight is about 10^{-3} sec. The lifetimes of excited atomic states are about 10^{-8} sec, so that most observations are restricted to ground levels or to metastable excited levels. On the other hand, due to the long lifetime of the states under observation it is possible to determine their energies with a very high precision. The half-width of the transition is limited by the length of the region in which the transition is observed. The consequent small half-width means that very precise measurements can be made and important second order effects studied.

An atom or a molecule has a magnetic moment which depends on its state and on the strength of the applied magnetic field. The force on a magnetic dipole is the product of the component of the dipole moment in the field direction and the gradient of the field. Thus, the path which an atom takes through a magnetic field depends on both the magnitude of the field and the gradient of the field. The fields and field gradients of two magnets may be arranged so that a particle leaving a source on the axis of the system has a net deflection of zero at the position of a detector for a given state of the particle. If, however, the particle absorbs or radiates energy somewhere on its path, so that the state changes, the net deflection will no longer be zero and the transition can be observed as a decrease in the intensity of the beam.

17. Atomic beam magnetic resonance method. The various experiments using atomic or molecular beams to study atomic and nuclear properties are refinements and modifications of the classic experiments of STERN and GERLACH. The developments are fully outlined in two books by FRASER [4]. The non-resonance methods for atomic and molecular beams are now almost entirely superseded by the more precise resonance methods. Fig. 6 is a schematic diagram of an apparatus for the resonance method. More complete discussions are given by KELLOGG and MILLMAN [8], RAMSEY [14], [15], KUSCH [11] and SMITH [17].

For refractory materials the source is a heated oven, while for normally diatomic gases, such as Cl_2, it may be a discharge tube to produce a beam of atoms. The defining slits for the beam are the source exit and the collimating slit. The deflection of a beam with the largest obtainable field gradients is only a few hundredths of a millimeter so the defining slits and the detector must be very narrow, usually about 0.01 mm in width. The fore slit and the separating slit do not define the beam, but serve to isolate the source chamber from the main chamber. In order to prevent scattering from the beam and to keep the detector stable the pressure in the main chamber should be 10^{-7} mm of mercury or lower. For this reason the source and fore chambers are separately pumped to remove the relatively large amounts of gas evolved in the source chamber.

The magnets A and B are shaped to produce inhomogeneous magnetic fields. If dH/dz is made very large, an adequate deflection can be supplied by relatively short A and B fields. A large field gradient means that the magnetic field is also large. In large magnetic fields F is no longer a good quantum number and the vectors $\boldsymbol{I}$ and $\boldsymbol{J}$ orient themselves independently in the field. The resulting moment of the atom is

$$\mu = -\left(m_J g_J + \frac{m_I g_I}{1836}\right)\mu_0.\qquad(17.1)$$

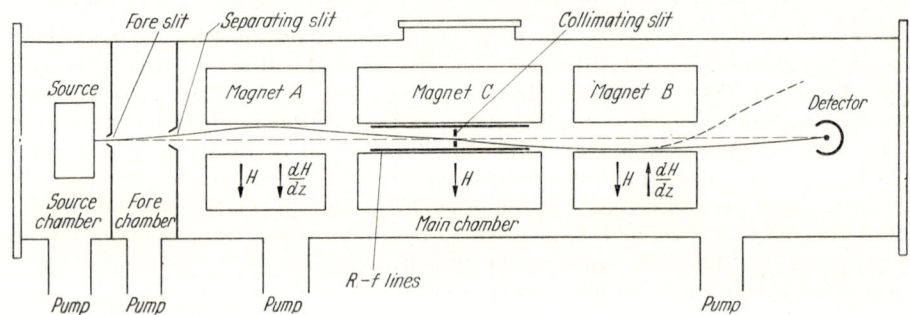

Fig. 6. Schematic diagram of the apparatus for the atomic beam magnetic resonance method.

Since the contribution from the nuclear magnetic moment is small, the deflection in the A and B fields will depend only on the value of m_J. Then, the beam intensity recorded at the detector will not change for transitions which leave the strong field value of m_J unchanged. The transitions which are detected by a decrease of the beam intensity are those which have strong field states with different values of m_J, independent of the field of the magnet C where the transitions actually take place.

Another arrangement is to make the A and B fields much longer so that an adequate deflection can be obtained with smaller values of the field gradient. Then, the field strengths of the A and B magnets may be in the region of intermediate field, or, in weak fields where F is still a good quantum number. The transitions which can now be detected are those in which weak or intermediate field states change. With suitable variations of the A and B magnets it is possible to arrange conditions under which all allowed transitions can be detected.

Magnet C is used to produce a homogeneous field in which transitions are induced by a radiofrequency field which is superimposed on the steady field. Absorption or induced emission will take place when the frequency of the applied field satisfies the BOHR rule

$$h\nu = E_n - E_m\qquad(17.2)$$

where ν is the frequency and E_n and E_m are the energies of two states of the atom in the magnetic field C. The field C must be as uniform as possible over the region in which transitions take place.

The selection rules for the transitions depend on the strength of the C field. For a weak field the ZEEMAN selection rules hold:

$$\Delta F = 0, \pm 1 \quad \text{and} \quad \Delta m_F = 0, \pm 1.\qquad(17.3)$$

For a strong field

$$\Delta m_I = 0, \pm 1 \quad \text{and} \quad \Delta m_J = 0, \pm 1. \tag{17.4}$$

The oscillating field is produced by a radiofrequency current in a "hairpin" shaped conductor in the C region.

A PIRANI gauge has been used to detect beams of hydrogen and other light gases. An ionization gauge has been used for the detection of heavy elements. Both these instruments measure the change in pressure in a chamber which is closed except for the narrow entrance slit for the beam. These detectors are relatively insensitive and have an inconvenientiy long response time. A more sensitive and stable detector with a rapid response utilizes the production of positive ions when atoms are evaporated from the surface of a heated metallic filament. If an atom strikes a heated tungsten filament it is evaporated as a positive ion provided the ionization potential of the atom is close to or less than the work function of the surface. This ion current is collected and amplified, and is proportional to the beam intensity. A method which can be universally electron bombardment and to detect the atoms of the beam in a applied[1] is to ionize the beam by mass spectrograph. This greatly extends the number of elements which can be detected but, due to the inefficiency of the ionization by electron bombardment, it is less sensitive and a strong source of atoms is required.

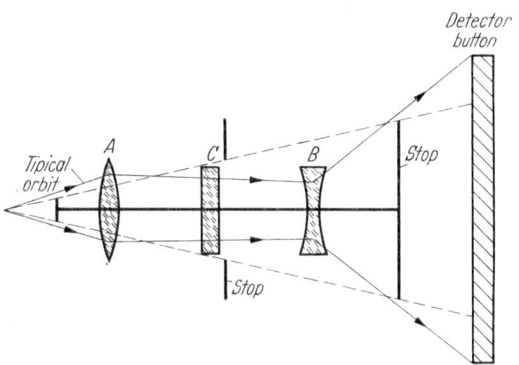

Fig. 7. Schematic diagram of the atom optics for atomic beam resonance method using focussing magnets and applied to radioactive nuclei.

An important modification of the method is to arrange the apparatus with the gradients of the A and B fields parallel to one another. In this case only those atoms which undergo a certain transition in the C field reach the detector. Atoms which do not undergo the transition are swept away. This restricts the number of observable transitions but has the advantage that it is possible to see lines arising from isotopes of small abundance in a beam which consists mostly of unwanted isotopes.

Excited states[2] of some alkali atoms have been studied with the atomic beam resonance method by irradiating the beam with resonance radiation in the C region. The alkali atoms are excited to their first $^2P_{\frac{1}{2}, \frac{3}{2}}$ levels and transitions between the various hyperfine states of these levels have been observed. The main application of these experiments is the determination of quadrupole moments from the departure from the interval rule in the $^2P_{\frac{3}{2}}$ level.

A recent modification[3], which can be applied to radioactive isotopes, replaces the conventional A and B magnets with six-pole focusing magnets. Magnet A, in Fig. 7, acts as a converging lens for atoms with positive values of m_J in the strong magnetic field, while magnet B acts as a diverging lens for atoms with negative values of m_J. The arrangement of the stops, shown schematically in Fig. 7, means that only those atoms which undergo a transition in the C field

[1] G. WESSEL and H. LEW: Phys. Rev. 92, 641 (1953).
[2] M. L. PERL, I. I. RABI and B. SENITSKY: Phys. Rev. 98, 611 (1955).
[3] A. LEMONICK, F. M. PIPKIN and D. R. HAMILTON: Rev. Sci. Instrum. 26, 1112 (1955).

which changes the sign of m_J follow the solid path in the diagram and reach the detector button. A single point on the resonance curve is obtained by exposing the detector button to the beam for a stated period, and then removing the button and determining the number of disintegrations per minute. The peak of the resonance curve, which consists of a plot of the frequency versus the counts per minute for a number of observations, gives the frequency of the transition. The obsverations are independent of the large number of non-radioactive atoms which are present in the beam.

18. Determination of nuclear spin. Optical experiments for the determination of nuclear spin require about a milligram of the isotope under study. The difficulty in obtaining this amount of radioactive material and the hazard involved in its use makes optical experiments impractical, except for nuclei with long half-lives. It is possible, however, to apply atomic beam techniques to radioactive nuclei, provided the apparatus is designed to conserve material and the detector does not record the large background of non-radioactive atoms which may be present.

Spin determinations with the atomic beam method depend on the measurement of the ZEEMAN effect of the hyperfine structure. The hyperfine structure separation, ΔW, is given by the interval rule [Eq. (2.16)]. Thus,

$$\Delta W = A_J F = h \Delta \nu \tag{18.1}$$

where A_J is the interval factor, F is the greater of the two F values belonging to the two adjacent hyperfine states which are under consideration, and, $\Delta \nu$ is the frequency corresponding to a transition between the two states. A magnetic field is weak if the ZEEMAN energy is small enough to be considered as a perturbation of the interaction between the electrons and the nuclear magnetic moment. Under these circumstances the total angular momentum F orients itself in the magnetic field and the ZEEMAN levels are specified by the values of the magnetic quantum number m_F.

In a strong magnetic field the coupling between the nuclear spin and the angular momentum of the electrons is broken down, and $\boldsymbol{I}$ and $\boldsymbol{J}$ orient themselves independently in the applied field. The energy of the system depends on the two magnetic quantum numbers, m_I and m_J. In both the strong and weak cases the calculation of the energy can be carried out in a relatively straight forward manner similar to that used for the PASCHEN-BACK and the ZEEMAN effects of the atomic fine structure (KOPFERMANN [9], p. 17; RAMSEY [14], p. 12). However, for the atomic beam resonance method intermediate field strengths are important. For this intermediate case the interaction between $\boldsymbol{I}$ and $\boldsymbol{J}$, and the interactions with the applied magnetic field must be treated as perturbations simultaneously and a secular equation can be obtained by a method similar to that used for the corresponding atomic fine structure case (CONDON and SHORTLEY [3]). Restricting the solution to the case where $J = \frac{1}{2}$ the energy of the states in a magnetic field is given by the BREIT-RABI[1] relation:

$$W(F, m) = -\frac{\Delta W}{2(2I+1)} - \frac{\mu_I}{I} H m \pm \frac{\Delta W}{2} \sqrt{1 + \frac{4m}{2I+1} x + x^2}. \tag{18.2}$$

ΔW is the hyperfine structure separation defined by (18.1). Since $J = \frac{1}{2}$, F is restricted to two values, and the positive sign in (18.2) is used for $F = I + \frac{1}{2}$

¹ G. BREIT and I. I. RABI: Phys. Rev. **38**, 2072 (1931). — MILLMAN, RABI and ZACHARIAS: Phys. Rev. **53**, 384 (1938).

while the negative sign applied to $F = I - \frac{1}{2}$. H is the applied magnetic field and x is a parameter depending on the field and given by

$$x = \frac{1}{\Delta W}\left(-\frac{\mu_J}{J} + \frac{\mu_I}{I}\right)H$$
$$= \frac{1}{\Delta W}\left(g_J + \frac{g_I}{1836}\right)\mu_0 H. \qquad (18.3)$$

For simplicity the magnetic quantum number, m_F, is written as m.

A plot of (18.2) is given in Fig. 8 for $I = \frac{3}{2}$. The main features of Fig. 8 are independent of I. All the states which have $F = I + \frac{1}{2}$ have positive effective

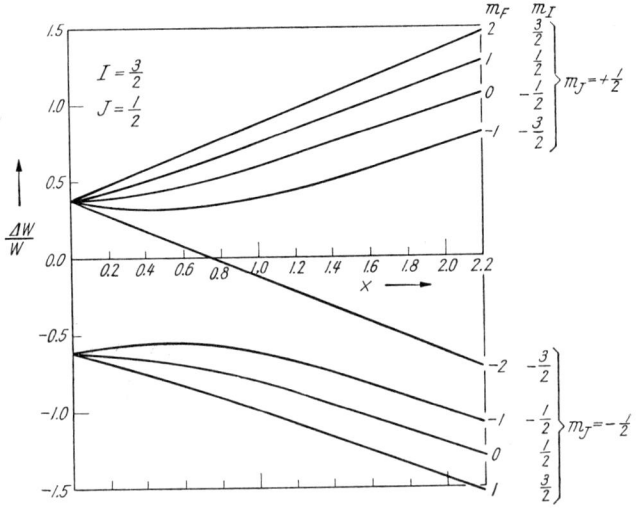

Fig. 8. The variation of the hyperfine structure states of an atom with a ground $^2S_{\frac{1}{2}}$ level in an external magnetic field. The nuclear spin is taken to be $\frac{3}{2}$ and the nuclear magnetic moment is assumed to be positive.

magnetic moments at high field strengths except the state with $m_F = -(I + \frac{1}{2})$. All the states with $F = I - \frac{1}{2}$, and the state with $m_F = -(I + \frac{1}{2})$ have negative effective magnetic moments at high field strengths. The crossing-over of the state with $m_F = -(I + \frac{1}{2})$ from the group of states belonging to $F = I + \frac{1}{2}$ at low field strengths to the group belonging to $F = I - \frac{1}{2}$ at high field intensity is well illustrated by an optical experiment summarized in Fig. 3. At high field strengths there are $2I + 1$ states corresponding to the various possible orientations of the nuclear spin, for each of the two values of m_J.

For small values of x ($\mu_0 H \ll \Delta W$) the terms in x^2 in (18.2) may be neglected. The second term in (18.2) $\left(\frac{\mu_I}{I} H m\right)$ may also be neglected due to the relatively small size of the nuclear magnetic moment. Then, the separation of any two adjacent m_F states in the ZEEMAN pattern of the hyperfine structure is

$$\Delta \nu = \frac{g_J \mu_0}{2h} \cdot \frac{H}{(I + \frac{1}{2})}. \qquad (18.4)$$

For a $^2S_{\frac{1}{2}}$ level $g_J = 2$ and substituting numerical values in (18.4), we obtain

$$\Delta \nu = 1.400 \frac{H}{I + \frac{1}{2}} \qquad (18.5)$$

where the final $\Delta\nu$ is in megacycles/sec. A measurement of this frequency in a known magnetic field determines the value of the nuclear spin.

The spins of four radioactive alkali nuclei have been measured by ZACHARIAS[1] and others with a conventional beam design, but with a long channel-shaped oven slit so that the atoms tend to leave the oven in a forward direction, and with a mass spectrograph incorporated into the apparatus following the hot tungsten wire detector to reduce the background of unwanted atoms. The magnetic field in the C region where the transitions take place can be measured by observing the resonance for a stable isotope with known properties, which in practice is present in the beam as a contamination. If $\Delta\nu_1$ is the resonance frequency for the isotope with known properties and $\Delta\nu_2$ is the frequency for the radioactive isotope we have

$$\frac{\Delta\nu_1}{\Delta\nu_2} = \frac{2I_2 + 1}{2I_1 + 1} . \tag{18.6}$$

The use of focusing magnets to replace the conventional A and B magnets (see Fig. 7) increases the beam intensity by a factor of twenty-five so that sources with a relatively small number of radioactive atoms can be employed. The spin of several radioactive elements[2] have been determined with this type of apparatus and the use of Eq. (18.6).

19. Determination of nuclear magnetic moments. The value of the hyperfine structure separation can be obtained by determining the resonant frequency for the atom in the C field at intermediate strength with the use of (18.2). More accurate observations of the splitting can be made once an approximate value is known by direct observation at very weak fields of the transitions $\Delta F = \pm 1$, $\Delta m_F = \pm 1, 0$. The nuclear magnetic moment may then be calculated with Eq. (4.7). If accurate values of the hyperfine structure separation and the nuclear magnetic moment for another isotope are known from other experiments the numerical factors in (4.7) can be determined from this data giving an increase in the accuracy of the determination of the moment of the new isotope.

A considerable part of the uncertainty of the determination of nuclear magnetic moments from optical experiments lies in the estimation of the numerical factors in (4.5) and (4.7) so that the large increase in accuracy of the value of the hyperfine structure splitting obtained by the atomic beam method does not always lead to a corresponding increase in the accuracy of the value of the nuclear moment. However, the increase in accuracy is very important for the study of second order effects. Such an effect is the so-called hyperfine structure anomaly already mentioned in Sect. 12. It has been observed that the ratios of the nuclear magnetic moments of Rb^{85} and Rb^{87} (and also K^{39} and K^{41}, and Ag^{107} and Ag^{109})[3,4] do not agree with the ratio of the two moments determined by the nuclear induction method.

BOHR and WEISSKOPF[5] have calculated the influence of the structure of the nucleus on the hyperfine splitting. The nuclear magnetic moment may be considered as arising from the spin and the orbital motion of the nucleons. The hyperfine structure anomaly will then depend on the values of g_S and g_L, the g-factors

[1] J. R. ZACHARIAS: Phys. Rev. **61**, 270 (1942). — L. DAVIS, D. E. NAGLE and J. R. ZACHARIAS: Phys. Rev. **76**, 1068 (1949).

[2] A. LEMONICK and F. M. PIPKIN: Phys. Rev. **95**, 1356 (1954). — CHRISTENSEN, HAMILTON, LEMONICK, PIPKIN and STROKE: Phys. Rev. **101**, 1389 (1956).

[3] S. A. OCHS, R. A. LOGAN and P. KUSCH: Phys. Rev. **78**, 184 (1950).

[4] P. B. SOGO and C. D. JEFFRIES: Phys. Rev. **93**, 174 (1954).

[5] A. BOHR and V. F. WEISSKOPF: Phys. Rev. **77**, 94 (1950). — A. BOHR: Phys. Rev. **81**, 331 (1951).

for the spin and orbital moments, and on g_I. It has been found that the assumption that $g_S = g$ (proton) and $g_L = 1$ for odd Z leads to agreement between theory and experiment. This experimental result supports the single particle model of the nucleus.

A similar discrepancy between the ratio of the hyperfine splittings for the ground states of hydrogen and deterium and the nuclear induction values of the proton and deuterium moments has been observed. The major part of this discrepancy has been explained[1] by considering the electron orbit in the region very close to the deuteron nucleus to be centered around the proton rather than around the centre of mass of the system. Then, for the region within the nucleus the neutron can be considered as a spherical distribution of magnetic moment outside the electron so that there is no interaction with the neutron in this region and the total hyperfine interaction is then reduced.

Very accurate measurements[2] of the hyperfine structure of the $^2P_{\frac{3}{2}}$ ground level of I^{127} and In^{115} have shown that the assumption of nuclear magnetic dipole and electric quadrupole moments is not sufficient to explain the observed intervals. The intervals can be fitted to the observed data if an interaction arising from a nuclear magnetic octupole moment is included in the calculations.

C. Molecular spectra.

20. Symmetry properties of homonuclear diatomic molecules. An important method for the determination of nuclear spins is the study of the band spectra or the rotational RAMAN effect of homonuclear diatomic molecules. The method is of special importance since it can lead to an unambiguous assignment of a zero nuclear spin. A homonuclear diatomic molecule is one in which the two nuclei are identical, for example, H_2^1 and O_2^{16}, but not $O^{16} O^{18}$. If, in the rotational spectra of such diatomic molecules alternate rotational lines are missing, a zero nuclear spin is indicated. On the other hand, alternating intensities in the rotational lines indicate a spin other than zero. These effects can be explained by a consideration of the symmetry properties of the complete eigenfunctions of homonuclear diatomic molecules.

The SCHRÖDINGER equation for a diatomic molecule takes the form

$$\frac{1}{m} \sum_i V_i^2 \psi + \sum_k \frac{1}{M_k} V_k^2 \psi + \frac{8\pi^2}{h^2} (E - V) \psi = 0 \tag{20.1}$$

where m is the electron mass, M_k is the nuclear mass and V^2 is the LAPLACE operator $\frac{\partial^2}{\partial x^2} + \frac{\partial^2}{\partial y^2} + \frac{\partial^2}{\partial z^2}$. The operator V_i^2 contains the co-ordinates of the electrons while V_k^2 contains the nuclear co-ordinates.

An important symmetry operation, known as an *inversion*, is the reflection of all the particles at the origin of the co-ordinates. For cartesian co-ordinates this means the x_i, y_i, z_i are replaced by $-x_i, -y_i, -z_i$. In (20.1) the potential depends only on the distances relative to the nuclei. Thus, a change in sign of the cartesian co-ordinates (an inversion) leaves the equation unchanged. Then, the eigenfunctions, ψ, which are solutions of (20.1) must either remain unchanged, or, since $-\psi$ is also a solution of (20.1), just change sign when an inversion is carried out. If the eigenfunction remains unchanged for an inversion of the co-ordinates, the parity of the state of the molecule is said to be positive, but if the eigenfunction changes sign the parity of the state is said to be negative.

[1] A. BOHR: Phys. Rev. **73**, 1109 (1948).
[2] V. JACCARINO, J. G. KING, R. A. SATTEN and H. H. STROKE: Phys. Rev. **94**, 1798 (1954). — P. KUSCH and T. G. ECK: Phys. Rev. **94**, 1799 (1954).

The division of the rotational levels of a diatomic molecule into positive and negative terms is quite general and does not depend on the approximations involved when the eigenfunction is separated into electronic, vibrational and rotational parts during the solution of (20.1). However, the separation of the eigenfunction into these three parts is needed for the purpose of assigning the parity to the individual terms.

A diatomic molecule in its ground state can be considered to be formed by two atoms which are both in their ground states. If, however, one of the atoms forming the molecule is in an excited state the resulting molecule is also in an excited electronic state. The energy differences between the various excidet electronic states and the ground electronic state of the molecule are large compared to the other molecular energies. The two nuclei of the diatomic molecule can vibrate along the internuclear axis of the molecule and the different frequencies of vibration correspond to different vibrational energy levels, so that each electronic energy level has a number of vibrational levels associated with it. Usually the vibrational energies are much smaller than the electronic energies. The molecule can also be considered as a rotator with the axis of rotation perpendicular to the internuclear line resulting in a number of rotational energy levels for each vibrational energy level. The rotational energies are in turn much smaller than the vibrational energies and are given by

$$E_{rot} = B J (J + 1) \tag{20.2}$$

where B is the rotational constant and J is the rotational quantum number which can take any integral value $0, 1, 2, \ldots$.

Then, for a diatomic molecule the total energy can be expressed as

$$E_0 = E_{el} + E_{vib} + E_{rot} \tag{20.3}$$

where E_{el} is the electronic energy, E_{vib} is the vibrational energy and E_{rot} is the rotational energy of the molecule, and $E_{el} \gg E_{vib} \gg E_{rot}$. Eq. (20.1) can be conveniently solved in the form

$$\psi = \psi_{el} \psi_{vib} \psi_{rot} \tag{20.4}$$

where ψ_{el}, ψ_{vib} and ψ_{rot} are the electronic, vibrational and rotational parts of the eigenfunction [5].

The eigenfunctions, ψ_{rot}, are those of the rigid rotator and are identical to the angular parts of the eigenfunctions of the hydrogen atom. They are expressed in terms of the so-called surface harmonics and contain only the angular co-ordinates of the nuclei. For the rotational eigenfunctions, an inversion just means a reversal of the direction. Then, the mathematical properties of the surface harmonics show that ψ_{rot} remains unaltered for even values of J, but changes sign for odd values of J when an inversion takes place.

The vibrational eigenfunctions, ψ_{vib}, depend only on the magnitude of the distance between the two nuclei so that an inversion always leaves ψ_{vib} unchanged and the vibrational eigenfunctions are symmetric in the nuclei.

The electronic eigenfunctions, ψ_{el}, contain the co-ordinates of all the electrons in the molecule. In the approximation (20.4), ψ_{el} is derived with the assumption that the nuclei are stationary so that the nuclear co-ordinates enter only as parameters. Now, for any diatomic molecule any plane through the line joining the nuclei is a plane of symmetry for the nuclei. If the discussion is limited to non-degenerate Σ states[1] of the molecule, the internuclear line is also an axis of symmetry for the electrons, and for a reflection in any plane which passes

---

[1] A Σ state is defined as one where the component (Λ) of the orbital angular momentum of the electrons along the internuclear axis is zero.

through the line joining the nuclei the eigenfunction must remain unchanged, or, just change sign. If the electronic eigenfunction remains unchanged for this reflection the state is known as a Σ^+ state, while an eigenfunction which changes sign is described as a Σ^- state. In diatomic molecules where the two nuclei have the same charge, but are not necessarily identical, there is also a centre of symmetry in the electric field at the midpoint of the line joining the two nuclei, and as a consequence of this the electronic eigenfunctions have an additional symmetry property. If the electronic eigenfunctions remain unchanged for a reflection of all the *electrons* at the centre of symmetry they are said to describe even (gerade) states while if the sign of the function changes for an inversion of the electrons it describes an odd (ungerade) state. Thus, there are four possible symmetry combinations for the electronic eigenfunctions of Σ states, Σ_g^+, Σ_g^-, Σ_u^+ and Σ_u^-.

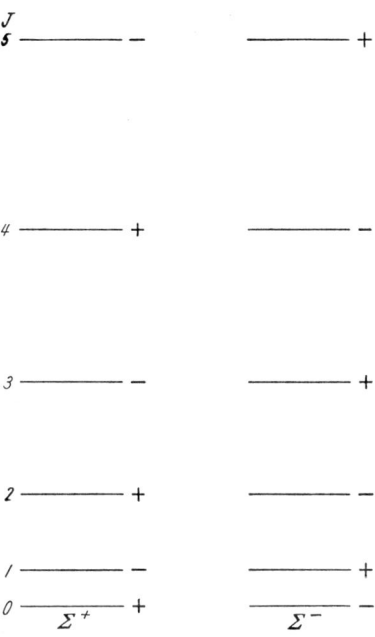

As already described, the vibrational eigenfunctions are always unchanged, while the rotational eigenfunctions remain unchanged for even J, but change sign for odd J during an inversion. The behaviour of the electronic eigenfunctions of a homonuclear diatomic molecule for an inversion may be determined by first rotating the molecule through 180° around an axis through the centre of symmetry at right angles to the line joining the two nuclei, and then, reflecting the electrons in the plane of symmetry passing through the internuclear line and perpendicular to the axis of this rotation. These two operations are equivalent to an inversion. The rotation through 180° does not change ψ_{el} since the electron co-ordinates are meas-

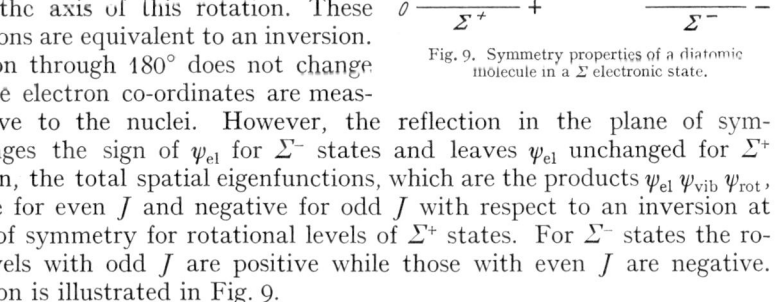

Fig. 9. Symmetry properties of a diatomic molecule in a Σ electronic state.

ured relative to the nuclei. However, the reflection in the plane of symmetry changes the sign of ψ_{el} for Σ^- states and leaves ψ_{el} unchanged for Σ^+ states. Then, the total spatial eigenfunctions, which are the products $\psi_{el}\,\psi_{vib}\,\psi_{rot}$, are positive for even J and negative for odd J with respect to an inversion at the centre of symmetry for rotational levels of Σ^+ states. For Σ^- states the rotational levels with odd J are positive while those with even J are negative. The situation is illustrated in Fig. 9.

The ground states of most known homonuclear diatomic molecules are in fact $^1\Sigma$ states. In excited states of molecules the component (Λ) of the orbital angular momentum along the internuclear axis may be different from zero so that states other than Σ states occur. Since Λ may take either of two directions with no change in the energy of the molecule such levels are doubly degenerate. This degeneracy, and multiplicities other than unity which arise when the total resultant spin angular momentum of the electrons is not zero, complicate the symmetry properties. A more complete discussion with references to original papers is given by HERZBERG [5]. For simplicity the discussions here are limited to $^1\Sigma$ states.

The symmetry properties which have been considered above apply also to homonuclear diatomic molecules. Now, if considerations are restricted to the homonuclear case, where the two nuclei are identical, a further element of symmetry enters, and for a homonuclear diatomic molecule (20.1) becomes fully symmetric in the co-ordinates of the two nuclei. Then, an exchange of the nuclei leaves the equation unchanged, and the solutions of the equation obtained by exchanging the nuclear co-ordinates are the same as those of the original equation, or, since $-\psi$ is also a solution of the original equation the sign of the solutions may change. In either case the square of the eigenfunction is not changed so that the probability density does not alter when the two nuclei are exchanged. If the sign of the eigenfunction remains unchanged the state is said to be sym-

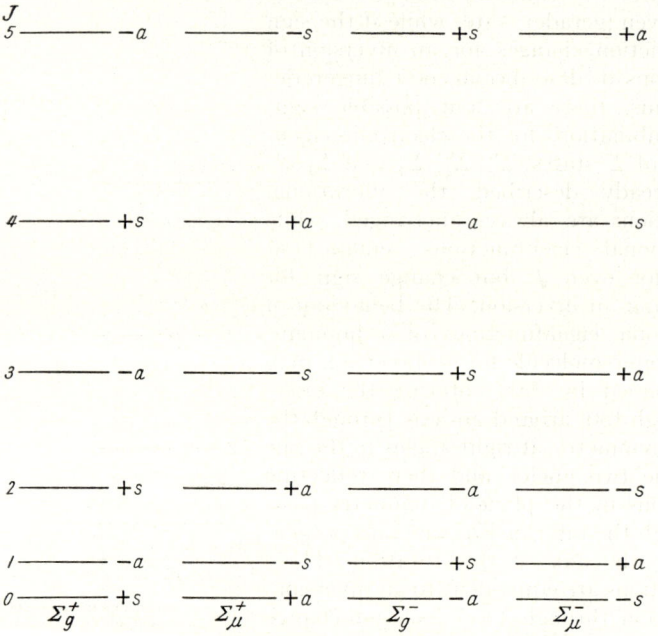

Fig. 10. Symmetry properties of a homonuclear diatomic molecule for the rotational levels of Σ electronic states. The $+$-sign means a positive rotational level while the $-$-sign indicates a negative rotational level. s means that the level is symmetric and a means that the level is antisymmetric.

metric for the exchange of the two nuclei, and if the sign of the eigenfunction changes the state is said to be antisymmetric. The above discussion has implicitly assumed non-degenerate states and has ignored nuclear spin. The result, however, does apply to degenerate levels. The influence of nuclear spin will be considered in Sect. 22.

The symmetry properties for the exchange of identical particles are more familiarly known for the exchange of two electrons in an atom. The SCHRÖDINGER equation for an atom is fully symmetric in the spatial co-ordinates of the electrons. Then, with the same reasoning as in the molecular case, the eigenfunction for the spatial co-ordinates must either remain the same or just change sign for the exchange of the spatial co-ordinates of any two electrons. According to the PAULI principle, which was discovered as a result of the study of atomic spectra, no two electrons may have identical sets of the four quantum numbers n, l, m_l and m_s. In quantum mechanics this means that only those states, whose complete eigenfunctions (including electron spin) are antisymmetric for the exchange of two electrons, occur in nature.

In order to determine whether the rotational levels are symmetric or anti-symmetric all the particles are reflected at the origin and then the electrons only are reflected back to their original positions. These two operations are equivalent to an exchange of the two nuclei. The reflection of all the particles changes the sign of the eigenfunction for negative levels and leaves the eigenfunction unchanged for positive levels. In the second operation the electrons alone are reflected at the origin and the sign of the eigenfunction changes for odd electronic states ad remains unchanged for even electronic states. Thus, positive rotational levels are symmetric in the nuclei for even electronic states such as Σ_g, and are antisymmetric for odd electronic states such as Σ_u. The opposite is true for negative rotational levels. It has already been shown that for Σ^+ states the rotational levels are positive for even J and negative for odd J, and for Σ^- states the opposite is true. The resulting symmetry properties of $^1\Sigma$ states are summarized in Fig. 10. The various symmetry operations and their results are collected in Table 4.

Table 4. *Symmetry properties of Σ states.*

Operation	Result
An inversion. All particles are reflected at the origin. $x_i,\ y_i,\ z_i \rightarrow -x_i,\ -y_i,\ -z_i$	$\psi = \psi_{el}\,\psi_{vib}\,\psi_{rot}$ $\psi \rightarrow \psi$ defines positive $(+)$ $\psi \rightarrow -\psi$ defines negative $(-)$
An inversion.	$\psi_{rot} \rightarrow \psi_{rot}$ for even J $\psi_{rot} \rightarrow -\psi_{rot}$ for odd J
An inversion.	$\psi_{vib} \rightarrow \psi_{vib}$
For any diatomic molecule and a reflection in any plane through the internuclear line.	$\psi_{el} \rightarrow \psi_{el}$ defines Σ^+ $\psi_{el} \rightarrow -\psi_{el}$ defines Σ^-
For a diatomic molecule with nuclei having the same charge and a reflection of all the electrons at the centre of symmetry of the electric field.	$\psi_{el} \rightarrow \psi_{el}$ defines Σ_g (even) $\psi_{el} \rightarrow -\psi_{el}$ defines Σ_u (odd)
For a homonuclear diatomic molecule. A rotation of $180°$ about an axis perpendicular to internuclear line then a reflection in the plane through the internuclear line = an inversion.	$\Sigma^+ \begin{cases} J \text{ even } + \\ J \text{ odd } - \end{cases}$ $\Sigma^- \begin{cases} J \text{ even } - \\ J \text{ odd } + \end{cases}$
For homonuclear diatomic molecules and an exchange of the co-ordinates of the nuclei.	$\psi \rightarrow \psi$ defines symmetric (s) $\psi \rightarrow -\psi$ defines antisymmetric (a)
For a homonuclear diatomic molecule. An inversion followed by a reflection of the electrons to their original position = exchange of the nuclei.	$\Sigma_g^+ \begin{cases} J \text{ even } +s \\ J \text{ odd } -a \end{cases}$ $\Sigma_u^+ \begin{cases} J \text{ even } +a \\ J \text{ odd } -s \end{cases}$ $\Sigma_g^- \begin{cases} J \text{ even } -a \\ J \text{ odd } +s \end{cases}$ $\Sigma_u^- \begin{cases} J \text{ even } -s \\ J \text{ odd } +a \end{cases}$

21. Homonuclear diatomic molecules with zero nuclear spin. The probability that a transition with dipole radiation will occur, depends on the value of the square of matrix elements of the form

$$R_x^{mn} = \int \psi_m^* \psi_n \left(\Sigma\, e_i\, x_i\right) d\tau \qquad (21.1)$$

where ψ_m and ψ_n are the eigenfunctions of the two levels involved in the transition and $\Sigma e_i x_i$ is the x-component of the dipole moment. The value of this definite integral must remain invariant for an inversion or any other transformation of co-ordinates. Since $\Sigma e_i x_i$ changes sign for an inversion, $\psi_m \psi_n$ must also change sign. Thus, transitions with dipole radiation can occur from positive to negative terms, but transitions between terms with the same parity are forbidden. On the other hand the selection rules for the RAMAN effect are derived from a consideration of the matrix elements of the polarizability of the molecule and the opposite selection rule holds. Thus, for the RAMAN effect positive terms combine with positive terms while changes in parity are forbidden. Dipole or RAMAN transitions are the only ones of importance in molecular spectra.

The probability of a transition of any type occuring depends on integrals of the form $\int \psi_m M \psi_n d\tau$ where, in the case of dipole radiation, M stands for the dipole moment. The value of this integral must be invariant for an exchange of the two nuclei in homonuclear molecules. The x component of the dipole moment, $\Sigma e_i x_i$, is symmetric in the two nuclei. Then, $\psi_m \psi_n$ must also be symmetric in the two nuclei or the integral would change sign when the nuclei are exchanged. This same result holds for quadrupole and all other types of radiation. The interaction for a molecule undergoing a collision is also symmetric in the two nuclei so that the condition that the two states must have the same symmetry properties in the two nuclei is quite general. Thus, for homonuclear diatomic molecules the only possible transitions are those between two levels which are either both symmetric, or, both antisymmetric in the nuclei. The prohibition of intercombinations is strict. A zero nuclear spin has been assumed in this discussion.

A molecule in a symmetric state cannot, then, with any type of transition go over into an antisymmetric state and a molecule in a symmetric state will always remain in a symmetric state. The same is true for antisymmetric states. Then, one would expect to find a particular kind of homonuclear diatomic molecules in either symmetric or antisymmetric states, but not in both kinds of states, so that every other rotational level for this molecule does not exist.

The prohibition of intercombinations taken with the selection rule for dipole radiation that the parity of the levels must change means that pure rotational and rotational-vibrational infrared spectra cannot occur with dipole radiation in homonuclear diatomic molecules. This is apart from the fact that the dipole moment of these molecules vanishes. Fig. 10 shows that the positive levels are all symmetric and the negative levels are all antisymmetric, or, that the opposite is true. Thus, transitions with dipole radiation cannot occur unless the electronic level changes.

In order to discuss transitions from one electronic level to another we may, for example, consider a molecule which has a Σ_g^+ ground electronic state and statistics such that only the symmetric rotational levels exist. Transitions from an excited electronic state such as Σ_u^+ are permitted by the selection rules and are observed. Dipole transitions are limited by the two symmetry rules already discussed and also by the selection rule $\Delta J = \pm 1$. In the ground electronic state only the rotational levels with even values of J occur while in the upper level only odd values of J appear. An examination of Fig. 10 shows that every other rotational line in the band is missing in comparison with the bands of a diatomic molecule where the two nuclei are different. An analysis of the spectrum permits a clear decision concerning the missing rotational levels so that the experiment provides unambiguous evidence that the nuclear spin is zero.

As already mentioned, the selection rules for the RAMAN effect differ from those for dipole radiation in that transitions occur between two positive terms or between two negative terms while changes of parity are forbidden. Changes in J are limited by the selection rule $\Delta J = 0, \pm 2$. This means that RAMAN transitions can be observed without change in electronic levels, since rotational levels for which J changes by 2 have the same symmetry properties. Since the levels are all symmetric or the levels are all antisymmetric only the rotational levels with even J, or only the rotational levels with odd J can occur in any electronic state of a homonuclear diatomic molecule. For the pure rotational RAMAN effect the molecules are all in their ground electronic and vibrational states and the RAMAN transitions will occur between the various rotational levels of the ground state. Since every other rotational level does not occur it is apparent that every other rotational line in the RAMAN spectrum will be missing. Again, an analysis of the spectrum (HERZBERG [5] p. 132) permits an unambiguous decision that the levels are missing so that an unambiguous assignement of zero spin for the nuclei of the molecule can be made.

The rotational RAMAN spectrum of O_2^{16} has been observed and an analysis shows that the even numbered rotational levels are missing[1]. This means that the nuclear spin of O^{16} is zero. An examination of the bands arising from electronic transitions confirms this value of the nuclear spin, since alternate rotational lines are missing. A study of the microwave absorption spectrum[2] of the O_2^{18} molecule has shown that alternate rotational levels are missing so that the spin of O^{18} is zero.

The bands from S_2 molecules can be expected to be similar to those of O_2 and this has been found to be the case. The fact that alternate lines in the bands are missing, shows conclusively that the nuclear spin of the most abundant naturally occurring isotope, S^{32}, is zero. Similar observations have been made for Se^{80}.

Helium does not form a stable diatomic molecule, but a number of band systems for the He_2^4 molecule can be observed. An analysis of these bands shows that alternate rotational levels are missing, proving that the nuclear spin of He^4 is zero.

The bands from homonuclear diatomic molecules formed from the naturally abundant C^{12} isotope and also from the important radioactive C^{14} nuclei both exhibit the missing rotational lines of diatomic molecules whose nuclei have zero spin. On the other hand, the bands from the C^{13} molecule do not have missing lines but show an alternation in intensity which is characteristic of non-zero nuclear spin.

It is usual to assume that nuclei with even charge and mass numbers have zero spin. However, many assignments of zero spin have been made from the fact that it has not been possible to observe splittings of energy levels when one would expect a separation with a reasonable value of the magnetic moment of the nucleus. An assignment of a zero spin from data of this type is not completely conclusive. Although there are no known exceptions to the rule that even-even nuclei have zero spin, some assignments of zero spin may have to be treated with reserve.

22. Homonuclear diatomic molecules with non-zero nuclear spin. Before considering the exchange of two nuclei in a homonuclear diatomic molecule when the nuclear spin is different from zero we will return to the exchange of two electrons in an atom, as introduced in Sect. 20, and further simplify the problem by limiting

---

[1] The ground state of O_2 is actually a $^3\Sigma_g^-$ state but the conclusion is not altered by this higher multiplicity.

[2] S. L. MILLER, A. JAVANS and C. H. TOWNES: Phys. Rev. **82**, 454 (1951).

the discussion to a helium atom where there are only two electrons. For the helium atom the SCHRÖDINGER equation is

$$\nabla_1^2 \psi + \nabla_2^2 \psi + \frac{8\pi^2 m}{h^2} \left(E + \frac{Z e^2}{r_1} + \frac{Z e^2}{r_2} - \frac{e^2}{r_{12}} \right) \psi = 0 \qquad (22.1)$$

where ∇_1^2 and ∇_2^2 are the LAPLACE operators for the first and second electrons, r_1 and r_2 are the distances between the first and second electrons and the nucleus and r_{12} is the distance between the two electrons. This equation is fully symmetric in the co-ordinates of the two electrons. An exchange of the electron co-ordinates then leaves the equation unaltered and the solutions of the equation are either symmetric or antisymmetric for an exchange of the two electrons.

A solution of (22.1) may be obtained by considering the electrostatic interaction between the two electrons, e^2/r_{12}, as a perturbation. The zero order of the eigenfunctions may then be taken as

$$\psi_0 = \varphi_n(x_1)\, \varphi_m(x_2) \qquad (22.2)$$

where x_1 and x_2 represent the position co-ordinates of the electrons and φ_n and φ_m are two hydrogenic eigenfunctions corresponding to the two sets of quantum numbers n and m. The zero order approximation may be symmetrized by taking the following linear combinations,

$$\psi^s = \frac{1}{\sqrt{2}} \left[\varphi_n(x_1)\, \varphi_m(x_2) + \varphi_n(x_2)\, \varphi_m(x_1) \right], \qquad (22.3)$$

$$\psi^a = \frac{1}{\sqrt{2}} \left[\varphi_n(x_1)\, \varphi_m(x_2) - \varphi_n(x_2)\, \varphi_m(x_1) \right] \qquad (22.4)$$

where ψ^s is symmetric and ψ^a is antisymmetric for the exchange of two electrons. This operation also removes the exchange degeneracy so that these two functions belong to two different energy levels. These symmetry properties must also be preserved in solutions of higher order. The eigenfunctions ψ^s and ψ^a contain only the position co-ordinates of the two electrons.

The arguments in Sect. 21 may be applied here to show that transitions between levels with differing symmetry are strictly forbidden. However, levels which have both types of symmetry are found in helium atoms so that some property of the electron which will enable transitions between levels which differ in their symmetry properties for the spatial co-ordinates has been neglected.

In Eq. (22.1) terms involving the electron spin have been ignored. If the spin-orbital interaction is now added to the wave equation, the extra term may be considered as a perturbation. Then, the total eigenfunction is a product of the co-ordinate eigenfunction, considered above, and a spin function which depends on the orientations of the electron spins. For helium the approximation involved here is a very good one since the spin-orbital interaction is very small.

The electron spin may take either of two orientations with respect to a fixed direction so that the spin function for each of the two electrons may take two values depending on whether the electron spin is parallel or antiparallel to the fixed direction. If the electron spin is parallel to the fixed direction the value of the spin function is χ^+ and if the spin is antiparallel to the fixed direction the value of the spin function is χ^-. In this discussion we are dealing with two electrons so that there are four possible combinations of spin directions. If the spins of both electrons are parallel to the fixed direction the spin function for the system is χ_1^s and if both electrons are antiparallel to the fixed direction the spin function is χ_2^s, where these functions are given by,

$$\left. \begin{aligned} \chi_1^s &= \chi^+(1) \cdot \chi^+(2) , \\ \chi_2^s &= \chi^-(1) \cdot \chi^-(2) . \end{aligned} \right\} \tag{22.5}$$

In these equations the arguments (1) and (2) refer to the first and second electrons respectively. These two functions do not alter if the two electrons are interchanged, so they are both symmetric as indicated by the superscript s. The other spin functions are $\chi^+(1) \cdot \chi^-(2)$ and $\chi^+(2) \cdot \chi^-(1)$ which apply when one of the electrons is parallel to the fixed direction and the other is antiparallel. These two functions are neither symmetric nor antisymmetric for an exchange of two electrons and must be symmetrized in a manner analogous to that used for Eqs. (22.3) and (22.4). This gives two more functions one of which is symmetric and the other antisymmetric as follows:

$$\left. \begin{aligned} \chi_3^s &= \frac{1}{\sqrt{2}} \left[\chi^+(1) \cdot \chi^-(2) + \chi^+(2) \cdot \chi^-(1) \right] , \\ \chi_4^a &= \frac{1}{\sqrt{2}} \left[\chi^+(1) \cdot \chi^-(2) - \chi^+(2) \cdot \chi^-(1) \right] . \end{aligned} \right\} \tag{22.6}$$

Thus, there is a total of three symmetric and one antisymmetric spin functions for a system of two electrons.

The PAULI exclusion principle is introduced through the requirement that the *complete* eigenfunction of the system must be antisymmetric for an exchange of two electrons. This means that only antisymmetric products of the eigenfunctions for the spatial co-ordinates and the spin functions are acceptable. The functions given by the symmetric products do not represent states which occur in nature. Then, the four functions $\psi^a \chi_1^s$, $\psi^a \chi_2^s$, $\psi^a \chi_3^s$, and $\psi^s \chi_4^a$, which are antisymmetric for the exchange of two electrons, represent observable states of the helium atom. The antisymmetric spatial eigenfunction appears three times, and the three products in which it occurs represent the triplet system of energy levels, while the single eigenfunction containing ψ^s represents the singlet system. In the approximation considered here the triplet levels are degenerate since the spin-orbital interaction has been neglected in (22.1). The triplet character of the terms which are antisymmetric in the spatial co-ordinates is expressed by a three-fold weight while the singlet terms are given a weight of unity. The prohibition of radiative transitions between symmetric and antisymmetric levels is still quite strict for helium since the intercombination lines are not observed although both types of levels can be excited.

The symmetry characteristics of the eigenfunctions for the spatial co-ordinates of a diatomic molecule have already been discussed in Sect. 20. We need now to discuss the exchange of two identical nuclei in a diatomic molecule when the nuclear spin is different from zero. The problem is analogous to that of the exchange of two electrons except that allowance must be made for nuclear spins of values $\frac{1}{2}, 1, \frac{3}{2}, 2, \ldots$, and the complete eigenfunctions may be either symmetric or antisymmetric for the exchange of the two nuclei depending on whether the nuclei follow BOSE-EINSTEIN or FERMI-DIRAC statistics.

For nuclei which have a spin $I = \frac{1}{2}$ the problem of the interchange of two nuclei is completely analogous to the exchange of two electrons. The complete eigenfunction for the molecule is made up of a product of the spatial co-ordinate function[1], ψ, and a nuclear spin function. For a molecule this approximation is a very good one since the coupling of the nuclear spin to the rest of the molecule

---

[1] Strictly the co-ordinate function ψ also contains co-ordinates for electron spin.

is quite weak. Just as in the case of the helium atom the value of the spin function depends only on the orientation of the spin with respect to a fixed direction. With $I = \frac{1}{2}$ the spin function for each nucleus can take only two values χ^+ and χ^- depending on whether the spin is parallel or antiparallel to the fixed direction. Then, if nuclei with $I = \frac{1}{2}$ follow FERMI-DIRAC statistics, the reasoning used above for the helium atom shows that the rotational levels with antisymmetric co-ordinate functions have a statistical weight of 3, while symmetric rotational levels have a statistical weight of 1, excluding the statistical weight due to J. Then, contrary to the case of the diatomic molecule where the nuclei have zero spin, rotational levels with both symmetric and antisymmetric co-ordinate eigenfunctions appear, but for $I = \frac{1}{2}$ the antisymmetric rotational levels appear three times as frequently as the symmetric ones.

In a diatomic molecule which contains nuclei with nuclear spin $I > \frac{1}{2}$ the projection M_I of the nuclear spin on a fixed direction can have $2I + 1$ different values. There are also $2I + 1$ different values of the spin function for a single nucleus, $\chi^{+I}, \chi^{+(I-1)}, \ldots, \chi^{-I}$, corresponding to the $2I + 1$ values of M_I. The spin function for the two nuclei can have $2I + 1$ symmetric functions made up of the products $\chi^{+I}(1) \cdot \chi^{+I}(2), \chi^{+(I-1)}(1) \cdot \chi^{+(I-1)}(2), \ldots, \chi^{-I}(1) \cdot \chi^{-I}(2)$. Other products are degenerate in pairs due to the exchange degeneracy. These degenerate pairs are symmetrized by taking linear combinations in the manner used in (22.6) to form functions represented by

$$\frac{1}{\sqrt{2}} \left[\chi^{+I}(1) \cdot \chi^{+(I-1)}(2) + \chi^{+I}(2) \cdot \chi^{+(I-1)}(1) \right]. \tag{22.7}$$

There are $I(2I + 1)$ such functions corresponding to the number of ways the $2I + 1$ χ-functions can be selected two at a time. The functions (22.7) are symmetric so that we have a total of $(I + 1)(2I + 1)$ symmetric spin functions. The functions (22.7) are made antisymmetric by changing the sum to a difference so that there are $I(2I + 1)$ possible antisymmetric spin functions.

Since the interaction between the nucleus and the rest of the molecule is very small compared to the other energies considered here the states represented by the symmetric spin functions (or the antisymmetric spin functions) are degenerate. As shown above the eigenfunction, ψ, for the spatial co-ordinates of the rotational levels is either symmetric or antisymmetric with respect to the exchange of the two nuclei. If the nuclei follow FERMI-DIRAC statistics the complete eigenfunction of the homonuclear diatomic molecule, formed by the product of ψ and the spin function for the two nuclei is antisymmetric. Then, rotational levels with antisymmetric ψ's are associated with symmetric spin functions and thus have a relative weight of $(I + 1)(2I + 1)$. Rotational levels with symmetric ψ's also appear but with a relative weight $I(2I + 1)$ from the antisymmetric spin functions.

On the other hand nuclei which follow BOSE-EINSTEIN statistics form molecules whose complete eigenfunction is symmetric for the exchange of the two nuclei. Then the rotational levels with antisymmetric ψ's have a relative weight of $I(2I + 1)$ while the rotational levels with symmetric ψ's have a weight of $(I + 1)(2I + 1)$. For either BOSE-EINSTEIN of FERMI-DIRAC statistics the ratio of the statistical weights is $(I + 1):I$. The symmetry of the eigenfunctions for a non-zero nuclear spin is summarized in Table 5.

Then, if a substance consists of homonuclear diatomic molecules containing nuclei whose spin is different from zero rotational levels with both antisymmetric and symmetric eigenfunctions for the spatial co-ordinates appear but with differing statistical weights. The selection rule prohibiting radiative transitions between levels with symmetric and antisymmetric co-ordinate eigenfunctions still holds strictly[1] for RAMAN and band spectra so that symmetric levels combine only with symmetric levels and antisymmetric levels combine only with antisymmetric levels. All the rotational lines will appear in the spectra but with an alternation in intensities superimposed on the usual intensity distribution.

Table 5. *Symmetry of the eigenfunction for non-zero spin.*

Co-ordinate function ψ		Nuclear spin function χ	Statistical weight	Total function $\chi\,\psi$	Nuclear statistics
sym.	{	sym.	$(I+1)(2I+1)$	sym.	BOSE
		anti.	$I(2I+1)$	anti.	FERMI
anti.	{	sym.	$(I+1)(2I+1)$	anti.	FERMI
		anti.	$I(2I+1)$	sym.	BOSE

The ratio of the intensity of the stronger to the weaker lines is $(I+1)/I$ so that a measurement of the intensity ratio determines the value of the nuclear spin.

For $I = \frac{1}{2}$ and 1 the values of this ratio are 3:1 and 2:1. These intensity ratios are relatively easy to determine experimentally. However, for large values of I

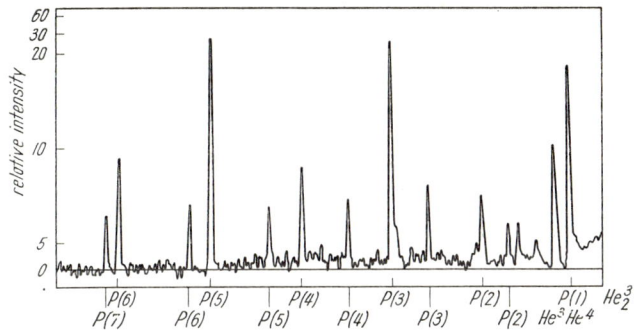

Fig. 11. The photometer curve of a portion of P branch of a band of the He_2^3 and the $\mathrm{He}^3\mathrm{He}^4$ molecules. The relative intensities are marked on the ordinate. The observed ratio of the intensity of alternate lines of the He_2^3 molecule is 2.8:1 which is close to the theoretical value of 3:1 expected for $I = \frac{1}{2}$ and much different from the ratio 1.67:1 which would apply for $I = \frac{3}{2}$. There is no regular variation in intensity for the rotational lines of the $\mathrm{He}^3\mathrm{He}^4$ molecule.

the ratio of the intensities is smaller and difficulties inherent in this method make it less reliable. In experiments care must be taken to ensure that the rotational lines under study are resolved from the lines of other bands, especially if the spectrum is complicated by the presence of two or more isotopes.

The spins of a number of nuclei have been measured by the method of band spectra. For H_2^1 an alternation in intensity corresponding to the value of the spin $I = \frac{1}{2}$ has been observed. For deuterium the alternation of intensities shows that the spin is equal to 1. The spin of the important He^3 nucleus[2] has been shown to be $\frac{1}{2}$ from the fact that a ratio of close to 3:1 was observed in the intensity of the rotational lines in a band spectrum as illustrated in Fig. 11.

The difficulty in making accurate measurements of this intensity ratio may be illustrated from the fact that band spectra measurements gave a value $I = \frac{5}{2}$

---

[1] If the interaction between the nuclear moments and the rest of the molecule is considered the probability of a transition between a symmetric and an antisymmetric level no longer vanishes. However, the probability is so small that the associated mean lifetime is measured in years in contrast to 10^{-8} sec for levels with permitted transitions.

[2] A. E. DOUGLAS and G. HERZBERG: Phys. Rev. **76**, 1529 (1949).

for the nuclear spin of the two chlorine isotopes while measurements from microwave spectroscopy show that the correct result is $I = \frac{3}{2}$. A number of other nuclear spins have been determined by this method, but unlike the case of zero nuclear spin and excepting He^3, the spins have been confirmed by other methods.

23. Nuclear statistics. The discussion in Sect. 22 shows that the statistics of nuclei may be determined from a study of the band spectra of homonuclear diatomic molecules. Important examples are furnished by the band spectra of hydrogen and deuterium. A theoretical discussion shows the ground electronic level of the stable H_2^1 molecule is a $^1\Sigma_g^+$ state [5]. Fig. 10 shows that for this electronic state all the rotational levels with even J are symmetric, and that the odd rotational levels are antisymmetric. Transitions to the ground electronic level can occur from the excited level $^1\Sigma_u^+$ where the symmetry of the even and odd J's is reversed. Observation shows that there is an alternation of intensity of 3:1, showing the proton spin is $\frac{1}{2}$, and also the transitions to the odd J values in the ground state are the stronger. Then, the antisymmetric levels have a statistical weight of 3 and from the discussion in Sect. 22 it follows that protons follow FERMI-DIRAC statistics, where the complete eigenfunction is antisymmetric. This result has important consequences in nuclear physics.

The electronic structure of heavy hydrogen or deuterium must be the same as that of ordinary hydrogen. The band spectra of the D_2 molecule, however, shows important differences from that of the H_2^1 molecule. In this case the ratio of the strong to the weak lines is 2:1 so that the nuclear spin is 1, but the stronger lines correspond to the symmetric rotational levels. This means that deuterium does not follow FERMI-DIRAC statistics, but that the complete eigenfunction is symmetric for the exchange of two deuterons which is characteristic of BOSE-EINSTEIN statistics. If deuterium is assumed to consist of a proton and a neutron it follows immediately that the neutron must follow FERMI-DIRAC statistics so that the exchange of a proton and a neutron at the same time leaves the sign of the complete eigenfunction unchanged.

From the historical point of view an important experiment[1] was the determination of the statistics of N^{14}. The rotation spectra of N_2^{14} showed an alternation in intensity the same as that described for deuterium so that the spin of N^{14} is 1 and this nucleus also follows BOSE-EINSTEIN statistics. If the nucleus consists of protons and electrons there would be 14 protons and 7 electrons, or, a total of 21 particles each of which follows FERMI-DIRAC statistics. Consequently the N^{14} nucleus should follow FERMI-DIRAC statistics. On the other hand, the proton-neutron hypothesis gives N^{14} an even number of particles, each of which follows FERMI-DIRAC statistics, so that the resulting nucleus is expected to follow BOSE-EINSTEIN statistics in agreement with observation.

The analysis of the band spectra or RAMAN effect of homonuclear diatomic molecules which contain nuclei with zero spin also enables one to determine nuclear statistics. For example, the spectra of O_2^{16} and S_3^{32}, which were mentioned in Sect. 21, show that the levels with even J values are missing. The configuration of the electrons in these molecules shows that the ground electronic level is a $^3\Sigma_g^-$ state. From Fig. 10 it can be seen that the levels which are present in the molecule are the symmetric ones, so that O^{16} and S^{32} follow BOSE-EINSTEIN statistics in agreement with the general rule that nuclei with even mass number follow this type of statistics. In some cases the assignment of the electron configuration may not be completely certain and the nuclear statistics are used to help specify the electronic configuration.

---

[1] W. HEITLER and G. HERZBERG: Naturwiss. **17**, 673 (1929). — F. RASETTI: Z. Physik **61**, 598 (1930).

General references.

[1] BETHE, H. A., and R. F. BACHER: Rev. Mod. Phys. **8**, 206 (1936). — This review article on nuclear theory includes a chapter on nuclear moments.

[2] BLIN-STOYLE, R. J.: Theories of Nuclear Moments. Rev. Mod. Phys. **28**, 75 (1956).

[3] CONDON, E. U., and G. H. SHORTLEY: Theory of Atomic Spectra. Cambridge: University Press 1953. — This book gives a complete and authoritative account of atomic spectra based on a quantum mechanical treatment of the nuclear-atom model.

[4] FRASER, F. G. J.: Molecular Rays. New York: MacMillan & Co. 1931. — Molecular beams. New York: Chemical Publishing Co. 1938. — These book describe technique for non-resonance molecular beam experiments.

[5] HERZBERG, G.: Molecular Spectra and Molecular Structure I. Spectra of Diatomic Molecules, Chap. III, IV and V. New York: D. van Nostrand, Inc. 1950. — This book gives an authoritative account of molecular spectra and contains many references to original papers.

[6] HUGHES, D. S., and C. ECKART: Phys. Rev. **36**, 694 (1930), calculate isotope shifts due to mass effects in lithium. — BARTLETT, J. H., and J. J. GIBBONS: Phys. Rev. **44**, 538 (1933), and VINTI, J. P.: Phys. Rev. **56**, 1120 (1939), extend the calculation to neon and magnesium respectively.

[7] JACKSON, D. A., and H. KUHN: Proc. Roy. Soc. Lond., Ser. A **148**, 335 (1935). — This paper is one of a series describing atomic beams used in absorption for hyperfine structure studies.

[8] KELLOGG, J. B. M., and S. MILLMAN: Rev. Mod. Phys. **18**, 323 (1946). — This paper reviews the molecular beam magnetic resonance experiments with atomic and molecular beams.

[9] KOPFERMANN, H.: Kernmomente, 2. Aufl. Stuttgart: Akademische Verlagsgesellschaft 1956. — This book gives a complete review of the theory and experiments for the determination of nuclear moments.

[10] KUHN, H.: Rep. Phys. Soc. Progr. Phys. **14**, 64 (1951) reviews interferometric technique. For aluminium films see BURRIDGE, KUHN and PERY: Proc. Phys. Soc. Lond. B **66**, 963 (1953).

[11] KUSCH, P., and V. W. HUGHES: Handbuch der Physik, Vol. XXXVII Atomic and Molecular Beams. Berlin: Springer 1957.

[12] MEISSNER, K. W.: Rev. Mod. Phys. **14**, 68 (1942) and CRAWFORD, SCHAWLOW, KELLY and GRAY: Canad. J. Res. A **28**, 558 (1950), describe the construction of atomic beam sources for optical spectroscopy.

[13] PAULING, L., and S. GOUDSMIT: The structure of Line Spectra. New York: McGraw-Hill, Inc. 1930. — Atomic spectra is discussed largely from the point of view of the vector model.

[14] RAMSEY, N. F.: Nuclear Moments. New York: John Wiley & Sons, Inc. 1953. — Theory and experimental determination of nuclear moments are discussed fully with emphasis on non-optical methods.

[15] RAMSEY, N. F.: Molecular Beams. Oxford: Oxford University Press 1956.

[16] ROSENTHAL, J. E., and G. BREIT: Phys. Rev. **41**, 459 (1932) do theoretical calculations. Extensions of the theory and comparison with experiments are given by CRAWFORD, M. F., and A. L. SCHAWLOW: Phys. Rev. **41**, 459 (1949), BRIX, P., and H. KOPFERMANN: Festschrift Akad. Wiss. Göttingen, Math.-physik. Kl. **17** (1951), and HUMBACH, W.: Z. Physik **133**, 589 (1952).

[17] SMITH, K. F.: Molecular Beams. London: Methuen & Co. 1955.

[18] TOLANSKY, S.: High Resolution Spectroscopy. London: Methuen & Co. 1947. — This monograph gives an account of the technique for optical hyperfine structure experiments.

[19] TOWNES, C. H.: Handbuch der Physik, Determination of Nuclear Quadrupole Moments. In this volume.

Isotope Shifts.

By

LAWRENCE WILETS.

With 10 Figures.

Introduction.

The atomic electron provides a useful and sensitive probe for the investigation of certain nuclear properties. Spectroscopic studies of hyperfine structure have yielded valuable information about nuclear spins, magnetic dipole moments and electrical quadrupole moments. Isotope shifts provide a tool for investigating the nuclear charge distribution.

The nuclear charge Z determines the principal character of the spectrum of an atom. Various other properties of the nucleus, however, cause small modifications of the spectrum as compared with the spectrum associated with an idealized, fixed, point nucleus. In addition to the hyperfine structure (hfs) splittings associated with a single (odd N and/or Z) isotope, there appear displacements among the *centers of gravity* of the hfs patterns of the various isotopes. Such isotope shifts were anticipated on the basis of the early Bohr model because of the finite nuclear mass, but the appearance of isotope shifts among the very heavy elements could be explained only in terms of an electron-nuclear interaction which deviates from COULOMB's $-Ze^2/r$ law. The investigation of isotope shifts among heavy elements, therefore, adds to our knowledge of nuclear *structure*.

Isotope shifts fall into the categories of mass effects and field effects. The mass effects are conveniently subdivided into normal and specific mass effects. The normal mass effect is exactly calculable while the specific mass effect is as complete—and only as complete—as our knowledge of atomic structure. Mass effects only are observed among light elements. Among heavy elements, the nuclear field effect dominates isotope shifts. The electrostatic potential due to a finite nucleus is higher inside the nucleus than the $-Ze^2/r$ potential of a point charge. Two nuclei differing in neutron number give rise to different potentials. The simple assumptions that nuclei are spherical and that their radii follow an $A^{\frac{1}{3}}$ law leave unexplained the following "anomalies":

(i) The ratios of observed to predicted shifts fluctuate widely, but in a manner correlated with nuclear magic numbers.

(ii) The observed isotope shifts are in the mean smaller than predicted.

(iii) The centroid of an odd isotope invariably lies closer (or at least as close) to the line of its lighter even-even neighbor than to the line of its heavier neighbor. This is referred to as even-odd staggering.

The fluctuations of the isotope shifts can be accounted for on the basis of nuclear deformations. The smallness of the shifts is consistent with "reasonable" estimates of nuclear compressibility and nuclear surface phenomena. Explanations for even-odd staggering in terms of nuclear deformation will be discussed.

A. Nuclear mass effects.

1. Normal mass effect. The non-relativistic Hamiltonian for an atom with n electrons is of the form

$$H = \frac{1}{2m} \sum_{j=1}^{n} \boldsymbol{p}_{0j}^2 + \frac{1}{2M} \boldsymbol{P}_0^2 + V(\boldsymbol{r}_{0j} - \boldsymbol{R}_0), \tag{1.1}$$

where $\boldsymbol{P}_0$ is the nuclear momentum and $\boldsymbol{p}_{0j}$ the electronic momenta relative to a *fixed* coordinate system, M and m are the nuclear and electronic masses, and V is the potential energy. The nuclear coordinate is designated $\boldsymbol{R}_0$ and the electronic coordinates $\boldsymbol{r}_{0j}$.

The following transformation leads to coordinates for the center of mass and electronic coordinates *relative to the nucleus:*

$$\boldsymbol{R} = \frac{M \boldsymbol{R}_0 + \sum_j m \boldsymbol{r}_{0j}}{M + nm}, \qquad \boldsymbol{r}_j = \boldsymbol{r}_{0j} - \boldsymbol{R}_0. \tag{1.2}$$

In terms of the new coordinates, the x-component of momentum of the center of mass is expressed as $(\boldsymbol{P})_x = -i\hbar\, \partial/\partial X$. The quantities $(\boldsymbol{p}_j)_x = -i\hbar\, \partial/\partial x_j$ are the x-components of electronic momentum *relative to the center of mass* even though x_j is measured relative to the *moving* nucleus.

In a coordinate system where the center of mass is at rest ($\boldsymbol{P} = 0$), the Hamiltonian assumes the form

$$H = \frac{1}{2m} \sum_j \boldsymbol{p}_j^2 + \frac{1}{2M} \left(\sum_j \boldsymbol{p}_j \right)^2 + V(\boldsymbol{r}_j). \tag{1.3}$$

Comparison with (1.1) leads to a simple interpretation of the terms in (1.3). The first term on the right-hand side of (1.3) is just the kinetic energy of the electrons ($\boldsymbol{p}_{0j} = \boldsymbol{p}_j$). The second term represents the recoil kinetic energy of the nucleus ($\boldsymbol{P}_0 = -\sum \boldsymbol{p}_j$).

It is convenient to expand the second term on the right-hand side of (1.3) and group the squared momenta with the first term:

$$H = \frac{1}{2} \left(\frac{1}{m} + \frac{1}{M} \right) \sum_j \boldsymbol{p}_j^2 + \frac{1}{2M} \sum_{j \neq k} \boldsymbol{p}_j \cdot \boldsymbol{p}_k + V. \tag{1.4}$$

The reduced mass μ is introduced by $1/\mu = 1/m + 1/M$.

The formal replacement of m by μ in the term for the electronic kinetic energy gives rise to the well-known normal mass effect. This is the only mass effect that occurs in hydrogen and one-electron ions.

The advantage of this grouping lies in the ease with which the energy dependence upon μ can be determined. Consider the Schrödinger equation

$$\left[-\frac{\hbar^2}{2\mu} \sum_j \nabla_j^2 + V(\boldsymbol{r}_1 \ldots \boldsymbol{r}_n) - W(\mu) \right] \psi = 0. \tag{1.5}$$

Let the $\boldsymbol{r}_j$ be formally replaced by $\boldsymbol{r}_j' = (\mu/m)\, \boldsymbol{r}_j$. The potential energy, if magnetic effects are ignored, scales inversely as a length:

$$V = -\sum_j \frac{Z e^2}{r_j} + \sum_{j<k} \frac{e^2}{r_{jk}}, \tag{1.6}$$

and hence

$$V(\boldsymbol{r}) = \frac{\mu}{m}\, V(\boldsymbol{r}').\tag{1.7}$$

The Schrödinger equation now becomes

$$\left[-\frac{\hbar^2}{2m}\sum_j V_j'^2 + V(\boldsymbol{r}') - \frac{m}{\mu}\, W(\mu)\right]\psi = 0.\tag{1.8}$$

The form (1.8) is the same as (1.5), and thus the energy levels depend upon the reduced mass according to

$$\frac{W(\mu)}{W(m)} = \frac{\mu}{m}.\tag{1.9}$$

Atomic energies scale directly and lengths inversely as the (reduced) electronic mass.

The normal mass effect was first observed in hydrogen[1]. In agreement with expectation, D_α is displaced by 1.79 Å to the blue of H_α (λ 6562.79), or about 0.027%.

2. Specific mass effects [*10*], [*1*]. The second term on the right-hand side of (1.4) gives rise to the so-called specific mass effect. Because of the occurrence of the nuclear mass in the denominator, this term is small compared with the rest of the Hamiltonian and may be treated as a perturbation.

VINTI[2] has given an expression for the mass term which does not involve derivatives:

$$\frac{1}{2M}\left\langle\left(\sum_j \boldsymbol{p}_j\right)^2\right\rangle = \frac{Z e^2}{2}\frac{m}{M}\left\langle\sum_{j\neq k}\frac{\boldsymbol{r}_j\cdot\boldsymbol{r}_k}{r_k^3}\right\rangle.\tag{2.1}$$

The equality is valid, however, only when the expectation value is taken with respect to solutions of SCHRÖDINGER's equation. In general, one has available only trial wave functions which are not exact solutions of SCHRÖDINGER's equation so that (2.1) is then not exact. VINTI's form is of particular utility when a table of values is given for the atomic wave function, rather than an analytic expression.

In the Hartree central field approximation, where the atomic wave function is written as a determinant of single particle functions (no configuration mixing), the specific mass operator

$$H' = \frac{1}{2m}\sum_{j\neq k}\boldsymbol{p}_j\cdot\boldsymbol{p}_k\tag{2.2}$$

has contributions only from exchange terms [*1*]. Thus for[3]

$$\Psi = (n!)^{-\frac{1}{2}}\,\|\,\varphi_j(\boldsymbol{r}_k)\,\|,\tag{2.3}$$

one obtains

$$\Delta W \equiv (\Psi, H'\Psi) = -\frac{1}{2M}\sum_{j\neq k}|\,(\varphi_j,\,\boldsymbol{p}\,\varphi_k)\,|^2.\tag{2.4}$$

The selection rules for $(\varphi_j,\,\boldsymbol{p}\,\varphi_k)\neq 0$ are the same as those for dipole radiation, namely, $l_j - l_k = \pm 1$, $m_{lj} - m_{lk} = 0,\ \pm 1$, and $m_{sj} - m_{sk} = 0$.

---

[1] UREY, BRICKWEDDE and MURPHY: Phys. Rev. **40**, 1 (1932).

[2] J. P. VINTI: Phys. Rev. **58**, 882 (1940).

[3] A determinant wave function of the type (2.3) generally will not be an eigenfunction of the operators $\boldsymbol{L}^2$, $\boldsymbol{S}^2$ and $\boldsymbol{J}^2$, although $M_l = \sum m_l$ and $M_s = \sum m_s$ may be good quantum numbers. Linear combinations of such determinants with various combinations of the m_l and m_s are required such that $\boldsymbol{L}^2$, $\boldsymbol{S}^2$ (in L-S coupling) and $\boldsymbol{J}^2$ are constants of the motion. These considerations do not affect the results quoted here.

Terms in (2.4) which arise between closed shell electrons and valence electrons will be independent of the m_l and m_s of the valence electron after summing over all the closed shell electrons. Thus the net contribution of closed shell electrons is to shift all members of a given multiplet equally. In complex atoms, therefore, a sensitive test of the theory is the relative shifts among various members of a multiplet.

The use of Hartree-type wave functions to calculate mass effects has achieved only moderate success. In the cases of neon[1] and magnesium[2], for example, qualitative agreement was obtained, but the magnitudes of the calculated shifts were generally too small.

The inadequacies of the Hartree-type wave functions appear to arise in large measure from the absence of electron-electron correlations (configuration mixing). The influence of correlations is sometimes referred to as the polarization mass effect to distinguish it from the exchange mass effect of (2.4).

Two-electron spectra provide the simplest examples for studying the specific mass effect. The configurations $ns\,n's$ cannot exhibit the exchange mass effect because of the selection rule $l_j - l_k = \pm 1$; any specific mass effect would thus be classified as polarization mass effect. The ground state of helium, $(1s)^2\,{}^1S_0$, is a case in point.

In Table 1 is given a list of energy upper bounds and specific mass energies obtained from various Hylleraas-type trial wave functions. The best value for the specific mass energy is $4.79\,\mathrm{cm}^{-1}$, which is well outside the $\pm 2\,\mathrm{cm}^{-1}$ accuracy of the HERZBERG and ZBINDEN[9] experiment. In this experiment, the absolute energy displacement is observed rather than isotope shifts. An accurate calculation is required in order to isolate relativistic effects[10] which are less than $1\,\mathrm{cm}^{-1}$ (although individual contributions are of the order of $20\,\mathrm{cm}^{-1}$).

Calculations[11] are also available for other two electron spectra from H^- to Mg^{10+}, but the prospect of observing specific mass effects in the ground states of these ions appears somewhat remote.

Table 1. *Specific mass displacements in the ground state of He from Hylleraas type of wave functions. Energies are given in atomic units. The specific mass displacement is given by* $(m/M)\,\langle \boldsymbol{p_1}\cdot\boldsymbol{p_2}\rangle$.

No. of parameters	$\langle H \rangle$	$\langle \boldsymbol{p_1}\cdot\boldsymbol{p_2}\rangle$
1	-2.848	$-0.$
3	$-2.9024\,{}^3$	$-0.17764\,{}^7$
6	$-2.90324\,{}^3$	$-0.16437\,{}^{7,\,8}$
10	$-2.903603\,{}^4$	$-0.15905\,{}^7$
14	$-2.9037009\,{}^5$	$-0.15923\,{}^7$
18	$-2.9037063\,{}^5$	$-0.15916\,{}^7$
20	$-2.9037179\,{}^6$	$-0.159078\,{}^6$

[1] J. H. BARTLETT jr. and J. J. GIBBONS jr.: Phys. Rev. **44**, 538 (1933).

[2] J. P. VINTI: Phys. Rev. **56**, 1120 (1939).

[3] E. A. HYLLERAAS: Z. Physik **54**, 347 (1929); **65**, 209 (1930).

[4] S. CHANDRASEKHAR, D. ELBERT and G. HERZBERG: Phys. Rev. **91**, 1172 (1953).

[5] S. CHANDRASEKHAR and G. HERZBERG: Phys. Rev. **98**, 1050 (1955).

[6] J. F. HART and G. HERZBERG: Phys. Rev. **106**, 79 (1957).

[7] L. WILETS and I. J. CHERRY: Phys. Rev. **103**, 112 (1956).

[8] H. A. BETHE: Handbuch der Physik, second edit., Vol. XXIV, Part 1, p. 375. Berlin: Springer 1933. — BETHE'S value is in error, as first noted by EDMONDS and WILETS, and KINOSHITA (both unpublished). This has been corrected by BETHE and SALPETER: Encyclopedia of Physics, Vol. XXXV.

[9] G. HERZBERG and R. ZBINDEN: Unpublished; referred to by CHANDRASEKHAR and HERZBERG: Phys. Rev. **98**, 1050 (1955).

[10] T. KINOSHITA: Phys. Rev. **105**, 1490 (1957). — These effects are also discussed by CHANDRASEKHAR and HERZBERG: Phys. Rev. **98**, 1050 (1955), where further references will be found.

[11] J. F. HART and G. HERZBERG: Phys. Rev. **106**, 79 (1957).

Shifts in excited states of helium have been investigated by various experimenters[1-4]. Comparison with theory[5] is generally satisfactory. Table 2 shows the comparison for the $1s\,2s\,^3S_1$ and $^1S_0$ levels.

Table 2. *Specific mass shifts in the* He I *configurations* $1s\,2s$.
The values labelled "observation" have the normal mass shift subtracted out.

	$^3S_1$	$^1S_0$
Observation		
FRED et al.[3]	$0.141\,K$	$0.099\,K$
BRADLEY and KUHN[4]	0.10 ± 0.02	0.10 ± 0.02
Calculations by STONE[5] using:		
Huang functions	0.159	0.076
Corrected Hylleraas-Undheim functions . .	0.149	0.0758

In the intermediate mass region, CRAWFORD's students[6-9] have attempted to isolate mass effects from field effects. Energy displacements due to field effects depend primarily on the electronic configuration, while specific mass effects depend also on the multiplet. For two electrons outside closed shells in the configuration $nsnp$, they express the difference between singlet and triplet specific mass displacements (for a given isotope) in the form

$$\Delta W(\text{singlet-triplet}) = -2R\,\frac{\mu}{M}\,k\,(ns,np)\,, \tag{2.5}$$

where R is RYDBERG's constant and $k\,(ns,\,np)$ is a parameter [compare Eq. (2.4)]. Experimental values have been obtained for magnesium[7], $k\,(3s,\,3p) = -0.14$; zinc[6], $k\,(4s,\,4p) = -0.18$; and cadmium[8], $k\,(5s,\,5p) = -0.10$. The parameter does not appear to be a strongly varying function of n. This is good evidence that the specific mass effect, like the normal mass effect, varies as $1/A$ for the energy displacements or as $1/A^2$ for the isotope shifts and is thus negligible for heavy elements.

B. Field effects for spherical, incompressible nuclei.

I. Nuclear volume effect.

3. Finite charge distribution. RACAH [*12*] and ROSENTHAL and BREIT [*14*] investigated the effect of the extended charge distribution in the nucleus. The probability density of an s-electron in the neighborhood of a point charge in the Dirac theory is given by

$$P(r) = \frac{2\,(2\varrho + 1)}{[\Gamma(2\varrho + 1)]^2}\,\Psi^2(0)\left(\frac{2Zr}{a_{\mathrm{H}}}\right)^{2\varrho - 2}\,, \tag{3.1}$$

where $\varrho = (1 - Z^2\alpha^2)^{\frac{1}{2}}$, $\Psi(0)$ is the nonrelativistic Schrödinger wave function at $r = 0$, and a_{H} is the first Bohr radius in hydrogen. Relativistic effects are important

---

[1] A. ANDREW and W. W. CARTER: Phys. Rev. **74**, 838 (1948).

[2] T. E. MANNING: Phys. Rev. **76**, 173 (1949).

[3] Private communication to MACK and ARROE [*11*]. The line shifts are reported by M. FRED, F. S. TOMKINS and J. K. BRODY: Phys. Rev. **75**, 1772 (1949); M. FRED, F. S. TOMKINS, J. K. BRODY and M. HAMERMESH: Phys. Rev. **82**, 406 (1951).

[4] L. C. BRADLEY III and H. KUHN: Nature, Lond. **162**, 412 (1948). — Proc. Roy. Soc. Lond., Ser. A **209**, 325 (1951).

[5] A. P. STONE: Nature, Lond. **176**, 130 (1955). — Proc. Phys. Soc. Lond. A **68**, 1152 (1955).

[6] W. M. GRAY: Doctoral Thesis, Univ. of Toronto 1949.

[7] F. M. KELLY: Doctoral Thesis, Univ. of Toronto 1949.

[8] G. R. HANES: Doctoral Thesis, Univ. of Toronto 1955.

[9] A. W. SMITH: Doctoral Thesis, Univ. of Toronto 1955.

because they lead to large density in the neighborhood of the nuclear charge. Normalization with respect to the nonrelativistic Schrödinger wave function is possible because only a very small portion of the electronic wave function is affected by relativistic considerations. The $p_{\frac{1}{2}}$-electron density becomes infinite at $r=0$ like the $s_{\frac{1}{2}}$-electron density, but with a smaller ($\sim Z^2\alpha^2$) coefficient; this may be regarded as arising from relativistic mixing of $p_{\frac{1}{2}}$- and $s_{\frac{1}{2}}$-states. The wave functions of all other electrons are zero at the origin. The principal effects of nuclear structure therefore appear in the energy levels of the s-electrons.

If one considers a nucleus which is not a point charge, but rather is characterized by a spherically symmetric charge distribution $f(r)$ and the corresponding

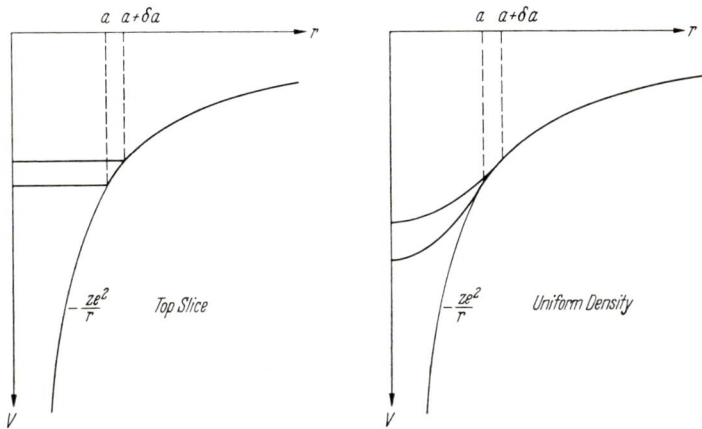

Fig. 1. Two models for the electron-nuclear potential [15]. Model A: BREIT's top slice. All nucleons are concentrated in a shell at the surface. Model B: Nucleons are distributed uniformly through the nucleus. The displacement ΔE is due to the difference between the actual potential (heavy lines) and the Coulomb potential (light line). The isotope shift $\delta \Delta E$ is due to the difference in the potentials between isotopes of radius a and $a+\delta a$.

electrostatic potential $V(r)$, the first-order perturbation energy relative to a point nucleus arising from the volume effect is given by

$$\Delta E = \int P(r)\left[V(r) + \frac{Ze^2}{r}\right] d\tau, \tag{3.2}$$

where $d\tau$ is the volume element. The interaction energy between the nucleus and an electron may also be computed by considering the nucleus in the field of the electron. Using the s-electron density given by (3.1), one obtains

$$\Delta E = F(Z)\, \Psi^2(0)\, R_1^{2\varrho}, \tag{3.3}$$

where

$$F(Z) = \frac{12\pi Z e^2(\varrho+1)}{[\Gamma(2\varrho+1)]^2\,\varrho\,(2\varrho+1)\,(2\varrho+3)}\left(\frac{2Z}{a_{\mathrm{H}}}\right)^{2\varrho-2}, \tag{3.4}$$

$$R_1 \equiv [(1+2\varrho/3)\,\langle r^{2\varrho}\rangle]^{\frac{1}{2\varrho}}, \tag{3.5}$$

$$\langle r^{2\varrho}\rangle = \int f(r)\, r^{2\varrho}\,\frac{d\tau}{Z}, \tag{3.6}$$

and $f(r)$ is normalized so that $\int f(r)\,d\tau = Z$. It can be seen that the perturbation energy depends on $\langle r^{2\varrho}\rangle$ and not on further details of the nuclear charge distribution. For evaluating the perturbation energy, an arbitrary charge distribution is equivalent to a uniform charge distribution contained within a sphere of radius R_1. If the charge were located on the surface of a sphere of radius R' (BREIT's "top slice"), we would have $R_1 = [1+2\varrho/3]\,R'$. The potentials corresponding to the uniform and "top slice" models are shown in Fig. 1.

The perturbation energy shift between two isotopes differing by δR_1 in the equivalent radius is thus

$$\delta \Delta E = 2\varrho F(Z) \Psi^2(0) R_1^2 \varrho \frac{\delta R_1}{R_1} . \tag{3.7}$$

The sum of the perturbation energies over all s-electrons in the atom would give the first-order displacement of the energy level of the atom, but this total displacement is not directly observable in practice. ΔE for an s-valence electron, for example, gives the displacement in the energy of a photon emitted during the transition of the electron from a p-state to an s-state (neglecting the smaller

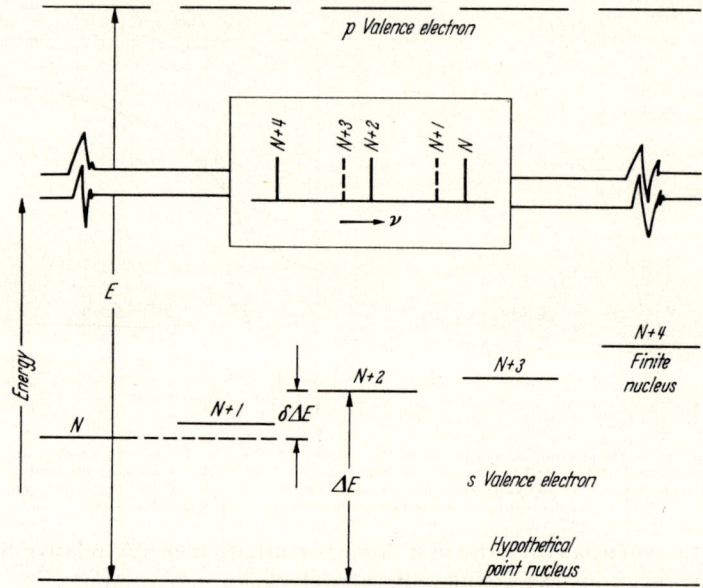

Fig. 2. Schematic representation of isotope displacements and shifts [15]. The transition is assumed to involve the change of a valence electron from a p state to an s state. Only the valence electron displacements are considered, and p-state shifts are assumed negligible. For a point nucleus, the s-states for all isotopes lie at the same energy. For a realistic nucleus, the s-states lie at progressively higher energies for heavier isotopes. Typically, the centroid of an odd-N isotope level lies closer to its lower even-N neighbor than to its heavier even-N neighbor (even-odd staggering, Z even). The energy scale is distorted, since $(\delta \Delta E)/(\Delta E)$ is the order of A^{-1}. In the inset is shown the appearance on a "photographic plate" of the isotope shift splittings.

displacement of the p-state). What is generally observed is the difference in ΔE between the lines of two different isotopes. We shall use "displacement" to mean the perturbation energy ΔE, and isotope shift ($\delta \Delta E$) to mean the difference in ΔE between two isotopes of the same element. Fig. 2 gives a schematic indication of the energy shifts involved.

Because the deviation from the unperturbed $1/r$ potential is large inside the nucleus, the first-order perturbation theory is not sufficiently accurate in determining the energy shifts. ROSENTHAL and BREIT [14] have subjected the perturbation treatment to analysis and found that a factor as small as 0.5 may be required to correct the results predicted by the first-order calculations in some cases. BROCH[1] has investigated the shifts by a treatment which avoids perturbation theory and, in the simple case where $V(r) = 0$ for $r < R$, he shows that a factor $k \approx 2\varrho^2/(\varrho + 1)$ is needed to correct the first order results. CRAWFORD and SCHAWLOW [7] investigated the perturbation theory in the case of thallium ($Z = 81$)

---

[1] E. K. BROCH: Arch. Math. Naturvidenskab **48**, 25 (1945).

for both the uniform and top slice models. Using Rosenthal and Breit wave functions for the top slice model and a power series solution for the case of uniform charge density, they found that the first-order calculations must be corrected by factors of 0.74 and 0.75, respectively. BROCH's correction factor for the zero potential model is 0.72. The proximity of the correction factors for such different nuclear models is striking and indicates that BROCH's factor k is generally accurate to a few percent. It is interesting to note that BROCH's correction factor is independent of the radius.

II. Nuclear charge distributions.

Several independent experiments which yield information about nuclear charge distributions are now available. An analysis of these experiments is given elsewhere in the Encyclopedia[1], but it is well to discuss here relevant features of some of the experiments.

4. μ-mesic atoms. Radiations from μ-mesic atoms have been detected by CHANG[2] in cosmic ray studies, and more recently by FITCH and RAINWATER[3] working with artificially produced mesons. The mesons are captured in outer Bohr orbits and cascade toward the nucleus, transferring energy by radiative and Auger transitions. The mesons arrive in the lower atomic states, for the most part, with $l = n - 1$ and proceed to decay by radiative transitions with $\Delta l = \Delta n = -1$, to the ground state. The $2p \to 1s$ transition has been studied most thoroughly.

In the low quantum states, atomic electrons do not perturb the mesic wave function and the system may be treated as a hydrogen-like atom, with due regard to the characteristic effects of the mesic mass. Only electrical forces are considered, since the μ-meson interacts only weakly with nuclear matter. The meson is treated as a Dirac particle with spin $\frac{1}{2}$ and magnetic moment $e\hbar/(2m_\mu c)$. The mass m_μ of the μ-meson is 207 times the mass of the electron, and the Bohr orbits are proportionately smaller. In lead, for example, the first mesic Bohr orbit is 3.12×10^{-13} cm compared with a nuclear radius of about 7×10^{-13} cm.

For quite light elements the nuclear volume effect is small and may be treated by perturbation theory; the energy displacement is proportional to $\langle r^2\varrho \rangle$ and is given by Eq. (3.3) with a_H replaced by the first mesic Bohr radius. But for medium weight and heavy nuclei the volume effect becomes very large and perturbation theory is no longer valid. In lead, for example, the finite extension of the nucleus accounts for the reduction of the $1s$-state energy from 21.3 Mev for a point nucleus to 10.1 Mev, and a reduction of the fine structure splitting of the $2p$ doublet from 0.55 Mev to 0.20 Mev.

A single spectral line yields only a single parameter of the nuclear charge distribution, but this parameter cannot be expressed simply in terms of some moment of the charge distribution. The $2p \to 1s$ transitions have been measured very precisely and give the most accurate single parameter. This parameter may be expressed in terms of an equivalent radius, i.e., the radius of a uniform charge distribution which leads to the same transition energy. The best value for heavy elements (especially lead) is[1]

$$R_\mu = 1.17 \times 10^{-13} A^{\frac{1}{3}} \text{ cm.} \tag{4.1}$$

This is not yet the number to be identified with R_1.

[1] D. L. HILL: Encyclopedia of Physics, Vol. XXXIX, p. 178. Heidelberg: Springer. 1957.
[2] W. Y. CHANG: Rev. Mod. Phys. **21**, 166 (1949). — Phys. Rev. **75**, 1315 (1949).
[3] V. F. FITCH and J. RAINWATER: Phys. Rev. **92**, 789 (1953).

5. Electron scattering [9]. High-energy electron scattering experiments are capable of yielding more details of the charge distribution than mesic atom spectroscopy. This is possible since a differential cross section contains information (phase shifts) of many angular momenta. In practice two parameters—radius and surface thickness—have been determined for "reasonable" distribution functions. Moreover, the scattering data do tend to limit the choice of distributions; in particular HILL, FREEMAN and FORD[1] find that the best fit to the data is obtained with little or no concentration of charge toward the surface.

The electronic scattering data of HOFSTADTER and collaborators [9] have been analyzed by several groups, of which we mention two in particular. RAVENHALL and YENNIE [13] use a density distribution function of the form

$$f(x) = f(0) [1 + e^{\varkappa(x-1)}]^{-1}, \tag{5.1}$$

where $x = r/R_c$ is the radius in units of the half-density radius R_c. HILL, FREEMAN and FORD[1] use the functional form

$$f(x) = f(0) \times \begin{cases} 1 - \tfrac{1}{2} e^{n(x-1)}, & x < 1; \\ \tfrac{1}{2} e^{-n(x-1)}, & x > 1; \end{cases} \tag{5.2}$$

where x has the same significance as in (5.1). For n and $\varkappa$ large compared with unity, the two functional forms are very similar if one chooses $\varkappa = (\pi/\sqrt{6})\, n$.

The various moments of the two charge distributions can be expressed to orders n^{-4} and $e^{-n} n^{-3}$ by the relationship:

$$\left[\frac{p+3}{3} \langle r^p \rangle\right]^{1/p} \approx R_c \left[1 + \frac{(p+5)}{n^2}\right], \tag{5.3}$$

where n can be replaced by $(\sqrt{6}/\pi)\, \varkappa$. This gives the desired relationship for the equivalent radius:

$$R_1 \approx R_c \left[1 + \frac{2\varrho + 5}{n^2}\right]. \tag{5.4}$$

Both RAVENHALL and YENNIE, and HILL, FREEMAN and FORD obtain substantially the same half-density radius. In the vicinity of Pb, an average between their values is

$$R_c = 1.11 \times 10^{-13}\, A^{\frac{1}{3}}\, \text{cm}. \tag{5.5}$$

The accuracy is the order of one percent. The quantity $R_c/\varkappa$ (or R_c/n) is proportional to the surface thickness and is found to be rather independent of the mass number, A. The former authors quote for $R_c/\varkappa$ a value of 0.45×10^{-13} cm $[\pi R_c/(\sqrt{6}\,\varkappa) = 0.58 \times 10^{-13}$ cm], while the latter authors quote for R_c/n a value of 0.66×10^{-13} cm; the errors here are the order of ten percent. We adopt the value

$$\frac{R_c}{n} = 0.63 \times 10^{-13}\, \text{cm}, \tag{5.6}$$

which corresponds to a 90 to 10% fall-off distance of 2.0×10^{-13} cm. Another quantity which is frequently calculated from electron scattering is R_2, the root mean square radius [Eq. (5.3) with $p = 2$]. This quantity is of some interest to us since R_1 does not differ greatly from the rms radius [$1.5 < p = 2\varrho < 2$]. An average value from electron scattering for heavy nuclei is

$$R_2 \approx 1.19 \times 10^{-13}\, A^{\frac{1}{3}}\, \text{cm}. \tag{5.7}$$

---

[1] HILL, FREEMAN and FORD: Unpublished; cited by K. W. FORD and D. L. HILL [8].

The charge distribution deduced from electron scattering yields an equivalent radius in agreement with mesic atom experiments.

6. X-ray fine structure and isotope shifts. The innermost electronic orbits experience relatively large energy displacements due to the finite extension of the nuclear charge. This effect is pronounced in the case of the $2p_{\frac{3}{2}} - 2p_{\frac{1}{2}}$ fine structure doublet of heavy elements. The $p_{\frac{1}{2}}$ electronic density is singular (for a point charge) at the nucleus while the $p_{\frac{3}{2}}$ density is negligible within the nuclear dimensions.

In many respects the x-ray fine structure experiments come very close to measuring the same effects as optical isotope shifts. There remain, however, radiative corrections (e.g. Houston-Lamb shift) which must be cleared up before direct comparisons can be made[1-4]. The difficulty in estimating these corrections is due to the relatively large size of the expansion parameter $Z\alpha$. More detailed calculations must be made before definitive statements can be made about the X-ray fine structure.

Estimates of x-ray isotope shifts have been made by IGO and WERTHEIM[5]. Their calculations take into account screening of the inner electrons by the others and the change in screening of the outer electrons when a K or L shell electron is removed (cf. Sect. 16). They estimate the shifts in Mo and U to be one-half and one-twentieth, respectively, of the experimental resolution of ROGOSA and SCHWARZ[6], who had previously failed to observe shifts. Further experiments would be desirable since it seems likely that the atomic effects included in these calculations are more accurate than those in the optical region.

III. Nuclear polarization.

7. Excitation of the nucleus by atomic electrons. BREIT, ARFKEN, and CLENDENIN[7] have investigated the possible polarization of the nucleus by atomic electrons. The polarization acts in the direction to increase the binding of the electrons and is thus contrary in sense to the volume effect. Only the monopole effect appears to be large enough to contribute appreciably to the isotope shifts.

The polarization of the nucleus may be interpreted in terms of admixtures to the nuclear ground state of low-lying excited levels, in such a way as to give greater concentration of protons toward the center of the nucleus where the electron probability density is greatest. The size of the effect depends upon the number and spacing of the low-lying excited levels of the nucleus, and might thus be expected to be greater for odd nuclei than for even ones. BREIT *et al.* were looking for an effect half as great as the volume effect in order to account for even-odd staggering. Reasonable approximations appeared to give no more than twenty percent of the needed one-half and rather extreme departures from the central field approximations for an individual nucleon seemed necessary to account for the observed effect. While the polarization may account in some part for the even-odd staggering, it probably is not important in explaining the small average value of the even shifts.

---

[1] A. L. SCHAWLOW and C. H. TOWNES: Science, Lancaster, Pa. **115**, 284 (1952). — Phys. Rev. **100**, 1273 (1955).

[2] C. H. TOWNES: Phys. Rev. **94**, 773 (1954).

[3] N. M. KROLL: Phys. Rev. **94**, 747 (1954).

[4] E. WICHMANN and N. M. KROLL: Phys. Rev. **96**, 232 (1954).

[5] G. IGO and M. S. WERTHEIM: Phys. Rev. **95**, 1097 (1954).

[6] G. L. ROGOSA and G. SCHWARZ: Phys. Rev. **92**, 1434 (1953).

[7] G. BREIT, G. B. ARFKEN and W. W. CLENDENIN: Phys. Rev. **78**, 390 (1950). — Other nuclear effects are also discussed.

IV. Non-Coulombic forces.

8. Electron-nucleon interaction [*16*]. The interaction between an electron and a nucleon does not follow COULOMB's $-Ze^2/r$ law at small distances. Deviations from the Coulomb law have been interpreted as arising from a finite extension of the nucleon (other interpretations are possible), plus the effect of the anomalous magnetic moment of the nucleon (i.e., deviations from the moment of a Dirac particle). Both effects can be described in terms of either an extended charge distribution *or* the addition of a short range potential to the Coulomb potential.

In the case of the neutron, there appears a short range potential which, according to recent experiments[1-2] has a value of about 4 kev for a square well of radius $r_e = 2.82 \times 10^{-13}$ cm. (Only the volume integral of the potential is measured.)

As a charge distribution, the neutron appears with positive charge toward the center and negative charge farther out. While the net charge is zero, the integral of r^2 over the charge distribution is finite and negative:

$$\langle r^2 \rangle_n = -(0.35 \times 10^{-13} \text{ cm})^2. \tag{8.1}$$

This is in complete agreement with FOLDY's[3] theoretical calculation based on the anomalous magnetic moment of the neutron alone.

The non-Coulombic part of the electron-proton potential appears repulsive. An effective charge distribution based on electron scattering experiments[4] yields

$$\langle r^2 \rangle_p = [(0.77 \pm 0.10) \times 10^{-13} \text{ cm}]^2. \tag{8.2}$$

The non-Coulombic forces need not concern us further for two reasons. Firstly, the interaction is quite small for isotope shift considerations[5], contributing less than two percent. Secondly, the estimates of nuclear charge distributions with which we are primarily concerned come from experiments where the electron is used as a probe (e.g. electron scattering): any charge distribution associated with the individual nucleons is already included.

C. Nuclear deformation effect.

Large nuclear deformations are known to occur in various regions of the periodic table. The magnitudes of the deformations can be determined by optical hfs quadrupole measurements in the case of odd nuclei, and by nuclear γ-ray lifetimes or Coulomb excitation measurements in general. The significance of these various measurements with respect to "intrinsic" nuclear deformation is somewhat model-dependent. An understanding of the role of the intrinsic deformation in nuclear physics is supplied by the BOHR-MOTTELSON unified model, a description of which is given elsewhere in the Encyclopedia [*3*]. Pertinent details of the model will be discussed in Sects. 10 to 12.

I. Contribution to isotope shifts.

9. Theory [*4*], [*15*]. Nuclei which are non-spherical in shape will, after averaging over angles, appear more extended radially than spherical nuclei of the same

---

¹ HUGHES, HARVEY, GOLDBERG and STAFNE: Phys. Rev. **90**, 497 (1953).
² MELKONIAN, RUSTAD and HAVENS: Bull. Amer. Phys. Soc. **1**, 62 (1956).
³ L. L. FOLDY: Phys. Rev. **83**, 688 (1951).
⁴ R. W. MCALLISTER and R. HOFSTADTER: Phys. Rev. **102**, 851 (1956). — E. E. CHAMBERS and R. HOFSTADTER: Phys. Rev. **103**, 1454 (1956).
⁵ L. WILETS and L. C. BRADLEY III: Phys. Rev. **82**, 285 (1951).

volume. This effect is illustrated in Fig. 3 for a uniformly charged spheroidal nucleus. The energy displacement ΔE is increased over that of a spherical nucleus. The effect on ΔE is small, but the differential effect measured by the isotope shift may be relatively large since the nuclear deformation may change by a large fraction of itself from one isotope to another.

Restricting consideration to axially symmetric deformations of second order (approximate spheroids), the equation for the nuclear surface may be written

$$R(\vartheta) = a_0[1 + \alpha P_2(\cos\vartheta)], \tag{9.1}$$

where α may be >0 (prolate) or <0 (oblate). The α defined by (9.1) is related to the Bohr deformation parameter β by the relationship $\beta = \sqrt{4\pi/5}\,|\alpha|$.

In order to isolate the deformation effect from other effects, it is assumed here that the nucleus has constant volume $4\pi a^3/3$ and uniform charge density

$$f(r) = \begin{cases} \dfrac{3Z}{4\pi a^3}, & r < R(\vartheta); \\ 0, & r > R(\vartheta). \end{cases} \tag{9.2}$$

The constant volume assumption relates the parameters a_0 and a appearing in (9.1) and (9.2):

$$a_0 = a[1 + \tfrac{3}{5}\alpha^2 + \tfrac{2}{35}\alpha^3]^{-\frac{1}{3}}. \tag{9.3}$$

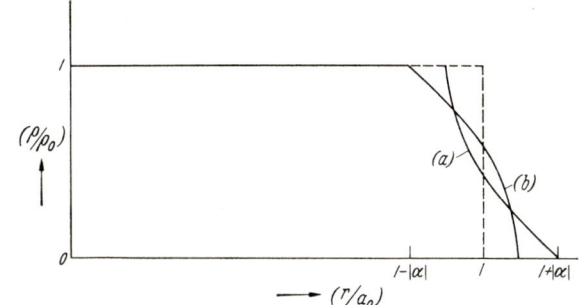

Fig. 3. Charge density of a deformed nucleus averaged over angles [15]. Dashed curve: hypothetical uniform density for a spherical nucleus of radius a_0. Curve (a): charge density averaged over angles for a prolate ($\alpha > 0$) spheroid with maximum radius $a_0(1+\alpha)$. Curve (b): charge density averaged over angles for an oblate ($\alpha < 0$) spheroid with maximum radius $a_0(1 + \tfrac{1}{2}|\alpha|)$.

Introducing the notation $m = $ minimum radius and $M = $ maximum radius, i.e.

$$m = a_0 \times \begin{cases} 1 - \tfrac{1}{2}\alpha, & \alpha > 0; \\ 1 - |\alpha|, & \alpha < 0; \end{cases} \tag{9.4}$$

$$M = a_0 \times \begin{cases} 1 + \alpha, & \alpha > 0; \\ 1 + \tfrac{1}{2}|\alpha|, & \alpha < 0; \end{cases} \tag{9.5}$$

one obtains for the charge density averaged over angles in the region $m < r < M$;

$$f(r) = f(0) \times \begin{cases} 1 - \sqrt{(r-m)/(M-m)}, & \alpha > 0; \\ \sqrt{(M-r)/(M-m)}, & \alpha < 0. \end{cases} \tag{9.6}$$

It is these espressions which are plotted in Fig. 3.

The contribution of deformation to the energy displacement can be calculated using

$$\Delta E = \tfrac{1}{5} F(Z)\,\varrho\,(2\varrho + 3)\,a^{2\varrho}\,\Psi^2(0)\,\alpha^2\,[1 + \tfrac{2}{21}(2\varrho + 3)\,\alpha + \cdots]. \tag{9.7}$$

This expression, to leading order, was first given by BRIX and KOPFERMANN [4]. The ratio of this displacement to that of the normal volume effect is, to order α^2,

$$\frac{\Delta E_\alpha}{\Delta E_v} = \varrho\,\frac{(2\varrho + 3)}{5}\,\alpha^2, \tag{9.8}$$

and is generally less than 0.1.

The ratio of isotope shift due to deformation to the shift due to the normal volume effect (3.7) is

$$\frac{(\delta\Delta E)_\alpha}{(\delta\Delta E)_v} = \frac{3}{10}(2\varrho + 3)\, A \left(\frac{\delta(\alpha^2)}{\delta N}\right)_z [1 + (2\varrho + 3)\bar{\alpha}/7 + \cdots], \qquad (9.9)$$

where $\bar{\alpha}$ is the mean value of α. Since corrections to the first order perturbation energy are insensitive to the shape of the density distribution, the ratio (9.9) should remain very nearly the same if higher-order corrections to perturbation theory are applied. The ratio can be greater than unity in particular cases and of either sign.

To lowest order the isotope shifts depend on changes in α^2 and not on the sign of α. For axially unsymmetric shapes, the nuclear surface can be described in terms of BOHR'S expansion

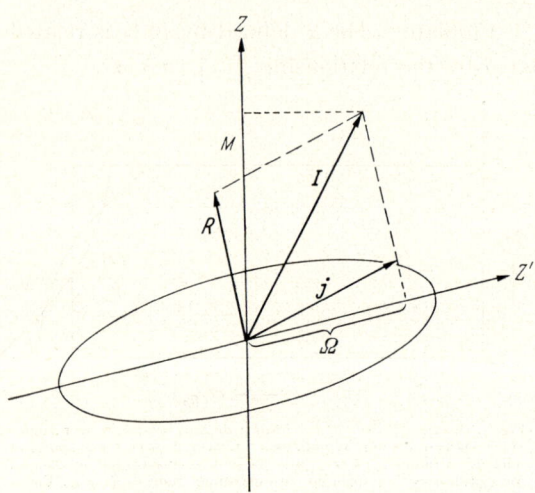

$$\left.\begin{array}{l} R(\vartheta, \varphi) \\ = a_0 [1 + \Sigma_\mu a_\mu Y_{2\mu}(\vartheta, \varphi)], \end{array}\right\} \quad (9.10)$$

where $\Sigma |a_\mu|^2 = \beta^2$ is to be identified with $(4\pi/5)\,\alpha^2$ in (9.9). For nuclei which vibrate, β may not be a constant of the motion; what is significant is $\langle\beta^2\rangle$. In many cases $\langle\beta^2\rangle$ can be large while the spectroscopic quadrupole moment (see below), which depends on $\langle\beta\rangle$, may be very small. Samarium may be an example [15].

Fig. 4. Vector model of a rotating nucleus [A. BOHR and B. R. MOTTELSON: Dan. Mat.-Fys. Medd. 27, No. 16 (1953)]. The nucleus is depicted as a prolate spheroid. The angular momentum j of the odd nucleon precesses about the nuclear symmetry axis Z' with constant projection Ω. The nuclear core has an angular momentum vector R normal to the symmetry axis. R and Ω (or more precisely, j) add together to form the total nuclear angular momentum vector I, which is rigorously a constant of the motion. The entire figure precesses about I. For the lowest rotational state, $|I| = \sqrt{\Omega(\Omega+1)}$. For an even-even nucleus, $\Omega = 0$ and $R = I$.

Although (9.9) was derived above for the case of a uniformly charged nucleus, it is valid for more general distributions (see FORD and HILL [8], Eq. 58).

II. Measurements of nuclear deformation.

10. Optical quadrupole moments [3]. Optical hfs quadrupole studies measure the quantity

$$Q = \langle r^2 (3\cos^2\vartheta - 1)\rangle_{M=I}. \qquad (10.1)$$

This is related to the projection of the intrinsic quadrupole moment Q_0 along the nuclear spin, and always $Q_0 > Q$. In the strong-coupling limit of the Bohr-Mottelson model, the nucleus is described by quantum numbers I, M, and K, where K is the projection of angular momentum along the nuclear symmetry axis; for the nuclear ground state, $K = I$. The nuclear symmetry axis precesses about the total angular momentum vector I (see Fig. 4). This model then gives the relationship

$$Q = \frac{I(2I-1)}{(I+1)(2I+3)}\, Q_0. \qquad (10.2)$$

The spectroscopic quadrupole moment Q vanishes, of course, for I equal to either 0 or $\frac{1}{2}$ and approaches Q_0 for large I. Q_0, however, need not be zero for $I=0$ or $\frac{1}{2}$. The relationship between Q_0 and α for a uniformly charged nucleus of mean radius a_0 [Eq. (9.1)] is given by

$$\alpha = \frac{5}{6} \frac{Q_0}{Z a_0^2}. \tag{10.3}$$

The strong coupling approximation appears to be valid in the regions $90 \leq N \lesssim 112$ and[1] $Z \gtrsim 88$.

11. γ-transitions and Coulomb excitation [3]. Nuclear transitions which proceed by $E2$ radiation or excitation measure a transition quadrupole moment which is frequently related to the ground state (static) quadrupole moment. The relationship is particularly pronounced when the nuclear excitation can be described in terms of collective oscillations (e.g. rotations or vibrations).

Transition quadrupole moments have been measured by both γ-ray lifetimes and Coulomb excitations; the results are in good agreement. The former method is probably the most accurate available for measuring any kind of quadrupole moments; the accuracy available there is about ten percent at present.

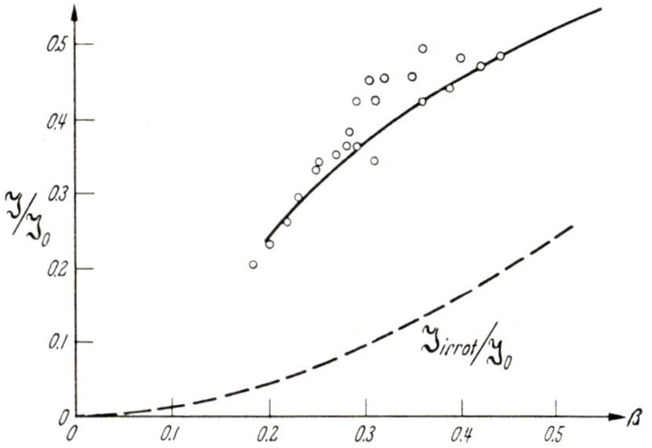

12. Nuclear energy levels [3]. There is a strong correlation between the energy of the first

Fig. 5. Moments of inertia obtained from rotational spectra [Eq. (12.1)] [A. Bohr and B. R. Mottelson: Dan. Mat.-Fys. Medd. **30**, No. 1 (1955)]. The ratio of observed moments of intertia to the solid body moment of intertia $\mathfrak{I}_0$ is plotted as a function of nuclear deformation β obtained mainly from Coulomb excitation. The solid curve is a two-nucleon model fit to the data. The broken curve is the irrotational moment of inertia, $\mathfrak{I}_{\text{irrot}} = \dfrac{45}{16\,\pi} \mathfrak{I}_0\,\beta^2$.

excited states of even-even nuclei and intrinsic deformation. The low-lying energy levels of strongly deformed nuclei (even or odd) are described by the rotational spectrum

$$E_{\text{rot}} = \frac{\hbar^2}{2\mathfrak{I}} \left[I(I+1) - I_0(I_0+1) \right], \quad (K \neq \tfrac{1}{2}). \tag{12.1}$$

In the early development of the collective model, a hydrodynamical model of irrotational flow was assumed in which the moment of inertia $\mathfrak{I}$ was given by

$$\mathfrak{I}_{\text{irrot}} \approx \frac{45}{16\pi} \mathfrak{I}_0 \beta^2 \tag{12.2}$$

where $\mathfrak{I}_0$ is the rigid body moment of inertia, and $\beta^2 = (4\pi/5)\,\alpha^2$. The irrotational flow assumption is now known to be incorrect, and the so-called "cranking model[2]" for nucleons in a nuclear potential plus residual interparticle forces leads[3] to moments of inertia intermediate between $\mathfrak{I}_{\text{irrot}}$ and $\mathfrak{I}_0$. While simple models show general correlations between deformations and moments of inertia, the precise relationship does depend on details of nuclear configurations (cf. Fig. 5). For the purpose of comparison with isotope shifts, it is at present better to use quadrupole moments measured by $E2$ transitions than those derived from energy level data.

[1] G. Scharff-Goldhaber: Phys. Rev. **103**, 837 (1956).
[2] D. R. Inglis: Phys. Rev. **96**, 1059 (1954); **103**, 1186 (1956).
[3] A. Bohr and B. R. Mottelson: Dan. Mat.-Fys. Medd. **30**, No. 1 (1955).

D. Nuclear compressibility and surface phenomena.

One of the consequences of nuclear saturation is that we are permitted to write the expression for the effective radius in the *approximate* form

$$R = r A^{\frac{1}{3}}$$

with r a constant. Thus we are tempted to write in Eq. (3.7)

$$\frac{\delta R}{R} = \frac{1}{3A}.$$

But in fact the nuclear radius is a function not of A alone, but rather of N and Z. The constant r above is a mean value for those nuclei which lie about the stable valley on which measurements are made. What is significant for isotope shifts is the change in nuclear radius with neutron number and this will generally be smaller than that given by the $A^{\frac{1}{3}}$ law. Two effects are given in the following section.

13. Compressibility. Let E_0 be the energy and R_0 the radius of the nucleus in the absence of Coulomb forces. In the presence of the Coulomb forces the energy as a function of radius is given by[1]

$$E = E_0 + E_c \left(\frac{R_0}{R}\right) + \frac{1}{2} E_0'' \left(\frac{R - R_0}{R_0}\right)^2 + \cdots, \tag{13.1}$$

where $E_0'' = [R^2 \partial^2 E_0 / \partial R^2]_{R_0}$ and E_c is the Coulomb energy at R_0. The equilibrium radius is thus given approximately by

$$R = R_0 (1 + E_c / E_0''), \tag{13.2}$$

with $R_0 = r_0 A^{\frac{1}{3}}$. The dependence of E_c and E_0'' on A and Z is given by

$$E_c = \frac{3 Z^2 e^2}{5 r_0 A^{\frac{1}{3}}} \tag{13.3}$$

and

$$E_0'' = K A. \tag{13.4}$$

The dependence of E_0'' on A follows from the property of nuclear saturation. From Eqs. (13.2) to (13.4) can be derived the quantities

$$\frac{1}{R} \frac{\partial R}{\partial N} = \frac{1}{3A} \left[1 - \frac{4 E_c}{E_0'' + E_c}\right], \tag{13.5}$$

$$\frac{1}{R} \frac{\partial R}{\partial Z} = \frac{1}{3A} \left[1 + \frac{4 E_c}{E_0'' + E_c} \left(\frac{3}{2} \frac{A}{Z} - 1\right)\right], \tag{13.6}$$

and

$$\frac{1}{R} \frac{dR}{dA} = \frac{1}{3A} \left[1 + \frac{4 E_c}{E_0'' + E_c} \left(\frac{3}{2} \frac{A}{Z} \frac{dZ}{dA} - 1\right)\right]. \tag{13.7}$$

The quantity $[R^{-1} \partial R / \partial N] \delta N$ is to be identified with the quantity $(\delta R_1 / R_1)$ in (3.7). Thus isotope shifts should be reduced by the factor $[1 - 4 E_c / (E_0'' + E_c)]$.

The electron scattering and mesic atom experiments are not sufficiently accurate at present to determine nuclear compressibility—*i.e.*, the nuclear density appears to be constant within experimental error. [This permits one, however, to place a lower limit of 50 Mev on the compressibility constant K.] BRUECKNER and GAMMEL[2] have obtained a theoretical estimate of the compressibility which gives $K \approx 187$ Mev. The semiempirical theory of BERG and WILETS [2] yields a value

[1] E. FEENBERG: Phys. Rev. **59**, 149 (1941).

[2] K. A. BRUECKNER and J. L. GAMMEL: Phys. Rev. **105**, 1679 (1957); **109**, 1023 (1958).

of $K \sim 200$ Mev. To these theoretical estimates must be added the effect of surface tension which varies with A in such a manner that $[2]$ the K of Eq. (13.4) should be identified with $K + K_s A^{-\frac{1}{3}}$ where $K_s \sim 200$ Mev.

From (13.5) it is seen that the effect of compressibility is to reduce the isotope shifts. While the theoretical estimates of K indicate that the effect is of some importance, it may not be sufficient to account for the smallness of the observed isotope shifts.

14. Surface phenomena. There exist nuclear effects other than compressibility which result in a decrease in the ratio $R_1^{-1} \partial R_1 / \partial N$. One of these is related to the difference in neutron and proton distributions. This effect is illustrated by semi-empirical calculations $[2b]$ on nuclear density distributions. For heavy nuclei ($N \approx 1.4Z$), the neutron and proton distributions appear similar; for a greater neutron excess the proton distribution would be smaller than that of the neutrons and for $N = Z$ the protons would extend farther than the neutrons. The phenomenon is related to surface thickness. The effect is clearly in the direction to reduce the magnitude of isotope shifts. While quantitative estimates are difficult, the effect appears to be large. An order of magnitude estimate can be obtained as follows:

A calculation has been made $[2b]$ for a nucleus with $Z = 70$ and $N = 130$, neglecting the Coulomb field. The protons were found to lie inside the neutrons by 20% of the surface thickness (the surface thickness being about $\frac{1}{3}$ of the nuclear radius). A nucleus with equal proton and neutron numbers would have identical proton and neutron density distributions in the absence of Coulomb forces. If the effect were linear, one would find the ratio $R^{-1} \partial R / \partial N$ diminished by 35%. In fact it is not linear, and specific Coulomb effects modify the results. A change in ratio of somewhat less than 35% is to be expected.

E. Atomic considerations.

The nuclear and atomic effects in isotope shifts are clearly separable, but the computation of the atomic effects is far from trivial. The problems are essentially: Calculation of $\Psi^2(0)$ for the transition electron; contribution of the p-electron; effect of the change in screening of the transition electron.

15. Calculation of $\Psi^2(0)$. FERMI and SEGRÈ[1] have obtained an expression for the s-electron Schrödinger density at $r = 0$, $\Psi^2(0)$, in the case of alkali metals:

$$\Psi^2(0) = \frac{Z Z_0^2}{\pi a_H^3 n_0^3} \left[1 - \frac{d\sigma}{dn} \right], \tag{15.1}$$

where Z_0 is the effective charge in the outer region; n_0 is the effective quantum number; and σ is the quantum defect: $n_0 = n - \sigma$.

While the Fermi-Segrè formula (15.1) was derived for alkali metals, it appears to be of far wider validity than originally expected[2]; it provides a semiempirical method for determining $\Psi^2(0)$ whenever the spectrum can be fitted to a series with a smoothly varying quantum defect σ.

A direct check on the validity of the Fermi-Segrè formula is available in those cases where the nuclear magnetic moment has been measured both by optical hfs and by nuclear induction methods. In the case of optical spectra the hfs splittings for an s-electron are proportional to the product of $\Psi^2(0)$ and the nuclear magnetic moment, while nuclear induction yields the magnetic moment directly. Comparisons made in this way indicate that the Fermi-Segrè formula is quite reliable.

[1] E. FERMI and E. SEGRÈ: Z. Physik **82**, 729 (1933).
[2] L. L. FOLDY: Phys. Rev. (to be published).

16. Contribution of p-electron $[14]$, $[7]$. Isotope shifts are generally observed in transitions of the type $p \to s$. While the shift of $p_{\frac{3}{2}}$-electron is negligible, ROSENTHAL and BREIT found that the $p_{\frac{1}{2}}$ shift may be of some consequence. For a $6p_{\frac{1}{2}}$-electron they estimated the shift to be about $\frac{1}{20}$ of that for a $6s$-electron. For complex configurations of the type sp, s^2p, etc., the j's of the individual electrons are not good quantum numbers in L-S coupling. Thus a p-electron will generally be a mixture of $p_{\frac{1}{2}}$- and $p_{\frac{3}{2}}$-electrons and will contribute to the isotope shifts in proportion to the amount of $p_{\frac{1}{2}}$-state present.

17. Screening effects $[7]$. During a transition of the valence electron of the form $p \to s$, the screening of the inner electron changes, and hence their wave functions also change. This represents one type of correlation among the electrons. The change in the wave functions of the inner electrons has the effect of a transition and so contributes to the isotope shifts.

The screening effect can be calculated as follows: Let p represent the fraction of time the valence electron lies inside the "orbit" of the inner ns electron in question. During this fraction of time the effective charge of the electron is reduced by one unit and [cf. Eq. (15.1)] $\Psi_n^2(0)$ is reduced by the fraction

$$[Z_0^2 - (Z_0 - 1)^2]/Z_0^2 \approx 2/Z_0.$$

Table 3. *Isotope shift screening correction for a transition involing a 6s-electron. From* CRAWFORD *and* SCHAWLOW *([7], Table 4).*

Electron	$\Psi^2(0) \cdot \pi\, a_H^3$ (HARTREE)	p	Screening correction per ns^2 as a fraction of $(\delta \Delta W)_{6s}$
$1s$	5.06×10^5	2.034×10^{-5}	0.016
$2s$	5.56×10^4	1.912×10^{-4}	0.017
$3s$	1.23×10^4	9.010×10^{-4}	0.020
$4s$	3.06×10^3	3.846×10^{-3}	0.028
$5s$	6.10×10^2	2.090×10^{-2}	0.079
$6s$	$3.2\ \times 10^1$		—

total 0.16

Since the contributions of these inner electrons vary as the change in $\Psi_n^2(0)$, the contribution of each electron, measured in units of the valence electron ($_0$) shift, is given by

$$\frac{(\delta \Delta E)_n}{(\delta \Delta E)_0} = \frac{\Psi_n^2(0)}{\Psi_0^2(0)} \frac{2}{Z_0} p. \tag{17.1}$$

The quantity p is obtained by evaluating the integral

$$p = (4\pi)^2 \int_0^\infty \Psi_n^2(r')\, r'^2\, dr' \int_0^{r'} \Psi_0^2(r)\, r^2\, dr, \tag{17.2}$$

CRAWFORD and SCHAWLOW $[7]$ evaluated p and $(\delta \Delta E)_n/(\delta \Delta E)_0$ for all of the inner s-electrons in mercury using Hartree wave functions. The results are given in Table 3. The total of sixteen percent should be regarded as an upper limit since errors are present in the Hartree wave functions and in the assumption that n_0 [Eq. (15.1)] is independent of Z_0.

The screening effect is in the direction of reducing the theoretical value of the shifts.

F. Comparison with experiment.

CRAWFORD and SCHAWLOW $[7]$ have compared in detail the experimental results in the isoelectronic sequence Hg II, Tl III and Pb IV with the theory of the volume effect. They included all of the atomic effects discussed in the previous section. For incompressible spherical nuclei of equivalent radii $R_1 = 1.19 \times 10^{-13}\, A^{\frac{1}{3}}$ cm [they assumed $1.5 \times 10^{-13}\, A^{\frac{1}{3}}$ cm], the ratio of experimental to theoretical values for the shifts was found to be 0.74.

BRIX and KOPFERMANN [5], and HUMBACH[1] have systematically analyzed the experimental data from a large number of elements in a manner similar to that of CRAWFORD and SCHAWLOW. The plot in Fig. 6, which is taken from BRIX and KOPFERMANN [5] except where noted in the caption, gives the ratio S of experimental to theoretical values of the shifts for even-N nuclei. [The theoretical values assume incompressible, spherical nuclei of equivalent radius $R_1 = 1.19 \times 10^{-13} A^{\frac{1}{3}}$ cm]. The accuracy of the theoretical volume shifts is probably better than 20%.

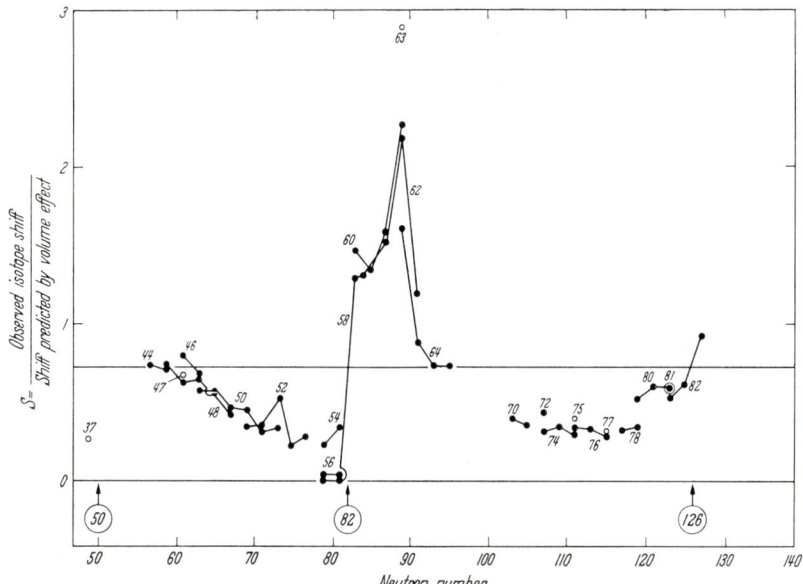

Fig. 6. Ratio, S, of experimental to theoretical isotope shifts. Only even-N shifts are given, plotted at the average neutron number. Even-Z shifts are given by solid circles, odd-Z by open circles. — The theoretical shifts have been calculated by W. HUMBACH [Z. Physik 133, 589 (1952)] and BRIX and KOPFERMANN [5], and are here based on $R_1 = 1.19 \times 10^{-13} A^{\frac{1}{3}}$ cm. References to the experimental values are given by MACK and ARROE [11] and BRIX and KOPFERMANN [6]. — This kind of plot was first used by BRIX and KOPFERMANN [5]. See also H. KOPFERMANN: Kernmomente, 2. Aufl. p. 167. Frankfurt: Akademische Verlagsgesellschaft 1956.

18. Fluctuations in S. WILETS, HILL and FORD [15] have shown that the observed fluctuations in S can be described in terms of changes in nuclear deformation according to Eq. (9.9). Estimates of α were obtained from the energies of first excited levels of even-even nuclei using Eqs. (12.1) and (12.2). The deformations so derived were reduced by the empirical factor 3 in α^2 in order to obtain agreement with spectroscopically measured quadrupole moments. The agreement between the experimental and theoretical shifts obtained in this manner was satisfactory and established definitively the role of deformation in isotope shifts.

An alternative approach is to derive nuclear deformations from isotope shifts. This can be done in a straightforward manner if the tentative assumption is made that α is a function of N alone. While this is not strictly true, there is a considerable amount of evidence that, in the rare earth region, nuclear deformations depend more strongly on N than Z. Then (9.9) can be rearranged (to order α^2),

$$\delta(\alpha^2) = \frac{10}{3(2\varrho + 3)A} \frac{(\delta \Delta E)_\alpha}{(\delta \Delta E)_v} \delta N, \qquad (18.1)$$

---

[1] W. HUMBACH: Z. Physik 133, 589 (1952).

and integrated (summed). Another assumption which is made is that fluctuations about the mean value $\overline{S}$ of the ratio S is due to deformation only. The mean value is chosen so that the deformation vanishes at magic neutron numbers (i.e. $N = 82$ and 126).

There is one point which requires further discussion. The mean value of S turns out to be less than unity. If we can ascribe this anamolously low value of $\overline{S}$ to nuclear compressibility, surface phenomena, or some analogous nuclear property, then the quantity $(\delta \Delta E)_\alpha/(\delta \Delta E)_v$ in (18.1) should be identified with $S - \overline{S}$. If, however, there has been some systematic error or omission in the

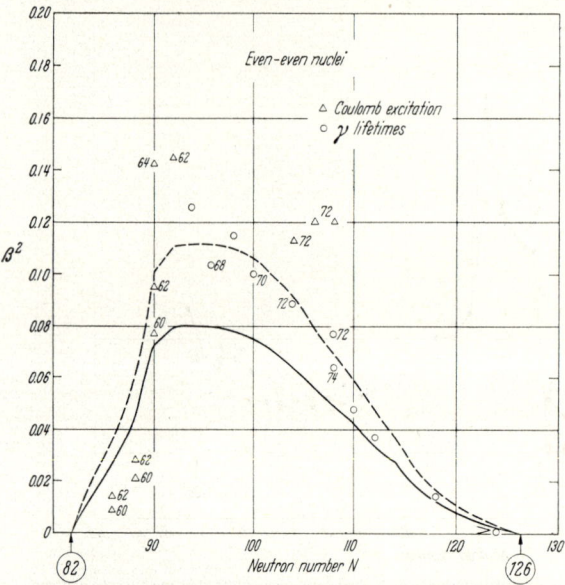

calculation of atomic effects so that, in fact, $\overline{S}$ should be unity, then the experimental shifts should be rescaled so that $(\delta \Delta E)_\alpha/(\delta \Delta E)_v$ is identified with $(S - \overline{S})/\overline{S}$. The latter integrated curve would be above the former by the constant factor $\overline{S}^{-1}$. Deformations ($\beta^2 = 4\pi \alpha^2/5$) derived according to both prescriptions are given in Fig. 7 along with the experimental (transition) values of β^2 determined from Coulomb excitations and γ-ray lifetimes.

The solid curve in Fig. 7 $[(\delta \Delta E)_\alpha/(\delta \Delta E)_v = S - \overline{S}]$ lies below the experimental values of β^2. This can be accounted for by reconsidering the assumption that α (or β) is a function of N alone. One can see from the experimental points that there is some dependence on Z as well, and the dependence is in the right

Fig. 7. Nuclear deformations deduced from isotope shifts and $E2$ transitions. The curves are obtained from fluctuations in the ratio S according to the procedure outlined in Sect. 18. The solid curve is obtained assuming $(\delta \Delta E)_\alpha/(\delta \Delta E)_v = S - \overline{S}$ with $\overline{S} = 0.72$. The broken curve is obtained by renormalizing the ratios S so that $\overline{S} = 1$. The triangles are deduced from Coulomb excitation and the circles from γ-lifetimes. [N. P. HEYDENBERG and G. M. TEMMER: Ann. Rev. Nucl. Sci. 6, 77 (1956)]. All data are for even-even nuclei.

direction. Thus the apparent better agreement of the broken curve ($\overline{S} \to 1$) is not particularly significant.

Properly, of course, one should compare the slope of the integrated curve with the changes in β^2 for constant Z (which is where we began). The conclusion to be drawn is that there are no apparent discrepancies between the observed fluctuations and the explanation in terms of nuclear deformations. Furthermore, isotope shifts appear to be a reliable guide in comparing the deformations of various isotopes of the same element.

19. Smallness of $\overline{S}$. While it is possible that the small mean value of the ratio, $\overline{S}$, is due to systematic errors in the analysis of atomic effects, it seems reasonable to attribute [15] a part of the discrepancy to nuclear compressibility. Let us, for the moment, relate the anomaly in the shifts to nuclear compressibility alone according to (3.7) and (13.5):

$$1 - \frac{4 E_c}{E_0'' + E_c} = \overline{S}, \tag{19.1}$$

with $E_c = 3Z^2 e^2/(5R)$ and $E_0'' = KA$. The left hand side of (19.1) varies roughly as $A^{\frac{2}{3}}$; in the region $82 \leq N \leq 126$, we will assume the mean values:

$$\left.\begin{aligned} \bar{S} &= 0.72, \\ \bar{Z} &= 70, \\ \bar{A} &= 174, \\ R_1 &= 1.19 \times 10^{-13} A^{\frac{1}{3}} \text{cm}. \end{aligned}\right\} \qquad (19.2)$$

This gives for K a value of 48 Mev, which is significantly lower than the estimates discussed in Sect. 13. It would appear that only about $\frac{1}{4}$ of the difference between $\bar{S}$ and unity can be attributed to compressibility. There thus remains a discrepancy of 15 to 20% in $\bar{S}$ which can reasonably be attributed to differences between neutron and proton distributions as a function of neutron number, as discussed in Sect. 14.

20. Even-odd staggering. There now exists a considerable amount of data available concerning even-odd staggering (cf. Table 4). In all cases of staggering

Table 4. *Even-odd staggering.*

Odd-isotope (A)	$\dfrac{(\nu_A - \nu_{A-1})}{(\nu_{A+1} - \nu_{A-1})}$	Odd-isotope (A)	$\dfrac{(\nu_A - \nu_{A-1})}{(\nu_{A+1} - \nu_{A-1})}$	Odd-isotope (A)	$\dfrac{(\nu_A - \nu_{A-1})}{(\nu_{A+1} - \nu_{A-1})}$
$_{48}\text{Cd}_{103}^{111}$	0.10	$_{62}\text{Sm}_{85}^{147}$	≥ 0.4 [1]	$_{80}\text{Hg}_{119}^{199}$	0.15
$_{48}\text{Cd}_{105}^{113}$	0.15	$_{62}\text{Sm}_{87}^{149}$	< 0.3	$_{80}\text{Hg}_{121}^{201}$	0.30
$_{52}\text{Te}_{71}^{123}$	0.0	$_{70}\text{Yb}_{101}^{171}$	0.4 [2]	$_{82}\text{Pb}_{125}^{207}$	0.38
$_{52}\text{Te}_{73}^{125}$	0.1	$_{70}\text{Yb}_{103}^{173}$	0.50	$_{92}\text{U}_{141}^{233}$	0.34 [2]
$_{56}\text{Ba}_{79}^{135}$	-1	W_{107}	0.5	$_{92}\text{U}_{143}^{235}$	0.25
$_{56}\text{Ba}_{81}^{137}$	-1	$_{78}\text{Pt}_{117}^{195}$	0.44		

[1] $1 - (\nu_{A+1} - \nu_A)/2(\nu_{A+1} - \nu_{A-3})$.
[2] $1 - (\nu_{A+1} - \nu_A)/(\nu_{A+2} - \nu_A)$.

References to the experimental data are given by DRIX and KOPFERMANN [6] and by MACK and ARROE [11] with the following additions:
Cd: H. G. KUHN and S. A. RAMSDEN: Proc. Roy. Soc. Lond., Ser. A 237, 485 (1956).
Ba: Additional interpretation received from O. H. ARROE: Private communication.
Pb: F. E. GEIGER: Phys. Rev. (to be published).

reported, the spectral line due to the odd isotope lies closer (or at least as close) to the lighter even-even neighbor than to the heavier neighbor. In some cases the line of the odd isotope lies beyond the lighter neighbor. Several explanations of the origin of the staggering have been proposed:

α) *Nucleus polarization.* This has been discussed in Sect. 7. While in the right direction, the effect appears to be too small to account alone for the staggering.

β) *Nuclear deformation.* Several mechanisms exist through which nuclear deformations may produce even-odd staggering. One of these [15] is based on the point that isotope shifts depend upon the *square* of the deformation.

In the simple strong coupling theory of the unified model, nucleons fill into the nucleus in pairs with $\Omega = \pm |\Omega|$ (Ω is the projection of angular momentum of a given particle along the nuclear symmetry axis). The deformation is approximately linear in the number of nucleons with a given j and $|\Omega|$, and in the simple theory is given by [1-3]

$$\alpha = \text{const} \sum_i \frac{j_i(j_i + 1) - 3\Omega_i^2}{4j_i(j_i + 1)}. \tag{20.1}$$

Isotope shifts depend on α^2, however, and α^2 is generally less for an odd isotope than is the mean value of α^2 for the even-even neighbors. This is illustrated in Fig. 8 for a shell of $j = \frac{9}{2}$ particles. The effect is most pronounced when the nuclear deformation is changing rapidly, for example near closed shells. Even-odd staggering, however, is also observed in regions where the deformation is not changing rapidly.

A more detailed study of the independent particle strong coupling theory[4] shows that the order of filling of nucleonic orbits may differ between an odd nucleon and the next pair. A manifestation of this may be the recurrence of the same spin in series of odd nuclei differing by two in A. Such considerations could lead to staggering of either sense, although the predominance appears fortuitously to favor the observed sense.

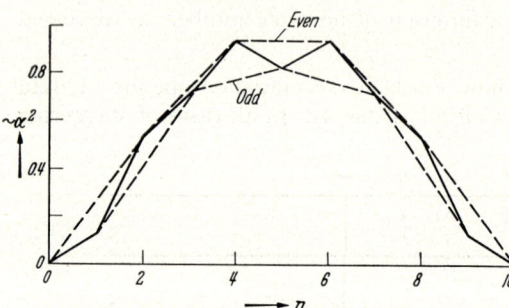

Fig. 8. Nuclear deformations in a shell of $j = \frac{9}{2}$ particles [15]. Plotted vertically is the quantity

$$\left[\sum_i \frac{j(j+1) - 3\Omega_i^2}{4j(j+1)} \right]^2,$$

which in the simple strong coupling theory is proportional to the square of the nuclear deformation, α^2. Plotted horizontally is the number of nucleons in the shell. The value of α^2 for n odd is less than the average value of α^2 for the neighboring n even points. This may afford an explanation of even-odd staggering.

The independent particle model is modified by considerations of "residual" two body interactions. These interactions not only affect the order of orbit filling but also are responsible for configuration mixing, just as in atomic structure. Calculations[5] indicate that for strong coupling those orbitals which favor pair overlap also favor large deformations. This indicates a larger average deformation for even-even nuclei than for odd-A nuclei, which is the direction of observed staggering. The effect is probably reversed when strong coupling breaks down, since then paired nucleons will tend to couple together to $J = 0$, thus favoring spherical symmetry.

Near closed shell neutron numbers 82 and 126, nuclear deformations are derived in Fig. 9 on the assumption that the staggering results from deformation effects. Where deformations are not large enough to lead to strong coupling, the values of $\langle \alpha^2 \rangle$ are to be interpreted as resulting from softness to oscillation.

γ) *Non-interaction of the odd neutron.* The initial success of the Schmidt model led some authors[6-8] to suggest that an odd nucleon does not interact strongly with the nuclear core, and hence does not cause so great an expansion

[1] J. RAINWATER: Phys. Rev. 79, 432 (1950).
[2] A. BOHR: Dan. Mat.-Fys. Medd. 26, No. 14 (1952).
[3] K. W. FORD: Phys. Rev. 90, 29 (1953).
[4] Cf. S. G. NILSSON: Dan. Mat.-Fys. Medd. 29, No. 16 (1955). — B. R. MOTTELSON and S. G. NILSSON: Phys. Rev. 99, 1615 (1955).
[5] D. M. CHASE and L. WILETS: Unpublished.
[6] M. FIERZ: Göttinger Nachr. 3, 1 (1947).
[7] P. BRIX and H. KOPFERMANN: Göttinger Nachr. 2, 31 (1947).
[8] G. BREIT: Phys. Rev. 78, 470 (1950); 79, 891 (1950).

of the core as a nucleon pair. While this model is very appealing, one might also argue that the larger orbit of the odd neutron should polarize the core and expand the proton distribution more than a tightly bound pair.

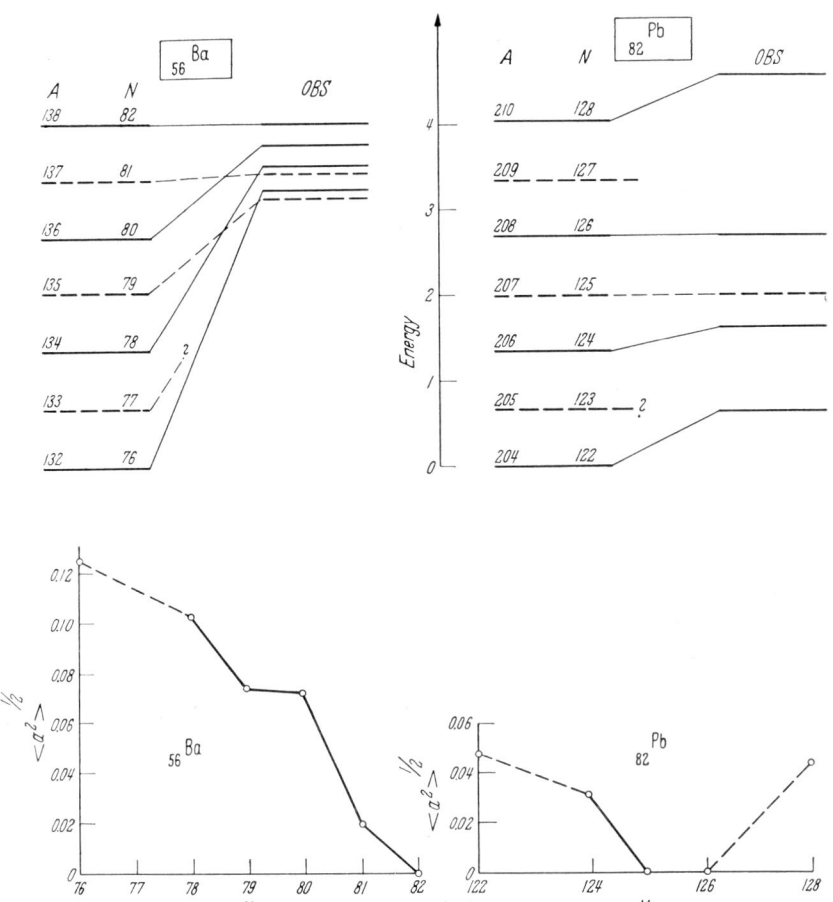

Fig. 9. Deformations derived from isotope shifts near closed neutron shells 82 and 126 (after WILETS, HILL and FORD [15]). The figure illustrates that even-odd staggering can be accounted for on the basis of reasonable variation in deformation. The unit of energy is the theoretical shift for a change in mass number by one unit and $R_1 = 1.19 \times 10^{-13} A^{\frac{1}{3}}$ cm. The left column for each element gives the *presumed* levels in the absence of deformation (i.e. levels spaced by $\bar{S} \approx 0.72$). The difference between the left column and observation is used to derive the deformations given at the bottom of the figure, assuming the difference is due to deformation alone. References for the experimental data are: Ba: O. H. ARROE: Doctoral Thesis, University of Copenhagen 1951; and private communication. — Pb: MANNING, ANDERSON and WATSON: Phys. Rev. 78, 417 (1950). — K. MURAKAWA and S. SUWA: J. Phys. Soc. Japan 5, 382 (1950). — F. GEIGER: Phys. Rev. 79, 212 (1950) and Phys. Rev. to be published. — BRIX, VON BUTTLAR, HOUTERMANS and KOPFERMANN: Z. Physik 133, 192 (1952).

Even-odd staggering presents a challenging problem to nuclear theory. Light should be shed on the subject when further experiments such as electron scattering and Coulomb excitation are performed on separated isotopes.

21. Application to spectral analysis. ZEEMAN and hyperfine structure patterns are very useful in spectral analysis, particularly in the identification of terms and J-values. Isotope shift data can be equally useful in the assignment of electron configurations, which frequently defy identification even after the terms are known.

In the case of erbium, a method of limiting possible configurations has been proposed[1]. This is schematized in Fig. 10. Transitions of the kind $p \to s$ lead to the usual positive shifts while the $s \to p$ transitions give rise to negative shifts. Through configuration mixing, two electron transitions of the kind $dp \to s^2$ are possible and should lead to doubly-sized positive shifts, which were believed to have been observed.

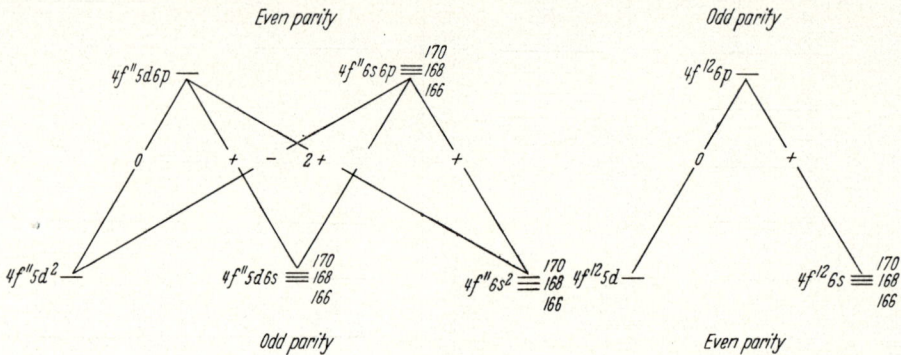

Fig. 10. Proposed electron configurations and transitions in Er II [L. WILETS and L. C. BRADLEY III: Phys. Rev. 87, 1018 (1952)]. The symbols $(-, 0, +, 2 +)$ indicate the relative size of the isotope shifts in the corresponding transitions.

G. Conclusion.

22. Atomic isotope shifts provide a useful tool for nuclear research. Most directly, in heavy nuclei they measure *changes* in the quantity $\langle r^{2\varrho} \rangle$; for a given element, ratios in the changes in $\langle r^{2\varrho} \rangle$ are given directly and irrespective of atomic considerations.

The variation with A in the ratio S of experimental to theoretical shifts for even-even nuclei appears to be accounted for on the basis of nuclear deformation while the smallness of the mean value of the shifts appears to be consistent with concepts of nuclear compressibility and surface phenomenon. Even-odd staggering, however, still presents a challenging problem to nuclear theorists, and it may be that several effects contribute.

Bibliography.

[1] BARTLETT jr., J. H., and J. J. GIBBONS jr.: Isotope Shifts in Neon. Phys. Rev. **44**, 538 (1933). — Contains a formulation of the specific mass problem for Hartree-type wavefunctions.
[2] (a) BERG, R. A., and L. WILETS: Nuclear Surface Effects. Phys. Rev. **101**, 201 (1956). — (b) L.WILETS: Neutron and Proton Densities in Nuclei. Phys. Rev. **101**, 1805 (1956). — L.WILETS: Rev. Mod. Phys. **30**, April (1958). — Contains a review of the two previous papers and corrects a numerical error which appeared in [2a] which affects the estimate of nuclear compressibility.
[3] S. MOSZKOWSKI: Encyclopedia of Physics, Vol. XXXIX. — BOHR, A.: The Coupling of Nuclear Surface Oscillations to the Motion of Individual Nucleons. Dan. Mat.-Fys. Medd. **26**, No. 14 (1952). — BOHR, A., and B. R. MOTTELSON: Collective and Individual Particle Aspects of Nuclear Structure. Dan. Mat.-Fys. Medd. **27**, No. 16 (1953).
[4] BRIX, P., and H. KOPFERMANN: Nachr. Ges. Wiss. Göttingen **1947**, 31. — Z. Physik **126**, 344 (1949). — Phys. Rev. **85**, 1050 (1952). — The authors here first correlated the large value of the samarium isotope shift with nuclear quadrupole moments.
[5] BRIX, P., and H. KOPFERMANN: Neuere Ergebnisse zum Isotopieverschiebungs-Effekt in den Atomspektren. Festschr. Akad. Wiss. Göttingen, Math.-physikal. Kl., p. 17, 1951.

---

[1] L. WILETS and L. C. BRADLEY III: Phys. Rev. **87**, 1018 (1952).

[6] BRIX, P., and H. KOPFERMANN: LANDOLT-BÖRNSTEIN, Zahlenwerte und Funktionen aus Physik, Chemie, Astronomie, Geophysik und Technik, 6. Aufl., Bd. 1, Teil 5, p. 1. Berlin: Springer 1952.

[7] CRAWFORD, M. F., and A. L. SCHAWLOW: Electron-Nuclear Potential Fields from Hyperfine Structure. Phys. Rev. 76, 1310 (1949). — The first complete treatment of atomic effects is given here.

[8] FORD, K. W., and D. L. HILL: The Charge Distribution in the Nucleus. Ann. Rev. Nucl. Sci. 5, 25 (1955).

[9] HOFSTADTER, R.: Electron Scattering and Nuclear Structure. Rev. Mod. Phys. 28, 214 (1956). — Further references are given in this paper.

[10] HUGHES, D. S., and C. ECKART: Phys. Rev. 36, 694 (1930). — The classic paper on nuclear motion.

[11] MACK, J. E., and O. H. ARROE: Isotope Shift in Atomic Spectra. Ann. Rev. Nucl. Sci. 6, 117 (1956). — Contains an extensive survey of the literature.

[12] RACAH, G.: Nuovo Cim. 8, 178 (1931). — Nature 129, 723 (1932). — The first correct formulation of field effects including relativity. An earlier nonrelativistic investigation by J. H. BARTLETT jr. [Nature, Lond. 128, 408 (1931)] contains an error.

[13] RAVENHALL, D. G., and D. R. YENNIE: Phys. Rev. 96, 239 (1954).

[14] ROSENTHAL, J. E., and G. BREIT: Phys. Rev. 41, 459 (1932). — BREIT, G.: Phys. Rev. 42, 348 (1932). — A formulation of the field effects including relativity done independently of RACAH and in considerable detail.

[15] WILETS, L., D. L. HILL and K. W. FORD: Isotope Shift Anomalies and Nuclear Structure. Phys. Rev. 91, 1488 (1953). — A survey of isotope shifts, introducing several new nuclear effects.

[16] YENNIE, D. R., M. M. LEVY and D. G. RAVENHALL: Electromagnetic Structure of Nucleons. Rev. Mod. Phys. 29, 144 (1957). — See also April 1958 issue of Reviews of Modern Physics for several articles on the subject.

Kernmagnetische Hochfrequenz-Spektroskopie.

Von

G. LAUKIEN.

Mit 110 Figuren.

Einführung.

Die Feinstruktur der optischen Linienspektren konnte durch die quantenmechanische Beschreibung der Wechselwirkungen zwischen den Bahn- und Eigenmomenten der Hüllenelektronen erklärt werden. Es lag daher nahe, zur Deutung der Hyperfeinstruktur von Atomspektren und ebenso zur Erklärung des Intensitätswechsels in Molekülbanden (neben dem Isotopieverschiebungs-Effekt) Wechselwirkungen zwischen Kern- und Hüllenmomenten zu vermuten. Die zuerst von PAULI[1] ausgesprochene Hypothese, daß manche Kerne ein mechanisches Impulsmoment und ein magnetisches Dipolmoment besitzen, erwies sich in der Folge als außerordentlich glücklich. Sie ermöglichte es, zahlreiche unerklärt gebliebene spektroskopische Beobachtungen zu verstehen, wie vor allem BACK, GOUDSMIT[2] und KOPFERMANN[3] gezeigt haben. Später konnten SCHÜLER und SCHMIDT[4] auch die Existenz elektrischer Kernquadrupolmomente experimentell nachweisen. Die Berechnungsverfahren hierzu[5] sowie die Einführung des allgemeinen Begriffs der Multipolmomente verdanken wir CASIMIR.

Mit der Deutung der optisch-spektroskopischen Beobachtungen wurde es erstmals möglich, die genannten Kerneigenschaften experimentell zu bestimmen[6]. Fast alle bekannten Kern-Drehimpulse sind zuerst optisch aus der Hyperfeinstruktur der Atomspektren und aus den alternierenden Intensitäten der Spektren von Molekülen mit identischen Kernen[7] gemessen worden. Auch der größere Teil der magnetischen Kerndipol- und der elektrischen Kernquadrupol-Momente ist zunächst aus der Hyperfeinstruktur-Aufspaltung mit einer Genauigkeit von 1 bis 10% ermittelt worden. Zahlreiche theoretische Arbeiten, z.B. die Berechnungen des Schalenmodells[8], und ebenso fast alle späteren mit den Hilfsmitteln der Hochfrequenz-Spektroskopie ausgeführten experimentellen Arbeiten haben sich wesentlich auf diese Resultate der optischen Spektroskopie gestützt (vgl. hierzu die Tabelle im Anhang).

Direkt bestätigt wurde die Existenz von Kernmomenten durch Molekularstrahl-Experimente. Solche Versuche sind zuerst von STERN und GERLACH[9] zum Nachweis der Richtungsquantelung der Impulsmomente von Atomhüllen

[1] W. PAULI: Naturwiss. **12**, 741 (1924).
[2] E. BACK u. S. GOUDSMIT: Z. Physik **43**, 321 (1927); **47**, 174 (1928).
[3] H. KOPFERMANN: Naturwiss. **19**, 400 (1931); Z. Physik **75**, 363 (1932).
[4] H. SCHÜLER u. TH. SCHMIDT: Z. Physik **94**, 457 (1935); **98**, 430 (1935).
[5] H. CASIMIR: Physica, Haag **2**, 719 (1935).
[6] Vgl. hierzu den Artikel von KELLY in diesem Bande.
[7] F. HUND: Z. Physik **42**, 93 (1927).
[8] O. HAXEL, H. JENSEN u. H. SUESS: Phys. Rev. **75**, 1766 (1949). — Z. Physik **128**, 295 (1950). — M. GOEPPERT-MAYER: Phys. Rev. **75**, 1969 (1949).
[9] O. STERN: Z. Physik **7**, 249 (1921). — W. GERLACH u. O. STERN: Ann. Physik **74**, 673 (1924).

in Magnetfeldern ausgeführt worden. Der erste geglückte Versuch, ein Kernmoment, und zwar das magnetische Dipolmoment des Protons, durch einen Strahlablenkungs-Versuch zu messen, ist von FRISCH, ESTERMANN und STERN[1] veröffentlicht worden. Das Ergebnis dieser Messung, welche eindeutig bewies, daß das magnetische Moment des Protons ungefähr 2,5mal größer als ein BOHRsches Kernmagneton ist, war verglichen mit den heutigen Meßresultaten sehr ungenau. Für die weitere Entwicklung der Kerntheorie war es jedoch von grundlegender Bedeutung, denn es zeigte eindeutig, daß sich die von der Elektronenhülle her bekannten Vorstellungen über magnetische Momente nicht unmittelbar auf Kerne übertragen lassen.

Den für die weitere Entwicklung entscheidenden meßtechnischen Fortschritt das sog. Molekularstrahl-Resonanzverfahren, verdanken wir RABI, MILLMAN, KUSCH und ZACHARIAS[2]. Angeregt durch eine Arbeit von GORTER[3], haben die genannten Forscher die Erfahrungen der physikalischen Experimentiertechnik, Strahl-Erzeugung, -Ablenkung und -Anzeige mit den von der Elektrotechnik in den vorangegangenen Jahrzehnten entwickelten Hilfsmitteln der Radiotechnik kombiniert. Das von ihnen entwickelte Verfahren kann man wie folgt beschreiben. Untersucht wird in ihm ein fein ausgeblendeter Strahl von Molekülen, die zwei entgegengesetzt inhomogene und ein dazwischenliegendes homogenes Magnetfeld durchfliegen. Strahlt man in dem mittleren Feld Hochfrequenzquanten ein, deren Größe gleich dem Energieunterschied zweier möglicher Orientierungen ist (Resonanz), dann wird ein Teil dieser Quanten absorbiert. Infolgedessen durchläuft ein Teil der Moleküle bzw. derer Kerne das dritte inhomogene Feld mit einer anderen Orientierung als das erste und trifft die Öffnung einer Beobachtungsblende und damit den Detektor nicht mehr (sog. flop out-Methode; bei einer Abänderung des Verfahrens, der sog. flop in-Methode[4], ordnet man die Felder und Blenden übrigens gerade so an, daß der Detektor nur in Resonanz von den Molekülen getroffen wird). Variiert man die Größe der Energiequanten, also ihre Frequenz, so ändert sich die Strahlintensität an der Beobachtungsstelle, wenn sich die Frequenz einer Orientierungs-Resonanzfrequenz des Moleküls oder eines seiner Kerne nähert, bzw. sich von dieser entfernt. Damit wird die Messung der kernmagnetischen Momente auf eine Frequenzbestimmung in den Kurz- und Mittelwellenbereichen der Hochfrequenztechnik zurückgeführt. Es ist also zu bemerken, daß mit der Molekularstrahl-Resonanzmethode nicht nur die Ergebnisse der optischen Kernmomentenforschung bestätigt und zum Teil erheblich präzisiert wurden, sondern daß sie auch als Methode die sog. Hochfrequenz-Spektroskopie begründete[5].

In konsequenter Fortführung dieser Entwicklung wurden vor 10 Jahren unabhängig voneinander zwei Verfahren entwickelt und unter den Namen *Kerninduktion*[6] und *Kernresonanzabsorption*[7] veröffentlicht. Bei beiden Verfahren werden, genau wie bei der Molekularstrahl-Resonanzmethode, erzwungene Übergänge zwischen den Energieniveaus der Kerndipoleinstellungen in einem äußeren Magnetfeld beobachtet. Ebenso erfolgt bei ihnen die Anregung aber auch

---

[1] R. FRISCH u. O. STERN: Z. Physik **85**, 4 (1933). — I. ESTERMANN u. O. STERN: Z. Physik **85**, 17 (1933).

[2] I. I. RABI, J. R. ZACHARIAS, S. MILLMAN u. P. KUSCH: Phys. Rev. **53**, 318 (1938). — I. I. RABI, S. MILLMAN, P. KUSCH u. J. R. ZACHARIAS: Phys. Rev. **55**, 526 (1939).

[3] C. GORTER: Physica, Haag **3**, 995 (1936).

[4] J. ZACHARIAS: Phys. Rev. **61**, 270 (1942).

[5] Vgl. hierzu auch die Artikel von P. KUSCH in Bd. XXXVII und von W. GORDY in Bd. XXVIII dieses Handbuches.

[6] F. BLOCH, W. W. HANSEN u. M. PACKARD: Phys. Rev. **69**, 127, 680 (1946).

[7] E. M. PURCELL, H. C. TORREY u. R. V. POUND: Phys. Rev. **69**, 37 (1946).

die Beobachtung der Übergänge mit elektronischen Hilfsmitteln. *Dagegen wird nicht mehr mit den einzelnen Molekülen eines Strahles experimentiert, sondern es werden kompakte Materiestücke in festem oder flüssigem Zustand untersucht*[1]. Die beiden Methoden unterscheiden sich voneinander nur in ihrer elektronischen Anordnung zur Resonanzfeststellung. Die Unterschiede in den Bezeichnungen rühren daher, daß ein Teil der Forscher vor allem die Sprache der Quantenphysik benutzte und infolgedessen von Absorptionsvorgängen sprach, während die andere Arbeitsgruppe Begriffe der klassischen Physik bevorzugte und somit die Beobachtungen durch Präzessionsbewegungen und Induktionseffekte erklärte.

Mit dem Beginn der Messungen von kernmagnetischen Dipolmomenten und später auch von kernelektrischen Quadrupolmomenten[2] an Mengen eng aneinander liegender Atome oder Moleküle wurde zugleich ein neuer physikalischer Problemkreis erschlossen. Bei der Verwendung von Flüssigkeiten und Festkörpern als Untersuchungssubstanz beeinflussen sowohl Wechselwirkungen zwischen den Kernen als auch zwischen diesen und den Hüllen benachbarter, insbesondere paramagnetischer Atome den Verlauf der Experimente. Umgekehrt lassen sich aus den Ergebnissen derartiger Versuche Aussagen über die Struktur der Probesubstanz gewinnen.

Um ein kernmagnetisches Experiment mit Erfolg ausführen zu können, muß man zunächst die Größe dieser Kopplungen wenigstens ungefähr richtig erraten. Ein Beispiel soll dies belegen. Sind die Wechselwirkungen zwischen den Kernen und ihrer Umgebung sehr klein (die sie beschreibende Relaxationszeit also sehr lang), dann besagt dies, daß Energieübergänge zwischen dem Freiheitsgrad der Kernmomentorientierung und allen übrigen Freiheitsgraden der Substanz nur sehr langsam erfolgen. Es genügt daher schon eine schwache Einstrahlung von Energiequanten der Resonanzfrequenz, um das Kernspinsystem vom thermischen Gleichgewichtszustand in den Zustand der Sättigung zu überführen. Ein gesättigtes System kann aber nicht absorbieren, somit kann ein Absorptionseffekt auch nicht beobachtet werden. In einer früheren Arbeit mußten die Verfasser[3] ein negatives Ergebnis ihrer Versuche mitteilen, da sie diesen Umstand bei der Auswahl der Probesubstanz nicht genügend berücksichtigen konnten.

Versucht man die zahlreichen Veröffentlichungen aus dem Gebiet der kernmagnetischen Forschung zu ordnen, dann kann man dabei einmal von den experimentellen Methoden ausgehen, die inzwischen entwickelt wurden. Bei der Mehrzahl dieser Verfahren regt man ebenso wie bei den zwei ursprünglichen Anordnungen die Kernresonanz durch eine kontinuierliche Einstrahlung an und beobachtet sie in irgendeiner Brückenschaltung. Klassisch gesprochen werden dabei meistens erzwungene Präzessionsbewegungen beobachtet. Weitere Möglichkeiten zur Resonanzstellensuche ergeben sich aus der Verwendung von schwach schwingenden Oszillatoren[4], Sperrschwingern und Oszillatoren in Pendelrückkopplungs-Schaltungen[5]. Die sog. freien Präzessionen werden in ihrer reinen Form mit impulstechnischen Hilfsmitteln angeregt und beobachtet[6]. Auch

---

[1] Zum Teil wurden auch komprimierte Gase untersucht, siehe z.B. E. M. Purcell, R. V. Pound u. N. Bloembergen: Phys Rev. **70**, 986 (1946).

[2] H. G. Dehmelt u. H. Krüger: Naturwiss. **37**, 111 (1950). — R. V. Pound: Phys. Rev. **79**, 685 (1950). Vgl. hierzu auch den Artikel von C. Townes in diesem Bande.

[3] C. J. Gorter u. L. F. J. Broer: Physica, Haag **9**, 591 (1942); s. auch: G. J. Gorter: Physica, Haag **17**, 169 (1951).

[4] R. V. Pound u. W. D. Knight: Rev. Sci. Instrum. **21**, 219 (1950). — H. W. Knoebel u. E. L. Hahn: Rev. Sci. Instrum. **22**, 904 (1951). — H. A. Thomas: Electronics, N. Y. **25**, 114 (1952).

[5] A. Roberts: Rev. Sci. Instrum. **18**, 845 (1947).

[6] H. C. Torrey: Phys. Rev. **75**, 1326 (1949); **76**, 1059 (1949); E. L. Hahn: Phys. Rev. **76**, 145 (1949); **77**, 297 (1950); **80**, 580 (1950).

die Untersuchung der Hyperfeinstruktur hüllen-paramagnetischer Resonanzen liefert Meßergebnisse zur Erforschung der Kernmomente[1]. Interessante Effekte findet man ferner, wenn man Elektronen- und Kernresonanzen gleichzeitig anregt[2].

Zum anderen kann man die Ergebnisse nach der physikalischen Fragestellung ordnen, die den Arbeiten zugrunde lag. Ein großer Teil der Veröffentlichungen befaßt sich natürlich mit der ursprünglichen kernphysikalischen Aufgabe, nämlich der möglichst genauen Bestimmung der magnetischen Kern-Dipolmomente. In allen kernmagnetischen Resonanzversuchen wird gemäß:

$$\omega = \gamma H \begin{cases} \omega = \text{LARMOR-Kreisfrequenz,} \\ H = \text{Magnetische Feldstärke,} \\ \gamma = \text{Gyromagnetisches Verhältnis,} \\ = \text{Magnetisches Moment/Impulsmoment} \end{cases}$$

das gyromagnetische Verhältnis aus einer Frequenzmessung und einer Magnetfeldmessung ermittelt. Das Impulsmoment kennt man fast immer aus optischen Daten, meist aus Hyperfeinstruktur-Termanalysen. Frequenzmessungen können sehr präzise ausgeführt werden, genaue Magnetfeldmessungen sind dagegen schwierig. In den meisten Arbeiten werden daher nur Relativmessungen ausgeführt. Dabei bestimmt man die Resonanzfrequenzen verschiedener Kerne, meist in dem gleichen, konstant gehaltenen Magnetfeld und vergleicht sie mit der bekannten Frequenz eines Standardkernes, etwa der des Protons. Absolutwerte des magnetischen Moments solcher Standardkerne sind auf verschiedene Art ermittelt worden. Einmal hat man versucht, den Betrag der magnetischen Resonanzfeldstärke möglichst genau direkt zu messen[3]. Weiter kann man aber auch die Kernresonanzfrequenz der Standardkerne mit deren Zyklotron-Umlauffrequenz[4] oder mit der Zyklotron-Umlauffrequenz der Elektronen[5] im selben Magnetfeld vergleichen. Auch auf diesem Wege lassen sich die kernmagnetischen Messungen an das System der physikalischen Fundamentalkonstanten-Bestimmungen anschließen.

Alle so ermittelten Resonanzfrequenzen müssen jedoch noch korrigiert werden. Das Magnetfeld am Ort eines Kernes setzt sich im allgemeinen aus dem starken äußeren Feld und einem kleinen Zusatzfeld der Elektronenhüllen zusammen. So hängen z.B. die kernmagnetischen Resonanzfrequenzen von der Art der Molekülbindung ab, in welcher die Kerne enthalten sind[6]. Die allgemeine Berechnung solcher Korrekturfelder ist schwierig[7]. Die Genauigkeit bei der Bestimmung kernmagnetischer Momente wird mit durch diesen Umstand begrenzt. Auf Grund solcher kernmagnetischen Feinstrukturuntersuchungen lassen sich jedoch umgekehrt Aussagen über die Probesubstanz gewinnen. Bei Versuchen dieser Art wird also nicht mehr nach der Größe der Kernmomente gefragt, sondern es wird ihre Kenntnis dazu benutzt, Kerne als Prüfsonden — bildlich gesprochen als Tastköpfe — für die inneren Felder zu verwenden. Dasselbe geschieht

[1] B. BLEANEY u. K. STEVENS: Progr. Phys. 16, 108 (1953).

[2] A. W. OVERHAUSER: Phys. Rev. 92, 411 (1953).

[3] H. A. THOMAS, R. L. DRISCALL u. J. A. HIPPLE: J. Res. Nat. Bur. Stand. 44, 569 (1950).

[4] H. SOMMER, H. A. THOMAS u. J. A. HIPPLE: Phys. Rev. 80, 487 (1950); 82, 697 (1951). — F. BLOCH u. C. D. JEFFRIES: Phys. Rev. 80, 305 (1950). — C. D. JEFFRIES: Phys. Rev. 81, 1040 (1951).

[5] J. H. GARDNER u. E. M. PURCELL: Phys. Rev. 76, 1262 (1949). — J. H. GARDNER: Phys. Rev. 83, 996 (1951).

[6] W. G. PROCTOR u. F. C. YU: Phys. Rev. 77, 717 (1950). — W. C. DICKINSON: Phys. Rev. 77, 736 (1950).

[7] Vgl. W. LAMB: Phys. Rev. 60, 817 (1941). — W. C. DICKINSON: Phys. Rev. 80, 563 (1950). — N. F. RAMSEY: Phys. Rev. 77, 567 (1950); 78, 339, 699 (1950); 86, 243 (1952).

bei der Untersuchung von Relaxations- oder Diffusionsvorgängen in kernmagnetischen Versuchsanordnungen. Auch hier wird nach den magnetischen Kopplungen der Substanz gefragt, während die angewendeten kernphysikalischen Verfahren zum Teil der experimentellen Methode werden.

Wollen wir demnach das Gebiet der kernmagnetischen Hochfrequenz-Spektroskopie nach seiner physikalischen Fragestellung einordnen, dann haben wir es nicht nur als Teilgebiet der Kernphysik, sondern auch als Teilgebiet des Magnetismus zu betrachten. Übrigens verläuft auch seine theoretische Behandlung in mancher Hinsicht ähnlich zu derjenigen des Atomparamagnetismus[1]. Der erste erfolgreiche Versuch zum direkten Nachweis kernmagnetischer Effekte war sogar eine rein statische Suszeptibilitätsmessung an festem Wasserstoff[2]. Auch wurde schon früh der Gedanke diskutiert, durch adiabatische Entmagnetisierung von Kernmomenten, analog zu den erfolgreichen Versuchen mit paramagnetischen Molekülen, die bisherige Temperaturgrenze von 10^{-3} °K experimentell und meßtechnisch zu unterschreiten[3].

Auch hier soll die Behandlung des Problemkreises als Teilgebiet des Magnetismus am Anfang stehen. Dementsprechend soll zunächst von Systemen paramagnetischer Kerne gesprochen werden.

A. Physikalische Grundlagen.

I. Statik und Dynamik eines Systems paramagnetischer Kerne.

1. Statische resultierende Kernmomente. Wir betrachten ein System von N gleichen Kernen, die ein endliches Impulsmoment der Größe $\sqrt{(I+1)}\,I\cdot\hbar$ (I ganz- oder halbzahlig) besitzen. Mit einem solchen Drehimpuls ist stets ein den Kern umgebendes rotations-symmetrisches Magnetfeld verbunden. Dieses Feld kann durch ein zum Kernspin paralleles oder antiparalleles magnetisches Dipolmoment beschrieben werden, da nach den bisherigen experimentellen Ergebnissen die höheren Glieder einer Multipolentwicklung des magnetischen Kernfeldes vernachlässigbar sind[4].

Das System befinde sich in einem homogenen Magnetfeld der Feldstärke H_0. Setzt man voraus, daß die Wechselwirkungen der Kerne untereinander und ebenso ihre Wechselwirkung mit sonstigen atomaren oder molekularen Feldern vernachlässigbar klein sind, dann hat jeder Kern infolge der Richtungsquantelung genau $2I+1$ mögliche Orientierungen zur Feldrichtung, deren Komponenten in Feldrichtung sich um ganzzahlige Vielfache von $\hbar$ unterscheiden (Fig. 1). Auf Grund der Kopplungen zwischen den Dipolmomenten und dem äußeren Feld entspricht jeder dieser Orientierungen ein Energiezustand. Diesen Zuständen können magnetische Quantenzahlen m zugeordnet werden. Die Energie eines solchen Niveaus m ist:

$$E_m = -\frac{m}{I}\,\mu\,H_0. \tag{1.1}$$

[1] Vgl. etwa J. H. VAN VLECK: Theory of Electric and Magnetic Susceptibilities. Oxford: Clarendon Press 1932. — I. WALLER: Z. Physik **79**, 370 (1932). — R. DE L. KRONIG: Physica, Haag **6**, 33 (1939). — J. H. VAN VLECK: Phys. Rev. **57**, 426 (1940). — G. J. GORTER: Paramagnetic Relaxation. Leiden: Elsevier 1937.

[2] B. G. LASAREW u. L. W. SCHUBNIKOW: Phys. Z. Sowjet. **11**, 445 (1937).

[3] E. TELLER u. W. HEITLER: Proc. Roy. Soc. Lond. **155**, 629 (1936).

[4] Neuerdings ist es gelungen, die Existenz von kernmagnetischen Oktopol-Momenten am Beispiel von Gallium-Isotopen experimentell zu beweisen; vgl. hierzu R. T. DALY u. J. H. HOLLOWAY: Phys. Rev. **96**, 539 (1954).

Dabei ist μ die maximale Komponente des magnetischen Moments in Feldrichtung. Die wirkliche Größe des Dipolmoments eines einzelnen Kernes ist nach der Quantentheorie: $\sqrt{(I+1)/I} \cdot \mu$. Aber da nur seine maximale Komponente im Experiment beobachtet werden kann, soll im weiteren, dem allgemeinen Sprachgebrauch folgend, die Komponente μ einfach als das magnetische Moment des Kernes bezeichnet werden. Einander folgende Zustände unterscheiden sich gemäß (1.1) um äquidistante Energiebeträge. Der Zustand des gesamten Systems läßt sich durch die Besetzungszahlen n_m der $2I+1$ Energiestufen beschreiben.

Zur Berechnung dieser Besetzungszahlen muß man voraussetzen, daß das Kernspinsystem thermischen Kontakt mit seiner Umgebung hat, und zwar soll es mit ihr im thermodynamischen Gleichgewicht sein. Mit der zuvor genannten Voraussetzung — vernachlässigbare Wechselwirkungen — sollten Störungen der Energieniveaus ausgeschlossen werden. Um widerspruchsfrei zu bleiben, muß man also einen Kompromiß schließen und wenigstens schwache Kopplungen zulassen. In der Thermodynamik der reversiblen Prozesse wird jedoch über die Einstellzeit eines Gleichgewichtszustandes nichts ausgesagt, sie kann insbesondere beliebig lang sein. Die Allgemeinheit der Betrachtung wird also nicht eingeschränkt, wenn man sehr kleine Wechselwirkungen doch zuläßt.

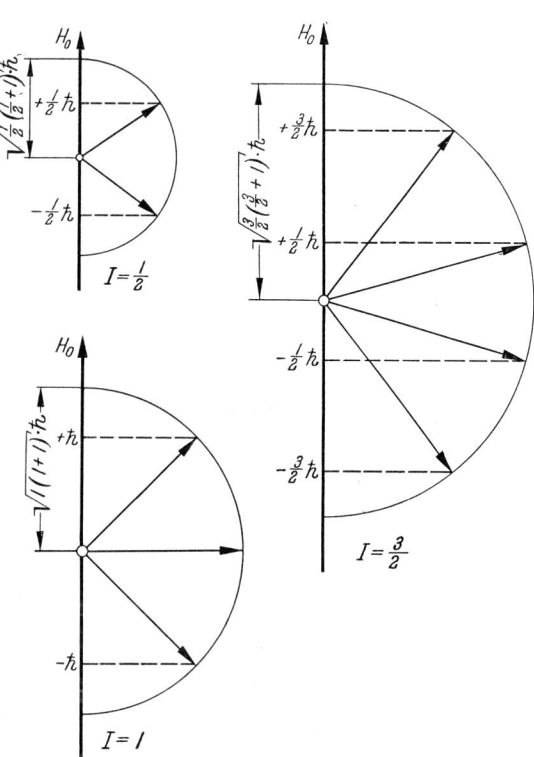

Fig. 1. Schematische Darstellung der möglichen Orientierungen von Kern-Drehimpulsen in einem Magnetfeld.

Im Gleichgewicht bei der Temperatur T sind die Besetzungszahlen von Energiezuständen proportional zum BOLTZMANN-Faktor $e^{-E_m/kT}$. Es ist also mit $\eta = \dfrac{\mu H_0}{I\,kT}$:

$$n_m \sim e^{m\eta} \quad \text{und} \quad N = \sum_{-I}^{I} n_m = C \sum_{-I}^{I} e^{m\eta}. \tag{1.2}$$

Die Summe läßt sich ausrechnen, es ist:

$$\sum_{-I}^{I} e^{m\eta} = e^{I\eta}\left(1 + e^{-\eta} + \cdots + e^{-2I\eta}\right) = e^{I\eta}\left(1 - e^{-(2I+1)\eta}\right)\sum_{0}^{\infty} e^{-m\eta},$$

worin

$$\sum_{0}^{\infty} e^{-m\eta} = \sum_{0}^{\infty}\left(e^{-\eta}\right)^m = \frac{1}{1-e^{-\eta}};$$

somit ist:

$$\sum_{-I}^{I} e^{m\eta} = \frac{e^{I\eta} - e^{-(I+1)\eta}}{1-e^{-\eta}} = \frac{e^{(I+\frac{1}{2})\eta} - e^{-(I+\frac{1}{2})\eta}}{e^{\eta/2} - e^{-\eta/2}} = \frac{\operatorname{Sin}\{(I+\frac{1}{2})\eta\}}{\operatorname{Sin}\{\eta/2\}} = Z(\eta). \tag{1.3}$$

Für die Besetzungszahlen gilt also:

$$n_m = \frac{N}{Z(\eta)} e^{+m\eta}.$$ (1.4)

Die unteren Energieniveaus sind stärker besetzt. Infolgedessen hat das System ein statisches magnetisches Moment M_0, und es ist:

$$M_0 = \sum_{-I}^{I} n_m \frac{m\mu}{I} = -\frac{1}{H_0} \sum_{-I}^{I} n_m E_m.$$

Die statische Gesamtenergie E_0 des Systems ist:

$$E_0 = \sum_{-I}^{I} n_m E_m = -\frac{N}{Z(\eta)} \eta\, kT \sum_{-I}^{I} m\, e^{m\eta}.$$

Beachtet man:

$$\sum_{-I}^{I} m\, e^{m\eta} = \frac{d}{d\eta} \sum_{-I}^{I} e^{m\eta} = \frac{dZ}{d\eta},$$

dann ist:

$$E_0 = -\frac{N\,kT\eta}{Z} \frac{dZ}{d\eta} = -N\,kT\eta \frac{d}{d\eta} \ln Z.$$

Rechnet man den Differentialquotienten aus, dann erhält man abschließend:

$$E_0 = -N\,kT\eta \frac{I \cdot \mathrm{Sin}\{(I+1)\eta\} - (I+1)\,\mathrm{Sin}\{I\eta\}}{2 \cdot \mathrm{Sin}\left\{\dfrac{\eta}{2}\right\} \mathrm{Sin}\left\{\dfrac{(2I+1)\eta}{2}\right\}}.$$ (1.5)

Die Größe des resultierenden Momentes beschreibt in Abhängigkeit von der Feldstärke, der Temperatur, dem Spin, dem Kernmoment und der Kernzahl die Brillouinsche Funktion:

$$M_0 = N\mu \frac{I \cdot \mathrm{Sin}\{(I+1)\eta\} - (I+1)\,\mathrm{Sin}\{I\eta\}}{2I\, \mathrm{Sin}\left\{\dfrac{\eta}{2}\right\} \mathrm{Sin}\left\{\dfrac{(2I+1)\eta}{2}\right\}}.$$ (1.6)

Ist $\eta \gg 1$, strebt also $H_0 \to \infty$ oder $T \to 0$, dann nähert sich M_0 seinem Sättigungswert $N\mu$. Bei kernparamagnetischen Untersuchungen ist jedoch fast immer die Energiedifferenz zweier benachbarter Orientierungen sehr klein gegen kT, also $\eta \ll 1$. Man kann dann die Brillouinsche Funktion durch Potenzreihenentwicklung nach η vereinfachen und erhält:

$$M_0 = \frac{N\mu^2}{3\,kT} \cdot \frac{I+1}{I} \cdot H_0.$$ (1.7)

Die statische kern-paramagnetische Suszeptibilität ist demgemäß:

$$\chi_0 = \frac{N\mu^2}{3\,kT} \cdot \frac{I+1}{I}.$$ (1.8)

Die klassische Theorie des Paramagnetismus liefert die Langevinsche Formel:

$$\chi_0 = \frac{N\mu^2}{3\,kT}.$$

Die Proportionalität zwischen χ_0 und $1/T$ wurde zuerst experimentell aufgefunden und ist als Curiesches Gesetz bekannt.

Wie erwähnt, muß man in den magnetischen Resonanzexperimenten zur Messung der Dipolmemente die optisch, oder aus der Hyperfeinstruktur der paramagnetischen Resonanz ermittelten Größen der Kern-Spins kennen. Man kann den Spin I aber auch aus den Resonanzbeobachtungen bestimmen, wenn es gelingt, den Betrag (1.7) des statischen Momentes direkt zu ermitteln, oder die Beträge (1.7) der statischen Momente zweier Kernarten, von denen der Spin der einen bekannt ist, miteinander zu vergleichen. Praktisch ist dies jedoch schwierig, da man zuvor sämtliche anderen Daten des Experimentes und der Substanzen kennen muß, vgl. hierzu Ziff. 20.

Einige Zahlenbeispiele sollen die Betrachtungen abschließen. Die statische paramagnetische Suszeptibilität von festem Wasserstoff ist nach (1.8) im cgs-System $\chi_0 = 7 \cdot 10^{-8}/T$ (T in $^\circ$K). Die diamagnetische Hüllensuszeptibilität von Festkörpern ist etwa 10^{-6} bis 10^{-7}. Die kernparamagnetische Suszeptibilität des festen Wasserstoffs hat also erst bei Temperaturen unterhalb 1° K etwa die gleiche Größenordnung. Für Kerne, deren Spin $I = \frac{1}{2}$ ist (z.B. Protonen) folgt aus (1.7), daß nur der Bruchteil $\mu H_0/kT$ der N Kerne der Substanz das resultierende Moment bildet. Der Quotient $\mu H_0/kT$ hat aber bei Zimmertemperatur und in Feldern von einigen tausend Gauß ungefähr die Größe 10^{-6}.

Für alle im folgenden beschriebenen Experimente und Überlegungen ist die Existenz eines resultierenden magnetischen Moments eine notwendige Voraussetzung. Alle kernmagnetischen Versuche haben als übereinstimmendes Merkmal, daß in ihnen Bewegungen solcher resultierender Momente angeregt und beobachtet werden. Es soll daher als nächstes die Dynamik solcher Kernmagnetisierungs-Vektoren besprochen werden.

2. Die Differentialgleichungen der kernmagnetischen Präzessionsbewegungen. Auf die Kerne eines Systems sollen äußere magnetische Felder einwirken. Solche Felder lassen sich in Abhängigkeit vom Ort und der Zeit durch eine Funktion $\boldsymbol{H}_a(\boldsymbol{r}, t)$ beschreiben, die mit der Versuchsanordnung vorgegeben ist. Weiter wirken auf die Kerne sog. innere Felder $\boldsymbol{H}_i(\boldsymbol{r}, t)$ ein. Mit diesem Begriff erfaßt man sämtliche magnetischen Wechselwirkungen. Jedoch erfolgt die Änderung der inneren Felder auf Grund der Wärmebewegung von Ort zu Ort und von Zeitpunkt zu Zeitpunkt statistisch ungeordnet. Weder besteht in einem bestimmten Zeitpunkt ein Funktionszusammenhang zwischen den Feldstärken an verschiedenen Orten, noch besteht in einem Ortspunkt ein solcher Zusammenhang zwischen den Feldstärken zu verschiedenen Zeiten. Der Problemkreis ist also in seiner allgemeinen Form schwierig zu behandeln und soll deshalb unterteilt werden.

Zunächst mögen nur äußere Felder und nur eine paramagnetische Kernart vorhanden sein. Die Felder sollen sich zwar zeitabhängig ändern können, jedoch sollen sie räumlich streng homogen sein $[\boldsymbol{H} = \boldsymbol{H}_a(t)]$.

Bezeichnet man mit $\boldsymbol{M}$ das resultierende magnetische Moment pro cm³ und mit $\boldsymbol{A}$ das resultierende Impulsmoment pro cm³, dann ist[1] das auf das Kernspinsystem einwirkende Drehmoment $\boldsymbol{T}$:

$$\boldsymbol{T} = \boldsymbol{M} \times \boldsymbol{H}.$$

Gleich diesem ist die zeitliche Änderung des Impulsmomentes $\boldsymbol{A}$:

$$\boldsymbol{T} = \frac{d\boldsymbol{A}}{dt}.$$

Auf Grund der Parallelität der Kernmomente $\boldsymbol{\mu}$ und $\boldsymbol{a}$ $\left(|\boldsymbol{\mu}| = \sqrt{\dfrac{I+1}{I}} \cdot \mu, \right.$ $\left. |\boldsymbol{a}| = \sqrt{(I+1)\,I} \cdot \hbar\right)$ kann man für einen einzelnen Kern schreiben:

$$\mu = \gamma\, a \quad \text{bzw.} \quad \mu = \gamma\, I\, \hbar \qquad (\gamma = \text{gyromagnetisches Verhältnis}). \tag{2.1}$$

---

[1] F. Bloch: Phys. Rev. **70**, 460 (1946).

Damit gilt aber auch für die resultierenden Momente:

$$M = \gamma A.$$

Zusammengefaßt ergeben diese Beziehungen für die Bewegungen des Magnetisierungsvektors M eine oft als Grundgleichung des Kernmagnetismus bezeichnete Differentialgleichung:

$$\frac{d M}{d t} = \gamma \left[M \times H \right]. \tag{2.2}$$

Ihre klassische Ableitung ist zulässig, denn die Erwartungswerte oder Beobachtungsgrößen eines aus hinreichend vielen Teilen (in unserem Fall also Kernen) bestehenden Systems stimmen mit den aus den klassisch-mechanischen Gesetzen folgenden entsprechenden Größen überein[1].

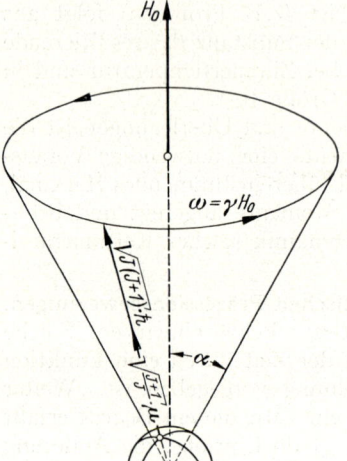

Diese erstaunlich einfache Gleichung zeigt, daß das resultierende Moment sich ähnlich wie ein mechanischer makroskopischer Kreisel unter der Einwirkung von Kräften verhält, nur daß dessen Bewegungsablauf im allgemeinen sehr viel komplizierter ist.

Eine triviale Lösung von (2.2) ist sofort zu erkennen. Ist das Magnetfeld konstant ($H = H_0$), dann präzessiert M um seine Richtung (Fig. 2). Die Präzessionsfrequenz (LARMOR-Frequenz) ist unabhängig vom Öffnungswinkel des Präzessionskegels:

$$\omega_0 = \frac{d \varphi}{d t} = \frac{d M}{M d t} = \gamma H_0. \tag{2.3}$$

Fig. 2. LARMOR-Präzession eines kernmagnetischen Dipolmoments in einem Magnetfeld.

Das gleiche Ergebnis soll zur Veranschaulichung auch noch auf zwei anderen Wegen für Einzelkerne direkt abgeleitet werden. Bezeichnet man mit α den Winkel zwischen der Achse der Kernmomente a und μ und dem Feld H_0, dann versucht ein Drehmoment vom Betrag $|\mu| \cdot H_0 \cdot \sin\alpha$ die Kerne in die Feldrichtung zu drehen. Als Kreisel weichen sie diesem Zwang seitwärts aus, präzessieren also mit einer gewissen Frequenz ω um die Feldrichtung. Um diesen Bewegungszustand zu erhalten, muß ein Kreiselmoment vom Betrag $|a| \cdot \omega \cdot \sin\alpha$ auf die Kernkreisel einwirken. Gleichsetzung ergibt:

$$\omega = \frac{|\mu|}{|a|} H_0 = \gamma H_0 = \omega_0.$$

Weiter unterscheiden sich quantentheoretisch gemäß (1.1) benachbarte mögliche Orientierungen um äquidistante Energiebeträge $\mu H_0/I$. Setzt man diesen Energiebetrag gleich $h\nu_0 = \hbar\omega_0$, dann erhält man wiederum $\omega_0 = \gamma H_0$.

Als nächstes soll der Einfluß statistisch schwankender innerer Felder auf den Bewegungsablauf behandelt werden. Bei kernmagnetischen Experimenten ist die durch sie verursachte Kopplung zwischen dem Freiheitsgrad der magnetischen Orientierung und allen übrigen Freiheitsgraden der Probe sehr von der Art der untersuchten Substanz abhängig und darf im allgemeinen nicht vernachlässigt werden. Im Rahmen einer dynamischen Betrachtung muß man also irreversible Annäherungsvorgänge an den Zustand des thermodynamischen Gleichgewichts mitberücksichtigen.

---

[1] Siehe z. B. R. K. WANGSNESS u. F. BLOCH: Phys. Rev. **89**, 728 (1953).

Ein Beispiel soll dies verdeutlichen. Befinden sich Atomkerne, etwa die in einem cm³ Wasser enthaltenen Protonen in einem von äußeren Feldern freien Raum, dann können diese natürlich kein resultierendes Moment haben. Zwar ist es möglich, daß in irgendeinem Zeitpunkt zufällig eine größere Anzahl Kerne gleich ausgerichtet sind. Im Zeitmittel haben jedoch die in wechselnden Richtungen so entstehenden und verschwindenden Magnetisierungsvektoren den Betrag Null. Schaltet man nun ein konstantes, homogenes Magnetfeld ein, dann wird zwar die Entartung der kernmagnetischen Energieniveaus aufgehoben, die Besetzungszahlen der Niveaus sind jedoch zunächst noch gleich groß. Das Kernspinsystem befindet sich im Zustand der Gleichverteilung und hat dementsprechend die Temperatur ∞. Handelt es sich wie im Beispiel um Protonen, deren Spin $I = \frac{1}{2}$ ist, dann sind $2I + 1 = 2$ Richtungen möglich (vgl. Fig. 1). Der Energieunterschied der zugehörigen Niveaus ist $\mu H_0/I = 2\mu H_0$. Dieser Energiebetrag muß bei jedem Umklappvorgang von anderen Energiespeichern aufgenommen oder abgegeben werden. Solche Energiespeicher sind alle übrigen Freiheitsgrade der Probe. Das System strebt nun vom Zustand der Gleichverteilung zum Zustand der Gleichgewichtsverteilung. Dabei wird die Substanz wärmer, während sich das Kernspinsystem abkühlt, bis beider Temperaturen gleich groß sind. Den umgekehrten Vorgang nutzt man bei der adiabatischen Entmagnetisierung aus. Dort hat der magnetische Orientierungsfreiheitsgrad nach dem Abschalten des Feldes die niedrigere Temperatur und entzieht den übrigen Freiheitsgraden Wärme[1].

Einen Kernresonanzversuch könnte man demnach im Prinzip auch kalorimetrisch ausführen[2]. Strahlt man ständig Hochfrequenzquanten der Resonanzfrequenz (2.3) in eine Substanz ein, dann wird dem Kernspinsystem Energie zugeführt. Damit werden bei den Wechselwirkungen innerhalb der Substanz die Übergänge zu den energetisch niedrigeren Niveaus häufiger. Der Resonanzeffekt wäre also auch an der Zunahme der Probentemperatur zu erkennen.

Durch statistische Wechselwirkungen verursachte irreversible Annäherungsvorgänge an einen Gleichgewichtszustand gehorchen im allgemeinen einer Exponentialfunktion. Man kann infolgedessen phänomenologisch die Änderung der Komponente des Magnetisierungsvektors in Feldrichtung (sie sei mit der z-Richtung eines kartesischen Koordinatensystems identisch) durch eine Differentialgleichung der Form:

$$\dot{M}_z = - \frac{M_z - M_0}{T_1} \qquad (2.4)$$

beschreiben. Darin ist M_0 der statische Wert der Magnetisierung (1.7) und T_1 eine Konstante, die physikalisch die Bedeutung einer Relaxationszeit hat. In Erinnerung an ihre Verknüpfung mit Energieübergängen nennt man diese Zeit meistens *Spin-Gitter-Relaxationszeit*, wobei unter dem Begriff Gitter die Summe aller übrigen Freiheitsgrade der Substanz verstanden wird. Daneben ist auch der Name *longitudinale Relaxationszeit* gebräuchlich. Er weist auf die übliche geometrische Anordnung der Versuche hin.

Als nächstes sind die Ausgleichsvorgänge in der zum Magnetfeld transversalen Richtung zu besprechen. In klassischer Betrachtung verhält sich jeder Kern wie ein Kreisel und präzessiert um die Richtung des Magnetfeldes. In einer die Probe umgebenden Spule kann man jedoch diese Präzessionsbewegungen der einzelnen Kerne nicht beobachten. Die Öffnungswinkel der Präzessionskegel und vor allem die Phasen der Umlaufbewegungen sind von Kern zu Kern verschieden. Die

---

[1] Zur adiabatischen Entmagnetisierung vgl. den Artikel von D. DE KLERK in Bd. XV dieses Handbuches.

[2] C. J. GORTER versuchte dies schon 1936. Physica, Haag **3**, 995.

Bildung eines beobachtbaren, transversalen, resultierenden Momentes hat man sich also klassisch als erzwungene Synchronisierung (Phasenübereinstimmung) der individuellen Bewegungen vorzustellen. Notwendig für seine Existenz ist die Erhaltung der Phasenbeziehungen zwischen dem Teil der Kerne, aus deren Dipolmomenten es sich zusammensetzt. Stößt einer der „in Phase" präzessierenden Kerne mit einem beliebigen anderen zusammen, etwa auf Grund der BROWNschen Bewegung, dann kann seine Phasenbeziehung zu den übrigen, kohärent präzessierenden zerstört werden. Auch Wechselwirkungen dieser Art sind statistische Ereignisse. Man kann daher wenigstens versuchsweise annehmen, daß sich auch eine transversale Magnetisierung ihrem Gleichgewichtswert exponentiell nähert. Im thermodynamischen Gleichgewicht existieren aber keine transversalen Komponenten des resultierenden Momentes, wenn man über die äußeren Felder voraussetzt, daß deren transversale Komponenten H_x und H_y gegenüber ihrer longitudinalen H_z vernachlässigbar klein sind. Es gelten dann also die einfachen Differentialgleichungen:

$$\dot{M}_x = -\frac{M_x}{T_2}, \qquad \dot{M}_y = -\frac{M_y}{T_2}. \tag{2.5}$$

Die den Vorgang beschreibende Relaxationszeit T_2 wird *Spin-Spin-Relaxationszeit* und *transversale Relaxationszeit* genannt. Die Namen sind aus dem Vorangehenden zu verstehen.

Zu beachten ist, daß Phasenübergänge nicht notwendig mit Energieübergängen, also Orientierungsänderungen verbunden sind, während die umgekehrte Aussage sicher richtig ist. Damit ist auch die Einführung von zwei verschiedenen Relaxationszeiten gerechtfertigt.

Ergänzt man die Grundgleichung (2.2) um die Glieder (2.4) und (2.5) und schreibt sie zugleich in Koordinaten, dann erhält man mit $H_0 = H_z$ die bekannten BLOCHschen Gleichungen:

$$\left.\begin{aligned}
\dot{M}_x &= \gamma\,(M_y H_0 - M_z H_y) - \frac{1}{T_2}\,M_x, \\
\dot{M}_y &= \gamma\,(M_z H_x - M_x H_0) - \frac{1}{T_2}\,M_y, \\
\dot{M}_z &= \gamma\,(M_x H_y - M_y H_x) - \frac{1}{T_1}\,(M_z - M_0),
\end{aligned}\right\} \quad \begin{aligned} (M_0 &= \chi_0 H_0 = \chi_0 H_z \\ H_0 &= H_z \gg H_x, H_y). \end{aligned} \tag{2.6}$$

Diese phänomenologisch abgeleiteten Gleichungen haben sich zur Deutung und Beschreibung der bisherigen experimentellen Ergebnisse gut bewährt.

BLOEMBERGEN und PURCELL haben ähnliche Differentialgleichungen abgeleitet

$$\left.\begin{aligned}
\dot{M}_x &= \gamma\,(M_y H_z - M_z H_y) - \gamma^2 F\,(M_x - \chi_0 H_x), \\
\dot{M}_y &= \gamma\,(M_z H_x - M_x H_z) - \gamma^2 F\,(M_y - \chi_0 H_y), \\
\dot{M}_z &= \gamma\,(M_x H_y - M_y H_x) - \gamma^2 F\,(M_z - \chi_0 H_z).
\end{aligned}\right\} \tag{2.7}$$

Ihre Beziehungen sind das Ergebnis einer Untersuchung der mikroskopischen Einzelprozesse in Gasen und Flüssigkeiten und deren statistischer Zusammenfassung. Demgemäß ergibt sich aus ihrer Ableitung zugleich ein expliziter Ausdruck für die einzige Relaxationszeit, die in ihnen enthalten ist ($F \sim 1/T$). In Ziff. 3 soll darauf näher eingegangen werden.

Die BLOCHschen Gleichungen beschreiben auch Experimente zu deren Deutung zwei kernmagnetische Relaxationszeiten erforderlich sind. Eingeschränkt ist ihre Gültigkeit durch die Bedingung $H_0 = H_z \gg H_x$ und H_y. In ihnen wird also der Beitrag der transversalen Komponenten des äußeren Feldes zur Gleichgewichtsmagnetisierung vernachlässigt. In den Gln. (2.7) ist dagegen keine Richtung ausgezeichnet. Sie sind symmetrisch in allen drei Koordinaten. Es ist selbstverständlich leicht, auch in den Gln. (2.6) die transversale Gleichgewichtsmagnetisierung phänomenologisch mit zu berücksichtigen, jedoch war dies bisher nicht erforderlich. Interessanter sind Versuche, die Relaxationszeiten T_1 und T_2 mit anderen mikroskopischen und makroskopischen physikalischen Größen der untersuchten Substanz zu verknüpfen.

3. Zur Deutung der Relaxationsvorgänge. Der Verlauf eines kernmagnetischen Experimentes und damit die Forderungen, die an die experimentellen Versuchsanordnungen zu stellen sind, hängen wesentlich von den Größen der beiden Relaxationszeiten der Substanz ab, in der sich die zu untersuchenden Kerne befinden. Um mit Erfolg Resonanzfrequenzen messen zu können, muß man die Größenordnungen der Zeiten T_1 und T_2 zuvor entweder aus theoretischen Annahmen oder auf Grund von Vorversuchen ungefähr abschätzen. Dies wird verständlich, wenn man bedenkt, daß in den bisherigen Arbeiten Relaxationszeiten von 10 Std bis 10^{-5} sec gefunden worden sind[1,2]. Aus diesem Grunde soll im folgenden über den gegenwärtigen Stand der Deutung der Relaxationsmechanismen berichtet werden.

Die in den Problemkreis der Dämpfung und Linienbreite einführenden grundlegenden Arbeiten wurden vor allem von BLOEMBERGEN, PURCELL und POUND verfaßt[3]. Leider sind in der Literatur die Bezeichnungen nicht einheitlich. Wir werden folgende Nomenklatur verwenden. Es sei:

1. T_1 die Zeitkonstante der Einstellung der thermodynamischen Gleichgewichtsverteilung *in Feldrichtung*.

2. T_2 die Zeitkonstante der durch *statistisch* schwankende, innere Felder verursachten Abnahme eines zur Feldrichtung *transversalen resultierenden* Moments.

3. T_2' die Zeitkonstante der Abnahme eines *transversalen* resultierenden Moments auf Grund der Inhomogenität des *inneren* statischen Felds der Substanz.

4. T_2'' die Zeitkonstante der Abnahme eines *transversalen* resultierenden Moments ortsfester Kerne auf Grund der Inhomogenität des *äußeren* statischen Felds.

5. T_2''' die Zeitkonstante der Abnahme eines transversalen resultierenden Moments auf Grund von Diffusionsbewegungen der Kerne in dem inhomogenen statischen Feld.

An sich kann man in der Relaxationsphysik im Prinzip jeden einzelnen, an einem Relaxationsvorgang als mitbeteiligt erkannten Mechanismus durch eine eigene Relaxationszeit beschreiben. Erfolgt die Annäherung an den Gleichgewichtszustand bei allen solchen Prozessen einfach exponentiell, d.h. jeweils proportional zu e^{-t/T_r}, dann ist ihr gemeinsamer Einfluß auf den Versuchsverlauf proportional zu dem Produkt $e^{-t/T_I} e^{-t/T_{II}} \ldots e^{-t/T_r} \ldots$, und für die im Experiment beobachtbare Relaxationszeit T gilt:

$$\frac{1}{T} = \frac{1}{T_I} + \frac{1}{T_{II}} + \cdots + \frac{1}{T_r} + \cdots . \qquad (3.1)$$

[1] G. R. HOLZMAN, J. H. ANDERSON u. W. KOTH: Phys. Rev. **98**, 542 (1955).
[2] Zum Beispiel A. M. SACHS u. E. TURNER: Zit. von E. M. PURCELL Physica, Haag **17** 282 (1951).
[3] E. M. PURCELL: Phys. Rev. **69**, 681 (1946). — N. BLOEMBERGEN, E. M. PURCELL u. R. V. POUND: Nature, Lond. **160**, 475 (1947). — N. BLOEMBERGEN, E. M. PURCELL u. R. V. POUND: Phys. Rev. **73**, 679 (1948). — N. BLOEMBERGEN: Nuclear Magnetic Relaxation. Den Haag: Nijhoff 1948.

Die angegebene Gliederung ist also sicher nicht vollständig und muß in Anpassung an die experimentellen Ergebnisse ständig weiter unterteilt werden. Aber da die Entwicklung dieses Forschungszweigs noch in Fluß ist, soll die vorliegende summarische Unterteilung genügen.

Ein echter Relaxationsprozeß verläuft stets irreversibel. In diesem Sinne sind die Zeiten T_2' und T_2'' nur scheinbar Relaxationszeiten. Wie bei der Besprechung der freien Präzessionsbewegungen (Ziff. 4) gezeigt wird, handelt es sich bei der Abnahme des resultierenden Moments auf Grund von Feldinhomogenitäten um einen experimentell umkehrbaren, also reversiblen Vorgang. Dabei gehorcht die Amplitude des Magnetisierungsvektors häufig einer Exponentialfunktion der Form $e^{-\frac{1}{2}(\gamma B t)^2}$ (s. etwa S. 160, B = Halbwertsbreite der Inhomogenitätenverteilung), die man natürlich auch in der Form $e^{-(t/T_2'')^2}$ schreiben kann. Das einfache Additionsgesetz (3.1) gilt also nicht allgemein.

Bei der Ableitung der kernmagnetischen Differentialgleichungen in Ziff. 2 wurde ständig vorausgesetzt, daß alle äußeren Felder räumlich streng homogen sind. Diese Voraussetzung schränkt jedoch die Allgemeinheit der analytischen Behandlung nicht ein; denn man kann immer das Volumen, in dem sich die zu untersuchenden Kerne befinden, in infinitesimale Teilgebiete derart zerlegen, daß auf alle Kerne, die sich in einem solchen Teilvolumen befinden, die exakt gleiche Feldstärke einwirkt. Natürlich darf man sich nicht vorstellen, daß ein solches Gebiet zusammenhängend sein muß. Eine räumliche Unterteilung dieser Art entspricht einer Zerlegung des resultierenden Momentes aller Kerne in infinitesimale Momente. Man kann dann die Blochschen Differentialgleichungen für diese Teilmomente lösen und kennt damit deren Bewegungsablauf als Funktion der Zeit und als Funktion der zugehörigen Feldstärke. Den Betrag und die Richtung des Gesamtmoments erhält man in Abhängigkeit von der Zeit durch eine Integration der Koordinatengleichungen dieser infinitesimalen Momente, multipliziert mit einer die Feldverteilung beschreibenden Funktion, über alle vorkommenden Feldstärken (s. z.B. S. 160).

Im Prinzip unterscheidet sich der Einfluß der Inhomogenitäten des äußeren Felds auf den allein beobachtbaren resultierenden Bewegungsablauf nicht von dem der Inhomogenitäten eines statischen inneren Felds. In beiden Fällen kennt man die Verteilungsfunktion der Felder zunächst nicht. Experimentell kann man sie aus dem Verlauf der Resonanzkurve erschließen. Besteht diese nur aus einer Resonanzlinie, so ist, wie allgemein bei der Diskussion von Linienformen, die wichtigste Meßgröße die Halbwertsbreite dieser Linie. Deren Größe hängt aber von der Dämpfung der Kernpräzession, also von den echten Relaxationsmechanismen und von der Breite der Feldstreuung ab. Das letztere ist einleuchtend wenn man bedenkt, daß in einem inhomogenen Feld beim Resonanzdurchgang innerhalb eines gewissen Intervalls der Feldstärke immer ein Teil der Kerne mit der Senderfrequenz genau in Resonanz ist. So ist es zur Berechnung der Linienbreite oft erleichternd und in manchen Fällen näherungsweise auch berechtigt, die Inhomogenitätenwirkung einfach als zusätzlichen Relaxationseinfluß zu berücksichtigen.

Um die eigentlichen Relaxationsmechanismen zu verstehen muß man das Verhalten der Einzelkerne diskutieren. Zur Einführung in den Problemkreis sollen zunächst einige Modellvorstellungen qualitativ erläutert werden und zwar zuerst zum Verständnis der longitudinalen, durch T_1 charakterisierten sog. Spin-Gitter-Relaxation.

Bei jedem Umklappen eines Kerns ändert sich die zur Feldrichtung parallele z-Komponente seines Dipolmoments und damit seine Energie um den Betrag $\hbar\omega$, Gl. (2.3). Nimmt man an, daß die betrachteten Kerne kein elektrisches

Quadrupolmoment haben, dann ist die einzige Möglichkeit ihre z-Komponente zu ändern, ihre Anregung durch ein magnetisches Wechselfeld von ungefähr der LARMOR-Frequenz (erzwungene Emission und Absorption). Die Wahrscheinlichkeit der spontanen Emission von Quanten der Energie $\hbar\omega$ ist im Bereich der Radiowellen vernachlässigbar klein. Ein magnetisches Wechselfeld erzeugt z.B. jeder Kern (wenn er ein endliches magnetisches Dipolmoment besitzt) selbst in seiner unmittelbaren Umgebung, da er um die Feldrichtung präzessiert. Wirken zwei gleiche Kerne wechselseitig aufeinander ein, dann kann es vorkommen, daß der eine im Feld des anderen umorientiert wird. Ändert sich während der Wechselwirkung die Summe der übrigen Energiebeträge (Translation, Rotation usw.) der beiden Moleküle oder des Moleküls nicht, das die zwei betrachteten Kerne enthält, so muß die z-Komponente des anderen Kerns gleichzeitig um denselben Betrag zu- oder abnehmen, um den sich die z-Komponente des ersten vergrößert oder verkleinert hat. Das durch die Eigenpräzession der Kerne verursachte gleichzeitige und gegensinnige Umklappen zweier identischer Kerne ändert aber den Betrag der Magnetisierung nicht und kann demnach auch nicht zur longitudinalen Relaxation beitragen (wohl aber zur transversalen Relaxation, vgl. S. 136). Zu deren Erklärung muß man also andere Ursachen suchen, die magnetische Wechselfelder in der Substanz erzeugen können. Es ist naheliegend, an die BROWNschen Bewegungen der in den Substanzen enthaltenen magnetischen Dipole zu denken.

Statistisch ungeordnete Bewegungen von der Art der thermischen Bewegungen der Atome und Kerne innerhalb der Moleküle, oder der Moleküle in der Substanz lassen sich durch die Intensitätsfunktion ihres Bewegungsspektrums beschreiben, vgl. hierzu die Gl. (3.2) und (3.21). Hat diese Funktion, welche die mittlere Intensität der BROWNschen magnetischen Wechselfelder in der Substanz in Abhängigkeit von der Frequenz beschreibt, für die LARMOR-Frequenz der betrachteten Kerne einen endlichen, ausreichend großen Betrag, so können Energieübergänge zwischen den verschiedenen Bewegungs-Freiheitsgraden und dem magnetischen Orientierungs-Freiheitsgrad, also zwischen dem Spinsystem und dem Gitter der Substanz erfolgen. An Hand eines Beispiels kann man sich diesen Mechanismus der Spin-Gitter-Relaxation folgendermaßen verdeutlichen. Die Substanz sei flüssig und bestehe nur aus einer Sorte von Molekülen (destilliertes Wasser etwa). Von den in ihr enthaltenen Atomkernen besitzt mindestens die untersuchte Art, im Beispiel also die Protonen, ein endliches paramagnetisches Dipolmoment. Es ist möglich, daß sich auch noch andere Kernarten der Verbindung paramagnetisch verhalten. Weiter besitzen aber auch die Elektronenhüllen der Moleküle magnetische Momente, auch wenn man, wie bei allen kernmagnetischen Resonanzexperimenten nur mit hüllendiamagnetischen Verbindungen experimentiert. Durch die BROWNschen Bewegungen der Moleküle werden magnetische Wechselfelder erzeugt, deren Frequenz sich ständig, zeitlich und örtlich statistisch schwankend ändert. Anschaulich kann man dies als ein magnetisches Rauschen beschreiben. Stimmt die Frequenz der Bewegung der Dipole in irgend einem Orts- und Zeitpunkt mit der Frequenz der betrachteten Kernpräzession überein, so kann ein zufällig in diesem Zeitpunkt an diesem Ort befindlicher Kern umorientiert werden und dadurch nach den bekannten Regeln der klassischen Elektrodynamik entweder die Rauschbewegung verstärken oder dämpfen. (Erhitzung oder Abkühlung des Gitters.) Die quantitative Behandlung liefert für die Intensität $J(\omega)$ eines thermischen Rauschspektrums eine Funktion der Form [vgl. Gl. (3.21)]

$$J(\omega) = C \frac{\tau_c}{1 + \omega^2 \tau_c^2} . \tag{3.2}$$

Darin wird die thermische Bewegung durch einen Amplitudenfaktor C und durch eine sog. Korrelationszeit τ_c charakterisiert. In Flüssigkeiten ist τ_c ungefähr gleich der Zeit in der sich ein Molekül um den Winkel eins dreht, oder in dem es sich um ein Wegstück von der Größenordnung des halben Abstands von einem betrachteten Wechselwirkungs-Partner vorwärtsbewegt. Ungefähr so lange kann man, etwa bei der thermischen Translationsbewegung, von einer Korrelation zweier Nachbarmoleküle sprechen.

Korrelationszeiten lassen sich mit makroskopischen Größen verknüpfen[1]. In der Diffusionstheorie werden z.B. für Flüssigkeiten für die Korrelationszeiten $\tau_{c,\mathrm{rot}}$ und $\tau_{c,\mathrm{trans}}$ der Rotations- bzw. Translations-Diffusionsbewegung die Beziehungen abgeleitet:

$$\tau_{c,\mathrm{rot}} = \frac{a^2}{6 D_{\mathrm{rot}}} \quad \text{und} \quad \tau_{c,\mathrm{trans}} = \frac{r^2}{12 D_{\mathrm{trans}}}. \tag{3.3 a, b}$$

Darin sind a der Molekülradius, r der Abstand der wechselwirkenden Moleküle und D_{rot} bzw. D_{trans} Diffusionskonstanten. Für $r = a$ ist $\tau_{c,\mathrm{rot}}$ doppelt so groß wie $\tau_{c,\mathrm{trans}}$. Dies wird verständlich, wenn man bedenkt, daß sich bei der Translationsdiffusion die beiden betrachteten Moleküle bewegen.

Auf Grund der Stokes-Einstein Beziehungen für Rotations- und Translationsbewegungen gilt in Flüssigkeiten:

$$D_{\mathrm{rot}} = \frac{\mathsf{k} T}{8 \pi \eta a} \quad \text{und} \quad D_{\mathrm{trans}} = \frac{\mathsf{k} T}{6 \pi \eta a}. \tag{3.4 a, b}$$

Darin ist η die Viscosität, k die Boltzmannsche Konstante und T die absolute Temperatur. Die Zeiten τ_c sind also proportional zur Viskosität η und umgekehrt proportional zur Temperatur T der Substanz, wenn die Beziehungen (3.3) und (3.4) gültig sind.

Die Wahrscheinlichkeit W erzwungener Übergänge zwischen den kernmagnetischen Energieniveaus ist, wie auseinandergesetzt, proportional zu der Intensität (3.2) des Rauschens in dem die Larmor-Frequenz umgebenden infinitesimalen Spektrumsabschnitt ($W \sim J(\omega_L)$, bezüglich der Wirkung weiterer Spektrumsabschnitte vgl. S. 141). Andererseits besteht zwischen der Übergangswahrscheinlichkeit W und der Zeitkonstanten T eines einfachen Relaxationsvorgangs der Zusammenhang $T = 1/W$. Im Fall der Spin-Gitter-Relaxation gilt also $1/T_1 = 2W$, denn jeder einzelne Übergang verändert die Differenz der Besetzungszahlen um zwei.

Um einen qualitativen Vergleich der Modellvorstellung ,,Spin-Gitter-Relaxation durch Brownsche Bewegung'' mit der experimentellen Erfahrung zu ermöglichen, genügt es, die Form des Rauschspektrums (3.2) zu diskutieren. Untersucht man eine bestimmte Substanz im thermodynamischen Gleichgewicht bei einem bestimmten Druck und bei einer festen Temperatur, so sind auch die Viscosität η und die Korrelationszeiten τ_c konstante Größen. In Fig. 3 wurden Rauschintensitätsverteilungen für verschiedene Werte des Parameters τ_c aufgetragen. Wenn die Larmor-Frequenz ω klein gegen $1/\tau_c$ ist, hängt die Intensität $J(\omega)$ des relaxierende Übergänge induzierenden spektralen Anteils des magnetischen Rauschens nur wenig von ω ab. Dagegen verschwindet der relaxierende Frequenzanteil des Rauschens für $\omega \geq 1/\tau_c$ schnell. Andererseits ist die

---

[1] P. Debye: Polare Molekeln. Leipzig: S. Hirzel 1928. Polar Molecules. New York-Dover 1945. — S. Chandrasekhar: Rev. Mod. Phys. **15**, 1 (1943). — N. Bloembergen, E. M. Purcell u. R. V. Pound: Phys. Rev. **73**, 679 (1948). — N. Bloembergen: Nuclear Magnetic Relaxation. Den Haag: Nijhoff 1948. Vgl. auch T. A. Litovitz: J. Acoust. Soc. Amer. **29**, 648 (1957).

anfänglich konstante Intensität des Spektrums bei einem festen Amplitudenfaktor C um so größer, je länger τ_c ist. Erinnert man sich der Proportionalität zwischen $J(\omega)$ und $1/T_1$ einerseits und zwischen τ_c und η/T andererseits (setzt also die Beziehungen (3.3) und (3.4) als gültig voraus), dann kann man dies folgendermaßen interpretieren. Wenn η/T klein ist, wie etwa bei Flüssigkeiten $(\tau_{c,\,H_2O} \sim 10^{-12}\,\text{sec})^1$, dann ist in dem experimentell zugänglichen Frequenz- bzw. Feldstärkenbereich $\omega\,\tau_c \ll 1$ und damit T_1 praktisch unabhängig von der Meßfrequenz bzw. Feldstärke und um so größer, je kleiner η/T ist.

Mißt man dagegen die Relaxationszeit T_1 einer bestimmten Substanz bei einer festen Frequenz und ändert etwa die Temperatur, also $\eta/T \sim \tau_c$, dann ist ein Verlauf der in Fig. 4 wiedergegebenen Art zu erwarten. Für kleine Werte von η/T ist T_1 lang, weil sich die

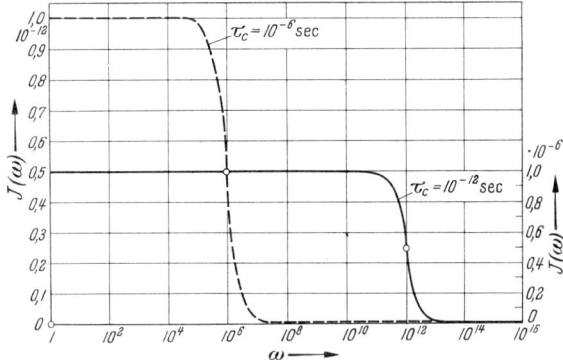

Fig. 3. Intensitätsverteilung des magnetischen Rausch-Spektrums (τ_c Korrelationszeit, ω Kreisfrequenz, die Konstante C in (3.2) wurde gleich eins gesetzt).

Intensität des Rauschspektrums über einen sehr breiten Frequenzbereich verteilt. Für große Werte von η/T ist T_1 ebenfalls lang; denn für die LARMOR-Frequenz ist dann die Bedingung $\omega \gg 1/\tau_c$ erfüllt; sie liegt also außerhalb des eigentlichen Rauschspektrums. Dazwischen existiert eine Temperatur, bei der T_1 ein Minimum hat, die Spin-Gitter-Kopplungen der Substanz somit maximal sind.

Als nächstes sollen einige Mechanismen der transversalen Relaxation besprochen werden. Ein transversales resultierendes Moment entsteht, wenn die in Feldrichtung im thermodynamischen Gleichgewicht vorhandene Überschußmagnetisierung irgendwie ganz oder teilweise aus dieser herausgedreht wird. Dadurch werden die Präzessionsbewegungen

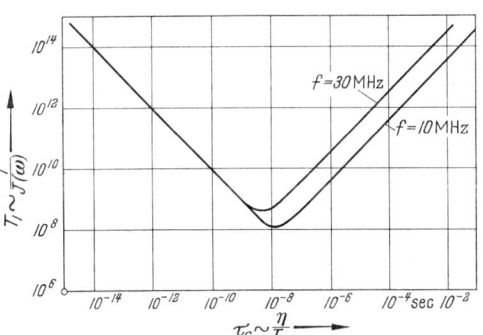

Fig. 4. Abhängigkeit der Spin-Gitter-Relaxationszeit von der Korrelationszeit (willkürliche Ordinateneinheiten).

der Kerne teilweise synchronisiert; sie erhalten übereinstimmende Phasen. Die damit entstandene resultierende Quermagnetisierung verschwindet wieder, wenn irgendwelche Vorgänge in der Substanz es dem Spinsystem erlauben, in den Zustand der statistischen Phasenverteilung zurückzukehren.

Wir wollen zunächst annehmen, daß in der Substanz keine thermischen Bewegungen stattfinden. Dann überlagern sich dem äußeren Feld, von dem angenommen werde, daß es streng homogen sei, die inneren statischen Felder der in der Substanz enthaltenen Elementarmagnete. Deren Beitrag zum Gesamtfeld

---

^{1} Auf Grund einer Messung der Dielektrizitätskonstanten von Wasser mit Mikrowellen findet man $\tau_D = 8 \cdot 10^{-12}\,\text{sec}$ ($\tau_c = \tau_D/3$). J. A. SAXTON: Meteorological Factors in Radio-Wave Propagation Conference. Phys. Soc. Lond., p. 278, 1946.

ist, verglichen mit der Feldstärke des äußeren Feldes, zwar sehr klein (hüllen-paramagnetische Atome oder Moleküle sollen in der Substanz nicht enthalten sein), jedoch wird das Feld inhomogen. Infolgedessen sind die LARMOR-Frequenzen der Kerne voneinander verschieden. Diese präzessieren mit verschiedener Geschwindig-keit und ein auf irgendeine Art entstandenes transversales Moment verschwindet schnell wieder. Betrachten wir dagegen die Elementarmagnete in der Substanz als thermisch so schnell bewegt, daß ihre Korrelationszeit τ_c sehr klein gegen die Zeit $1/\omega$ ist, in der sich die Kerne auf Grund ihrer LARMOR-Präzession um den Winkel eins drehen, dann mitteln sich die inneren statischen Felder fast voll-ständig aus. Die Präzessionsfrequenz schwankt in diesem Fall zwar statistisch, ihr Mittelwert ist aber für alle Kerne praktisch gleich der zur äußeren Feldstärke gehörenden LARMOR-Frequenz. In Flüssigkeiten und Gasen ist die obige Be-dingung in der Regel erfüllt.

Während die beschriebene, durch innere oder äußere Feldinhomogenitäten verursachte Momentabnahme, wie schon erwähnt, ein reversibler Vorgang ist, gilt dies für einen ähnlichen Mechanismus nicht. Ist die Versuchsdauer, z.B. die Zeit eines Resonanzdurchgangs, so lang, daß die Kerne während ihr größere Weg-stücke diffundieren können, dann gelangen sie in Gebiete, deren Feldstärken sich von denen an ihren Anfangsorten unterscheiden. Jedoch kann der dadurch aus-gelöste, ebenfalls indirekt durch die Feldinhomogenitäten verursachte Abnahme-prozeß der beobachteten transversalen Magnetisierung natürlich nicht im Experi-ment umgekehrt werden. Ist er allein für das Verschwinden eines transversalen Moments verantwortlich, so ist dessen Abklingkurve übrigens unter gewissen plausiblen Voraussetzungen eine Funktion der Form[1] $e^{-(t/T_2''')^3}$ also ebenfalls keine einfache Exponentialfunktion, obwohl es sich um einen irreversiblen Vorgang handelt.

Anschaulich ist die Vermutung naheliegend, daß sich bei jedem durch Spin-Gitter-Wechselwirkungen verursachten Übergang eines Kerns in ein benachbartes Niveau, zugleich die Phase der Präzessionsbewegung dieses Kerns sprunghaft ändert. Der besprochene Spin-Gitter-Relaxationsmechanismus trägt also auch zur Spin-Spin-Relaxation bei. Allerdings ist zu beachten, daß sein Beitrag zur trans-versalen Phasen-Übergangswahrscheinlichkeit $W = 1/T_2$ gleich $1/2 T_1$ ist; denn es ändert sich bei einem Umklappvorgang dieser Art zwar die Differenz der Niveau-Besetzungszahlen um zwei, aber es springt nur die Phase des einen umklappenden Kernes. Würde dieser Mechanismus als einziger für den transversalen Phasen-ausgleich sorgen, dann wäre die Spin-Spin-Relaxationszeit T_2 doppelt so groß wie die longitudinale Zeit T_1, vgl. hierzu jedoch S. 147.

Auch eine weitere Art von Spin-Spin-Wechselwirkungen wurde bereits be-schrieben, das durch die magnetischen Wechselfelder der LARMOR-Präzession ver-ursachte gleichzeitige und gegensinnige Umklappen zweier Kerne. Dabei ändern sich vermutlich die Phasen beider Kerne unstetig. Als Spin-Spin-Relaxation wirkt ein solcher Prozeß aber nur, wenn wenigstens einer der Kerne anfänglich kohärent präzessierte. Von der Wahrscheinlichkeit eines solchen Phasenübergangs ist an-zunehmen, daß sie proportional zu der Zeit ist, in der zwei Kerne aufeinander ein-wirken können, also proportional zu der Korrelationszeit τ_c.

Einen der beschriebenen Theorie entsprechenden Relaxationsverlauf[2] zeigt Fig. 5. Solange η/T und damit τ_c klein sind, stimmen die gemessenen Werte von T_1 überein und nehmen umgekehrt proportional zu η/T ab. Für größere Werte von η/T wächst T_1 aber wieder, und zwar liegt das Minimum um so tiefer, je

---

[1] H. Y. CARR u. E. M. PURCELL: Phys. Rev. **94**, 630 (1954).
[2] N. BLOEMBERGEN, E. M. PURCELL u. R. V. POUND: Phys. Rev. **73**, 679 (1948).

kleiner die Meßfrequenz ist. Die Spin-Spin-Relaxationszeit T_2 nimmt dagegen innerhalb des Meßbereichs ständig ungefähr umgekehrt proportional zu η/T ab.

Von besonderer Bedeutung sind im Gebiet der kernmagnetischen Relaxation die sog. paramagnetischen Katalysatoren. Man versteht darunter einen Zusatz von hüllenparamagnetischen Fremdatomen, meist Ionen, zu der untersuchten Substanz. Da Hüllen-Dipolmomente etwa tausendmal größer sind als Kernmomente, verkürzen schon geringe Beimischungen solcher Fremdatome die Relaxationszeiten der Substanzen erheblich. In Flüssigkeiten wächst $1/T_1$ proportional mit dem Quadrat der Dipolmomente[1] und proportional mit deren Konzentration[2],

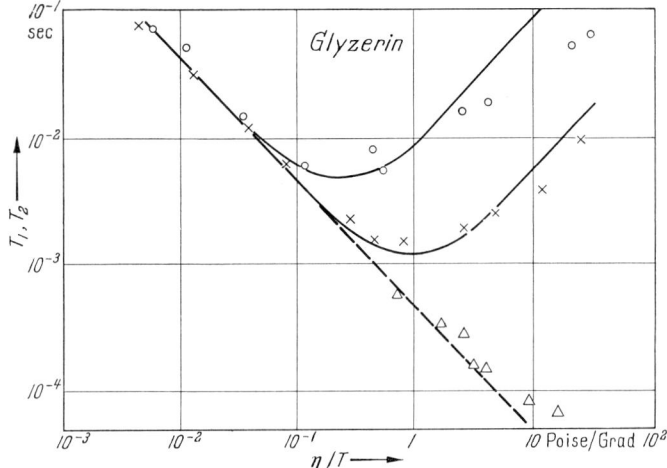

Fig. 5. Messungen der Relaxationszeiten T_1 und T_2 der in Glyzerin enthaltenen Protonen in Abhängigkeit von dem Verhältnis der Viskosität zur absoluten Temperatur. (o T_1 gemessen bei 29 MHz, × T_1 gemessen bei 4,8 MHz, △ T_2 gemessen bei 29 MHz.)

vgl. (3.29). Durch entsprechend dosierte Zusätze paramagnetischer Ionen lassen sich also die Relaxationszeiten, vor allem von Flüssigkeiten, bequem in einen für die experimentellen Beobachtungen günstigen Bereich verschieben.

Die voranstehende anschaulich-qualitative Beschreibung der kernmagnetischen Relaxationsvorgänge ist unvollständig. Einmal ist sie vor allem für Flüssigkeiten begründet worden, und zum anderen wurde als einziger Kopplungsmechanismus nur die bekannte klassische sog. direkte magnetische Wechselwirkung zwischen Dipolen berücksichtigt. Es ist aber naheliegend, daß die Theorie der kernmagnetischen Relaxationsphänomene, allgemeiner gesagt, die Theorie der Wechselwirkungen von Atomkernen mit den sie enthaltenden Substanzen, das vom Aggregatzustand abhängende Verhalten der Substanzen, also die Erfahrungen und Vorstellungen der Gasphysik, der Flüssigkeitsphysik und der Festkörperphysik mindestens teilweise widerspiegeln muß. Am Ende dieser Ziffer sollen deshalb die Besonderheiten des Relaxationsverhaltens der Gase und Festkörper in zwei speziellen Abschnitten diskutiert werden. Ebenso ist leicht einzusehen, daß außer der einfachen magnetischen Dipol-Dipol-Kopplung wegen der Existenz von höheren Kernmomenten noch weitere, ebenfalls klassisch-anschaulich verständliche Kopplungsmechanismen zur Relaxation beitragen können.

[1] R. L. CONGER u. P. W. SELWOOD: J. Chem. Phys. 20, 383 (1952).
[2] N. BLOEMBERGEN, E. M. PURCELL u. R. V. POUND: Phys. Rev. 73, 679 (1948). — G. LAUKIEN u. J. SCHLÜTER: Z. Physik 146, 113 (1956). — R. HAUSSER u. G. LAUKIEN: Arch. d. Sci. 10, 235 (1957).

Kerne, welche endliche elektrische Quadrupolmomente besitzen, sind mit dem Gitter der Substanz z.B. auch elektrisch gekoppelt. Bekanntlich wirkt auf ein Quadrupolmoment in einem elektrischen Feld ein Drehmoment ein, dessen Betrag proportional zum Betrag des Feldgradienten des elektrischen Felds ist. Die elektrischen Relaxations-Wechselwirkungen werden also durch die thermischen statistischen Schwankungen des Feldgradienten des inneren elektrischen Felds vermittelt. Ein schönes Beispiel für diesen Relaxationsmechanismus ist die Messung der Zeit T_1 einer Mischung von normalem und schwerem Wasser (je 50%). Dabei wurde gefunden, daß die Spin-Gitter-Relaxationszeit der Protonen 3 sec ist, während die der Deuteronen nur 0,5 sec beträgt[1]. Die Deuteronen haben aber ein endliches, wenn auch kleines Quadrupolmoment und die H_2O- und D_2O-Moleküle elektrische Dipolmomente.

Haben die untersuchten Kerne höhere magnetische Momente, etwa Oktopole, dann können auch diese im Prinzip zur Relaxation beitragen, da auf Grund der thermischen Bewegung der Substanz auch die zweite räumliche Ableitung des inneren magnetischen Felds statistisch schwankt. (Das auf Oktopolmomente in einem magnetischen Feld einwirkende Drehmoment ist proportional zum Betrag der zweiten Ableitung des Felds.) Jedoch konnte ein solcher Einfluß bis jetzt experimentell nicht nachgewiesen werden.

Neben den klassisch erklärbaren elektromagnetischen Wechselwirkungen zwischen Momenten und deren Feldern tragen möglicherweise auch nur quantentheoretisch verständliche Kopplungsmechanismen zur kernmagnetischen Relaxation merklich bei. Vor der Besprechung eines ersten auf nichtklassischen Wechselwirkungen fußenden Deutungsversuchs kernmagnetischer Relaxationszeitmessungen (vgl. S. 147) soll jedoch zuerst die quantitative Formulierung der Vorstellungen der kernmagnetischen Relaxationstheorie[2] skizziert werden und zwar im wesentlichen am Beispiel der thermisch ausreichend schnell bewegten Substanzen und unter der einschränkenden Voraussetzung, daß die Kerne nur über direkte Dipol-Dipol-Kopplungen mit dem Gitter der Substanz wechselwirken können.

Aus der elementaren Physik ist bekannt, daß ein Dipol mit dem magnetischen Moment μ_i an einem Ort, der mit dem Dipol durch einen Vektor r_{is} verbunden ist, ein magnetisches Feld

$$H_i(r_{is}) = - \frac{\mu_i}{r_{is}^3} + 3 \frac{(\mu_i\, r_{is}) \cdot r_{is}}{r_{is}^5} \qquad (3.5)$$

erzeugt. Ein zweiter Dipol mit dem magnetischen Moment μ_s hat am Ort r_{is} die potentielle Energie $V_{is} = -H_i \cdot \mu_s$. Bei atomaren oder nuklearen magnetischen Dipolen sind die Momentvektoren parallel oder antiparallel zu den Spinvektoren, es ist also etwa $\mu_i = \gamma_i \hbar I$ und $\mu_s = \gamma_s \hbar S$ (γ_i und γ_s seien die zugehörigen gyromagnetischen Verhältnisse). Die Wechselwirkungs-Energie zweier beliebiger Dipole im Abstand r_{is} ist demnach

$$V_{is} = \gamma_i \gamma_s \hbar^2 r_{is}^{-3} \{(IS) - 3(I\,e_{is})(S\,e_s e_{is})\}, \qquad (3.6)$$

---

[1] W. A. ANDERSON u. J. T. ARNOLD: Phys. Rev. **94**, 497 (1954).

[2] N. BLOEMBERGEN, E. M. PURCELL u. R. V. POUND: Phys. Rev. **73**, 679 (1948). — N. BLOEMBERGEN: Nuclear Magnetic Relaxation. Den Haag: Nijhoff 1948. — R. K. WANGSNESS u. F. BLOCH: Phys. Rev. **89**, 728 (1953). — R. KUBO u. K. TOMITA: J. Phys. Soc. Japan **9**, 888 (1954). — I. SOLOMON: Phys. Rev. **99**, 559 (1955). — F. BLOCH: Phys. Rev. **102**, 104 (1956); **105**, 1206 (1957). — J. SEIDEN: J. Phys. Radium **18**, 274 (1957). Vgl. hierzu auch: I. WALLER: Z. Physik **79**, 370 (1932). — P. GÜTTINGER: Z. Physik **73**, 169 (1932). — E. MAJORANA: Nuovo Cim. **9**, 43 (1932) . — I. I. RABI: Phys. Rev. **51**, 652 (1937). — J. H. VAN VLECK Phys. Rev. **74**, 1168 (1948).

wenn man mit e_{is} einen Einheitsvektor parallel zu r_{is} bezeichnet. Die verwandte Symbolik I und S deutet an, daß in einem Teil der anschließenden Betrachtung I ein Kernspin und S ein Elektronenspin ist, jedoch soll sie den ebenfalls im weiteren diskutierten Fall identischer wechselwirkender Kernspins nicht ausschließen. Zur Ausrechnung der Gl. (3.6) ist es zweckmäßig zwei Koordinatensysteme einzuführen, ein kartesisches x, y, z-System in dem die Komponenten I_x, I_y, I_z und S_y, S_y, S_z der Spinvektoren I und S gemessen werden und ein Kugelkoordinaten r, φ, ϑ-System in dem die Länge r_{is} und die Orientierung φ_{is}, ϑ_{is} des Verbindungsvektors r_{is} gemessen werden. Beide Systeme sollen denselben Ursprung haben, ϑ_{is} sei der Winkel zwischen r_{is} und der z-Achse des Cartesischen Systems und φ_{is} der in der x, y-Ebene dieses Systems gemessene Orientierungswinkel von r_{is}. Das Ergebnis der Ausrechnung von (3.6) ist:

$$\left.\begin{aligned}
V_{is}=\gamma_i\gamma_s\hbar^2 r_{is}^{-3}\{&I_x S_x(1-3\sin^2\vartheta_{is}\cos^2\varphi_{is})+I_y S_y(1-3\sin^2\vartheta_{is}\sin^2\varphi_{is})+\\
&+I_z S_z(1-3\cos^2\vartheta_{is})-3(I_x S_y+I_y S_x)\sin^2\vartheta_{is}\sin\varphi_{is}\cos\varphi_{is}-\\
&-3(I_y S_z+I_z S_y)\sin\vartheta_{is}\cos\vartheta_{is}\sin\varphi_{is}-\\
&-3(I_z S_x+I_x S_z)\sin\vartheta_{is}\cos\vartheta_{is}\cos\varphi_{is}\}.
\end{aligned}\right\} \quad (3.7)$$

Ersetzt man $\cos\varphi_{is}$ und $\sin\varphi_{is}$ mit Hilfe der EULERschen Formeln $\cos\varphi = \frac{1}{2}(e^{i\varphi}+e^{-i\varphi})$ und $\sin\varphi = \frac{1}{2i}(e^{i\varphi}-e^{-i\varphi})$ und ordnet um, so erhält (3.7) die übersichtlichere Form

$$V_{is}=\gamma_i\gamma_s\hbar^2\{A+B+C+D+E+F\} \qquad (3.8)$$

mit:

$$A=[I_z S_z]\cdot[(1-3\cos^2\vartheta_{is})\,r_{is}^{-3}],$$

$$B=-\tfrac{1}{4}[(I_x-iI_y)(S_x+iS_y)+(I_x+iI_y)(S_x-iS_y)]\cdot[(1-3\cos^2\vartheta_{is})\,r_{is}^{-3}],$$

$$C=-\tfrac{3}{2}[(I_x+iI_y)S_z+(S_x+iS_y)I_z]\cdot[\sin\vartheta_{is}\cos\vartheta_{is}\,e^{-i\varphi_{is}}\,r_{is}^{-3}],$$

$$D=-\tfrac{3}{2}[(I_x-iI_y)S_z+(S_x-iS_y)I_z]\cdot[\sin\vartheta_{is}\cos\vartheta_{is}\,e^{+i\varphi_{is}}\,r_{is}^{-3}],$$

$$E=-\tfrac{3}{4}[(I_x+iI_y)(S_x+iS_y)]\cdot[\sin^2\vartheta_{is}\,e^{-2i\varphi_{is}}\,r_{is}^{-3}],$$

$$F=-\tfrac{3}{4}[(I_x-iI_y)(S_x-iS_y)]\cdot[\sin^2\vartheta_{is}\,e^{+2i\varphi_{is}}\,r_{is}^{-3}].$$

Die beiden Dipole μ_i und μ_s sollen sich in einem homogenen zur z-Achse parallelen Magnetfeld H befinden. Dann präzessieren sie mit den Kreisfrequenzen $\omega_i = \gamma_i H$ und $\omega_s = \gamma_s H$ um dessen Richtung, wenn die Wechselwirkungs-Energie (3.8) genügend klein gegen die Orientierungsenergien $-\mu_i H = -\gamma_i\hbar I_z H$ und $-\mu_s H = -\gamma\hbar S_z H$ der Dipole in diesem Feld ist. Bei den in der kernmagnetischen Spektroskopie üblicherweise verwandten Feldstärken ist diese Bedingung stets erfüllt.

In dem äußeren Magnetfeld sind die Spin-Komponenten I_z und S_z zeitunabhängig und die Spinkomponenten I_x, I_y und S_x, S_y zeitabhängig:

$$\left.\begin{aligned}
I_x(t)&=I_x\cos(\omega_i t)-I_y\sin(\omega_i t) &\text{bzw.}& &S_x(t)&=S_x\cos(\omega_s t)-S_y\sin(\omega_s t),\\
I_y(t)&=I_x\sin(\omega_i t)+I_y\cos(\omega_i t) & & &S_y(t)&=S_x\sin(\omega_s t)+S_y\cos(\omega_s t).
\end{aligned}\right\} \quad (3.9)$$

Die zeitunabhängigen, im Zeitpunkt $t=0$ vorhandenen Spinkomponenten wurden unverändert mit I_x, I_y, S_x und S_y bezeichnet.

Setzt man die Gl. (3.9) in die Gl. (3.8) ein und benutzt für die Spinkomponenten in der zur Feldrichtung transversalen Ebene die Abkürzungen

$$I^{\pm}=I_x\pm iI_y \quad\text{und}\quad S^{\pm}=S_x\pm iS_y \qquad (3.10)$$

und führt ferner für die von der Länge r_{is} und der Orientierung φ_{is}, ϑ_{is} der Verbindungsachse der Dipole abhängenden Faktoren der Glieder von (3.8) die Bezeichnungen

$$\left. \begin{aligned} f_0 &= (1 - 3\cos^2\vartheta_{is})\, r_{is}^{-3} \\ f_1^{\mp} &= \sin\vartheta_{is}\cos\vartheta_{is}\, e^{\mp i\varphi_{is}}\, r_{is}^{-3}, \\ f_2^{\mp} &= \sin^2\vartheta_{is}\, e^{\mp 2i\varphi_{is}}\, r_{is}^{-3}, \end{aligned} \right\} \qquad (3.11)$$

ein, so findet man für die Wechselwirkungs-Energie:

$$V_{is} = \gamma_i\gamma_s\hbar^2 (A_1 + B_1 + B_2 + C_1 + C_2 + D_1 + D_2 + E_1 + F_1) \qquad (3.12)$$

mit:

$$A_1 = I_z \cdot S_z \cdot f_0$$

$$B_1 = -\tfrac{1}{4} \cdot I^- \cdot S^+ \cdot f_0\, e^{i(\omega_s - \omega_i)t}, \qquad B_2 = -\tfrac{1}{4} \cdot I^+ \cdot S^- \cdot f_0 \cdot e^{-i(\omega_s - \omega_i)t},$$

$$C_1 = -\tfrac{3}{2} \cdot I^+ \cdot S_z \cdot f_1^- \cdot e^{i\omega_i t}, \qquad C_2 = -\tfrac{3}{2} \cdot S^+ \cdot I_z \cdot f_1^- \cdot e^{i\omega_s t},$$

$$D_1 = -\tfrac{3}{2} \cdot I^- \cdot S_z \cdot f_1^+ \cdot e^{-i\omega_i t}, \qquad D_2 = -\tfrac{3}{2} \cdot S^- \cdot I_z \cdot f_1^+ \cdot e^{-i\omega_s t},$$

$$E_1 = -\tfrac{3}{4} \cdot I^+ \cdot S^+ \cdot f_2^- \cdot e^{i(\omega_i + \omega_s)t}, \qquad F_1 = -\tfrac{3}{4} \cdot I^- \cdot S^- \cdot f_2^+ \cdot e^{-i(\omega_i + \omega_s)t}.$$

Für identische Kerndipole folgt aus (3.12) die etwas einfachere Beziehung $(\boldsymbol{I} = \boldsymbol{I}_i, \, \boldsymbol{S} = \boldsymbol{I}_j)$:

$$V_{ij} = \gamma^2\hbar^2 (A' + B' + C' + D' + E' + F') \qquad (3.13)$$

mit:

$$A' = I_{i,z} \cdot I_{j,z} \cdot f_0, \qquad\qquad\qquad B' = -\tfrac{1}{4}(I_i^- I_j^+ + I_i^+ I_j^-) \cdot f_0,$$

$$C' = -\tfrac{3}{2} \cdot (I_i^+ I_{j,z} + I_j^+ I_{i,z})\, f_1^- \, e^{i\omega t}, \qquad D' = -\tfrac{3}{2} \cdot (I_i^- I_{j,z} + I_j^- I_{i,z}) \cdot f_1^+ \cdot e^{-i\omega t},$$

$$E' = -\tfrac{3}{4} \cdot I_i^+ \cdot I_j^+ \, f_2^- \cdot e^{i2\omega t}, \qquad\qquad F' = -\tfrac{3}{4} \cdot I_i^- \cdot I_j^- \cdot f_2^+ \cdot e^{-i2\omega t}.$$

Die Wechselwirkungs-Energien (3.12) und (3.13) sind Störungen der Orientierungs-Energien der Dipole in dem äußeren Feld, welche auf Grund der Richtungsquantelung nur $(2I + 1)$ bzw. $(2S + 1)$ diskrete Energieeigenwerte E_r besitzen. In der Diracschen Störungstheorie wird gezeigt, daß für die Übergangswahrscheinlichkeit eines quantenmechanischen Systems unter dem Einfluß einer Störung von einem anfänglichen Energiezustand E_0 zu einem Energiezustand E_n zur Zeit t die Gleichung

$$W_n = |a_n(t)|^2 = a_n a_n^* \qquad (3.14)$$

gilt. Dieser Beziehung liegt der Ansatz zugrunde, daß sich die Eigenfunktion $\Psi(\boldsymbol{r}, t)$ des gestörten Zustands als Summe der zu den Energieeigenwerten E_r gehörenden Eigenfunktionen $\Psi_r(\boldsymbol{r}, t)$ des ungestörten Zustands darstellen läßt:

$$\Psi(\boldsymbol{r}, t) = \sum_r a_r \Psi_r(\boldsymbol{r}, t). \qquad (3.15)$$

Für die durch (3.15) definierten Koeffizienten a_n findet man die näherungsweise gültige Beziehung

$$a_n(t) = \frac{1}{i\hbar} \int_0^t (n\,|\,V\,|\,0)\, e^{-\frac{i(E_0 - E_n)t}{\hbar}} \cdot dt \qquad (3.16)$$

in der die übliche Abkürzung

$$(n\,|\,V\,|\,m) = \int \psi_n^*(\boldsymbol{r})\, V\, \psi_m(\boldsymbol{r})\, d\tau \qquad (3.17)$$

benutzt wurde.

Setzt man die Wechselwirkungs-Energien (3.12) oder (3.13) in die Gl. (3.16) ein, so findet man, daß nur die zeitunabhängigen, sog. säkularen Terme endliche Übergangswahrscheinlichkeiten ergeben. Damit wird die weitere Diskussion zu einer Untersuchung der Zeitabhängigkeit der Glieder $A_1, B_1, \ldots, A', B', \ldots, F'$. Ferner zeigt es sich, daß von den zueinander konjugiert komplexen Termen einer die Wahrscheinlichkeit der Umorientierung unter Absorption eines Quants und der andere die Wahrscheinlichkeit der Umorientierung unter Emission eines Quants beschreibt.

In der Darstellungsweise der Gl. (3.12) und (3.13) ist die von der Präzessionsbewegung der Dipole herrührende Zeitabhängigkeit abgetrennt worden. Man übersieht deshalb sofort, daß in einer innerlich völlig bewegungslosen Substanz, in der alle Radiusvektoren r_{is} dem Betrag und der Richtung nach ideal starr sind, nur die Glieder A_1, A' und B' säkular sind.

Zur Erklärung der großen Linienbreiten der kernmagnetischen Resonanzlinien von Festkörpern genügt es fast immer diese Terme zu betrachten (vgl. hierzu Ziff. 28). Ursache der durch die Säkularität der Glieder A_1 und A' bewirkten Linienverbreiterung ist die auch anschaulich leicht übersehbare Inhomogenisierung des äußeren Felds H durch die ihm in festen Körpern überlagerten Felder der starren Dipole. Der Term B ist nur im Fall identischer Dipole säkular, er beschreibt die bereits diskutierte wechselseitige Spin-Umorientierung zweier Dipole.

In thermisch bewegten Substanzen sind die Koordinaten r_{is}, φ_{is} und ϑ_{is} Funktionen der Zeit, denn sie ändern sich fortwährend statistisch schwankend. Dementsprechend tragen die Glieder A_1, A' und B' um so weniger zur Linienbreite bei, je geringer die spektrale Intensität der Funktion $f_0(r_{is}, \vartheta_{is})$ in der Nachbarschaft der Frequenz 0 ist, je kleiner also ihre säkularen Anteile sind. Dagegen erhalten alle übrigen, in einer starren Substanz rein periodisch von der Zeit abhängenden Glieder auf Grund der thermischen Bewegung endliche säkulare Anteile. Man übersieht sofort, welche Frequenzbereiche, des thermischen Rauschens diese Säkularisierung bewirken, und damit zur Relaxation beitragen.

Im Fall verschiedener Dipole, der wichtigste hierzu gehörende Fall ist derjenige der Wechselwirkung zwischen Kern- und Elektronendipolen[1], erfolgt die durch die Zeit T_1 charakterisierte Spin-Gitter-Relaxation des Kernspins I gemäß (3.12) um so schneller, je größer die Intensität des Rauschens in den Frequenzintervallen um die Frequenzen $\omega_i, \omega_s - \omega_i$ und $\omega_s + \omega_i$ ist. Neben dem direkten Übergang eines Quants $\hbar\omega_i$ der beobachteten LARMOR-Frequenz in oder aus dem magnetischen Rauschen wirken demnach auch die Übergänge von Quanten $\hbar(\omega_s - \omega_i)$ und $\hbar(\omega_s + \omega_i)$ der Differenz und der Summe der LARMOR-Frequenzen der beiden nach (3.12) wechselwirkenden Dipole relaxierend. Anschaulich bedeutet dies einfach, daß von der thermischen Rauschbewegung der Dipole gegeneinander in dem zweitgenannten Fall ein Energiebetrag aufgenommen oder abgegeben wird, der gleich der Energie ist, welche bei der gleichzeitigen Umorientierung der beiden wechselwirkenden Dipole abgegeben oder verbraucht wird, während in dem erstgenannten Fall mit der Umorientierung des einen

---

1 N. BLOEMBERGEN, E. M. PURCELL u. R. V. POUND: Phys. Rev. **73**, 679 (1948). — Nuclear magnetic relaxation, Diss., Utrecht 1948. — R. L. CONGER u. P. W. SELWOOD: J. Chem. Phys. **20**, 383 (1952). — J. R. ZIMMERMAN: J. Chem. Phys. **21**, 1605 (1953); **22**, 950 (1954). — B. M. KOZYREV u. A. I. RIVKIND: ZETP, USSR **27**, 69 (1954). — A. I. RIVKIND: DAN, USSR **102**, 1107 (1955). — A. BLOOM: J. Chem. Phys. **25**, 793 (1956). — G. LAUKIEN u. J. SCHLÜTER: Z. Physik **146**, 113 (1956). — L. O. MORGAN, A. W. NOLLE u. a.: J. Chem. Phys. **25**, 206 (1956); **26**, 642 (1957). — S. BROERSMA: J. Chem. Phys. **27**, 481 (1957). — R. HAUSSER u. G. LAUKIEN: Arch. d. Sci. **10**, 235 (1957). — H. PFEIFER, H. WINKLER u. G. EBERT: Ann. Phys. (6) **20**, 322 (1957).

Dipols gleichzeitig und gegensinnig eine Umorientierung des anderen Dipols erfolgt. Wegen der vermuteten Verknüpfung von Energieübergängen und Phasensprüngen, tragen die zu den aufgezählten Frequenzen gehörenden Spin-Gitter-Übergangswahrscheinlichkeiten auch zur Spin-Spin-Relaxation oder anders ausgedrückt zur Phasenübergangswahrscheinlichkeit bei, jedoch, wie bereits auseinandergesetzt, nur mit der Hälfte ihres Betrags. Ebenfalls dieser Verknüpfung wegen bewirkt auch der Term C_2 von (3.12) einen Phasenausgleich, denn es ist anzunehmen, daß sich bei jeder Umorientierung eines Elektronenspins S durch die Rauschfrequenz ω_s auch die Präzessionsphase des wechselwirkenden Kerndipols um einen Betrag statistisch schwankender Größe ändert. Zu der Spin-Spin-Relaxation T_2 tragen also insgesamt fünf Frequenzen des Rausch-spektrums bei, nämlich die Frequenzintervalle um 0, ω_i, $\omega_i - \omega_s$, ω_s, und $\omega_s + \omega_i$. Sind die wechselwirkenden Dipole identisch, so sind entsprechend die relaxie-renden Frequenzen der direkten Dipol-Dipol-Wechselwirkung nach (3.13) 0, ω und 2ω.

Der nächste Schritt der theoretischen Beschreibung der kernmagnetischen Relaxationserscheinungen ist die Analyse der Zeitabhängigkeit $f(t)$ der Orts-funktionen (3.11), d.h. die Bestimmung ihrer spektralen Intensität $J(\nu)$ in Funktion der Frequenz. Dazu soll vorausgesetzt werden, daß die Bewegungen der Dipole rein statistisch, also vollständig zufällig verlaufen sollen.

In der Statistik bezeichnet man den Mittelwert

$$K(\tau) = \overline{f(t)\, f^*(t+\tau)} \tag{3.18}$$

einer Funktion $f(t)$ als deren Korrelationsfunktion. Wang und Uhlenbeck[1] haben gezeigt, daß zwischen der Intensitätsfunktion $J(\nu)$ und der Korrelations-funktion $K(\tau)$ eines statistischen Bewegungsverlaufs der Zusammenhang

$$J(\nu) = \int_{-\infty}^{\infty} K(\tau)\, e^{i\omega\tau} d\tau \quad \text{und} \quad K(\tau) = \int_{-\infty}^{\infty} J(\nu)\, e^{-i\omega\tau} d\tau \tag{3.19a, b}$$

besteht, die Funktionen $J(\nu)$ und $K(\tau)$ sind also zueinander Fourier-transformiert.

Von der Funktion $K(\tau)$ weiß man auf Grund der Voraussetzung über den rein zufälligen Charakter der Bewegung, daß sie gerade sein muß $\big(K(\tau) = K(-\tau)\big)$. Bei der Mittelwertsbildung $\overline{f(t)\, f^*(t+\tau)}$ ist nämlich die Lage des Zeitnullpunkts ohne Bedeutung und deshalb auch keine Zeitrichtung bevorzugt. Daraus ergibt sich weiter, daß $K(\tau)$ reell sein muß, denn mit Hilfe der Zeitachsentransformation $t = t' - \tau$ findet man $K^*(\tau) = \overline{f^*(t)\, f(t+\tau)} = \overline{f^*(t'-\tau)\, f(t')} = K(-\tau) = K(\tau)$. Ferner folgt unmittelbar aus der Definition (3.18) und der genannten Voraussetzung, daß $K(0)$ der Maximalwert der Funktion $K(\tau)$ ist, und daß $K(\tau)$ für $\tau \to \pm\infty$ gegen 0 konvergiert.

Unter den verschiedenen möglichen Vermutungen über die Form der gesuchten Korrelationsfunktion zeichnet sich der Ansatz

$$K(\tau) = K(0)\, e^{-\frac{|\tau|}{\tau_c}} \tag{3.20}$$

durch besondere Einfachheit aus. Er besitzt die aufgezählten Eigenschaften, zu seiner Rechtfertigung bedarf er jedoch ebenso wie alle anderen Ansätze der experimentellen Bestätigung. Die Konstante τ_c in (3.20) hat die Dimension einer Zeit und ist mit der bereits eingeführten Korrelationszeit identisch.

---

[1] M. Ch. Wang u. G. E. Uhlenbeck: Rev. Mod. Phys. **17**, 323 (1945).

Setzt man den Ansatz (3.20) in (3.19a) ein, so erhält man

$$J(\nu) = \overline{f(t)\, f^*(t)} \int\limits_{-\infty}^{\infty} e^{i\omega\tau - \frac{|\tau|}{\tau_c}}\, d\tau = 2\cdot \overline{f(t)\, f^*(t)}\, \frac{\tau_c}{1+\omega^2 \tau_c^2}. \qquad (3.21)$$

In Fig. 3 wurde der Verlauf dieser Intensitätsfunktion in Abhängigkeit von der Frequenz für zwei Werte von τ_c bereits dargestellt.

Damit sind die Grundlagen zur Berechnung der Übergangswahrscheinlichkeiten nach (3.14), (3.16) und (3.17) vollständig. Setzt man in (3.16) die Gln. (3.12) und in die Gln. (3.12) die spektralen Intensitäten (3.21) ein, so findet man[1] für die resultierende Spin-Gitter-Übergangswahrscheinlichkeit der drei relaxierenden Rauschfrequenzen ω_i, $\omega_s - \omega_i$ und $\omega_s + \omega_i$ für den Fall der Wechselwirkung Kernmoment-Elektronenmoment den Ausdruck

$$2 W_1 = \frac{1}{T_1} = \gamma_i^2 \gamma_s^2 \hbar^2 S(S+1) \left\{ \frac{1}{12} J_0(\omega_s - \omega_i) + \frac{3}{2} J_1(\omega_i) + \frac{3}{4} J_2(\omega_s + \omega_i) \right\}. \qquad (3.22)$$

Entsprechend findet man für den Sonderfall identischer Kerne die Beziehung[1,2]

$$2 W_1 = \frac{1}{T_1} = \frac{9}{8} \gamma^4 \hbar^2 \{ J_1(\omega) + J_2(2\omega) \}. \qquad (3.23)$$

Die Gl. (3.23) folgt nicht aus der Gl. (3.22) für $\omega_s = \omega_i$, da die Wechselwirkung Kernspin—Elektronenspin, wie bei der Ableitung von (3.22) vorausgesetzt, keinen wesentlichen Beitrag zur Elektronenspin-Relaxation liefert.

Die angeschriebenen Übergangswahrscheinlichkeiten bzw. Relaxationszeiten wurden in der bisherigen Betrachtung nur durch einen, nach der klassischen Beziehung (3.5) direkt Dipol-Dipol-gekoppelt wechselwirkenden Nachbarn verursacht. Die gesamte Substanz, welche den betrachteten Dipol enthält, bzw. in der Sprache der kernmagnetischen Theorie deren Gitter, wird in dem voranstehenden Formalismus nur durch diesen einen Wechselwirkungsparameter repräsentiert. In Wirklichkeit wechselwirkt aber jeder Kern mit allen ihn umgebenden Dipolen, wenn auch der Betrag dieser Wechselwirkungen mit zunehmendem Dipolabstand sehr schnell und zwar mit der dritten Potenz des Abstands abnimmt, vgl. hierzu (3.12). Die Allgemeinheit der bisherigen Betrachtung wird jedoch durch diesen Umstand nicht eingeschränkt, denn die von verschiedenen Wechselwirkungs-Partnern herrührenden Beiträge zur Übergangswahrscheinlichkeit addieren sich einfach, und man muß dementsprechend zur Berechnung der resultierenden Übergangswahrscheinlichkeit bzw. Relaxationszeit nur über alle Dipole der Substanz summieren.

Bei der praktischen Ausführung von Relaxationsberechnungen nach dem skizzierten Schema stößt man in der Regel auf mehrere Schwierigkeiten. In grober Näherung kann man die auftretenden Probleme in vier Gruppen einteilen.

a) Zunächst ist zu prüfen, welche Kopplungsarten für das Relaxationsgeschehen verantwortlich sind. Unter diesem Gesichtspunkt ist der einfachste Fall die Berechnung der Relaxationszeit einer Kernart mit $I = \frac{1}{2}$ in einer Substanz, in welcher die beobachteten Kerne mit allen Nachbarn nur über direkte magnetische Dipol-Dipol-Kopplungen wechselwirken.

b) Weiter ist zu klären, welche Wechselwirkungs-Partner in der Substanz enthalten sind. Am kürzesten wird die Rechnung, wenn nur eine Art von Partnern berücksichtigt werden muß, etwa weil überhaupt nur eine Kernart der

[1] I. Solomon: Phys. Rev. **99**, 559 (1955).
[2] R. Kubo u. K. Tomita: J. Phys. Soc. Japan **9**, 888 (1954).

Substanz endliche magnetische Dipolmomente besitzt (z.B. Wasser), oder weil die Relaxationsbeiträge der betrachteten Partner alle anderen Beiträge bei weitem überwiegen (z.B. paramagnetische Ionen und Flüssigkeiten). Enthält die Substanz mehrere Dipolarten von vergleichbarer Momentstärke, so wächst zwar das Volumen der Rechnungen erheblich, nicht aber ihr Schwierigkeitsgrad.

c) Zur Ausrechnung der zeitlichen Mittelwerte $\overline{f(t) \cdot f^*(t)}$ der Ortsfunktion $f(t)$ muß man den zeitlich mittleren Ordnungszustand der Umgebung des betrachteten Kerns, d.h. aber der ganzen Substanz kennen. In Flüssigkeiten kennt man im allgemeinen die unmittelbare Kernumgebung, nämlich die Struktur und den Zustand des Moleküls in dem der betrachtete Kern enthalten ist. Dagegen bestehen häufig Unklarheiten über die zeitlich mittlere Mikrostruktur der Moleküle innerhalb der Flüssigkeiten. Angefangen von dem Fragenkreis der zeitlich mittleren Komplexgrößen (z.B. der Hydratmolekülzahl in verschiedenen Lösungen) bis zur Frage nach der Existenz kristallähnlicher Flüssigkeitsstrukturen verhindern in vielen Fällen noch ungelöste Probleme der Flüssigkeitsphysik die Berechnung der kernmagnetischen Relaxationszeiten. Es ist daher nicht verwunderlich, daß man häufig den umgekehrten Weg geht und versucht aus kernmagnetischen Messungen Aussagen über die Strukturen der betrachteten Flüssigkeiten zu gewinnen.

d) Die Korrelationszeiten der Rauschspektren von Wechselwirkungs-Partnern, welche verschieden weit von dem betrachteten Kern entfernt sind, sind bereits in einer im zeitlichen Mittel völlig homogenen thermisch bewegten Dipolverteilung voneinander verschieden, vgl. (3.3 b). Darüber hinaus hat die Existenz von Mikrostrukturen stets eine weitere, mehr oder weniger komplizierte Differenzierung des Spektrums der Korrelationszeiten der Substanz zur Folge. Zur Verdeutlichung dieses Sachverhalts denke man etwa an eine wäßrige Lösung paramagnetischer Ionen, in der z.B. neben den normalen Platzwechselvorgängen der nichtgebundenen Wassermoleküle die Austauschvorgänge zwischen den freien Molekülen und den in Hydratkomplexen gebundenen Molekülen gesondert betrachtet werden müssen. Meistens weiß man über die Bewegungen in Flüssigkeiten und damit über das den inneren Bewegungszustand beschreibende Korrelationszeiten-Spektrum noch wesentlich weniger als über die zeitliche mittlere Struktur der Flüssigkeiten. Diese Schwierigkeit ist sicher einer der Gründe, welche zur Zeit die Anwendbarkeit der kernmagnetischen Theorie begrenzen. Der Gedanke liegt nahe, das Studium kernmagnetischer Relaxationsphänomene nicht nur als Hilfsmittel statischer Substanzstrukturen (vgl. auch Ziff. 28), sondern auch als Informationsquelle über das dynamische Substanzverhalten zu benutzen. Zum Beispiel kann man aus der Abhängigkeit der Relaxationszeiten von der Meßfrequenz auf die Korrelationszeiten-Verteilung der Substanz in Funktion ihrer Temperatur und ihres Drucks experimentell schließen.

Als Beispiel für die Anwendung der Gln. (3.22) und (3.23) soll der Rechnungsgang zur Bestimmung der Spin-Gitter-Relaxationszeit von Wasser erläutert werden. In der Verbindung Wasser (H_2O^{16}) haben nur die Protonen endliche magnetische Dipolmomente. Je zwei Protonen sind praktisch starr verkoppelt. Bezeichnet man den konstanten Abstand dieser beiden nächsten, sog. intramolekularen Wechselwirkungspartner mit r_0, so findet man für ihre Ortsfunktionen (3.11) durch eine Mittelwertsbildung über alle räumlichen Richtungen (die ja alle gleich wahrscheinlich sind), d.h. durch eine Integration über die Einheitskugel

$$\overline{f(t)\, f^*(t)} = \frac{1}{4\pi} \iiint f\, f^*\, d\Omega, \qquad (3.24)$$

die zeitlichen Mittelwerte

$$\overline{f_0 f_0^*} = \tfrac{4}{5} r_0^{-6}, \qquad \overline{f_1 f_1^*} = \tfrac{2}{15} r_0^{-6}, \qquad \overline{f_2 f_2^*} = \tfrac{8}{15} r_0^{-6}, \tag{3.25}$$

(3.21) und (3.25) in (3.23) eingesetzt, liefert direkt den intramolekularen Beitrag zur Spin-Gitter-Relaxationszeit des Wassers.

Das Produkt Meßfrequenz · Korrelationszeit ist in Wasser sehr klein gegen eins ($\tau_c \approx 10^{-12}$ sec). Dadurch vereinfacht sich der Rechnungsgang erheblich. Das Ergebnis ist:

$$2 W_{\text{intra}} = \frac{1}{T_{1,\text{intra}}} = \frac{3}{2} \gamma^4 \hbar^2 r_0^{-6} \cdot \tau_{c,\text{rot}}. \tag{3.26}$$

$\tau_{c,\text{rot}}$ kann man nach (3.3 a) berechnen, γ, $\hbar$ und r_0 sind sehr viel genauer bekannt, als es für die vorliegende Rechnung erforderlich wäre (vgl. Ziff. 19). Die intermolekularen Beiträge zur Spin-Gitter-Übergangswahrscheinlichkeit lassen sich ganz ähnlich berechnen. Da man die Abstände r der Protonen verschiedener Moleküle als groß gegen den intramolekularen Protonenabstand r_0 betrachten kann, sind bei der Berechnung der intermolekularen Übergangswahrscheinlichkeit die Rotationsbewegungen der Moleküle, welche die betrachteten Kerne enthalten, gegenüber ihrer translatorischen Diffusionsbewegung im Rahmen dieser Näherung vernachlässigbar. Aus der Theorie der BROWNschen Bewegung folgt, vgl. (3.3 b), daß die Korrelationszeit $\tau_{c,\text{trans}}$ proportional mit dem Quadrat des Abstands r wächst. Trotzdem kann man auch hier die Gültigkeit der Ungleichung $\omega \tau_{c,\text{trans}} \ll 1$ voraussetzen, denn die Relaxationsbeiträge der weiter entfernten Dipole sind wegen ihrer Proportionalität zur r^{-6} vernachlässigbar klein. (Die intermolekulare Übergangswahrscheinlichkeit aller Dipole im Abstand r nimmt allerdings nur proportional zu r^{-2} ab, vgl. (3.28). Aus (3.21) und (3.23) folgt damit, wenn man in (3.25) r_0 durch r ersetzt, für den Beitrag der beiden Protonen des j-ten Moleküls

$$2 W_{\text{inter}} = \frac{1}{T_{1,\text{inter}}} = 4 \cdot \gamma^4 \hbar^2 r^{-6} \cdot \tau_{c,\text{trans}}, \tag{3.27}$$

wenn es sich um ein Orthomolekül handelt ($I = 1$) handelt und der Beitrag Null, wenn es ein Wassermolekül im Parazustand ($I = 0$) ist.

Bezeichnet man mit N die Zahl der Moleküle pro cm³, so enthält eine Kugelschale, welche den betrachteten Kern im Abstand $r, r + dr$ umschließt $4 \pi r^2 \cdot N \cdot dr$ Moleküle. Von diesen sind 75% unter Normalverhältnissen im Orthozustand wie in der Quantentheorie gezeigt wird. Damit ergibt sich für die gesamte intermolekulare Übergangswahrscheinlichkeit ($a = $ Molekülradius)

$$2 W_{\text{inter}} = \frac{1}{T_{1,\text{inter}}} = 3 \gamma^4 \hbar^2 N \int_{2a}^{\infty} \frac{\tau_{c,\text{trans}}}{r^6} \cdot 4 \pi r^2 \, dr$$

und mit (3.3 b)

$$2 W_{\text{inter}} = \frac{1}{T_{1,\text{inter}}} = \frac{\gamma^4 \hbar^2 N \pi}{D} \int_{2a}^{\infty} \frac{dr}{r^2} = \frac{\pi}{2} \cdot \frac{\gamma^4 \hbar^2 N}{a D}. \tag{3.28}$$

Die gesamte Spin-Gitter-Übergangswahrscheinlichkeit ist gleich der Summe der intramolekularen und der intermolekularen Übergangswahrscheinlichkeiten.

Für die Spin-Spin-Relaxationszeit T_2 findet man[1] für nicht zu große Werte von τ_c im Falle der Wechselwirkung nichtidentischer Dipole die Gleichung

$$\left. \begin{aligned} \frac{1}{T_2} &= \gamma_i^2 \gamma_s^2 \hbar^2 S(S+1) \times \\ &\times \left\{ \frac{1}{6} J_0(0) + \frac{1}{24} J_0(\omega_s - \omega_i) + \frac{3}{2} J_1(\omega_s) + \frac{3}{4} J_1(\omega_i) + \frac{3}{8} J_2(\omega_s + \omega_i) \right\} \end{aligned} \right\} \tag{3.29}$$

[1] I. SOLOMON: Phys. Rev. 99, 559 (1955).

und für identische Wechselwirkungs-Partner die einfachere Beziehung[1,2]

$$\frac{1}{T_2} = \frac{9}{8}\,\gamma^4\,\hbar^2\left\{\frac{1}{4}\,J_0(0) + \frac{5}{2}\,J_1(\omega) + \frac{1}{4}\,J_2(2\omega)\right\}. \tag{3.30}$$

Auch die Gl. (3.30) folgt nicht aus der Gl. (3.29) für $\omega_i = \omega_s$, vgl. die Bemerkung nach Gl. (3.23).

Die Formeln (3.29) und (3.30) enthalten neben den durch die Rauschfrequenzen $\omega_i, \omega_s, \omega_s - \omega_i, \omega_s + \omega_i$ und $\omega, 2\omega$ bewirkten Phasen-Übergangswahrscheinlichkeiten auch den durch T_2' beschriebenen Einfluß des Rauschspektrums-Abschnitts um die Frequenz 0, also den Einfluß des statischen Anteils der inneren Felder. Um diesen Beitrag anschaulich zu verstehen, betrachtet man das von einem beliebigen Kern k mit dem Spin I am Ort eines zweiten betrachteten Kerns j erzeugte magnetische Feld. Parallel zur Richtung des äußeren Felds $\boldsymbol{H}$ (parallel zur Richtung dieses Felds verlaufe wie üblich die z-Achse des Koordinatensystems) hat es den Betrag

$$H_k' = \gamma\,\hbar\,I_z|(3\cos^2\vartheta_{jk} - 1)|\cdot r_{jk}^{-3} = \gamma\,\hbar\,I_z\cdot|f_{0,k}|.$$

Das Quadrat dieses Felds hat gemittelt über die $2I+1$ möglichen und durch Quantenzahlen m_k charakterisierten Orientierungen des Spins k die Größe

$$H_k'^2 = \gamma^2\,\hbar^2\cdot\frac{f_{0,k}^2}{2I+1}\sum_{-I}^{+I} m_k^2 = \frac{1}{3}\,I(I+1)\,\gamma^2\,\hbar^2\,f_{0,k}^2. \tag{3.31}$$

Wegen der thermischen Bewegung der Dipole k und j mittelt sich der größte Teil des Felds H_k' zeitlich aus. Nur der spektrale Intensitätsanteil $\int_{-\delta\nu}^{+\delta\nu} J_{0,k}(\nu)\,d\nu$ von $f_{0,k}$ dessen zeitliche Änderung ausreichend langsam erfolgt, trägt zu der durch T_2' beschriebenen statischen Linienbreite bei. Summiert man die spektralen Intensitäten aller Nachbardipole:

$$J_0(\nu) = \sum J_{0,k}(\nu), \tag{3.32}$$

so findet man für das Quadrat des gesamten quasistatischen Felds die Beziehung

$$\overline{H'^2} = \frac{1}{3}I(I+1)\,\gamma^2\,\hbar^2\int_{-\delta\nu}^{+\delta\nu} J_0(\nu)\,d\nu. \tag{3.33}$$

(Im Falle identischer Kerne muß wegen der Existenz des im starren Gitter ebenfalls säkularen Terms B', vgl. (3.13), in (3.33) noch ein Faktor $(\frac{3}{2})^2$ eingefügt werden, vgl. auch Ziff. 28).

Es ist anzunehmen, daß die Grenze $\delta\nu$ bis zu welcher Rauschfrequenzen als genügend nahe bei 0 betrachtet werden dürfen, von der gleichen Größenordnung wie die gesuchte Linienbreite ist. Aus (3.33) folgt also mit $\delta\omega^2 = \gamma^2\overline{H'^2}$ und $\delta\omega = 2\pi\,\delta\nu$

$$(\delta\nu)^2 = \frac{1}{12\pi^2}\cdot I(I+1)\,\gamma^4\,\hbar^2\int_{-\delta\nu}^{+\delta\nu} J_0(\nu)\,d\nu, \tag{3.34}$$

eine Bestimmungsgleichung für die durch die quasistationären Felder bewirkte Linienbreite $\delta\nu$. Zwischen der Linienbreite $\delta\nu$ und der Relaxationszeit T_2' besteht der Zusammenhang $\delta\nu = \frac{1}{\pi\,T_2'}$, die Beziehung (3.34) ist also auch eine Bestimmungsgleichung für T_2'. Da zur expliziten Berechnung der Spin-Spin-Relaxationszeit

---

[1] I. SOLOMON: Phys. Rev. 99, 559 (1955).
[2] R. KUBO u. K. TOMITA: J. Phys. Soc. Japan 9, 888 (1954).

die Gln. (3.29) und (3.30) zur Verfügung stehen, soll hier von einer Berechnung der Relaxationszeit T_2' nach (3.34) abgesehen werden. Die exakte Theorie liefert für (3.34) einen etwas anderen Ausdruck[1].

Vergleicht man die theoretische Spin-Gitter-Relaxationszeit (3.22) und die zugehörige Formel (3.29) der theoretischen Spin-Spin-Relaxationszeit (Wechselwirkung Kernmoment—Elektronenmoment), so kommt man zu dem Schluß, daß für $\omega_i \tau_c \ll 1$ und $\omega_s \tau_c \ll 1$ T_1 ungefähr gleich T_2 ist. Für $\omega_i \tau_c \ll 1$ und $\omega_s \tau_c = 1$ ist $T_1/T_2 = 1,04$ und für $\omega_i \tau_c \ll 1$ und $\omega_s \tau_c \gg 1$ erreicht T_1/T_2 den Wert $1,17$.

Einige experimentelle Ergebnisse[2,3,4], beobachtet an wäßrigen Lösungen paramagnetischer Ionen haben jedoch wesentlich größere T_1/T_2-Verhältnisse erbracht und widersprechen damit den genannten theoretischen Resultaten.

Auf der Suche nach einer Erklärung für diese Diskrepanz schlug BLOEMBERGEN[5] vor, als weiteren Kopplungsmechanismus neben der direkten Dipol-Dipol-Kopplung, auch eine Spin-Spin-Austauschwechselwirkung, d.h. eine elektronengekoppelte Dipol-Dipol-Wechselwirkung von der Form $A\,\boldsymbol{I} \cdot \boldsymbol{S}$ in Erwägung zu ziehen. Bezüglich der modellmäßigen Erklärung dieses Kopplungsmechanismus sei auf die Ausführungen in Ziff. 27 verwiesen. Seine wesentlichste Eigenschaft ist, daß er auch bei ständiger statistischer Umorientierung der Verbindungslinie der beiden wechselwirkenden Dipole eine von der Geschwindigkeit der BROWNschen Drehbewegung unabhängige statische Kopplung der beiden Dipole ergibt. Infolgedessen liefert bei kurzen Korrelationszeiten der Wechselwirkungsterm $A I_z S_z$ den Hauptbeitrag der elektronengekoppelten Spin-Spin-Wechselwirkung zum transversalen Relaxationsmechanismus. [Auch im Fall der direkten Spin-Spin-Kopplung beschreibt dieser Term die statische Wechselwirkung, vgl. (3.12) und (3.13).] Für die Größe der auf der elektronengekoppelten Spin-Spin-Wechselwirkungen beruhenden transversalen Relaxationszeit findet BLOEMBERGEN die Beziehung

$$\frac{1}{T_2} = \frac{1}{3} S(S+1) \hbar^{-2} A^2 \tau \cdot p. \tag{3.35}$$

Darin ist p die Wahrscheinlichkeit, daß ein Wassermolekül an ein paramagnetisches Ion angrenzt, d.h. praktisch, daß es sich in dessen Hydrathülle befindet. τ ist die mittlere Dauer der Wechselwirkung. Dementsprechend ist τ entweder gleich τ_h, der mittleren Aufenthaltsdauer eines Wassermoleküls in der Hydrathülle oder gleich τ_s, der Spin-Gitter-Relaxationszeit des Elektronenspins des paramagnetischen Ions, je nachdem welche der beiden Zeiten kürzer ist. Man darf erwarten, daß die Korrelationszeit τ der elektronengekoppelten Spin-Spin-Wechselwirkung länger ist, als die durch die Rotationsgeschwindigkeit der Wassermoleküle bestimmte Korrelationszeit τ_c der direkten Dipol-Dipol-Kopplung. Genau dies ist aber die Voraussetzung, welche es gemäß den Ausführungen zu Fig. 5 erlaubt, das Zustandekommen von T_1/T_2 Verhältnissen wesentlich über eins zu verstehen. Der Beitrag der elektronengekoppelten Spin-Spin-Wechselwirkung zur Spin-Gitter-Übergangswahrscheinlichkeit $W_1 = \dfrac{1}{2T_1}$ ist nach BLOEMBERGEN[5] gegenüber dem der direkten Wechselwirkung vernachlässigbar.

Weitere Versuche, T_1/T_2-Verhältnisse größer als eins in Flüssigkeiten zu erklären, wurden von KUBO und TOMITA[1] (nicht-isotrope Umgebung der Kerne) und von NOLLE und MORGAN[6] (Hydrat-Effekt) diskutiert.

[1] R. KUBO u. K. TOMITA: J. Phys. Soc. Japan 9, 888 (1954).
[2] J. R. ZIMMERMAN: J. Chem. Phys. 21, 1605 (1953); 22, 950 (1954).
[3] A. I. RIVKIND: DAN, USSR 102, 1107 (1955).
[4] G. LAUKIEN u. J. SCHLÜTER: Z. Physik 146, 113 (1956).
[5] N. BLOEMBERGEN: J. Chem. Phys. 27, 572 (1957).
[6] A. W. NOLLE u. L. O. MORGAN: J. Chem. Phys. 26, 642 (1957).

Der größere Teil der in dieser Ziffer wiedergegebenen Erläuterungen kernmagnetischer Relaxationsphänomene bezieht sich auf Flüssigkeiten. Einmal ist das Relaxationsverhalten flüssiger Substanzen relativ am genauesten untersucht worden, und zum anderen sind flüssige Substanzen zur Lösung der eigentlichen kernphysikalischen Aufgabe der kernmagnetischen Spektroskopie, der Messung aller Kerndipolmomente, besonders geeignet. Ergänzend soll deshalb kurz auf das Relaxationsverhalten der Gase und Festkörper eingegangen werden.

Gase verhalten sich ähnlich wie Flüssigkeiten niedriger Viskosität. Die Energie- und Phasenänderung der Kerne werden ebenfalls durch das in der Probe auf Grund der BROWNschen Translations- und Rotationsbewegung der Moleküle zeitlich und örtlich schwankende innere Magnetfeld vermittelt. Haben die Kerne ein endliches elektrisches Quadrupolmoment, so ist das Kernspin-System auch durch das schwankende innere elektrische Feld der Substanz mit den übrigen Freiheitsgraden der Substanz gekoppelt. Zum Beispiel fanden BRUN, OESER, STAUB und TELSCHOW[1], daß die Relaxationszeit T_1 des Edelgasisotops Xe^{129} ($I = \frac{1}{2}$, $Q = 0$) sehr viel länger als diejenige des Isotops Xe^{131} ($I = \frac{3}{2}$, $Q = -0,15$) ist. Auffallend ist, daß die Relaxationszeiten der mehratomigen Gase allgemein sehr viel kleiner sind als die der einatomigen, vgl. unter anderen[2-4]. So hat z.B. die Spin-Gitter-Relaxationszeit T_1 des Wasserstoffgases[2] bei 10 Atmosphären eine Größe von etwa $^1/_{100}$ sec. Die inneren Kopplungen der mehratomigen Gase werden demnach vor allem durch intramolekulare Magnetfelder bewirkt. Das Magnetfeld am Ort eines der Protonen des H_2-Moleküls hat z.B. zwei molekülinnere Quellen, einmal das andere Proton innerhalb des Moleküls und zweitens die Rotationsbewegung des Moleküls. Stoßen zwei Moleküle infolge ihrer BROWNschen Translationsbewegung zusammen, so können sich sowohl ihre Rotationsquantenzahlen als auch ihre ebenfalls gequantelten räumlichen Lagen bezüglich der Richtung des äußeren Feldes ändern. Die Annahme, daß sich bei jedem Zusammenstoß mindestens eine dieser Größen ändert, führt zu dem Ergebnis, daß die Korrelationszeit τ_c dieselbe Größenordnung wie die mittlere Zeit zwischen zwei Stößen hat (z.B. in H_2-Gas bei 10 Atmosphären und 290° K ungefähr 10^{-11} sec). Demnach sollten T_1 und T_2 proportional mit dem Druck p des Gases wachsen, denn zwischen den genannten Größen bestehen die Proportionalitätsbeziehungen $\tau_c \sim 1/p$ und $T_{1,2} \sim 1/\tau_c$. PACKARD und WEAVER[4] konnten diesen Zusammenhang für H_2-Gas im Intervall 0,3 bis 40 Atmosphären experimentell bestätigen.

Charakteristisch für das Relaxationsverhalten typischer Festkörper sind die in der Regel recht langen Spin-Gitter-Relaxationszeiten einerseits und die verhältnismäßig großen Linienbreiten der kernmagnetischen Resonanzlinien andererseits. Beide Erscheinungen beruhen auf der gleichen Ursache. In Festkörpern haben die Kerndipole fast keine Translations- und fast keine Rotations-Bewegungsfreiheit. Infolgedessen sind die statischen Wechselwirkungsfelder sehr viel größer als in Gasen und Flüssigkeiten. Dagegen besitzen die für die Spin-Gitter-Relaxationszeit verantwortlichen dynamischen Felder sehr viel geringere spektrale Intensitäten als in Gasen und Flüssigkeiten. In Kristallen findet man Linienbreiten, welche umgerechnet in Relaxationszeiten T_2-Werte der Größenordnung 10^{-6} ergeben, während die Spin-Gitter-Relaxationszeiten der gleichen Festkörper bis zu 10^{+4} sec betragen können. Andererseits gibt es auch äußerlich feste Körper, deren Relaxationsverhalten demjenigen der Flüssigkeiten sehr ähnlich ist. Eine der interessantesten Lehren des Studiums kernmagnetischer

[1] E. BRUN, J. OESER, H. H. STAUB u. C. G. TELSCHOW: Phys. Rev. **93**, 904 (1954).
[2] E. M. PURCELL, R. V. POUND u. N. BLOEMBERGEN: Phys. Rev. **70**, 988 (1946).
[3] F. VERBRUGGE u. R. L. HENRY: Phys. Rev. **83**, 211 (1951).
[4] M. E. PACKARD u. H. E. WEAVER: Phys. Rev. **88**, 163 (1952).

Spektren ist gerade die Erfahrung, daß zwischen dem festen und dem flüssigen Zustand zahlreiche Übergangsformen existieren, z.B. freie Beweglichkeiten eines Teils der Partikel eines Festkörpers in diesem, freie Rotationsmöglichkeiten von Atom- oder Molekülgruppen in einem Festkörper usw. (vgl. hierzu Ziff. 28).

In wirklich starren Festkörpern haben die Partikel nur eine Bewegungsmöglichkeit, die der Schwingungen. Alle Versuche, die experimentell beobachteten Spin-Gitter-Relaxationszeiten von Kernen mit dem Spin $I = \frac{1}{2}$ (keine Quadrupolwechselwirkungen) in diamagnetischen Elektronenhüllen (keine Wechselwirkungen mit ungesättigten Elektronenmomenten) allein durch deren Gitter-Temperaturbewegung zu erklären, haben zu große T_1-Werte ergeben[1-3].

Man nimmt deshalb an, vgl. hierzu Arbeiten von ROLLIN und HATTON[4,5] und von BLOEMBERGEN[6], daß der Einfluß paramagnetischer Verunreinigungen auf den Relaxationsverlauf praktisch nie vernachlässigt werden kann. Kein wirklicher Festkörper ist ganz frei von nicht abgesättigten Elektronenmomenten, sei es in der Form von paramagnetischen Ionen, von F-Zentren oder von anderen Störstellenarten. Als Beleg dieser Vorstellungen hat BLOEMBERGEN gezeigt, daß in einem Aluminium-Alaun-Kristall, in dem ein Teil der diamagnetischen Al^{+++}-Ionen durch paramagnetische Cr^{+++}-Ionen ersetzt wurde, die Spin-Gitter-Relaxationszeit der Protonen ungefähr umgekehrt proportional zur Konzentration der Chrom-Verunreinigung ist.

Zur Erklärung des auch in kleinster Konzentration (bis 10^{-6}) so überraschend wirksamen Relaxationseinflusses freier Elektronenmomente nimmt man an, daß ein an irgendeiner Stelle des Festkörpers vorhandener Orientierungsenergie-Überschuß mittels des paarweisen Spinorientierungs-Austausches (vgl. S. 136) im Gitter umherdiffundiert. Kommt die Überschußenergie in die Nachbarschaft eines unabgesättigten Elektronenmoments, so kann sie sich infolge der dort viel stärkeren magnetischen Wechselwirkungen mit entsprechend größerer Wahrscheinlichkeit in BROWNsche Bewegungsenergie des Gitters umwandeln.

In Festkörpern, welche bewegliche Partikel oder Partikelgruppen enthalten, werden die Relaxationsvorgänge vor allem durch deren Felder bewirkt. Die Beschreibung der Relaxationsprozesse kann in solchen Fällen ganz analog zu dem bei der Behandlung der Flüssigkeiten entwickelten Formalismus erfolgen. Handelt es sich etwa um statistisch schwankend verlaufende Umorientierungen einer Molekülgruppe, so kann man für die Korrelationszeit dieses Prozesses

$$\tau_c = \tau_0 \, e^{\frac{E}{kT}} \tag{3.36}$$

ansetzen. Darin ist E die Höhe des bei dem Prozeß zu überwindenden Potentialwalls. Wie bei der Erklärung der Fig. 4 auseinandergesetzt, erreicht der Beitrag des zu einer Korrelationszeit τ_c gehörenden Relaxationsmechanismus seinen Maximalwert, wenn das Produkt $\omega \, T_c$ (Beobachtungskreisfrequenz · Korrelationszeit) ungefähr gleich eins ist, vgl. hierzu[7,8]. Die zugehörige Relaxationszeit T_1 wächst also sowohl beim Übergang zu sehr kleinen als auch beim Übergang zu sehr großen Beobachtungsfrequenzen. Leichter läßt sich allerdings experimentell

[1] J. WALLER: Z. Physik **79**, 370 (1932).
[2] E. M. PURCELL: Physica, Haag **17**, 282 (1951).
[3] G. R. CHUZISCHWILI: ZETP, USSR **22**, 382 (1952).
[4] B. V. ROLLIN u. J. HATTON: Phys. Rev. **74**, 346 (1948).
[5] J. HATTON u. B. V. ROLLIN: Proc. Roy. Soc. Lond., Ser. A **199**, 222 (1949).
[6] N. BLOEMBERGEN: Physica, Haag **15**, 386 (1949).
[7] A. H. COOKE u. L. E. DRAIN: Proc. Phys. Soc. Lond. A **65**, 894 (1952).
[8] E. R. ANDREW u. R. G. EADES: Proc. Roy. Soc. Lond., Ser. A **216**, 398; **218.** 537 (1953).

die Substanztemperatur T und damit die Korrelationszeit τ_c über große Meßbereiche variieren.

Außer der geschilderten Temperaturabhängigkeit der Spin-Gitter-Relaxationszeit gemäß (3.36) und (3.2) beobachtet man beim Erwärmen einer Substanz häufig, daß sich von bestimmten Temperaturschwellen an die Linienbreiten der kernmagnetischen Resonanzlinien plötzlich stark verengen. Zur Erklärung dieses Effekts liegt die Annahme nahe, daß bei diesen Temperaturen Teile des Festkörpers plötzlich beweglich werden, daß also bildlich gesprochen, eingefrorene Bewegungsmöglichkeiten bei bestimmten Sprungtemperaturen auftauen[1,2].

Kerne, welche einen Spin $I \geq \frac{1}{2}$ besitzen, haben der experimentellen Erfahrung nach stets ein endliches elektrisches Quadrupolmoment. Zwischen der Wechselwirkung eines Dipolmoments mit dem Feld des magnetischen Rauschens und der Wechselwirkung eines Quadrupolmomentes mit dem Feldgradienten des elektrischen Rauschens besteht kein prinzipieller Unterschied, wohl aber ein Unterschied in der Größenordnung. Wie auf S. 324 gezeigt wird, verhalten sich die elektrischen und die magnetischen Wechselwirkungs-Energien etwa wie $e^2 Q/\mu^2$. In allen untersuchten Substanzen ist dieser Quotient wesentlich größer als eins. Da die Übergangswahrscheinlichkeit proportional mit dem Quadrat der Wechselwirkungs-Energie wächst, kommt man zu dem mit der experimentellen Erfahrung übereinstimmenden Ergebnis[3], daß für Kerne mit $I \geq \frac{1}{2}$ die Quadrupolrelaxation die Dipolrelaxation überwiegt.

Unter den Festkörpern sind die Metalle durch die in ihnen frei existierenden Leitungselektronen ausgezeichnet. Das Vorhandensein dieser Elektronen ermöglicht es den in Metallen enthaltenen Kernen auf einem weiteren Weg zu relaxieren[4-9]. Die bisher besprochenen Relaxationsmechanismen tragen zwar ebenfalls zur Gesamtrelaxation bei, allerdings erweist sich ihr Beitrag gegenüber dem der Wechselwirkung Kern—Leitungselektronen meist als vernachlässigbar. Die Wahrscheinlichkeit eines Energieübergangs von dem Freiheitsgrad der Kernorientierung zu den Freiheitsgraden der kinetischen Energie der Elektronen ist auf Grund der Beweglichkeit der Elektronen und auf Grund ihrer Dipolfelder der Anschauung nach sehr groß. Zu beachten ist jedoch, daß die Leitungselektronen die zugehörigen Energieübergänge nur ausführen können, wenn ihnen unbesetzte Energieniveaus im Abstand $\hbar\omega$ von ihrem anfänglichen Niveau offen stehen. Diese Bedingung begrenzt die Wirksamkeit der Kernrelaxation mittels der Leitungselektronen erheblich, denn die FERMI-DIRAC-Verteilung der Elektronen in Metallen enthält nur an ihrem oberen, zur Grenzenergie E_0 gehörenden Rand über eine Saumbreite der Größenordnung kT Leerstellen, während alle tieferliegenden Energieniveaus lückenlos besetzt sind (Entartung des Elektronengases). Die Orientierungsenergien der Kerne $\hbar\omega$ sind unter praktischen Versuchsbedingungen sehr viel kleiner als kT, infolgedessen kann nur ein Bruchteil der Größenordnung kT/E_0 der Leitungselektronen zur Kernrelaxation beitragen. Die

[1] N. L. ALPERT: Phys. Rev. **72**, 637 (1947); **75**, 398 (1949).
[2] H. S. GUTOWSKY u. G. E. PAKE: J. Chem. Phys. **18**, 162 (1950).
[3] R. V. POUND: Phys. Rev. **79**, 685 (1950).
[4] W. HEITLER u. E. TELLER: Proc. Roy. Soc. Lond., Ser. A **155**, 629 (1936).
[5] B. V. ROLLIN u. J. HATTON: Phys. Rev. **74**, 346 (1948). — Proc. Roy. Soc. Lond., Ser. A **199**, 222 (1949).
[6] N. BLOEMBERGEN: Physica, Haag **15**, 588 (1949).
[7] N. J. POULIS: Physica, Haag **16**, 373 (1950).
[8] J. KORRINGA: Physica, Haag **16**, 601 (1950).
[9] R. E. NORBERG u. C. P. SLICHTER: Phys. Rev. **83**, 1074 (1951).

Berechnung des Beitrags der Elektronen im Leitungsband zur Übergangswahrscheinlichkeit der Kerne zeigt, in Übereinstimmung mit der voranstehenden anschaulichen Beschreibung, daß dieser proportional zur Temperatur wächst. Die zugehörige Relaxationszeit nimmt dementsprechend, wie experimentell bestätigt, umgekehrt proportional zur Temperatur ab.

II. Präzessionsbewegungen in einem konstanten Magnetfeld.

4. Freie Präzessionsbewegungen. In Analogie zu den Bezeichnungen „erzwungene Schwingung" und „freie Schwingung" kann man bei den bekannten kontinuierlichen, Kerninduktion und Kernresonanzabsorption genannten Verfahren von einer Beobachtung der „erzwungenen Präzessionsbewegung" sprechen, während man eine Anregung durch Impulse als Anstoß „freier Präzessionsbewegungen" des Kern-Spin-Systems beschreiben kann.

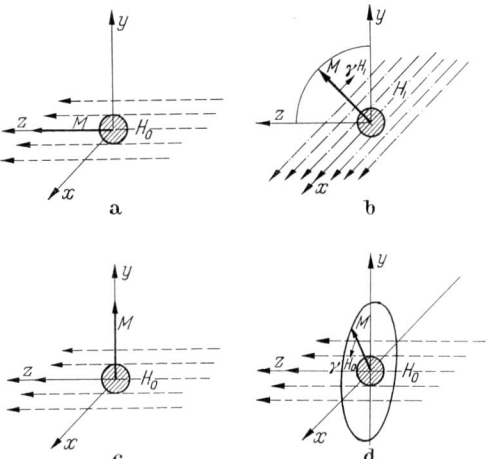

Fig. 6 a–d. Anregung freier Präzessionsbewegungen in einem Feld H_0 durch ein kurzzeitig eingeschaltetes starkes transversales Magnetfeld H_1 (Gedankenexperiment, $H_1 \gg H_0$).

Ein einfaches Gedankenexperiment soll dies verdeutlichen. Befindet sich eine zu untersuchende Substanz genügend lange in einem konstanten Magnetfeld $H_0 = H_z$ (Fig. 6a), dann entsteht ein makroskopisches kernmagnetisches Moment M, vgl. Gl. (1.7). Schaltet man nach der Einstellung des thermischen Gleichgewichts im Feld H_0, senkrecht zu diesem, ein zweites ebenfalls konstantes, aber vielfach stärkeres Feld $H_1 = H_x$ kurzzeitig ein (Fig. 6b), dann kann man während seiner Einschaltdauer, die im übrigen klein gegen die Relaxationszeiten T_1 und T_2 sei, das Feld H_0 vernachlässigen. Der zuvor entstandene Magnetisierungsvektor M präzessiert um die Richtung des neuen Feldes H_1 mit der LARMOR-Frequenz $\omega_1 = \gamma H_1$, Gl. (2.3). Wird nach einer Zeit t_1 das Feld H_1 wieder abgeschaltet, dann hat sich M um einen Winkel $\alpha_1 = \gamma H_1 t_1$ gedreht. Bemißt man t_1 so, daß $\alpha_1 = 90°$ wird (Fig. 6c), dann liegt nach dem Gleichfeldimpuls das Moment M senkrecht zum ursprünglichen Feld H_0. Es setzt seine Präzessionsbewegung um dessen Richtung fort (Fig. 6d), und zwar jetzt mit der Kreisfrequenz $\omega_0 = \gamma H_0$. Umschließt man die Probesubstanz mit einer Spule, deren Achse in der x, y-Ebene liegt, dann induziert der frei rotierende Magnetisierungsvektor in ihr eine Wechselspannung der LARMOR-Frequenz ω_0. Die Messung dieser Frequenz würde den Betrag des gyromagnetischen Verhältnisses der Kerne liefern, und die Beobachtung des Amplitudenabfalles des freien Präzessionssignales in einem ausreichend homogenen Feld gäbe Aufschluß über das transversale Relaxationsverhalten der Probe. Damit ist das Verfahren der Anregung und Beobachtung freier Präzessionsbewegungen im Prinzip schon beschrieben.

Gleichfeldimpulse solch hoher Feldstärke können mit den derzeitigen technischen Hilfsmitteln leider nicht erzeugt werden. Man kann aber das beschriebene Experiment mit Hochfrequenzimpulsen beliebig kleiner Amplitude in fast der gleichen Form ausführen[1].

---

[1] E. L. HAHN: Phys. Rev. **77**, 297 (1950).

Dazu legt man an die im Gedankenexperiment nur zur Beobachtung der Präzessionen benutzte Spule eine Wechselspannung an, deren Frequenz gleich der LARMOR-Frequenz ist. Diese erzeugt in der Spule (Fig. 7) ein oszillierendes Magnetfeld, dessen Maximalfeldstärke (zur Vereinfachung der Schreibweise) die Größe $2H_1$ haben soll. Der Ursprung eines ortsfesten Koordinatensystems x, y, z liege in dem Mittelpunkt der Spule und der zu untersuchenden Substanz. In diesem System hat das oszillierende Feld die Komponenten $H_{1x} = 2H_1 \cos \omega_0 t$, $H_{1y} = H_{1z} = 0$. Ein solches Feld läßt sich in zwei gegeneinander rotierende Felder der Amplitude H_1 aufspalten[1]. Ein zweites Koordinatensystem x', y', z', dessen

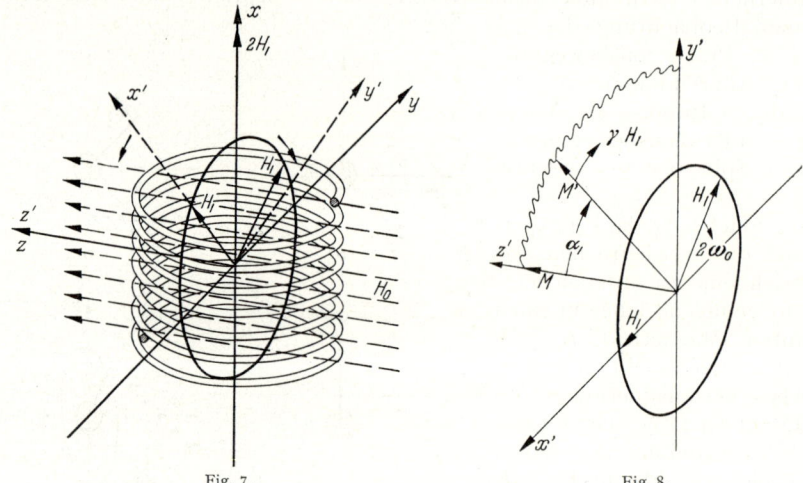

Fig. 7.   Fig. 8.

Fig. 7. Zerlegung des oszillierenden Feldes $2H_1 \cos(\omega t)$ einer Spule in zwei mit der Frequenz ω rotierende Felder. Die Richtung des einen Feldes ist parallel zur x'-Achse eines mitrotierenden Koordinaten-Systems. Das andere Feld rotiert gegensinnig, und zwar vom x', y', z'-System aus gesehen mit der Kreisfrequenz 2ω.

Fig. 8. Einfluß der beiden rotierenden Feldanteile des oszillierenden Spulenfeldes auf den im mitrotierenden Koordinaten-System zu beobachtenden Bewegungsablauf. Um den im x', y', z'-System konstanten Feldanteil H_1 präzessiert der Magnetisierungsvektor mit der Kreisfrequenz γH_1. Dieser Bewegung überlagert sich — verursacht durch den mit der Frequenz 2ω rotierenden Feldanteil — eine schwache Nutationsbewegung.

Ursprung mit dem des ortsfesten übereinstimmt, soll mit der Kreisfrequenz ω_0 um die mit der z-Achse identische z'-Achse rotieren[2]. In diesem System wird eines der im ortsfesten System rotierenden Felder zum konstanten Feld. Das sich gegensinnig drehende Feld rotiert dagegen im bewegten Koordinatensystem mit der doppelten Kreisfrequenz $2\omega_0$. Unter dem alleinigen Einfluß des ersten Feldanteils bewegt sich ein makroskopisches Moment in dem zweiten Koordinatensystem nun ebenso wie in dem Gedankenexperiment unter dem Einfluß des dort auch im ortsfesten System konstanten Feldes H_1. Wir müssen jedoch nicht mehr fordern, daß H_1 sehr viel größer als H_0 ist, um das Feld H_0 während der Einschaltdauer des Feldes H_1 vernachlässigen zu können. Da dem mitrotierenden Koordinatensystem gerade vorgeschrieben wurde, daß es mit der LARMOR-Frequenz ω_0 rotiert, kann in diesem keine Präzessionsbewegung mehr beobachtet werden. Das Feld H_0 ist aus ihm eliminiert, und die Bedingung $H_1 \gg H_0$ somit immer erfüllt.

Weiter ist der Einfluß des gegensinnig rotierenden Feldanteils auf den Bewegungsablauf des makroskopischen Momentes abzuschätzen. Hierzu muß man

---

[1] Vgl. hierzu F. BLOCH u. A. SIEGERT: Phys. Rev. 57, 522 (1940).
[2] Zur Verwendung mitrotierender Koordinatensysteme s. auch: F. BLOCH: Phys. Rev. 70, 460 (1946). — I. I. RABI, N. F. RAMSEY u. J. SCHWINGER: Rev. Mod. Phys. 26, 167 (1954).

verlangen, daß seine Umlaufdauer $1/\nu_0 = 2\pi/\omega_0$ klein gegen die Einschaltdauer t_1 des Wechselfeldes ist. Jede vom rotierenden Feld verursachte Bewegung wird dann rückgängig gemacht durch die nach einer halben Periode erfolgende umgekehrte Bewegung (Fig. 8), da das rotierende Feld dann gerade in der x', y'-Ebene in die entgegengesetzte Richtung weist. Die genannte Bedingung ist aber bereits erfüllt, wenn das hochfrequente Feld wenigstens einige Perioden lang eingeschaltet bleibt. Der vom konstanten Feldanteil verursachten Bewegung der Spitze des resultierenden Moments auf einer Kreisbahn wird also vom rotierenden Feldanteil nur eine vernachlässigbare Nutationsbewegung überlagert[1].

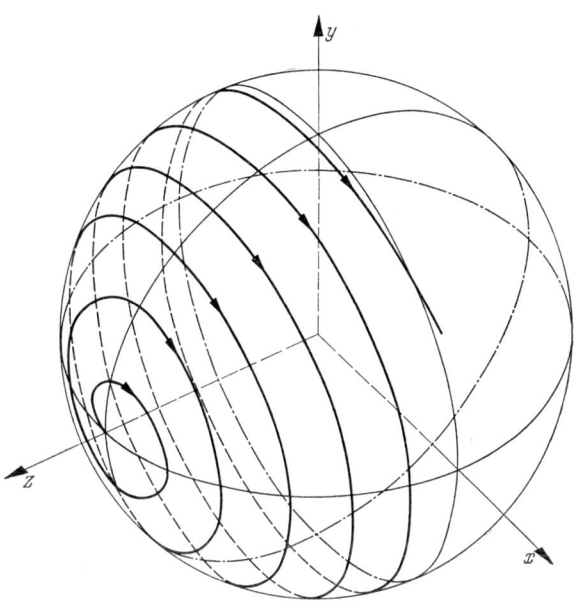

Damit ist bewiesen, daß sich ein Magnetisierungsvektor ebenso mit Hochfrequenzimpulsen kleiner Amplitude drehen läßt, wie mit kurzzeitig eingeschalteten starken Magnetfeldern. Der Drehwinkel ist ebenfalls $\alpha_1 = \gamma H_1 t_1$. Fordert man, daß er einen bestimmten Betrag hat, dann kann man sowohl die Amplitude $2H_1$ des Impulses als auch seine Dauer t_1 entsprechend einstellen.

Nach einem 90°-Impuls hat das resultierende Kernmoment im rotierenden System die Richtung der positiven y'-Achse, in dem ortsfesten System dagegen die Richtung eines beliebigen, der Beobachtungsebene angehörenden Halbstrahls. Seine Anfangslage nach dem Impuls und damit die Anfangsphase der von ihm induzierten Spannung interessiert aber im allgemeinen nicht. Den vom ortsfesten System aus zu beobachtenden Verlauf der Drehbewegung gibt Fig. 9 wieder.

Fig. 9. Bahnkurve der Spitze des Magnetisierungsvektors nach dem Einschalten eines Wechselfeldes der Resonanzfrequenz (beobachtet in einem ortsfesten Koordinaten-System, $T_1 = T_2 = \infty$).

Durch Spin-Spin-Wechselwirkungen werden die für die Existenz eines transversalen Moments notwendigen Phasenbeziehungen allmählich zerstört. Dementsprechend erwartet man, daß die induzierte Spannung exponentiell abnimmt. Aber dieser Effekt wird, wenn die transversale Relaxationszeit zu lang ist, durch die immer vorhandene Inhomogenität eines wirklichen Magnetfeldes verdeckt. Jede Probe enthält gleichzeitig Kerne, die in einem etwas stärkeren Feld und Kerne, die in einem etwas schwächeren Feld sind. Ebenso sind die zugehörigen LARMOR-Frequenzen größer, bzw. kleiner als ihr Mittelwert. Während der zur mittleren LARMOR-Frequenz gehörende infinitesimale Teil der Kerne im mitrotierenden Koordinatensystem stillsteht, eilen die anderen infinitesimalen Momente in diesem System vor, bzw. nach (Fig. 10c, d). Dieses Auseinanderfächern

---

[1] Ein rotierendes Hochfrequenzfeld läßt sich auch direkt herstellen (wesentlich für absolute Vorzeichenmessungen des magnetischen Moments) vgl. hierzu Ziff. 21, sowie eine Arbeit von H. H. STAUB u. E. H. ROGERS: Helv. phys. Acta **23**, 63 (1950).

hat ebenfalls zur Folge, daß das transversale Moment abnimmt und zwar um so schneller je inhomogener das Magnetfeld ist. Jedoch bleiben die Phasen-

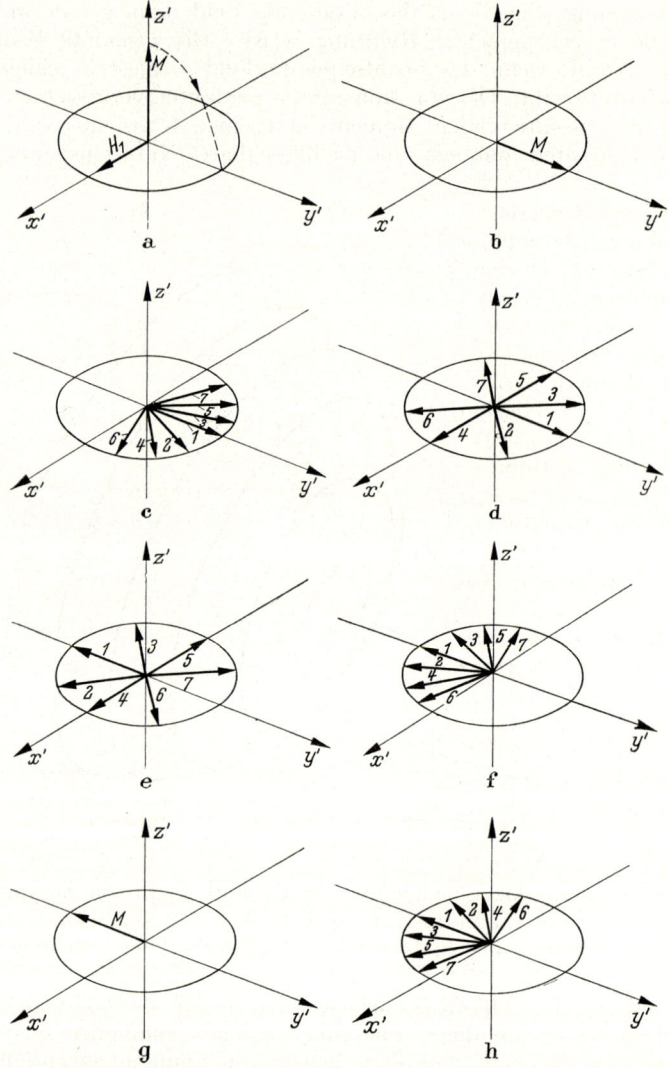

Fig. 10 a–h. Anregung freier Präzessionsbewegungen durch eine 90°–180°-Impulsfolge in einem inhomogenen Magnetfeld. $T_1 = T_2 = \infty$; H_1 wirksame Komponente des hochfrequenten Feldes; M resultierendes Moment der Kerne; 1 infinitesimales Moment, dessen LARMOR-Frequenz mit der Koordinatenfrequenz übereinstimmt; 3, 5, 7 infinitesimale Momente, deren LARMOR-Frequenz größer ist; 2, 4, 6 infinitesimale Momente, deren LARMOR-Frequenz kleiner ist. (a) Nach der Einstellung des thermodynamischen Gleichgewichtes ist parallel zur Feldrichtung (z'-Richtung) eine resultierende Kernmagnetisierung M entstanden. Durch einen 90°-Impuls wird diese in die y'-Richtung gedreht (b). Die infinitesimalen Momente fächern auseinander, das Kernsignal verschwindet (c, d). Durch einen zweiten 180°-Impuls wird der Fächer umgedreht (e). Die infinitesimalen Momente setzen ihre Präzessionsbewegung im ursprünglichen Sinn fort (f) und vereinigen sich erneut in der negativen y'-Richtung (g). Anschließend fächern die infinitesimalen Momente wiederum auseinander, und das Signal verschwindet (h).

beziehungen zwischen den Kernen dabei erhalten, auch wenn sie zunächst nicht mehr beobachtbar sind. Schaltet man t_2 sec nach dem ersten 90°-Impuls einen zweiten 180°-Impuls ein, dann wird durch diesen der Fächer der infinitesimalen Anteile (Fig. 10d) einfach umgedreht (Fig. 10e). Nach dem zweiten Impuls kann zunächst kein transversales Moment beobachtet werden; sämtliche infinitesimalen Momente präzessieren aber auch nach ihm mit dem ursprünglichen Drehsinn innerhalb der x', y'-Ebene weiter und nähern sich somit wieder einander (Fig. 10f). Nach ungefähr der gleichen Zeit t_2, die ihnen zwischen den Impulsen zum Auseinanderfächern zur Verfügung stand, sind sie wieder vereint (Fig. 10g) und zwar nun in der Richtung der negativen y'-Achse. Für den Induktionsvorgang ist es aber ohne Belang, in welche Richtung das resultierende Moment in der x', y'-Ebene zeigt. Vom Zeitpunkt $2t_2$ an trennen sich die Momente wieder, ebenso wie nach dem ersten Impuls (Fig. 10h). E. L. HAHN hat diesen von ihm Spin-Echo[1] genannten Effekt entdeckt. Allerdings hat er auch als zweiten

Impuls einen 90°-Impuls verwandt. Die Wiederherstellung des makroskopischen Momentes ist in diesem Fall nicht vollständig. Es handelt sich um einen anschaulich schwieriger zu verstehenden Sonderfall.

Zusammenfassend wird der zeitliche Spannungsverlauf an der die Probe umgebenden Spule in Fig. 11 wiedergegeben. Die Darstellung ist jedoch in verschiedener Hinsicht idealisiert. Einmal ist ihr Maßstab falsch. Die von außen angelegte Spannung ist in der Regel einige hundert Volt. Die von den Kernen induzierte Spannung hat die Größenordnung einiger zehn μ-Volt, ist also etwa 10^6mal kleiner. Weiter wurde stillschweigend vor-

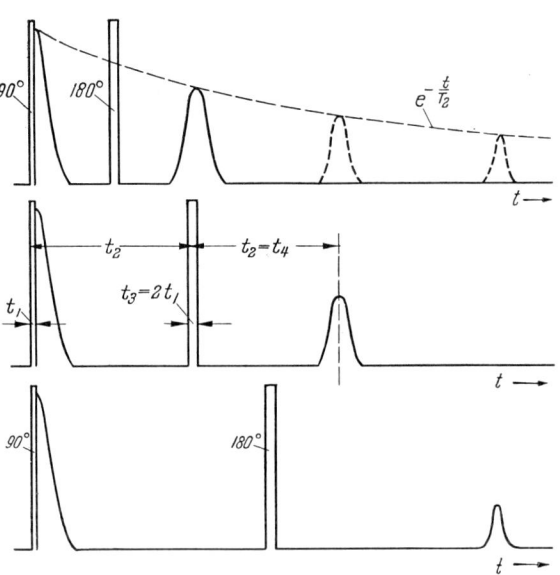

Fig. 11. Schematische Darstellung einer Messung der Spin-Spin-Relaxationszeit T_2 aus einer mit zunehmendem Impulsabstand registrierten Folge von Spin-Echos.

ausgesetzt, daß die Feldverteilung durch eine Fehlerfunktion beschrieben werden kann. Darauf soll später eingegangen werden. In Analogie zur Interferenz von Schwingungen kann man den ganzen Vorgang auch als „kernmagnetische Interferenz" bezeichnen.

Die Maximalamplitude des Echos muß natürlich kleiner als die Amplitude des Abfallsignales nach dem ersten Impuls sein, da in der zwischen ihnen vergangenen Zeit $2t_2$ ein Teil der Kerne auf Grund von Wechselwirkungen inkohärent weiter präzessiert. Vergrößert man den Abstand der beiden Impulse und damit auch den des Echos (Fig. 11), dann nimmt die Echoamplitude um so schneller ab, je stärker die inneren Kopplungen des Kernspin-Systems sind. So liefert die Umhüllende einer mit zunehmendem Impulsabstand registrierten Folge von Interferenzsignalen direkt die Spin-Spin-Relaxationszeit T_2. Eine weitere Voraussetzung ist, daß Diffusionsvorgäng einnerhalb der Substanz vernachlässigt werden können (s. auch S. 136).

---

[1] E. L. HAHN: Phys. Rev. 80, 580 (1950). Siehe auch T. P. DAS u. A. K. SAHA: Phys. Rev. 93, 749 (1954); H. Y. CARR u. E. M. PURCELL: Phys. Rev. 94, 630 (1954); G. LAUKIEN Z. Natorforsch. 11a, 222 (1956) u. a.

Andere Impulskombinationen erlauben die direkte Bestimmung der longitudinalen Relaxationszeit T_1. Fig. 12 zeigt ein Beispiel. Ein erster 180°-Impuls dreht den Magnetisierungsvektor in die zur Feldrichtung entgegengesetzte negative z'-Richtung. Auf Grund der Spin-Gitter-Kopplungen verschwindet die

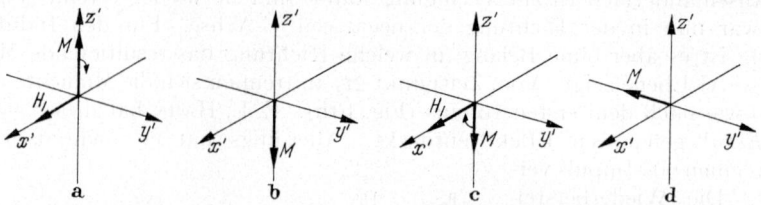

Fig. 12 a—d. Drehung des resultierenden Momentes durch eine 180°—90°-Impulsfolge. (a) Nach der Einstellung des thermodynamischen Gleichgewichts ist parallel zur Feldrichtung ein resultierendes Kernmoment M entstanden. Durch einen 180°-Impuls wird es in die zur Feldrichtung antiparallele negative z'-Richtung gedreht (b). Auf Grund der Spin-Gitter-Wechselwirkungen wird M kleiner (c). Durch einen 90°-Impuls wird der in diesem Zeitpunkt noch vorhandene Restvektor M in die Beobachtungsebene gedreht (d). Ist der zwischen den Zeitpunkten (b) und (c) vergangene Zeitabschnitt größer als in der Zeichnung angenommen wurde, dann ist M inzwischen verschwunden bzw. in der positiven z'-Richtung (Feldrichtung) erneut entstanden.

Magnetisierung in dieser Richtung allmählich und entsteht danach in der Feldrichtung, also der positiven z'-Richtung von neuem. Schaltet man in verschiedenen Abständen hinter dem 180°-Impuls einen zweiten 90°-Impuls ein, dann ist

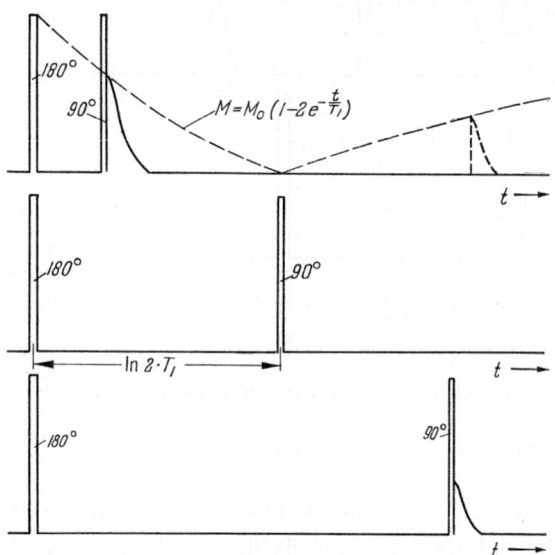

Fig. 13. Schematische Darstellung einer Messung der Spin-Gitter-Relaxationszeit T_1 aus einer mit zunehmendem Impulsabstand registrierten Folge der abklingenden Kernsignale nach dem 2. Impuls eines 180°—90°-Programms.

die Amplitude des Abfallsignals hinter diesem Impuls vom zeitlichen Abstand der beiden Impulse abhängig. Ist dieser Abstand sehr klein oder sehr groß, dann erreicht die Signalamplitude ihren maximalen Grenzwert. Dazwischen existiert ein Abstand, bei dem das Kernsignal völlig verschwindet. Mißt man diesen Abstand (Fig. 13), oder beobachtet den Verlauf des Amplitudenanstiegs, dann ist damit auch T_1 bestimmt[1].

Nach dieser anschaulichen Einführung in das Gebiet der freien Präzessionsbewegungen soll jetzt deren quantitative Behandlung besprochen werden. Experimentell beobachtbar sind nur Bewegungen des resultierenden Moments aller Kerne einer Probesubstanz. Zu deren Berechnung muß man jedoch von dem Bewegungsablauf eines infinitesimalen Anteils dieses Moments ausgehen. Unter einem solchen Anteil soll das resultierende Moment all jener Kerne verstanden werden, auf welche die nach Richtung und Betrag gleichen magnetischen Felder einwirken. Der Verlauf der Präzession folgt aus den kernmagnetischen Differentialgleichungen (2.6). Dabei wird der Rechengang sehr übersichtlich, wenn die

---

[1] Vgl. hierzu auch G. LAUKIEN: Z. Naturforsch. **11**a, 266 (1956).

Impulsbreiten t_ν klein gegen die Relaxationszeiten T_1 und T_2 sind, und wenn im Experiment der Anregungsvorgang (Impulseinstrahlung) und der Beobachtungsvorgang (Induktion einer Wechselspannung durch den frei präzessierenden Kernmagnetisierungsvektor) voneinander zeitlich getrennt sind. Man kann dann während der Anregungszeit die Relaxationsglieder der Differentialgleichungen vernachlässigen, und für den Versuchsablauf sind nur die Präzessionsglieder von (2.6) maßgebend. In den anschließenden Beobachtungszeitabschnitten läßt sich aber mit dem Kunstgriff des mitrotierenden Koordinatensystems die freie Präzessionsbewegung des Magnetisierungsvektors im wesentlichen aus der Darstellung eliminieren. In dieser Zeit wird der Versuchsablauf also vor allem durch die Relaxationsglieder von (2.6) beschrieben.

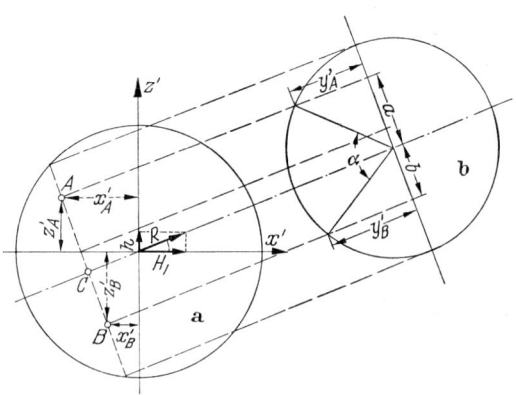

Ein Moment befinde sich, den Experimenten entsprechend in einem Gleichfeld der Feldstärke H, dem zeitweise senkrecht dazu ein mit der Frequenz ω rotierendes zweites Feld H_1 überlagert wird. Die Bahnkurve des Moments soll in einem System x', y', z' bestimmt werden, das mit der gleichen Frequenz wie das Feld H_1, also mit der Senderfrequenz rotiert. Jedoch soll nicht verlangt werden, daß die LARMOR-Frequenz der Kerne mit der Sender- bzw. mit der damit identischen Koordinatenfrequenz auch nur näherungsweise übereinstimmt. Im rotierenden System

Fig. 14 a u. b. Drehung eines Moments beliebiger Anfangslage durch einen Impuls. H_1 Hochfrequenzfeld; h Resonanzabstand; R Resultante; A, B Anfangslage bzw. Endlage der Spitze des Moments; C Mittelpunkt des in Fig. 14a gestrichelt und in Fig. 14b ausgezogen gezeichneten Präzessionskreises auf der Kugel.

beeinflußt außer H_1 noch ein weiteres Feld, es sei h genannt, den Bewegungsablauf. Seine Feldstärke ist gleich der im mitrotierenden System beobachtbaren Differenz $H - \dfrac{\omega}{\gamma}$. Auf das Moment wirkt während der Einschaltdauer des Feldes H_1 die Resultante beider Felder $R = \sqrt{H_1^2 + h^2}$ (Fig. 14), und um deren Richtung präzessiert es in dieser Zeit mit der Winkelgeschwindigkeit γR. Die Bewegung beginne in einem beliebigen Punkt A der Einheitskugel[1]. Nach t sec hat sich das Moment auf dem in Fig. 14a gestrichelt eingezeichneten Kreis um den Winkel $\alpha = \gamma \sqrt{H_1^2 + h^2} \cdot t$ gedreht (Fig. 14b). Seine Spitze befindet sich jetzt in dem Punkt B. Nennt man den Mittelpunkt des Präzessionskreises C und die in der x', z'-Ebene gemessenen Abstände zwischen ihm und den Punkten A und B: a und b, dann sind die gesuchten Koordinaten von B:

$$x'_B = x'_C - \frac{h}{\sqrt{H_1^2 + h^2}} \cdot b; \qquad y'_B = \sin\alpha \cdot a + \cos\alpha \cdot y'_A; \qquad z'_B = z'_C + \frac{H_1}{\sqrt{H_1^2 + h^2}} b.$$

Aus der geometrischen Anordnung folgt:

$$a = \frac{1}{\sqrt{1 + (h/H_1)^2}} \left(z'_A - \frac{h}{H_1} x'_A \right); \qquad b = \cos\alpha \cdot a - \sin\alpha \cdot y'_A;$$

$$x'_C = \frac{h/H_1}{1 + (h/H_1)^2} \left(z'_A + \frac{H_1}{h} x'_A \right); \qquad z'_C = \frac{(h/H_1)^2}{1 + (h/H_1)^2} \left(z'_A + \frac{H_1}{h} x'_A \right).$$

---

[1] G. LAUKNɪE Z. Naturforsch. **11**a, 222 (1956).

Kürzt man ab: $u = h/H_1$, dann besteht zwischen den Anfangs- und Endkoordinaten der Zusammenhang:

$$x'_B = \frac{u}{1+u^2}\left[\left(z'_A + \frac{1}{u}\,x'_A\right) + \sqrt{1+u^2}\,y'_A \sin\alpha - (z'_A - u\,x'_A)\cos\alpha\right],$$

$$y'_B = y'_A \cos\alpha + \frac{z'_A - u\,x'_A}{\sqrt{1+u^2}}\sin\alpha\,,$$

$$z'_B = \frac{1}{1+u^2}\left[u^2\left(z'_A + \frac{1}{u}\,x'_A\right) - \sqrt{1+u^2}\,y'_A \sin\alpha + (z'_A - u\,x'_A)\cos\alpha\right].$$

$$(4.1)$$

Die Anfangsrichtung des Moments ist bei allen Versuchen die zum Gleichfeld H parallele positive z'-Richtung ($z'_0 = 1$, $x'_0 = y'_0 = 0$). Nach einem ersten Impuls der Dauer t_1 sind demnach die Endkoordinaten, geschrieben als Funktionen des Drehwinkels in Resonanz $\alpha_1 = \gamma H_1 t_1$ und des relativen Resonanzabstandes $u = h/H_1$:

$$x'_1 = \frac{u}{1+u^2}\left(1 - \cos\{\alpha_1\sqrt{1+u^2}\}\right), \qquad y'_1 = \frac{1}{\sqrt{1+u^2}}\sin\{\alpha_1\sqrt{1+u^2}\},$$

$$z'_1 = \frac{1}{1+u^2}\left(u^2 + \cos\{\alpha_1\sqrt{1+u^2}\}\right).$$

Nach dem Impuls setzt das Moment seine Präzession um die Richtung des Feldes h fort. Außerdem nimmt sein Betrag auf Grund der Dämpfungsglieder der Differentialgleichungen (2.6) ab. Es ist (t gleich der vom Impulsende gerechneten Zeit):

$$x' = \left[\frac{u}{1+u^2}\left(1 - \cos\{\alpha_1\sqrt{1+u^2}\}\right)\cos(\gamma h t) + \frac{1}{\sqrt{1+u^2}}\sin\{\alpha_1\sqrt{1+u^2}\}\sin(\gamma h t)\right]e^{-\frac{t}{T_2}},$$

$$y' = \left[-\frac{u}{1+u^2}\left(1 - \cos\{\alpha_1\sqrt{1+u^2}\}\right)\sin(\gamma h t) + \frac{1}{\sqrt{1+u^2}}\sin\{\alpha_1\sqrt{1+u^2}\}\cos(\gamma h t)\right]e^{-\frac{t}{T_2}},$$

$$z' = 1 - \left(1 - \left[\frac{u^2}{1+u^2} + \frac{1}{1+u^2}\cos\{\alpha_1\sqrt{1+u^2}\}\right]\right)e^{-\frac{t}{T_1}}.$$

$$(4.2)$$

Dabei sind die in den eckigen Klammern eingeschlossenen Ausdrücke die nicht relaxierten Komponenten.

Entsprechend findet man für die Komponenten des Moments nach einem zweiten Impuls (t gleich der vom Ende des zweiten Impulses an gerechneten Zeit, $t_2 =$ Abstand der beiden Impulse, $\alpha_3 = \gamma H_1 t_3 =$ Drehwinkel des zweiten Impulses):

$$x' = e^{-\frac{t_2+t}{T_2}}\left[+f_1\cos\{\gamma h(t_2+t)\} + f_2\sin\{\gamma h(t_2+t)\} + \right.$$
$$\left. + f_3\cos\{\gamma h(t_2-t)\} + f_4\sin\{\gamma h(t_2-t)\}\right] +$$
$$+ e^{-\frac{t}{T_2}}\left(1 - f_5 e^{-\frac{t_2}{T_1}}\right)\left[f_6\sin\{\gamma h t\} + f_7\cos\{\gamma h t\}\right],$$

$$y' = e^{-\frac{t_2+t}{T_2}}\left[-f_1\sin\{\gamma h(t_2+t)\} + f_2\cos\{\gamma h(t_2+t)\} + \right.$$
$$\left. + f_3\sin\{\gamma h(t_2-t)\} - f_4\cos\{\gamma h(t_2-t)\}\right] +$$
$$+ e^{-\frac{t}{T_2}}\left(1 - f_5 e^{-\frac{t_2}{T_1}}\right)\left[f_6\cos\{\gamma h t\} - f_7\sin\{\gamma h t\}\right].$$

$$(4.3)$$

Darin sind die zur Abkürzung eingeführten Ausdrücke f_ν Funktionen der Anregungsdaten u, $a_1 = \alpha_1 \sqrt{1+u^2}$ und $a_3 = \alpha_3 \sqrt{1+u^2}$:

$$\left.\begin{aligned} f_1 &= \frac{u}{1+u^2}\left[\cos a_3 - \cos(a_1+a_3) + \frac{1}{2(1+u^2)}(1-\cos a_1)(1-\cos a_3)\right], \\ f_2 &= \frac{1}{(1+u^2)^{\frac{1}{2}}}\left[-\sin a_3 + \sin(a_1+a_3) + \right. \\ &\quad + \left.\frac{1}{2(1+u^2)}\left\{\sin a_1(1-\cos a_3) + 2\sin a_3(1-\cos a_1)\right\}\right], \\ f_3 &= \frac{u}{2(1+u^2)^2}(1-\cos a_1)(1-\cos a_3); \quad f_4 = \frac{1}{2(1+u^2)^{\frac{3}{2}}}\sin a_1(1-\cos a_3), \\ f_5 &= \frac{1}{1+u^2}(1-\cos a_1); \quad f_6 = \frac{1}{(1+u^2)^{\frac{1}{2}}}\sin a_3; \quad f_7 = \frac{u}{1+u^2}(1-\cos a_3). \end{aligned}\right\} \quad (4.4)$$

Die Gln. (4.3) gehen über in (4.2), wenn entweder die Relaxationszeiten T_1 und T_2 sehr kurz sind, oder der Abstand der beiden Impulse t_2 sehr lang ist. Das Kernspin-System befindet sich dann im Zeitpunkt t_2 bereits wieder im thermischen Gleichgewicht.

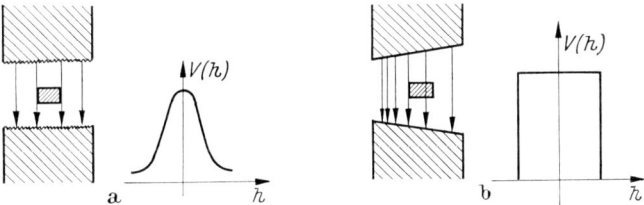

Fig. 15a u. b. Beispiele von Inhomogenitätsarten. a Die Polschuhe sind feinkörnig rauh, die Feldverteilung $V(h)$ gehorcht einer Fehlerfunktion. b Die Feldstärke nimmt in einer Richtung linear zu, die Feldverteilung $V(h)$ hat die Form eines Rechtecks.

Würde man den Betrag H bzw. h des Gleichfeldes in jedem Punkt der untersuchten Substanz kennen, dann könnte man daraus exakt berechnen, welcher Bruchteil $V(h)\,dh$ der Kerne in dem Feldintervall $(h, h+dh)$ liegt. Jedoch kann man über die Form der Feldverteilung $V(h)$ im allgemeinen nichts aussagen, sondern muß sich mit versuchsweisen Annahmen begnügen. Naheliegend ist die Annahme, daß die Feldinhomogenität nur durch zufällige Ursachen hervorgerufen wird (z.B. durch die Rauhigkeit von Polschuhoberflächen). Sie ist zugleich die notwendige und hinreichende Bedingung dafür, daß die Verteilungsfunktion die Form der Fehlerfunktion hat (Fig. 15a):

$$V(h) = \frac{1}{\sqrt{2\pi}\,B}\,e^{-\frac{(h-\bar{h})^2}{2B^2}}. \quad (4.5)$$

Dabei soll unter B die Halbwertsbreite der Feldstreuung — definiert durch $B^2 = \overline{(\bar{h}-h)^2}$ — verstanden werden, während mit $\bar{h}$ ihr quadratischer Mittelwert bezeichnet werden soll. Interessant sind weiter alle Annahmen, die von magnetostatisch, also aus der geometrischen Anordnung berechenbaren Feldverteilungen ausgehen. Nimmt die Stärke des konstanten Feldes z.B. in einer Richtung linear zu, dann wird die Feldverteilung durch eine Rechteckfunktion beschrieben (Fig. 15b).

Allgemein erhält man die Komponenten des resultierenden Momentes durch den Ansatz

$$M_{x'} = M_0 \int_{-\infty}^{\infty} V(h) \cdot x' \cdot dh, \quad M_{y'} = M_0 \int_{-\infty}^{\infty} V(h) \cdot y' \cdot dh. \quad (4.6)$$

Darin ist M_0 das statische magnetische Moment (1.7). Im mitrotierenden System sind auch negative Feldstärken möglich, die untere Integrationsgrenze ist also $-\infty$. Außerdem muß für jede Verteilungsfunktion $V(h)$ die Normierungsbedingungen $\int_{-\infty}^{\infty} V(h)\, dh = 1$ erfüllt sein.

Setzt man für x' und y' die Gl. (4.2) und für $V(h)$ den Ansatz (4.5) ein, dann erhält man mit $\bar{u} = \bar{h}/H_1$ (mittlerer, relativer Resonanzabstand) und $B' = B/H_1$ (relative Halbwertsbreite) für das resultierende Moment nach dem ersten Impuls die Komponentengleichungen:

$$\begin{matrix} M_{x'} \\ M_{y'} \end{matrix} = \frac{M_0\, e^{-\frac{t}{T_2}}}{\sqrt{2\pi}\, B'} \left[\pm \int_{-\infty}^{\infty} e^{-\frac{(u-\bar{u})^2}{2B'^2}} \begin{matrix} \cos \\ \sin \end{matrix} (\gamma h t)\left(1 - \cos\{\alpha_1 \sqrt{1+u^2}\}\right) \frac{u}{1+u^2}\, du + \right.$$

$$\left. + \int_{-\infty}^{\infty} e^{-\frac{(u-\bar{u})^2}{2B'^2}} \begin{matrix} \sin \\ \cos \end{matrix} (\gamma h t) \sin\{\alpha_1 \sqrt{1+u^2}\} \frac{1}{\sqrt{1+u^2}}\, du \right].$$

Leider sind diese Integrale nicht in geschlossener Form lösbar. Führt man als neue Variable $z = u - \bar{u}$ ein und entwickelt die Funktionen g_1 und g_2:

$$g_1 = \left(1 - \cos\{\alpha_1 \sqrt{1 + \bar{u}^2 + 2\bar{u}z + z^2}\}\right) \frac{z + \bar{u}}{1 + \bar{u}^2 + 2\bar{u}z + z^2},$$

$$g_2 = \sin\{\alpha_1 \sqrt{1 + \bar{u}^2 + 2\bar{u}z + z^2}\} \frac{1}{\sqrt{1 + \bar{u}^2 + 2\bar{u}z + z^2}}.$$

in TAYLORsche Reihen um $z = 0$ dann erhält man Integrale der Form:

$$\int_{-\infty}^{\infty} e^{-\lambda z^2} \cos(v z)\, z^n\, dz \quad \text{und} \quad \int_{-\infty}^{\infty} e^{-\lambda z^2} \sin(v z)\, z^n\, dz.$$

Deren Lösungen sind für $n = 0$, bzw. für $n = 1$:

$$\int_{-\infty}^{\infty} e^{-\lambda z^2} \cos(v z)\, dz = \sqrt{\frac{\pi}{\lambda}}\, e^{-\frac{v^2}{4\lambda}} \quad \text{und} \quad \int_{-\infty}^{\infty} e^{-\lambda z^2} \sin(v z)\, z\, dz = \frac{\sqrt{\pi}}{2} \cdot \frac{v}{\lambda^{\frac{3}{2}}}\, e^{-\frac{v^2}{4\lambda}}. \quad (4.7)$$

Weiter haben die linken Integrale für ungerade n und die rechten für gerade n den Wert Null, da ihre Integranden dann antisymmetrische Funktionen sind. Für beliebige Werte von n kann man Rekursionsformeln ableiten:

$$\left. \begin{aligned} \int_{-\infty}^{\infty} e^{-\lambda z^2} \begin{matrix} \cos \\ \sin \end{matrix} (v z)\, z^n\, dz &= \frac{n-1}{2\lambda} \int_{-\infty}^{\infty} e^{-\lambda z^2} \begin{matrix} \cos \\ \sin \end{matrix} (v z)\, z^{n-2}\, dz \mp \\ &\mp \frac{v}{2\lambda} \int_{-\infty}^{\infty} e^{-\lambda z^2} \begin{matrix} \sin \\ \cos \end{matrix} (v z)\, z^{n-1}\, dz. \end{aligned} \right\} \quad (4.8)$$

Damit erhält man für die Komponenten des resultierenden Magnetisierungsvektors Reihendarstellungen:

$$\left. \begin{aligned} \begin{matrix} M_{x'} \\ M_{y'} \end{matrix} = M_0 \Big\{ & e^{-\frac{t}{T_2}} e^{-\frac{1}{2}(\gamma B t)^2} \Big[\Big(\pm \begin{matrix} \cos \\ \sin \end{matrix} \{\gamma \bar{h} t\} g_{10} + \begin{matrix} \sin \\ \cos \end{matrix} \{\gamma \bar{h} t\} g_{20} \Big) + \\ & + \frac{B}{H_1} \gamma B t \Big(-\begin{matrix} \sin \\ \cos \end{matrix} \{\gamma \bar{h} t\} g_{11} \pm \begin{matrix} \cos \\ \sin \end{matrix} \{\gamma \bar{h} t\} g_{21} \Big) + \\ & + \Big(\frac{B}{H_1} \Big)^2 \big(1 - (\gamma B t)^2\big) \Big(\pm \begin{matrix} \cos \\ \sin \end{matrix} \{\gamma \bar{h} t\} g_{12} + \begin{matrix} \sin \\ \cos \end{matrix} \{\gamma \bar{h} t\} g_{22} + \cdots \Big) \Big] \Big\}, \end{aligned} \right\} \quad (4.9)$$

in denen die Konstanten $g_{1\nu}$ und $g_{2\nu}$ Koeffizienten der Reihen g_1 und g_2 sind.

Berücksichtigt man nur die ersten Glieder von (4.9), dann ist der dadurch verursachte Fehler glücklicherweise klein. Der Faktor $e^{-\frac{z^2}{2B'^2}}$ sorgt dafür, daß der Integrand für große Werte von z rasch verschwindet. Ferner streben auch die Funktionen g_1 und g_2 gegen Null für $z \to \infty$. Zu beachten ist, daß die Größe des Fehlers nur von der relativen Halbwertsbreite B/H_1 der Feldinhomogenitäten abhängt, nicht aber vom Resonanzabstand, also von der Differenz der Impulsträger- und der mittleren Präzessionsfrequenz. Gemäß der Definition der Variablen $z = u - \bar{u} = (h - \bar{h})/H_1$ ist eine Näherung für kleine Werte von z um so genauer, je weniger die Feldstärken h um ihren Mittelwert $\bar{h}$ streuen. Außerdem ist sie um so besser, je größer die Impulsfeldstärke H_1 ist. Aus der Winkelbeziehung $\alpha_\nu = \gamma H_1 t_\nu$ folgt aber, daß für vorgegebene Winkelwerte α_ν die zugehörigen Impulsdauern dann sehr klein werden.

Die Berechnung der Bewegungsgleichungen des resultierenden Moments nach einem zweiten Impuls kann von den Beziehungen (4.3) und (4.5) ausgehend, ebenso erfolgen. Ist $f_{\nu\lambda}$ der λ-te Koeffizient der Entwicklung der Funktion f_ν (4.4), dann gilt:

$$
\begin{aligned}
\left.\begin{array}{l} M_{x'} \\ M_{y'} \end{array}\right\} = M_0 \Bigg\{ & e^{-\frac{t_2+t}{T_2}} e^{-\frac{1}{2}\gamma^2 B^2 (t_2+t)^2} \times \\
& \times \left[\left(\begin{array}{c}\cos\\\sin\end{array}\{\gamma\bar{h}(t_2+t)\} f_{10} + \begin{array}{c}\sin\\\cos\end{array}\{\gamma\bar{h}(t_2+t)\} f_{20} \right) + \frac{B}{H_1}(\cdots) + \cdots \right] + \\
& + e^{-\frac{t}{T_2}} e^{-\frac{1}{2}\gamma^2 B^2 t^2} \times \\
& \times \left[\left(1 - f_{50}\, e^{-\frac{t_2}{T_1}}\right)\left(\begin{array}{c}\sin\\\cos\end{array}\{\gamma\bar{h}t\} f_{60} \pm \begin{array}{c}\cos\\\sin\end{array}\{\gamma\bar{h}t\} f_{70} \right) + \frac{B}{H_1}(\cdots) + \cdots \right] + \\
& + e^{-\frac{t_2+t}{T_2}} e^{-\frac{1}{2}\gamma^2 B^2 (t_2-t)^2} \times \\
& \times \left[\left(+ \begin{array}{c}\cos\\\sin\end{array}\{\gamma\bar{h}(t_2-t)\} f_{30} \pm \begin{array}{c}\sin\\\cos\end{array}\{\gamma\bar{h}(t_2-t)\} f_{40} \right) + \frac{B}{H_1}(\cdots) + \cdots \right] \Bigg\}.
\end{aligned} \tag{4.10}
$$

Die Bewegungen des beobachtbaren Moments werden in (4.10) durch eine Summe von drei Gliedern beschrieben. Ist der Abstand t_2 der beiden Impulse genügend groß, dann kann der Beitrag des ersten Gliedes zum resultierenden Moment und damit zur induzierten Spannung vernachlässigt werden. Und in der Tat stellt man ja im Experiment den Abstand der Impulse meistens so ein, daß die verschiedenen kernmagnetischen Signale deutlich zeitlich getrennt sind. Damit liefert aber die in dem ersten Glied beschriebene Umwandlung der Restamplitude des Abfallsignales nach dem ersten Impuls nur einen Beitrag zur Signalspannung, der unterhalb der experimentellen Beobachtungsgrenze, nämlich der Rauschspannung des Empfängers liegt. Das zweite Glied verschwindet mit dem Faktor $e^{-\frac{1}{2}\gamma^2 B^2 t^2}$ und beschreibt demgemäß die Form eines unmittelbar hinter dem zweiten Impuls entstehenden Kernsignals. Das dritte zu $e^{-\frac{1}{2}\gamma^2 B^2 (t_2-t)^2}$ proportionale Glied erreicht seinen Maximalwert ungefähr zur Zeit $t = t_2$ und gibt die Umhüllende des t_2 sec später beobachteten Echos wieder.

Allgemein hat das beobachtbare Moment den Betrag $M_{x',y'} = \sqrt{M_{x'}^2 + M_{y'}^2}$. In Resonanz ist im mitrotierenden Koordinatensystem $\bar{h} = \bar{u} = 0$. In der hier verwendeten Bezeichnungsweise war die Anfangsrichtung des resultierenden Moments die positive z'-Richtung, während das Impulsfeld H_1 parallel zur positiven x'-Achse liegt. Jede von ihm verursachte Bewegung des mittleren infinitesimalen

Moments verläuft also in der z', y'-Ebene. Die Bewegungen der übrigen infinitesimalen Momente können dagegen in einer beliebigen Raumrichtung enden. Jedoch gehört zu jedem Moment außerhalb des Mittelwertes bei symmetrischen Feldverteilungen ein anderes, dem Betrage nach gleichgroßes, dessen Bewegung unter dem Einfluß derselben Felder in der in Bezug auf die z', y'-Ebene symmetrischen Richtung endet. Die infinitesimalen x'-Komponenten heben sich somit bei einer symmetrischen Feldverteilung in Resonanz exakt gegenseitig auf, und es gilt der erleichternde Zusammenhang $M_{x',y'} = M_{y'}$.

Aus (4.9) folgt damit für das Abfallsignal nach dem ersten Impuls in Resonanz:

$$M_{x',y'} = M_0\, e^{-\frac{1}{2}(\gamma Bt)^2}\, e^{-\frac{t}{T_2}} \left[g_{20} - \frac{B}{H_1}\gamma Bt\, g_{11} + \left(\frac{B}{H_1}\right)^2 (1-(\gamma Bt)^2)\, g_{22} + \cdots \right] \quad (4.11)$$

aus (4.10) für das Abfallsignal nach dem zweiten Impuls:

$$M_{x',y'} = M_0\, e^{-\frac{1}{2}(\gamma Bt)^2}\, e^{-\frac{t_2}{T_2}} \Bigg[\left(1 - f_{50}\, e^{-\frac{t_2}{T_1}}\right) f_{60} - \frac{B}{H_1}\gamma Bt \left(\left(1 - f_{50}\, e^{-\frac{t_2}{T_1}}\right) f_{71} - \right.$$
$$\left. - f_{51} f_{70}\, e^{-\frac{t_2}{T_1}}\right) + \left(\frac{B}{H_1}\right)^2 (1-(\gamma Bt)^2)\left(\left(1 - f_{50}\, e^{-\frac{t_2}{T_1}}\right) f_{62} - f_{51} f_{61}\, e^{-\frac{t_2}{T_1}}\right) + \cdots \Bigg] \quad (4.12)$$

und ebenfalls aus (4.10) für das Echo in Resonanz:

$$M_{x',y'} = M_0\, e^{-\frac{1}{2}(\gamma B(t_2-t))^2}\, e^{-\frac{(t_2+t)}{T_2}} \left[-f_{40} + \frac{B}{H_1}\gamma Bt\, f_{31} - \left(\frac{B}{H_1}\right)^2 (1-(\gamma Bt)^2)\, f_{42} + \cdots \right]. \quad (4.13)$$

Sind die Einschaltzeiten t_ν der Impulse so kurz und entsprechend ihre Feldstärke so groß, daß man während ihnen nicht nur Relaxationseffekte ($t_\nu \ll T_1, T_2$), sondern auch Inhomogenitätseffekte vernachlässigen kann ($B/H_1 \ll 1$), dann erhalten die Beziehungen die einfache Form:

$$M_{x',y'} = M_0 \sin\alpha_1\, e^{-\frac{1}{2}(\gamma Bt)^2} \cdot e^{-\frac{t}{T_2}}, \quad (4.14)$$

$$M_{x',y'} = M_0 \left(1 - (1-\cos\alpha_1)\, e^{-\frac{t_2}{T_1}}\right) \sin\alpha_3\, e^{-\frac{1}{2}(\gamma Bt)^2}\, e^{-\frac{t}{T_2}}, \quad (4.15)$$

$$M_{x',y'} = -M_0\, \tfrac{1}{2} \sin\alpha_1 (1-\cos\alpha_3)\, e^{-\frac{1}{2}(\gamma B(t_2-t))^2}\, e^{-\frac{t_2+t}{T_2}}. \quad (4.16)$$

Die damit angeschriebenen ersten Näherungen entsprechen der in dieser Ziffer eingangs besprochenen anschaulichen Beschreibung der Präzessionen. Sie zeigen wie die Amplituden der drei Kernsignale mit der Zunahme der Drehwinkel periodisch größer und kleiner werden.

In allen Näherungen ist die Zeit t nur in dem dimensionslosen Produkt γBt enthalten. Die in Feldern verschiedener Inhomogenität entstehenden Signale gehen also exakt durch eine lineare Zeitachsen-Transformation ineinander über. Es ist daher zweckmäßig, das Produkt γBt als unabhängige Variable zu verwenden. Dies ist in Fig. 16 geschehen. Parameter der dort in dritter Näherung wiedergegebenen theoretischen Signalformen ist die relative Halbwertsbreite B/H_1. Dabei wurde angenommen, daß die Relaxationszeiten T_1 und T_2 vernachlässigbar groß sind. Ist das Verhältnis B/H_1 nicht ausreichend klein gegen eins, dann sind auch die aus den Gln. (4.14), (4.15) und (4.16) folgenden anschaulichen Aussagen über die Amplituden-Extrema der Kernsignale als Funktion der Drehwinkel α_ν nur näherungsweise gültig. Die Anfangsamplitude des Signals nach dem ersten Impuls wird z.B. für $B/H_1 = 0.25$ zum Maximum, wenn $\alpha_1 = 87{,}3°$ ist.

Zum Vergleich soll noch die Formel des resultierenden Moments für den Fall der linearen Zunahme des konstanten Felds (Fig. 15 b) angeschrieben werden.

Mit B, der halben Breite des Rechtecks, gilt für das Abfallsignal nach einem ersten Impuls in Resonanz in erster Näherung:

$$M_{x',y'} = M_0 \sin \alpha_1 \frac{\sin(\gamma B t)}{\gamma B t}. \qquad (4.17)$$

Fig. 17 zeigt die in diesem Fall zu erwartende Form der Kernsignale. Ihre Umhüllende ist mit der Kreisfrequenz γB moduliert. Beobachtet wird nur der Betrag des resultierenden Moments.

Hochfrequenz-Rechteckimpulse enthalten ein mehr oder weniger breites Frequenzspektrum. So ist es einleuchtend, daß sich mit ihnen auch Kernsignale anregen lassen, deren mittlere LARMOR-Frequenz sich von der Trägerfrequenz der Impulse um einen gewissen Betrag unterscheidet. Bei der Ableitung der Gln. (4.9) und (4.10) wurde dies bereits berücksichtigt. Die Zusammenfassung der dort angegebenen Komponentengleichungen liefert für den transversalen Magnetisierungsvektor nach dem ersten Impuls in erster Näherung in und außerhalb Resonanz:

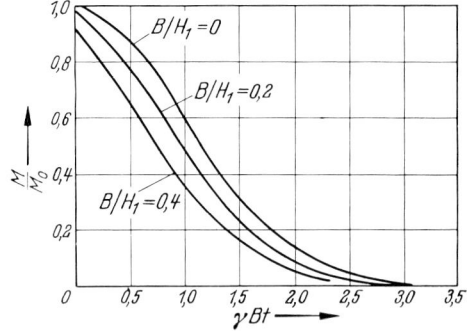

Fig. 16. Abhängigkeit der Signalform von der Halbwertsbreite B der Feldverteilung, wenn diese einer Fehlerfunktion gehorcht. 90°-Impuls; B/H_1 relative Halbwertsbreite; berechnet in dritter Näherung.

$$\left.\begin{aligned} M_{x',y'} &= M_0\, e^{-\frac{1}{2}(\gamma B t)^2} e^{-\frac{t}{T_2}} \sqrt{g_{10}^2 + g_{20}^2} \\ &= M_0 \frac{1}{1+\bar{u}^2} \sqrt{2\bar{u}^2(1-\cos\{\alpha_1\sqrt{1+\bar{u}^2}\}) + \sin^2\{\alpha_1\sqrt{1+\bar{u}^2}\}}\, e^{-\frac{1}{2}(\gamma B t)^2} e^{-\frac{t}{T_2}}. \end{aligned}\right\} \quad (4.18)$$

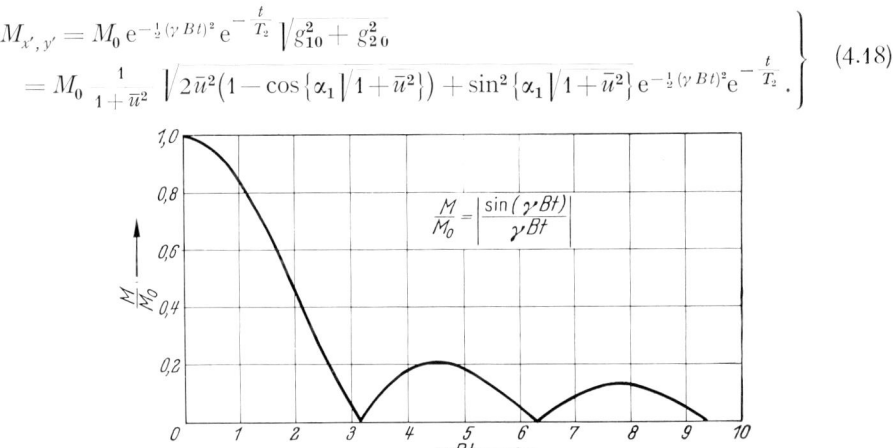

Fig. 17. Form des abklingenden Kernsignals, wenn die Feldverteilung einer Rechteckfunktion gehorcht. 90°-Impuls; B halbe Breite des Rechtecks; berechnet in erster Näherung.

Die Form des Kernsignals unterscheidet sich außerhalb Resonanz nicht von derjenigen in Resonanz. Nur der vor den zeitabhängigen Exponentialfunktionen stehende Amplitudenfaktor ist verglichen mit (4.14) komplizierter geworden. In Fig. 18 wurde dieser Amplitudenfaktor als Funktion des mittleren relativen Resonanzabstandes $\bar{u} = \bar{h}/H_1$ aufgetragen. Kurvenparameter ist der Resonanzdrehwinkel α_1. Die Signalamplitude oszilliert beiderseits außerhalb Resonanz um so schneller, je größer α_1 ist. Von Bedeutung ist, daß sie nur vom relativen, nicht aber vom absoluten Resonanzabstand abhängt. Mit der Anwendung stärkerer Amplituden H_1 des Impulsfeldes werden demnach automatisch immer weiter entfernte Resonanzfrequenzen angeregt.

11*

Übrigens läßt sich das zur Einführung besprochene Gedankenexperiment stetig in die Betrachtung einbeziehen. Ist H_1 im Grenzfall sogar wesentlich größer als die Stärke H_0 des konstanten Feldes, dann werden auf Grund der Winkel-

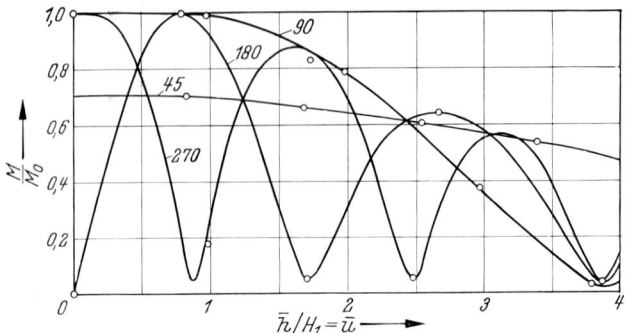

Fig. 18. Abhängigkeit der Amplitude des abklingenden Kernsignals vom Resonanzabstand. Die ausgezogenen Kurven wurden für Impulse berechnet, die in Resonanz den Magnetisierungsvektor um 45°, 90°, 180° bzw. 270° drehen, $\bar{u}$ ist der mittlere relative Resonanzabstand; die eingezeichneten Kreise umschließen Meßwerte.

beziehungen $\alpha_\nu = \gamma H_1 t_\nu$ für vorgegebene Winkel α_ν die zugehörigen Impulsdauern klein gegen die Schwingungsdauer der LARMOR-Präzession. Aus dem hochfrequenten Impulsfeld ist damit aber ein Gleichfeldimpuls geworden, mit dem sich sämtliche Resonanzfrequenzen anregen lassen.

Die periodische Zu- und Abnahme der Signalamplitude bei der Veränderung des Resonanzabstands ist auch anschaulich zu verstehen (Fig. 19). Außerhalb Resonanz präzessiert das resultierende Moment während der Anregung um die Resultante R. Die gestrichelt eingezeichneten Präzessionskreise seiner Spitze liegen senkrecht zu dieser Resultanten und parallel zur y-Achse. Eine 90°-Drehung des Magnetisierungsvektors würde in den hinteren Schnittpunkten der Präzessionskreise mit den y', R-Ebenen enden. Die zu beobachtende Signalamplitude ist gleich dem Betrag der x', y'-Komponente des resultierenden Momentes. Durch die Drehung des Präzessionskreises um den Winkel β (tan $\beta = \bar{h}/H_1$) wäre diese Komponente also schon in Fig. 19b merklich kleiner geworden. Die Abnahme wird jedoch zunächst dadurch kompensiert, daß auch der Winkel α außer-

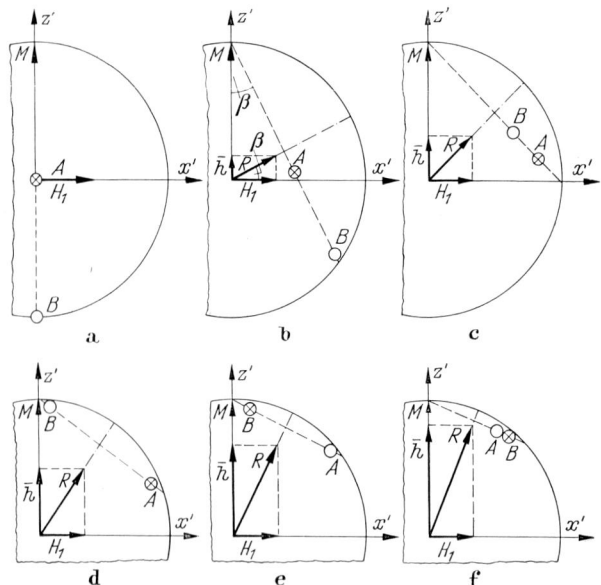

Fig. 19 a—f. Zur Erklärung der Signalanregung außerhalb Resonanz. $\bar{h}$ mittlerer Resonanzabstand; H_1 Stärke des mitrotierenden Feldes. Die Präzessionsbewegung des Magnetisierungsvektors (seine Spitze beschreibt die gestrichelt eingezeichneten Kreise) endet in den Punkten A bzw. B, wenn der Impuls in Resonanz (Fig. 19a) das resultierende Moment um 90° bzw. 180° dreht.

halb Resonanz wächst. Aus $\alpha = \alpha_1 \sqrt{1 + (\bar{h}/H_1)^2}$ folgt z.B. für Fig. 19b ein Drehwinkel von 101°, d.h. der Magnetisierungsvektor wird bis in den nahezu in der x', y'-Ebene liegenden Punkt A des Präzessionskreises gedrängt. Das Kreuz in dem Kreis um Punkt A erinnert daran, daß der Vektor in die Zeichenebene hineinzeigt. In Fig. 19d wird α größer als $180°$, und von da an trägt jede weitere Zunahme des Winkels α ebenfalls zur Momentabnahme bei. Regt man mit einem $180°$-Impuls an, dann zeigt der Magnetisierungsvektor nach der Präzession in die Punkte B der Fig. 19. Außerhalb Resonanz sorgen zunächst die größer werdende Kreisneigung und die Zunahme von α gleichsinnig für ein schnelles Anwachsen der Signalamplitude. Sie erreicht ihren Maximalwert zwischen den Fig. 19b und c. Zwischen den Fig. 19d und e verschwindet die Amplitude wiederum und nähert sich in Fig. 19f ihrem zweiten, nun aber kleineren Maximalwert.

Kurz sollen noch die durch einen zweiten Impuls außerhalb Resonanz angeregten Kernsignale erwähnt werden. Die Gleichung des Echos ist in erster Näherung in und außerhalb Resonanz:

$$\left. \begin{aligned} M_{x',y'} &= M_0\, e^{-\frac{1}{2}\gamma^2 B^2(t_2-t)^2}\, e^{-\frac{t_2+t}{T_2}}\, \sqrt{f_{30}^2 + f_{40}^2} \\ &= M_0\, \frac{1-\cos\{\alpha_3\sqrt{1+\bar{u}^2}\}}{2(1+\bar{u}^2)^2}\, \sqrt{2\bar{u}^2\left(1-\cos\{\alpha_1\sqrt{1+\bar{u}^2}\}\right)+\sin^2\{\alpha_1\sqrt{1+\bar{u}^2}\}}\, e^{-\frac{1}{2}\gamma^2 B^2(t_2-t)^2}\, e^{-\frac{t_2+t}{T_2}}. \end{aligned} \right\} \quad (4.19)$$

In Fig. 20 wurde die Echoamplitude im Zeitpunkt $t=t_2$ als Funktion des relativen Resonanzabstandes aufgetragen[1]. Vergrößert man den Resonanzabstand, dann entstehen und verschwinden periodisch weitere Interferenzsignale, sog. Seitenbandechos. Allerdings nimmt deren Amplitude im Gegensatz zu (4.18) umgekehrt proportional zur 3. Potenz des relativen Resonanzabstandes ab. So ist es zu verstehen, daß weit außerhalb Resonanz zwar immer noch Abfallsignale, aber keine Echos mehr beobachtet werden können.

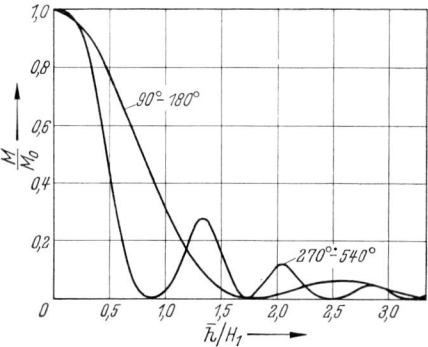

Fig. 20. Abhängigkeit der Amplitude des Echos vom Resonanzabstand (Seitenbandecho). Die Kurven wurden für eine $90°-180°$-, bzw. für eine $270°-540°$-Impulsfolge berechnet.

5. **Erzwungene Präzessionsbewegungen.** Bei der Besprechung des Anregungsvorgangs freier Präzessionsbewegungen wurde in Ziff. 4 zunächst verlangt, daß in den kurzen Impulseinstrahlungszeiten sowohl Relaxationseffekte, als auch der Einfluß der Inhomogenitäten auf den Bewegungsablauf vernachlässigt werden können. Die anschauliche Beschreibung der Bewegungen des Magnetisierungsvektors wird besonders übersichtlich, wenn beide Bedingungen erfüllt sind. Auf die zweite Voraussetzung wurde in der in Ziff. 4 anschließend besprochenen quantitativen Darstellung verzichtet. Im folgenden wird der umgekehrte Fall behandelt, Inhomogenitätseffekte sollen also vernachlässigt und Relaxationseffekte berücksichtigt werden. Praktisch sprechen wir damit über die Impulsanregung von Substanzen mit starken inneren Kopplungen. Vor allem aber soll in dieser Ziffer verdeutlicht werden, auf welche Weise die anfänglich immer instationäre Präzessionsbewegung zu stationären Endlagen des Magnetisierungsvektors führt. In Erinnerung an die Analogie zum erzwungen schwingenden gedämpften Oszillator, befassen wir uns also jetzt mit Einschwingvorgängen[2]. Dagegen entsprachen die freien Präzessionsbewegungen dem Ausschwingvorgang eines gedämpften Oszillators.

Während der Anregungszeit, gleichgültig ob sie nun zeitlich befristet oder unendlich lang ist, finden immer erzwungene Präzessionsbewegungen statt. In

---

[1] G. LAUKIEN: Dtsch. Physiker-Tagung Hamburg: Physik, Verh. **5**, 170 (1954).
[2] H. C. TORREY: Phys. Rev. **76**, 1059 (1949).

der bisherigen Darstellung war dies allerdings kaum zu bemerken. Die Kunstgriffe, Einführung eines mitrotierenden Koordinatensystems, Berücksichtigung der Feldinhomogenitäten durch eine nachträgliche Integration und die Vernachlässigung der Relaxationseffekte führten dazu, daß von den kernmagnetischen Differentialgleichungen (2.6) während der Anregung nur die triviale Lösung (2.3) benutzt wurde.

In Resonanz haben die Gln. (2.6) in einem mit der Resonanzfrequenz rotierenden System die einfache Form:

$$\left.\begin{array}{l} \dot{x}' = \qquad\qquad - \dfrac{1}{T_2}\, x' \\[3mm] \dot{y}' = \quad \gamma\, H_1\, z' - \dfrac{1}{T_2}\, y' \\[3mm] \dot{z}' = -\gamma\, H_1\, y' - \dfrac{1}{T_1}\, z' + \dfrac{1}{T_1} \end{array}\quad \begin{array}{l} \text{mit } M = (x',\, y',\, z') \\[3mm] \text{und } M_0 = (0,\, 0,\, 1). \end{array}\right\} \qquad (5.1)$$

Die erste Gleichung kann gesondert gelöst werden. Sie hat gemäß der Anfangsbedingung $x_0' = 0$ die Lösung $x' = 0$. Die stationäre Lösung des Gleichungssystems (5.1) erhält man, wenn man $\dot{y}' = \dot{z}' = 0$ setzt:

$$y_\infty' = \frac{\gamma\, H_1\, T_2}{1 + \gamma^2\, H_1^2\, T_1 T_2}\,; \qquad z_\infty' = \frac{1}{1 + \gamma^2\, H_1^2\, T_1 T_2}\,. \qquad (5.2)$$

Die charakteristische Gleichung des homogenen Systems:

$$\left.\begin{array}{l} \dot{y}' = -\dfrac{1}{T_2}\, y' + \gamma\, H_1\, z', \\[3mm] \dot{z}' = -\gamma\, H_1\, y' - \dfrac{1}{T_1}\, z' \end{array}\right\} \qquad (5.3)$$

lautet:

$$\begin{vmatrix} \omega - \dfrac{1}{T_2} & \gamma\, H_1 \\[3mm] -\gamma\, H_1 & \omega - \dfrac{1}{T_1} \end{vmatrix} = 0,$$

bzw.:

$$\omega^2 - \left(\frac{1}{T_1} + \frac{1}{T_2}\right)\omega + \frac{1}{T_1 T_2} + \gamma^2\, H_1^2 = 0.$$

Sie hat die Lösungen[1]:

a) falls gilt: $\dfrac{1}{2}\left(\dfrac{1}{T_2} - \dfrac{1}{T_1}\right) < \gamma\, H_1$,

$$\omega_{1,2} = \frac{1}{2}\left(\frac{1}{T_1} + \frac{1}{T_2}\right) \pm i\, \sqrt{\gamma^2\, H_1^2 - \frac{1}{4}\left(\frac{1}{T_2} - \frac{1}{T_1}\right)^2} = \omega' \pm i\, \omega'', \qquad (5.4\text{a})$$

b) falls gilt: $\dfrac{1}{2}\left(\dfrac{1}{T_2} - \dfrac{1}{T_1}\right) = \gamma\, H_1$,

$$\omega_{1,2} = \frac{1}{2}\left(\frac{1}{T_1} + \frac{1}{T_2}\right) = \omega', \qquad (5.4\text{b})$$

c) falls gilt: $\dfrac{1}{2}\left(\dfrac{1}{T_2} - \dfrac{1}{T_1}\right) > \gamma\, H_1$,

$$\omega_{1,2} = \frac{1}{2}\left(\frac{1}{T_1} + \frac{1}{T_2}\right) \pm \sqrt{\frac{1}{4}\left(\frac{1}{T_2} - \frac{1}{T_1}\right)^2 - \gamma^2\, H_1^2} = \omega' \pm \omega''. \qquad (5.4\text{c})$$

---

[1] G. J. KRÜGER u. G. LAUKIEN: Z. Physik **145**, 456 (1956).

Damit erhält man die Lösungen des Systems (5.1):

a) $\quad y' = y'_\infty + e^{-\omega' t} \left(A' \cos (\omega'' t) + A'' \sin (\omega'' t)\right),$
$\quad\quad z' = z'_\infty + e^{-\omega' t} \left(B' \cos (\omega'' t) + B'' \sin (\omega'' t)\right),$ (5.5a)

b) $\quad y' = y'_\infty + e^{-\omega' t} \left(A' + A'' t\right),$
$\quad\quad z' = z'_\infty + e^{-\omega' t} \left(B' + B'' t\right),$ (5.5b)

c) $\quad y' = y'_\infty + e^{-\omega' t} \left(A' \text{Cos} (\omega'' t) + A'' \text{Sin} (\omega'' t)\right),$
$\quad\quad z' = z'_\infty + e^{-\omega' t} \left(B' \text{Cos} (\omega'' t) + B'' \text{Sin} (\omega'' t)\right).$ (5.5c)

Setzt man die Lösungen (5.5) ohne die stationären Glieder y'_∞ und z'_∞ in (5.3) ein und betrachtet speziell den Zeitpunkt $t=0$, dann erhält man für die Koeffizienten A'' und B'' die Beziehungen:

a), c) $\quad A'' = \frac{1}{\omega''}\left[\left(\omega' - \frac{1}{T_2}\right)A' + \gamma H_1 B'\right], \quad B'' = \frac{1}{\omega''}\left[\left(\omega' - \frac{1}{T_1}\right)B' - \gamma H_1 A'\right]$ (5.6a, c)

b) $\quad A'' = \left(\omega' - \frac{1}{T_2}\right)A' + \gamma H_1 B', \quad B'' = \left(\omega' - \frac{1}{T_1}\right)B' - \gamma H_1 A'.$ (5.6b)

Die Koeffizienten A' und B' folgen aus den Anfangsbedingungen $y'_0 = 0$ und $z'_0 = 1$:

$$A' = -y'_\infty = -\frac{\gamma H_1 T_2}{1 + \gamma^2 H_1^2 T_1 T_2}; \quad B' = 1 - z'_\infty = \frac{\gamma^2 H_1^2 T_1 T_2}{1 + \gamma^2 H_1^2 T_1 T_2}. \quad (5.7)$$

Setzt man die Ausdrücke (5.6) und (5.7) für die Integrationskonstanten in (5.5) ein und schreibt zur Abkürzung

$$\tau = \gamma H_1 t, \quad \tau_1 = \gamma H_1 T_1 \quad \text{und} \quad \tau_2 = \gamma H_1 T_2,$$

so erhält man statt (5.5a) und (5.5c)

$$y' = \frac{\tau_2}{1 + \tau_1 \tau_2} \left\{ 1 - e^{-\frac{1}{2}\left(\frac{1}{\tau_1} + \frac{1}{\tau_2}\right)\tau} \times \right.$$

$$\left. \times \left[\begin{matrix} \cos \\ \text{Cos} \end{matrix} \sqrt{\pm 1 \mp \frac{1}{4}\left(\frac{1}{\tau_1} - \frac{1}{\tau_2}\right)^2}\, \tau - \frac{\tau_1 - \frac{1}{2}\left(\frac{1}{\tau_1} - \frac{1}{\tau_2}\right)}{\sqrt{\pm 1 \mp \frac{1}{4}\left(\frac{1}{\tau_1} - \frac{1}{\tau_2}\right)^2}} \begin{matrix} \sin \\ \text{Sin} \end{matrix} \sqrt{\pm 1 \mp \frac{1}{4}\left(\frac{1}{\tau_1} - \frac{1}{\tau_2}\right)^2}\, \tau \right] \right\}.$$

$$z' = \frac{1}{1 + \tau_1 \tau_2} \left\{ 1 + \tau_1 \tau_2\, e^{-\frac{1}{2}\left(\frac{1}{\tau_1} + \frac{1}{\tau_2}\right)\tau} \times \right.$$

$$\left. \times \left[\begin{matrix} \cos \\ \text{Cos} \end{matrix} \sqrt{\pm 1 \mp \frac{1}{4}\left(\frac{1}{\tau_1} - \frac{1}{\tau_2}\right)^2}\, \tau + \frac{\frac{1}{2}\left(\frac{1}{\tau_1} + \frac{1}{\tau_2}\right)}{\sqrt{\pm 1 \mp \frac{1}{4}\left(\frac{1}{\tau_1} - \frac{1}{\tau_2}\right)^2}} \begin{matrix} \sin \\ \text{Sin} \end{matrix} \sqrt{\pm 1 \mp \frac{1}{4}\left(\frac{1}{\tau_1} - \frac{1}{\tau_2}\right)^2}\, \tau \right] \right\}$$

(5.8a, c)

und statt (5.5b)

$$y' = \frac{\tau_2}{1 + \tau_1 \tau_2} \left\{ 1 - e^{-\frac{1}{2}\left(\frac{1}{\tau_1} + \frac{1}{\tau_2}\right)\tau} \left[1 - \left(\tau_1 - \frac{1}{2}\left(\frac{1}{\tau_1} - \frac{1}{\tau_2}\right)\right)\tau \right] \right\},$$

$$z' = \frac{1}{1 + \tau_1 \tau_2} \left\{ 1 + \tau_1 \tau_2\, e^{-\frac{1}{2}\left(\frac{1}{\tau_1} + \frac{1}{\tau_2}\right)\tau} \left[1 + \frac{1}{2}\left(\frac{1}{\tau_1} + \frac{1}{\tau_2}\right)\tau \right] \right\}.$$

(5.8b)

Diese Lösungen sollen nun diskutiert werden. Mit den obigen Abkürzungen lauten die stationären Lösungen (5.2):

$$y'_\infty = \frac{\tau_2}{1 + \tau_1 \tau_2}; \quad z'_\infty = \frac{1}{1 + \tau_1 \tau_2}.$$

Einsetzen der zweiten Gleichung in die erste liefert mit $T_1/T_2 = \tau_1/\tau_2 = n^2$

$$\frac{y'_\infty}{z'_\infty} = \tau_2 \quad \text{bzw.} \quad \frac{1}{z'_\infty} = 1 + \frac{\tau_1}{\tau_2}\tau_2^2 = 1 + n^2\left(\frac{y'_\infty}{z'_\infty}\right)^2.$$

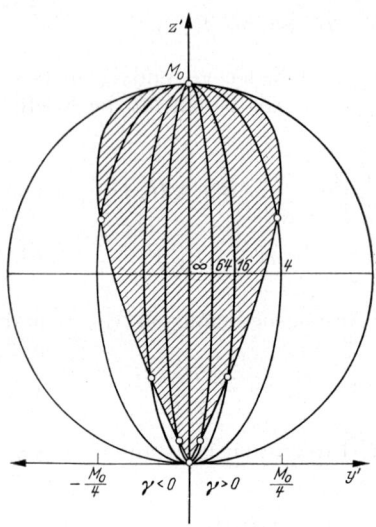

Fig. 21. Stationäre Lagen der Spitze des Magnetisierungsvektors in Resonanz. Ellipsenparameter: $T_1/T_2 = \tau_1/\tau_2 = 1, 4, 16, 64$ und ∞. Aperiodische Bewegungen enden innerhalb der Parabeln (graues Gebiet), periodische außerhalb.

Somit gilt:

$$\frac{y'^2_\infty}{\left(\frac{1}{2n}\right)^2} + \frac{\left(z'_\infty - \frac{1}{2}\right)^2}{\left(\frac{1}{2}\right)^2} = 1. \quad (5.9)$$

Diese Gleichung stellt eine Ellipsenschar mit den Halbachsen $\frac{1}{2}$ und $1/(2n)$ dar. In Fig. 21 wurden solche Ellipsen für $n = 1, 2, 4, 8$ und ∞, also für $T_1/T_2 = 1, 4, 16, 64$ und ∞ gezeichnet. Für $n = 1$ entarten die Ellipsen zu einem Kreis und für $n \to \infty$ gehen sie in das der z'-Achse angehörende Geradenstück $(0,1)$ über. Zu den oberen Scheitelpunkten ($y'_\infty = 0$, $z'_\infty = 1$) gehören die Relaxationszeiten $T_1 = T_2 = 0$ und zu den unteren Scheitelpunkten ($y'_\infty = 0$, $z'_\infty = 0$) die Zeiten $T_1 = T_2 = \infty$. In einem dieser Ellipsenpunkte endet immer die Bewegung der Spitze des resultierenden Momentes nach unendlich langer Anregungszeit. Der aperiodische Grenzfall ist bestimmt durch $\frac{1}{4}(1/\tau_1 - 1/\tau_2)^2 = 1$. Mit $\tau_2 \leq \tau_1$ gilt also für ihn: $\tau_2 = (n^2 - 1)/2n^2$ und $\tau_1 = (n^2 - 1)/2$ und damit:

$$y'_\infty = \frac{2(n^2 - 1)}{(n^2 + 1)^2}, \quad z'_\infty = \frac{4n^2}{(n^2 + 1)^2}.$$

Eliminiert man den Parameter n^2, dann erhält man:

$$(z'_\infty - y'_\infty)^2 = z'_\infty - 2y'_\infty, \quad (5.10)$$

die Gleichung zweier Parabeln (Fig. 21), deren Achsen mit der y'- und der z'-Achse den Winkel $45°$ einschließen. Diese Parabeln schneiden die z'-Achse in den Punkten 0 und 1 und haben dort die Steigungen $dz/dy = 2$ bzw. $= 0$. Für den aperiodischen Grenzfall galt: $T_2 = (n^2 - 1)/2n^2\gamma H_1$. Die Präzessionsbewegung ist periodisch, wenn T_2 größer als dieser Quotient, und aperiodisch, wenn T_2 kleiner als dieser Quotient ist. Bei allen Punkten innerhalb der Parabeln in Fig. 21 stellt sich der stationäre Zustand also nach einer aperiodischen Bewegung und bei allen Punkten außerhalb der Parabeln nach einer periodischen Bewegung ein. Ist speziell $n^2 = 1$, dann gilt im aperiodischen Grenzfall $T_2 = 0$, d.h. die Bewegung ist immer periodisch. Rechts von der z'-Achse ist im übrigen $\gamma > 0$ und links von ihr $\gamma < 0$.

Es sollen nun Grenzfälle des Relaxationsverhaltens betrachtet werden. Streben T_1 und $T_2 \to \infty$, dann gilt $(1/T_2 - 1/T_1)^2 \to 0$. Es genügt also, den periodischen Fall a) zu betrachten. Die Gln. (5.8a) gehen über in:

$$y' = -\sin\{\gamma H_1 t\}, \quad z' = \cos\{\gamma H_1 t\},$$

d.h., das Moment rotiert mit der Kreisfrequenz γH_1 um die x'-Achse in der y', z'-Ebene. Der Betrag des Momentes bleibt konstant eins. Diese anschaulich leicht zu übersehende Lösung wurde in Ziff. 4 benutzt. Die zugehörige, in Fig. 21 enthaltene stationäre Lösung — der Punkt 0, 0 — wird erst nach unendlich langer Zeit erreicht.

Im Falle starker Relaxation streben T_1 und $T_2 \to 0$. Die Gln. (5.8) gehen dann über in $y' = 0$ und $z' = 1$. Man kann dies leicht verstehen. Ein aus der z'-Richtung herausgedrehtes Moment verschwindet in den transversalen Richtungen sehr schnell, während es sich in der Feldrichtung sofort wieder neu bildet.

Ist $n^2 = 1$ $(T_1 = T_2 = T)$[1], die Relaxation also isotrop, dann sind die Bewegungen des Momentes, wie gezeigt, immer periodisch. Die Gln. (5.8a) erhalten die einfache Form:

$$y' = y'_\infty \left\{ 1 - e^{-\frac{t}{T}} \left[\cos\{\gamma H_1 t\} - \gamma H_1 T \sin\{\gamma H_1 t\} \right] \right\},$$

$$z' = z'_\infty \left\{ 1 + \gamma^2 H_1^2 T^2 e^{-\frac{t}{T}} \left[\cos\{\gamma H_1 t\} + \frac{1}{\gamma H_1 T} \sin\{\gamma H_1 t\} \right] \right\}$$

oder:

$$y' = y'_\infty - y'_\infty \sqrt{1 + \gamma^2 H_1^2 T^2} \, e^{-\frac{t}{T}} \cos\{\gamma H_1 t + \varphi_{y'}\} \quad \text{mit} \quad \tan \varphi_{y'} = \gamma H_1 T,$$

$$z' = z'_\infty + y'_\infty \sqrt{1 + \gamma^2 H_1^2 T^2} \, e^{-\frac{t}{T}} \{\sin \gamma H_1 t + \varphi_{z'}\} \quad \text{mit} \quad \tan \varphi_{z'} = \gamma H_1 T.$$

Quadrieren und Addieren ergibt

$$(z' - z'_\infty)^2 + (y' - y'_\infty)^2 = \frac{\gamma^2 H_1^2 T^2}{1 + \gamma^2 H_1^2 T^2} \, e^{-\frac{2t}{T}}. \tag{5.11}$$

Die Bahnkurve der Vektorspitze besteht demnach aus einer Folge von mit der Kreisfrequenz γH_1 durchlaufenen Kreisen um den stationären Endwert y'_∞, z'_∞, deren Radien mit $e^{-t/T}$ abnehmen. Das Moment beschreibt also eine logarithmische Spirale um die Endlage. Für $T = \infty$ beschreibt der Vektor um den Nullpunkt einen Kreis vom Radius 1. Man kann zeigen, daß auch für $0 < T < \infty$ und für $T_1 > T_2$ die Spitze des resultierenden Momentes niemals das Innere dieses Kreises verläßt.

Nach diesen Vorbereitungen sollen nun die allgemeinen Fälle diskutiert werden. Aus den Gln. (5.8a, c) kann man die sinus- und cosinus- bzw. Sinus- und Cosinusglieder eliminieren. Man erhält in beiden Fällen:

$$(y' - y'_\infty)^2 - \left(\frac{1}{\tau_2} - \frac{1}{\tau_1}\right)(y' - y'_\infty)(z' - z'_\infty) + (z' - z'_\infty) - \frac{y'^2_\infty}{z'_\infty} \frac{\tau_1}{\tau_2} e^{-2\omega' t} = 0, \tag{5.12}$$

die Gleichung einer Kegelschnittschar. Dreht man das Koordinatensystem um $45°$ und verschiebt den Nullpunkt in die stationäre Endlage (y'_∞, z'_∞) des Magnetisierungsvektors, dann geht (5.12) über in $(y' \to \eta, z' \to \zeta)$:

$$\pm \frac{\eta^2}{\dfrac{\tau_1 \cdot y'_\infty \cdot e^{-2\omega' t}}{\pm 1 \mp \dfrac{1}{2}\left(\dfrac{1}{\tau_2} - \dfrac{1}{\tau_1}\right)}} + \frac{\zeta^2}{\dfrac{\tau_1 \cdot y'_\infty \cdot e^{-2\omega' t}}{1 + \dfrac{1}{2}\left(\dfrac{1}{\tau_2} - \dfrac{1}{\tau_1}\right)}} = 1. \tag{5.13}$$

Darin sollen die oberen Vorzeichen gelten, wenn (a) $\gamma H_1 > \frac{1}{2}\left(\frac{1}{T_2} - \frac{1}{T_1}\right)$ und die unteren, wenn (c) $\gamma H_1 < \frac{1}{2}\left(\frac{1}{T_2} - \frac{1}{T_1}\right)$ ist.

---

[1] Die absolute Temperatur T kommt in diesem Abschnitt nicht vor, Symbolverwechslungen sind also wohl nicht möglich.

Fall a). Periodische Einschwingvorgänge. Die Gl. (5.13) beschreibt Ellipsen, deren Halbachsen mit der Zeitkonstanten $\frac{1}{\omega'} = 2\,\frac{T_1 T_2}{T_1 + T_2}$ exponentiell abnehmen. Der Magnetisierungsvektor setzt sich zusammen aus einem konstanten, den Nullpunkt des y', z'-Systems mit dem stationären Endpunkt y'_∞, z'_∞ verbindenden Vektor und aus einem um diesen Endpunkt mit periodisch schwankender Winkelgeschwindigkeit rotierenden Vektor abnehmenden Betrages. Seine Spitze beschreibt eine logarithmische Spirale, die jedoch, dem Achsenverhältnis der Ellipse entsprechend, affin verzerrt ist. Fig. 22 zeigt ein Beispiel eines solchen Bewegungsverlaufes. Für $T_1 = T_2$ werden die Ellipsen zu Kreisen und die Winkelgeschwindigkeit konstant gleich γH_1.

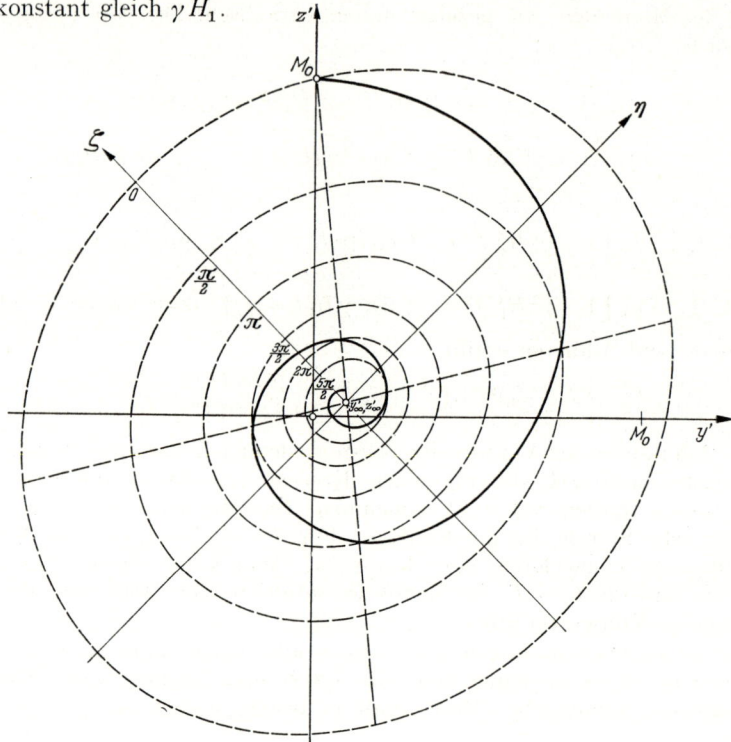

Fig. 22. Bahnkurve einer periodischen Bewegung der Spitze des Magnetisierungsvektors in Resonanz $(\frac{1}{4}\,(1/T_1 - 1/T_2)^2 < \gamma^2 H_1^2)$. Es wurde angenommen: $\gamma H_1 T_2 = 2,4$; $T_1/T_2 = 4$. Daraus folgt: $y'_\infty = 0,1 \cdot M_0$; $z'_\infty = 0,044 \cdot M_0$; $\omega' = 0,625/T_2$; $\omega'' = 2,36/T_2$ Ellipsenparameter: $\omega'' t = 2,36 \cdot t/T_2 = 0$, $\pi/2$, π, $3\pi/2$, 2π, $5\pi/2$.

Fall c). Aperiodische Einschwingvorgänge. Die Gl. (5.13) beschreibt Hyperbeln, deren Halbachsen ebenfalls mit der Zeitkonstanten $1/\omega'$ exponentiell abnehmen. Man kann sich die Bewegung des um den stationären Endpunkt rotierenden Vektors wie folgt vorstellen (Fig. 23). Einmal wandert seine Spitze im Punkt (0,1) der ersten Hyperbel beginnend den rechten Hyperbelästen entlang. Gleichzeitig nimmt aber sein Betrag sehr schnell ab. Der Vektor beschreibt also nur einen kurzen Bogen, an dessen Ende (also nach unendlich langer Zeit) seine Richtung mit derjenigen der Hyperbelasymptoten übereinstimmt. Sein Betrag ist in diesem Zeitpunkt aber schon Null geworden.

Fall b). Der aperiodische Grenzfall. Eliminiert man aus (5.8b) das zur Zeit t proportionale Glied, dann erhält man die Gleichung:

$$(y' - y'_\infty) - (z' - z'_\infty) + \frac{1}{2}\left(\frac{\tau_1}{\tau_2} + 1\right) y'_\infty \, \mathrm{e}^{-\omega' t} = 0. \tag{5.14}$$

Sie stellt eine Geradenschar dar, deren Steigungswinkel $45°$ beträgt. Die erste dieser Geraden geht durch den Anfangspunkt $(0, 1)$ der Bewegung, die letzte durch den stationären Endpunkt. Der Bewegungsverlauf läßt sich ebenso wie der im Falle c) besprochene beschreiben. Fig. 24 zeigt ein Beispiel. Die Tangente der Bahnkurve ist in allen drei Fällen im Zeitpunkt 0, also im Ortspunkt $(0,1)$

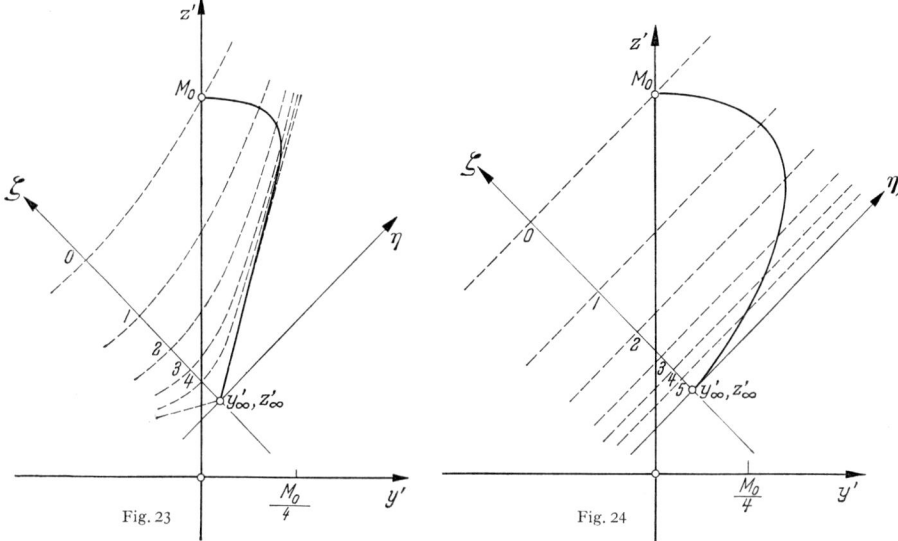

Fig. 23.   Fig. 24

Fig. 23. Bahnkurve einer aperiodischen Bewegung der Spitze des Magnetisierungsvektors in Resonanz ($\frac{1}{4}(1/T_1 - 1/T_2)^2 > \gamma^2 H_1^2$). Es wurde angenommen: $\gamma H_1 T_2 = 0,25$; $T_1/T_2 = 64$. Daraus folgt: $y'_\infty = 0,05 \cdot M_0$; $z'_\infty = 0,2 \cdot M_0$; $\omega' = 0,508/T_2$; $\omega'' = 0,424/T_2$. Hyperbel-Parameter: $t/T_2 = 0, 1, 2, 3, 4$.

Fig. 24. Bahnkurve der Spitze des Magnetisierungsvektors im aperiodischen Grenzfall in Resonanz ($\frac{1}{4}(1/T_1 - 1/T_2)^2 = \gamma^2 H_1^2$). Es wurde angenommen: $\gamma H_1 T_2 = 0,468$; $T_1/T_2 = 16$. Daraus folgt: $y'_\infty = 0,104 \cdot M_0$; $z'_\infty = 0,221 \cdot M_0$; $\omega' = 0,531/T_2$. Parameter der Geraden: $t/T_2 = 0, 1, 2, 3, 4, 5$.

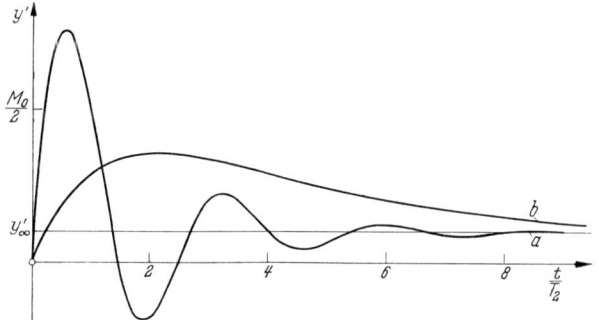

Fig. 25. Verlauf der y'-Komponente des Magnetisierungsvektors. a Periodische Bewegung: Daten siehe Fig 22. b Aperiodischer Grenzfall: Daten siehe Fig. 24.

horizontal. Ihre Tangente im Augenblick der Einmündung in einen dem aperiodischen Grenzfall angehörenden stationären Endwert hat stets den Steigungswinkel $45°$.

Beobachtbar ist im Versuch nur die y'-Komponente des Magnetisierungsvektors (eine resultierende x'-Komponente existiert in Resonanz nicht). Deren Verlauf ist daher in Fig. 25 gezeichnet worden. H. C. TORREY[1] hat dem Fall periodischer Anregung entsprechende Kurven experimentell erhalten.

---

[1] H. C. TORREY: Phys. Rev. 76, 1059 (1949).

III. Präzessionsbewegungen in einem modulierten Magnetfeld.

6. Der langsame adiabatische Resonanzdurchgang. Bei den meisten Verfahren zur Untersuchung kernmagnetischer Präzessionsbewegungen werden die Kernsignale gleichzeitig angeregt und beobachtet. Um die von dem rotierenden Magnetisierungsvektor induzierte Signalspannung leicht verstärken zu können, moduliert man ihre Amplitude niederfrequent, meist mit einer Frequenz von etwa 30 Hz. Grundsätzlich sind drei Modulationsarten möglich. Man kann die Feldstärke H_1, also die Amplitude des Senders — etwa in Resonanz — modulieren, man kann seine Frequenz, also den Resonanzabstand, periodisch variieren, und man kann drittens der Feldstärke H_0 ein niederfrequent moduliertes Feld überlagern und damit ebenfalls den Resonanzabstand periodisch ändern. Aus technischen Gründen zieht man eine der beiden zuletzt genannten, im übrigen physikalisch äquivalenten Methoden vor.

Aus Ziff. 5 wissen wir: Ändern sich in einem Versuch von einem bestimmten Zeitpunkt an weder die Amplitude H_1, noch die Frequenz des Hochfrequenzgenerators, noch die Feldstärke H_0 des Magneten, dann stellt sich theoretisch nach unendlich langer Dauer, praktisch aber nach einer endlichen Zeit, der konstante stationäre Endwert des Magnetisierungsvektors ein. Die Zeit Δt, die bis zur Einstellung des stationären Zustandes vergeht, ist um so kleiner, je weniger sich der neue stationäre Wert von dem vorangegangenen unterscheidet, je kleiner also die Änderung des Resonanzabstandes $\Delta(\omega_0 - \omega) = \gamma \Delta(H_0 - H)$ ist. Ferner wissen wir aus den Gln. (5.8), daß eine instationäre Bewegung um so schneller verschwindet, je größer ihre Dämpfung, je größer also die Linienbreite beim Resonanzdurchgang ist. Eine genauere Untersuchung zeigt[1]: Erfolgt die Änderung des Resonanzabstandes so langsam, daß dabei stets die Ungleichung:

$$\frac{d\omega}{dt} \ll (\Delta\omega_h)^2 \qquad \text{bzw.} \qquad \frac{dH}{dt} \ll |\gamma| (\Delta H_h)^2 \qquad (6.1)$$

($\Delta\omega_h$ = Linienhalbwertsbreite in Frequenzeinheiten, ΔH_h = Linienhalbwertsbreite in Feldstärkeeinheiten) erfüllt bleibt, dann wird die Präzessionsbewegung des resultierenden Moments zu einem stetigen Durchlaufen stationärer Endwerte, und man spricht von einem adiabatischen Resonanzdurchgang. Darüber hinaus bezeichnet man einen solchen Resonanzdurchgang als langsam adiabatisch, wenn im Experiment die Modulationsfrequenz ω_m so klein gewählt wird, daß die beiden kernmagnetischen Relaxationszeiten klein gegen die Dauer eines Resonanzdurchgangs sind:

$$\frac{1}{\omega_m} \gg T_1, T_2 . \qquad (6.2)$$

Die stetig durchlaufenen stationären Lagen folgen unter diesen Bedingungen aus den Differentialgleichungen (2.6) der gedämpften Präzessionsbewegungen.

In Gl. (5.2) wurden die Komponenten des stationären Magnetisierungsvektors im Resonanzfall als Funktion der experimentellen Daten berechnet. Dies soll jetzt verallgemeinert werden. Dazu setzen wir in den BLOCHschen Differentialgleichungen

$$H_x = H_1 \cos\omega t \qquad H_y = -H_1 \sin\omega t \qquad (\text{es sei } \gamma > 0) \qquad (6.3)$$

und transformieren wiederum in ein Koordinatensystem x', y', z', das im Drehsinn der LARMOR-Präzession (ω_L), aber mit der Frequenz ω des Feldes H_1 mitrotiert. Dabei soll die positive x'-Achse parallel zur Richtung des Feldes liegen. Zwischen den Magnetisierungskomponenten im raumfesten System x, y, z und

---

[1] B. A. JACOBSOHN u. R. K. WANGSNESS: Phys. Rev. **73**, 942 (1948).

im rotierenden System x', y', z' gelten dann die Gleichungen

$$\left.\begin{aligned} M_{x'} &= M_x \cos \omega t - M_y \sin \omega t, \\ M_{y'} &= M_x \sin \omega t + M_y \cos \omega t \end{aligned}\right\} \text{ bzw. } \left\{\begin{aligned} M_x &= M_{x'} \cos \omega t + M_{y'} \sin \omega t, \\ M_y &= -M_{x'} \sin \omega t + M_{y'} \cos \omega t \end{aligned}\right\} \quad (6.4\,\text{a, b})$$

$$(M_z = M_{z'}).$$

Die Gln. (2.6) haben im x', y', z'-System die Form:

$$\left.\begin{aligned} \dot{M}_{x'} &= + (\omega_L - \omega)\, M_{y'} - \frac{M_{x'}}{T_2}, \\ \dot{M}_{y'} &= - (\omega_L - \omega)\, M_{x'} - \frac{M_{y'}}{T_2} + \omega_1 M_{z'}, \\ \dot{M}_{z'} &= - \omega_1 M_{y'} - \frac{1}{T_1}(M_{z'} - M_0). \end{aligned}\right\} \quad (6.5)$$

Im stationären Zustand ist $\dot{M}_{x'} = \dot{M}_{y'} = \dot{M}_{z'} = 0$. Daraus folgen die stationären Lösungen[1] ($\omega_1 = \gamma H_1$):

$$M_{x'\infty} = \frac{\gamma\, H_1 T_2^2\, \Delta\omega\, M_0}{1 + \gamma^2 H_1^2 T_1 T_2 + (\Delta\omega)^2 T_2^2}, \quad (6.6\,\text{a})$$

$$M_{y'\infty} = \frac{\gamma\, H_1 T_2\, M_0}{1 + \gamma^2 H_1^2 T_1 T_2 + (\Delta\omega)^2 T_2^2}, \quad (\Delta\omega = \omega_L - \omega) \quad (6.6\,\text{b})$$

$$M_{z'\infty} = \frac{(1 + (\Delta\omega)^2 T_2^2)\, M_0}{1 + \gamma^2 H_1^2 T_1 T_2 + (\Delta\omega)^2 T_2^2}, \quad (6.6\,\text{c})$$

und mit Hilfe von (6.4) im raumfesten System:

$$\left.\begin{aligned} M_{x\infty} &= \frac{\frac{1}{2}\chi_0 \omega_0 T_2 (2H_1 \Delta\omega\, T_2 \cos \omega t + 2H_1 \sin \omega t)}{1 + \gamma^2 H_1^2 T_1 T_2 + (\Delta\omega)^2 T_2^2}, \\ M_{y\infty} &= \frac{\frac{1}{2}\chi_0 \omega_0 T_2 (2H_1 \cos \omega t - 2H_1 \Delta\omega\, T_2 \sin \omega t)}{1 + \gamma^2 H_1^2 T_1 T_2 + (\Delta\omega)^2 T_2^2}, \end{aligned}\right\} \begin{aligned} &(M_0 = \chi_0 H_0, \\ &\;\omega_0 = \gamma H_0). \end{aligned} \quad (6.7\,\text{a, b})$$

Formal kann man die stationäre Magnetisierung (6.7) auch durch die Einführung einer komplexen, sog. dynamischen Suszeptibilität

$$\chi = \chi' - i\chi'' \quad (6.8)$$

beschreiben. Beachtet man, daß das anregende Feld tatsächlich linear in Richtung der x-Achse polarisiert ist, und die maximale Amplitude $2H_1$ hat, so findet man für die Komponenten von χ:

$$\left.\begin{aligned} \chi' &= \frac{1}{2}\chi_0 \omega_0 T_2 \frac{\Delta\omega\, T_2}{1 + \gamma^2 H_1^2 T_1 T_2 + (\Delta\omega)^2 T_2^2} = \frac{1}{2H_1} M_{x'\infty}, \\ \chi'' &= \frac{1}{2}\chi_0 \omega_0 T_2 \frac{1}{1 + \gamma^2 H_1^2 T_1 T_2 + (\Delta\omega)^2 T_2^2} = \frac{1}{2H_1} M_{y'\infty}. \end{aligned}\right\} \quad (6.9\,\text{a, b})$$

Von einem Faktor $1/(2H_1)$ abgesehen sind also die der Beobachtungsebene angehörenden Komponenten des stationären Magnetisierungsvektors im mitrotierenden Koordinatensystem identisch mit den Real- und Imaginäranteilen der komplexen Suszeptibilität.

Der Verlauf eines Resonanzdurchgang-Experimentes hängt davon ab, ob das Produkt $\gamma^2 H_1^2 T_1 T_2 \lessgtr 1$ ist. Dieser Ausdruck ist offenbar ein Maß dafür, ob man die Kernsignale mit einer schwachen, mittleren oder starken Hochfrequenzfeldstärke H_1 anregt. Genauer gesagt, gibt er an, welchen Sättigungsgrad das Kernspin-System erreicht. Die Sättigung und damit die Temperatur eines solchen

---

[1] F. Bloch: Phys. Rev. **70**, 460 (1946).

Systems sind um so höher, je größer die anregende Feldstärke, je größer das gyromagnetische Verhältnis der Kerne und je schwächer seine Kopplungen zu den übrigen Substanzfreiheitsgraden sind. Um so mehr geht das System vom Zustand der Gleichgewichtsverteilung in den unendlich hoher Spin-Temperatur zuzuordnenden Zustand der Gleichverteilung über.

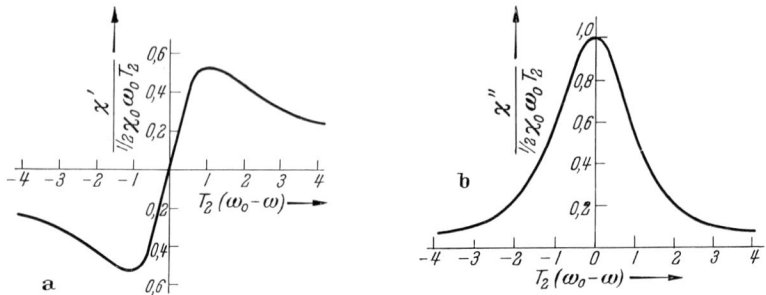

Fig. 26 a u. b. Abhängigkeit der Dispersion (a) und Absorption (b) einer kernmagnetischen Substanz vom Resonanzabstand; $\gamma^2 H_1^2 T_1 T_2 \ll 1$; berechnet nach Gl. (6.10a, b).

Wir betrachten zunächst den Fall schwacher Hochfrequenzfeldstärke $\gamma^2 H_1^2 T_1 T_2 \ll 1$. Die Gln. (6.9a, b) erhalten damit die einfachere Form:

$$\begin{aligned}
\chi' &= \frac{1}{2H_1} M_{x'\infty} = \frac{1}{2}\chi_0\omega_0 T_2\left[\frac{\Delta\omega T_2}{1+(\Delta\omega)^2 T_2^2}\right], \\
\chi'' &= \frac{1}{2H_1} M_{y'\infty} = \frac{1}{2}\chi_0\omega_0 T_2\left[\frac{1}{1+(\Delta\omega)^2 T_2^2}\right].
\end{aligned}$$
(6.10a, b)

In den Fig. 26a und b wurden die durch (6.10a, b) beschriebenen Kurven aufgetragen[1]. Sie sind mit den Dispersions- bzw. Absorptionskurven eines gedämpften und erzwungen angeregten Oszillators identisch. Die Fig. 27a und b zeigen experimentelle Registrierungen[2].

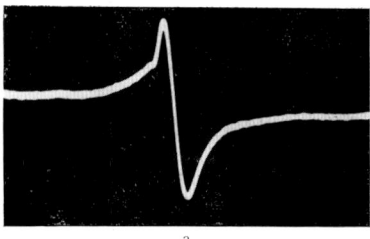
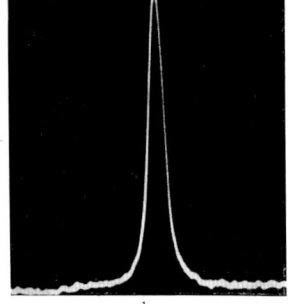

Fig. 27 a u. b. Protonenresonanz beobachtet in einer Eisennitratlösung. a Dispersionskurve. b Absorptionskurve (N. BLOEMBERGEN, E. M. PURCELL und R. V. POUND).

Der Imaginärteil χ'' der Suszeptibilität und damit die Absorption bzw. die zum Hochfrequenzfeld H_1 senkrechte y'-Komponente des stationären Magnetisierungsvektors erreichen ihre Maximalwerte $\frac{1}{2}\chi_0\omega_0 T_2$ bzw. $\gamma H_1 T_2 M_0$ in Resonanz, also für $\omega_L = \omega$. Die Halbwertsbreite $\Delta\omega_h$ eines beim Resonanzdurchgang erhaltenen Absorptionskernsignals folgt aus

$$\frac{1}{1+\left(\frac{\Delta\omega_h}{2}\right)^2 T_2^2} = \frac{1}{2} \quad \text{zu} \quad \Delta\omega_h = \frac{2}{T_2}.$$
(6.11)

[1] G. E. PAKE: Amer. J. Phys. 18, 438 (1950).
[2] N. BLOEMBERGEN, E. M. PURCELL u. R. V. POUND: Phys. Rev. 73, 679 (1948).

Die Extrema der Dispersionskurve (Fig. 26a) erhält man, wenn man die eckige Klammer von (6.10a) differenziert. Ihre Abszissen sind $\Delta\omega = \pm 1/T_2$, sie stimmen also mit den Halbwertsabszissen der Absorptionskurve überein. Der Betrag der Dispersionsextrema ist $\pm\frac{1}{4}\chi_0\omega_0 T_2$ bzw. $\pm\frac{1}{2}\gamma H_1 T_2 M_0$.

Kann das Produkt $\gamma^2 H_1^2 T_1 T_2$ nicht vernachlässigt werden, oder wird es gar größer als eins, dann verschieben sich die Extrema von $\chi'(\sim M_{x'\infty})$ und entsprechend die Halbwertsbreiten von $\chi''(\sim M_{y'\infty})$ nach außen. Die Halbwertsbreite ist dann auch von der Hochfrequenzfeldstärke abhängig. Aus (6.9) erhält man:

$$\Delta\omega_h = \frac{2}{T_2}\sqrt{1 + \gamma^2 H_1^2 T_1 T_2}. \qquad (6.12)$$

Im Absorptionsmaximum ist

$$\chi'' = \frac{\frac{1}{2}\chi_0\omega_0 T_2}{1 + \gamma^2 H_1^2 T_1 T_2} \quad \text{bzw.} \quad M_{y'\infty} = \frac{\gamma H_1 T_2 M_0}{1 + \gamma^2 H_1^2 T_1 T_2}. \qquad (6.13\,\text{a, b})$$

In der Literatur wird der Bruch $s = 1/(1 + \gamma^2 H_1^2 T_1 T_2)$ übrigens als Sättigungsfaktor bezeichnet[1]. Der Abstand der Extrema der Dispersionskurve von der Resonanzstelle ist ebenfalls $\frac{1}{T_2}\sqrt{1 + \gamma^2 H_1^2 T_1 T_2}$, ihre Höhe dagegen:

$$\chi' = \pm\frac{\frac{1}{2}\chi_0\omega_0 T_2}{2\sqrt{1 + \gamma^2 H_1^2 T_1 T_2}} \quad \text{bzw.} \quad M_{x'\infty} = \pm\frac{\gamma H_1 T_2 M_0}{2\sqrt{1 + \gamma^2 H_1^2 T_1 T_2}}. \qquad (6.14\,\text{a, b})$$

Ist $\gamma^2 H_1^2 T_1 T_2 \gg 1$, dann wird die Halbwertsbreite des Absorptionssignals bzw. der Abstand der Dispersionsextrema:

$$\Delta\omega_h = 2\gamma H_1\sqrt{\frac{T_1}{T_2}}. \qquad (6.15)$$

Wir fragen jetzt nach der für die Beobachtung der Kernsignale günstigsten Hochfrequenzfeldstärke H_1. Die größten Suszeptibilitätsänderungen erhält man beim Resonanzdurchgang gemäß (6.9), wenn $\gamma^2 H_1^2 T_1 T_2$ verschwindend klein ist, H_1 also gegen Null strebt. Beobachtet wird aber immer die von den Kernen induzierte Spannung, und diese ist proportional zu dem Betrag des stationären resultierenden Momentes in der x', y'-Ebene und damit nach Gl. (6.6) proportional zu H_1. Differenzieren von (6.13b) nach H_1 zeigt, daß die zur Beobachtung von Absorptionssignalen optimale Hochfrequenzfeldstärke den Betrag

$$H_{1,\text{opt}} = \frac{1}{\gamma\sqrt{T_1 T_2}} \qquad (\gamma^2 H_{1,\text{opt}}^2 T_1 T_2 = 1) \qquad (6.16)$$

hat. Das resultierende Moment in diesem Fall (die x'-Komponente verschwindet in Resonanz)

$$M_{x',y'}^{\text{opt}} = \frac{1}{2} M_0\sqrt{\frac{T_2}{T_1}} \qquad (6.17)$$

ist unter der Voraussetzung $T_1 \geq T_2$ am größten, wenn $T_1 = T_2$ ist.

Die Gl. (6.14b) zeigt, daß die Disperionsextrema für $H_1 \to \infty$ sich asymptotisch monoton den Grenzwerten $\pm\frac{1}{2}M_0\sqrt{T_2/T_1}$ nähern. Zur Beobachtung des Dispersionsverlaufes ist es also zweckmäßig, die Kernsignale mit einem möglichst starken Feld H_1 anzuregen. Gleichzeitig wächst dann aber auch in (6.15) der Abstand der Dispersionsextrema, während die y'-Komponente des stationären Magnetisierungsvektors immer mehr vernachlässigt werden kann.

---

[1] Manche Autoren bezeichnen das Produkt $\gamma^2 H_1^2 T_1 T_2$ als Sättigungsfaktor.

Als nächstes soll der Verlauf der Präzessionsbewegung beim Resonanzdurchgang besprochen werden. Aus (6.10) erhält man das beobachtbare transversale Moment, wenn $\gamma^2 H_1^2 T_1 T_2 \ll 1$ ist:

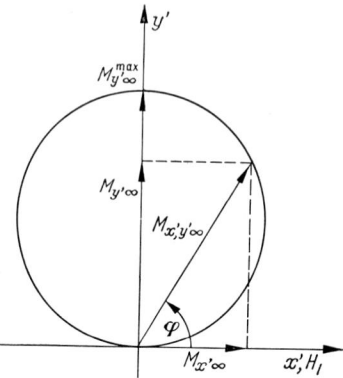

$$M_{x',y'\infty} = \sqrt{M_{x'\infty}^2 + M_{y'\infty}^2} = \frac{\gamma H_1 T_2 M_0}{\sqrt{1 + (\Delta\omega)^2 T_2^2}}.$$

In diesem Fall ist der Maximalwert von $M_{x'y'\infty}$ gleich $M_{y'\infty}^{\max} = \gamma H_1 T_2 M_0$. Bezeichnet man mit φ den Winkel zwischen dem Magnetisierungsvektor $M_{x',y'\infty}$ und der zum rotierenden Feld H_1 stets parallelen x'-Richtung (Fig. 28), dann ist:

$$\sin\varphi = \frac{M_{y'\infty}}{M_{x',y'\infty}} = \frac{1}{\sqrt{1 + (\Delta\omega)^2 T_2^2}}.$$

Man kann für $M_{x',y'\infty}$ also auch schreiben:

$$M_{x',y'\infty} = M_{y'\infty}^{\max} \sin\varphi. \tag{6.18}$$

Fig. 28. Verlauf der Präzessionsbewegung des resultierenden transversalen Magnetisierungsvektors bei einem langsamen Resonanzdurchgang. $\gamma^2 H_1^2 T_1 T_2 \ll 1$; φ Phasenverschiebung zwischen H_1 und $M_{x',y'\infty}$.

Weit außerhalb Resonanz ist die transversale Magnetisierung verschwindend klein und die Phasendifferenz zwischen ihr und dem anregenden Feld Null. In Resonanz erreicht sie ihren maximalen Betrag, und die Phasendifferenz wird 90°. Nach dem Resonanzdurchgang nimmt die Magnetisierung wieder ab, während die Phasendifferenz bis 180° anwächst. Auch in dieser Beziehung verhält sich ein schwach angeregter Magnetisierungsvektor wie ein gezwungen angeregter gedämpfter Oszillator.

Für beliebige Feldstärken H_1 folgt aus (6.9), wenn man den Resonanzabstand $\Delta\omega$ eliminiert:

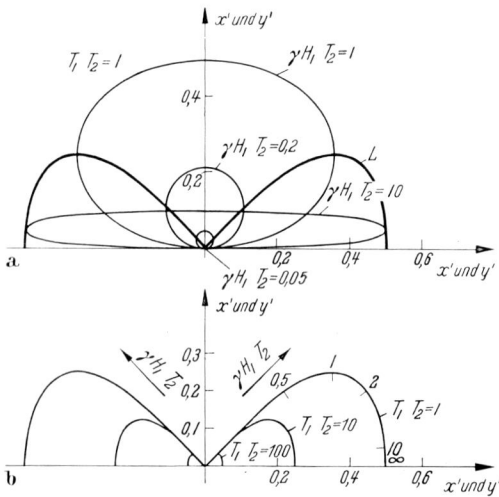

$$\left. \frac{M_{x'\infty}^2}{\left(\frac{\gamma H_1 T_2 M_0}{2\sqrt{1 + \gamma^2 H_1^2 T_1 T_2}}\right)^2} + \frac{\left(M_{y'\infty} - \frac{\gamma H_1 T_2 M_0}{2(1 + \gamma^2 H_1^2 T_1 T_2)}\right)^2}{\left(\frac{\gamma H_1 T_2 M_0}{2(1 + \gamma^2 H_1^2 T_1 T_2)}\right)^2} = 1 \right\} \tag{6.19}$$

die Gleichung einer Ellipsenschar. In Fig. 29a wurden solche Ellipsen für $T_1 = T_2$ gezeichnet. Die als Leitkurve der Ellipsenschar bezeichnete, in Fig. 29a stark ausgezogene, Kurve verbindet

Fig. 29a u. b. Bahnkurven der Spitze des Magnetisierungsvektors beim langsamen Resonanzdurchgang, wenn das Produkt $\gamma^2 H_1^2 T_1 T_2$ nicht vernachlässigt werden kann. a) $T_1/T_2 = \tau_1/\tau_2 = n^2 = 1$. b) Lagen der Ellipsenscheitel, berechnet nach (6.19).

alle äußeren Ellipsenscheitel miteinander. Scharparameter ist das Produkt $\gamma H_1 T_2$, praktisch also die Feldstärke H_1. Fig. 29b zeigt die Leitkurven der Ellipsenscharen $T_1/T_2 = 1$, 10 und 100. Ebenso wie in Fig. 28 ist in Resonanz die Phasenverschiebung zwischen dem Feld und der Magnetisierung stets 90°. Der gemeinsame Berührungspunkt aller Ellipsen, der Koordinatennullpunkt, beschreibt den Zustand des Kernspin-Systems weit außerhalb Resonanz.

Mit Hilfe der Beziehung (6.6c) kann man übrigens die Temperatur eines durch ein Hochfrequenzfeld der Feldstärke H_1 angeregten Kernspin-Systems bestimmen. Im dynamischen Gleichgewicht, wenn also die Absorptions- und die Relaxations-Elementarprozesse in dem System gleich häufig erfolgen, hat nämlich die Komponente des Magnetisierungsvektors in Richtung des äußeren Felds (also parallel zur z-Achse) nach (6.6c) den Betrag

$$M_{z,\,\text{dynam.-stationär}} = M_{z'} {}_\infty = \frac{1 + (\Delta\omega)^2 T_2^2}{1 + \gamma^2 H_1^2 T_1 T_2 + (\Delta\omega)^2 T_2^2} M_0. \qquad (6.20)$$

Der Betrag der statischen Magnetisierung (ohne Anregung durch ein Feld H_1) nimmt umgekehrt proportional mit der Temperatur der Substanz ab, in welcher das Kernspin-System enthalten ist und zwar gilt nach (1.7)

$$M_{z,\,\text{statisch}} = \frac{N\mu^2}{3K}\, \frac{I+1}{I}\, \frac{H_0}{T_{\text{Subst}}}. \qquad (6.21)$$

In dem bei der Ableitung von (6.21) vorausgesetzten thermischen BOLTZMANN-Gleichgewicht sind die Temperaturen aller Freiheitsgrade der Substanz gleich groß, insbesondere ist die Temperatur des Kernspin-Systems T_{Spin} gleich der Temperatur T_{Subst} aller anderen Freiheitsgrade der Substanz. Dagegen ist die Temperatur eines durch Absorption von Quanten der LARMOR-Frequenz angeregten Kernspin-Systems höher als die Temperatur der Substanz, in welcher das System enthalten ist, denn der Orientierungs-Freiheitsgrad der Kerne enthält mehr Energie als im statischen Gleichgewicht. Durch Gl. (6.21) sind Temperatur und Magnetisierung umkehrbar eindeutig verknüpft. Mann kann daher die Magnetisierung als Maß der Kernspin-Temperatur verwenden und Gl. (6.21) zur Definitionsgleichung der Temperatur des kernmagnetischen Freiheitsgrads verallgemeinern:

$$T_{\text{Spin}} = \frac{N\mu^2}{3K}\, \frac{I+1}{I}\, \frac{H_0}{M_z} = \frac{M_0}{M_z} \cdot T_{\text{Subst}}. \qquad (6.22)$$

Aus den Gln. (6.20) und (6.22) folgt für das Verhältnis Kernspin-Temperatur zu Substanz-Temperatur in Resonanz ($\Delta\omega = 0$) z.B. die Beziehung:

$$T_{\text{Spin}}/T_{\text{Subst.}} = 1 + \gamma^2 H_1^2 T_1 T_2. \qquad (6.23)$$

Im Fall der optimalen Anregung [vgl. (6.16)] ist demnach die Kernspin-Temperatur doppelt so groß wie die Substanz-Temperatur.

7. Der schnelle adiabatische Resonanzdurchgang. Man bezeichnet einen Resonanzdurchgang als schnell, wenn seine Dauer klein gegen die beiden Relaxationszeiten T_1 und T_2 ist. Soll er zugleich adiabatisch, also als stetige Folge stationärer Lagen des Magnetisierungsvektors verlaufen, dann muß man im Experiment dafür sorgen, daß die Ungleichung (6.1) stets erfüllt wird. Die zeitliche Änderung des Resonanzabstandes $d\omega/dt$ ist näherungsweise proportional zur Modulationsfrequenz ω_m; denn die Ableitung der Modulationsfunktion $\cos\omega_m t$ hat an deren Nullstellen den Betrag ω_m. Als Definitionsgleichung eines adiabatischen schnellen Resonanzdurchganges kann man somit die Bedingungen anschreiben:

$$\frac{d\omega}{dt} \ll (\Delta\omega_h)^2 \quad \text{und} \quad \frac{1}{\omega_m} \ll T_1, T_2. \qquad (7.1)$$

In dem zuvor besprochenen Fall des langsamen Resonanzdurchganges wurde $T_1, T_2 \ll 1/\omega_m$ vorausgesetzt. In der Regel ist damit aber auch die Ungleichung (6.1) mit erfüllt.

In dem jetzt zu besprechenden Fall zwingt die doppelte Ungleichung (7.1) zur Anwendung starker Hochfrequenzamplituden H_1 — die Linienbreite $\Delta\omega_h$ wird dadurch größer [Gl. (6.15)] — und beschränkt außerdem die Anwendbarkeit des Verfahrens auf Substanzen mit langen Relaxationszeiten. Um schnell angeregte Präzessionsbewegungen dieser Art diskutieren zu können, sollen noch auf einem weiteren Weg quasistationäre Lösungen der BLOCHschen Differentialgleichungen abgeleitet werden. Dazu setzen wir in (2.6) $H_x = H_1\cos\omega t$ und $H_y = -H_1\sin\omega t$ ein (es sei $\gamma > 0$) und erhalten:

$$\dot{M}_x = \gamma\,(M_y H_0 + M_z H_1 \sin\omega\,t) - \frac{1}{T_2}\,M_x, \tag{7.2a}$$

$$\dot{M}_y = \gamma\,(M_z H_1 \cos\omega\,t - M_x H_0) - \frac{1}{T_2}\,M_y, \tag{7.2b}$$

$$\dot{M}_z = -\gamma\,(M_x H_1 \sin\omega\,t + M_y H_1 \cos\omega\,t) - \frac{1}{T_1}\,(M_z - M_0). \tag{7.2c}$$

Für die Lösungen dieses Systems kann man ansetzen:

$$M_x = M_{x,y}(t)\cos\omega\,t, \qquad M_y = -M_{x,y}(t)\sin\omega\,t. \tag{7.3}$$

Das gesamte resultierende Moment ist:

$$M^2 = M_{x,y}^2 + M_z^2,$$

folglich gilt:

$$M_{x,y} = \frac{M}{\sqrt{1 + \left(\frac{M_z}{M_{x,y}}\right)^2}} \quad \text{bzw.} \quad M_z = \frac{M}{\sqrt{1 + \left(\frac{M_{x,y}}{M_z}\right)^2}}. \tag{7.4}$$

Setzt man (7.3) in die Differentialgleichungen (7.2a, b) ein, dann findet man:

$$\left[\dot{M}_{x,y} + \frac{1}{T_2}M_{x,y}\right]\frac{\cos\omega t}{\gamma H_1} + \left[\frac{H_0 - \frac{\omega}{\gamma}}{H_1}M_{x,y} - M_z\right]\sin\omega\,t = 0,$$

$$-\left[\dot{M}_{x,y} + \frac{1}{T_2}M_{x,y}\right]\frac{\sin\omega t}{\gamma H_1} + \left[\frac{H_0 - \frac{\omega}{\gamma}}{H_1}M_{x,y} - M_z\right]\cos\omega\,t = 0.$$

Es ist also:

$$\frac{M_z}{M_{x,y}} = \frac{H_0 - \frac{\omega}{\gamma}}{H_1} = \frac{\Delta\omega}{\gamma H_1} = u \tag{7.5a}$$

und

$$\dot{M}_{x,y} + \frac{1}{T_2}M_{x,y} = 0. \tag{7.5b}$$

Ferner ergibt sich durch Einsetzen von (7.3) in (7.2c):

$$\dot{M}_z + \frac{1}{T_1}M_z = \frac{1}{T_1}M_0. \tag{7.6}$$

Aus (7.4) und (7.5) folgt:

$$M_{x,y} = M_{x'} = \frac{M}{\sqrt{1+u^2}}, \qquad M_z = M_{z'} = \frac{uM}{\sqrt{1+u^2}}. \tag{7.7}$$

Differenziert man (7.4) und setzt (7.5b) und (7.6) ein, dann erhält man für M die Differentialgleichung:

$$\dot{M} + \frac{\dfrac{T_1}{T_2} + u^2}{T_1(1+u^2)} M - \frac{u}{T_1\sqrt{1+u^2}} M_0 = 0. \tag{7.8}$$

Die von F. BLOCH angegebene Lösung dieser Gleichung lautet[1]:

$$M(t) = \frac{1}{T_1} \int_{-\infty}^{t} M_0(t') \frac{u(t')}{\sqrt{1+u^2(t')}} e^{-[g(t)-g(t')]} dt'. \tag{7.9}$$

Darin ist $M_0(t') = \chi_0 H_0(t')$ und:

$$g(t) - g(t') = \frac{1}{T_1} \int_{t'}^{t} \frac{\dfrac{T_1}{T_2} + u^2(t'')}{1 + u^2(t'')} dt''. \tag{7.10}$$

Zu beachten ist, daß $M(t)$ sowohl positiv als auch negativ sein kann, je nach dem ob $u(t)$ vor dem Resonanzdurchgang positiv oder negativ war. Hatte das konstante Feld zuvor längere Zeit den Betrag $H \neq H_0$, wobei $|u| \gg 1$ war, und findet der Resonanzdurchgang in einer Zeit statt, die klein gegen die Spin-Gitter-Relaxationszeit T_1 ist, dann stammt der Hauptbeitrag zum Integral (7.9) aus der Zeit vor dem Durchgang, als M_0 den konstanten Wert $\chi_0 H$ hatte. Oft ist auch noch die Ungleichung $|u|^2 \gg T_1/T_2$ erfüllt. In diesem Fall hat das Integral (7.10) einfach den Betrag:

$$g(t) - g(t') = \frac{t - t'}{T_1},$$

und es ist:

$$M(t) = \pm \frac{\chi_0 H_1}{T_1} \int_{-\infty}^{t} e^{-\frac{(t-t')}{T_1}} dt' = \pm \chi_0 H = \pm M_0.$$

Das Pluszeichen gilt, wenn vor dem Resonanzdurchgang das Feld H stärker und das Minuszeichen, wenn vor dem Durchgang H schwächer als das Resonanzfeld war. Damit erhält die Lösung im mitrotierenden System die einfache Form:

$$M_{x'} = \pm \frac{M_0}{\sqrt{1+u^2}}, \qquad M_{y'} = 0, \qquad M_{z'} = \frac{u M_0}{\sqrt{1+u^2}}. \tag{7.11}$$

Dieses spezielle Integral der BLOCHschen Differentialgleichungen (2.6) bzw. (6.5) bzw. (7.2) hätte man auch durch eine Berechnung der stationären Lösungen der Grundgleichung (2.2) direkt gewinnen können. Die Gl. (7.11) sind also nichts anderes als Lösungen der genannten Differentialgleichungen für den analytisch einfachsten Sonderfall, daß sowohl die zeitlichen Ableitungen $\dot{M}_x$, $\dot{M}_y$ und $\dot{M}_z$ als auch die Relaxationsglieder M_x/T_2, M_y/T_2 und $(M_z - M_0)/T_1$ vernachlässigbar sind. Die erste Vereinfachung besagt, daß der Resonanzdurchgang adiabatisch, d.h. stationär erfolgen soll und ist erlaubt, wenn die linke Seite der Ungleichung (7.1) erfüllt ist [vgl. hierzu Fig. 31 sowie die folgende Ableitung der Gl. (7.13)]. Die zweite Vereinfachung

$$\gamma H_0, \gamma H_1 \gg \frac{1}{T_1}, \frac{1}{T_2} \tag{7.12}$$

besagt, daß die Präzessionsglieder die Relaxationsglieder weitaus überwiegen, daß also der Resonanzdurchgang schnell erfolgen soll. Praktisch ist immer

---

[1] F. BLOCH: Phys. Rev. **70**, 460 (1946).

$H_0 > H_1$ und $T_1 \geq T_2$. Demnach sind alle Ungleichungen (7.12) erfüllt, wenn $\gamma H_1 \gg 1/T_2$ ist, und diese Bedingung ist wiederum im allgemeinen milder, als die in der rechten Seite von (7.1) bereits aufgestellte Forderung.

Der Verlauf (7.11) von $M_{x'}$ beim schnellen adiabatischen Resonanzdurchgang beschreibt die Form des im Experiment zu beobachtenden und in Fig. 30 dargestellten Kernsignals. Zu beachten ist, daß zwischen dem zur x'-Achse parallelen anregenden Feld H_1 und dem transversalen Kernfeld $M_{x'}$, keine Phasenverschiebung besteht. Es handelt sich also um ein Dispersionssignal, obwohl seine Form Ähnlichkeit mit der des Absorptionssignals (Fig. 26b) beim langsamen, adiabatischen Resonanzdurchgang hat[1]. Bei diesem haben wir ferner unterschieden zwischen starker und schwacher Anregung. Beim schnellen, adiabatischen Resonanzdurchgang muß dagegen nach Ungleichung (7.1) bzw. (7.12) die Anregungsfeldstärke H_1 immer groß sein.

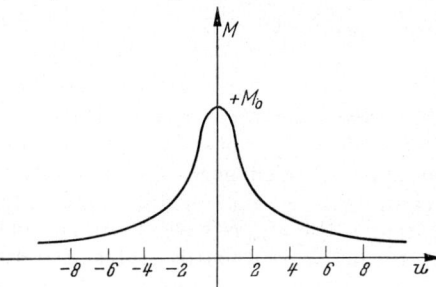

Fig. 30. Form der Kernsignale beim schnellen adiabatischen Resonanzdurchgang. Berechnet aus (7.11).

Wir wollen uns nun den Präzessionsverlauf, der zu den stationären Lösungen (7.11) führt, verdeutlichen. Weit außerhalb Resonanz ist $M_{x'} = M_{y'} = 0$ und $M_{z'} = M_0$, so als ob gar kein Hochfrequenzfeld vorhanden wäre. Geht man mit der Feldstärke H sehr schnell in Resonanz (H_0), oder besser, schaltet man erst in Resonanz das hochfrequente Feld H_1 plötzlich ein, dann präzessiert der Vektor M_0 um H_1, also um die x'-Achse, und seine Spitze beschreibt einen Kreis in der y', z'-Ebene (Fig. 31). Dies ist der uns von der Besprechung der freien Präzessionsbewegungen her bekannte Fall. Bei der Lösung (7.11) wurde aber gerade vorausgesetzt, daß der Resonanzdurchgang adiabatisch langsam, wenn auch im Vergleich zu den Relaxationszeiten T_1 und T_2 schnell erfolgt. Nähert man sich daher unter diesen Bedingungen der Resonanzstelle, dann beschreibt die Spitze des Magnetisierungsvektors einen in der x', z'-Ebene liegenden Großkreis auf der in der Fig. 31 abgebildeten Kugel vom Radius M_0.

Anschaulich versteht man das Zustandekommen dieser Präzessionsbewegung wie folgt. Wir ersetzen die stetige Annäherung des Magnetfeldes an die Resonanzfeldstärke durch eine Folge infinitesimal kleiner momentaner Feldstärkeänderungen, zwischen denen die Feldstärke infinitesimal kurze Zeiten konstant bleibt (in Fig. 31 wurde übrigens angenommen, daß die anfängliche Feldstärke H größer als die Resonanzfeldstärke H_0 war). In dem der Betrachtung zugrunde liegenden mitrotierenden System x', y', z' wirkt außer dem Hochfrequenzfeld H_1 ein Differenzfeld $h = H - \dfrac{\omega}{\gamma}$ auf den Magnetisierungsvektor M_0 ein. Es habe in Fig. 31 außerhalb Resonanz zunächst etwa den Betrag h_8. Im stationären Zustand liegt das resultierende Moment M_0 parallel zur Resultante R_8 beider Felder. Zur Zeit $t = 0$ werde die Feldstärke h_8 plötzlich auf den Betrag h_7 vermindert. Das Moment präzessiert nun um die Resultante R_7 mit der Winkel-

---

[1] Untersucht man nacheinander mit derselben experimentellen Anordnung Substanzen, deren Relaxationszeiten immer kürzer sind, dann geht tatsächlich die anfänglich beobachtete Signalform des schnellen Resonanzdurchgangs (Fig. 30) über in die des Dispersionssignals beim langsamen Resonanzdurchgang (Fig. 26a). Die Beobachtung dieser Signalformänderung kann zur T_1-Messung ausgenutzt werden: L. E. DRAIN: Proc. Phys. Soc. Lond. A **62**, 301 (1949). — S. D. GVOZDO' ER u. A. A. MAGAZANIK: J. exp. theor. Phys. USSR. **20**, 705 (1950).

geschwindigkeit γR_7. Die Schnittkurve des Präzessionskegels mit der Kugel vom Radius M_0 ist der linke Kreis in Fig. 31. Der Öffnungswinkel des Kegels hat die Größe $\Delta \alpha_7$. Wird nach einer Zeit Δt die Feldstärke h wiederum verkleinert, ihr Betrag sei jetzt h_6, dann zeigt in diesem Augenblick M_0 im Zeitmittel in die Richtung von R_7 und zwar um so genauer, je kleiner der Öffnungswinkel $\Delta \alpha_7$ und je größer der Präzessionswinkel $\gamma R_7 \Delta t$ ist, wenn also die Ungleichung $\dfrac{\Delta \alpha_7}{\gamma R_7 \Delta t} \ll 1$ erfüllt ist. Das Moment setzt seine Präzessionsbewegung um die

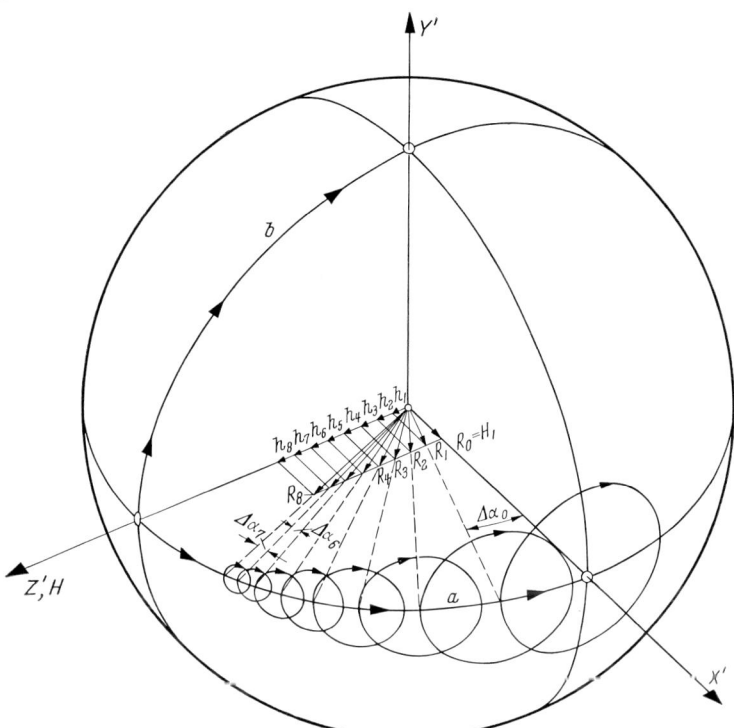

Fig. 31. Zur Erklärung des Verlaufs der Präzessionsbewegung beim schnellen adiabatischen Resonanzdurchgang. Außerhalb Resonanz liege das resultierende Moment zunächst parallel zur Resultante R_8. Der Resonanzabstand ist nacheinander h_8, h_7, ..., h_1, 0; $\Delta \alpha_\nu$ ist der Öffnungswinkel des Präzessionskegels um R_ν. Die Spitze des resultierenden Momentes wandert bei der Annäherung an Resonanz dem in der x', z'-Ebene liegenden Kreis a entlang. Wird in Resonanz das Feld H_1 plötzlich eingeschaltet, dann beschreibt der Magnetisierungsvektor den in der y', z'-Ebene liegenden Kreis b (Impulsanregung).

Richtung von R_6 fort und beschreibt auf der M_0-Kugel den zweiten Kreis von links. Der Öffnungswinkel hat jetzt die Größe $\Delta \alpha_6 > \Delta \alpha_7$ und die Präzessionsgeschwindigkeit den Betrag $\gamma R_6 < \gamma R_7$. Soll der obige Gedankengang in derselben Weise fortgesetzt werden können, dann muß auch die Ungleichung $\dfrac{\Delta \alpha_6}{\gamma R_6 \Delta t} \ll 1$ erfüllt sein. Mit dem letzten Schritt der Resonanzannäherung verschwindet das Differenzfeld h völlig. Die Folge $\Delta \alpha_\nu$ strebt also gegen $\Delta \alpha_0 = \Delta h / H_1$ und die Folge R_ν gegen $R_0 = H_1$. Die letzte der zu befriedigenden Ungleichungen ist also:

$$\frac{\Delta \alpha_0}{\gamma H_1 \Delta t} \ll 1 \quad \text{bzw.} \quad \frac{\Delta h}{\Delta t \cdot H_1} \ll \gamma H_1, \tag{7.13}$$

und diese schließt alle vorangegangenen ein.

Beim Übergang von Präzessionskreis zu Präzessionskreis wurde angenommen, daß der Magnetisierungsvektor während einer plötzlichen Feldänderung $h_\nu \to h_{\nu-1}$ im Zeitmittel in die Richtung R_ν zeigt. In Wirklichkeit hat er natürlich in diesem Augenblick die Richtung einer der Mantellinien des Kegels um R_ν. Ist die Zahl N der Feldsprünge aber genügend groß, und sind damit die Feldstärkeänderungen $\varDelta h$ ausreichend klein, dann mitteln sich solche Momentanabweichungen heraus. Im Grenzübergang geht also der beschriebene Bewegungsverlauf über in die durch die Lösung (7.11) dargestellte einfache Präzession um die y'-Achse in der x', z'-Ebene.

Die Halbwertsbreite $\varDelta H_h$ der Resonanzlinie beim schnellen Resonanzdurchgang folgt aus (7.11) zu $\varDelta H_h = \sqrt{3} H_1$. Allgemein war die Bedingung eines adiabatischen Resonanzdurchgangs (6.1) $dH/dt \ll |\gamma| (\varDelta H_h)^2$. Sie lautet also hier:

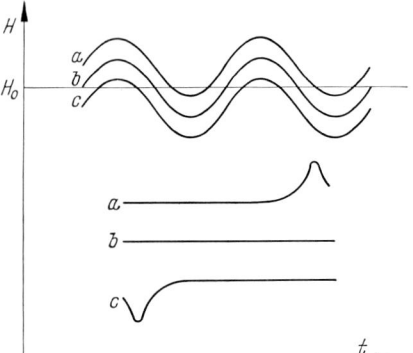

$dH/dt \ll |\gamma| H_1^2$. Dieses Ergebnis stimmt mit dem oben anschaulich abgeleiteten [Gl. (7.13)] überein.

Nach dem Resonanzdurchgang wird das Differenzfeld h negativ und der Magnetisierungsvektor setzt seine Präzessionsbewegung in der gleichen Weise in Richtung der negativen z'-Richtung fort. Am Ende des Resonanzdurchganges liegt also M_0 ungefähr antiparallel zur Feldrichtung. Durchläuft man nun von diesen Anfangsbedingungen ($M_0 \downarrow\uparrow z'$, $H < H_0$) ausgehend die Resonanzstelle wiederum nach oben, dann begleitet auch hierbei der Vektor M_0 den Vektor R, und beide sind in jedem Zeitpunkt parallel zueinander.

Fig. 32. Verlauf der Kernsignale in einem adiabatisch-schnell modulierten Feld. Die Kernsignale haben die Form a, b, c, wenn der Mittelwert des Feldes H größer, gleichgroß bzw. kleiner als die Resonanzfeldstärke H_0 ist (F. BLOCH, W.W. HANSEN und M. E. PACKARD).

Liegt dagegen vor einem Resonanzdurchgang von unten ($H < H_0 \to H > H_0$) M_0 parallel zur Feldrichtung ($M_0 \uparrow\uparrow z$), dann folgt auch hierbei der Magnetisierungsvektor M_0 der Bewegung der Resultanten R. Beide Vektoren sind dann aber stets antiparallel zueinander. Daraus folgt, daß in Resonanz M_0 in die Richtung der negativen x'-Achse zeigt. Erfolgt der Resonanzdurchgang also unter diesen Anfangsbedingungen, dann besteht zwischen dem anregenden Feld H_1 und der beobachtbaren transversalen Komponente $M_{x'}$ von M_0 stets eine Phasendifferenz von 180°, d.h. die Magnetisierung $M_{x'}$ ist nicht wie oben zum Feld H_1 zu addieren, sondern muß von diesem abgezogen werden. Während im ersten Fall ($M_0 \uparrow\uparrow z'$, $H > H_0 \to H < H_0$; $M_0 \downarrow\uparrow z'$, $H < H_0 \to H > H_0$) die Umhüllende des Spannungsverlaufs an der die Substanz umgebenden Spule bei einem periodisch wiederholten Resonanzdurchgang die in Fig. 32a wiedergegebene Form hat, wird sie im zweiten Fall ($M \uparrow\uparrow z'$, $H < H_0 \to H > H_0$; $M \uparrow\downarrow z'$, $H > H_0 \to H < H_0$) durch Fig. 32c beschrieben[1].

8. Nichtadiabatische Bewegungen. In den adiabatisch verlaufenden kernmagnetischen Experimenten werden nur Präzessionsbewegungen zwischen stetig aneinander anschließenden stationären Lagen des Magnetisierungsvektors beobachtet. Im Gegensatz dazu beobachtet man bei der Untersuchung der mit Impulsen angeregten freien Präzessionsbewegungen nur instationäre Annäherungsvorgänge, zum Teil während der Impulsanregung, meist aber in dem sich daran anschließenden Zeitabschnitt des Abklingens der Präzessionsbewegung. Die

---

[1] F. BLOCH, W. W. HANSEN u. M. PACKARD: Phys. Rev. **70**, 474 (1946).

Kernsignale werden dabei überhaupt erst auf Grund ihrer Modulation durch den nichtstationären Verlauf der Bewegungen der Beobachtung zugänglich, während die stationären Endlagen des Magnetisierungsvektors als konstante Größen keinen Beitrag zur niederfrequenten Signalmodulation liefern. Dagegen rührt die Signalmodulation bei den kontinuierlichen adiabatischen Experimenten nur von der durch die Variation der Magnetfeldstärke bzw. der Senderfrequenz bewirkten Änderung des stationären Zustands her.

Selbstverständlich ist es leicht, Experimente anzugeben, die zwischen diesen beiden Grenzfällen liegen, bei denen also die Kernsignale sowohl durch die Wanderung der stationären Endlagen beim Resonanzdurchgang, als auch durch instationäre Annäherungsbewegungen moduliert werden. Physikalisch liefern solche gemischten Experimente an sich keine neuen Erkenntnisse oder Meßdaten, aber da zur Aufklärung der zahlreichen dabei möglichen Effekte doch eine ganze Anzahl Arbeiten erforderlich war[1], sollen wenigstens einige Beispiele erwähnt werden.

Erfolgt ein Resonanzdurchgang zu schnell um völlig adiabatisch verlaufen zu können, dann wird er dies doch am Anfang tun. Laut Voraussetzung ist die in Resonanz zu befriedigende Ungleichung (7.13) nicht erfüllt, sei es weil die Hochfrequenzamplitude H_1 zu klein oder die Geschwindigkeit des Durchgangs zu groß ist. Der zweite Fall braucht uns nicht zu interessieren; denn er entspricht praktisch der Anregung durch einen Hochfrequenzimpuls, der ja auch eine endliche Anstiegszeit hat. Im ersten Fall gelten aber im allgemeinen bis zu einer gewissen Grenze ν in der unmittelbaren Nachbarschaft der Resonanzstelle immer noch die Ungleichungen $\dfrac{\varDelta \alpha_\nu}{\gamma\, R_\nu \varDelta t} \ll 1$. Daraus folgt aber, daß die Annäherung an die Resonanzstelle, wie in Ziff. 7 (S. 180) beschrieben, zunächst noch adiabatisch erfolgt. Unmittelbar vor der Resonanzstelle wird sich der Magnetisierungsvektor aber von der Richtung des resultierenden Feldes R entfernen; damit wird die Bewegung kompliziert instationär. Uns interessiert nur, daß das resultierende Moment jetzt im allgemeinen eine transversale Komponente hat, die auch beim Verlassen der Resonanzstelle wenigstens zum Teil erhalten bleibt. Diese transversale Komponente induziert in der die Probe umgebenden Spule eine Wechselspannung, deren Frequenz gemäß $\omega_L = \gamma H$ der jeweiligen Feldstärke $H(t)$ proportional ist, sich also beim Verlassen der Resonanzstelle stetig ändert. In allen experimentellen Anordnungen, die mit einem kontinuierlich arbeitenden Generator betrieben werden, wird diese Spannung jedoch mit der Senderspannung der Frequenz $\omega_0 = \gamma H_0$ gemischt. Demnach ist zu erwarten, daß das abklingende Kernsignal mit einer zunehmenden Differenzfrequenz moduliert ist. Nimmt man an, daß die Resonanzstelle im Zeitpunkt $t = 0$ passiert wird, und daß in diesem Augenblick keine Phasendifferenz zwischen der Senderspannung und der von den Kernen induzierten Spannung besteht, dann ist die Phasendifferenz $\varphi(t)$ zwischen den beiden Spannungen in irgendeinem späteren Zeitpunkt t:

$$\varphi(t) = \gamma \int_0^t \big(H(t) - H_0\big)\, dt. \qquad (8.1)$$

¹ B. A. JACOBSOHN u. R. K. WANGSNESS: Phys. Rev. 73, 942 (1948). — L. GIULOTTO: Nuovo Cim. 5, 498 (1948). — A. BOLLE u. G. ZANOTELLI: R. C. Accad. 40, 27 (1948). — E. E. SALPETER: Proc. Phys. Soc. Lond. A 63, 337 (1950). — R. GABILLARD: C. R. Acad. Sci., Paris 232, 1477 (1951). — M. SOUTIF u. R. GABILLARD: Physica, Haag 17, 319 (1951). — M. SOUTIF: Rev. Sci., Paris 89, 203 (1951). — G. J. BÉNÉ, P. M. DENIS u. R. C. EXTERMANN: Arch. Sci., Genève 3, 452 (1950); 4, 212, 266 (1951); 5, 32, 406 (1952). — C. R. Acad. Sci., Paris 231, 1294 (1950). — Helv. phys. Acta 22, 388 (1949); 24, 304 (1951). — Physica, Haag 17, 308 (1951); gemeinsam mit H. J. BONHOMME: Helv. phys. Acta 26, 435 (1953). — K. TAYLOR: Nature, Lond. 172, 722 (1953). — C. MANUS, R. MERCIER, P. DENIS, G. BÉNÉ u. R. EXTERMANN: C. R. Acad. Sci., Paris 238, 1315 (1954).

Gilt $H(t) = H_0 + \dot{H}(0) \cdot t$, ändert sich $H(t)$ also linear — näherungsweise ist dies auch bei einer sinusförmigen Modulation in der Nachbarschaft der Resonanzdurchgänge erfüllt —, dann ist:

$$\varphi(t) = \tfrac{1}{2}\gamma \dot{H}(0)\, t^2.$$

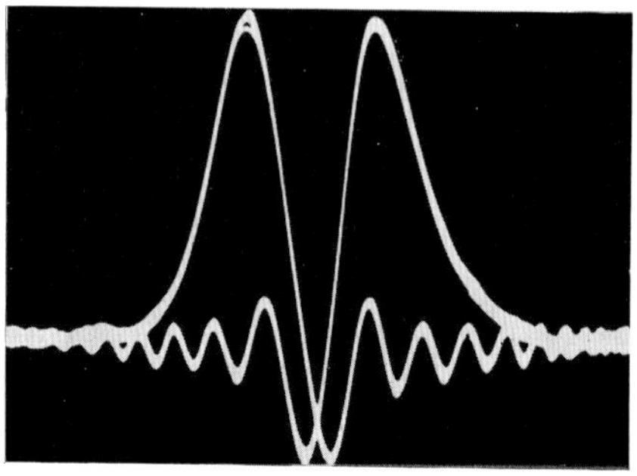

In einem ausreichend homogenen Magnetfeld wird der Signalabfall nur durch die Spin-Spin-Relaxationszeit T_2 bestimmt. Man erwartet somit ein Signal der Form:

$$\cos\left\{\tfrac{1}{2}\gamma \dot{H}(0)\, t^2\right\} e^{-\frac{t}{T_2}}.$$

Fig. 33. Nicht-adiabatischer Verlauf einer in Wasser bei 29 MHz beobachteten Protonenresonanz. Die Modulation des Feldes und die Zeitablenkung des Oszillographen wurden mit dem gleichen Generator sinusförmig moduliert. Die beiden einander überlagerten Kurven gehören zu dem Vor- bzw. zu dem Rücklauf des Elektronenstrahls der Oszillographen-Röhre. Nach jedem Resonanzdurchgang entstehen Schwebungen des Signals (wiggles). (N. BLOEMBERGEN, E. M. PURCELL und R.V. POUND.)

Ein experimentelles Beispiel[1] zeigt Fig. 33. Es bestätigt, daß die sog. „wiggles" oder Schwebungen erst nach dem Resonanzdurchgang beobachtet werden können. Ist die wahre, durch T_2 bestimmte Linienbreite sehr schmal, dann wird der Abklingvorgang ebenso wie bei den Impulsexperimenten durch die Feldinhomogenität beschleunigt. Die zu Gebieten verschiedener Feldstärke gehörenden infinitesimalen Momente fächern auseinander, und das Signal verschwindet schneller.

Im nächsten Beispiel soll vorausgesetzt werden, daß die Zeit eines einfachen Resonanzdurchganges $t_D = \pi/\omega_m$ (ω_m = Modulations-Kreisfrequenz) vergleichbar mit T_2 oder kleiner als T_2 ist. Außerdem soll zunächst angenommen werden, daß das Feld völlig homogen ist. Am Ende eines Resonanzdurchganges ist dann die Signalspannung und damit auch die Schwebung noch nicht abgeklungen (Fig. 34). Während der nachfolgenden rückläufigen erneuten Annäherung an die Resonanzstelle nimmt ihre Amplitude weiter exponentiell ab und wird schließlich beim Resonanzdurchgang wiederum irgendwie verstärkt. Diesen Gedankengang kann man fortsetzen. Zusammengefaßt, stellt sich ebenso wie bei der Anregung durch eine unendlich lange

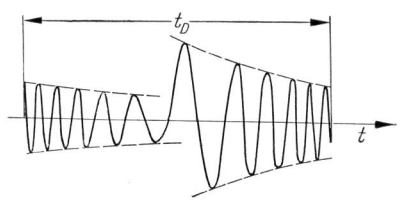

Fig. 34. Form des Kernsignals bei einem nicht-adiabatischen Resonanzdurchgang, wenn das Feld ausreichend homogen und die Zeit t_D eines Durchgangs kleiner als die Relaxationszeiten der Substanz ist.

Impulsfolge[2] ein stationärer Zustand ein[3], und das Verhältnis der Signalamplitude unmittelbar vor der Resonanzstelle zu derjenigen direkt nach ihr ist gleich e^{-t_D/T_2}.

Läßt man die Homogenitätsvoraussetzung fallen, dann verschwindet das Schwebungssignal schneller und zwar im allgemeinen schon vor den Umkehrpunkten der Feldmodulation. Aber auch hierbei hat die periodisch wiederholte Anregung zur Folge, daß der Auseinanderfächerungs-Vorgang zum Teil umgekehrt wird. Unter gewissen Bedingungen kann dann ebenfalls aus dem oben genannten Amplitudenverhältnis auf T_2 geschlossen werden und zwar unab-

[1] N. BLOEMBERGEN, E. M. PURCELL u. R. V. POUND: Phys. Rev. 73, 679 (1948).
[2] H. C. TORREY: Phys. Rev. 85, 365 (1952).
[3] J. S. GOODEN: Nature, Lond. 165, 1014 (1950). — R. GABILLARD: C. R. Acad. Sci., Paris 232, 1551 (1951); 233, 39, 307 (1951). — Phys. Rev. 85, 694 (1952). — Rev. Sci., Paris 90, 307 (1952).

hängig von der Feldinhomogenität. Wie man sieht, werden bei allen solchen Experimenten erzwungene und freie Präzessionsbewegungen mehr oder weniger gleichzeitig beobachtet und damit die Vorgänge entsprechend kompliziert.

B. Experimentelle Technik.

I. Brückenschaltungen.

9. Anregung und Beobachtung mit einer Spule. Präzessionsbewegungen eines resultierenden Kernmagnetisierungs-Vektors lassen sich in einem konstanten oder niederfrequent modulierten Magnetfeld durch ein transversales oszillierendes

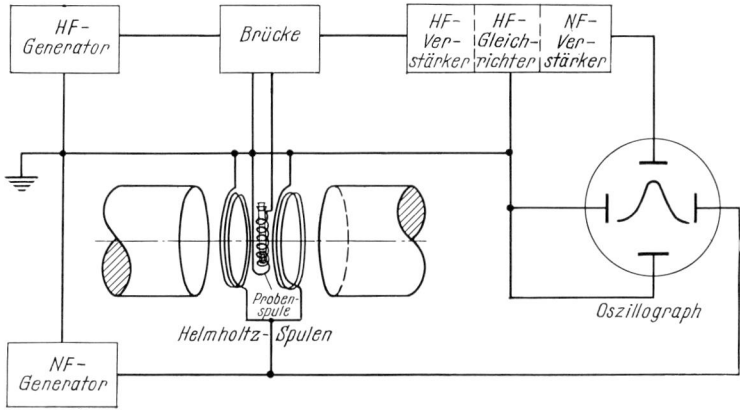

Fig. 35. Prinzip-Schaltbild einer Apparatur zur Beobachtung erzwungener kernmagnetischer Präzessionsbewegungen.

Hochfrequenzfeld der ungefähren LARMOR-Frequenz anregen. Dem oszillierenden Feld überlagert sich das Feld der synchron präzessierenden Kerne. Damit wird aber die an der Anregungs- und Beobachtungsspule liegende Senderspannung verstärkt oder geschwächt und außerdem in ihrer Phasenlage verändert. Fig. 35 zeigt das Prinzipschaltbild einer Apparatur zur Beobachtung solcher Spannungsänderungen. Ein Niederfrequenzgenerator speist zur Modulation des konstanten Feldes ein HELMHOLTZ-Spulenpaar und steuert zugleich die Zeitablenkung eines Oszillographen. Der Resonanzabstand, also die Differenz zwischen der Sender- und der LARMOR-Frequenz und die Oszillographenabszisse ändern sich somit periodisch synchron. Der Hochfrequenzgenerator liefert die an der Probenspule liegende Wechselspannung. Diese Spule ist ein Glied einer Brückenschaltung. Der Hochfrequenzverstärker empfängt die von den Kernen induzierte Signalspannung vollständig, dagegen von der Senderspannung nur den nichtkompensierten Teil.

In Ziff. 6 wurde gezeigt, daß sich die Rückwirkung erzwungen synchron präzessierender Kerne auf das anregende Wechselfeld auch als dynamische Suszeptibilität der untersuchten Substanz beschreiben läßt. Beide Betrachtungsarten sind äquivalent. Zur Besprechung der Beobachtungsverfahren, welche Brückenschaltungen benutzen, ist es jedoch bequemer, sich der zweiten, hochfrequenztechnischen Ausdrucksweise zu bedienen. Dies soll im folgenden geschehen.

Die Induktivität der Probenspule ist $(1 + 4\pi\chi) L$, wenn man mit L die Induktivität der leeren Spule und mit χ die dynamische Suszeptibilität der Kerne bezeichnet[1]. Zur Erhöhung der Empfindlichkeit gegenüber Induktivitätsänderungen ergänzt man die Spule durch einen Kondensator der Kapazität C zu einem Parallelschwingkreis der ungefähren Kreisfrequenz $\omega = \dfrac{1}{\sqrt{LC}}$. Vor dem Einbringen der Probe soll der Kreis durch den Verlustwiderstand R_v gedämpft sein (Fig. 36a). Seine Güte ist also $Q = R_v/\omega L$. Mit der Probe ist der Leitwert des Kreises $(G_v = 1/R_v)$:

$$G = G_v + i\left(\omega C - \frac{1}{\omega L(1 + 4\pi\chi)}\right).$$

Der Schwingkreis wird stets auf Resonanz[2] abgestimmt. Berücksichtigt man, daß $|4\pi\chi| \ll 1$ ist, so folgt [vgl. (6.9)]:

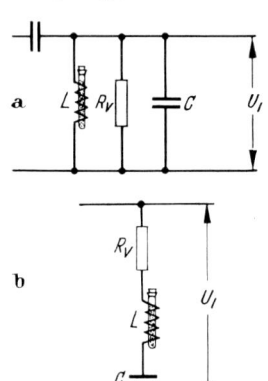

$$\left.\begin{aligned} G &= G_v(1 + 4\pi Q\chi \cdot i) \\ &= G_v + \frac{4\pi}{\omega L}\chi \cdot i = G_v + G_a + i\,G_d. \end{aligned}\right\} \quad (9.1)$$

Mit

$$G_a = \frac{4\pi}{\omega L}\cdot\chi'' = 4\pi G_v Q\chi'' = \text{Absorptionsleitwert}$$

und

$$G_d = \frac{4\pi}{\omega L}\cdot\chi' = 4\pi G_v Q\chi' = \text{Dispersionsleitwert}.$$

Fig. 36a u. b. a Parallel-Schwingkreis. b Serien-Schwingkreis zur Beobachtung von Präzessionsbewegungen.

Beim Resonanzdurchgang ändert sich χ und damit G. Die zugehörige Spannungsänderung des Kreises ist am größten, wenn der durch ihn fließende Strom J konstant bleibt. Zur Erfüllung dieser Forderung schaltet man zwischen den Hochfrequenzgenerator und den Kreis einen hohen komplexen Widerstand $R_0 \left(|R_0| \gg \dfrac{1}{|G|}\right)$, z.B. eine kleine Koppelkapazität oder einen großen OHMschen Widerstand. Nennt man die Spannung weit außerhalb Resonanz U_1 $(\chi = 0,\ U_1 = J/G_v)$, dann ist die am Kreis liegende Spannung U in der Nachbarschaft der Resonanzstelle $(|4\pi Q\chi| \ll 1)$:

$$U = U_1(1 - 4\pi Q\chi \cdot i) = U_1 + U_{\chi''} + i\,U_{\chi'} \qquad (9.2)$$

mit

$$U_{\chi''} = -4\pi U_1 Q\chi'' = \text{Absorptionsspannung}$$

und

$$U_{\chi'} = -4\pi U_1 Q\chi' = \text{Dispersionsspannung}.$$

Ähnlich sind die Verhältnisse, wenn man den Resonanzdurchgang in einem Serienschwingkreis beobachtet (Fig. 36b). Sein Widerstand ist außerhalb Resonanz $(\omega L \neq 1/\omega C)$:

$$R = R_v + i\left(\omega L(1 + 4\pi\chi) - \frac{1}{\omega C}\right)$$

---

[1] In der Regel ist die Probenspule nicht vollständig mit der untersuchten Substanz gefüllt. Diesen Umstand kann man durch die Einführung eines Füllfaktors f $(f \leqq 1)$ berücksichtigen. Genauer hat demnach die Induktivität der Spule die Größe $(1 + 4\pi f\chi)L$. Da sich jedoch der Faktor f erforderlichenfalls leicht in allen folgenden Beziehungen nachträglich einfügen läßt, kann zur Vereinfachung der Darstellung im weiteren $f = 1$ vorausgesetzt werden.

[2] Zu beachten ist, daß der Begriff Resonanz in zwei Bedeutungen verwandt wird. Kreisresonanz liegt vor, wenn zwischen der Frequenz des anregenden Senders und den Schwingkreisdaten L, C die Beziehung $\omega = 1/\sqrt{LC}$ erfüllt ist; Kernresonanz dagegen, wenn die Senderfrequenz ω mit der LARMOR-Frequenz γH der Kerne übereinstimmt.

und in Kreisresonanz in der Nachbarschaft der Kernresonanzstelle (die Güte eines Serienschwingkreises ist $Q = \omega L/R_v$):

$$R = R_v(1 + 4\pi Q \chi \cdot i) = R_v + 4\pi \omega L \chi \cdot i = R_v + R_a + i R_d \qquad (9.3)$$

mit

$$R_a = 4\pi \omega L \chi'' = 4\pi R_v Q \chi'' = \text{Absorptionswiderstand}$$

und

$$R_d = 4\pi \omega L \chi' = 4\pi R_v Q \chi' = \text{Dispersionswiderstand}.$$

In Resonanz erreicht der zusätzliche reelle Kernresonanz-Absorptionswiderstand R_a seinen maximalen Betrag. Der Dispersions-(Blind-)Widerstand R_d verschwindet in Resonanz, und ist auf der einen Seite der Resonanzstelle kapazitiv und auf der anderen induktiv.

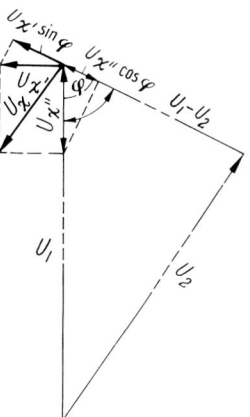

Fig. 37. Elektrotechnisches Vektordiagramm einer Brückenschaltung. U_1 Ausgangsspannung des Meßzweiges; U_2 Ausgangsspannung des Blindzweiges (Kompensationsspannung); U_χ Kernsignalspannung.

Die am Serienkreis liegende Spannung ist:

$$U = U_1(1 + 4\pi Q \chi \cdot i) = U_1 + U_{\chi''} + i U_{\chi'} \qquad (9.4)$$

mit

$$U_{\chi''} = 4\pi U_1 Q \chi'' = \text{Absorptionsspannung}$$

und

$$U_{\chi'} = 4\pi U_1 Q \chi' = \text{Dispersionsspannung}.$$

In beiden Fällen werden die Amplitude und die Phase der hochfrequenten Spannung U_1 beim periodischen Resonanzdurchgang niederfrequent moduliert. Allerdings ist der Modulationsgrad sehr klein, z. B. $1:10^6$. Um eine Hochfrequenzverstärkung der niederfrequenten Modulation, also der Kernsignale zu ermöglichen, muß man die Spannung U_1 teilweise kompensieren. Technisch kann dies auf sehr verschiedene Arten erfolgen. Verfahrensunterschiede sollen uns jedoch zunächst noch nicht interessieren. Im elektrotechnischen Vektordiagramm läßt sich jede derartige vollständige oder teilweise Kompensationsart durch die Überlagerung einer zweiten Spannung U_2 beschreiben. Die dem Empfänger angebotene Gesamtspannung ist also:

$$U = U_1 - U_2 + U_{\chi''} + i U_{\chi'}.$$

Setzt man voraus, daß die Kompensation nur teilweise erfolgt:

$$|U_1|, |U_2| \gg |U_1 - U_2| \gg |U_\chi| \qquad (9.5)$$

und bezeichnet mit φ den Phasenwinkel zwischen U_1 und $U_1 - U_2$ (Fig. 37), dann hat in guter Näherung die am Empfängereingang liegende Spannung die Phase von $U_1 - U_2$ und den Betrag:

$$|U| = |U_1 - U_2| + \cos \varphi \cdot U_{\chi''} - \sin \varphi \cdot U_{\chi'}.$$

[In Fig. 37 wurde $U_{\chi''} < 0$ und $U_{\chi'} < 0$ angenommen.] Für einen Serienschwingkreis gilt somit:

$$|U| = |U_1 - U_2| + 4\pi Q |U_1| (\cos \varphi \cdot \chi'' - \sin \varphi \cdot \chi'). \qquad (9.6)$$

und für einen Parallelschwingkreis:

$$|U| = |U_1 - U_2| + 4\pi Q |U_1| (-\cos \varphi \cdot \chi'' + \sin \varphi \cdot \chi'). \qquad (9.7)$$

Die Form des zu beobachtenden Kernsignals wird im wesentlichen durch die Größe des Phasenwinkels φ, also durch den Brückenabgleich bestimmt. In zwei Sonderfällen erhält man gut auswertbare Signalformen:

a) $\varphi = 0, \pi$: Gemäß der Ungleichung (9.5) sind in diesem Fall die Amplituden der Brückenzweige verstimmt, während ihre Phasen abgeglichen sind (Fig. 38a). Die Ausgangsspannung der Brücke

$$|U| = |U_1 - U_2| \pm 4\pi Q |U_1| \chi'' \quad \text{bzw.} \quad |U| = |U_1 - U_2| \mp 4\pi Q |U_1| \chi''$$

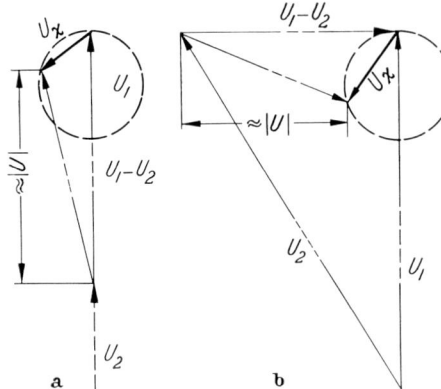

ist proportional zu χ'' moduliert, und das Kernsignal hat die Form eines reinen Absorptionssignals (Fig. 26b und 27b).

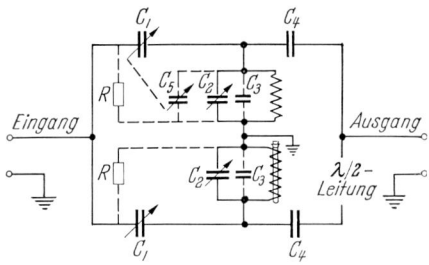

Fig. 38a u. b. Diagramm zur Abhängigkeit der Signalform von der Art des Brückengleichgewichts. a Im Phasengleichgewicht entsteht ein Absorptionssignal. b Im Amplitudengleichgewicht entsteht ein Dispersionssignal. Zu bemerken ist, daß die Bahnkurve von U_χ nur dann ein Kreis ist, wenn $\gamma^2 H_1^2 T_1 T_2 \ll 1$ ist (vgl. Fig. 28).

Fig. 39. Brückenschaltung von BLOEMBERGEN, PURCELL und POUND. Die Phase eines Brückenzweiges wird durch eine $\lambda/2$-Leitung um 180° gedreht. Die ausgezogenen Linien zeigen das Prinzipschaltbild, die praktisch verwandte Schaltung enthält auch die gestrichelt eingezeichneten Leitungen.

b) $\varphi = \pi/2, 3\pi/2$: Praktisch befindet sich die Brücke in diesem Fall im Amplitudengleichgewicht, da nach (9.5) $|U_1|, |U_2| \gg |U_1 - U_2|$ sein soll (vgl. Fig. 38b). Die Ausgangsspannung der Brücke

$$|U| = |U_1 - U_2| \mp 4\pi Q |U_1| \chi' \quad \text{bzw.} \quad |U| = |U_1 - U_2| \pm 4\pi Q |U_1| \chi'$$

ist proportional zu χ' moduliert, und das Kernsignal hat die Form eines reinen Dispersionssignals (Fig. 26a und 27a).

Hat φ einen beliebigen Zwischenwert, oder ist die Ungleichung (9.5) nicht erfüllt, dann entstehen gemischte, unsymmetrische Signalformen, aus denen sich weder die Absorption noch die Dispersion der Substanz ermitteln lassen. Aus diesem Grunde vergrößert man in Brückenschaltungen den ursprünglichen Modulationsgrad nur auf etwa $1:10^3$. Um die Kernsignale gut beobachten zu können, muß man sie deshalb nach der durch Empfängerübersteuerung begrenzten Hochfrequenzverstärkung noch ausreichend niederfrequent verstärken.

10. Verfahren zur Regelung der Kompensationsspannung. Die Kompensationsspannung U_2 kann auf sehr verschiedene Arten mit der Anregungs- und Signalspannung $U_1 + U_\chi$ gemischt werden. Wohl am bekanntesten ist das von BLOEMBERGEN, PURCELL und POUND angegebene Verfahren[1] geworden. Fig. 39 zeigt ausgezogen das Prinzipschaltbild ihrer Brückenanordnung und insgesamt die wirklich verwandte Schaltung. Der einseitig geerdete Eingang der Brücke ist mit dem Hochfrequenzgenerator verbunden, und der ebenfalls unsymmetrisch

---

[1] N. BLOEMBERGEN, E. M. PURCELL u. R. V. POUND: Phys. Rev. **73**, 679 (1948).

geerdete Ausgang mit dem Verstärker. Die beiden Brückenzweige bestehen jeweils aus zwei Koppelkondensatoren (C_1, C_4) und einem Schwingkreis. Die Induktivität des Meßkreises befindet sich zwischen den Polschuhen eines Magneten und enthält die zu untersuchende Substanz. Oft ist es zur Erleichterung des Abgleichs zweckmäßig, auch in die Spule des zweiten, sog. Blindkreises ein Reagenzglas mit der gleichen Probesubstanz zu stecken. Damit vermeidet man eine zu starke, nicht mehr ausregelbare Verstimmung der beiden Kreise, wie sie z. B. durch eine hohe Dielektrizitätskonstante der Probe verursacht werden kann.

U_1 und U_2 sollen voneinander subtrahiert werden. Die beiden Spannungen müssen also im wesentlichen — zur Beobachtung der Absorption exakt — um 180° verschiedene Phasen haben. In der Schaltung der Fig. 39 befindet sich deshalb im Meßbrückenzweig eine Leitung der Länge $\lambda/2$. Selbstverständlich kann man dieses Verzögerungskabel auch an einer anderen Stelle der Brücke einfügen. Zur exakten Justierung des Phasen- bzw. Amplitudengleichgewichts der Brücke sind die im übrigen kleinen (s. Ziff. 9, S. 186) Eingangs-Koppelkondensatoren C_1 variabel. Die Widerstände R dienen zur Anpassung der Brücke an den Innenwiderstand des Eingangskabels. Zur Feinabstimmung der Kreise ist es zweckmäßig, die Schwingkreiskapazität aufzuteilen. Sie besteht deshalb in beiden Kreisen aus einem keramischen Kondensator fester Kapazität (C_3) und aus einem kleinen Trimmer (C_2). Um das Phasen- bzw. das Amplitudengleichgewicht der Brücke orthogonal, d. h. unabhängig voneinander justieren zu können, ist die Achse des einen Koppelkondensators C_1 mechanisch gegensinnig mit der Achse eines zusätzlichen Trimmers C_5 derart verbunden, daß die gesamte wirksame Schwingkreiskapazität von einer Änderung der Ankopplung unabhängig wird. Damit wird es möglich, das Amplitudengleichgewicht der Brücke zu ändern, ohne ihr Phasengleichgewicht zu stören. Die Ausgangs-Koppelkondensatoren C_4 dienen zur Anpassung der Brücke an den Verstärker. Im allgemeinen schaltet man vor den eigentlichen, meist handelsüblichen Hochfrequenz-Schmalbandverstärker noch einen besonders rauscharmen Vorverstärker, z. B. eine Cascodeschaltung. Die ausgangsseitige Anpassung hat zum Ziel, den Beitrag der Brücke zum Eingangsrauschen des Empfängers möglichst niedrig zu halten[1]. BLOEMBERGEN hat ein Verfahren angegeben[2], nach dem sich mit Hilfe von Rauschdioden die optimalen Werte der Kondensatoren C_4 experimentell ermitteln lassen.

Ein wesentliches Kriterium zur Beurteilung des Wertes einer Brückenschaltung ist natürlich das maximal erzielbare Signal-Rauschverhältnis. Zum Rauschen liefert aber der Hochfrequenzgenerator einen Beitrag, da auch seine Ausgangsamplitude statistisch schwankt. Es ist ein Vorzug der Brückenschaltungen, daß sich bei der Mischung der Spannungen ihrer beiden Zweige diese Schwankungen gegenseitig kompensieren. Zwar soll die Kompensation von U_1 und U_2 nicht vollständig sein, vgl. (9.5), jedoch wird durch die Spannungskompensation um etwa einen Faktor 1000 auch die Generator-Rauschspannung im gleichen Verhältnis vermindert und kann dann im allgemeinen gegenüber den anderen Rausch-Spannungsquellen vernachlässigt werden.

Die Grundgedanken dieses Verfahrens zur Beobachtung kernmagnetischer Resonanzen sind schon in der ersten zu diesem Thema von PURCELL, TORREY und POUND veröffentlichten Arbeit[3] zu finden. Allerdings haben die Verfasser anfänglich zur Anregung und Beobachtung der Signale, anstatt normaler aus

---

[1] Siehe z. B. K. FRÄNZ: Hochfrequenztechn. **59**, 105, 143 (1942).
[2] N. BLOEMBERGEN: Nuclear Magnetic Relaxation. The Hague: Nijhoff.
[3] E. M. PURCELL, H. C. TORREY u. R. V. POUND: Phys. Rev. **69**, 37 (1946).

Spulen und Kondensatoren bestehenden Schwingkreisen, einen mit der Probe-substanz (Paraffin) gefüllten Topfkreis verwandt. Dies hat sich jedoch in dem vorliegend zur Diskussion stehenden Frequenzgebiet der Kurzwellen-Spektro-skopie nicht bewährt.

Ein praktischer Vorzug der beschriebenen und auch von vielen anderen Autoren[1,2] benutzten Brückenschaltung ist, daß beide Schwingkreise und der Ein- und Ausgang der Brücke einseitig geerdet sind. Damit wird die Lösung aller Abschirmprobleme erleichtert, insbesondere kann die Probenspule in ein geschlossenes Kästchen, meist Probenkopf genannt, eingebaut werden. Weiter sind auch die zu justierenden Schwingkreis-Kondensatoren einseitig geerdet und somit bequem (keine „Handempfindlichkeit") zu regeln. Ein Nachteil der Anordnung ist dagegen, daß zur Abstimmung zu viele, teilweise aufeinander rückwirkende Parameter zur Verfügung stehen. Insbesondere ist es schwierig, das Phasengleichgewicht unabhängig vom Amplitudengleichgewicht zu ändern. Ferner beträgt die für das $\lambda/2$-Verzögerungskabel

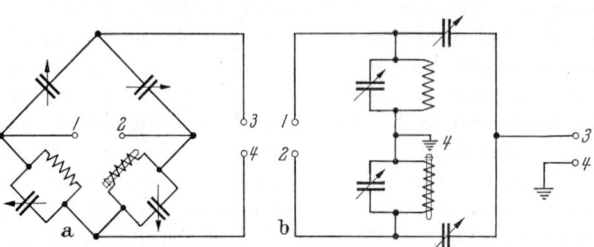

Fig. 40 a u. b.   Wheatstone-Bücken. a Symmetrischer Ein- und Ausgang.
b Symmetrischer Eingang, geerdeter Ausgang.

erforderliche Länge im vorliegenden Kurzwellengebiet immerhin einige Meter. Geringe, durch mechanische Störungen verursachte Längenänderungen haben aber bereits eine merkliche Verstimmung der Brücke zur Folge. Solche auch in den übrigen Teilen der Brücken störende Erschütterungen ändern übrigens in erster Näherung nur das Phasengleichgewicht, nicht aber das Amplituden-gleichgewicht. Zur Feststellung der Resonanzstelle ist es daher meist bequemer, die Resonanzabsorption zu beobachten.

Die Schaltung der Fig. 39 soll noch mit der wohl am besten bekannten WHEAT-STONE-Brücke verglichen werden. Fig. 40a zeigt eine solche Brücke in der geläufigen Darstellungsweise. Ein Zweig der Brücke besteht aus zwei variablen Kondensatoren und der andere aus zwei Schwingkreisen. Eine der beiden Spulen enthält die zu untersuchende Substanz. Wahlweise kann man, wie bei jeder Brücke, sowohl die Punkte 1, 2 als auch 3, 4 als Ein- bzw. Ausgang benutzen. Nur einer dieser Punkte kann aber geerdet werden, da man sonst einen Zweig der Brücke kurzschließen würde. Am anderen Punktpaar muß dann die Span-nung symmetrisch angelegt oder abgenommen werden. In Fig. 40b wurde der Punkt 4 geerdet und die Punkte 1, 2 als Eingang angenommen. Man erkennt sofort, daß die Prinzipschaltung 39 aus 40b durch Hinzufügen eines die Phase eines Zweiges um 180° drehenden Gliedes hervorgeht, wenn man wünscht, daß der Brückeneingang ebenfalls einseitig geerdet werden kann. Die Schaltung Fig. 40b ist übrigens auch praktisch zur Beobachtung der Resonanzabsorption verwandt worden[3], allerdings nur bei niedrigen Frequenzen.

Zur Phasendrehung kann man natürlich anstatt des Verzögerungskabels auch andere elektronische Schaltelemente verwenden. Naheliegend ist es, Übertrager mit passend gewähltem Wicklungssinn zu benützen. Dies haben PAKE[4] und

¹ W. C. DICKINSON: Phys. Rev. **81**, 717 (1951).
² E. R. ANDREW u. R. G. EADES: Proc. Roy. Soc. Lond., Ser. A **216**, 398 (1953).
³ R. M. BROWN: Phys. Rev. **78**, 530 (1950).
⁴ G. E. PAKE: J. Chem. Phys. **16**, 327 (1948).

TORREY[1] getan. Fig. 41 zeigt die von ihnen verwandte Brückenschaltung. In der Anordnung von PAKE befindet sich der Ausgang auf der Seite A und der Eingang auf der Seite B, während TORREY die Brücke umgekehrt schaltete. Durch den Umformer wird die Hochfrequenzspannung den beiden Zweigen mit entgegengesetzter Phase entnommen, bzw. beide Zweige werden mit entgegengesetzter Phase gespeist. Die Brücke wird dadurch symmetrischer und unempfindlicher gegen mechanische Störungen. Außerdem beeinflussen Schwankungen der Generatorfrequenz ein einmal justiertes Phasengleichgewicht weniger. Die Symmetrierung und Anpassung des Übertragers ist natürlich mühevoll, insbesondere müssen Streukapazitäten so weit als möglich vermieden werden, sie wird jedoch durch eine recht stabile Anordnung belohnt. Für die eingangs- bzw. ausgangs-

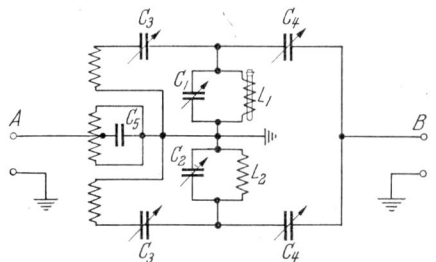

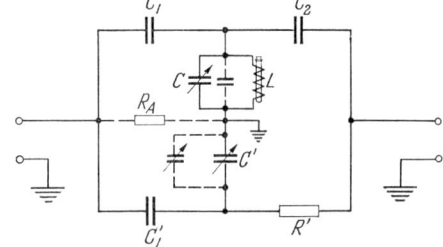

Fig. 41. Brückenschaltungen von PAKE und TORREY. Die Hochfrequenzspannung wird den beiden Brückenzweigen durch einen Übertrager mit entgegengesetzter Phase entnommen bzw. zugeführt. In der Anordnung von PAKE ist A der Ausgang und B der Eingang, in der Anordnung von TORREY wird die Brücke umgekehrt benutzt.

Fig. 42. Unsymmetrische Doppel-T-Brücke von ANDERSON. Mit Hilfe der Kondensatoren C bzw. C' kann das Phasen- bzw. das Amplitudengleichgewicht der Brücke orthogonal justiert werden. Die ausgezogenen Linien zeigen das Prinzipschaltbild, die praktisch verwandte Schaltung enthält auch die gestrichelt eingezeichneten Leitungen.

seitige Anpassung der Brücke durch die Koppelkondensatoren C_3 und C_4 gilt das bereits Gesagte.

Auf Grund einer Arbeit von TUTTLE[2] über die sog. Doppel-T-Brücken hat ANDERSON einen weiteren Typ von Brückenschaltungen erstmals zur Beobachtung paramagnetischer Kernresonanzen mit Erfolg benutzt[3]. Man versteht in der Vierpoltheorie unter einem T-Glied eine Kombination von drei miteinander verbundenen elektronischen Schaltelementen, von denen außerdem das erste geerdet, das zweite mit dem Eingang des Vierpols und das dritte mit dessen Ausgang verbunden ist. Betrachtet man auch einen Parallel-Schwingkreis als ein solches Schaltelement, dann sind die bisher besprochenen Anordnungen im Kern ebenso wie die in Fig. 42 dargestellte Schaltung Doppel-T-Brücken. Jedoch hat ANDERSON bewußt auf eine symmetrische Brückenanordnung verzichtet und damit die Möglichkeit gewonnen, unter Ausnutzung der reichen Erfahrungen mit T-Gliedern, ohne ein zusätzliches phasendrehendes Schaltelement auszukommen. Die Berechnung der Brücke Fig. 42 zeigt, daß sie im Amplitudengleichgewicht ist, wenn die Bedingung:

$$\omega^2 C_1 C_2 \left(1 + \frac{C'}{C_1'}\right) R' R_p = 1$$

erfüllt ist. Darin ist ω die Frequenz des Generators und R_p der äquivalente Nebenschluß-Widerstand der Probenspule bei dieser Frequenz. Phasengleichgewicht liegt dagegen vor, wenn die Beziehung:

$$C + C_1 + C_2 \left(1 + \frac{C_1}{C_1'}\right) = \frac{1}{\omega^2 L}$$

[1] H. C. TORREY: Phys. Rev. 76, 1059 (1949).
[2] W. N. TUTTLE: Proc. Inst. Radio Engrs. 28, 23 (1940).
[3] H. L. ANDERSON: Phys. Rev. 76, 1460 (1949).

befriedigt wird. Von den beiden variablen und zur Erleichterung der Abstimmung unterteilten Kondensatoren C und C' beeinflußt der eine nur das Amplituden- und der andere nur das Phasengleichgewicht. Die beiden Gleichgewichte können somit unabhängig voneinander justiert werden, die Brücke ist also exakt orthogonal. Ein weiterer Vorteil ist, daß sowohl die Probenspule, als auch die beiden abzugleichenden Kondensatoren ebenso wie der Ein- und Ausgang einseitig geerdet sind. Ein Nachteil ist jedoch, daß in den obigen Gleichungen die Generatorfrequenz ω enthalten ist. Diese Frequenz muß deshalb möglichst stabil sein, es ist daher zweckmäßig, einen quarzgesteuerten Oszillator zu verwenden. Der Widerstand R_a ist der Abschlußwiderstand des Eingangskabels.

Unabhängig haben übrigens Grivet, Soutif, Buyle and Gabillard ähnliche Schaltungen vorgeschlagen[1-3]. Eine dieser Anordnungen[1] unterscheidet sich

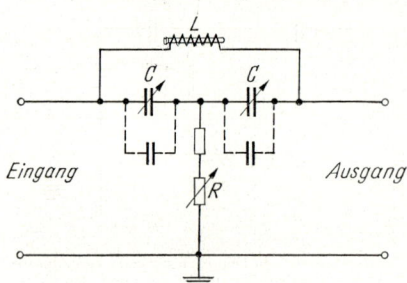

Fig. 43. Überbrücktes T-Glied von Waring, Spencer und Custer. In der ähnlichen Schaltung von Grivet, Soutif und Buyle ist die Probenspule ein Teil des T-Gliedes.

wesentlich von der besprochenen[4]. Anstatt zweier T-Glieder kann man nämlich zur Beobachtung der Resonanz auch ein überbrücktes T-Glied benutzen. In der ersten Schaltung dieser Art war die Probenspule zugleich das innere Schaltelement eines T-Gliedes, während die Kompensationsspannung U_2 über einen variablen Ohmschen Widerstand zu dem Brückenausgang geleitet wurde. Die Phase der Generatorspannung wird also nur in dem T-Glied gedreht. Eine andere von Waring, Spencer und Custer angegebene Anordnung[5] dieses Typs zeigt Fig. 43. Hier wurde das T-Glied mit der Probenspule überbrückt. Diese Schaltung ist jedoch nicht orthogonal. Außerdem sind weder die Probenspule noch die abzugleichenden Kondensatoren geerdet, die Abschirmung des Eingangs vom Ausgang wird dadurch schwierig.

Ein ganz anderer Weg zur Erzeugung der Kompensationsspannung U_2 ist von Rollin bereits 1946 eingeschlagen worden[6]. Läßt man die erste Stufe eines Hochfrequenzverstärkers im C-Betrieb arbeiten, sperrt also seine erste Röhre durch eine ausreichend negative Gittervorspannung, dann verstärkt dieser von einer ihm angebotenen schwach modulierten Hochfrequenzspannung nur den die Sperrspannung überwiegenden Teil, während der größte Teil des nichtmodulierten Trägerfrequenzpegels einfach abgeschnitten wird. Der Modulationsgrad des empfangenen Teils der Eingangsspannung wird dadurch wesentlich erhöht, und diese kann somit hochfrequent verstärkt werden. An die Stelle der bisher besprochenen Brückenschaltungen tritt bei der Anwendung dieses Prinzips zur Messung kernmagnetischer Resonanzen ein einfacher Schwingkreis, etwa von der in Fig. 36a dargestellten Art, dessen kleiner Eingangs-Koppelkondensator, wie auf S. 186 besprochen, die Impedanz des Hochfrequenzgenerators so erhöht, daß dieser zu einer Quelle konstanten Stroms wird. Zweifellos ist dies die übersichtlichste und technisch einfachste Art, kernmagnetische Resonanzen zu beobachten. Es leuchtet sofort ein, daß die auf diesem Wege der Spannung $U_1 + U_\chi$ beigemischte fiktive Kompensationsspannung U_2 stets dieselbe Phase wie U_1 hat. Die

[1] P. Grivet, M. Soutif u. M. Buyle: C. R. Acad. Sci. Paris **229**, 113 (1949).
[2] M. Soutif: Rev. Sci. Paris **89**, 203 (1951).
[3] P. Grivet, M. Soutif u. R. Gabillard: Physica, Haag **17**, 420 (1951).
[4] Siehe Fußnote 3, S. 191.
[5] C. E. Waring, R. H. Spencer u. R. L. Custer: Rev. Sci. Instrum. **23**, 497 (1952).
[6] B. V. Rollin: Nature, Lond. **158**, 669 (1946). — Rep. Progr. Phys. **12**, 22 (1949). — J. Hatton u. B. V. Rollin: Proc. Roy. Soc. Lond., Ser. A **199**, 222 (1949).

Kernsignale sind deshalb immer Absorptionssignale und können auch durch Frequenzschwankungen des Generators nicht gestört werden. Dagegen wird das Amplitudenrauschen des Senders mit den Kernsignalen verstärkt, während es in den zuvor beschriebenen Brücken kompensiert wurde. Das Signal-Rauschverhältnis dieser Beobachtungsart ist also erheblich schlechter. Aus diesem Grunde haben THOMAS und HUNTOON von dem obigen Prinzip ausgehend, eine sog. Amplitudenbrücke[1] (Fig. 44) entwickelt. Am Ausgang des in der Abbildung oberen Meßkreises werden die Kernsignale und mit ihnen das Generatorrauschen

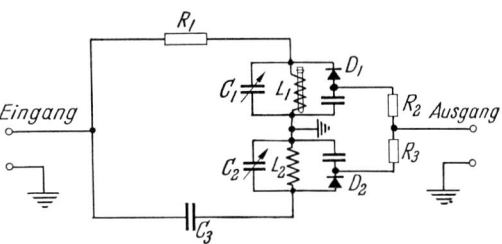

sofort durch eine Diode gleichgerichtet. Auf Hochfrequenzverstärkung wird also ganz verzichtet. Das identische Generatorrauschen des zweiten, in der Abbildung unteren, symmetrischen Blind-Brückenzweiges wird gegenphasig gleichgerichtet. Bei der Mischung beider Spannungen ändern sich die Kernsignale nicht, dagegen werden der Rausch und eine eventuelle Netzfrequenzmodulation (Brumm) des Generators kompensiert.

Fig. 44. Amplitudenbrücke von THOMAS und HUNTOON. Die Hochfrequenzspannung wird sofort gleichgerichtet, der Blindkreis dient zur Kompensation des Generatorrauschens.

Ebenso wie bei der Kompensation durch anfängliche C-Verstärkung erhält man unabhängig von Phasenschwankungen stets ein Absorptionssignal. Leider wird mit den Gleichrichterdioden eine neue

Rauschquelle in die Schaltung eingeführt. Um diesen Nachteil zu vermindern, ist es zweckmäßig, anstatt Kristalldioden Röhrendioden zu verwenden. Der Verzicht auf Hochfrequenzverstärkung und damit auf ein rauscharmes schmalbandiges Arbeiten wirkt sich auf das Signal-Rausch-Verhältnis natürlich nachteilig aus.

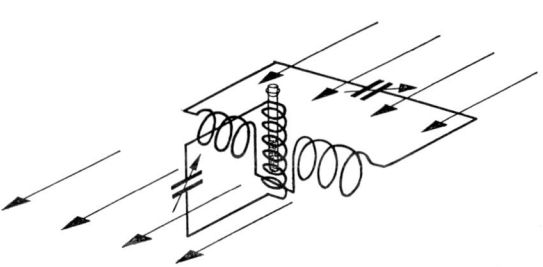

Fig. 45. Anordnung eines Zweispulensystems im Magnetfeld.

Brauchbar ist diese Brückenschaltung also nur, wenn die Kernsignale sehr stark sind. Dann ist sie allerdings wegen ihrer einfachen Handhabung und auf Grund ihrer geringen Störanfälligkeit zu empfehlen.

11. Anregung und Beobachtung mit zwei gekreuzten Spulen („Kerninduktion"). BLOCH und seine Mitarbeiter HANSEN und PACKARD haben in dem von ihnen entdeckten Verfahren[2,3] zur Beobachtung kernmagnetischer Präzessionsbewegungen — sie nannten es *Kerninduktion* — ein System von zwei gekreuzten Spulen (Fig. 45) benutzt. In deren Mitte befindet sich die zu untersuchende feste oder flüssige Substanz. Beide Spulen liegen in der zur Richtung des konstanten Feldes transversalen Ebene. Außerdem stehen auch ihre Achsen aufeinander senkrecht. Durch eine, meist die äußere Spule fließt der Generatorstrom, sie dient also zur Anregung der Präzessionsbewegung. In der zweiten, mit dem Empfänger verbundenen Spule wird die Bewegung beobachtet. Zur Vergrößerung des Schwingkreis-Stromes bzw. zur Erhöhung der Empfindlichkeit gegenüber Induktivitätsänderungen ergänzt man beide Spulen durch Kondensatoren zu Schwingkreisen

[1] H. A. THOMAS u. R. D. HUNTOON: Rev. Sci. Instrum. 20, 516 (1949).
[2] F. BLOCH, W. W. HANSEN u. M. PACKARD: Phys. Rev. 69, 127 (1946); 70, 474 (1946).
[3] F. BLOCH: Phys. Rev. 70, 460 (1946).

derselben Resonanzfrequenz. Wir wollen zunächst davon ausgehen, daß die Kreise elektrisch exakt voneinander entkoppelt sind, insbesondere soll die Gegeninduktivität der beiden Spulen Null sein. Aus Ziff. 6 wissen wir, daß etwa beim langsamen adiabatischen Resonanzdurchgang eine Präzessionsbewegung des resultierenden Magnetisierungsvektors angeregt wird. Dessen transversale Komponenten wurden in Gl. (6.7) berechnet. In jeder Spule, deren Achse in der zum Gleichfeld senkrechten Ebene liegt, induziert der erzwungen rotierende Magnetisierungsvektor eine Wechselspannung der LARMOR-Frequenz.

Auf Grund der Voraussetzung idealer Entkopplung liegt nur diese Spannung am Beobachtungskreis. Ihre Größe findet man wie folgt. Die Achse der Beobachtungsspule habe in einem ortsfesten Koordinatensystem die Richtung der positiven y-Achse. Die Induktion B der Komponente $M_{y\infty}$, Gl. (6.7b), des Magnetisierungsvektors ist $B_y = 4\pi M_{y\infty}$. Der Gesamtfluß Φ durch die Spule, die aus N Windungen bestehen und die Querschnittsfläche F haben möge, hat also den Betrag:

$$\Phi = N F B_y = 4\pi N F M_y.$$

In der Spule wird eine Spannung der Größe

$$U = -\frac{1}{c}\frac{d\Phi}{dt} = -\frac{4\pi}{c} N F \frac{dM_y}{dt}$$

induziert. Der Betrag dieser Spannung hängt stark von der Gleichfeldstärke, von den Relaxationszeiten und von den apparativen Daten ab. Bei den meisten Versuchen ist die induzierte Spannung etwa 10^{-3} bis 10^{-6} Volt. Sie kann natürlich kleiner und auch noch größer sein. Verstärkt und gleichgerichtet kann man ihre Abhängigkeit vom Resonanzabstand z.B. auf einem Oszillographenschirm verfolgen und somit die Resonanzfrequenz der Kerne ermitteln.

Aus der voranstehenden Darstellung könnte man den Eindruck gewinnen, daß sich diese Beobachtungsart grundsätzlich von den in Ziff. 10 besprochenen Brückenschaltungen unterscheidet. Dies ist jedoch nicht der Fall. Die Signalspannung eines ideal entkoppelten Zweispulensystems stimmt mit der von einer vollkommen abgeglichenen Brücke gelieferten Spannung überein. Beide Justierungen sind aber gleich schwierig. Um eindeutige Signalformen zu gewinnen, ist es sogar zweckmäßig, wie in Ziff. 9 ausgeführt wurde, nicht vollständig zu kompensieren $(|U_1 - U_2| \gg U_x)$, bzw. bei der vorliegenden Anordnung einen Teil der Generatorspannung mit der Signalspannung zu mischen.

Eine Überlagerung beider Spannungen im Beobachtungskreis stellt sich aber bei zwei gekreuzten Spulen ganz von selbst ein. Unsymmetrien in den Spulen und Streukapazitäten zwischen den beiden Kreisen sorgen dafür, daß stets ein Teil der Senderspannung auch auf die Empfangsspule übertragen wird. Selbst wenn man diese technisch unvermeidlichen Kopplungen ausschließen könnte, dürfte bei der geometrischen Senkrechtjustierung der beiden Spulenachsen die Winkelabweichung höchstens 0,2″ betragen, wenn die vom Beobachtungskreis aufgenommene Spannung 10^6 mal kleiner sein soll als die des Anregungskreises[1,2]. Um definiert beobachten zu können, muß man außerdem die Amplitude und Phase der Überlagerungsspannung regeln. BLOCH, HANSEN und PACKARD haben deshalb folgenden Kunstgriff angewandt. Nach der geometrischen Grobjustierung der Spulenkörper regeln sie die Richtung des Feldes der Generatorspule. Dazu befinden sich an den Enden dieser Spule ein oder zwei um ihre Achse drehbar

[1] L. GIULOTTO, A. GIGLI u. P. SILLANO: Nuovo Cim. 4, 201 (1947).
[2] L. GIULOTTO u. A. GIGLI: Nuovo Cim. 4, 275 (1947).

angeordnete Leitbleche (Trimmer, Paddles), z.B. halbkreisförmige Kupferscheiben. Der hochfrequente magnetische Fluß induziert in ihnen Wechselspannungen und damit Wirbelströme. Je nach der Stellung der Trimmer wird die Richtung des Flusses geändert (Fig. 46). Auf diese Weise kann man die Gegeninduktivität der beiden Spulen tatsächlich sehr klein machen. Praktisch ist es jedoch wichtiger, daß man auch die Phase der übertragenen Restspannung, genau wie in den übrigen Brückenschaltungen regeln und somit wahlweise entweder die Dispersion oder die Absorption der untersuchten Substanz beobachten kann. Dies ist leicht einzusehen, die in den Leitblechen induzierten Wirbelströme und damit der Justierfluß haben sowohl eine Blind- als auch eine Wirkkomponente. Im Prinzip ist also die Zweispulen-Anordnung ebenfalls eine Brückenschaltung. Man kann sie etwa als eine Brücke beschreiben, die einen Übertrager benutzt (vgl. Fig. 41), in dessen Mitte sich die Probe befindet. Die beschriebene Feinjustierung ist natürlich mühevoll und kann nur durch die Beobachtung des Kernsignales selbst überprüft werden. Fig. 47 zeigt zwei praktische Ausführungen solcher Probenköpfe mit gekreuzten Spulen. Das optimal erreichbare Signal-Rausch-Verhältnis ist gleich dem der besten übrigen Brückenschaltungen, denn vom Generatorrauschen gelangt ebenfalls nur ein kleiner Rest zum Empfänger.

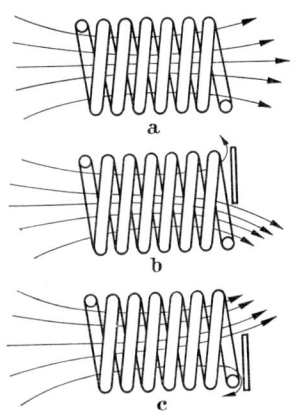

Fig. 46a — c. Zur Justierung des magnetischen Hochfrequenzflusses einer Spule mit Hilfe von Leitblechen nach BLOCH, HANSEN und PACKARD. a Normaler Verlauf des Flusses. b und c Änderung der Flußrichtung durch Leitbleche.

Ausdrücklich soll noch einmal darauf hingewiesen werden, daß die beiden Namen Kerninduktion und Kernresonanz-Absorption verschiedene Bezeichnungen des gleichen Experimentes sind. Dem Namen Kerninduktion liegt die Vorstellung zugrunde, daß angeregte Kernmomente synchron präzessieren und somit in einer Spule eine Wechselspannung induzieren. Dagegen weisen die Bezeichnungen Kern-Absorption bzw. -Dispersion auf die Änderung des Imaginär- bzw. Realteils der Kern-Suszeptibilität der Substanz in Resonanz hin. Beide Beschreibungsweisen sind aber einander äquivalent. Insbesondere ist zu betonen, daß das Präzessionsbild nicht fiktiv, sondern tatsächlich den Ablauf des Experimentes beschreibt. Den abschließenden Beweis zu dieser Frage haben die Beobachtungen der freien Präzessionsbewegung, also des Abklingvorgangs geliefert. Zum Teil ist es üblich geworden, von Kerninduktion nur dann zu sprechen, wenn eine Zweispulen-Anordnung benutzt wird; und von Absorption bzw. Dispersion, wenn die Kernsignale in einer Spule angeregt und beobachtet werden. Dies ist wie gezeigt wurde, nicht berechtigt.

Meßtechnisch kann man mit Hilfe gekreuzter Spulen auch das Vorzeichen des gyromagnetischen Verhältnisses γ relativ ermitteln, kann also feststellen, ob das Impulsmoment eines Kernes parallel oder antiparallel zu seinem magnetischen Moment liegt. Je nach dem Vorzeichen von γ rotiert der Magnetisierungsvektor im Feld im Rechts- oder Links-Sinn. Entsprechend wird das Vorzeichen der Signalspannung positiv oder negativ, bzw. anders ausgedrückt ihre Phase unterscheidet sich in den beiden Fällen um $180°$. Beobachtet man etwa das normale, auf dem Oszillographen-Bildschirm über der Grundlinie liegende Absorptionssignal einer Kernart mit positivem gyromagnetischen Verhältnis und daran anschließend mit der gleichen Justierung des Hochfrequenzfeldes und derselben elektronischen Anordnung eine zweite Kernart mit negativem γ, dann liegt deren

Absorptionssignal umgeklappt unter der Grundlinie. Der einzige Parameter, der bei einer solchen Relativbestimmung des Vorzeichens von γ geändert werden darf, ist die Stärke des konstanten Feldes, denn die Signale beider Kernarten müssen natürlich nacheinander in Resonanz, also bei derselben mit der Senderfrequenz übereinstimmenden Larmor-Frequenz verglichen werden. Werden die

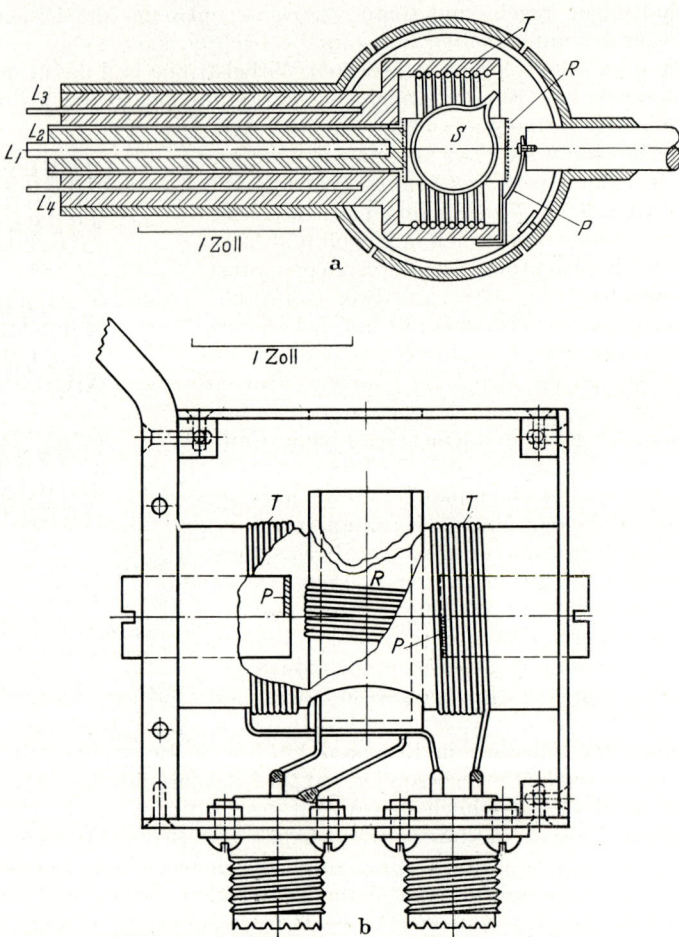

Fig. 47a u. b. Konstruktionsbeispiele von Probenköpfen mit gekreuzten Spulen. a Anordnung von Bloch, Hansen und Packard. T Senderspule, R Empfangsspule, S untersuchte Substanz, P Leitblech, $L_1 - L_4$ koaxiale Zuleitung zu den Spulen. Die Abschirmung ist geschlitzt, um durch die Feldmodulation verursachte Wirbelströme zu verhindern. b Anordnung von Proctor. Die Senderspule ist in ein Helmholtz-Spulenpaar aufgeteilt. An der Oberseite des Kopfes befindet sich eine Öffnung, durch welche die meist in einem Reagenzglas enthaltene Substanz in die Empfangsspule gesteckt wird. Dadurch können die Proben während einer Versuchsreihe von außen leicht ausgewechselt werden.

Kernsignale in einer Spule angeregt und beobachtet, dann kann das Vorzeichen von γ aus Symmetriegründen die Signalformen prinzipiell nicht beeinflussen.

Die Phasen- und Amplituden-Regelung der von der Sender- auf die Empfängerspule übertragenen Wechselspannung ist allein mit Hilfe der beschriebenen Feldtrimmer schwierig und unbefriedigend. Ebenso wie bei den übrigen Brückenschaltungen ist es wünschenswert, daß man das Phasen- und das Amplitudengleichgewicht wahlweise unabhängig voneinander einstellen kann. In der Folge haben deshalb die meisten Autoren, z.B. PACKARD[1], LEVINTHAL[2], PROCTOR[3] zusätzlich eine durch variable RC-Glieder regelbare Spannung direkt vom Sender zum Empfänger geleitet und dort mit der Spannung der Beobachtungsspule gemischt. Praktisch handelt es sich hierbei um eine teilweise Kompensation der direkt übertragenen Spannung, entsprechend den im letzten Paragraphen beschriebenen Methoden. GVOZDOVER und IEVSKAYA[4] haben sogar ganz auf Feldtrimmer verzichtet und die Überlagerungsspannung nur auf diesem Wege, also durch elektrische Schaltelemente geregelt. Dagegen hat WEAVER[5] die geometrische Feldjustierung verbessert. Die Einstellung des Absorptionsgleichgewichtes erfolgt in seiner Anordnung grob durch die mechanische Neigung einer Hälfte der Senderspule und damit des Senderfeldes mit Hilfe einer Mikrometerschraube und fein durch eine kleine verstellbare Kupfer-scheibe (Wirbelstromregelung). Zur Regelung des Dispersionsgleichgewichtes benutzt WEAVER, fußend auf eine Idee von F. WESTERN, eine Hilfsspule von der in Fig. 48 dargestellten Art. In einer Schleife senkrecht zur Achse der Senderspule wird durch deren Feld eine Spannung induziert. Der Stromkreis ist über eine zweite transversale Schleife und über einen Widerstand R geschlossen. R ist so groß, daß der Widerstand der gesamten die Empfänger- und Senderspule verkoppelnden Anordnung praktisch reell ist. Dreht man sie um die y-Achse, dann ändert sich zwar die induzierte Spannung und damit die Stromstärke nicht, wohl aber der Übertragungsgrad, also die Gegenkopplung, auf die zur x-Achse parallele Empfangsspule.

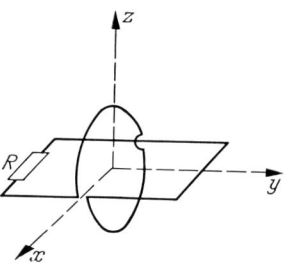

Fig. 48. Hilfsspule nach WESTERN zur Regelung der Kopplung zwischen zwei gekreuzten Spulen. Die Achse der Senderspule liegt parallel zur x-Achse und die der Empfangsspule parallel zur y-Achse. Der OHMsche Widerstand R ist so groß, daß der Gesamtwiderstand der Hilfsspule praktisch reell ist.

Eine weitere Verbesserung des Signal-Rausch-Verhältnisses hat BAKER vorgeschlagen. Um Schwankungen des Hochfrequenzgenerators als Rausch- und Brummquelle ganz auszuschließen, muß man mit einer völlig abgeglichenen Brücke arbeiten, bzw. im Fall des Zweispulensystems durch geometrische oder elektrische Gegenkopplung dafür sorgen, daß die Beobachtungsspule ganz von der Senderspule entkoppelt ist. Im Hochfrequenzempfänger wird dann nur das reine Kernsignal verstärkt. Um eindeutige Signale justieren zu können, mischt man unmittelbar vor der Gleichrichtung die verstärkten Kernsignale mit einer vom Generator abgezweigten Vergleichsspannung variabler Phase und Amplitude, die man vorher beschnitten und damit von Rausch- und Brummspannungen gereinigt hat[6]. Alle Zweispulen-Systeme müssen mechanisch sehr stabil ausgeführt werden, da sonst jede Erschütterung ihre Justierung zerstören würde. In dieser Hinsicht sind sie unbequemer als die in Ziff. 10 besprochenen Brückenschaltungen.

12. Differentielle Abtastung und phasenempfindliche Gleichrichtung, statische Abtastung. Zur Verminderung des Rauschens, d.h. zur Verbesserung des Signal-Rausch-Verhältnisses gibt es im Prinzip zwei Möglichkeiten. Man kann die Intensität des Rauschspektrums, also die Leistung der verschiedenen Rauschgeneratoren

1 M. E. PACKARD: Rev. Sci. Instrum. **19**, 435 (1948).
2 E. C. LEVINTHAL: Phys. Rev. **78**, 204 (1950).
3 W. G. PROCTOR: Phys. Rev. **79**, 35 (1950).
4 S. D. GVOZDOVER u. N. M. IEVSKAYA: J. Exp. Theor. Phys. **25**, 435 (1953).
5 H. E. WEAVER: Phys. Rev. **89**, 923 (1953).
6 E. B. BAKER: Rev. Sci. Instrum. **25**, 390 (1954).

durch eine entsprechende Dimensionierung der Brücke, durch die Auswahl geeigneter Eingangsröhren des Verstärkers und durch eine möglichst vollständige Kompensation des Generatorrauschens so klein als möglich halten. Die schrittweise Entwicklung solcher technischer Verbesserungen ist naturgemäß mühevoll. Schneller und drastischer hat man dagegen Erfolg, wenn man den Abschnitt des Rauschspektrums verkleinert, der gemeinsam mit den Kernsignalen verstärkt und beobachtet wird. Der Betrag der Rauschspannung nimmt proportional mit der Wurzel einer solchen Verminderung der Beobachtungs-Bandbreite des Verstärkers ab.

In der bisher beschriebenen Beobachtungstechnik werden die beim Resonanzdurchgang angeregten Kernsignale (Fig. 35) ganz auf dem Bildschirm eines Oszillographen abgebildet. Das konstante Magnetfeld wird also mit einer Amplitude,

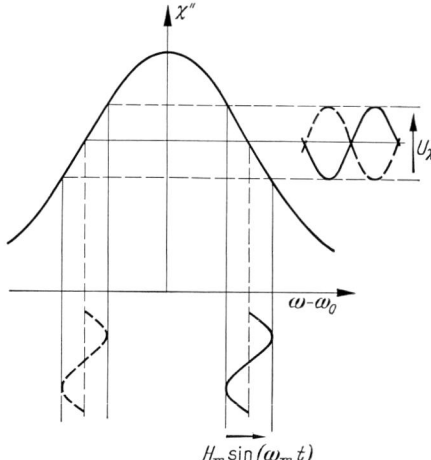

Fig. 49. Differentielle Abtastung einer Absorptionskurve. Die Phase der Kernsignalspannung ändert sich um 180°, wenn die Steigung der Signalkurve das Vorzeichen wechselt.

die groß gegen die Halbwertsbreite des Kernsignals ist, niederfrequent moduliert. Die notwendige Mindestbandbreite des Empfängers folgt bei dieser Beobachtungsart aus der Forderung nach formgetreuer Wiedergabe der Signalformen durch den Verstärker. Um Verzerrungen zu vermeiden, muß sie auf jeden Fall ein Vielfaches der reziproken Signaldauer betragen.

Will man demnach die Beobachtungsbandbreite um Größenordnungen herabsetzen, dann muß man zunächst die Beobachtungstechnik abändern. Eine Möglichkeit hierzu ist, die Amplitude der Modulation des konstanten Feldes drastisch zu verkleinern. Dies bedeutet, daß innerhalb einer Modulationsperiode nur ein Ausschnitt der Resonanzlinie beobachtet wird. Setzt man voraus, daß dieser Abschnitt klein gegen die Halbwertsbreite, etwa des Absorptionssignals ist (Fig. 49), dann kann man jedes derartige Stück der Kernsignalkurve durch ein Geradenstück approximieren. Die zu beobachtende Signalspannung ist dann praktisch sinusförmig. Ihre Amplitude ist proportional zur Steigung des untersuchten Kurvenstücks, während ihre Frequenz mit derjenigen der Modulation übereinstimmt. Ferner ändert sich die Phase dieser Wechselspannung um 180°, wenn die Ableitung der Signalkurve ihr Vorzeichen wechselt. Zur Untersuchung des gesamten Kernsignals ändert man entweder die Senderfrequenz oder die Magnetfeldstärke langsam monoton und registriert während des Resonanzdurchgangs die Amplitude und Phase der Signalspannung. Auf diesem Wege erhält man den Verlauf der ersten Ableitung der Signalkurve, man bezeichnet deshalb das voranstehend beschriebene Beobachtungsverfahren als differentielle Abtastung.

Da stets nur dieselbe Modulationsfrequenz zu verstärken ist, kann der an den Hochfrequenzempfänger anschließende Niederfrequenzverstärker beliebig schmalbandig sein. Zur Anwendung des obigen Abtastverfahrens ist jede kontinuierlich arbeitende Apparatur von der Art der Fig. 35 brauchbar. Man muß nur die Amplitude des Modulationsgenerators entsprechend verkleinern und die Bandbreite des Niederfrequenz-Verstärkers soweit als möglich herabsetzen. Zusätzlich benötigt man noch einen Niederfrequenz-Demodulator, dessen Ausgangs-Gleichspannung proportional zur Amplitude der niederfrequenten Eingangs-Wechselspannung ist. Außerdem soll das Vorzeichen dieser Gleichspannung von der

Phasenlage der Signalspannung abhängen, es soll also z. B. positiv sein, wenn die Signalkurve steigt und negativ, wenn sie fällt. Zur Feststellung der Signalphase muß diese an irgend einer Stelle des Niederfrequenzteils der Schaltung mit der Phase der Modulationsspannung verglichen werden. Anschließend an den Niederfrequenz-Gleichrichter wird die Signalgleichspannung, meist zuvor durch ein Gleichspannungs-Röhrenvoltmeter verstärkt und entkoppelt, von irgendeinem Schreiber aufgezeichnet, dessen Papiergeschwindigkeit proportional zur Geschwindigkeit des Resonanzdurchganges, also etwa des langsamen Feldanstiegs ist. Den in Fig. 35 eingezeichneten Oszillographen benötigt man bei der differentiellen Abtastung an sich nicht mehr, behält ihn jedoch im allgemeinen zur Überwachung der Apparatur bei.

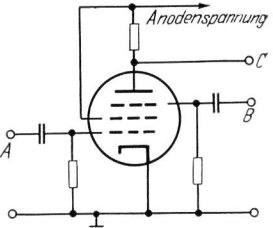

Glücklicherweise sind in der Hochfrequenztechnik schon seit langem Anordnungen bekannt[1], die, wie hier gefordert, extrem schmalbandig, linear und phasenempfindlich verstärken und gleichrichten. Berichtet haben darüber u. a. CHANCE[2] und GEYGER[3].

Die Arbeitsweise solcher phasenempfindlicher Verstärker (,,Lock-in") versteht man leicht an Hand der Schaltung[4] Fig. 50. An dem Gitter einer Pentode (A) liegt die Signalspannung und an deren Bremsgitter (B) die Vergleichspannung. Die letztere wird über einen Phasen- und Amplitudenregler vom Modulationsgenerator abgezweigt. Sie ist im allgemeinen exakt sinusförmig. Die am Steuergitter liegende Spannung hat zwar dieselbe Grundfrequenz, enthält jedoch auch

Fig. 50. Schaltung einer phasenempfindlich verstärkenden Pentode. Am Gitter A liegt die Signal-Eingangsspannung, am Bremsgitter B die vom Modulationsgenerator abgezweigte Vergleichsspannung und an der Anode C die phasenempfindlich verstärkte Ausgangsspannung.

Oberwellen, denn die Signalkurve kann ja an keiner Stelle exakt durch ein Geradenstück ersetzt werden. Der durch die Röhre fließende Wechselstrom und damit die Wechselspannung am Punkt C ist somit proportional zu dem Produkt:

$$(a + b \sin(\omega_m t)) \cdot \sum_{n=1}^{\infty} c_n \sin(n\omega_m t + \varphi_n). \qquad (12.1)$$

Darin ist die rechtsstehende Summe die in eine FOURIER-Reihe entwickelte Signalspannung und die linke Klammer die von der Vergleichsspannung abhängige Röhrensteilheit (multiplikative Mischung). Für die effektive (zeitlich mittlere) Spannung $\overline{U}$ gilt somit:

$$\overline{U} \sim \left[a \sum_{n=1}^{\infty} c_n \int_0^{2\pi/\omega_m} \sin(n\omega_m t + \varphi_n)\, dt + b \sum_{n=1}^{\infty} c_n \int_0^{2\pi/\omega_m} \sin(\omega_m t) \sin(n\omega_m t + \varphi_n)\, dt \right].$$

Das erste Integral verschwindet für sämtliche Werte von n und das zweite für alle $n > 1$. Es ist also:

$$\overline{U} \sim c_1 \int_0^{2\pi/\omega_m} \sin^2(\omega_m t) \cos \varphi_1 \, dt \sim c_1 \cos \varphi_1. \qquad (12.2)$$

Die mittlere Ausgangsspannung des phasenempfindlichen Verstärkers ist demnach proportional zum FOURIER-Koeffizienten c_1 der Grundschwingung der Signalspannung, unabhängig von eventuell vorhandenen Oberwellen und phasenempfindlich, denn ihr Vorzeichen wird durch die Phasendifferenz φ_1 bestimmt.

[1] C. H. WALTER: Z. techn. Phys. **13**, 363, 436 (1932).
[2] B. CHANCE: Waveforms, p. 511. New York: McGraw-Hill 1949.
[3] W. GEYGER: Arch. elektr. Übertragung **3**, 165 (1949).
[4] E. R. ANDREW: Nuclear Magnetic Resonance, p. 44. Cambridge: University Press 1955.

Im Experiment sorgt man durch eine passende Justierung der Vergleichs-spannung dafür, daß ihre Phasendifferenz φ_1 zur Signalspannung für eine an-steigende Signalkurve Null wird. Damit wird die mittlere Ausgangsspannung positiv, bzw. für eine fallende Signalkurve negativ, außerdem wird der Betrag der mittleren Spannung in beiden Fällen maximal. Eine derartige phasenempfind-liche Schaltung verstärkt, gemittelt über eine Periode der Signalspannung (effektiv), von dem gesamten ihr am Eingang angebotenen Rauschspektrum nur den Frequenzabschnitt unmittelbar um die Frequenz der Vergleichs- und Signal-spannung. Allerdings schwankt die Phasenlage dieser aus dem Rauschpegel aus-gesonderten Spannung statistisch, und mit ihr verändert sich auch der Betrag der resultierenden effektiven Ausgangsspannung ständig unregelmäßig. Um diese Rauschschwankungen soweit als möglich auszugleichen, sorgt man dafür, daß die Registriervorrichtung (Röhrenvoltmeter, Schreiber) auf Änderungen der mittleren Spannung nur sehr langsam anspricht. Man macht also ihre Zeitkon-stante t_R, etwa durch groß dimensionierte RC-Glieder sehr hoch. Die Bandbreite des gesamten Verstärkers wird damit auf den Betrag $1/t_R$ reduziert. Arbeitet die Registriervorrichtung z.B. mit einer Zeitkonstante von 500 sec, dann ist die Bandbreite $^2/_{1000}$ Hz, und nur dieser schmale Ausschnitt des Rauschspektrums stört die Beobachtung der Signalspannung. Die Bandbreite der Vorverstärker ging in die voranstehende Betrachtung nicht ein. Trotzdem ist es bei der An-wendung der phasenempfindlichen Verstärkung und Gleichrichtung zweckmäßig, auch diese möglichst klein zu halten. Eine zu große am Eingang liegende Rausch-spannung kann die Pentode übersteuern, und damit wird der obige Ansatz falsch. Außerdem ist es klar, daß mit Rücksicht auf die große Zeitkonstante der Beob-achtungsvorrichtung die Geschwindigkeit des Resonanzdurchgangs sehr klein sein muß. Nur dann kann der Schreiber den Änderungen der Signalspannungs-amplitude folgen.

Eine einfache Schaltung von der Art der Fig. 50 ist jedoch zur praktischen Anwendung nicht geeignet. Selbst wenn der Rauschpegel am Eingang A ver-nachlässigbar klein wäre, würde die Ausgangsspannung am Punkt C doch un-regelmäßig schwanken. Weder die Versorgungsspannungen der Pentode (Anode, Heizung), noch ihre elektrischen Daten (Steilheit, Innenwiderstand u. a.), noch die Amplitude der Vergleichsspannung können in einem sich über mehrere Stunden erstreckenden Experiment als ausreichend stabil vorausgesetzt werden.

In der ersten zur Registrierung kernmagnetischer Resonanzen mit Erfolg angewandten phasenempfindlichen Anordnung[1] hat daher DICKE eine Gegen-taktschaltung benutzt (Fig. 51). Ein Teil der genannten Störquellen kompen-sieren sich in dieser Schaltung in erster Näherung. Ihr Eingang und ihr Ausgang sind symmetrisch. Die Vergleichsspannung liegt gleichphasig an den beiden Bremsgittern. Die Wirkungsweise der Anordnung stimmt im übrigen mit der von Fig. 50 überein.

Eine Schaltung besonders hoher Stabilität hat SCHUSTER entwickelt[2] (Fig. 52). Er verwendet eine Pentode, in deren Anodenleitung sich eine Doppel-triode befindet. Die beiden Triodensysteme werden von der Vergleichsspannung gesteuert, und zwar sind sie alternierend gesperrt bzw. stromführend. Der durch die Signaleingangsspannung gesteuerte Anodenstrom der Pentode fließt also im Rhythmus der Vergleichsspannung abwechselnd durch die Widerstände R_1 bzw. R_2. Die Widerstände R_3 und R_4 dienen zur Begrenzung der Triodengitterströme. Diese Gitterströme fließen direkt zu den Kathoden der Trioden, sie beeinflussen

¹ R. H. DICKE: Rev. Sci. Instrum. 17, 268 (1946); s. auch N. BLOEMBERGEN: Nuclear Magnetic Relaxation. The Hague: Nijhoff.
² N. A. SCHUSTER: Rev. Sci. Instrum. 22, 254 (1951).

also den Anodenstrom der Pentode nicht. Die Kondensatoren C_1 und C_2 verhindern, daß die Trioden-Anoden in den Sperrperioden wieder das Potential $+B$ erhalten, sie schließen den Wechselstromanteil der beiden Anodenstromzweige kurz. Der an den Fußpunkten von R_1 und R_2 liegende Ausgang der Schaltung liefert damit eine Gleichspannung.

Zur Stabilität der Schaltung trägt wesentlich bei, daß der Innenwiderstand einer Pentode ($\sim 10^6$ Ohm) immer sehr viel größer ist, als der einer Triode ($\sim 10^4$ Ohm). Schwankt etwa die Amplitude der Vergleichsspannung oder der Innenwiderstand eines der im Gegentakt arbeitenden Triodensysteme um einen gewissen Wert, dann ist die zugehörige Anodenstromänderung 100mal kleiner als in der Schaltung Fig. 51. Ein weiterer Vorteil ist, daß die Signalspannung

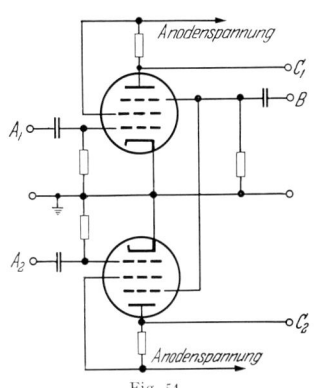

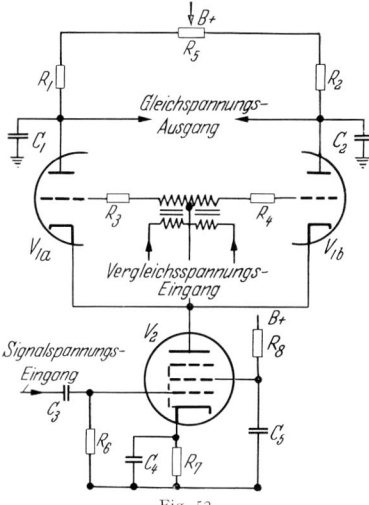

Fig. 51.   Fig. 52.

Fig. 51. Phasenempfindliche Gegentakt-Verstärkerschaltung von DICKE. Die Eingangsspannung ist symmetrisch, liegt also mit 180°-Phasendifferenz an den Gittern (A_1, A_2) der zwei Pentoden. Die Vergleichsspannung steuert gleichphasig deren Bremsgitter (B). Der durch die Anoden (C_1, C_2) beider Röhren gebildete Ausgang ist symmetrisch und liefert die Differenzspannung der beiden im Gegentakt arbeitenden Hälften der Schaltung.

Fig. 52. Phasenempfindlicher Verstärker und Gleichrichter von SCHUSTER. Die Signaleingangsspannung liegt am Gitter einer verstärkenden Pentode. In deren Anodenleitung befindet sich eine Doppeltriode, die gitterseitig über einen Übertrager durch die Vergleichsspannung gesteuert wird, also ein synchronisierter elektronischer Schalter. Dadurch werden in Funktion der Phasenlage der Signalspannung die Filter R_1C_1 und R_2C_2 verschieden aufgeladen. Der durch sie gebildete Ausgang liefert unmittelbar die Differenz-Gleichspannung beider Stromzweige und damit die effektive Signalspannung.

nur eine Röhre steuert. Steilheitsschwankungen dieser Röhre beeinflussen aber die Stromstärke in beiden Stromzweigen gleichermaßen und wirken sich somit auf die Messung der Differenzspannung in erster Näherung nicht aus. Entsprechend wird auch die Empfindlichkeit gegenüber Änderungen der Versorgungsspannung reduziert.

Die beiden Ausgänge des beschriebenen phasenempfindlichen Gleichrichters kann man über zwei sehr große Widerstände und einen zwischen diesen liegenden großen Kondensator verbinden. Die Zeitkonstante dieses RC-Gliedes bestimmt dann die Bandbreite des Verstärkers. Um registrieren zu können, muß man solch hohe Impedanzen natürlich transformieren. Es ist empfehlenswert, hierzu zwischen den Gleichspannungsausgang und den Schreibereingang zwei Kathodenfolger zu schalten.

Zu beachten ist, daß in der Anordnung von SCHUSTER die beiden Stromzweige periodisch gesperrt werden, obwohl die Schaltspannung sinusförmig ist. Jedoch werden deren positive Halbwellen durch Gitterstrombegrenzung abgeschnitten, während sich ihre negativen zum Teil in den Sperrbereichen der Trioden befinden. Von einer multiplikativen Mischung kann man also nicht mehr sprechen, denn die Steilheit der Triodensysteme ändert sich unstetig. In einer von Cox angegebenen Schaltung[1] (Fig. 53) eines phasenempfindlichen Gleichrichters wird die Vergleichsspannung sogar zunächst in eine Rechteckwellenspannung umgeformt. Diese Spannung liegt an den Anoden zweier Dioden, deren Kathoden über zwei Widerstände R_1 und R_2 mit dem Signaleingang verbunden sind. Das Gleichspannungspotential der Rechteckwelle ist so gelegt, daß die Dioden in den unteren Halbwellen gesperrt sind und in den oberen eine leitende Verbindung zur Erde bilden. Die Anordnung arbeitet somit als synchronisierter Gleichrichter. Die Differenz der Ladespannungen der beiden Filter R_1C_1 und R_2C_2 kann direkt mit einem Röhrenvoltmeter gemessen werden.

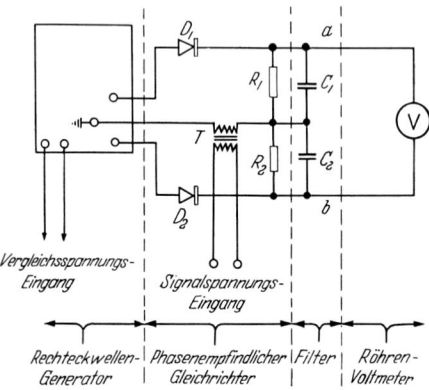

Fig. 53. Phasenempfindliche Gleichrichterschaltung von Cox. Durch eine Rechteckwellen-Vergleichsspannung werden die beiden Dioden D_1 und D_2 abwechselnd gesperrt bzw. zur Erde kurzgeschlossen. Die Aufladung der beiden Filter R_1C_1 und R_2C_2 durch die Signalspannung wird damit phasenempfindlich gesteuert. Die zeitlich mittlere Differenzspannung kann direkt mit einem Röhrenvoltmeter gemessen werden.

Für einen phasenempfindlichen Gleichrichter der zuletzt beschriebenen Art gilt natürlich der Ansatz (12.1) nicht mehr. Die Vergleichsspannung enthält Oberwellen und damit verschwinden die Integrale nicht mehr für $n > 1$. Es werden also auch höhere Harmonische der Grundschwingung der Signalspannung und damit auch des Rauschpegels mitregistriert. Praktisch stört dies jedoch nicht, denn man kann die Amplitude dieser Oberwellen bereits im Vorverstärker beliebig schwächen, wenn man dessen Bandbreite ausreichend klein macht.

Ergänzend sollen noch einige Anordnungen zur Registrierung kernmagnetischer Signale besprochen werden, die zwar nicht zu der voranstehend behandelten Verfahrenstechnik, wohl aber zu dem darin enthaltenen physikalischen Problem kreis gehören. Um sehr schwache im Rauschpegel verschwindende Resonanzlinien beobachten zu können, haben BLOCH und GARBER eine normale Apparatur, entsprechend dem Prinzipschaltbild der Fig. 35 benutzt, jedoch die Registrierung der Kernsignale photographisch verbessert[2]. Ist die Verstärkungsziffer des Verstärkers genügend groß, dann geht die normalerweise auf dem Oszillographenschirmbild zu beobachtende gerade Grundlinie in eine statistisch schwankende Kurve über, die den jeweiligen Verlauf der Rauschspannung wiedergibt. Visuell kann man schwache Signale innerhalb dieser sich laufend verändernden Kurve nicht mehr erkennen. BLOCH und GARBER haben das Schirmbild deshalb photographiert, und dabei die Belichtungszeit so lang gewählt (einige 100 sec), daß in jeder Aufnahme mehrere tausend Ablenkperioden des Oszillographen-Bildpunktes registriert werden. Auf der Photographie überlagern sich dann alle Einzelkurven zu einem mehr oder weniger diffus begrenztem Band, dessen Breite vom Verstärkungsgrad abhängt. Die Schwärzung des Bildes ist in der Mitte des Bandes am dichtesten und wird nach beiden Seiten kleiner, denn die Aufenthaltswahrscheinlichkeit des Bildpunktes ist natürlich in der Nähe der Grundlinie am größten. Dies gilt jedoch nicht für die Stelle des Bandes, an der bei jedem Resonanzdurchgang das stets gleiche Kernsignal angeregt wurde. Dort nähert sich die Kurve maximaler Schwärzung, etwa bei einem

---

[1] H. L. Cox: Rev. Sci. Instrum. **24**, 307 (1953).
[2] F. BLOCH u. D. H. GARBER: Phys. Rev. **76**, 585 (1949).

Absorptionssignal seinem oberen Rand. Um die Schwärzungsunterschiede deutlich hervorzuheben, muß man Filmmaterial mit hohem Kontrast benutzen, außerdem kann man die Registrierungen noch hart kopieren und zur Auswertung ein empfindliches Densitometer verwenden.

Bei diesem Verfahren werden, ebenso wie bei den zuvor besprochenen trägen elektrischen Registriervorrichtungen, die Amplituden vieler Kernsignale zur Heraushebung aus dem Rauschpegel addiert. Zu beachten ist jedoch, daß die Analogie auch physikalisch nicht vollkommen ist. In den elektrischen Verfahren erfolgt die Integration der Rauschspannungs-Amplituden algebraisch, Spannungsschwankungen entgegengesetzter Phase heben sich also gegenseitig auf. Bei der photographischen Methode werden dagegen die Beträge aller Rauschspannungs-Amplituden summiert, denn auslöschend kann man bekanntlich nicht belichten. Sie entspricht damit der bekannten, zur Herabsetzung von Fehlergrenzen angewandten quadratischen Mittelwertsbildung von vielen Einzel-Meßergebnissen.

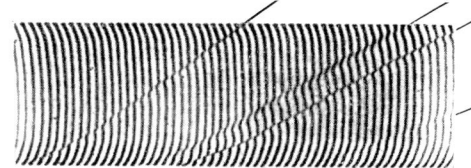

Fig. 54. Registrierung der Kernresonanzlinien von Br⁷⁹, Br⁸¹, Cu⁶³ und Cu⁶⁵ nach der Methode von Ross und JOHNSON. Die Linien der beiden Kupferisotope (Cu⁶³: 886 Gauß, Cu⁶⁵: 827 Gauß) rühren von den in dem Kupferdraht der Empfangsspule enthaltenen Kernen her. Die untersuchte Probe ist eine Lösung von Lithiumbromid. Die Magnetfeldstärke wächst von rechts nach links, Meßfrequenz 1 MHz.

Eine technische Variante des photographischen Registrierprinzips wurde von Ross und JOHNSON entwickelt[1]. In ihrer Anordnung wird nach jedem Resonanzdurchgang die Magnetfeldstärke um einen geringen Betrag erhöht und gleichzeitig der Film in der Kamera ein kleines Stück weiter bewegt. Die Kernsignale bilden dann in dem durch die aufeinanderfolgenden Belichtungen erzeugten Grundlinienraster eine schräge Spur. Ein Beispiel einer solchen Registrierung zeigt Fig. 54. Ein Vorzug dieser Anordnung ist, daß mit der Modulationsfrequenz korrelierte Rauschextrema gerade Spuren bilden und somit nicht stören.

Ein weiteres Verfahren zur summierenden Registrierung kernmagnetischer Interferenzen hat SURYAN angegeben[2]. Er bedient sich des bekannten im Magnetophon zur Schallspeicherung angewandten Prinzips der magnetischen Registrierung. Dazu wird die Ausgangsamplitude des Verstärkers auf einem hohlen Stahlzylinder aufgezeichnet. Dessen Achse ist mit der Achse eines Wechselstromgenerators verbunden, der die Modulationsspannung des Gleichfeldes liefert. Der Zylinder rotiert also synchron mit der Feldmodulation, jedoch wird er während der Registrierung langsam achsenparallel verschoben. Zur Einprägung der Verstärkerspannung wird, ebenso wie zur Schallregistrierung auf Tonbändern, ein kleiner magnetischer Übertragerkopf verwendet. Dadurch wird der Spannungsverlauf auf einer engen Spirale des Zylindermantels aufgezeichnet. Zur Abtastung benutzt man dagegen einen Kopf, der den Zylinder längs einer ganzen Mantellinie bedeckt. Damit löscht sich der größte Teil der Rauschspannungen bei der Abnahme gegenseitig aus, denn deren Vorzeichen wurde ja mitregistriert. Die Amplituden der Kernsignale addieren sich aber, da die Signalspannung längs aller Mantellinien des Zylinders übereinstimmt. Der Spannungsverlauf der Summe aller vorangegangenen Resonanzdurchgänge läßt sich wiederum auf einen Oszillographen mit synchronisierter Zeitablenkung sichtbar machen. Auf dessen Bildschirm kann man auf diese Weise verfolgen, wie während des Versuches allmählich das Kernsignal aus dem Rauschpegel hervorwächst. Physikalisch ist die

[1] I. M. Ross u. F. B. JOHNSON: Nature, Lond. **167**, 286 (1951).
[2] G. SURYAN: Phys. Rev. **80**, 119 (1950).

Analogie zwischen diesem magnetischen und den voranstehend beschriebenen elektrischen Registrierverfahren vollständig.

Ein gemeinsames Kennzeichen aller in diesem Abschnitt bisher diskutierten Registrierverfahren ist die niederfrequente Modulation (z.B. mit 30 Hz) des Resonanzabstands, also der Frequenz oder der Feldstärke um Beträge, die größer (Beobachtung der ganzen Resonanzkurve) oder sehr viel kleiner (differentielle Abtastung der Resonanzkurve) als die Linienbreite sind. Unter den Beobachtungsarten der kontinuierlich angeregten Kernresonanz besitzt das Verfahren der Beobachtung des langsamen $\left(\frac{1}{\omega_m} \gg T_1, T_2\right)$ adiabatischen Resonanzdurchgangs, wie in den Ziff. 6 bis 8 ausgeführt, seiner Übersichtlichkeit wegen wesentliche Vorteile. Die Relaxationszeiten einer Anzahl Substanzen sind jedoch so groß (z.B. in vielen Flüssigkeiten einige Sekunden), daß man entweder im schnellen Resonanzdurchgang beobachten, oder den Resonanzabstand ausreichend langsam unmoduliert ändern muß (statische bzw. quasistatische Abtastung der Resonanzkurve). Der zweite Weg ist vorzuziehen, da er vor allem in der hochauflösenden kernmagnetischen Molekül-Spektroskopie, vgl. Ziff. 27, zu genaueren Aussagen über die Spektrenstrukturen führt. In einer Hinsicht ist er jedoch technisch schwieriger: Die durch die Änderung der Kernsuszeptibilität mit dem Resonanzabstand bewirkte Änderung der Ausgangsspannung der Brücke, d.h. der Modulationsgrad der Signal-Trägerfrequenz, muß nämlich, wie bereits erläutert [vgl. (9.5)], klein sein, wenn man definierte Signalformen erhalten will. Dementsprechend muß man bei der statischen Abtastung, wenn man die Kompensationsbzw. Überlagerungs-Spannung, so wie bisher beschrieben, in der Brücke mit der Signalspannung mischt, die nach der Hochfrequenz-Verstärkung gleichgerichtete Signalspannung mit Hilfe eines Gleichstromverstärkers weiter verstärken. Gleichstromverstärker haben aber im allgemeinen ungünstigere Eigenschaften (Stabilität, Eigenrausch) als Hochfrequenz-Verstärker. ARNOLD[1] hat deshalb ein etwas abweichendes Verfahren angewandt. Man kann nämlich auch zunächst mit einer möglichst gut abgeglichenen Brücke, bzw. mit einem möglichst vollständig entkoppelten Kreuzspulen-System die Kernsignale beobachten, und die so viel stärker modulierte Signalspannung erst nach der Hochfrequenz-Verstärkung mit der Kompensationsspannung mischen.

II. Empfindliche Oszillatoren.

13. Resonanzbeobachtung in schwach schwingenden Oszillatorschaltungen. POUND[2] und ROBERTS[3] haben erstmals eine weitere Gruppe von Verfahren zur kernmagnetischen Resonanzbeobachtung erprobt. Ebenso wie in den Brückenschaltungen befindet sich die zu untersuchende Probe im Inneren der Spule eines Schwingkreises. In diesem Kreis (im allgemeinen einem Parallel-Schwingkreis) werden in irgendeiner Oszillatorschaltung Schwingungen angeregt, deren Frequenz und deren Amplitude durch den Probenkreis selbst bestimmt werden. Die Spule samt der Probe befindet sich ebenso wie in der Anordnung der Fig. 35 in einem modulierten Magnetfeld. Weit außerhalb Kernresonanz schwingt der Oszillator mit der aus den Kreisdaten folgenden Resonanzfrequenz $\omega = \sqrt{\dfrac{1}{LC}}$. Die Amplitude seiner Schwingungen wird unter anderem durch die Güte des Kreises bestimmt. Auf Grund der Modulation des Magnetfeldes ändert sich in

---

[1] J. T. ARNOLD: Phys. Rev. **102**, 136 (1956).
[2] R. V. POUND: Phys. Rev. **72**, 527 (1947).
[3] A. ROBERTS: Rev. Sci. Instrum. **18**, 845 (1947).

der Nähe der Kernresonanzstelle die kernmagnetische Suszeptibilität χ [Gl. (6.9)] der in der Spule enthaltenen Probe periodisch. Die Änderung ihres Realteils χ' verursacht eine Änderung der Induktivität und damit der Frequenz der in dem Kreis erregten Schwingungen. Gleichzeitig ändert sich aber auch der Imaginärteil χ'' der Suszeptibilität und somit die Kreisgüte. Die Amplitude der Schwingungen des Oszillators wird also durch χ'' und ihre Frequenz durch χ' gesteuert. Benutzt man demgemäß zur Beobachtung der Oszillator-Ausgangsspannung einen zum Empfang amplitudenmodulierter Schwingungen vorgesehenen Verstärker und Gleichrichter (AM-Empfänger), dann erhält man auf dem Oszillographen-Bildschirm stets ein Absorptionssignal. Verwendet man dagegen einen zum Empfang frequenzmodulierter Schwingungen geeigneten Verstärker und Gleichrichter (FM-Empfänger), dann erhält man beim Resonanzdurchgang auf dem Bildschirm immer ein Dispersionssignal der untersuchten Kerne. Die Art der Signalbeobachtung wird damit durch die Wahl des Empfängersystems eindeutig festgelegt. Justierprobleme, wie bei den Brückenschaltungen, gibt es bei den vorliegenden Verfahren also nicht. Allerdings wird dieser Vorzug dadurch vermindert, daß die gleichzeitige Änderung der Absorption und Dispersion der Substanz eine zur Kernresonanzfrequenz symmetrische Deformation der Signale zur Folge hat. Deren Formen stimmen also nur in erster Näherung mit den in Fig. 26a und b wiedergegebenen überein. Die Frequenz eines serienbedämpften Kreises hängt z. B. geringfügig von seiner Dämpfung ab, ebenso wird auch die Güte eines Schwingkreises durch Induktivitätsänderungen beeinflußt.

Grundsätzlich können zur Beobachtung von Kernresonanzlinien nach dem voranstehend beschriebenen Prinzip alle Oszillatortypen benutzt werden. Aus praktischen Gründen ist es jedoch zweckmäßig, keine Anordnung zu verwenden, in denen mehrere Kreise, wie z. B. beim HUTH-KÜHN-Oszillator, enthalten sind. Unbequem sind ferner alle Anordnungen, in denen die Probenspule zur Rückkopplung noch zusätzlich angezapft wird, wie z. B. beim HARTLEY-Oszillator. Wesentlich beschränkt wird jedoch die Brauchbarkeit der meisten bekannten Standardschaltungen durch einen anderen Umstand. Aus mehreren Gründen dürfen in dem die Probe enthaltenden Kreis nur sehr schwache Schwingungen angeregt werden:

a) Die Feldstärke H_1 des in der Probenspule angeregten magnetischen Hochfrequenzfeldes muß ausreichend klein sein, damit das Kernspin-System nicht gesättigt wird (vgl. S. 175). Zur Untersuchung von Substanzen mit besonders langen Spin-Gitter-Relaxationszeiten T_1, in denen der Energieaustausch zwischen dem Freiheitsgrad der Kernorientierung und den übrigen Freiheitsgraden nur sehr langsam erfolgt, sind die mit empfindlichen Oszillatoren arbeitenden Verfahren also weniger gut geeignet.

b) Weiter reagiert ein Röhren-Oszillator auf Suszeptibilitäts-Änderungen im frequenzbestimmenden Schwingkreis um so empfindlicher, je kleiner die Amplitude der in ihm erregten Schwingungen ist [vgl. Gl. (13.8), S. 208].

c) Der Modulationsgrad der Ausgangsspannung des Oszillators ist beim Kernresonanzdurchgang um so größer, je schwächer der Oszillator schwingt. Um so leichter ist es, schmalbandig hochfrequent und damit rauscharm zu verstärken, ohne den Empfänger zu übersteuern.

Die Erzeugung elektromagnetischer Schwingungen in den bekannten Oszillatorschaltungen läßt sich anschaulich wie folgt beschreiben. Jede verstärkende Schaltung beginnt zu schwingen, wird also zum Generator, wenn ein ausreichend großer Bruchteil von ihrer Ausgangsspannung abgezweigt und phasenrichtig rückgekoppelt ihrem Eingang zugeführt wird. Genauer gesagt muß im Augenblick des Schwingeinsatzes die Ungleichung $vr > 1$ erfüllt sein, wenn man mit v den Verstärkungsgrad und mit r den Rückkopplungsfaktor bezeichnet. Im stationären Schwingzustand geht diese Beziehung in die Gleichung $vr = 1$ über. Meistens bleibt r in der Einschwingzeit konstant, während v auf Grund der Kennlinienkrümmung mit der Zunahme der Schwingungsamplitude abnimmt. Denselben Sachverhalt kann man aber formal auch auf eine andere Art beschreiben. Ein solcher Rückkopplungs-Oszillator verhält sich nämlich zugleich wie ein sog. negativer Widerstand. Schaltet man einen solchen negativen Widerstand R_n etwa parallel zu einem Schwingkreis, dessen Resonanz-Widerstand R_d immer positiv und reell ist, so werden in dem Kreis genau dann Schwingungen erregt, wenn der parallelgeschaltete negative Leitwert $1/R_n$ dem Betrage nach größer ist als der positive Leitwert $1/R_d$ des Schwingkreises. Die im Augenblick des Schwingeinsatzes zu befriedigende Ungleichung lautet also:

$$\frac{1}{R_n} + \frac{1}{R_d} < 0.$$

Nach einer gewissen Zeit wird die Schwingung stationär, und es gilt $1/|R_n| = 1/|R_d|$. Damit die Ungleichung zur Gleichung werden kann, muß mindestens einer der beiden Leitwerte von der Amplitude der erregten Schwingungen abhängen. Im allgemeinen wird der Betrag von $1/R_n$ mit der Zunahme der Schwingungsamplitude kleiner. Beim kernmagnetischen Resonanzexperiment ändert sich der Gesamtdämpfungs-Widerstand R_d des Probenkreises periodisch. Man sieht sofort, daß sich dadurch auch der Schwingungs-Gleichgewichtszustand, also der Oszillatorpegel verändert.

Die voranstehende Beschreibungsart erlaubt somit besonders bequem, die Rückwirkung einer sich beim Resonanzdurchgang ändernden kernmagnetischen Suszeptibilität auf einen Oszillator zu verstehen. Sie ist aber auch allgemeiner als die zuerst genannte. Eine ganze Gruppe von Schwingungserzeugern läßt sich zwanglos nur auf diese Weise beschreiben. Wohl am bekanntesten ist aus der elementaren Physik die Anregung elektrischer (und auch akustischer) Schwingungen mit Hilfe eines Lichtbogens. Dessen Strom-Spannungs-Charakteristik ist fallend, sein Widerstand also negativ, da die Zahl der den Ladungstransport besorgenden Elektronen und Ionen mit der Stromstärke stark steigt bzw. fällt. Schaltet man einen solchen Bogen parallel zu einem Schwingkreis, dann werden in diesem Schwingungen zunehmender Amplitude erregt, bis schließlich im Gleichgewichtszustand der durch die Krümmung der Bogencharakteristik bestimmte und über eine Periode der Schwingung gemittelte negative Widerstand des Bogens gleich dem positiven Resonanz-Dämpfungswiderstand des Kreises ist. Nach dem gleichen Prinzip kann man aber auch in Röhrenschaltungen Schwingungen erzeugen. Es hat sich gezeigt, daß Oszillatoren dieser Art zur Untersuchung kernmagnetischer Resonanzlinien besonders geeignet sind, da man mit ihrer Hilfe sehr schwache und von den Daten des Probenkreises empfindlich abhängende Schwingungen erzeugen kann. Einen zusammenfassenden Bericht über die Erregung von Schwingungen mit Hilfe solcher negativer Widerstände im engeren Sinn hat Herold[1] veröffentlicht.

Als Beispiel soll zunächst die von Knoebel und Hahn erstmals für kernmagnetische Messungen[2] benutzte Transitronschaltung[3-5] besprochen werden. Wir wollen als erstes das Zustandekommen des negativen Widerstandes in ihrer Anordnung verstehen. Fig. 55 zeigt eine Schaltung zur Aufnahme von Transitron-

[1] E. W. Herold: Proc. Inst. Radio Engrs. 23, 1201 (1935).
[2] H. W. Knoebel u. E. L. Hahn: Rev. Sci. Instrum. 22, 904 (1951).
[3] K. C. van Ryn: The Numan's Oscillator. Wireless Engr. 2, 134 (1924).
[4] C. Brunetti: Proc. Inst. Radio Engrs. 25, 1595 (1937); 27, 88 (1939).
[5] P. G. Sulzer: Proc. Inst. Radio Engrs. 38, 540 (1950).

Kennlinien. An allen Elektroden der Pentode liegen Gleichspannungen. Die Größe des gesamten durch die Röhre fließenden Stroms, also des Kathodenstroms, hängt im wesentlichen von der Steuergitter-Vorspannung und von der Schirmgitter-spannung ab. Die direkt am Bremsgitter lie-gende Spannungsquelle Ug_3 ist variabel. Wird ihre Ausgangsspannung um einen kleinen Be-trag erhöht, dann steigt die Schirmgitterspan-nung und mit ihr der Kathodenstrom gering-fügig an. Wesentlicher ist aber, daß durch die gleichzeitige Zunahme der Bremsgitter-spannung die Stromverteilung in der Röhre verändert wird, und zwar wächst der Anoden-strom auf Kosten des Schirmgitterstroms.

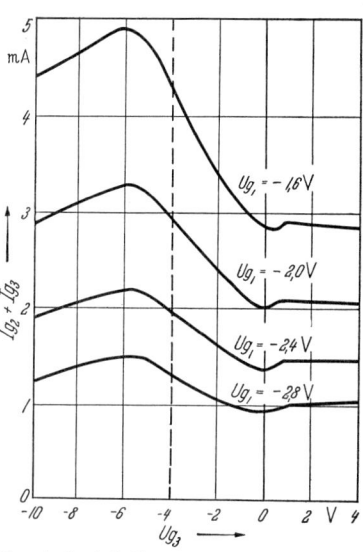

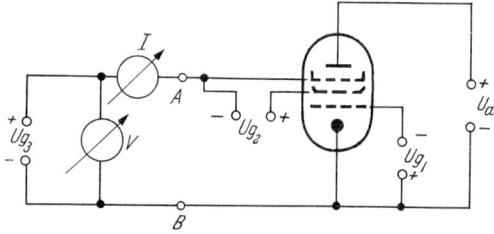

Fig. 55. Schaltung zur Aufnahme statischer Transitronkennlinien. Gemessen wird die Summe des Schirmgitter- und Bremsgitterstroms in Funktion der Spannung U_{g3}. Parameter der Kennlinien ist die Größe der Spannung U_{g1}. Diese beiden Spannungsquellen sind also variabel. Bei der Aufnahme der Kennlinien der Pentode 6AS6 (Fig. 56) war z.B $U_a = U_{g2} = 90$ V.

Fig. 56. Statische Transitron-Kennlinien der Pen-tode 6AS6. Im Bereich der Steuergitter-Vor-spannung −2,8 bis −1,6 Volt liegt der Betrag des an den Wendepunkten der Kurven gemes-senen negativen Widerstands zwischen 1900 und 7400 Ohm.

Der durch das Meßinstrument I fließende Schirmgitterstrom wird also infolge der mit dem Voltmeter V gemessenen Spannungserhöhung zwischen den beiden Punkten A und B kleiner. Entsprechend verursacht eine Spannungsabnahme zwischen den Punkten A und B eine Zunahme des Stroms I. Damit ist ge-zeigt, daß sich der in der Fig. 55 rechts von den Punkten A und B befindende Teil der Schaltung, innerhalb eines ge-wissen Spannungsintervalls, tatsächlich wie ein negativer Widerstand verhält. Fig. 56 zeigt eine Schar solcher fallender Kennlinien, die mit der Anordnung von Fig. 55 für verschiedene Steuergitter-Spannungen aufgenommen wurden. Zur Erregung hochfrequenter Schwingungen haben KNOEBEL und HAHN die Schal-tung der Fig. 57 benutzt. Die Steuergit-ter-Spannung ist regelbar. Man kann also

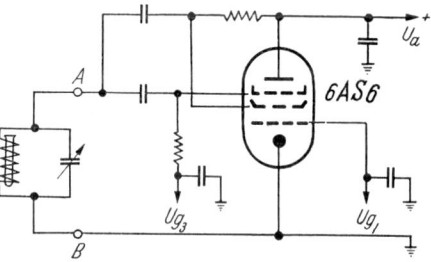

Fig. 57. Transitron-Schaltung nach KNOEBEL und HAHN. Der Empfänger bzw. Gleichrichter wird zwischen den Punkten A und B der Schaltung angeschlossen. Die beiden an der Röhre 6AS6 liegenden Regelspannungen dienen zur Auswahl der Arbeitskennlinie (U_{g1}) und zur Einstellung des Arbeitspunktes auf diesen Kennlinien (U_{g3}).

die Steilheit des fallenden Kennlinienabschnitts und damit die Empfindlichkeit des Oszillators ändern. Die Regelspannung am Bremsgitter dient zur Einstellung des Arbeitspunktes auf der gewählten Kennlinie. In der Regel ist der Wendepunkt dieser Kurve der günstigste Arbeitspunkt.

Stellt man den Verlauf einer solchen Kennlinie durch eine Potenzreihe um den Wendepunkt dar,

$$J = -\alpha U + \beta U^2 + \gamma U^3 + \cdots \qquad (13.1)$$

dann verschwinden in dieser Reihe auf Grund der Antisymmetrie der Kurve die zu den geraden Potenzen gehörenden Koeffizienten. Ferner ist die Steigung der Kennlinie am Arbeitspunkt, d.h. der dortige Leitwert, negativ. Die höheren Glieder der Reihe kann man im allgemeinen vernachlässigen. In der Nachbarschaft des Wendepunktes ist also der negative Widerstand $R_n = U/I = -1/(\alpha - \gamma U^2)$. Die Spannung zwischen den Punkten A und B (Fig. 57) gehorcht der Gleichung

$$\frac{U}{R_n} + \frac{U}{R_d} + C\frac{dU}{dt} + \frac{1}{L}\int U\, dt = 0, \tag{13.2}$$

wenn man mit C und L die Kapazität bzw. Induktivität des Probenkreises und mit R_d seinen Parallel-Verlustwiderstand bezeichnet. Mit den Abkürzungen

$$\alpha' = \frac{1}{C}\left(\alpha - \frac{1}{R_d}\right), \quad \gamma' = \frac{3}{C}\gamma \quad \text{und} \quad \omega_0^2 = \frac{1}{LC} \tag{13.3}$$

geht diese Gleichung durch Differenzieren über in:

$$\frac{d^2 U}{dt^2} - (\alpha' - \gamma' U^2)\frac{dU}{dt} + \omega_0^2 U = 0. \tag{13.4}$$

Transformiert man die Koordinaten durch:

$$U = \sqrt{\frac{|\alpha'|}{\gamma'}}\, v \quad \text{und} \quad t = \frac{w}{\omega_0} \quad \text{und setzt} \quad \varepsilon = \frac{\alpha'}{\omega_0} \tag{13.5}$$

dann erhält man die sog. van der Polsche Differentialgleichung:

$$\frac{d^2 v}{dw^2} - \varepsilon(1 - v^2)\frac{dv}{dw} + v = 0. \tag{13.6}$$

Diese Gleichung wurde ausführlich untersucht[1]. Sie hat für $|\varepsilon| \ll 1$ eine periodische Lösung. Nach dieser hat die Schwingungsamplitude im stationären Zustand den konstanten Endwert:

$$\hat{U} = 2\sqrt{\frac{\alpha - \dfrac{1}{R_d}}{3\gamma}}. \tag{13.7}$$

Der resultierende Leitwert $G_r = -\alpha + \dfrac{1}{R_d}$ ändert sich beim Resonanzdurchgang. Entsprechend steigt oder fällt auch die Ausgangsspannng des Oszillators. Dessen Empfindlichkeit gegenüber Leitwertsänderungen ist:

$$\frac{d\hat{U}}{d|G_r|} = \frac{2}{3\gamma}\frac{1}{\hat{U}}. \tag{13.8}$$

Sie ist also um so höher, je kleiner der Ausgangsspannungspegel des Oszillators ist (vgl. S. 205). Für Sättigungsuntersuchungen zur Bestimmung der Spin-Gitter-Relaxationszeit T_1 ist die vorliegende Beobachtungsart also nicht geeignet.

Oft sind die Kernsignale so stark (z.B. die Protonensignale elektrolytischer Lösungen), daß sich ihre hochfrequente Verstärkung erübrigt. Man kann dann unmittelbar die Ausgangsspannung des empfindlichen Oszillators, d.h. die in Fig. 57 zwischen A und B auftretende Spannung, gleichrichten. Eine solche Anordnung nennt man in der amerikanisch-englischen Literatur einen Autodyn-Detektor. Die Schaltung zeichnet sich dadurch aus, daß die ganze Resonanz-beobachtungs-Apparatur nur einen abzustimmenden Schwingkreis enthält. Damit

---

[1] B. van der Pol: Jb. drahtl. Telegr. **28**, 178 (1926); **29**, 114 (1927). — Proc. Inst. Radio Engrs. **22**, 1051 (1934).

wird es praktisch möglich, anstatt der sonst üblichen Modulation des konstanten Feldes, die Senderfrequenz periodisch zu variieren, z.B. mit Hilfe eines mechanisch niederfrequent schwingenden Kondensators. Die Transitron-Detektor-Schaltung ist also für Magnetfeldmessungen besonders geeignet (vgl. auch Ziff. 18). Hinzu kommt noch, daß sich mit fast der gleichen Anordnung Frequenzen von etwa 1 bis 100 MHz erregen lassen. Der Frequenzbereich wird nach oben nur durch die Laufzeit der Elektronen in der Röhre begrenzt, praktisch allerdings auch durch die Frequenzabhängigkeit der Drosseln.

POUND und KNIGHT[1] haben zur Kernresonanzbeobachtung eine kathodengekoppelte empfindliche Oszillatorschaltung entwickelt. Fig. 58 zeigt das Prinzipschaltbild des verwandten Oszillators und Fig. 59 die Gesamtanordnung. Ebenso wie beim Transitron hat eine Spannungserhöhung am Punkt A der Anordnung Fig. 58 zur Folge, daß das an ihm liegende Potential noch mehr steigt, und zwar verursacht der Spannungsanstieg am Gitter des linken Triodensystems zunächst eine Stromzunahme in diesem System und damit eine Erhöhung des gemeinsamen Kathodenpotentials beider Röhrenhälften. Dies hat zur Folge, daß der Anodenstrom des rechten Systems und mit diesem der Spannungsabfall an seinem Anodenwiderstand R_A abnimmt. Die daraus resultierende Anodenspannungs-Vergrößerung wird über einen Koppelkondensator C_k auf den Schwingkreis übertragen. Der Strom-Spannungs-

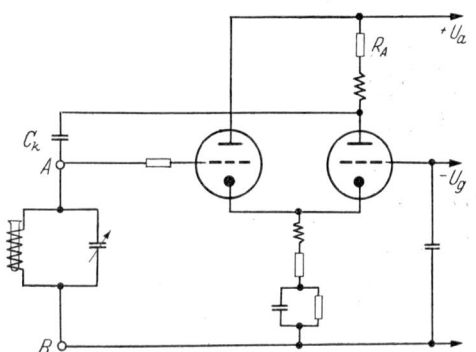

Fig. 58. Prinzipschaltbild des kathodengekoppelten Oszillators von POUND und KNIGHT. Der Spannungspegel des Oszillators wird durch die am Steuergitter der rechten Triode liegende Spannung geregelt. Das Signal wird zwischen A und B abgenommen. Die Induktivitäten in der Kathodenleitung und in der rechten Anodenleitung sind erforderlich, damit die Ausgangsspannung des Oszillators bei höheren Frequenzen nicht zu stark abfällt.

zusammenhang zwischen den Punkten A und B wird also ebenfalls durch eine fallende Charakteristik beschrieben. Die durch das Kernresonanzsignal modulierte Oszillatorspannung wird bei der vorliegenden Anordnung (Fig. 59) zur Verbesserung des Signal-Rauschverhältnisses zunächst hochfrequent verstärkt. Die Drehkondensatoren des Probenkreises und des Resonanzverstärkers sind mechanisch gekoppelt, so daß das Durchlaßband des letzteren mit der Oszillatorfrequenz im Gleichlauf verschoben werden kann.

Variiert man die Oszillatorfrequenz, so ändert sich auch der Resonanzwiderstand des Probenkreises. Damit steigt oder fällt die Senderamplitude, und wie wir wissen auch seine Empfindlichkeit gegenüber Suszeptibilitätsänderungen. Um diese unerwünschte Frequenzabhängigkeit zu eliminieren, haben POUND und KNIGHT einen Kunstgriff angewandt. Sie koppeln einen Teil der nach der Resonanzverstärkerstufe gleichgerichteten Oszillatorspannung als Regelspannung auf das Gitter des rechten Triodensystems zurück. Natürlich dürfen durch diese Amplitudenstabilisierung des Senders die Kernsignale selbst nicht ausgeregelt werden. Der Amplituden-Schwundausgleich erfolgt deshalb mit einer Zeitkonstante, die verglichen mit der Dauer der Kernsignale groß ist, und zwar beträgt die Ansprechzeit der Regelung in der vorliegenden Anordnung etwa 2 sec.

---

[1] R. V. POUND u. W. D. KNIGHT: Rev. Sci. Instrum. 21, 219 (1950). — Vgl. auch G. D. WATKINS u. R. V. POUND: Phys. Rev. 82, 343 (1951) sowie R. V. POUND: Progr. Nucl. Phys. 2, 21 (1952).

Nach der Niederfrequenzverstärkung kann die Signalspannung wie bei den Brückenschaltungen extrem schmalbandig verstärkt und phasenempfindlich (vgl. Ziff. 12) gleichgerichtet werden. Das optimale Signal-Rauschverhältnis solcher empfindlicher Oszillatorschaltungen erreicht dann etwa den gleichen Wert wie das der Brückenschaltungen.

Von Thomas[1] wurde ein kathodengekoppelter Gegentaktoszillator[2] angegeben, dessen Frequenz quarzstabilisiert ist. Er verwandte diese Anordnung zur Stabilisierung des Magnetfeldes eines Zyklotrons (vgl. Ziff. 18).

Zahlreiche weitere Schaltungen zur Beobachtung der magnetischen Kernresonanz arbeiten ebenfalls mit empfindlichen Oszillatoren. Zu nennen sind

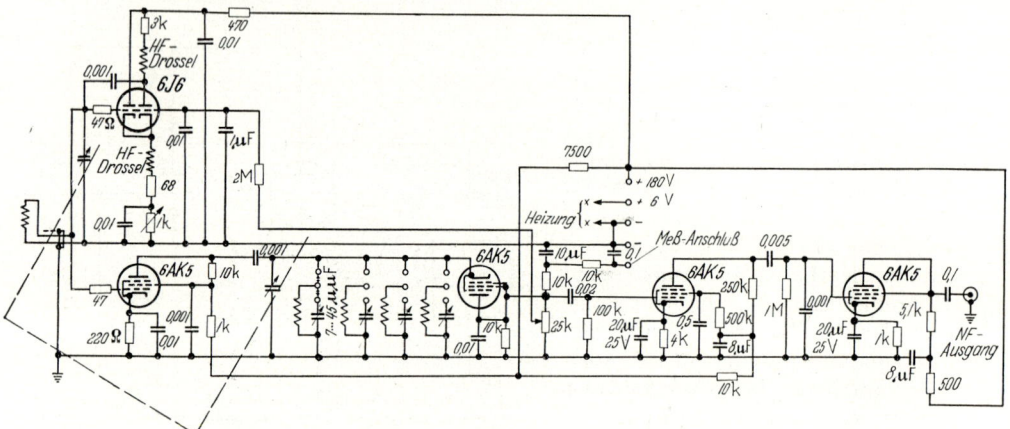

Fig. 59. Schaltung des Spektrometers von Pound und Knight. Der Ausgang des schwach schwingenden Oszillators (6 J 6, vgl. Fig. 58) ist (abgesehen von einem kleinen Schutzwiderstand) direkt mit dem Steuergitter einer hochfrequent verstärkenden Pentode (6AK 5, links unten) verbunden. Deren Ausgangsspannung wird mit Hilfe einer weiteren als Diode geschalteten Röhre 6AK6 gleichgerichtet. Ein Teil dieser gleichgerichteten Spannung wird über ein RC-Glied großer Zeitkonstante als Regelspannung auf das rechte Triodensystem des Oszillators zurückgekoppelt, um dessen Amplitude frequenzunabhängig zu stabilisieren. Die beiden weiteren, ebenfalls mit 6AK 5-Pentoden arbeitenden Stufen (rechts unten) dienen zur niederfrequenten Verstärkung.

unter anderem die Anordnungen von Hopkins[3], der übrigens einen rückgekoppelten Oszillator benutzt hat; von Gabillard und Soutif[4]; von Poulis[5]; von Kakiuchi, Shono, Komatsu und Kigoshi[6]; von Volkoff, Petsch und Smellie[7], von Pekárek und Urbanec[8]; von Gindsberg und Beers[9]; von Gutowsky, Meyer und McClure[10]. Bezüglich der Einzelheiten muß auf die Originalarbeiten verwiesen werden. Über die theoretische Behandlung solcher Schaltungen wurde ferner von Soutif, Grivet und Gabillard[11-13] berichtet.

14. Resonanzbeobachtung in Pendelrückkopplungs-Schaltungen. In den Pendelrückkopplungs-Verfahren (super-regenerative methods) beobachtet man ebenfalls

[1] H. A. Thomas: Electronics 25, 114 (1952).
[2] H. J. Reich: Proc. Inst. Radio Engrs. 25, 1387 (1934).
[3] N. J. Hopkins: Rev. Sci. Instrum. 20, 401 (1949).
[4] R. Gabillard u. M. Soutif: C. R. Acad. Sci. Paris 230, 1754 (1950).
[5] N. J. Poulis: Physica, Haag 17, 392 (1951).
[6] Y. Kakiuchi, H. Shono, H. Komatsu u. K. Kigoshi: J. Phys. Soc. Japan 7, 102 (1952).
[7] G. M. Volkoff, H. E. Petsch u. D. W. L. Smellie: Canad. J. Phys. 30, 270 (1952).
[8] L. Pekárek u. J. Urbanec: Czech. J. Phys. 1, 78 (1952).
[9] J. Gindsberg u. Y. Beers: Rev. Sci. Instrum. 24, 632 (1953).
[10] H. S. Gutowsky, L. H. Meyer u. R. E. McClure: Rev. Sci. Instrum. 24, 644 (1953).
[11] P. Grivet, M. Soutif u. R. Gabillard: C. R. Acad. Sci., Paris 229, 27 (1949).
[12] M. Soutif u. R. Gabillard: C. R. Acad. Sci., Paris 230, 2012 (1950).
[13] M. Soutif: Rev. Sci., Paris 89, 203 (1951).

die Rückwirkung der magnetischen Kernresonanz auf den Schwingzustand eines Oszillators, dessen frequenzbestimmendes Element der Probenkreis ist. Jedoch wird dieser Oszillator, meist verwendet man einen der gebräuchlichen rückgekoppelten Generatoren, periodisch durch eine Steuerspannung ein- und ausgeschaltet, er arbeitet also pendelnd in einer Art Impulstrieb. Gemessen wird der zeitliche Mittelwert der gleichgerichteten Oszillatorspannung. Eine solche Anordnung steht somit zwischen den zuvor besprochenen kontinuierlichen Oszillatorschaltungen und den im folgenden Abschnitt zu behandelnden Impulsverfahren.

ROBERTS[1] hat dieses aus der Anfangszeit der Rundfunktechnik wohl bekannte Prinzip erstmals zur kernmagnetischen Resonanzbeobachtung benutzt. Einen zusammenfassenden Bericht über Pendelrückkopplungs-Empfänger hat WHITEHEAD[2] veröffentlicht. Fig. 60 zeigt die von ROBERTS verwandte Schaltung. Die Oszillatorröhre wird anodenseitig getastet, und zwar mit einer Pendelfrequenz deren Größenordnung einige kHz bis einige 10 kHz beträgt. Es ist nicht erforderlich, daß der Empfänger direkt galvanisch, kapazitiv oder induktiv mit dem Probenkreis verbunden ist. In jeder Pendelperiode wird zweimal der empfindlichste Arbeitsbereich des Oszillators, nämlich derjenige verschwindend kleiner Amplitude überstrichen (vgl. S. 205). So ist es einleuchtend, daß das vorliegende

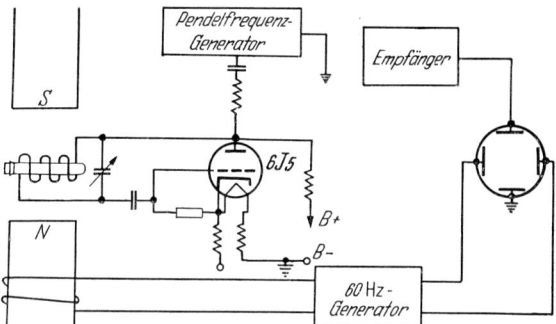

Fig. 60. Schaltung der Pendelrückkopplungs-Apparatur von ROBERTS. Die Anodenspannung des Oszillators ist mit der Pendelfrequenz moduliert. Die abgestrahlten Signale werden in einem Empfänger verstärkt und hochfrequent gleichgerichtet. Nach der Aussiebung der Pendelfrequenz können sie auf dem Bildschirm eines Oszillographen beobachtet werden, dessen Ablenkfrequenz wie üblich mit der Modulationsfrequenz des Magnetfeldes übereinstimmt.

Prinzip auf Grund seiner hohen Empfindlichkeit zur Feststellung von Resonanzstellen gut geeignet ist. Ein wesentlicher Nachteil des Verfahrens folgt jedoch daraus, daß die Feldstärke H_1 des die Präzessionsbewegung anregenden Feldes zeitlich inkonstant ist. Definierte Linienformen, die zur Ermittlung der Absorption bzw. der Dispersion der untersuchten Substanz ausgewertet werden könnten, erhält man daher nicht.

Um die Arbeitsweise der Pendelrückkopplungs-Schaltungen zu verstehen, muß man streng zwischen ihren verschiedenen möglichen Betriebsarten unterscheiden, und zwar kann man diese wie folgt gliedern: (α) Inkohärenter Betrieb, (β) präzessionskohärenter Betrieb, (γ) schwingungskohärenter Betrieb.

In allen drei Fällen wird der Oszillator periodisch an- und abgeschaltet, und zwar meistens durch einen Sinuswellen-Generator, dessen Spannung an einer Steuer-Elektrode der Schwingröhre liegt. Zur Vereinfachung der Betrachtung wollen wir jedoch annehmen, daß die Steuerspannung die Form einer Rechteckwelle hat. Die Schwingröhre kann formal als ein zum Probenkreis parallel geschalteter Widerstand betrachtet werden. In den Einschalt-Zeitabschnitten ist dieser Widerstand R_n negativ, d.h. anfachend (genauer gesagt ist der Eingangsleitwert der Röhre negativ), und dem Betrage nach größer als der

---

[1] A. ROBERTS: Phys. Rev. **72**, 182 (1947). — Rev. Sci. Instrum. **18**, 845 (1947).
[2] J. R. WHITEHEAD: Super-regenerative Receivers. Cambridge: University Press 1950.

positive Resonanz-Dämpfungswiderstand R_d des Probenkreises. Die Schwingungs-amplitude wächst infolgedessen exponentiell an, und zwar mit der Zeitkon-stanten $2CR_E$, wenn man mit C die Kapazität des Probenkreises und mit R_E den resultierenden Widerstand der Schaltung während der Einschaltperiode $\left(R_E = \dfrac{R_n \cdot R_d}{R_n + R_d}\right)$ bezeichnet. Mit der Zunahme der Schwingamplitude wird $|R_n|$ kleiner. Im stationären Zustand ist schließlich $|R_n| = |R_d|$. In den Abschalt-Zeit-abschnitten ist der resultierende Widerstand R_A positiv $\left(R_A = \dfrac{R_p \cdot R_d}{R_p + R_d}; \ R_p =\right.$ positiver Eingangswiderstand der Röhre im gesperrten Zustand$\left.\right)$, und die Schwingung klingt mit der Zeitkonstanten $2CR_A$ exponentiell ab. In der nächsten Pendelperiode beginnt dann wieder der Einschwingvorgang usw. Wesentlich ist nun, von welchem Spannungspegel aus sich die Schwingungen dabei jeweils aufbauen.

α) *Im inkohärenten Betrieb* liegt im Augenblick des Schwingeinsatzes am Probenkreis nur die seinem Dämpfungswiderstand entsprechende Rausch-spannung. Die Schwingungen starten also vom Rauschpegel. Befindet sich die Induktivität des Schwingkreises samt der in ihr enthaltenen Probe in einem Magnetfeld der ungefähren LARMOR-Resonanzfeldstärke, dann absorbieren die Kerne Hochfrequenzquanten. Dem entspricht eine Abnahme des Resonanz-Dämpfungswiderstandes des Schwingkreises auf den Betrag R_d^*. Dies hat zur Folge, daß die Zeitkonstante des Einschwingvorganges $2CR_E^*$ größer wird, der Oszillator also langsamer anschwingt. Dagegen wird der Widerstand R_A^* und mit ihm die Zeitkonstante $2CR_A^*$ kleiner. In den Ausschaltperioden klingen also die Schwingungen des Oszillators schneller ab. Fig. 61 zeigt, daß beide Kernresonanz-effekte zur Verminderung der zeitlich mittleren Oszillatoramplitude beitragen. Auf Grund der Modulation des Magnetfeldes wird somit der zeitliche Mittelwert der an den Vertikalplatten des Oszillographen liegenden gleichgerichteten Signal-spannung periodisch größer und kleiner und erreicht seinen Minimalwert direkt in Kernresonanz. Um übersichtliche Verhältnisse zu gewinnen, muß im übrigen zwischen der Senderfrequenz ω_s, der Pendelfrequenz ω_p und der Modulations-frequenz ω_m die Ungleichung

$$\omega_s \gg \omega_p \gg \omega_m$$

gelten. Zur zeitlichen Mittelung der Oszillator-Ausgangsspannung siebt man die Pendelfrequenz mit einem Filter aus, dessen Zeitkonstante gegen die Modulations-periode genügend klein ist.

β) *Die präzessionskohärente Betriebsart* einer Pendelrückkopplungs-Schaltung unterscheidet sich außerhalb Kernresonanz nicht von der voranstehend be-schriebenen. Die Oszillatorschwingungen sind also bereits unter den Rausch-pegel abgeklungen, wenn der zum Schwingkreis parallel geschaltete Röhren-widerstand wieder negativ wird. Befindet sich nun in der Probenspule eine Substanz, deren Spin-Spin-Relaxationszeit T_2 von derselben Größenordnung wie die Pendelperiode oder gar größer als diese ist, dann wird in jeder Ein-schaltperiode durch den in dieser Zeit in die Probe eingestrahlten Hochfrequenz-impuls eine freie Präzessionsbewegung des Kernspinsystems angeregt (vgl. Ziff. 4), die zu Beginn der nächsten Einschaltperiode noch nicht abgeklungen ist. Diese induziert in der Schwingkreis-Induktivität eine Wechselspannung der LARMOR-Fre-quenz. Infolge der Resonanzüberhöhung ist die resultierende Spannung am Schwing-kreis gleich dem Produkt aus induzierter Spannung und Kreisgüte Q. Von dieser, meist wesentlich über dem Rauschpegel liegenden Spannung aus startet demgemäß der Oszillator in der folgenden Schwingperiode. Damit wird aber in Kernresonanz die mittlere Hochfrequenzleistung des Oszillators größer als außerhalb Kernresonanz, falls dieser Effekt die beiden zuvor besprochenen und sicher stets auch vorhandenen überwiegt. Im einzelnen hängt die sich beim Resonanzdurchgang ausbildende Signalform bei dieser Betriebsart in recht komplizierter Weise von den Daten der Apparatur und der Substanz ab. Insbesondere sorgen Interferenzen zwischen den nacheinander angeregten kernmagnetischen Momenten und zwischen den zu verschiedenen Feldstärken des (praktisch immer etwas inhomogenen) Ma-

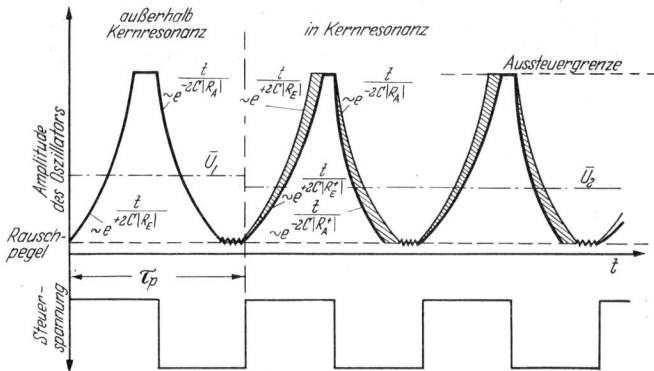

Fig. 61. Inkohärente Betriebsart eines Pendelrückkopplungs-Oszillators. In dem linken Teil der Abbildung ist der zeitliche Verlauf der Amplitude des Oszillators in Funktion der negativen Steuerspannung weit außerhalb Kernresonanz dargestellt. Der rechte Teil zeigt den gleichen Verlauf in Kernresonanz. Am Anfang jeder positiven Halbwelle aller Pendelperioden τ_p schwingt der Oszillator an. Zur Vereinfachung wurde ebenso wie in den folgenden Fig. 62 und 63 angenommen, daß der dem Schwingkreis parallel geschaltete negative anfachende Widerstand R_n während dem Einschwingvorgang konstant bleibt und dann an der Aussteuergrenze unstetig gleich dem Resonanz-Dämpfungswiderstand R_d des angekoppelten Schwingkreises wird. In Wirklichkeit erfolgt dieser Übergang natürlich stetig. Auf Grund dieser Voraussetzung wächst die Oszillatoramplitude exponentiell, und zwar außerhalb Kernresonanz mit der Zeitkonstanten $+2C\,|R_E|$ $\left(R_E = \dfrac{R_n\,R_d}{R_n+R_d}\right)$ und in Kernresonanz mit der Zeitkonstanten $+2C\,|R_E^*|\left(R_E^* = \dfrac{R_n\,R_d^*}{R_n+R_d^*}\,;\ R_d^* < R_d\right)$, also langsamer. In den negativen Halbwellen der Pendelperioden schwingt der Oszillator aus, und zwar in Kernresonanz mit der Zeitkonstante $-2C\,|R_A|$ $\left(R_A = \dfrac{R_p\,R_d}{R_p+R_d}\right)$ und außerhalb Kernresonanz mit der Zeitkonstante $-2C\,|R_A^*|$ $\left(R_A^* = \dfrac{R_p\cdot R_d^*}{R_p+R_d^*}\,;\ R_d^* < R_d\right)$, also schneller. In Kernresonanz nimmt somit die zeitlich mittlere Ausgangsspannung $\bar U_1$ des Oszillators außerhalb Kernresonanz auf den Betrag $\bar U_2$ ab. — Zur Verdeutlichung wurden die Größenverhältnisse natürlich verzerrt gezeichnet, der Abstand zwischen $\bar U_1$ und $\bar U_2$ ist in Wirklichkeit viel kleiner, außerdem ist die Rauschspannung um viele Größenordnungen kleiner als etwa die Oszillatorspannung an der Aussteuergrenze. Da die Hochfrequenz-Schwingungen sich in jeder Pendelperiode erneut vom Rauschpegel aus aufbauen, streuen die Phasen aller Wellenzüge statistisch, und man nennt daher diese Betriebsart inkohärent.

gnetfeldes gehörenden infinitesimalen Momenten (vgl. Ziff. 4) dafür, daß die Signalformen meist auch zeitlich nicht stabil sind. Trotzdem läßt sich die Resonanzstelle selbst natürlich identifizieren. In Fig. 62 wurde zur Veranschaulichung angenommen, daß die Zunahme der Oszillatorleistung auf Grund der freien Präzessionsbewegung größer ist als deren durch die Resonanzabsorption verursachte Verminderung. Außerdem wurde zur Vereinfachung vorausgesetzt, daß die Anregung der freien Präzessionsbewegung einen stationären Zustand erreicht hat, daß also Interferenzeffekte nicht stören.

$\gamma)$ *Die schwingungskohärente Betriebsart* ist dagegen in ihrer reinen Form wieder leichter zu übersehen. Dazu muß man voraussetzen, daß die durch die freie Präzessionsbewegung des Kernspinsystems induzierte Signalspannung vor jedem neuen Schwingeinsatz im wesentlichen abgeklungen ist. Die Spin-Spin-Relaxationszeit der Substanz muß also ausreichend klein gegen die Pendelperiode sein. Jedoch sollen die HF-Schwingungen des Oszillators selbst zu diesem Zeitpunkt noch eine Amplitude haben, die merklich über dem Rauschpegel liegt (praktisch etwa 100 bis 1000-fach). Fig. 63 zeigt den Spannungsverlauf des Oszillators unter diesen Bedingungen. In Kernresonanz wachsen auf Grund

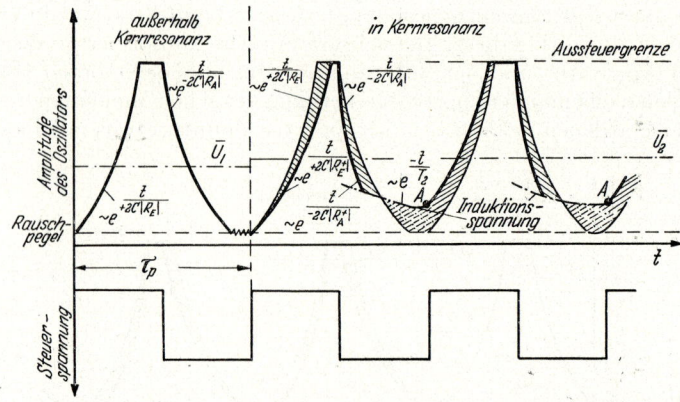

Fig. 62. Präzessionskohärente Betriebsart eines Pendelrückkopplungs-Oszillators. Außerhalb Kernresonanz verläuft die Oszillatoramplitude ebenso wie in Fig. 61 (bezüglich der allgemeinen Voraussetzungen vergleiche den dortigen Abbildungstext). Auf Grund der Annahme, daß die Spin-Spin-Relaxationszeit T_2 der untersuchten Substanz größer oder von der gleichen Größenordnung wie die Pendelperiode τ_p ist, startet der Oszillator jedoch zu Beginn jeder positiven Halbwelle der Steuerspannung von dem durch die freie Präzessionsbewegung des Kern-Spin-Systems am Schwingkreis liegenden Spannungspegel (strichpunktiert eingezeichnet), also von den Punkten A der Abbildung. Die ,,Induktionsspannung'' ist das Produkt aus der durch die Präzessionsbewegung in der Spule induzierten Spannung und der Kreisgüte. In der obigen Darstellung wurde angenommen, daß die Zunahme der mittleren Ausgangsspannung auf Grund dieses Induktionseffektes (ansteigend schraffiert gezeichnet) die durch die Absorptionseffekte bewirkte Abnahme (fallend schraffiert gezeichnet) überwiegt. Das durch den Verlauf der Induktionsspannung begrenzte Flächenstück wurde ansteigend gestrichelt schraffiert. Es trägt zwar ebenfalls zur Erhöhung der mittleren abgestrahlten Oszillatorleistung bei, die Größe dieses Beitrags wird jedoch in der obigen Darstellung auf Grund der Verzerrung des Ordinatenmaßstabs stark übertrieben. Stationäre Verhältnisse vorausgesetzt, besteht zwischen der Oszillatorspannung und der durch sie angeregten Kernsignalspannung stets dieselbe feste Phasendifferenz. Die Phasen der einzelnen Wellenzüge stimmen also bei dieser Betriebsart zwar im allgemeinen nicht überein, haben aber konstante Abstände. Um Verwechslungen auszuschließen, ist es daher empfehlenswert, diese Betriebsart als präzessionskohärent zu bezeichnen.

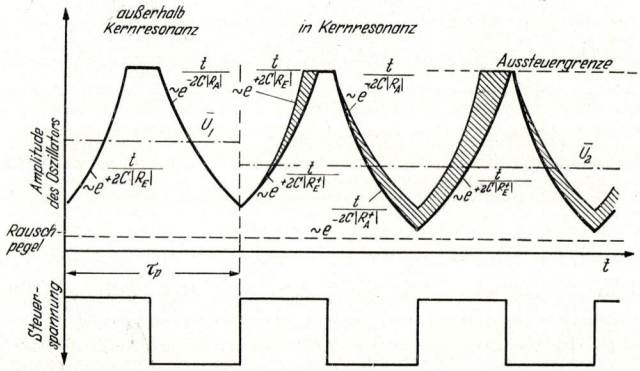

Fig. 63. Schwingungskohärente Betriebsart eines Pendelrückkopplungs-Oszillators. Bei dieser Betriebsart klingen die Oszillatorschwingungen niemals ganz ab. Dies hat zur Folge, daß in Kernresonanz von den beiden Absorptionseffekten (vgl. den Text der Fig. 61) vor allem der zweite (schnelleres Abklingen der Oszillatorschwingung auf Grund der Resonanzabsorption) eine starke Verminderung der mittleren Oszillatorleistung verursacht. Alle Wellenzüge haben stetig aneinander anschließende Phasen, die Schwingungen sind also kohärent. Zur Unterscheidung von der in der vorigen Abbildung beschriebenen Betriebsart ist es empfehlenswert, die obige als schwingungskohärent zu bezeichnen[1].

der Resonanzabsorption (vgl. auch inkohärenter Betrieb) die Schwingungen langsamer an, vor allem aber klingen sie in den Abschaltperioden schneller aus. Dies hat zur Folge, daß der Schwingungseinsatz von einem niedrigeren Pegel aus

---

[1] In der hochfrequenztechnischen Literatur wird diese Betriebsart z. T. logarithmischkohärent genannt.

beginnt. Von der Höhe dieses Pegels hängt aber die zeitlich mittlere Oszillatorleistung sehr empfindlich ab (vgl. hierzu auch die Arbeiten von KOHN[1], FRINK[2]
und DEHMELT[3]). Diese Betriebsart dürfte daher empfindlicher als die inkohärente
und definierter als die präzessionskohärente sein. Allerdings ist bei kernmagnetischen Experimenten die Relaxationszeit T_2 oft so groß, daß sich die reine
schwingungskohärente Betriebsart nicht verwirklichen läßt. Man erhält dann
wiederum mehr oder weniger kompliziert gemischte Betriebszustände.

Eine Hochfrequenz-Impulsfolge der Trägerfrequenz ω_s und der Wiederholungsfrequenz ω_p enthält außer der Zentralfrequenz ω_s beiderseits noch Serien
von Seitenbandfrequenzen $\omega = \omega_s \pm n\,\omega_p$ (n ganzzahlig). Erfüllt eine dieser
Frequenzen die Kernresonanzbe-
dingung $\omega = \gamma H$, dann werden
ebenfalls Kernsignale angeregt.
Beim Resonanzdurchgang erhält
man also um das mittlere Signal
eine Anzahl äquidistanter Seitenbandsignale. Zur Identifizierung
des mittleren Signals muß man
nur die Pendelfrequenz etwas ver-
ändern. Auf dem Oszillographenschirm bleibt dann das Haupt-
signal ortsfest, während alle anderen zusammenrücken oder auseinander wandern. Übrigens kann
man die Seitenbandsignale als
Frequenzmarken bei der Vermessung einer unbekannten Linie
benutzen. Man sieht sofort ein,
daß die Pendelfrequenz größer
sein muß als die in Frequenz-

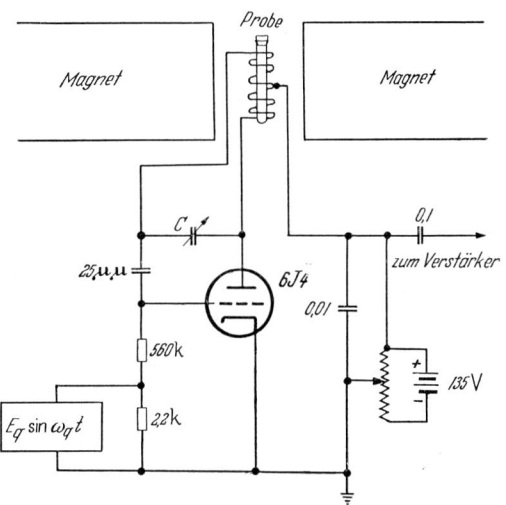

Fig. 64. Pendelrückkopplungs-Oszillator nach WILLIAMS.

einheiten gemessene Linienbreite der untersuchten Substanz, damit die Probe
jeweils nur durch eine diskrete Frequenz des Linienspektrums der Impulsfolge angeregt wird.

Das Pendelrückkopplungs-Verfahren wurde vor allem von WILLIAMS[4] und seinen Mitarbeitern ZIMMERMANN[5], CHAMBERS[6], SHERIFF[7] und YU TING[8] mit
Erfolg zur Messung kernmagnetischer Resonanzlinien benutzt. Fig. 64 zeigt
eine für den Frequenzbereich von 5 bis 10 MHz von WILLIAMS angegebene
Schaltung.

Die Ausgangsspannung eines pendelnden Oszillators kann man, ebenso wie
bei den mit Hilfe schwach schwingender Oszillatoren arbeitenden Verfahren,
entweder direkt gleichrichten oder zur Erhöhung der Empfindlichkeit zunächst
noch hochfrequent schmalbandig verstärken. Im ersten Fall hat man wiederum
nur einen abzustimmenden Schwingkreis und kann somit leicht zur Frequenz-
modulation übergehen. Solche Super-Regenerativ-Detektoren können deshalb

[1] H. KOHN: Jb. drahtl. Telegr. 37, 51, 98 (1937).
[2] F. W. FRINK: Proc. Inst. Radio Engrs. 26, 76 (1938).
[3] H. G. DEHMELT: Z. Physik 130, 356 (1951).
[4] D. WILLIAMS: Physica, Haag 17, 454 (1951).
[5] J. R. ZIMMERMANN u. D. WILLIAMS: Phys. Rev. 76, 350 (1949).
[6] W. H. CHAMBERS u. D. WILLIAMS: Phys. Rev. 76, 638 (1949).
[7] R. E. SHERIFF u. D. WILLIAMS: Phys. Rev. 82, 651 (1951).
[8] Y. TING u. D. WILLIAMS: Phys. Rev. 89, 595 (1953).

bequem zur Messung von Magnetfeldern benutzt werden. Außerdem eignen sie sich besonders gut zur Messung von Quadrupolmomenten[1,2], deren Frequenzen sich nicht von außen durch ein Magnetfeld modulieren lassen.

Zahlreiche weitere Autoren haben ebenfalls das Pendelrückkopplungs-Verfahren zur kernmagnetischen Resonanzbeobachtung angewandt. Zu nennen sind unter anderen Grivet und Soutif[3]; Kojima, Ogawa und Torizuka[4]; Suryan[5] sowie Gutowsky, Meyer und McClure[6]. Bezüglich der technischen Einzelheiten ihrer Apparaturen muß auf die Originalarbeiten verwiesen werden. Auch bei diesem Beobachtungsverfahren läßt sich zur Erhöhung des Signal-Rauschverhältnisses das Lock-in-Prinzip (vgl. Ziff. 12) anwenden[7].

III. Impulstechnische Schaltungen.

15. Zeitlich getrennte Anregung und Beobachtung. Zur Beobachtung kernmagnetischer Resonanzen moduliert man in den kontinuierlichen Verfahren das konstante Feld niederfrequent $(H = H_0 + H_m \cos \omega_m t; \; H_m \ll H_0)$, um bequem elektronisch verstärken zu können, jedoch so langsam, daß die Präzessionsbewegung zu einem stetigen Durchlaufen der zu den jeweiligen Feldstärken gehörenden stationären Lagen des Magnetisierungsvektors wird. Voraussetzung eines solchen sog. adiabatischen Präzessionsverlaufs ist die Erfüllung der Ungleichung $\left|\dfrac{dH}{dt}\right| \ll |\gamma| (\Delta H)^2$ ($\Delta H =$ Linienbreite, vgl. Ziff. 6). Dagegen wird die Signalmodulation bei den Impulsexperimenten nur durch den instationären Bewegungsverlauf verursacht, unabhängig davon, ob man den Anregungsvorgang (vgl. Ziff. 16) oder die in dieser Ziffer zu besprechende Beobachtungstechnik des Abklingens der Präzessionsbewegung betrachtet. Alle impulstechnischen kernmagnetischen Untersuchungen werden also in einem konstanten Feld mit einer festen Trägerfrequenz ausgeführt. Die nichtadiabatischen Resonanzdurchgang-Experimente (vgl. Ziff. 8) nehmen eine Mittelstellung ein. In ihnen wird die Signalamplitude sowohl durch die periodische Veränderung des Resonanzabstandes als auch durch den instationären Verlauf der Präzessionsbewegung moduliert. Ist übrigens bei dieser Beobachtungsart im Grenzfall die Ungleichung $\left|\dfrac{dH}{dt}\right| \gg |\gamma| (\Delta H)^2$ erfüllt, dann geht der Bewegungsverlauf in den durch eine äquidistante Impulsfolge angeregten über. Die ersten erfolgreichen Versuche, paramagnetische Kernresonanzen impulsförmig anzuregen, wurden von Torrey[8,9] und Hahn[10,11] veröffentlicht. Auf die Arbeiten des erstgenannten Autors soll erst in Ziff. 16 eingegangen werden. Fast alle späteren Publikationen aus dem vorliegenden Arbeitsgebiet schließen an eine grundlegende Arbeit von Hahn[11] an, der erstmals konsequent den Anregungs- und Beobachtungsvorgang der kernmagnetischen Präzessionsbewegungen zeitlich getrennt hat.

[1] H. G. Dehmelt u. H. Krüger: Z. Physik **129**, 401 (1951).
[2] H. G. Dehmelt: Z. Physik **130**, 356 (1951).
[3] P. Grivet u. M. Soutif: C. R. Acad. Sci. Paris **228**, 1852 (1949). — M. Soutif: Rev. Sci. Paris **89**, 203 (1951).
[4] S. Kojima, S. Ogawa u. K. Torizuka: Sci. of Light, Tokyo **1**, 101 (1951).
[5] G. Suryan: J. Indian Inst. Sci. A **35**, 25 (1953).
[6] H. S. Gutowsky, L. H. Meyer u. R. E. McClure: Rev. Sci. Instrum. **24**, 644 (1953).
[7] H. Krüger u. U. Meyer-Berkhout: Z. Physik **132**, 171 (1952).
[8] H. C. Torrey: Phys. Rev. **75**, 1326 (1949).
[9] H. C. Torrey: Phys. Rev. **76**, 1059 (1949).
[10] E. L. Hahn: Phys. Rev. **76**, 145 (1949); **77**, 297 (1950).
[11] E. L. Hahn: Phys. Rev. **80**, 580 (1950).

In Fig. 65 ist das Prinzipschaltbild einer Impulsapparatur[1] wiedergegeben. Die ganze Anordnung wird von einem Impulsgruppen-Generator gesteuert. Solche Generatoren liefern Einzelimpulse, sowie Gruppen von zwei, drei oder mehr Rechteckimpulsen, mit denen der eigentliche Hochfrequenzsender getastet wird. Gestartet wird jedes derartige Impulsprogramm durch einen Triggerimpuls, der zugleich zur Synchronisation der Zeitablenkung des Oszillographen benutzt wird. Meistens verwendet man als Triggergenerator einen Sperrschwinger, dessen Wiederholungsfrequenz etwa in dem Bereich 1 bis 100 Hz variabel ist. Es ist zweckmäßig, die Horizontalablenkung des Kathodenstrahls nicht im synchronisierten Freilauf, sondern nach dem Prinzip der wiederholten einmaligen Ablenkung arbeiten zu lassen. Jede einzelne Strahlablenkung wird dabei durch einen Triggerimpuls ausgelöst, und der Strahlrücklauf verdunkelt. Die Ablenkgeschwindigkeit des Strahls, also der Zeitmaßstab der Abszisse des Schirmbildes, läßt sich an guten Oszillographen unabhängig von der Triggerfrequenz einstellen. Dieses Verfahren ermöglicht die Beobachtung eines von den Impulsgruppendaten unabhängigen stehenden Bildes auf dem Oszillographenschirm auch dann, wenn die Trigger-

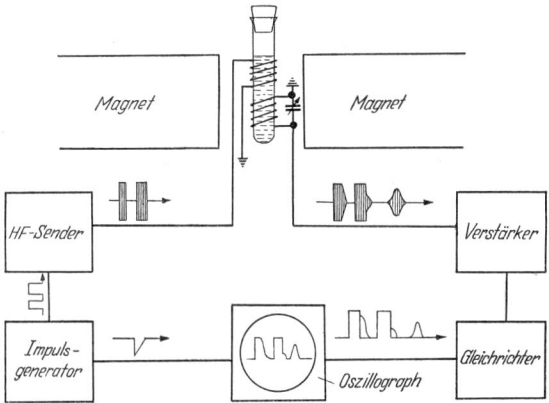

Fig. 65. Blockschaltbild einer Impulsapparatur zur Beobachtung freier kernmagnetischer Präzessionsbewegungen. In Wirklichkeit sind die Senderimpulse vor dem Verstärkereingang viele Größenordnungen höher als die Kernsignale.

frequenz kleiner als etwa 25 Hz ist, wenn also das Bild flimmert oder gar nur in mehr oder weniger großen Abständen erscheint. In diesem Fall ist es empfehlenswert, eine Kathodenstrahlröhre mit einer lang nachleuchtenden Bildschicht zu verwenden.

Die Wahl der Impulsprogrammfrequenz, also der Versuchs-Wiederholungsfrequenz, folgt aus den Relaxationsdaten der zu untersuchenden Substanz. Vielfachinterferenzen der von aufeinanderfolgenden Impulsgruppen angeregten Präzessionsbewegungen stören die Beobachtung. Um definierte Verhältnisse zu bekommen, muß deshalb die Dauer einer Triggerperiode τ groß gegen die Spin-Gitter-Relaxationszeit T_1 sein (für T_2 ist diese Ungleichung dann erst recht erfüllt). Gilt diese Beziehung, dann befindet sich das Kernspinsystem vor jeder neuen Einstrahlung von Hochfrequenzquanten wieder im thermischen Gleichgewicht, und die nacheinander angeregten Bewegungsabläufe sind identisch. Zur Erfüllung der Ungleichung $\tau \gg T_1$ muß man deshalb bei der Untersuchung von Substanzen mit sehr langem T_1 zur einmaligen Tastung übergehen. Praktisch ist es natürlich bequemer, ein konstantes Bild zu beobachten. Außerdem nimmt mit der Wiederholungsfrequenz die Lichtintensität zu, der Versuchsablauf läßt sich also auch einfacher photographisch registrieren. Man muß somit einen Kompromiß schließen, meist genügt es, wenn τ etwa 50mal größer als T_1 ist.

Als Impuls-Hochfrequenzsender können zwei wesentlich verschiedene Typen verwandt werden. Einmal kommen Anordnungen in Frage, in denen der Oszillator selbst getastet wird. Diese werden besonders einfach, wenn man einen

[1] G. LAUKIEN: Z. Naturforsch. 11a, 222 (1956).

Leistungsoszillator benutzt, also keine Verstärkerstufen mehr benötigt. Die Phasen der einzelnen Hochfrequenz-Wellenzüge streuen in diesem Fall natürlich statistisch, und man nennt deshalb diese Impulsbetriebsart *inkohärent*. Ein bekannter Nachteil der direkt getasteten Oszillatoren ist ihre schlechte Frequenzkonstanz. Von Vorteil ist dagegen, daß sich auf diese Weise hohe Senderleistungen und damit starke Hochfrequenzfelder (H_1) erzeugen lassen, deren zeitlicher Verlauf in guter Annäherung rechteckförmig ist. Steile An- und Abstiegsflanken der Impulse, sowie große Feldstärken H_1 (vgl. Ziff. 4) erleichtern die theoretiche Deutung des experimentell beobachteten Verlaufs der Präzessionsbewegung. Benutzt man dagegen einen schwach belasteten und dementsprechend frequenzstabilen Oszillator, dann muß man sehr breitbandig verstärken, damit die Impulsflanken nicht zu flach werden. Sollen z.B. die Impulsflanken nicht länger als $2 \cdot 10^{-7}$ sec sein, dann muß die Bandbreite des Gesamtverstärkers mindestens 5 MHz betragen. Die Bandbreite der einzelnen Stufen muß dementsprechend noch größer sein, damit sinkt deren Spannungsverstärkungs-Ziffer auf etwa 3 bis 4, der elektronische Aufwand wird also insgesamt recht beträchtlich.

Lohnend ist ein derartiges Vorgehen deshalb nur, wenn man eine extreme Konstanz der Impuls-Trägerfrequenz benötigt, oder zum *kohärenten* Impulsbetrieb übergehen will. In diesem Fall erzeugt man die hochfrequenten Schwingungen natürlich in einem quarzgesteuerten, kontinuierlich arbeitenden und schwach belasteten Oszillator. Dessen Ausgangsspannung wird dann in den Verstärkerstufen impulsförmig moduliert. Soll die Ausgangsspannung des gesamten Senders in den Impulspausen verschwindend klein sein, dann müssen mehrere seiner Verstärkerstufen im C-Betrieb arbeiten. Auf diese Weise läßt sich zwar die direkt verstärkte Oszillatorspannung leicht abschneiden, zu beachten ist jedoch, daß über die Gitter-Anodenkapazität der Röhren immer ein gewisser Spannungsrest zur nächsten Röhre weiter geleitet wird. Der Abschwächungsfaktor für die Pausenspannung beträgt in praktischen Fällen etwa 100 pro gesperrter Röhre. Unkontrollierte Streueinstrahlungen müssen selbstverständlich vermieden werden. Erleichtert wird die Lösung dieses Problems, wenn man einen Teil der Verstärker zugleich als Frequenzvervielfacher schaltet. Damit bleibt die Kohärenz zwischen den Impuls-Trägerschwingungen nach wie vor erhalten, die Oszillatorspannung hat jedoch eine wesentlich niedrigere Frequenz als die Senderausgangsspannung, und Einstreuungen werden dadurch weniger wirksam. Zur Lösung mancher Probleme ist es jedoch gerade günstig, wenn man in den Impulspausen der Kernsignalspannung einen Teil der Oszillatorspannung als Normalfrequenz beimischt. Darauf soll später (S. 227) eingegangen werden.

Das eigentliche Beobachtungselement einer Impulsapparatur ist sehr einfach beschaffen. Meist genügt es, einen einzigen Schwingkreis zu benutzen, der sowohl an den HF-Sender wie an den Verstärker-Eingang angepaßt ist (Leistungs-Anpassung). Ferner ist gleichzeitig eine Rausch-Anpassung an den Empfänger anzustreben. Die Probenspule dient also während der Impulssendezeiten zur Anregung der Kern-Präzessionsbewegung und in den Sendepausen zu deren Beobachtung. Ein Nachteil dieser einfachen Anordnung ist, daß die Senderspannung durch den Empfänger-Eingangswiderstand gedämpft wird. Außerdem liegt während der Senderimpulse deren volle Spannung am Gitter der ersten Röhre des Verstärkers. Dort können also Überschläge auftreten. Beide Gründe begrenzen die Höhe der Senderspannung und ergeben damit zugleich für einen festen gewünschten Drehwinkel des Impulses (z.B. 90°) eine untere Grenze für die Impulslängen. Stört dies, dann ist es zweckmäßig, einen Probenkopf mit gekreuzten Spulen zu verwenden. Die durch die geometrische Senkrechtjustierung der Spulen (oder eine andere geeignete Brückenschaltung) erzielbare Verminderung (ohne Feldtrimmer oder ähnliche Hilfsmittel) der auf den Empfänger übertragenen Senderspannung reicht im allgemeinen aus, um die angeführte technische Schwierigkeit zu vermeiden. Bei der Verwendung gekreuzter Spulen ist folgendes zu beachten: da im allgemeinen die Empfänger-Spule mit einem Kondensator auf die LARMOR-Frequenz der Kernpräzession abgestimmt ist (selektiver Empfänger-Eingang) scheint die Wirkung der gekreuzten Spulen aufgehoben. Denn nicht nur das Kernsignal, sondern auch die einstreuende Senderspannung unterliegt der Resonanzüberhöhung. Bei einer Entkopplung der beiden Spulen von $k = U_{Sender}/U_{Empfänger} \sim \frac{1}{100}$ und einer Kreisgüte $R \sim 100$ des Verstärker-Eingangskreises ist die am Empfänger-Eingang liegende Spannung gleich der Senderimpulsspannung.

Man hat nun jedoch die Möglichkeit, sekundärseitig die Senderspannung stark zu beschneiden, ohne den Ausgang des HF-Senders dadurch wesentlich zu belasten (wegen der geringen Kopplung). Eine geeignete Begrenzer-Schaltung ist z.B. durch zwei geringfügig vorgespannte Dioden gegeben, die entgegengesetzt gepolt parallel zum Empfänger-Eingang liegen. Die Kernsignalspannung wird durch die Anordnung nicht bedämpft, solange sie die Vorspannung der Dioden nicht überschreitet.

Schwierig zu vereinen sind die Forderungen, die an den Empfänger einer Impulsapparatur zu stellen sind. Am Eingang des Verstärkers liegen Senderimpulse, deren Spannung etwa 10^{+2} bis 10^{+3} Volt beträgt und Kernsignalspannungen, z.B. der Größenordnung 10^{-5} Volt. Am Ausgang sollen beide Spannungen zur Erleichterung der oszillographischen Beobachtung von der gleichen Größenordnung sein. Dementsprechend muß der Empfänger die Signalspannung verstärken und die Senderspannung begrenzen. Weiter soll die am Empfängerausgang liegende Rauschspannung möglichst klein sein, damit man auch schwache Signale beobachten kann. Die notwendige Mindestbreite des Verstärkers folgt aus der Forderung nach formgetreuer Wiedergabe der Umhüllenden der Kernsignale. Da die Senderimpulse begrenzt werden müssen, ist es unabhängig von der Bandbreite des Verstärkers prinzipiell unmöglich, deren Form unverändert zum Ausgang weiterzuleiten. Vor allem aber wird der Empfänger durch die Senderimpulse übersteuert. Sollen auch rasch abklingende Signale beobachtet werden, so muß er seine volle Empfindlichkeit nach jedem Impuls möglichst schnell wiedergewinnen. Zur Erfüllung dieser letzten Forderung, eines hohen zeitlichen Auflösungsvermögens also, darf der Verstärker nur RC-Glieder mit kleinen Zeitkonstanten und Kreise mit starker Dämpfung enthalten. Dies sind typische Merkmale von Breitbandempfängern, bei denen außerdem die Kreise gegeneinander verstimmt und verschieden stark gedämpft werden[1]. Proportional zur Wurzel der so erhaltenen großen Bandbreite steigt aber die Rauschspannung und vermindert sich dementsprechend die Empfindlichkeit des Empfängers. Man hat also nach einem von der Art der zu beobachtenden Signale abhängenden optimalen Kompromiß zu suchen.

Im folgenden sollen nun einige zur kernmagnetischen Resonanzbeobachtung geeignete Impulsgenerator-, Hochfrequenzsender- und Empfängerschaltungen erläutert werden.

α) *Triggerung, einmalige Tastung.* Gute kommerzielle Oszillographen und Impulsgeneratoren enthalten im allgemeinen einen Triggergenerator, dessen untere Grenzfrequenz jedoch für kernmagnetische Untersuchungen oft zu groß ist. In Fig. 66 ist deshalb eine Sperrschwinger-Schaltung wiedergegeben, die auch im Bereich von wenigen Hertz zufriedenstellend arbeitet. Zur einmaligen Tastung kann man im Prinzip jede Morsetaste benutzen und mit deren Hilfe

---

[1] Vgl. z.B. R. FELDTKELLER: Einführung in die Theorie der Hochfrequenz-Bandfilter, 4. Aufl. Stuttgart: S. Hirzel 1953.

irgendeinen Stromkreis unterbrechen. Jedoch ist die Flankensteilheit eines auf diese Weise erzeugten Spannungsimpulses wenig definiert. Es ist daher empfehlenswert, in einem besonderen, entsprechend dimensionierten Tastgerät mit diesem Impuls zunächst einen monostabilen Multivibrator anzustoßen, dessen Zeitkonstante so lang ist, daß er durch das Abheben der Taste nicht mehr gestört werden kann. Mit der Endflanke dieses Rechteckimpulses läßt sich ein zweiter monostabiler Multivibrator sehr kurzer Zeitkonstante auslösen. Der so erzeugte kurze steile Impuls kann dann, über einen Kathodenverstärker entkoppelt, als Triggerimpuls verwendet werden.

β) *Impulsgruppen-Generator.* Fig. 67 zeigt die Schaltung eines solchen Geräts, das im Prinzip Gruppen mit beliebig vielen Impulsen liefern kann. Es besteht aus einer Kette monostabiler Multivibratoren, sowie aus Verstärkungs-, Begrenzungs- und Entkopplungs-Stufen. Gezeichnet wurden vier Multivibratoren. Ein dem Eingang des Geräts zugeleiteter positiver Triggerimpuls wird zunächst differenziert. Die Anstiegsflanke des Impulses liefert so eine positive Spannungsspitze und seine Endflanke eine negative. Die letztere wird über eine Diode zu dem ersten Multivibrator geleitet und sperrt dort das normalerweise stromführende rechte Triodensystem. Dadurch kippt der Multivibrator und liefert an der Anode des rechten Systems einen positiven Rechteckimpuls, dessen Dauer durch die Zeitkonstante des eingezeichneten variablen RC-Glieds bestimmt wird. Dieser Impuls wird zur zweiten Stufe geleitet. Dort löst, wiederum nach einer Differen-

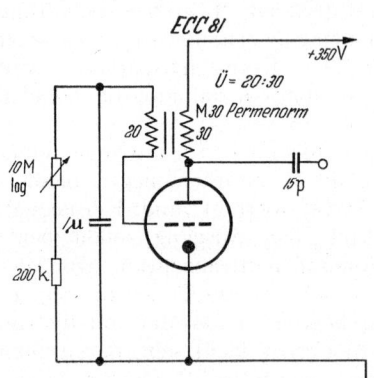

Fig. 66. Schaltung eines als Trigger-Generators geeigneten Sperrschwingers. Beide Systeme der Doppeltriode ECC 81 sind parallel geschaltet.

tiation, seine Endflanke den Kippvorgang des zweiten Multivibrators aus usw. Auf diese Weise erhält man eine Kette von zeitlich aneinander anschließenden Rechteckimpulsen. Jeden zweiten Impuls führt man, über besondere Kathodenverstärker entkoppelt, gemeinsamen Verstärker- und Begrenzungs-Stufen zu. Auf diese Weise erhält man eine Folge von je nach der Verstärkungsart positiven oder negativen Rechteckimpulsen variabler übereinstimmender Ausgangsspannung. Sämtliche Zeitabstände zwischen den Impulsen (bestimmt durch die Zeitkonstanten des ersten, dritten, fünften Multivibrators, usw.), sowie die Breiten aller Impulse (bestimmt durch die Zeitkonstanten des zweiten, vierten, sechsten Multivibrators, usw.) können unabhängig voneinander geändert werden. In der wiedergegebenen Multivibratorkette wurden übrigens zwei verschiedene Schaltungstypen dargestellt. Die angegebenen Gleichspannungspotentiale ergaben sich aus einer speziellen Problemstellung.

γ) *Getastete Oszillatoren.* Einen sehr einfachen Oszillator dieser Art zeigt Fig. 68. Eine Doppeltriode (ECC 81) schwingt in einer HUTH-KÜHN-Gegentaktschaltung. Am Gitter liegt eine negative Sperrspannung. Die Röhre wird durch positive Impulse geöffnet. Die Leistungsabgabe dieses Oszillators reicht z.B. aus, das Kernspin-System von Protonen mit einem Impuls der Dauer 10^{-5} sec um 90° zu drehen.

Fig. 69 gibt eine von HAHN[1] angegebene Schaltung wieder. Sie besteht aus einem Multivibrator (6 J 6), einem Anodenverstärker (6 V 6) und einem als

---

[1] E. L. HAHN: Phys. Rev. **80**, 580 (1950).

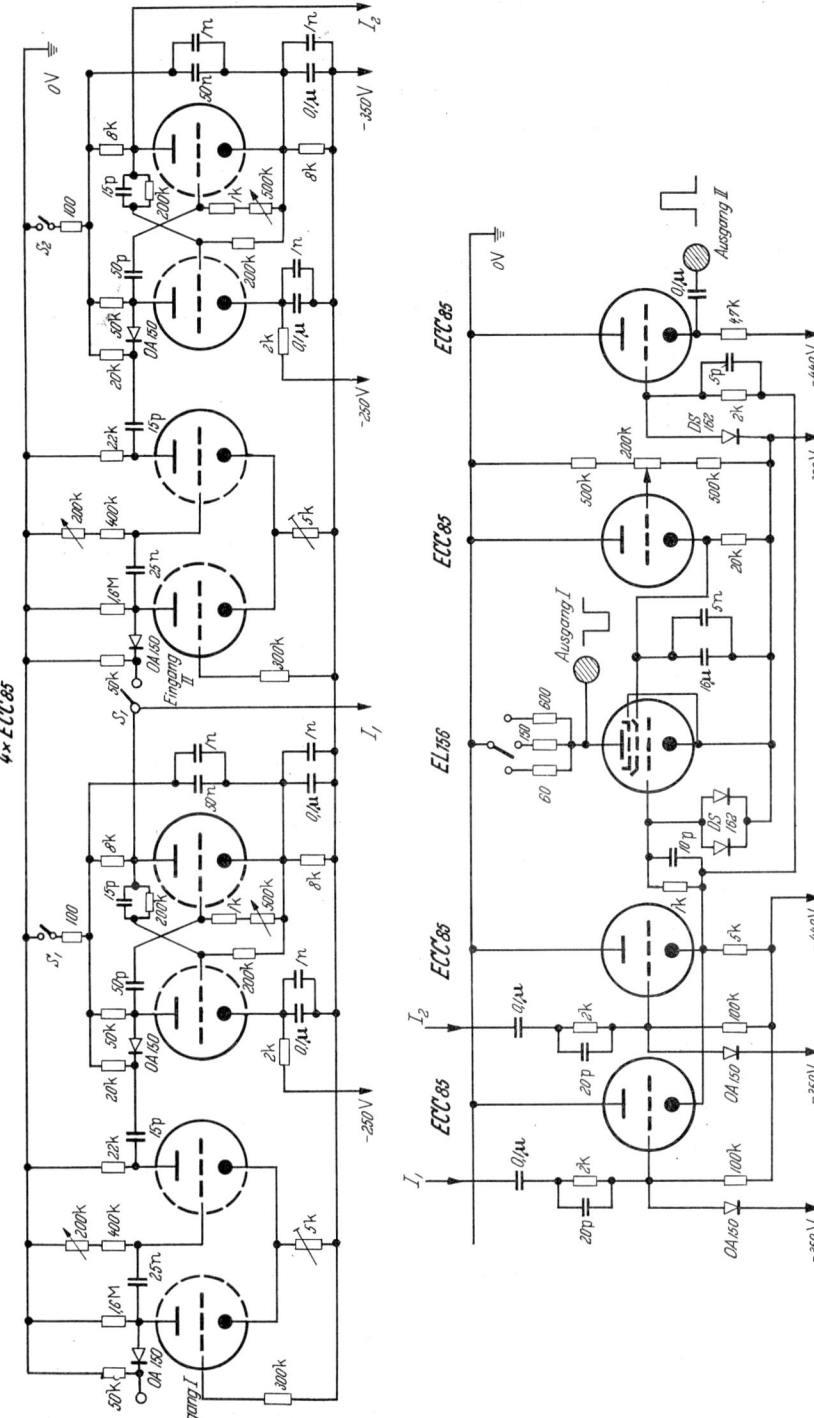

Fig. 67. Schaltung eines Impulsgruppen-Generators. Durch jeden den Eingängen I und II zugeführten Trigger-Impuls wird ein Programm parallel geschalteter Rechteckimpulse ausgelöst. Leitet man den Trigger-Impuls nur zu dem Eingang I und schließt den Schalter S_1, dann addieren sich alle Zeiten der Multivibratorkette und die Impulse sind in Serie geschaltet. Mit Hilfe der Schalter S_1 und S_2 können die Impulse einzeln abgeschaltet werden, man kann mit diesem Gerät also auch normale Impulsfolgen erzeugen. In den Verstärkerstufen werden jeweils beide Systeme der Doppeltrioden ECC85 parallel geschaltet. Die zweitletzte Röhre ECC85 dient zur Regelung der Schirmgitterspannung der EL156 und damit zur Feinregelung der Amplitude der negativen Impulse. Mit dem Anodenwiderstandsschalter der EL156 kann diese Amplitude grob geregelt werden.

Dreipunktschaltung ausgeführten Triodenoszillator (809). Jeder dem Eingang zugeführte positive Triggerimpuls wird in der ersten Stufe in einen negativen Rechteckimpuls umgeformt, dessen Breite sich mit Hilfe eines variablen Widerstandes (0,2 MΩ) regeln läßt. Alle mit dieser Anordnung erzeugten Hochfrequenz-Impulse

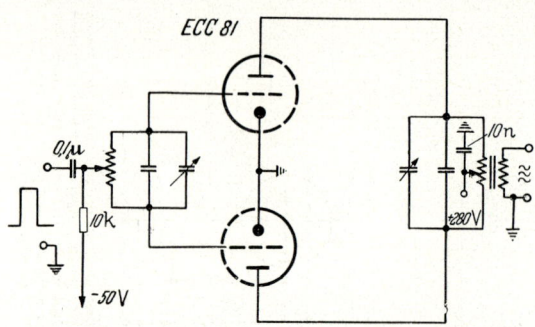

haben demnach die gleiche Breite. In den Impulspausen fließt durch die Röhre 6V6 und damit durch deren Anodenwiderstand Strom. Die am Gitter der Oszillatorröhre liegende Spannung ist auf Grund der Gleichspannungspotential-Anordnung also in dieser Zeit negativ. Während der Impulse verschwindet diese Sperrspannung, da dann die Röhre 6V6 gesperrt ist.

Fig. 68. Schaltung eines Gegentakt-Impulsoszillators. Der normalerweise durch die negative Vorspannung der Steuergitter gesperrte HUTH-KÜHN-Oszillator wird durch positive Rechteckimpulse geöffnet.

Eine weitere Anordnung haben BLOOM, HAHN und HERZOG kürzlich veröffentlicht[1]. In der Kathodenleitung (Fig. 70) eines Gegentaktoszillators (829B) befindet sich eine normalerweise gesperrte Röhre (6L6). Wird

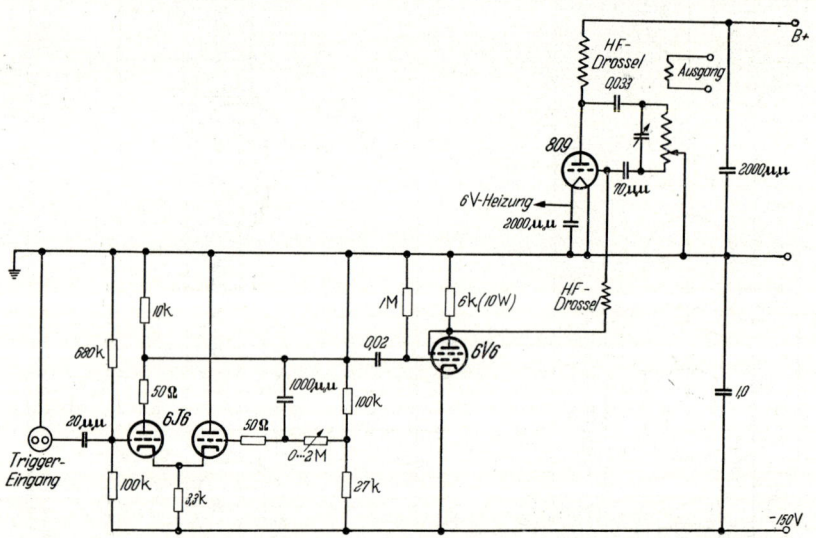

Fig. 69. Schaltung eines Impulsoszillators nach HAHN. Das Gerät enthält außer dem eigentlichen in einer Dreipunktschaltung schwingenden Oszillator (809) eine Anodenverstärkerstufe (6V6 oder 6L6) und einen Multivibrator (6J6), dessen Zeitkonstante die Breite der Impulse bestimmt. Gesteuert wird es durch positive Trigger-Impulse.

diese durch positive Impulse geöffnet, so schwingt der Oszillator. Die Eingangsröhre (6J6) der Schaltung dient zur Phasenumkehr (Polaritätswechsel) der in das Gerät einzukoppelnden negativen Steuerimpulse.

Fig. 71 zeigt die Schaltung des sorgfältig dimensionierten Hochfrequenz-Impulssenders, den BENEDEK[2], CARR[3] und PURCELL für ihre Versuche entwickelt

---

[1] M. BLOOM, E. L. HAHN u. B. HERZOG: Phys. Rev. 97, 1699 (1955).
[2] G. B. BENEDEK u. E. M. PURCELL: J. Chem. Phys. 22, 2003 (1954).
[3] H. Y. CARR: Phys. Rev. 88, 415 (1952); 94, 630 (1954).

haben. Der Oszillator (6 J 6) schwingt ebenso wie in der Schaltung der Fig. 68 in einer HUTH-KÜHN-Gegentaktschaltung, die ähnlich wie in der Schaltung Fig. 70 durch eine in der Kathodenleitung liegende Röhre ($\frac{1}{2}$ 6 J 6) getastet wird.

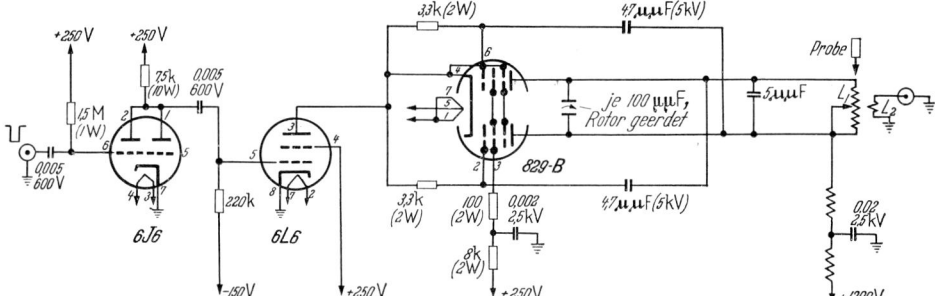

Fig. 70. Schaltung eines in der Kathodenleitung getasteten Gegentakt-Impulsoszillators nach BLOOM, HAHN und HERZOG. Die normalerweise gesperrte Röhre 6 L 6 liegt in der Kathodenleitung des Oszillators (829 B). Beide Systeme der Doppeltriode 6 J 6 sind parallel geschaltet und werden zur Phasendrehung der am Eingang liegenden positiven Rechteckimpulse benutzt.

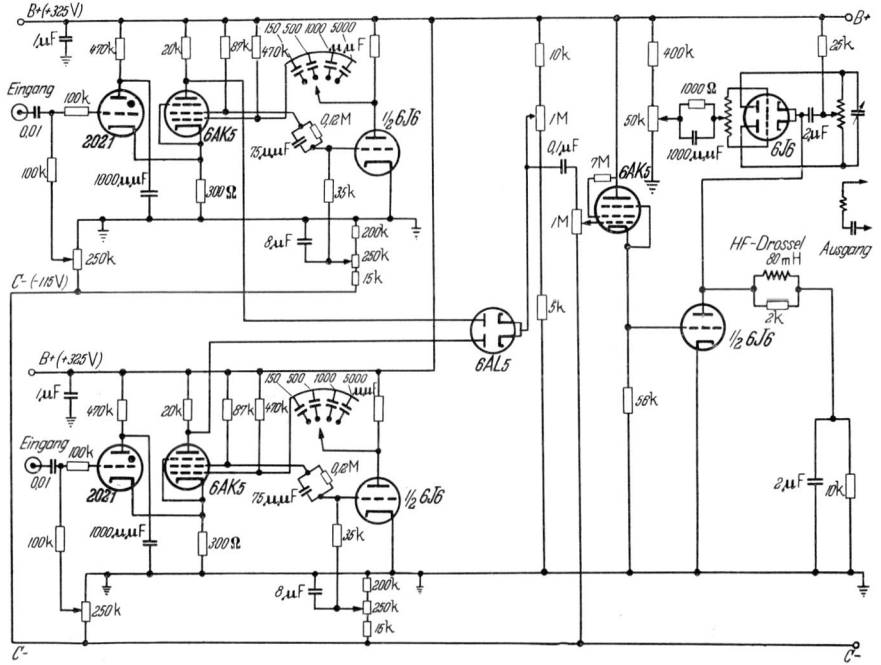

Fig. 71. Schaltung eines in der Kathodenleitung getasteten Gegentakt-Impulsoszillators nach BENEDEK, CARR und PURCELL. Von links nach rechts enthält der Sender folgende Stufen: Zwei voneinander unabhängige und einzeln fremd zu steuernde Multivibratoren, deren Zeitkonstanten die Impulslängen bestimmen. Sie bestehen aus den Röhren 2x(2021+6 AK 5 + 1/26 J 6). Über eine Doppeldiode (6 AL 5) und einen Kathodenfolger (6 AK 5) werden die Rechteckimpulse an das Gitter der in der Kathodenleitung des Oszillators (6 J 6) liegenden Röhre (1/26 J 6) geleitet und öffnen diese.

Auf der linken Seite des Schaltbildes befinden sich, übereinander gezeichnet, zwei identische monostabile Multivibratoren. In diese beiden Stufen werden zu verschiedenen Zeiten positive Triggerimpulse eingeleitet. Dadurch wird eine der Trioden 2021 geöffnet, und die ihr benachbarte Pentode 6 AK 5 über einen gemeinsamen Kathodenwiderstand geschlossen. Zur Verstärkung der Rückwirkung liegen die Gitter zweier weiterer Trioden ($\frac{1}{2}$ 6 J 6) an den Schirmgittern der Pentoden. Die an den Anoden der Pentoden erzeugten positiven Rechteck-impulse werden dann über eine Doppeldiode (6AL5) und einen Kathodenfolger (6AK5) der Taströhre zugeleitet. Mit dem beschriebenen Oszillator kann man also Gruppen von Impulsen mit zwei verschiedenen und voneinander unab-hängigen Breiten erzeugen.

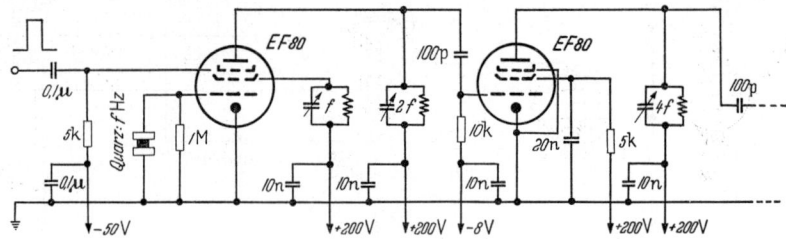

Fig. 72. Schaltung eines quarzgesteuerten kohärenten Impulssenders. Die erste Stufe enthält einen kontinuierlich schwin-genden Quarzoszillator, dessen Schaltelemente mit dem Schirmgitter, dem Steuergitter und der Kathode der Pentode verbunden sind. Impulsförmig moduliert wird das Bremsgitter. Die Frequenz im Anodenkreis ist verdoppelt. Die zweite Stufe beschneidet und verstärkt im C-Betrieb. Außerdem wird die Frequenz nochmals verdoppelt. Die weiteren, ebenfalls im C-Betrieb arbeitenden Stufen wurden nicht gezeichnet.

δ) *Erzeugung kohärenter Impulse.* Fig. 72 zeigt eine für diesen Zweck brauch-bare Schaltung. Das in der ersten Röhre enthaltene innere Triodensystem (be-stehend aus Schirmgitter, Steuergitter und Kathode) erzeugt die quarzgesteuerte

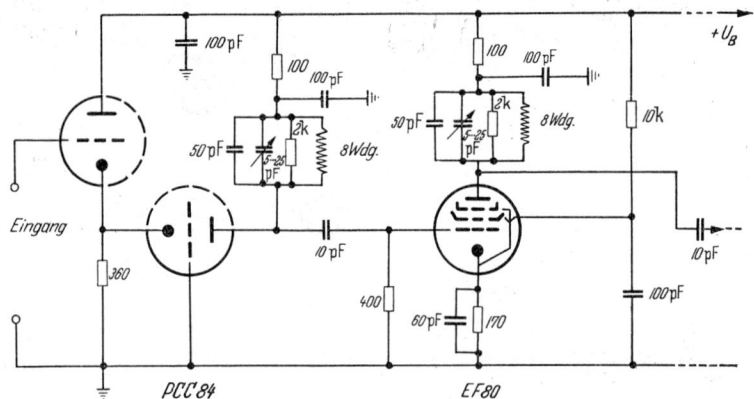

Fig. 73. Schaltung eines abgestimmten Empfängers zur Beobachtung freier Präzessionssignale. In der Eingangsstufe wird eine rauscharme Doppeltriode PCC84 benutzt. Das linke System dieser Röhre ist als Kathodenverstärker und das rechte als Gitterbasisverstärker geschaltet. Alle weiteren Stufen arbeiten als normale abgestimmte Verstärker.

kontinuierliche Oszillatorspannung. Am Bremsgitter, das normalerweise negativ gesperrt ist, liegen positive Impulse. In dem mit der Anode verbundenen zweiten LC-Kreis werden also impulsförmig modulierte Hochfrequenz-Schwingungen an-geregt. Außerdem ist dieser Kreis auf eine Oberschwingung der Quarz-Grund-frequenz abgestimmt (in Fig. 72 auf die zweite), die erste Röhre dient somit zugleich auch zur Frequenzvervielfachung. Alle folgenden Stufen des Senders (gezeichnet wurde nur eine) sind im C-Betrieb arbeitende Verstärker, in denen die Frequenz ebenfalls noch vervielfacht werden kann.

ε) *Empfänger.* Eine zur Beobachtung freier Kern-Präzessionsbewegungen geeignete Verstärkeranordnung ist in Fig. 73 dargestellt. Ihre erste Stufe — ver-wendet wurde eine besonders rauscharme UKW-Anfangsdoppeltriode (PCC84) — ist eine Anodenbasis-Gitterbasis-Schaltung. Diese Eingangsschaltung zeichnet sich wie die bekannte Cascode-Schaltung durch eine große Schwingsicherheit aus. Der große Eingangswiderstand, der den Probenkreis nur geringfügig bedämpft, spricht jedoch zugunsten der in Fig. 73 gezeigten Anordnung. Da andererseits infolge $R_K > \frac{1}{S}$ ($R_K =$ Kathodenwiderstand) die Leistungsanpassung der Anodenbasisstufe an das zweite Röhrensystem zufriedenstellend erreicht wird, ist die Rauschzahl des Verstärkers im wesentlichen durch das erste Röhrensystem gegeben und der Beitrag der folgenden Stufe nur gering. Zur Erzielung der optimalen Rauschzahl (Rausch- + Leistungsanpassung) ist jedoch die übliche Cascode-Schaltung vorzuziehen, da in diesem Fall die hier gezeigte Anordnung

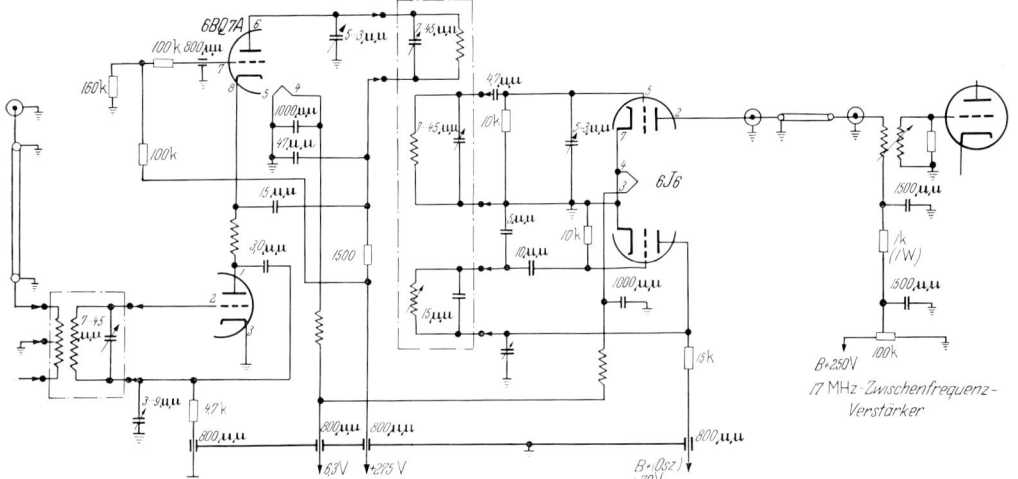

Fig. 74. Schaltung eines abgestimmten Überlagerungsempfängers nach BLOOM, HAHN und HERZOG. Zur rauscharmen Vorverstärkung wird eine Cascoden-Schaltung (6 BQ7A) benutzt. Das untere System der folgenden Doppeltriode 6 J6 dient zur Erzeugung der Überlagerungsfrequenz und das obere zu deren Mischung mit der Signalspannung. Daran anschließend wurde noch die erste Stufe des Zwischenfrequenz-Verstärkers gezeichnet.

ihre Stabilität verliert[1]. Durch die Anodenbasis-Gitterbasis-Kombination werden die übersteuernden Senderimpulse in Fig. 73 gut begrenzt. Für die negativen Halbwellen der Hochfrequenz-Schwingungen ist von einer gewissen Grenze an der Kathodenverstärker gesperrt (Gitterpotential negativ, Kathodenpotential null) und für die positiven Halbwellen gilt das gleiche für den Gitterbasisverstärker (Gitterpotential null, Kathodenpotential positiv). Alle folgenden Stufen sind normale abgestimmte Verstärkerschaltungen. Die Größe ihrer RC-Glieder und ihrer Kreisbedämpfungen ergeben sich aus der Art der zu untersuchenden Substanz, also aus der erforderlichen Bandbreite, der notwendigen Empfindlichkeit und dem zu fordernden zeitlichen Auflösungsvermögen der Apparatur.

Eine weitere geeignete Schaltung haben BLOOM, HAHN und HERZOG veröffentlicht[2]. Es handelt sich um einen entsprechend der Aufgabenstellung abgeänderten Fernseh-Überlagerungsempfänger (untersucht wurden mit diesem Verstärker übrigens Quadrupol-Resonanzfrequenzen). Fig. 74 zeigt die Schaltung des rauscharmen Vorverstärkers (Cascode, 6 BQ7A) und der Mischstufe (6 J6) des Gerätes. Zur Verbesserung des Signal-Rausch-Verhältnisses haben der die Überlagerungsfrequenz erzeugende Oszillator und der Vorverstärker voneinander

---

[1] G. E. VALLEY u. H. WALLMAN: Vacuum tube amplifiers. New York: McGraw-Hill Book Company 1948.
[2] Siehe Fußnote 1, S. 222.

unabhängige Anodenspannungsquellen. Die Abbildung zeigt noch die erste Stufe des bei 17 MHz arbeitenden Zwischenfrequenz-Verstärkers, dessen Bandbreite (in der Fernsehtechnik 6 MHz) durch die Herausnahme von Dämpfungswiderständen zur Verminderung des Rauschens reduziert wurde. Der Empfänger enthält so eine Empfindlichkeit von einem Mikrovolt (Signal-Rausch-Verhältnis: eins) und benötigt etwa 20 μsec, bis er nach starken Übersteuerungen durch Hochfrequenz-Impulse wieder voll empfangsbereit wird. Über das Problem der Erholungszeit (rise-time) von Impulsverstärkern hat auch Elmore[1] u. a. eine Arbeit publiziert.

Fast alle bekannt gewordenen impulstechnischen kernmagnetischen Untersuchungen befassen sich mit der Messung von Relaxationszeiten und Selbstdiffusionskonstanten, also mit der Bestimmung von Substanzeigenschaften, meist in Funktion der Konzentration, des Druckes, der Temperatur, der Viskosität bzw. anderer makroskopischer Größen. Zur Lösung von Fragen aus dieser Problemgruppe haben sich die Impulsverfahren besonders bewährt. Mit ihrer Hilfe kann man beide Relaxationszeiten, die Geschwindigkeit der Selbstdiffusion und den Verlauf der inneren bzw. äußeren inhomogenen Feldverteilung direkt und unabhängig voneinander ermitteln[2]. Die freie Präzessionsbewegung wird bei solchen Experimenten entweder durch Folgen äquidistanter und gleich großer Impulse[3], oder wie beschrieben durch Gruppen von gleich oder

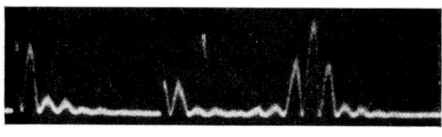

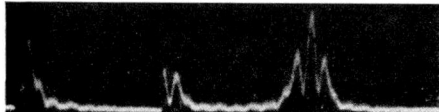

Fig. 75. Überlagerungssignale von F^{19}-Kernen, deren Resonanzfrequenzen sich auf Grund ihrer verschiedenen chemischen Bindung unterscheiden. A. Mischung von $CF_3CCl=CCl_2$ und $C_6H_4F_2$. B. Mischung von $CF_3CCl=CCl_2$ und $C_6H_3F_3$.

verschieden breiten Impulsen (z. B. durch $90°-90°-$ oder $90°-180°-180°-\cdots$ Impulsgruppen u. a., vgl. Ziff. 4 und die Originalarbeiten[4]) angeregt. Im ersten Fall muß man unterscheiden, ob die Abstände zwischen den Impulsen groß oder klein gegen die Relaxationszeiten sind, ob also nur einmalige, im übrigen zur Erzeugung eines stehenden Bildes auf dem Oszillographenschirm wiederholte Abklingbewegungen beobachtet werden, oder ob sich ein mehr oder weniger stationärer Zustand der periodisch angeregten untereinander interferierenden Präzessionsbewegungen einstellt.

Resonanzfrequenzmessungen lassen sich jedoch ebenfalls impulstechnisch ausführen. Ein Vorzug dieses Verfahrens ist evident. Hochfrequenzimpulse enthalten gemäß der bekannten Fourier-Zerlegung außer ihrer Trägerfrequenz noch Nachbarfrequenzen. Mit ihrer Hilfe können also unbekannte Resonanzlinien bequem angeregt werden. Damit ist aber zugleich ein wesentlicher Nachteil verknüpft. Eine exakte Messung der zur Resonanzstelle gehörenden Senderfrequenz, analog zu der bei den kontinuierlichen Verfahren angewandten Meßtechnik, ist prinzipiell unmöglich. Um genau zu messen, muß man die Frequenz der freien

[1] W. C. Elmore: Nucleonics 5, 48 (1949).

[2] E. L. Hahn: Phys. Rev. 80, 580 (1950). — H. Y. Carr u. E. M. Purcell: Phys. Rev. 88, 415 (1952); 94, 630 (1954). — G. Laukien: Z. Naturforsch. 11a, 266 (1956).

[3] Vgl. z. B. R. Bradford, C. Clay u. E. Strick: Phys. Rev. 84, 157 (1951). — E. Strick, R. Bradford, C. Clay u. A. Craft: Phys. Rev. 84, 363 (1951). — H. C. Torrey: Phys. Rev. 85, 365 (1952). — T. P. Das u. D. K. Roy: Phys. Rev. 98, 525 (1955).

[4] E. L. Hahn: Phys. Rev. 80, 580 (1950). — G. Laukien: Z. Naturforsch. 11a, 222 (1956). — H. Y. Carr u. E. M. Purcell: Phys. Rev. 88, 415, (1952); 94, 630 (1954). — T. P. Das u. A. K. Saha: Phys. Rev. 93, 749 (1954). — A. L. Bloom: Phys. Rev. 98, 1105 (1955).

Präzessionsbewegung selbst bestimmen. Zwei Aufgabenstellungen müssen dabei unterschieden werden.

Es ist technisch sehr einfach, den Frequenzabstand zweier nah benachbarter Resonanzlinien (z.B. den zweier Gruppen gleicher Kerne in verschiedenen chemischen Bindungszuständen zu ermitteln. Beide Kernspin-Systeme werden durch dieselben Impulse angeregt. In den anschließenden Beobachtungspausen präzessieren ihre resultierenden Magnetisierungsvektoren aber mit verschiedenen LARMOR-Frequenzen und mit deren Differenzfrequenz sind die Kernsignale

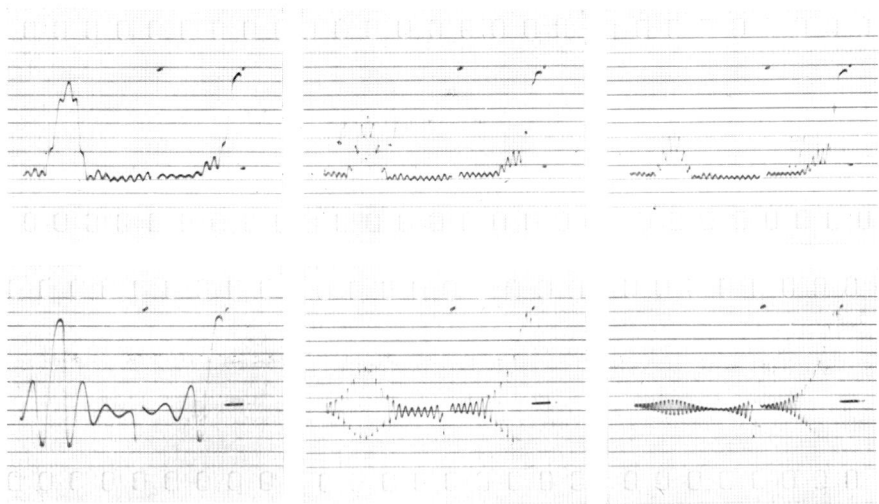

Fig. 76. Freie Präzessionssignale von Protonen, überlagert mit der Impulsträgerfrequenz. Zeitachse von rechts nach links. Die Frequenz des quarzgesteuerten Oszillators betrug 27,36 MHz. Die Signale wurden außerhalb Kernresonanz aufgenommen, und zwar ist der Resonanzabstand und mit ihm die Differenz der Oszillator- und der LARMOR-Frequenz in den rechten Abbildungen am größten. In den Registrierungen der oberen Reihe ist die maximale Signalspannung etwa zehnmal größer als die überlagerte Normalspannung, in den Aufnahmen der unteren Reihe ist die Amplitude der überlagerten Schwingungen fast ebenso groß wie die maximale Signalamplitude. — Die erste Justierart ist zu empfehlen, wenn man das Überlagerungsverfahren bei Relaxationszeitmessungen zur Kontrolle des Resonanzabstandes, bzw. zur Justierung der Resonanzeinstellung benutzt. Die zweite ist dagegen vorteilhaft, wenn man die Differenzfrequenz elektronisch messen will und dementsprechend zunächst verstärken muß. Bezüglich der Abhängigkeit der Signalamplitude vom Resonanzabstand vgl. Ziff. 4 (G. LAUKIEN).

moduliert[1]. Fig. 75 zeigt ein Beispiel solcher intern modulierter Kernsignale. Die Differenzfrequenz, also der Linienabstand kann dann z.B. oszillographisch durch Auszählen der Schwingungszahl pro Zeiteinheit gemessen werden.

Soll die Frequenz der freien Präzessionsbewegung dagegen möglichst exakt absolut gemessen werden, so muß man zu einem Überlagerungsverfahren übergehen[2]. Das zur Zeit genaueste Verfahren zur Ermittlung der Frequenz einer Schwingung ist deren Mischung mit einer benachbarten genau bekannten Normalfrequenz (etwa der eines Quarzoszillators). Damit wird die Messung der unbekannten Frequenz auf die Bestimmung der um einige Stellen kleineren Differenzfrequenz zurückgeführt, und um entsprechend viele Zehnerpotenzen kleiner ist der Meßfehler im Gesamtergebnis. Dieses Prinzip läßt sich im vorliegenden Fall praktisch wie folgt anwenden: Man benutzt einen kohärenten Impulssender, dessen Oszillator quarzgesteuert stetig durchschwingt (vgl. δ). In den

---

[1] E. L. HAHN: Phys. Rev. **80**, 580 (1950). — E. B. McNEIL, C. P. SLICHTER u. H. S. GUTOWSKY: Phys. Rev. **84**, 1245 (1951). — E. L. HAHN u. D. E. MAXWELL: Phys. Rev. **84**, 1246 (1951); **88**, 1070 (1952).
[2] G. LAUKIEN: Phys. Verh. **5**, 170 (1954).

15*

Impulspausen mischt man einen Teil von dessen Ausgangsspannung mit der Signalspannung. Dies ist sehr leicht, denn über die Gitter-Anoden-Kapazitäten der C-Verstärkerröhren wird auch in den Sperrzeiten eine schwache, von Stufe zu Stufe kleiner werdende Restspannung des Oszillators zum Probenkreis weitergeleitet und mischt sich dort mit der Signalspannung. Praktisch wünscht man sich jedoch, um einwandfrei justieren zu können, eine nach Amplitude und Phase regelbare Überlagerungsspannung. Es ist also zweckmäßiger, den direkt durchgelassenen Spannungsrest so zu vermindern, daß er unter dem Rauschpegel liegt und statt dessen von der Ausgangsspannung des Oszillators einen bequem regelbaren Teil abzuzweigen und diesen etwa in den Probenkreis oder gar erst im Empfänger zur Mischung mit der Signalspannung einzukoppeln. Fig. 76 zeigt einige auf diese Weise erhaltenen Kernsignal-Registrierungen, die sich untereinander im Überlagerungsgrad, also in dem Verhältnis Signalspannung zu Normalspannung unterscheiden. Die wiedergegebenen Kernsignale wurden außerhalb Resonanz photographiert. Die LARMOR-Frequenz der Kerne unterscheidet sich also mehr oder weniger von der mit der Impulsträgerfrequenz identischen Normalfrequenz. Mit dem beschriebenen Verfahren lassen sich auch weit auseinanderliegende Resonanzlinien vergleichen, wenn man verschiedene Oberwellen desselben Quarzes als Sender- und Normalfrequenz verwendet und dabei jeweils in demselben Magnetfeld die Differenzfrequenzen zu den entsprechenden LARMOR-Frequenzen bestimmt.

16. Gleichzeitige Anregung und Beobachtung. — Vergleich der kontinuierlichen und impulstechnischen Verfahren. Will man den Verlauf der Einschwingvorgänge erzwungener instationärer Präzessions-Bewegungen (vgl. KRÜGER und LAUKIEN[1] und Ziff. 5) experimentell beobachten, so muß man die in Ziff. 15

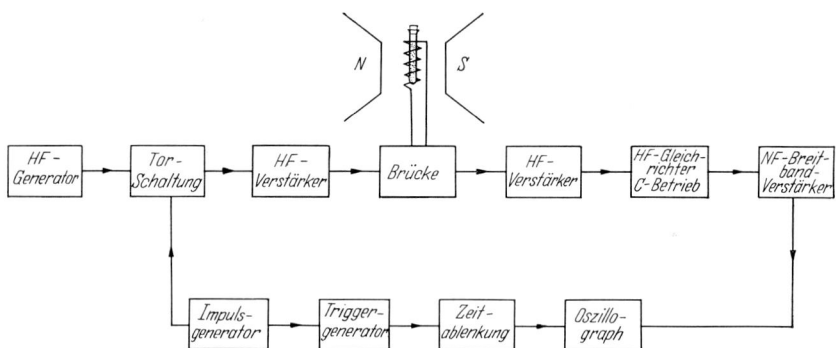

Fig. 77. Blockschaltbild einer Impuls-Brücken-Apparatur zur Beobachtung erzwungener kernmagnetischer Einschwingbewegungen.

besprochenen impulstechnischen Versuchs-Anordnungen durch Brückenschaltungen (vgl. Ziff. 10) ergänzen. Solche Versuche wurden von TORREY veröffentlicht[2]. Fig. 77 zeigt das Prinzip-Schaltbild seiner Apparatur. Der Oszillator schwingt elektronengekoppelt kontinuierlich. Seine Ausgangsspannung wird in einer durch einen Impulsgenerator gesteuerten Torschaltung impulsförmig moduliert. Diese besteht aus einer Doppel-T-Brücke (Fig. 78), deren einer Zweig als mittleres Element einen variablen Kondensator und symmetrisch dazu zwei feste OHMsche Widerstände enthält. Bei dem anderen Zweig ist es gerade umgekehrt, das mittlere einseitig geerdete Element ist also ein fester OHMscher Widerstand. Dieser Wider-

[1] G. KRÜGER u. G. LAUKIEN: Z. Physik **145**, 456 (1956).
[2] H. C. TORREY: Phys. Rev. **76**, 1059 (1949).

stand liegt zugleich in der Kathodenleitung einer gleichstromleitenden Triode, er bildet also mit dieser Röhre einen Kathodenverstärker. Damit setzt sich der dynamische Widerstand des unteren T-Gliedes aus zwei Parallelwiderständen zusammen, einem festen (500 Ω) und einem zweiten, dessen Betrag gleich der reziproken Triodensteilheit ist. In diesem Betriebszustand (Trioden-Gitterpotential null) wird die Brücke abgeglichen. Leitet man nun negative Impulse ausreichender Spannung an das Gitter der Triode, dann wird diese gesperrt, und im unteren Brückenzweig steigt der OHMsche Widerstand sprunghaft (von etwa 200 Ω) auf den Wert des Festwiderstands (500 Ω). Dadurch wird die Brücke verstimmt, und der Weg für die Hochfrequenz-Spannung zum Verstärker ist offen.

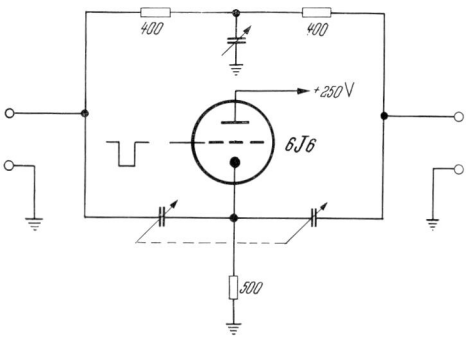

Fig. 78. Brückenschaltung zur impulsförmigen Modulation einer hochfrequenten Spannung. Beide Brückenzweige sind T-Glieder. Der OHMsche Widerstand des unteren Zweigs bildet zusammen mit einer gleichstromleitenden Triode einen Kathodenverstärker. Die abgeglichene Brücke läßt sich durch negative Impulse im Gitter der Triode verstimmen und damit öffnen.

Die eigentliche zur Kompensation der Hochfrequenz-Impulse und damit zur Beobachtung der Präzessionsbewegung des Kernspin-Systems während der Anregung benutzte Brücke wurde bereits besprochen (vgl. S. 191 und Fig. 41). Sie läßt sich bis auf etwa 40 bis 60 db abgleichen. Auf Grund der Absorption und Dispersion der Kernmomente in der Probenspule ist die von der Brücke abgegebene Restspannung der Hochfrequenz-Impulse amplitudenmoduliert. Diese Modulationsspannung verläuft proportional zur x'- bzw. zur y'-Komponente des Kernmagnetisierungsvektors, wenn entweder die Phasen oder die Amplituden der die beiden Brückenzweige durchlaufenden Spannungen vollkommen abgeglichen wurden (vgl. Ziff. 9).

Die modulierten Hochfrequenz-Impulse werden zunächst verstärkt und dann mit Hilfe einer Germanium-Diode gleichgerichtet. Der Gleichrichter ist negativ vorgespannt, um die Signalmodulation von dem relativ großen, durch den unvollständigen Brückenabgleich bedingten Spannungs-Grundpegel abzuschneiden. Ebenso wie in den früher besprochenen Verfahren werden die Signale auf einem Oszillographen beobachtet, dessen Zeitablenkung gleichzeitig mit dem Impulsgenerator getriggert wird.

Das beschriebene Verfahren ist ein Beispiel dafür, wie sich impulstechnische und brückentechnische Methoden miteinander kombinieren lassen. Von praktischer Bedeutung ist es in zweifacher Hinsicht. Einmal können, wie erwähnt, auf diese Weise Einschwingvorgänge untersucht und damit bequem die Frequenzen und Relaxationszeiten besonders stark gedämpfter Resonanzlinien gemessen werden, und zum anderen kann mit Hilfe solcher Impuls-Brücken-Schaltungen auch die Beobachtung der Ausschwing-Bewegungen von Kernspin-Systemen, also deren freier Präzessionsbewegung, erleichtert werden. Gelingt es, die Impulssender-Spannung soweit zu kompensieren, daß sie am Brückenausgang höchstens von der gleichen Größenordnung wie die Kernsignal-Spannung ist, so wird der Empfänger nicht mehr übersteuert, und damit seine Konstruktion wesentlich erleichtert. Von dieser Möglichkeit wurde jedoch bisher kein Gebrauch gemacht.

Zur Messung von Abständen kernmagnetischer Resonanzfrequenzen sind im Prinzip alle hochfrequenz-spektroskopischen Verfahren gleich gut geeignet. Ein Vergleich der äußerlich so verschiedenen kontinuierlichen und impulstechnischen

Verfahren soll dies noch belegen. Kontinuierlich angeregte Kernsignale sind auf dem Oszillographenschirm um so schmaler, je weniger die Resonanzlinie gedämpft ist, je größer also T_2 ist. Diese natürliche Linienbreite ($\Delta\omega_H = 2/T_2$, vgl. Gl. (6.11), bzw. $\Delta\omega_H = 2\sqrt{2}/T_2$, wenn man der Empfindlichkeit wegen mit der optimalen Feldstärke H_1 arbeiten muß, vgl. S. 175) kann durch apparative Einflüsse, z.B. durch eine zu starke Hochfrequenz-Einstrahlung [vgl. Gl. (6.12)] oder durch die Inhomogenität des äußeren Magnetfeldes, in dem die Probe untersucht wird, allgemein also durch technische Mängel, nur vergrößert werden. Die impulstechnisch angeregten freien Präzessionssignale verhalten sich auf dem Bildschirm umgekehrt. Sie klingen um so langsamer ab (proportional zu e^{-t/T_2}), sind also um so breiter, je länger T_2 ist [vgl. Gl. (4.10)]. Ist das Magnetfeld inhomogen, so werden die Signale dadurch verkürzt. Dementsprechend ist aber die Frequenz-Meßgenauigkeit bzw. das Auflösungsvermögen einer kontinuierlichen kernmangetischen Apparatur um so größer, je schmaler die zu vergleichenden Signale sind, während die Genauigkeit der Differenzfrequenz-Messung einer mit Impulsen angeregten abklingenden Schwingung proportional zu der zur Verfügung stehenden Beobachtungszeit[1], d.h. mit der Signallänge zunimmt. Das theoretisch höchste Auflösungsvermögen stimmt bei beiden Verfahren überein, und apparative Mängel, insbesondere die Inhomogenität des Magnetfeldes, vermindern es in beiden Fällen gleichermaßen.

Die zweite wesentliche Größe zur Charakterisierung der Leistungsfähigkeit einer kernmagnetischen Apparatur ist ihre Empfindlichkeit (elektronisch gesprochen, das mit ihr erzielbare Signal-Rausch-Verhältnis). Bei den kontinuierlichen Verfahren ist die optimale Signalspannung proportional zu $M_0\sqrt{T_2/T_1}$ [vgl. Gl. (6.17)]. Die Rauschspannung kann im Prinzip durch Verkleinern der Beobachtungs-Bandbreite beliebig vermindert werden (vgl. Ziff. 12). Bei den Impulsverfahren hängt dagegen die erforderliche Bandbreite von den Eigenschaften der zu untersuchenden Substanz ab, und zwar muß sie mindestens gleich dem Reziproken der Spin-Spin-Relaxationszeit sein, damit der Empfänger die mit der Zeitkonstanten T_2 abfallenden Signale verstärken kann. Proportional zur Wurzel der Bandbreite, d.h. proportional zu $\sqrt{1/T_2}$ steigt aber die Rauschspannung. Dagegen ist die optimale Signalspannung in diesem Fall proportional zu M_0, dem statischen resultierenden Moment der Kerne (vgl. Ziff. 1). Ein quantitativer Vergleich der in beiden Fällen erzielbaren optimalen Empfindlichkeit ist schwierig, da die Größen der Rauschspannungen außerdem natürlich noch in komplizierter Weise von den Eigenschaften des Probenkreises und der Verstärker-Eingangsschaltung u. a. abhängen. Sicher läßt sich jedoch folgendes sagen: Benötigt man eine hohe Empfindlichkeit, weil M_0 sehr klein ist, z.B. bei der Untersuchung von Kernen, die in der Probe nur in geringer Zahl enthalten sind, oder überhaupt nur in sehr kleinen Mengen zur Verfügung stehen (z.B. radioaktive Kerne), so sind bestimmt die kontinuierlichen Verfahren überlegen. Dagegen kann die Anwendung der Impulsmethode vorteilhaft sein, wenn das Verhältnis T_2/T_1 extrem klein ist. In Festkörpern findet man dies häufig, da in diesen, neben der eigentlichen durch T_2 charakterisierten Dämpfung auch inhomogene statische Felder innerhalb der Substanz zur Verbreiterung der Resonanzlinien beitragen. Andererseits lassen sich längere Relaxationszeiten gerade impulstechnisch gut messen, da sich mit diesem Verfahren der Einfluß der Feldinhomogenität eliminieren läßt.

Welches Verfahren man für eine bestimmte Meßaufgabe auswählt, hängt somit im wesentlichen von den Daten der zu untersuchenden Substanz ab, außerdem

---

[1] H. O. KNESER: Arch. elektr. Übertragung 2, 167 (1948).

in der Praxis noch von der Instituts-Tradition, auf Grund derer sich in der Regel zahlreiche, den experimentellen Aufbau erleichternde Erfahrungen in einer bestimmten Arbeitsrichtung ansammeln.

IV. Das Magnetfeld.

17. Räumliche Homogenisierung des Feldes. Wie bereits erwähnt, benötigt man für viele kernmagnetische Untersuchungen außerordentlich homogene Magnetfelder. Ein Zahlenbeispiel soll dies verdeutlichen. Die Spin-Spin-Relaxationszeit der in einer flüssigen Verbindung enthaltenen Protonen betrage etwa $^1/_{10}$ sec. Dies ist keineswegs ein ungewöhnlich großer Wert, in reinem Wasser ist T_2 sogar 30mal größer. Die natürliche Linienbreite der betrachteten Resonanzstelle ist somit in Frequenzeinheiten $\Delta \nu = 10$ Hz und in Feldstärkeeinheiten dementsprechend $\Delta H = \Delta \omega / \gamma \approx 2/1000$ Gauss. Sollen die im wesentlichen für die apparative Linienverbreiterung verantwortlichen Streuungen der Feldstärke über den Bereich des Probenvolumens gegenüber dem voranstehenden Betrag der natürlichen Linienbreite vernachlässigbar klein sein, etwa zur Auflösung sehr nah benachbarter Resonanzlinien oder um die Form der Resonanzkurve untersuchen zu können (zur Bestimmung der kernmagnetischen Absorption bzw. Dispersion), dann darf die Stärke des konstanten Feldes z.B. bei einer Meßfrequenz von 10 MHz ($\hat{=}$ 2300 Gauss) innerhalb eines Volumens von vielleicht 1 cm³ nur um Relativbeträge schwanken, die kleiner als 10^{-6} sind.

Untersucht man dieselbe Kernresonanzstelle mit einer niedrigeren Frequenz und damit auch in einem schwächeren Feld, so wird zwar die obige Forderung entsprechend vermindert, jedoch muß man dafür zwei andere Nachteile in Kauf nehmen. Die Amplitude der Kernsignale nimmt proportional zum Quadrat der konstanten Feldstärke H_0 zu bzw. ab. Einmal ist der Betrag der statischen resultierenden Magnetisierung M_0 proportional zu H_0 [vgl. Gl. (1.7)] und zum andern wächst auch die durch die präzessierende Magnetisierung induzierte Signalspannung mit der zur Feldstärke H_0 proportionalen LARMOR-Frequenz ω [vgl. Gl. (11.2)]. Zur Beobachtung kernmagnetischer Resonanzen bei niedrigen Feldstärken benötigt man also eine wesentlich höhere Empfindlichkeit der Apparatur. Darüber hinaus kann auch die Frequenz-Meßgenauigkeit bzw. das Auflösungsvermögen $\Delta \nu / \nu$ bei niedrigen Feldstärken kleiner sein, wenn die natürliche Linienbreite kontinuierlich angeregter Signale bzw. die natürliche Abklingdauer freier Präzessionssignale von dem Betrag der Feldstärke und der Meßfrequenz unabhängig ist, während der Linienabstand bzw. die Differenzfrequenz proportional mit der Meßfrequenz zunimmt.

Zur Erzeugung der erforderlichen starken homogenen Magnetfelder werden meist Elektromagnete verwendet, deren Polschuhdurchmesser 10 bis 40 cm beträgt. Die Größe des Polschuhabstands (1 bis 6 cm) hängt davon ab, wieviel Raum zur Unterbringung des Probenkopfes und der Modulationsspulen benötigt wird.

Häufig werden jedoch auch Permanentmagnete benutzt, die für viele Untersuchungen beachtliche Vorzüge aufweisen. Ihr Feld läßt sich mit Hilfsspulen im Luftspalt um etwa ± 25 Gauss variieren, insbesondere auch modulieren. Größere Feldänderungen sind nicht mehr ganz reversibel. Zur Untersuchung verschiedener Kerne muß man also die Senderfrequenz ändern. Dieser nicht sehr wesentliche Nachteil (Messungen der Relaxationszeiten einer bestimmten Substanz in Funktion der LARMOR-Frequenz sind natürlich nicht möglich)wird jedoch durch die zeitliche Stabilität des Feldes von Dauermagneten (vgl. Ziff. 18) und durch den Wegfall der Stromversorgungsanlage aufgewogen. Eine Anleitung zum Bau und zur Homogenisierung von Permanentmagneten ist in einer Arbeit von GUTOWSKY und HOFFMAN[1] zu finden, vgl. hierzu auch die Publikationen von SANFORD[2], PRIMAS und GÜNTHARD[3]. In einer Veröffentlichung von ADAMS[4] ist eine Liste von Materialien enthalten, die zur Fertigung von Dauermagneten geeignet sind.

Die Inhomogenitäten des in dem Luftspalt erzeugten Feldes haben drei Ursachen. Da der Polschuhdurchmesser eine endliche Größe hat, nimmt die Feldstärke von innen nach außen ab. Der Betrag dieser aus der Geometrie der Anordnung folgenden Inhomogenität ist um so kleiner, je größer das Verhältnis der Durchmesser der Polschuhe zu deren Abstand ist. Um den Einfluß der Oberflächenrauhigkeit der Polschuhflächen auf den Feldverlauf möglichst klein zu halten, muß die Bearbeitungsgüte der beiden planparallelen Flächen entsprechend hoch sein, es ist daher empfehlenswert, sie zu schleifen und zu polieren. Weiter

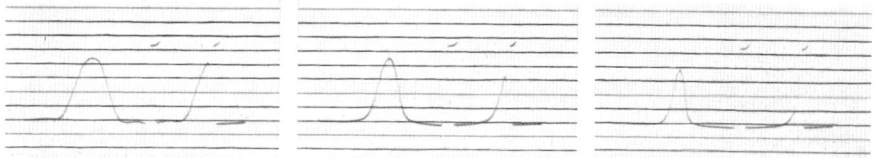

Fig. 79. Abhängigkeit der Echo-Halbwertsbreite von der Feldhomogenität. Zwischen der Halbwertsbreite B der Feldverteilung (in Gauß) und der Halbwertsbreite b der Interferenzsignale (in sec) gilt für Protonen der Zusammenhang: $B = 0,422 \cdot 10^{-4}/b$. Aus den obigen Registrierungen folgt: $B = 0,37$; $0,66$; $1,05$ Gauß. Meßfrequenz: 27,36 MHz, Probenform: zylindrisch, Probendurchmesser: 8 mm, Impulsdaten: $90° - 180°$, Zeitmaßstab: 1 mm $\triangleq 0,445 \cdot 10^{-4}$ sec. Die Messungen wurden zwischen den planparellen Polschuhen eines Magneten von 22 mm Polschuhabstand und 100 mm Polschuhdurchmesser ausgeführt. Der Probenabstand von der Feldmitte betrug während der Aufnahme der obigen Registrierungen 0 mm, 19 mm und 23 mm (G. LAUKIEN).

können auch Fehler in dem für die Polschuhe verwendeten Eisen (Einlagerungen, Lunker u.a.) Störungen im Feldverlauf verursachen. Aus diesem Grunde bevorzugt man feinkristalline, sorgfältig ausgeglühte Eisenlegierungen. Jedoch erfolgt die Herstellung der Legierungen zur Zeit noch rein empirisch, da mit den kernmagnetischen Experimenten überhaupt erstmals eine genügend empfindliche Methode zur Messung derartig kleiner Feldinhomogenitäten zur Verfügung steht. Übrigens wurde diese Schwierigkeit auch schon bei der Herstellung magnetischer Linsen für Elektronenmikroskope bemerkt. Auch hier streuen die Abbildungsfehler einer Serie identischer aus demselben Material gefertigter Linsen gleicher Bearbeitungsgüte auf Grund der Materialfehler recht erheblich. Zur Erhöhung des kernmagnetischen Auflösungsvermögens verschiebt man deshalb die Probe im Feld zwischen den Polschuhen und ermittelt so dessen homogenste Stelle empirisch aus den an verschiedenen Feldorten beobachteten Signalformen. Die Halbwertsbreite kontinuierlich angeregter Signale wird bekanntlich kleiner, wenn die Halbwertsbreite der Feldstärkestreuung innerhalb des Probevolumens abnimmt, während die Halbwertsbreite impulstechnisch angeregter Signale (vgl. Fig. 79) in diesem Fall größer wird. Diese Zu- bzw. Abnahme der Signalbreite endet, wenn der Inhomogenitäteneinfluß gegenüber der natürlichen durch T_2 charakterisierten Signaldämpfung vernachlässigbar klein wird. Außerdem kann man natürlich den Feldverlauf mit sehr kleinen Proben von Punkt zu Punkt

---

[1] H. S. GUTOWSKY u. C. J. HOFFMAN: J. Chem. Phys. **19**, 1259 (1951).
[2] R. L. SANFORD: Permanent Magnets. Circular of the Nat. Bur. of Standards C 448 U.S.G.P.O.
[3] H. PRIMAS u. Hs. H. GÜNTHARD: Helv. phys. Acta **30**, 297 (1957).
[4] E. ADAMS, W. M. HUBBARD u. A. M. SYELES: J. Appl. Phys. **23**, 1207 (1952).

ausmessen (BRUCE[1] benutzte z.B. Proben, deren Volumen 1 mm³ betrug), indem man die zugehörigen kernmagnetischen Resonanzfrequenzen miteinander vergleicht. Auf diese Weise kann man jede Messung einer magnetischen Feldstärke auf eine sehr viel genauer ausführbare Frequenzmessung zurückführen.

Besonders bei kleineren Magneten ist die überwiegende Ursache des inhomogenen Feldverlaufs der geometrische Randfehler. Verschiedene Autoren haben sich deshalb bemüht, diese Fehlerquelle soweit als möglich zu beseitigen. Wohl am bekanntesten ist das Verfahren, den radialen Feldstärkeabfall durch auf den Polschuhrändern angebrachte schmale Eisenringe — sog. shims — zu kompensieren. ROSE[2] hat die günstigsten Abmessungen solcher Ringe in erster Näherung für zylindrische Polschuhe berechnet. Fig. 80 gibt das nur vom Polschuhabstand abhängende Ergebnis wieder. Die an der Abszis-

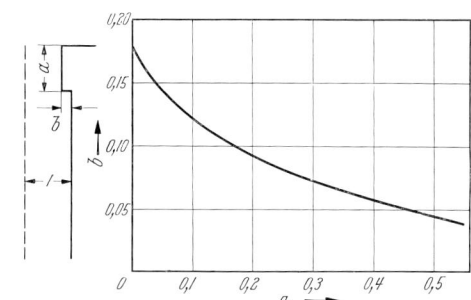

Fig. 80. Abmessungen der Shim-Ringe nach ROSE.

sen- und Ordinatenachse angeschriebenen Zahlenwerte sind mit dem halben Polschuhabstand zu multiplizieren. Fig. 81 zeigt Messungen von DOLEGA[3], der den Feldverlauf mit und ohne Shim-Ringe in einem Magneten von 10 cm

Polschuhdurchmesser mit der kernmagnetischen Resonanzmethode ausgemessen hat. ANDREW und RUSHWORTH[4] haben die günstigsten Shim-Ringabmessungen für konisch verjüngte Polschuhe berechnet. Vor allem bei Permanentmagneten — für die bezüglich ihrer Homogenisierung natürlich alles in dem vorliegenden Abschnitt besprochene ebenso gilt — kann man auf dieses Hilfsmittel zur Erhöhung der Feldstärke im Luftspalt meist nicht verzichten. Zur Herstellung von Dauermagneten geeignete Werkstoffe haben fast

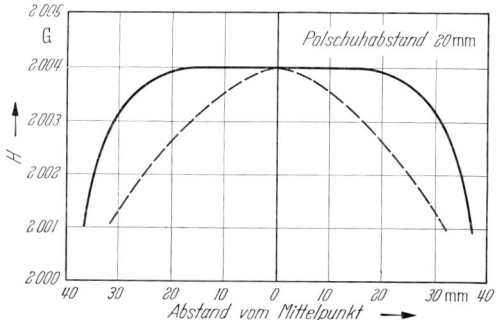

Fig. 81. Verbesserung der Feldhomogenität durch Shim-Ringe nach Messungen von DOLEGA. Die Abmessungen des Magneten und der Shim-Ringe sind: Polschuhdurchmesser = 100 mm, Polschuhabstand = 20 mm, $a = 2,5$ mm, $b = 0,8$ mm (vgl. Fig. 80).

immer auf Grund ihrer Vergütung ein recht inhomogenes Gefüge. Es ist daher beim Bau von Permanentmagneten empfehlenswert, die unmittelbar an den Luftspalt angrenzenden Teile der Polschuhe aus Weicheisen zu fertigen, wenn auch die Feldstärke im Luftspalt dadurch etwas kleiner wird.

WIMETT[5] hat zur Homogenisierung des Feldes die Polschuhflächen empirisch konkav geformt und außerdem zum weiteren Feldausgleich noch eine Anordnung von Kompensationsspulen benutzt. Es gelang ihm mit diesem Verfahren die

[1] C. R. BRUCE: Phys. Rev. 89, 896 (1953).
[2] M. E. ROSE: Phys. Rev. 53, 715 (1938).
[3] U. DOLEGA: Dipl.-Arb. Universität Leipzig 1952.
[4] E. R. ANDREW u. F. A. RUSHWORTH: Proc. Phys. Soc. Lond. B 65, 801 (1952).
[5] T. F. WIMETT: Phys. Rev. 91, 499 (1953).

relative Inhomogenität des Feldes innerhalb des Probevolumens so zu vermindern, daß sie kleiner als $3 \cdot 10^{-7}$ wurde.

Eine interessante Methode zur Verringerung des Einflusses der Feldstreuung auf das Auflösungsvermögen wurde von Bloch[1] vorgeschlagen und von Anderson und Arnold[2] erprobt. Dreht man eine Probe während ihrer kernmagnetischen Beobachtung im Magnetfeld mit einer hohen Rotationsgeschwindigkeit, so beschreiben alle in ihr enthaltenen Kerne Kreise. Auf diesen Bahnkurven schwankt die Stärke des Magnetfeldes, und damit präzessieren die Kerne zeitweise schneller und zeitweise langsamer. Diese Schwankungen mitteln sich aus, wenn zwischen der Dauer τ eines Umlaufs der Probe und der Halbwertsbreite ΔH der Feldstreuung die Ungleichung

$$\tau < \frac{2\pi}{|\gamma|\Delta H}$$

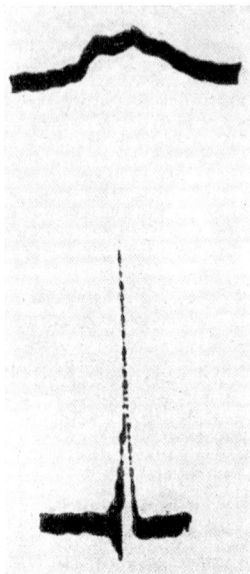

Fig. 82. Verkleinerung der apparativen Linienbreite durch Rotation. Beide Registrierungen zeigen die Protonenresonanzlinie von destilliertem Wasser. Jeder Resonanzdurchgang dauerte 30 sec. Aus der Halbwertsbreite des oberen Signals (ruhende Probe) folgt, daß die Feldinhomogenität $1,7 \times 10^{-3}$ Gauß beträgt. Das untere Signal wurde beobachtet, als die Probe mit der Geschwindigkeit 25 Umdrehungen/sec rotierte. Dadurch nimmt die wirksame Inhomogenität auf etwa 10^{-4} Gauß ab, gleichzeitig wächst die Signalamplitude um den Faktor 7.

erfüllt ist. Man versteht dies leicht, wenn man den Verlauf der Präzessionsbewegung in einem mit der Larmor-Frequenz rotierenden Koordinatensystem betrachtet. In einem solchen System schwankt die Larmor-Frequenz im wesentlichen innerhalb der Grenzen $-|\gamma|\Delta H$, $+|\gamma|\Delta H$. Die obige Ungleichung fordert also nur, daß die durch die Feldinhomogenität verursachte maximale Larmor-Umlaufsdauer kleiner als die Dauer einer Rotationsperiode ist. Dies hat zur Folge, daß sich die Phasenschwankungen der Präzessionsbewegung jedes einzelnen Kerns in jeder Periode völlig ausgleichen, sofern nur der betreffende Kern seine Bahnkurve nicht verläßt (z. B. auf Grund der Diffusionsbewegung). Alle auf einer bestimmten Kreisbahn befindlichen Kerne verhalten sich demnach im kernmagnetischen Experiment so, als ob die Feldstärke entlang dieses Kreises völlig homogen wäre. Zwischen den einzelnen Kreisbahnen erfolgt natürlich kein Feldausgleich. Trotzdem ist es den genannten Forschern mit dieser Methode gelungen, die Halbwertsbreite der Feldstärkestreuung um etwa einen Faktor 20 zu vermindern. Allerdings setzt das Verfahren voraus, daß das Feld schon vorher möglichst gut homogenisiert wurde, da sonst die gemäß der obigen Ungleichung erforderlichen Drehzahlen technisch nicht mehr erreicht werden können. Fig. 82 zeigt als Beispiel den Verlauf einer Resonanzlinie einer ruhenden Probe und darunter die Form derselben Linie, wenn die Probe ausreichend schnell rotiert.

Die bisher höchste Feldhomogenität und damit auch das größte kernmagnetische Auflösungsvermögen hat Arnold[3] erreicht. Zur Kompensation des Randfeldabfalls (etwa 2 Gauss in 2,5 cm Entfernung vom Polschuhmittelpunkt) befestigte er auf den Polschuhflächen ebene Kupferdrahtspiralen. Die durch diese Spulen fließenden Ströme erzeugen Magnetfelder, deren Richtung antiparallel zu der des Hauptfeldes ist. Sie sind in der Feldmitte am stärksten, denn dort liefern alle Windungen der Spiralen einen Beitrag. Nach außen nimmt die

[1] F. Bloch: Phys. Rev. **94**, 496 (1954).
[2] W. A. Anderson u. J. T. Arnold: Phys. Rev. **94**, 497 (1954).
[3] J. T. Arnold: Phys. Rev. **102**, 136 (1956).

Kompensationsfeldstärke von Windung zu Windung ab und verschwindet schließlich am Polschuhrand (current shimming, Spiralstrom-Homogenisierung, vgl. hierzu auch die Publikationen von H. PRIMAS und Hs. H. GÜNTHARD[1]). ARNOLD hat auf beiden Polschuhflächen je neun solcher Spiralen kreissymmetrisch verteilt. Jede Spule hat etwa 15 Windungen. Die günstigsten Stromstärken der einzelnen Spiralen (5 bis 45 mA) wurden empirisch ermittelt. Mit diesem Verfahren gelang es, die Halbwertsbreite der Feldverteilung in ausgewählten Feldgebieten bei einer Feldstärke von 7000 Gauss über die Probenabmessung von etwa 5 mm auf ein halbes Milligauss ($1:1,4 \cdot 10^7$) zu reduzieren. Das endgültige Auflösungsvermögen ($\frac{1}{2}$ Hertz bei 30 MHz!) erhielt der Verfasser mit Hilfe der bereits beschriebenen Rotations-Homogenisierung.

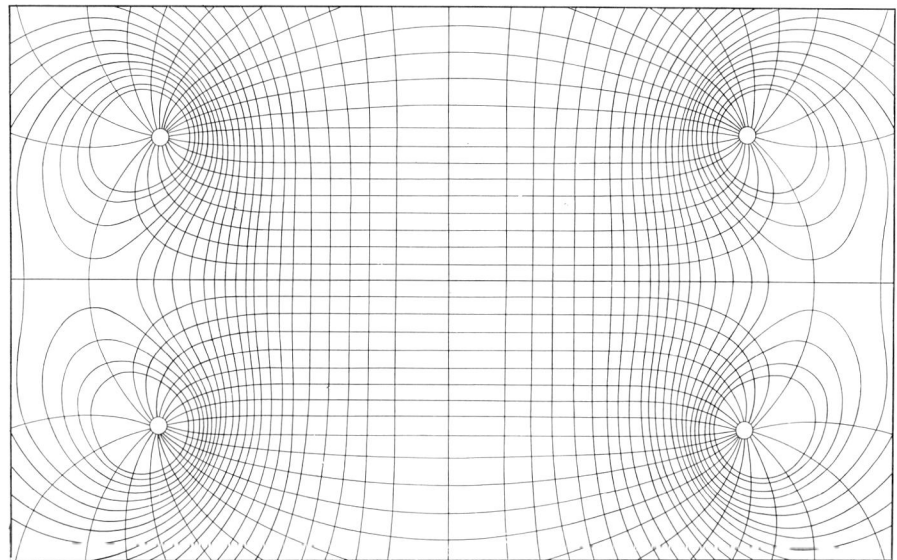

Fig. 83. Verlauf der magnetischen Feld- und Potentiallinien in einem HELMHOLTZ-Spulenpaar.

Gelegentlich wurden zur Erzeugung des konstanten Feldes H_0 auch Luftspulen benutzt. Zweckmäßig ist dies allerdings nur, wenn man bei niedrigen Feldstärken messen will. So haben z.B. MANUS, BÉNÉ, EXTERMANN und MERCIER[2] die Protonenresonanzlinie noch im Frequenzbereich von 2 bis 15 kHz beobachten können. PACKARD und VARIAN[3] haben mit einer Anordnung zur Beobachtung freier Präzessionsbewegungen das erdmagnetische Feld gemessen. In allen diesen Fällen ist es zweckmäßig, HELMHOLTZ-Spulen zu verwenden. Diese bestehen aus zwei koaxialen Spulen, in denen der Strom gleichsinnig fließt, und in denen der Abstand jedes Windungspaars gleich dem Windungsradius ist. In der Mitte des Spulenraums hat die Feldstärke den Betrag[4]: $H = 0,286 \cdot N \cdot I/R$ (H in Oersted, N = Anzahl der Windungen jeder Spule, I in Ampère, R in cm). Fig. 83 zeigt den Feldlinien- und den Potentiallinienverlauf eines solchen Windungspaars. Das Feld ist in einem großen Bereich innerhalb der Spulenmitte

[1] H. PRIMAS u. Hs. H. GÜNTHARD: Helv. phys. Acta **30**, 331 (1957).
[2] C. MANUS, G. BÉNÉ, R. EXTERMANN u. R. MERCIER: Helv. phys. Acta **28**, 616 (1955). — B. CAGNAC, C. MANUS, G. BÉNÉ u. R. EXTERMANN: Helv. phys. Acta **28**, 626 (1955).
[3] M. E. PACKARD u. R. VARIAN: Phys. Rev. **93**, 941 (1954).
[4] E. v. ANGERER u. H. EBERT: Technische Kunstgriffe bei physikalischen Untersuchungen, 8. Aufl. Braunschweig: Fr. Vieweg 1952.

praktisch homogen, vgl. hierzu auch BÉNÉ[1]. Aus diesem Grund benutzt man auch zur Modulation fast immer Spulen in HELMHOLTZ-Anordnung, denn die Homogenität des Gleichfeldes soll natürlich nicht durch das Modulationsfeld verschlechtert werden. Durch zusätzliche Kompensationsspulen läßt sich übrigens der Feldverlauf noch mehr homogenisieren, darüber haben LIN und KAUFMANN[2] berichtet.

18. Zeitliche Stabilisierung des Feldes. Schwankt die Stärke des Magnetfeldes während einer Resonanzfrequenzmessung, so wird dadurch, ebenso wie durch die räumliche Inhomogenität des Feldes, die Resonanzlinie apparativ verbreitert, und damit das Auflösungsvermögen vermindert. Es ist demnach nicht sinnvoll, zur Erhöhung der Resonanzfrequenz-Meßgenauigkeit einer kontinuierlichen Apparatur an eine der beiden komplementären apparativen Konstanten, der räumlichen Homogenität und der zeitlichen Stabilität des Magnetfeldes, größere Ansprüche als an die andere zu stellen. Allgemein gilt dies natürlich nicht. Sollen z.B. impulstechnisch sehr lange Relaxationszeiten gemessen werden, dann ist die Meßgenauigkeit von der räumlichen Inhomogenität des Feldes unabhängig, während zeitliche Feldschwankungen das Meßergebnis verfälschen. Andererseits stören zeitliche Feldschwankungen wenig, wenn der Abstand zweier nah benachbarter Resonanzlinien aus der Differenzfrequenz ihrer freien Präzessionsbewegungen ermittelt wird. Das Auflösungsvermögen wächst in diesem Fall im wesentlichen proportional mit der räumlichen Feldhomogenität. Welche Forderungen zur Lösung eines bestimmten Problems an das Magnetfeld zu stellen sind, hängt somit von der Meßaufgabe und von dem Meßverfahren ab.

Zur Charakterisierung der zeitlichen Stabilität eines Magnetfelds innerhalb einer fest vorgegebenen Beobachtungsdauer muß man kurzzeitige Feldstärkeschwankungen und monotone Feldstärkeänderungen getrennt betrachten. Zweckmäßigerweise beschreibt man Schwankungen, ebenso wie Inhomogenitäten, durch die Halbwertsbreite der zeitlichen Feldverteilung, eine stetige Zu- bzw. Abnahme der Feldstärke dagegen durch den Absolutbetrag der Feldänderung. Hervorgerufen werden die kurzzeitigen Änderungen vor allem durch Schwankungen der Versorgungsspannung des Magneten. Jede von einem Netzgerät oder von einem Umformer erzeugte Gleichspannung hat eine gewisse Restwelligkeit (Netzfrequenz nebst Oberwellen). Auch Batterien liefern keine völlig konstante Ausgangsspannung, da sich in ihnen z.B. unregelmäßig Blasen bilden. Daneben kann die Konstanz des Feldes auch durch mechanische Erschütterungen des Magneten gestört werden. Eine wesentliche Ursache der monotonen Feldstärkeänderungen ist die Inkonstanz der Magnettemperatur. Einmal sind sowohl die Suszeptibilität des Eisens als auch der Widerstand der Spulen temperaturabhängig, und zum anderen verursachen Temperaturänderungen auch Geometrieänderungen des Magneten (z.B. eine Vergrößerung oder Verkleinerung des Polschuhabstands).

Die Halbwertsbreite der kurzzeitigen Feldschwankungen ist von einer gewissen Beobachtungsdauer an unabhängig von der Größe dieses Zeitintervalls. Dagegen nimmt der Absolutbetrag der langwelligen Feldänderungen natürlich mit der Beobachtungsdauer zu. Dementsprechend wachsen aber auch die Anforderungen an die zeitliche Stabilität des Magnetfeldes mit der Zeit, die zur Ausführung eines Experiments erforderlich ist. In vielen Fällen folgt die Mindestgröße dieser Zeit aus praktischen Gesichtspunkten, wenn etwa in demselben Magnetfeld zwei verschiedene Messungen nacheinander ausgeführt werden sollen

---

[1] G. J. BÉNÉ: Helv. phys. Acta **24**, 367 (1951).
[2] S. T. LIN u. A. R. KAUFMANN: Rev. Mod. Phys. **25**, 182 (1953).

(vgl. z.B. Ziff. 20), und dementsprechend ein Teil der Apparatur umgebaut werden muß. Oft ist jedoch eine Verkürzung der Beobachtungsdauer auch grundsätzlich nicht möglich, da sowohl das Auflösungsvermögen als auch die Empfindlichkeit einer kernmagnetischen Apparatur von der zur Verfügung stehenden Meßzeit abhängen.

Bei allen kernmagnetischen Resonanzfrequenz-Messungen wird eine Gruppen-frequenz[1] bestimmt, denn die untersuchten Resonatoren schwingen stets gedämpft, und ihre Bewegung wird mit Hilfe eines weiteren Resonators (Proben-Schwingkreis) beobachtet. Für solche Messungen gilt die bekannte Unschärfe-relation[2]

$$\Delta \nu \gtrless \frac{1}{T_B}$$

wenn man mit T_B die zur Messung verfügbare Zeit und mit $\Delta \nu$ den prinzipiellen Fehler der Messung bezeichnet. Die maximale Dauer der Zeit T_B ergibt sich aus der natürlichen und eventuell auch aus der apparativen Dämpfung der Resonanz-linie. Gemäß der obigen Relation kann einer Beobachtung der erzwungenen bzw. der freien Präzessionsbewegung grundsätzlich keine Aussage über den Eigenwert der LARMOR-Frequenz der Kernresonatoren entnommen werden, deren Fehler kleiner als der Kehrwert der Beobachtungszeit ist. Jede verfahrens-mäßig bedingte zusätzliche Begrenzung der Meßzeit hat also eine Verminderung der Meßgenauigkeit zur Folge.

Praktisch ist es oft noch wichtiger, daß auch die Empfindlichkeit einer kern-magnetischen Apparatur mit der Meßdauer zunimmt. Es ist wohl bekannt, daß man bei der Beobachtung der kernmagnetischen Präzessionsbewegung in einer bestimmten Substanz derzeit das größte Signal-Rauschverhältnis erhält, wenn man die Resonanzlinie differentiell abtastet, und die Signalspannung phasen-empfindlich mit einer möglichst großen Zeitkonstanten gleichrichtet (vgl. Ziff. 12). Der Kehrwert der so erzielbaren extrem kleinen Beobachtungs-Bandbreiten ist gleich der Einschwingzeit des Empfängers. Dementsprechend muß eine solche Messung aber auch ausreichend langsam ausgeführt werden, damit die Registrier-vorrichtung den Änderungen der Signalspannung folgen kann. Dieser Erfahrung liegt ebenfalls ein prinzipieller Sachverhalt zugrunde. Zur Ermittlung einer Meß-größe (im vorliegenden Gebiet also immer der Signalspannung), der eine statistische Schwankung vielfach größeren Betrags (hier die Rauschspannung) überlagert ist, kann man stets die momentane Beobachtungsgröße über die Zeit integrieren. Die Meßgröße wächst dann proportional mit der Integrations- bzw. Beobachtungs-dauer, während die algebraische Summe der Schwankungen nur mit der Wurzel dieser Zeit größer wird.

Sehr empfindliche kernmagnetische Messungen erstrecken sich somit zwangs-läufig oft über viele Stunden und entsprechend hoch muß demgemäß auch die zeitliche Stabilität des Feldes sein. Zur Erzeugung solcher extrem konstanten Felder kann man sowohl Permanentmagnete, als auch Elektromagnete, deren Stromquellen zeitlich stabilisiert sind, benutzen.

In vielen Fällen sind Permanentmagnete natürlich besonders vorteilhaft. Über Konstruktionseinzelheiten von Dauermagneten haben unter anderen PAKE[3], POUND[4], GUTOWSKY und HOFFMAN[5], ANDREW und RUSHWORTH[6] sowie GU-

---

[1] H. O. KNESER: Arch. elektr. Übertragung **2**, 167 (1948).
[2] D. GABOR: Nature, Lond. **159**, 591 (1947).
[3] G. E. PAKE: J. Chem. Phys. **16**, 327 (1948).
[4] R. V. POUND: Phys. Rev. **79**, 685 (1950).
[5] H. S. GUTOWSKY u. C. J. HOFFMAN: J. Chem. Phys. **19**, 1259 (1951).
[6] E. R. ANDREW u. F. A. RUSHWORTH: Proc. Phys. Soc. Lond. B **65**, 801 (1952).

TOWSKY, MEYER und McCLURE[1] berichtet. Fig. 84 zeigt eine von ARNOLD[2] veröffentlichte Dauermagnetkonstruktion. Ein Polschuh kann hydraulisch verschoben werden. Fig. 85 zeigt den Verlauf der Feldstärke in Abhängigkeit vom Polschuhabstand. Ändert sich die Temperatur des Magneten um $1/100^\circ$ C, dann ändert sich seine Feldstärke um ungefähr $1/1000$ ihres Betrags. Da der Magnet eine hohe Wärmekapazität besitzt, folgt er jedoch Raumtemperaturänderungen nur sehr langsam. Seine thermische Zeitkonstante beträgt etwa 6 Std. Trotzdem

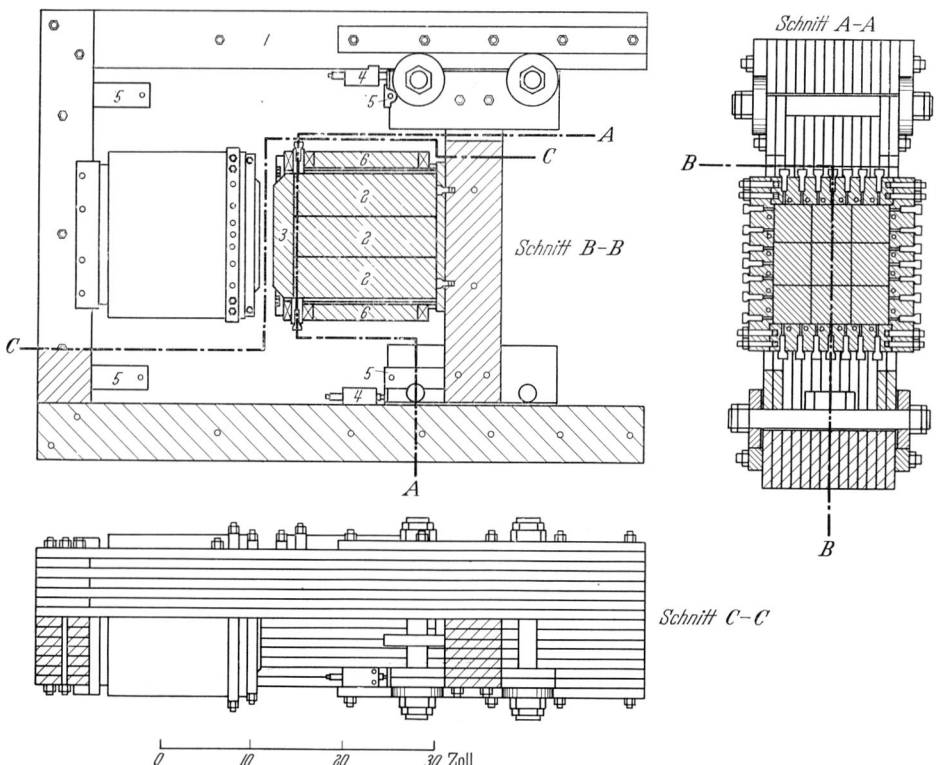

Fig. 84. Konstruktionszeichnung eines Permanentmagneten nach ARNOLD. 1. Stahl-Joch; 2. Dauermagnetblöcke (Alnico); 3. Polschuhkappen; 4. Druckschraubenhalterung; 5. Vorrichtungen zur hydraulischen Abstandsänderung; 6. Magnetisierungs- und Regelspulen.

hat schon eine Temperaturdifferenz von 2° C zwischen dem Magneten und der umgebenden Luft eine Feldstärkedrift von 10^{-4} Gauss/sec zur Folge. Diese Feldänderungsgeschwindigkeit ist immerhin noch 10mal größer als die niedrigste gewünschte Resonanzdurchgangsgeschwindigkeit. Es ist also unbedingt erforderlich, daß die Raumtemperatur möglichst konstant gehalten wird.

Zur Stabilisierung des Feldes eines Elektromagneten muß zunächst dessen Strom stabilisiert werden. Dazu vergleicht man die an einem im Magnetstromkreis befindlichen Normalwiderstand abfallende Spannung mit einer Normalspannung. Mit der verstärkten Differenzspannung wird, falls ein Umformer als Stromquelle dient, dessen Erregerwicklung gesteuert. Benutzt man als Stromquelle ein Netzgerät, dann kann man in bekannter Weise den Widerstand

[1] H. S. GUTOWSKY, L. H. MEYER u. R. E. McCLURE: Rev. Sci. Instrum. 24, 644 (1953).
[2] J. T. ARNOLD: Phys. Rev. 102, 136 (1956).

einer Anzahl im Stromkreis liegender Röhren zur Stromstabilisierung mit dieser Differenzspannung gegengekoppelt ändern. Als Faustregel gilt, daß sich Ströme unter 10 Amp besser mit Röhren stabilisieren lassen. Darüber wird meist ein geregeltes Motor-Generatoraggregat benutzt. Über allgemeine Probleme der Stromstabilisierung von Elektromagneten haben unter anderen HILL[1] und JERVIS[2] berichtet. Praktische Schalteinzelheiten wurden unter anderen von CARO und PARRY[3], SMITH[4], WILLS[5] sowie von KANDIAH und BROWN[6] mitgeteilt. SOMMERS, WEISS und HALPERN[7] gelang es mit einer reinen Stromstabilisierung eine zeitliche Stabilität von $1/10^6$ pro min zu erzielen. Sie benutzten zwei Kanäle zur Gegenkopplung. Über einen, zur Ausregelung niederfrequenter Feldänderungen bestimmten Kanal wurde die Erregerwicklung des Generators gesteuert. Gegen hochfrequente Schwankungen wurde der Magnet außerdem durch eine parallel geschaltete Röhrenstabilisation geschützt.

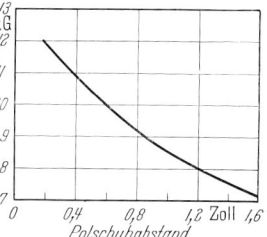

Fig. 85. Verlauf der Feldstärke des in Fig. 84 dargestellten Permanentmagneten in Abhängigkeit vom Polschuhabstand.

Auf eine besonders einfache Art, Felder von Elektromagneten gegen Spannungsschwankungen wirksam zu stabilisieren, hat SURYAN[8] hingewiesen. Oft benötigt man die mit einem bestimmten Magneten maximal erreichbare Feldstärke nicht. Man kann dann seine Spulen in zwei Gruppen gleicher Windungszahlen unterteilen. Eine dieser Spulengruppen schließt man direkt an die Spannungsquelle an, durch ihre Windungen fließt also der Magnetisierungsstrom. Durch die zweite Gruppe der sog. Kontrollspulen läßt man den Strom in umgekehrter Richtung fließen, kompensiert jedoch mit Hilfe einer in Serie geschalteten Normal-Spannungsquelle, z.B. einer Batterie ausreichender Konstanz und Kapazität, die Versorgungsspannung fast vollständig. Dies hat zur Folge, daß jede Schwankung der Versorgungsspannung in beiden Spulengruppen entgegengesetzte Stromschwankungen hervorruft. Unterscheiden sich die Widerstände der beiden Stromkreise um den Betrag ΔR und hat der Widerstand des Laststromkreises die Größe R, dann verursacht eine Änderung der Netzspannung um 1% nur eine Feldänderung vom Betrag:

$$\Delta H = \frac{\Delta R}{R^2} \cdot$$

Die Güte der Stabilisation hängt demnach nur vom Widerstandsabgleich des Last- und des Kontrollstromkreises ab, und dieser läßt sich im Prinzip mit Hilfe zusätzlicher Serienwiderstände beliebig vervollkommnen. Korrekt ist diese Aussage allerdings nur, wenn die beiden Spulengruppen auf dem Magneten symmetrisch verteilt sind. Übrigens ist es nicht empfehlenswert, die Versorgungsspannung mit einer Akkumulatoren-Batterie vollkommen zu kompensieren, da deren Lade- und Entladespannungen sich bekanntlich unterscheiden. Mit diesem Spannungs-Stabilisationsverfahren werden Stromstärkeänderungen auf Grund der Temperaturabhängigkeit des OHMschen Spulenwiderstands natürlich nicht

[1] W. R. HILL: Proc. Inst. Radio Engrs. **33**, 785 (1945).
[2] M. W. JERVIS: Electronic Eng. **26**, 429 (1954).
[3] D. E. CARO u. J. K. PARRY: J. Sci. Instrum. **26**, 375 (1949).
[4] H. G. SMITH: Electronic Eng. **24**, 173 (1952).
[5] M. S. WILLS: Electronic Eng. **25**, 202 (1953).
[6] K. KANDIAH u. D. E. BROWN: Proc. Inst. Electr. Engrs. **99**, 314 (1952).
[7] H. S. SOMMERS, P. R. WEISS u. W. HALPERN: Rev. Sci. Instrum. **20**, 244 (1949); **22**, 612 (1951).
[8] G. SURYAN: J. Sci. Instrum. **29**, 335 (1952).

erfaßt; da aber auch bei einer reinen Stromstabilisation die Temperaturab-hängigkeit der Eisensuszeptibilität und der Magnetgeometrie ebenfalls nicht ausregelbare Feldänderungen zur Folge hat, ist dies nicht sehr wichtig.

Alle oben beschriebenen Verfahren sind indirekte Feldstabilisationen. Zu einem völlig anderen Gegenkopplungsprinzip gelangt man, wenn man kern-magnetische Resonanzeffekte zur direkten Feldstabilisation ausnutzt. Eine solche Anordnung wurde, einer Anregung von BLOCH folgend, erstmals von PACKARD entwickelt und beschrieben[1]. Ihr Prinzip ist leicht zu verstehen. Zur Kontrolle der jeweiligen magnetischen Feldstärke beobachtet man die LARMOR-frequenz der in einer geeigneten Substanz enthaltenen Protonen. Der Proben-kreis ist Teil einer Brückenschaltung, die zur Beobachtung des Resonanzabsorp-

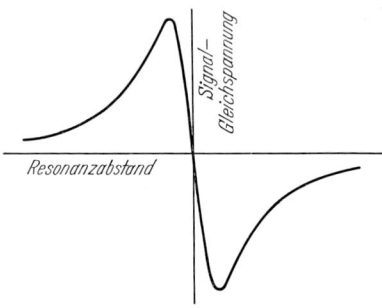

Fig. 86. Verlauf der phasenempfindlich gleichgerich-teten Signalspannung in Funktion des Resonanz-abstands bei der differentiellen Abtastung eines Absorptionssignals.

tionssignals justiert wurde (vgl. Ziff. 9). Die Absorptionskurve wird mit einer festen Senderfrequenz differentiell abgetastet, und die Kernsignalspannung phasenempfindlich gleichgerichtet (vgl. Ziff. 12). Der Verlauf der Signalspannung in Abhängigkeit vom Resonanzabstand wird also durch die erste Ableitung der Absorptionskurve beschrie-ben (vgl. Fig. 86). Demnach verschwindet die gleichgerichtete Signalspannung in Re-sonanz. In der unmittelbaren Nachbar-schaft der Resonanzstelle wächst der Betrag der Signalspannung proportional mit der Zunahme des Resonanzabstands. Dagegen ist das Vorzeichen der Signalspannung auf der einen Seite der Resonanzlinie positiv und auf der anderen Seite negativ. Man kann also die Magnetstromstärke unmittelbar mit der Kernsignalspannung regeln. Jede Abweichung des Feldes von seinem durch die feste Generator-frequenz bestimmten Sollwert erzeugt eine Steuerspannung, mit deren Hilfe sich die Stärke des Magnetstroms gegengekoppelt zur ursprünglichen Feldänderung regeln läßt. Die Feldstabilisation erfolgt um so genauer, je steiler das differen-zierte Kernsignal in Resonanz verläuft, je kleiner also die Linienbreite des eigent-lichen Absorptionssignals ist. Daraus folgt, daß sich ein räumlich homogenes Feld genauer stabilisieren läßt als ein inhomogenes Feld. Der Stabilisations-bereich, d.h. das mittlere lineare Kurvenstück der Fig. 86, wird natürlich mit der Abnahme der Linienbreite kleiner. Soll somit die grundsätzlich mögliche hohe Stabilisationsgenauigkeit der kernmagnetischen Verfahren voll ausgenützt wer-den, so muß der Feldschwankungsbereich schon vorher ausreichend eingeschränkt werden, z.B. durch eine zusätzliche Strom- und Temperaturstabilisation.

Zusammenfassend haben die kernmagnetischen Feldstabilisationsverfahren gegenüber den übrigen bekannten Methoden zwei prinzipielle Vorzüge. Einmal stabilisiert man die Feldstärke selbst und erfaßt damit alle Ursachen der Feld-änderung, und zum anderen wird die Feldstärke mit Hilfe eines Frequenznormals (der Generatorfrequenz) stabilisiert. Normalfrequenzen lassen sich aber wesent-lich genauer und zeitlich konstanter erzeugen als etwa Normalspannungen.

Fig. 87 zeigt das Schaltbild der Anordnung von PACKARD. Als Probesub-stanz verwandte er eine 0,1 molare wäßrige Lösung von Mangansulfat. Der Zu-satz der paramagnetischen Mangan-Ionen dient zur Verkürzung der beiden Relaxationszeiten T_1 und T_2. Die Linienbreite, also die effektive Zeit T_2 wird

---

[1] M. E. PACKARD: Rev. Sci. Instrum. **19**, 435 (1948).

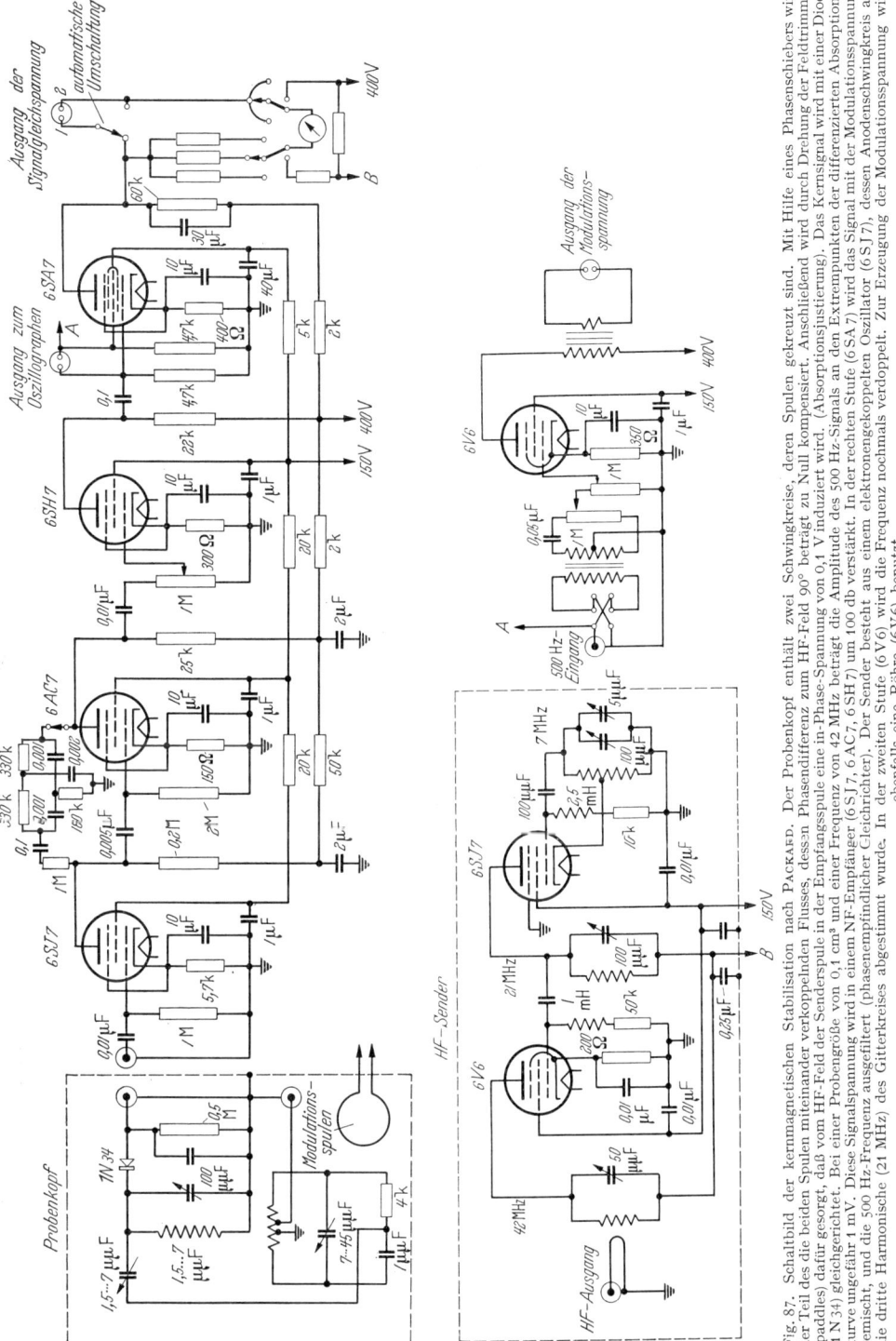

Fig. 87. Schaltbild der kernmagnetischen Stabilisation nach PACKARD. Der Probenkopf enthält zwei Schwingkreise, deren Spulen gekreuzt sind. Mit Hilfe eines Phasenschiebers wird der Teil des die beiden Spulen miteinander verkoppelnden Flusses, dessen Phasendifferenz zum HF-Feld 90° beträgt zu Null kompensiert. Anschließend wird durch Drehung der Feldtrimmer (paddles) dafür gesorgt, daß vom HF-Feld der Senderspule in der Empfangsspule eine in-Phase-Spannung von 0,1 V induziert wird. (Absorptionsjustierung). Das Kernsignal wird mit einer Diode (1 N 34) gleichgerichtet. Bei einer Probengröße von 0,1 cm³ und einer Frequenz von 42 MHz beträgt die Amplitude des 500 Hz-Signals an den Extrempunkten der differenzierten Absorptionskurve ungefähr 1 mV. Diese Signalspannung wird in einem NF-Empfänger (6 SJ7, 6 AC7, 6 SH7) um 100 db verstärkt. In der rechten Stufe (6 SA7) wird das Signal mit der Modulationsspannung gemischt, und die 500 Hz-Frequenz ausgefiltert (phasenempfindlicher Gleichrichter). Der Sender besteht aus einem elektronengekoppelten Oszillator (6 SJ7), dessen Anodenschwingkreis auf die dritte Harmonische (21 MHz) des Gitterkreises abgestimmt wurde. In der zweiten Stufe (6 V6) wird die Frequenz nochmals verdoppelt. Zur Erzeugung der Modulationsspannung wird ebenfalls eine Röhre (6 V6) benutzt.

dadurch praktisch nicht verändert, da sie sich im wesentlichen aus der räumlichen Feldinhomogenität ergibt, wohl aber das Verhältnis $T_{2,\text{eff}}/T_1$. Proportional zu dessen Wurzel wächst aber die Signalspannung (vgl. Ziff. 6). Als Brückenschaltung wurde ein System zweier gekreuzter Spulen verwendet (vgl. Ziff. 11). Die Modulationsfrequenz des Feldes betrug 500 Hz. Mit der Ausgangsspannung der Anordnung wurde das stromstabilisierte Versorgungsgerät der Magnetspulen zusätzlich gesteuert.

Schwankt das zu stabilisierende Feld um seinen Sollwert mit einer sehr niedrigen Frequenz, so kann man das Regelfeld stets so justieren, daß zwischen ihm und der Feldschwankung eine Phasendifferenz von 180° besteht (Gegenkopplung). In fast allen Teilen einer Regelanordnung findet aber eine Phasenverschiebung statt, deren Betrag von der Frequenz abhängt. So ändert sich z.B. das Verhältnis Feldstärke/Versorgungsspannung mit der Frequenz, der Widerstand der Magnetspulen ist komplex, und auch in allen Stufen der eigentlichen elektronischen Schaltung hängt die Größe der Phasendrehung der Steuerspannung von der Frequenz ab. Verursachen diese Phasenverschiebungen bei einer bestimmten Schwankungsfrequenz eine zusätzliche Phasendrehung um 180°, und ist die Verstärkung des Regelkreises bei dieser Frequenz ≥ 1, dann schwingt die ganze Anordnung. Um dies zu verhindern, muß man entweder komplizierte Anforderungen an den Amplituden- und Phasengang des Verstärkers in Funktion der Frequenz stellen, oder die Zeitkonstante des Magneten so groß machen, daß praktisch nur sehr langsame Feldänderungen ausgeregelt werden müssen.

Weitere Hinweise und Einzelheiten über Feldstabilisationsverfahren sind unter anderem in Veröffentlichungen von Thomas, Driscoll und Hipple[1], Thomas[2], Pound und Knight[3], Lindström[4] sowie in einer Arbeit von Knoebel und Hahn[5] mitgeteilt worden.

Die Aufgabe, ein Magnetfeld mit Hilfe einer Normalfrequenz zu stabilisieren, ist nur ein Sonderfall des allgemeineren Problems der Regeltechnik, eine Frequenz durch die Feldstärke eines Magneten oder die Feldstärke eines Magneten durch eine Frequenz zu steuern. Im Bereich der kernmagnetischen experimentellen Technik genügt es im allgemeinen, die spezielle Aufgabe der Konstant-Steuerung, also der Stabilisation zu lösen. In manchen anderen Gebieten der Physik, z.B. bei der Konstruktion von Synchrotrons, sind die Verknüpfungsvorschriften von Frequenzen und Feldstärken jedoch wesentlich komplizierter. Es lag daher nahe zu fragen, ob die kernmagnetischen experimentellen Erfahrungen zur Lösung solcher Probleme beitragen können. Allgemein läßt sich diese Frage nicht beantworten. Der Larmor-Präzessions-Effekt von Kern- und Elektronenmomenten verknüpft die Meßgröße Feldstärke und Frequenz durch eine Proportionalitätsbeziehung ($\omega = \gamma \cdot H$). Im Prinzip können Vorrichtungen zu seiner Messung demnach als Steuerelemente verwendet werden. Ob dies praktisch möglich ist, hängt im wesentlichen von der Größe des geforderten Regelbereichs ab. Bekanntlich ist der kernmagnetische Resonanzeffekt in schwachen Magnetfeldern, der kleinen Signalamplitude wegen, schwieriger nachweisbar als in starken Feldern. Um trotzdem ein günstiges Signal-Rausch-Verhältnis zu erzielen, muß man die Rauschspannung so weit wie möglich verkleinern. Andererseits zwingt aber die

---

[1] H. A. Thomas, R. L. Driscoll u. J. A. Hipple: J. Res. Nat. Bur. Stand. **44**, 569 (1950).

[2] H. A. Thomas: Electronics **25**, 114 (1952).

[3] R. V. Pound u. W. D. Knight: Rev. Sci. Instrum. **21**, 219 (1950).

[4] G. Lindström: Ark. Fys. **4**, 1 (1951).

[5] H. W. Knoebel u. E. L. Hahn: Rev. Sci. Instrum. **22**, 904 (1951).

Forderung eines großen Regelbereichs zu einer entsprechend großen Bandbreite der Signalverstärker und führt damit zwangsläufig zu einer Vergrößerung der Rauschspannung.

Unter den verschiedenen Versuchen geeignete Kompromißlösungen ausfindig zu machen (grundsätzlich kann man bei der Konstruktion der Steuerelemente von allen in diesem Abschnitt besprochenen Verfahren ausgehen), ist das Prinzip des von SCHMELZER[1] angegebenen Spin-Generators besonders interessant. Dieser besteht aus einem Breitbandverstärker und einem Paar gekreuzter Spulen (vgl. Ziff. 11). Die Empfängerspule samt der Probe ist an den Eingang des Verstärkers und die Anregungs- bzw. Senderspule an dessen Ausgang angeschlossen. Das frequenzbestimmende Element des Generators ist das Ensemble der präzessierenden Kernspins. Eine solche Anordnung kann sich unter Umständen selbst erregen, ebenso wie dies, wie aus der Akustik bekannt, bei einer Serienschaltung eines Mikrophons, eines Verstärkers und eines Lautsprechers möglich ist. Dazu muß die Schaltung ein frequenzbestimmendes Gebilde enthalten, und die Rückkopplungsbedingung muß sowohl für die Amplitude als auch die Phase der rückgekoppelten Spannung erfüllt sein. Im Falle des Spingenerators ist die erste Bedingung erfüllt. Die zweite Bedingung ist identisch mit der Forderung nach einem ausreichenden Signal-Rausch-Verhältnis und infolgedessen bei schwachen Feldstärken schwierig zu verwirklichen. Die dritte Bedingung kann zwar grundsätzlich bei jeder Frequenz durch eine entsprechende Justierung des Phasenwinkels eingehalten werden. Eine über einen großen Frequenzbereich von der Resonanzfrequenz unabhängige Einstellung ist jedoch schwierig. Man übersieht dies sofort, wenn man sich den Mechanismus der Rückkopplung des Spingenerators verdeutlicht. Vor dem Einschwingvorgang präzessieren die Kernspins inkohärent, also mit statistisch verteilten Phasen. Sie können demgemäß in der Beobachtungsspule keine Spannung induzieren. Im Verlauf der statistisch erfolgenden Phasenänderungen der einzelnen Kernspins kommt es aber immer wieder vor, daß ein kleiner Teil der Kernspins kurzzeitig kohärent präzessiert. Die von dem resultierenden Magnetisierungsvektor dieser Kernspins in der Beobachtungsspule induzierte Spannung wird verstärkt der Anregungsspule zugeführt. Sie verstärkt dort die ursprüngliche statistische transversale Magnetisierung nur dann, wenn die komplizierten Phasenbedingungen der Kernspin-Präzessionsbewegung erfüllt sind, vgl. hierzu die Abschnitte A II und A III dieses Artikels.

Eine endgültige Entscheidung, welches experimentelle Verfahren der kernmagnetischen Spektroskopie sich am besten für Magnetfeld-Frequenz-Steuerungsaufgaben eignet, ist bis jetzt noch nicht möglich.

C. Ergebnisse.

I. Absolutmessungen. — Experimente zur Bestimmung physikalischer Elementarkonstanten.

19. Prinzip der Absolutmessungen magnetischer Kerndipolmomente. Zur Messung der kernmagnetischen Dipolmomente bestimmt man bei allen hochfrequenzspektroskopischen Verfahren die in der Resonanzbedingung der LARMOR-Präzessionsbewegung

$$\omega = 2\pi\nu = \gamma H = \frac{\mu}{I\hbar}H \quad \text{bzw.} \quad h\nu = \frac{\mu}{I}H \tag{19.1}$$

---

[1] C. SCHMELZER: Lectures on the theory and design of an alternating-gradient proton-synchrotron, 26—28, Okt. 1953, S. 115—125, Genf. CERN.

16*

enthaltenen Parameter. Um den Betrag des Dipolmoments irgendeiner Kernart zu ermitteln, muß man somit im selben Experiment die Frequenz der Präzessionsbewegung und die Feldstärke des Gleichfeldes messen, in dem die Kerne präzessieren. Weiter müssen aber auch die Kern-Spinzahl I der untersuchten Kernart und der Betrag der PLANCKschen Konstante h bzw. $\hbar$ ausreichend genau bekannt sein.

Da die Kern-Spinzahlen I nur ganz- oder halbzahlig sein können, lassen sie sich fehlerfrei ermitteln. Die meisten Kenntnisse über die Spinzahlen der verschiedenen Kernarten verdanken wir früheren Arbeiten über die Struktur der optischen Spektren. Jedoch können auch diese Zahlen hochfrequenz-spektroskopisch gemessen werden. Darüber soll in der folgenden Ziffer berichtet werden.

Die Linienbreiten der kernmagnetischen Resonanzfrequenzen sind in vielen Verbindungen, vor allem im flüssigen und gasförmigen Aggregatzustand, außerordentlich klein. Hinzu kommt, daß Frequenzen bekanntlich sehr genau gemessen werden können. Beide Umstände haben zur Folge, daß sowohl die Absolutbeträge der LARMOR-Frequenzen, als auch die Abstände nahe benachbarter Kernresonanzlinien zu den experimentell am besten zugänglichen Meßgrößen der Physik gehören. Der relative Fehler $\Delta\nu/\nu$ solcher Frequenzbestimmungen ist bei den besten veröffentlichten Ergebnissen[1] kleiner als 10^{-7}.

Die wesentlich höhere Ungenauigkeit der eigentlichen Kerndipolmoment-Messungen hat zwei andere Ursachen. Vor allem kennen wir den Betrag der PLANCKschen Konstante h nur auf etwa vier Stellen genau[2]. Das eigentliche meßtechnische Problem des vorliegenden Arbeitsgebiets ist jedoch die Bestimmung des vierten Parameters der Gl. (19.1), nämlich der *Magnetfeldstärke*. Im Prinzip ist hierzu jedes in einem zeitlich konstanten Magnetfeld am Ort der Kernprobe ausgeführte Experiment geeignet, das es erlaubt, die dortige Feldstärke mit anderen Meßgrößen der Physik zu verknüpfen. Ausreichende Genauigkeit vorausgesetzt, liefern alle vergleichenden Experimente dieser Art nützliche Beiträge zum System der physikalischen Elementarkonstanten-Bestimmungen. Führt man speziell einen magnetomechanischen Versuch in demselben Feld aus, in dem zuvor die kernmagnetische Resonanzfrequenz gemessen wurde — etwa eine Bestimmung der auf einen stromdurchflossenen Leiter wirkenden Kraft —, dann kann man den Betrag des kernmagnetischen Moments direkt in cgs-Einheiten angeben, und man spricht von einer Absolutmessung. Zu beachten ist aber, daß jeder Vergleich der Resultate zweier Experimente verschiedener Art im selben Feld im weiteren Sinne als Absolutmessung bezeichnet werden muß. Beispiele hierzu werden wir in Ziff. 22 kennenlernen.

Der relative Fehler von Feldstärkenmessungen mit einer der bekannten Methoden (Flußmesser, Wismutspirale, HALL-Sonde, Feldmühle u. a.) beträgt bekanntlich etwa 10^{-3}. THOMAS, DRISCOLL und HIPPLE[3] (vgl. Ziff. 21) ist es jedoch gelungen, die Stärke des Magnetfelds am Ort einer Kernresonanzprobe mit einem relativen Fehler von nur 10^{-5} zu bestimmen. Auf Grund ihres Resultats kann z.B. in Zukunft überall jede Magnetfeldstärke mit der gleichen Genauigkeit absolut mit Hilfe einer kernmagnetischen Resonanzfrequenzmessung ermittelt werden.

---

[1] Vgl. z.B. J. T. ARNOLD: Phys. Rev. **102**, 136 (1956). — W. A. ANDERSON: Phys. Rev. **102**, 151 (1956).

[2] Derzeit bester Wert aus Ausgleich aller atomaren Konstanten: $h = (6,62517 \pm 0,00023) \times 10^{-27}$ ergsec, vgl. den Beitrag von DuMOND und COHEN in Bd. XXXV dieses Handbuches, S. 83.

[3] H. A. THOMAS, R. L. DRISCOLL u. J. A. HIPPLE: J. Res. Nat. Bur. Stand. **44**, 569 (1950).

Der weitaus überwiegende Teil der kernmagnetischen Veröffentlichungen befaßt sich mit *Relativmessungen*. Bei diesen Experimenten werden die Resonanzfrequenzen verschiedener Kernarten im selben Magnetfeld miteinander verglichen. Aus Gl. (19.1) folgt für das Verhältnis von Resonanzfrequenzen:

$$\frac{\omega_a}{\omega_b} = \frac{\gamma_a}{\gamma_b} = \frac{\mu_a I_b}{\mu_b I_a}.$$

Messungen des Verhältnisses der magnetischen Dipolmomente verschiedener Kerne haben also die gleiche hohe Genauigkeit wie Frequenzmessungen, sofern die Kern-Spinzahlen bekannt sind. Die Ergebnisse dieser Untersuchungen sollen jedoch erst im folgenden Abschnitt besprochen werden. Wesentlich ist zunächst nur, daß die Absolutbeträge der Dipolmomente aller Kerne sich einfach dadurch bestimmen lassen, daß man die Verhältnisse ihrer Resonanzfrequenzen im gleichen Feld bezogen auf gewisse Standardkerne (meist Protonen) ermittelt, deren Kerndipolmomente direkt gemessen wurden.

20. Hochfrequenz-spektroskopische Bestimmung von Kern-Spinzahlen. Wie schon erwähnt, wurden die meisten Kern-Spinzahlen mit anderen experimentellen Methoden (auf Grund der Hyperfeinstruktur der Linienspektren, aus der Struktur der Bandenspektren, aus Atomstrahlversuchen u. a.) ermittelt[1]. Einige dieser Zahlen wurden jedoch auch erstmals hochfrequenz-spektroskopisch gemessen. Außerdem wurden auf diese Weise zahlreiche frühere Ergebnisse nachgeprüft.

Zur relativen Messung der Spinzahlen zweier Kernarten, deren eine bekannt ist, kann man z.B. davon ausgehen, daß die kernmagnetische Signalspannung bei allen Beobachtungsarten proportional mit dem Betrag des statischen resultierenden Moments

$$M_0 = \frac{N \mu^2}{3 k T} \left(\frac{I + 1}{I} \right) H_0 \tag{20.1}$$

wächst. Beobachtet man also mit der gleichen elektronischen Anordnung die Signale zweier Kernarten, dann läßt sich aus deren Amplitudenquotienten das Verhältnis der Kern-Spinzahlen der verglichenen Kernarten errechnen, sofern man alle übrigen Daten des Experiments kennt. Einige Beispiele sollen dies zeigen. Sehr einfach wird der Vergleich, wenn man die Kernsignale mit derselben Senderfrequenz impulsförmig, und zwar mit 90°-Impulsen anregt. Zur Einstellung der verschiedenen Resonanzstellen kann man die Magnetfeldstärke H_0 ändern. Für jede Messung muß die Impulsbreite neu eingestellt werden. Da es sich hierbei aber um eine Maximumsjustierung des Signals handelt, ist dies unkritisch. An der übrigen elektronischen Justierung muß nichts geändert werden. Das Verhältnis der Anfangsspannungen U_0 der freien Präzessionssignale ist dann direkt gleich dem Quotienten der resultierenden Momente:

$$\frac{U_{0a}}{U_{0b}} = \frac{v_a N_a \mu_a^2 (I_a + 1) I_b H_{0a}}{v_b N_b \mu_b^2 (I_b + 1) I_a H_{0b}}.$$

Darin sind v_a und v_b die beiden Probevolumen, N_a und N_b die Kernzahlen pro cm³ und H_{0a} bzw. H_{0b} die beiden Resonanzfeldstärken. Mit $\mu = I \hbar \gamma$ und $H_0 = \omega_0/\gamma$ vereinfacht sich dieser Ausdruck zu:

$$I_a(I_a + 1) = I_b(I_b + 1) \frac{U_{0a} v_b N_b \gamma_b}{U_{0b} v_a N_a \gamma_a}. \tag{20.2}$$

Ist die Nukleonenzahl der Kernart a, deren Spinzahl bestimmt werden soll, ungerade, so ist diese Zahl sicher halbzahlig ($\frac{1}{2}, \frac{3}{2}, \frac{5}{2}$ usw.). Setzt man demgemäß

---

die Meßergebnisse auf der rechten Seite von (20.2) ein, so muß eine der Zahlen 0,75, 3,73, 8,75, ... herauskommen. Ist dagegen die Nukleonenzahl der Kernart gerade und damit die Spinzahl ganzzahlig, dann kann das Produkt $I(I+1)$ nur die Werte 2, 6, 12, ... annehmen. In beiden Fällen unterscheiden sich also die möglichen Meßergebnisse so stark voneinander, daß auch bei geringer Meßgenauigkeit stets eine einwandfreie Entscheidung über die Größe der unbekannten Spinzahl getroffen werden kann.

Das beschriebene Verfahren läßt sich jedoch in dieser einfachen Form nur anwenden, wenn mehrere Voraussetzungen erfüllt sind. Der Verstärker, der Gleichrichter und die Registriervorrichtung müssen eine lineare Charakteristik haben, damit das am Eingang liegende Signalspannungsverhältnis am Ausgang unverändert beobachtet werden kann. Es muß also darauf geachtet werden, daß insbesondere die Endstufen des Empfängers nicht übersteuert werden. Weiter müssen die Relaxationszeiten T_1 und T_2 der untersuchten Substanzen sehr groß gegen die Impulsdauer sein, damit sich während der Anregungszeit der Präzessionsbewegung die Beträge der resultierenden Magnetisierungsvektoren nicht ändern. Als letztes muß gefordert werden, daß die Impulsfeldstärke H_1 sehr groß ist verglichen mit den in Gauß (gemäß $\omega = \gamma H$) gemessenen Breiten der Larmor-Frequenzstreuungen um die Resonanzstellen der untersuchten Substanzen, unabhängig davon, ob es sich dabei um durch innere oder äußere Feldinhomogenitäten verursachte Linienverbreiterungen handelt ($B/H_1 \ll 1$, vgl. S. 160), oder ob eine der Linien sogar aufgespalten ist. Die beiden zuletzt genannten Bedingungen hängen natürlich eng zusammen. Sollen die Impulse z. B. zur Erfüllung der ersten sehr kurz sein, dann muß auch H_1 sehr groß sein, damit das resultierende Moment durch den Impuls stets um den gleichen 90°-Winkel gedreht wird und umgekehrt.

Ähnlich lassen sich Kern-Spinzahlen aus einem Vergleich kontinuierlich angeregter Kernsignale bestimmen. Befinden sich die Kerne der verglichenen Substanzen nacheinander am selben Ort in einem völlig homogenen Feld und beobachtet man z. B. die Resonanzabsorption, jeweils bei der gleichen Frequenz, so ist das Verhältnis der Signalspannungen gemäß (9.8) proportional zum Quotienten der Imaginärteile χ'' (6.9b) der dynamischen Kernsuszeptibilitäten:

$$\frac{U_{0a}}{U_{0b}} = \frac{\chi_a''}{\chi_b''} = \frac{\chi_{0a} T_{2a}(1 + \gamma_b^2 H_1^2 T_{1b} T_{2b} + (\Delta\omega_b)^2 T_{2b}^2)}{\chi_{0b} T_{2b}(1 + \gamma_a^2 H_1^2 T_{1a} T_{2a} + (\Delta\omega_a)^2 T_{2a}^2)}. \tag{20.3}$$

Zur Vereinfachung der Verhältnisse ist es empfehlenswert, die Kernspinsysteme nur schwach anzuregen ($\gamma^2 H_1^2 T_1 T_2 \ll 1$). Berücksichtigt man dies und setzt den Betrag der statischen Kernsuszeptibilität χ_0, Gl. (1.8), ein, so ist:

$$\frac{U_{0a}}{U_{0b}} = \frac{v_a N_a \mu_a^2 (I_a + 1) I_b T_{2a}(1 + (\Delta\omega_b)^2 T_{2b}^2)}{v_b N_b \mu_b^2 (I_b + 1) I_a T_{2b}(1 + (\Delta\omega_a)^2 T_{2a}^2)}. \tag{20.4}$$

Mißt man die Signalspannung beider Kernarten genau in Resonanz, so vereinfacht sich (20.4) mit $\mu = I\hbar\gamma$ zu:

$$I_a(I_a + 1) = I_b(I_b + 1) \frac{U_{0a} v_b N_b \gamma_b^2 T_{2b}}{U_{0b} v_a N_a \gamma_a^2 T_{2a}}. \tag{20.5}$$

Zur Bestimmung der Spinzahl der Kernart a muß man also in diesem Fall nicht nur die Spinzahl der Kernart b, die beiden Produkte vN und den Quotienten der gyromagnetischen Verhältnisse kennen, sondern außer dem Signalspannungs-Verhältnis auch noch den Quotienten der Spin-Spin-Relaxationszeiten T_{2a}/T_{2b}

messen. Gemäß Gl. (6.11) ist T_{2a}/T_{2b} gleich dem Quotienten der Halbwerts-breiten der Absorptionssignale der verglichenen Substanzen. ALLER und YU[1] haben mit diesem Verfahren erstmals den Spin des Sauerstoff-Isotops O^{17} ge-messen $[I(O^{17}) = \frac{5}{2}]$. Ebenso wie bei der zuvor beschriebenen Methode muß natür-lich auch hier die gesamte Empfangsanordnung eine lineare Charakteristik haben. Eingeschränkt wird die Anwendbarkeit des Verfahrens in der beschrie-benen Art vor allem aber durch die in der obigen Ableitung enthaltene Voraus-setzung, daß die Formen der Absorptionssignale nur durch Relaxationsvorgänge in den Substanzen bestimmt werden. Diese Bedingung ist offensichtlich nicht erfüllt, wenn das äußere Magnetfeld, in dem experimentiert wird, zu inhomogen ist, oder wenn dem äußeren ausreichend homogenen Feld substanzinnere statische inhomogene Felder überlagert werden, oder wenn die Resonanzlinie durch eine zusätzliche Kopplung aufgespalten ist, vgl. z.B. SANDS und PAKE[2] sowie WATKINS und POUND[3].

Unter diesen Umständen geht man am besten von dem für symmetrische Resonanzkurven gültigen Ansatz

$$U = k \cdot \chi_0 \cdot f(H - H_0) \qquad (20.6)$$

aus. Danach ist U gleich dem Produkt einer apparativen Konstante k, in die u. a. die Verstärkungszahl des Empfängers, die Abmessungen der Beobachtungs-spule und die Justierart der Brückenschaltung eingeht, multipliziert mit der statischen Suszeptibilität χ_0 und einer normierten Funktion $f(H - H_0)$, welche die Signalform in Funktion des Resonanzabstands beschreibt ($H_0 =$ Resonanz-feldstärke). Mit $\int\limits_0^\infty f(H - H_0)\, dH = 1$ folgt:

$$\int\limits_0^\infty U\, dH = k \cdot \chi_0.$$

Mißt man also das Integral der Signalspannung über alle Feldstärken zweier Kernarten beliebiger Singnalform bei derselben Frequenz mit der gleichen elek-tronischen Anordnung, so ist stets:

$$\frac{\int\limits_0^\infty U_a\, dH}{\int\limits_0^\infty U_b\, dH} = \frac{\chi_{0a}}{\chi_{0b}} = \frac{v_a N_a \gamma_a^2 I_a(I_a + 1)}{v_b N_b \gamma_b^2 I_b(I_b + 1)}. \qquad (20.7)$$

Im Prinzip kann man jede Kern-Spinzahl auf diesem Wege bestimmen. Praktisch ist dies jedoch oft sehr schwierig. Sind die Linien breit, dann werden die Signal-spannungen entsprechend schwach und liegen möglicherweise zum Teil unter dem Rauschpegel.

Mehrere Autoren haben zur Bestimmung von Spinzahlen den Verlauf der Kernsignale von Festkörpern untersucht. Auf diesem Wege gelang es u. a. GU-TOWSKY, MCCLURE und HOFFMAN[4] sicherzustellen, daß Be^9 den Spin $\frac{3}{2}$ hat. Sie verglichen die theoretisch zu erwartende Signalform der F^{19}-Resonanz mit den experimentell erhaltenen Absorptionssignalen von glasigem Berylliumfluorid (BeF_2) und von Kalium-Beryllium-Fluorid (K_2BeF_4). In festen Proben beträgt die Signalbreite auf Grund der Wechselwirkungen zwischen benachbarten kern-magnetischen Momenten oft mehrere Gauß. Die durch solche Kopplungen

---

[1] F. ALLER u. F.C. YU: Phys. Rev. **81**, 1067 (1951).
[2] R. H. SANDS u. G. E. PAKE: Phys. Rev. **89**, 896 (1953).
[3] G. D. WATKINS u. R. V. POUND: Phys. Rev. **89**, 658 (1953).
[4] H. S. GUTOWSKY, R. E. MCCLURE u. C. J. HOFFMAN: Phys. Rev. **81**, 635 (1951).

verursachte Signalverbreiterung ist um so stärker, je größer die Spinzahl I der Kerne ist, die auf die eigentlich untersuchten Kerne einwirken[1] (vgl. Ziff. 28).

Durch magnetische Wechselwirkungen zwischen den verschiedenen Kernen eines Moleküls können auch die Resonanzlinien flüssiger und gasförmiger Proben aufgespalten werden, vgl. z. B. die Arbeit von GUTOWSKY, McCALL und SLICHTER[2]. Die Zahl dieser sog. Hyperfeinstruktur-Linien (vgl. Ziff. 27) wächst mit der Kern-Spinzahl der wechselwirkenden Kerne. OGG und RAY[3] konnten so zeigen, daß Si^{29} die Spinzahl $\frac{1}{2}$ hat. Sie untersuchten die Form der Protonensignale in der tetraedisch-symmetrischen Verbindung SiH_4 im flüssigen Zustand und fanden außer der starken zentralen Resonanzlinie symmetrisch zu dieser im gleichen Abstand zwei schwache Linien. Natürliches Silizium besteht zu 4,7% aus dem Isotop Si^{29}, der Rest ist Si^{28} und Si^{30}. Die beiden letzten Isotope haben die Spinzahlen Null. Ein Amplitudenvergleich der drei Linien zeigt, daß die mittlere Linie den Molekülen $Si^{28} H_4$ und $Si^{30} H_4$ zuzuordnen ist, während die beiden äußeren zu den Protonen der Moleküle $Si^{29} H_4$ gehören. Da deren Resonanzlinie nur in zwei aufgespalten ist, muß Si^{29} den Spin $\frac{1}{2}$ (Linienzahl $= 2I + 1$) haben. Zwar wäre es möglich, daß in der starken mittleren Linie noch eine dritte $Si^{29} H_4$-Linie verborgen ist, daraus würde aber $I(Si^{29}) = 1$ folgen, und diese Möglichkeit kann man bei einem Kern mit ungerader Nukleonenzahl ausschließen. WILLIAMS, McCALL und GUTOWSKY[4] haben übrigens bei einer Untersuchung der Fluor-Resonanz in der Verbindung SiF_4 das gleiche Ergebnis erhalten.

Linienaufspaltungen können in Festkörpern auch noch andere Ursachen haben. Im Magnetfeld hat ein Kern, dessen Spinzahl I beträgt, $2I + 1$ mögliche Energieniveaus. Wenn deren Energieunterschiede äquidistant sind, beobachtet man nur eine Resonanzlinie. Kerne mit $I > \frac{1}{2}$ haben aber im allgemeinen ein endliches elektrisches Quadrupolmoment. Existiert am Ort eines solchen Kerns neben dem Magnetfeld noch ein inhomogenes elektrisches Feld, so hängt die Energie des Kerns nicht nur von der Lage seiner Spin- bzw. Dipolachse bezüglich der Magnetfeldrichtung ab, sondern auch von dem Winkel, den das elektrische Quadrupolmoment mit dem Gradienten des elektrischen Feldes einschließt. Dies hat zur Folge, daß die $2I$-Abstände zwischen den $2I + 1$ nach der Quantentheorie möglichen Orientierungen bzw. Energieniveaus ungleichmäßig werden. Damit spaltet die Resonanzlinie des Kerns in $2I$-Linien auf. Statische elektrische Feldgradienten, der für eine solche Aufspaltung erforderlichen Größe, wurden in Festkörpern beobachtet, deren Kristall-Symmetrie niedriger als kubisch ist. Da man im Experiment stets eine große Anzahl von Kernen derselben Art beobachtet, ist die Struktur der Resonanzkurve allerdings nur dann so leicht zu verstehen, wenn die elektrischen Feldgradienten an allen Kernorten gleich ausgerichtet sind. Zur Bestimmung von Kern-Spinzahlen aus Quadrupol-Aufspaltungen eignen sich also besonders Einkristalle (vgl. hierzu Ziff. 28).

21. Direkte Messung des magnetischen Dipolmoments der Protonen und des Vorzeichens ihres gyromagnetischen Verhältnisses. Von ganz besonderer Bedeutung ist für alle Zweige der kernmagnetischen Forschung, wie schon erwähnt, die Absolutmessung des gyromagnetischen Verhältnisses des Protons bzw. seines Dipolmoments in Einheiten des physikalischen cgs-Systems. Die bisher genaueste

---

[1] H. S. GUTOWSKY, G. B. KISTIAKOWSKY, G. E. PAKE u. E. M. PURCELL: J. Chem. Phys. **17**, 972 (1949).
[2] H. S. GUTOWSKY, D. W. McCALL u. C. P. SLICHTER: J. Chem. Phys. **21**, 279 (1953).
[3] R. A. OGG jr., u. J. D. RAY: J. Chem. Phys. **22**, 147 (1954).
[4] G. A. WILLIAMS, D. W. McCALL u. H. S. GUTOWSKY: Phys. Rev. **93**, 1428 (1954).

Messung dieser Größe verdanken wir THOMAS, DRISCOLL und HIPPLE[1,2] vom National Bureau of Standards (USA). In einer außerordentlich sorgfältig ausgeführten Arbeit haben die genannten Forscher mit Hilfe einer COTTON-Waage, vgl.[3-5], die Feldstärke eines Magneten am Ort einer Protonenprobe und zugleich die LARMOR-Resonanzfrequenz bestimmt. Der Fehler ihrer Messung ist kleiner als $2,5 \cdot 10^{-5}$. Bedenkt man, daß der Fehler einer mit normalen Laboratoriums-Hilfsmitteln ausgeführten Feldstärkemessung etwa $5 \cdot 10^{-3}$ beträgt, so wird offenbar, welche Präzision und auch welcher Aufwand erforderlich waren, um das vorliegende Ergebnis zu erhalten. Aus diesem Grunde sollen im folgenden Einzelheiten der zur Ausführung der Messung verwandten experimentellen Anordnung beschrieben werden.

Zur Messung der Kernresonanzfrequenz wurde die von THOMAS und HUNTOON[6] angegebene Amplitudenbrücke (vgl. Fig. 44) benutzt. Sie ist für den vorliegenden Zweck besonders geeignet, da sie gegen mechanische Störungen (microphonics) aller Art unempfindlich ist und ohne Justierschwierigkeiten ein reines Absorptionssignal liefert. Der einzige Nachteil der Amplitudenbrücke, ihr vergleichsweise schlechtes Signal-Rausch-Verhältnis, ist für das vorliegende Meßproblem bedeutungslos, da die Protonensignale bekanntlich besonders stark sind.

Zwei derartige Brückenschaltungen wurden benötigt. Beide befanden sich in kleinen Abschirmgehäusen außerhalb des Luftspalts des Magneten. Die eigentlichen Beobachtungsspulen mit den Proben befanden sich innerhalb des Luftspalts in den Enden von Messingrohren, die mit den Brückengehäusen fest verbunden waren. Die Hochfrequenzeingänge beider Brücken wurden mit demselben Quarzoszillator verbunden. Die Einstellung des Temperaturgleichgewichts des Oszillators dauerte etwa 3 Std. Danach schwankte seine Frequenz von ungefähr 20 MHz um weniger als $1 \cdot 10^{-6}$, wie mehrfach durch einen Vergleich mit der Normalfrequenz (WWV) des Nat. Bur. of Stand., überprüft wurde. Die beiden Brückenausgänge wurden mit verschiedenen Niederfrequenzverstärkern verbunden. (Die Hochfrequenz-Gleichrichtung der Signalspannungen erfolgt in den Brücken.)

Eine der beiden Resonanz-Meßanordnungen diente zur zeitlichen Stabilisierung der Magnetfeldstärke. Solche Schaltungen wurden in Ziff. 18 besprochen; es sollen deshalb hier nur einige Daten genannt werden. Die Zeitkonstante des Magneten betrug 2,7 sec. Am Probenort wurde das Magnetfeld durch kleine HELMHOLTZ-Spulen mit 800 Hz moduliert. Die Modulationsamplitude war kleiner als die Linienbreite des Absorptionssignals von etwa 0,3 Gauß. Die phasenempfindlich gleichgerichtete Signalspannung steuerte einen Kraftverstärker gegengekoppelt zur Feldänderung, der seinerseits zwei direkt hinter den Polschuhen des Magneten befindliche Hilfsspulen speiste. Durch die Stabilisierungsanordnung wurde die durch eine Änderung des Hauptstroms des Magneten um 2% verursachte Feldstärkenänderung auf 0,02 Gauß bei einer Gesamtfeldstärke von 4700 Gauß, also auf $\sim 4 \cdot 10^{-4}\%$ reduziert.

Die zweite Anordnung diente zur eigentlichen Messung der LARMOR-Frequenz und wurde außerdem zur relativen Messung der Feldverteilung zwischen den Polschuhen benutzt. Dies war zulässig, denn in der Gleichung $\Delta\omega = \gamma \Delta H$ war γ mindestens auf 1% genau bekannt, und mit der gleichen Genauigkeit konnten

[1] H. A. THOMAS, R. L. DRISCOLL u. J. A. HIPPLE: Phys. Rev. 75, 902 (1949); 78, 339, 787 (1950).
[2] H. A. THOMAS, R. L. DRISCOLL u. J. A. HIPPLE: J. Res. Nat. Bur. Stand. 44, 569 (1950).
[3] A. COTTON u. G. DUPOUY: Congr. internat. d'Electricité 3, Sect. 2, 208 (1932).
[4] F. A. SCOTT: Phys. Rev. 46, 633 (1934).
[5] G. H. BRIGGS u. A. F. A. HARPER: J. Sci. Instrum. 13, 119 (1936).
[6] H. A. THOMAS u. R. D. HUNTOON: Rev. Sci. Instrum. 20, 516 (1949).

somit die Feldstärkedifferenzen (einige Gauß innerhalb eines Kreises vom Radius 5 cm um die Polschuhmittelpunkte) gemessen werden. Zur Messung der Resonanzfrequenz in Funktion des Probenorts wurde auf einem Oszillographenschirm das ganze Absorptionssignal der zweiten Anordnung beobachtet. Das Feld in der Beobachtungsspule wurde ebenfalls durch kleine HELMHOLTZ-Spulen moduliert, und zwar mit der Frequenz 20 Hz. Zur Steuerung der Zeitablenkung des Oszillographen wurde die Modulations-Spannung verstärkt und direkt an dessen Horizontalplatten gelegt. Außerdem wurde sie in einer weiteren Schaltung in eine Rechteckspannung gleicher Frequenz und gleicher Phasenlage umgewandelt. Diese Rechtecke wurden differenziert und die so erhaltenen Impulse mit der Kernsignalspannung gemischt. Auf dem Bildschirm des Oszillographen dienten diese Impulse als Justiermarken. Verschiebungen des Resonanzsignals, verursacht durch eine Änderung der Feldstärke, konnten somit relativ zu diesen Marken mit guter Präzision beobachtet werden. Zur Messung der Feldverteilung wurde dem Gleichfeld in der Beobachtungsspule das Feld eines weiteren HELMHOLTZ-Spulenpaars überlagert. Änderte man die Stromstärke des durch diese Spulen fließenden Gleichstroms, so verschob sich das Absorptionssignal auf dem Bildschirm. Zur Eichung dieser Spulen wurde die Oszillatorfrequenz um einen kleinen bekannten Betrag geändert. Dadurch verschob sich der Signalort auf dem Bildschirm bezüglich der Justiermarken. Diese Verschiebung läßt sich rückgängig machen, wenn man durch die HELMHOLTZ-Spulen einen Strom bestimmten Betrags fließen läßt. Mißt man diesen, so kennt man die Eichkonstante c des Spulenpaars, und es ist $\Delta H = \dfrac{\Delta \omega}{\gamma} = \dfrac{c \cdot J}{\gamma}$. Das Ergebnis war $\dfrac{c}{\gamma} = (0{,}055 \pm 5\%)$ Gauß/mA.

Als Probensubstanz wurden in beiden Anordnungen schwach konzentrierte wäßrige Lösungen von Eisennitrat verwandt. Durch den Zusatz der paramagnetischen Ionen wird die Spin-Gitter-Relaxationszeit T_1 der Protonen verkürzt und damit die Signalamplitude erhöht [vgl. Gl. (6.17)], ohne daß sich die im wesentlichen durch die Feldinhomogenität bestimmte Linienbreite der Signale ändert, das wirksame T_2 bleibt also konstant. Nachteilig ist jedoch, daß sich auf Grund der paramagnetischen Suszeptibilität der Eisenionen auch die Feldstärke im Probeninnern geringfügig ändert. Glücklicherweise zeigte ein Vergleich der für die Messungen benutzten Proben mit Ölproben, daß dieser Effekt keine mit der verwandten Apparatur feststellbare Verschiebung des Resonanzsignals verursacht. Die Ergebnisse der Resonanzfrequenzmessung können also auf Grund dieses Fehlers höchstens um den Betrag $4 \cdot 10^{-6}$ falsch sein, denn mit dieser Genauigkeit ließen sich Frequenzverschiebungen des Absorptionssignals auf den Oszillographenschirm messen. Die hohe relative Genauigkeit wird verständlich, wenn man bedenkt, daß auf dem Schirmbild nur kleine Abweichungen von der sehr viel größeren Oszillatorfrequenz, also der eigentlichen Meßgröße beobachtet werden. In der gesamten Fehlerbilanz (vgl. Tabelle 1, S. 253) erscheint der Fehler vom Betrag $4 \cdot 10^{-6}$ noch ein drittes Mal, da auch die oben beschriebene Eichung der HELMHOLTZ-Spulen mit ihm behaftet ist.

Die zur Feldmessung angewandte Methode erfordert ein homogenes Magnetfeld beträchtlicher Ausdehnung. Es wurde deshalb ein Magnet mit großen Polschuhen benötigt. Dessen maximale Feldstärke ist aus zwei Gründen begrenzt. Die in seinem Luftspalt aufgehängte COTTON-Waage ist sehr empfindlich gegen Luftströme. Aus diesem Grunde müssen die Temperatur-Unterschiede zwischen den Magnetoberflächen und der Raumtemperatur möglichst klein sein, da sonst starke thermische Strömungen entstehen. Der Magnet muß also gut gekühlt werden und darf nur wenig Leistung aufnehmen. Außerdem soll der Streufluß von den Polschuhen zum Joch gering sein, damit einmal das Feld im Luftspalt möglichst homogen und zum anderen das Streufeld am oberen Ende der Waagenspule möglichst klein wird. Es zeigt sich, daß über 5000 Gauß Luftspalt-Feldstärke das Joch teilweise gesättigt war. Wählt man andererseits die Feldstärke zu klein, dann wird das Signal-Rausch-Verhältnis schlecht und auch die Genauigkeit der Kraftmessung vermindert sich. Alle Messungen wurden deshalb in einem Luftspaltfeld von etwa 4700 Gauß ausgeführt. Der Temperaturanstieg der heißesten Stelle der Wicklungen betrug $30°$ C bei einem Kühlwasserfluß von 1,6 Liter/min. Zur Homogenisierung des Felds im Luftspalt wurden auf einem Polschuh (bei späteren Kontrollmessungen auf beiden Polschuhen) Nickel-Shim-Bleche befestigt.

Zur Messung der Feldstärke im Luftspalt wurde eine lange rechteckige Spule benutzt. Als Spulenkörper diente eine 70 cm lange, 10 cm breite und 7 mm dicke Glasplatte, in deren Kanten parallele Nuten zur festen Lagerung der neun Windungen der Spule geschnitten wurden. Der Übergang der Windungen von Nut zu Nut erfolgte im oberen Teil der Spule, also außerhalb des Magnetfelds. Die exakte Herstellung der Glasplatte war ein besonderes technisches Problem, zu dessen Lösung spezielle Werkzeugvorrichtungen entwickelt werden mußten[1,2]. Beim Wickeln wurde darauf geachtet, daß die Spannung des verwandten sauerstoff-freien Kupferdrahts ($\varnothing = 0,56$ mm) stets konstant blieb. Die fertige Spule wurde in einem temperaturkonstanten Raum sorgfältig vermessen. Trotzdem lieferte der Fehler dieser Messung zum gesamten Fehler einen Beitrag von $10 \cdot 10^{-6}$. Hinzu kommt ein Fehler vom Betrag $5 \cdot 10^{-6}$, verursacht durch die Unbestimmtheit des benutzten Längennormals.

Die Spule wurde an einer analytischen Waage vertikal so aufgehängt, daß die unteren horizontalen Leiter sich in der Mitte zwischen den Polschuhen befanden. Gemessen wurde die in vertikaler Richtung auf diese Leiter wirkende Kraft, wenn durch sie Strom floß. Die auf die vertikalen Drähte der Spule wirkenden Kräfte heben sich, abgesehen von zu berücksichtigenden Korrekturbeträgen auf. Das restliche Streufeld des Magneten in der Umgebung der oberen horizontalen Leiter der Spule wurde durch das Feld eines weiteren HELMHOLTZ-Spulenpaars neutralisiert. Die Genauigkeit, mit der die Kompensation des etwa 3 Gauß großen Streufeldes erfolgte, war $\pm 0,02$ Gauß. Dementsprechend war der Beitrag dieser Fehlerquelle zur Unbestimmtheit des Endergebnisses der Messung $5 \cdot 10^{-6}$. Die Feldneutralisation wurde mit einer leichtbeweglich aufgehängten Magnetnadel kontrolliert. Die Kraftmessung liefert somit den Betrag der mittleren Feldstärke entlang den unteren Leitern. Um von diesem Wert auf die Feldstärke am Ort der Protonenprobe, direkt unterhalb der Mitte der Waagenspule, zu schließen, muß dementsprechend die Verteilung der Feldstärke in diesem Gebiet ermittelt werden. Das zur Messung der Feldverteilung angewandte Verfahren wurde bereits besprochen. Die Messungen ergaben, daß die mittlere Feldstärke $\bar{H}$ um 0,04 Gauß niedriger ist als die Feldstärke am Probenort. Der Fehlerbeitrag dieser Messung zum Gesamtergebnis betrug $10 \cdot 10^{-6}$. Mit der kernmagnetischen Methode konnte die Feldverteilung in vertikaler Richtung entlang der Rechteckspule nur bis etwa 7 cm über den unteren Spulendrähten bestimmt werden. Darüber wird das Feld zu inhomogen, dementsprechend die Linienbreite des Absorptionssignals zu groß und seine Amplitude zu schwach. Oberhalb dieser Grenze wurde die Feldstärke deshalb mit einer kleinen Spule bestimmt, die durch einen Synchronmotor angetrieben mit 1800 U/min rotierte.

---

[1] Siehe Fußnote 2, S. 249.
[2] H. L. CURTIS, C. MOON u. C. M. SPARKS: J. Res. Nat. Bur. Stand. **21**, 375 (1938).

Diese Messung war erforderlich, um die auf die vertikalen Drähte wirkenden Kräfte zu berechnen, deren nicht kompensierter Rest, wie oben erwähnt, bei der Berechnung der gesamten an der Spule angreifenden Kraft als Korrektur berücksichtigt werden mußte.

Als Spannungsquelle des durch die Rechteckspule fließenden Stroms diente eine 12 V-Batterie. Die genaue Stromstärke wurde in einer Kompensationsschaltung unter Verwendung eines Galvanometers, eines Normalwiderstands und einer Normalspannungsquelle bestimmt. Die beiden Normale befanden sich in Thermostaten. Ein in der Schaltung enthaltenes Potentiometer wurde stets so justiert (möglicher Fehler $1 \cdot 10^{-6}$), daß die Stromstärke konstant den Betrag 0,101 725 9 A hatte. Der Gesamtfehler der Stromstärkemessung war kleiner als $3 \cdot 10^{-6}$. Der Spulenwiderstand betrug bei $25°$ C 1,1006 Ohm. Die jeweilige Temperatur der Spule konnte aus ihrem Widerstand mit einer Genauigkeit von $0,1°$ C berechnet werden. Sie mußte bekannt sein, da die Spulenabmessungen von der Temperatur abhängen. Der Widerstand der Spule wurde durch einen Vergleich des Potentialabfalls an ihr mit dem Spannungsabfall an einem vom gleichen Strom durchflossenen Normalwiderstand ($R = 1$ Ohm) ermittelt. Die in Ampere gemessene Stromstärke mußte dann noch in absolute Einheiten umgerechnet werden. Der Umrechnungsfaktor läßt sich als Produkt $k \cdot \sqrt{g_1}$ schreiben. Darin ist k eine Konstante, die von CURTIS, DRISCOLL und CRITCHFIELD[1] mit einer Genauigkeit von $10 \cdot 10^{-6}$ gemessen wurde und g_1 der Wert der Erdbeschleunigung bei dieser Messung. Der Kraftmessung in der vorliegenden Arbeit, in welche die Unbestimmtheit des Betrags der Erdbeschleunigung natürlich auch eingeht, lag ein etwas anderer, 1948 im Nat. Bur. of Stand. bestimmter Wert dieser Konstanten ($g = 980,08$ cm/sec²) zugrunde. In der Formel zur Berechnung der mittleren Feldstärke $\bar{H}$ steht im Zähler g und im Nenner $\sqrt{g_1}$ (Kraft $\sim$ Feldstärke $\cdot$ Strom). Die Unbestimmtheit des Quotienten $g/\sqrt{g_1}$ ist aber nur halb so groß wie der eigentliche Fehler bei der Bestimmung von g ($5 \cdot 10^{-6}$), sie betrug also höchstens $3 \cdot 10^{-6}$.

Zur Messung der an den stromführenden Leitern angreifenden Kraft wurde die Waage zunächst bei eingeschaltetem Strom ins Gleichgewicht gebracht. Dann wurde die Stromrichtung umgepolt und auf die Waagschale eine mit einem Fehler von $1 \cdot 10^{-6}$ bekannte Masse aufgelegt. Deren Größe wurde so bemessen, daß die Waage auch nach der Umpolung wiederum ungefähr im Gleichgewicht war. Anschließend wurde die zum endgültigen Ausgleich erforderliche Differenzmasse bestimmt. Das Verfahren der Stromumkehr hat mehrere Vorteile, u. a. bleibt die Spulentemperatur und damit das Gewicht der Spule konstant, während die Meßgröße verdoppelt wird. Die Größe der Masse wurde nach der obigen Anpassung bei $25°$ C Raumtemperatur bestimmt. Da das Gewicht der von ihr verdrängten Luftmenge von der Temperatur abhängt, mußte bei jeder Einzelmessung eine Temperaturkorrektur des Massenwertes vorgenommen werden.

Tabelle 1 zeigt eine Zusammenstellung der im Text erwähnten Einzelfehler. Die Wurzel aus der Summe ihrer Quadrate ergibt die Unbestimmtheit der gesamten Messung. Diese betrug $22 \cdot 10^{-6}$. Aus den Ergebnissen der Feldstärkemessung am Probenort [$H_0 = (4697,56 \pm 0,05$ Gauß] und der Resonanzfrequenzmessung [$\nu_0 = (20\,001\,075 \pm 20)$ Hz] folgt für das gyromagnetische Verhältnis der Protonen in der untersuchten Eisennitratlösung:

$$\gamma_{p\text{-Eisennitratlösung}} = (2,675\,23 \pm 0,000\,06) \cdot 10^4 \sec^{-1} \text{Gauß}^{-1}. \tag{21.1}$$

---

[1] R. W. CURTIS, R. L. DRISCOLL u. C. L. CRITCHFIELD: J. Res. Nat. Bur. Stand. **28**, 133 (1942).

Tabelle 1. *Zusammenstellung der möglichen Einzelfehler bei der Absolutbestimmung des gyromagnetischen Verhältnisses der Protonen durch* THOMAS, DRISCOLL *und* HIPPLE.

Fehlerquelle	Unsicherheit $(\pm \cdot 10^{-6})$
Platin-Iridium-Masse	1
Präzision der Wägung	9
Unbestimmtheit der Erdbeschleunigungskonstanten $(g/\sqrt{g_1})$	3
Unbestimmtheit des Umrechnungsfaktors in absolute Ampère	10
Messung der Stromstärke der Rechteckspule	3
Spulenlänge	10
Unsicherheit des Längenstandardmaßes	5
Fehler bei der Messung der Feldverteilung	10
Fehler bei der Neutralisation des Streufelds im Gebiet der oberen Leiter der Rechteckspule	5
Justierung der Resonanzspitze auf dem Oszillographenschirm	4
Eichung der zur Messung der Feldverteilung benutzten HELMHOLTZ-Spulen	4
Eventuelle Resonanzverschiebung durch die in der Probe enthaltenen Fe^{+++}-Ionen	4
Unsicherheit der Frequenzmessung	1

Da die Eisennitratlösung keine geeignete Standardsubstanz ist, hat THOMAS in einer späteren Arbeit[1] die Resonanzfrequenzen der Protonensignale dieser Lösung und der Protonensignale von destilliertem (und vor allem von gelöstem Sauerstoff freiem) Wasser mit der Resonanzfrequenz der im molekularen Wasserstoff (H_2-Gas unter einem Druck von 40 atü) enthaltenen Protonen verglichen. Die Korrekturgrößen betrugen:

$$\Delta H/H \text{ (Eisennitratlösung)} = +1,3 \cdot 10^{-6}, \qquad \Delta H/H \text{ (H}_2\text{O)} = -0,6 \cdot 10^{-6}$$

aus $\left| \dfrac{\Delta H}{H} \right| = \left| \dfrac{\Delta \gamma}{\gamma} \right|$ und (21.1) folgt:

$$\gamma_{p\text{-H}_2\text{O}} = (2,675\,23_5 \pm 0,00006) \cdot 10^4 \, \text{sec}^{-1} \, \text{Gauß}^{-1}, \tag{21.2}$$

$$\gamma_{p\text{-H}_2} = (2,675\,23_3 \pm 0,00006) \cdot 10^4 \, \text{sec}^{-1} \, \text{Gauß}^{-1}. \tag{21.3}$$

Die Unterschiede zwischen den Werten (21.1), (21.2) und (21.3) liegen also weit innerhalb der Fehlergrenzen. Von RAMSEY[2] wurde der diamagnetische Abschirmungseffekt der Hüllenelektronen des H_2-Moleküls berechnet. Sein Ergebnis ist $\Delta H/H = 26,8 \cdot 10^{-6}$. Korrigiert man das Ergebnis (21.3) dementsprechend, so folgt für das gyromagnetische Verhältnis des freien Protons:

$$\gamma_p = (2,675\,30_5 \pm 0,00006) \cdot 10^4 \, \text{sec}^{-1} \, \text{Gauß}^{-1}. \tag{21.4}$$

Außerdem ergibt sich für die diamagnetische Abschirmung der Protonen im H_2O-Molekül $\Delta H/H = 26,2 \cdot 10^{-6}$.

Das magnetische Moment des Protons ist auf Grund der Unbestimmtheit des Werts der PLANCKschen Konstante weniger genau als sein gyromagnetisches Verhältnis bekannt. Benutzt man den von DuMOND und COHEN[3] ermittelten Wert $h = (6,6234 \pm 0,0011) \cdot 10^{-27} \, \text{erg} \cdot \text{sec}$, so folgt:

$$\mu_p = (1,4100 \pm 0,0002) \cdot 10^{-23} \, \text{dyn} \cdot \text{cm} \cdot \text{Gauß}^{-1}. \tag{21.5}$$

---

[1] H. A. THOMAS: Phys. Rev. **80**, 901 (1950).
[2] N. F. RAMSEY: Phys. Rev. **77**, 567; **78**, 339, 699 (1950).
[3] J. W. M. DuMOND u. E. R. COHEN: Rev. Mod. Phys. **20**, 82 (1948).

Gegenwärtig wird von Kirchner und Wilhelmy[1,2] mit einer anderen experimentellen Anordnung das gyromagnetische Verhältnis des Protons erneut bestimmt. Ihr derzeitiges Ergebnis[3]

$$\gamma_{p\text{-Eisenchloridlösung}} = (2,675\,49 \pm 0,000\,17) \cdot 10^4 \sec^{-1} \text{Gauß}^{-1} \qquad (21.6)$$
$$\text{(diamagnetisch nicht korrigiert)}$$

unterscheidet sich von (21.1), (21.2) und (21.3) um 0,1 Promille. Leider sind die Ursachen dieser dem Betrage nach geringfügigen Differenz bisher nicht bekannt. Um ihre möglichen Gründe diskutieren zu können, sollen deshalb die wesentlichen Unterschiede zwischen den beiden Verfahren kurz besprochen werden

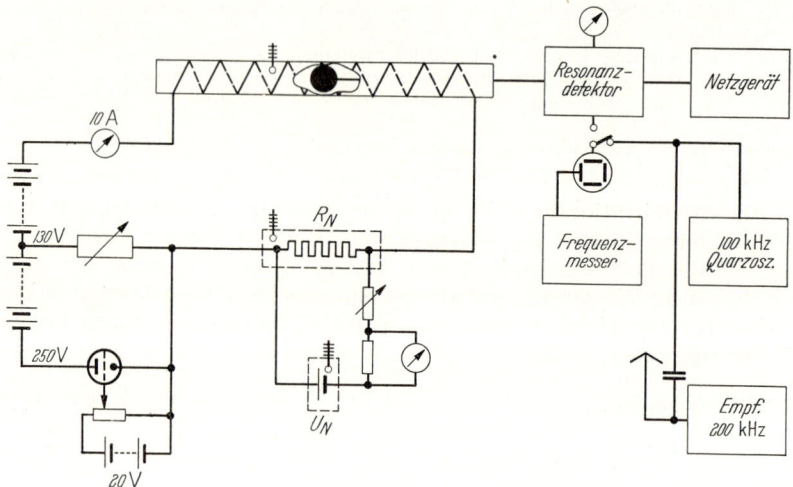

Fig. 88. Meßanordnung von Kirchner und Wilhelmy.

Der Fehler der Feldstärkemessung von Thomas, Driscoll und Hipple war etwa zehnmal größer als der Fehler ihrer Larmor-Frequenz-Messung. Bei jeder Neubestimmung mußte demzufolge versucht werden, die Zahl der mit der Feldmessung verknüpften Fehlerquellen zu vermindern. Kirchner und Wilhelmy haben deshalb das Magnetfeld in einer eisenfreien Spule erzeugt und seine Stärke aus den Spulendimensionen und dem Betrag des durchfließenden Stroms berechnet. Das Verfahren der Feldmessung ist also, verglichen mit dem zuvor beschriebenen, wesentlich einfacher. Jedoch wachsen die Anforderungen an die Kernresonanzapparatur erheblich. Die obere Grenze der Feldstärke eines Präzisions-Solenoids beträgt etwa 100 Gauß. Die kernmagnetische Signalspannung nimmt ungefähr proportional mit dem Quadrat der Feldstärke ab. In einem solch schwachen Feld wird demnach das Signal-Rausch-Verhältnis wesentlich niedriger als in Feldern von einigen 1000 Gauß, in denen sonst experimentiert wird.

Fig. 88 zeigt das Schaltbild der gesamten Meßanordnung von Kirchner und Wilhelmy. Das Solenoid hat eine Länge von 100 cm und einen Durchmesser von 6 cm. Als Spulenkörper wurde ein doppelwandiges Messingrohr verwandt, auf dessen Außenseite ein Gewinde zur Aufnahme der 796 Windungen eingeschnitten war. Der Raum zwischen den Wänden wird von Kühlwasser

---

[1] F. Kirchner u. W. Wilhelmy: Z. Naturforsch. 10a, 657 (1955).
[2] W. Wilhelmy: Dissertation an der Universität Köln 1955.
[3] W. Wilhelmy: Ann. Physik 6, 329 (1957).

durchflossen. Ein Ende der Spule ist mit dem Messingrohr verlötet, so daß der Strom am anderen Ende durch eine bifilare Leitung zugeführt werden kann. Dadurch werden Störfelder vermieden. Die Stärke des Spulenstroms (etwa 10 A) wird mittels eines Potentiometers und einer elektronischen Feinregelung so justiert, daß sein Spannungsabfall an einem Normalwiderstand gleich der Spannung eines Normalelements ist. Das im Kompensationskreis liegende Spiegelgalvanometer muß also Stromlosigkeit anzeigen. Der Normalwiderstand und das Normalelement (Typ Weston) befanden sich während der Messung in Thermostaten. Beide Schaltelemente wurden bei der Physikalisch-Technischen Bundesanstalt geeicht. Die an den Resonanzdetektor angeschlossene Beobachtungsspule der kernmagnetischen Präzessionsbewegung, samt der in ihr enthaltenen Protonenprobe (10 cm³ wäßrige Eisen-III-Chlorid-Lösung mit 10^{17} Fe^{+++}/cm³), war im Zentrum der Feldspule. Für die dortige Feldstärke gilt die Beziehung:

$$H = \frac{0.4 \cdot \pi \cdot n}{\sqrt{L^2 + D^2}} J = c \cdot J \ [\text{Gauß}], \tag{21.7}$$

wenn man mit n die Gesamtwindungszahl, mit L die Spulenlänge und mit D den Spulendurchmesser bezeichnet. Die Spulendaten wurden mit einem Mikroskop und mit einem in der Physikalisch-Technischen Bundesanstalt geeichten Glasmaßstab bestimmt. Beachtet werden mußte jedoch, daß die Windungsdichte auf der Spule nicht ganz konstant ist. Der eigentliche gemessene Wert $L_{20°C} =$ (99,470 $\pm$ 0,002) cm mußte dementsprechend korrigiert werden. Die effektive Spulenlänge betrug $L_{20°C} =$ (99,468 $\pm$ 0,002) cm, ihr Durchmesser war $D_{20°C} =$ (6,100 $\pm$ 0,002) cm. Die Spannung des Normalelements betrug 1,01864 abs.V, der Normalwiderstand hatte die Größe 0,100150 abs.Ω. Damit folgte für den Spulenstrom 10,1711 abs.A und für die Feldstärke $H = 102,102$ Gauß. Eine Berechnung der Feldverteilung ergab, daß die Inhomogenität innerhalb des Probevolumens kleiner als $\pm 2 \cdot 10^{-6}$ war. In ihrer Fehlerbilanz schätzen die Verfasser, daß die möglichen Einzelfehler höchstens

$$\text{Normalelement} \quad \pm 1 \cdot 10^{-5}$$
$$\text{Normalwiderstand} \ \pm 1 \cdot 10^{-5}$$
$$\text{Spulendimensionen} \ \pm 2 \cdot 10^{-5}$$

betragen haben können.

Dem Eigenfeld der Spule ist stets das erdmagnetische Feld überlagert. Die Spule wurde deshalb so aufgestellt, daß ihre Achse dieselbe Richtung wie das Erdfeld hat. Dadurch addieren bzw. subtrahieren sich die beiden Felder in Abhängigkeit von der Richtung des Spulenstroms. Bestimmt man in beiden Fällen die Resonanzfrequenz, so ist die zum eigentlichen Spulenfeld gehörige Resonanzfrequenz gleich dem arithmetischen Mittelwert der beiden gemessenen Frequenzen. Die Orientierung der Spule im Erdfeld ist nicht kritisch. Selbst bei einer Fehljustierung von 10° beträgt der Fehler nur $2 \cdot 10^{-7}$.

Der Resonanzdetektor (Fig. 89) besteht im wesentlichen aus einem Oszillator (Dreipunktschaltung mit kathodenseitiger Rückkopplung), dessen Schwingspule die Protonenprobe enthält. Der Schwingkreis enthält neben Festkondensatoren einen Drehkondensator mit dem eine Feinabstimmung im Intervall zwischen 430 und 440 kHz möglich ist. R_2 ist ein hochohmiger Gitterableitwiderstand, der Oszillator ist also ein Sperrschwinger, d.h. seine Hochfrequenzspannung ist impulsförmig moduliert, und zwar sind die Impulse etwa 0,3 msec und die Pausen ungefähr 2,5 msec lang. Die Impulsfolgefrequenz wird in erster Näherung durch die Zeitkonstante $R_2 C_2$ bestimmt. Die hochfrequenten Schwingungen der einzelnen

Impulse sind inkohärent. Deswegen wurde zur Messung der Frequenz des Oszillators ein kleinerer Gitterableitwiderstand R_1 eingeschaltet. Auf Grund einer besonderen Untersuchung glauben die Verfasser, daß diese Umschaltung keine Veränderung der Oszillatorfrequenz bewirkt. In Kernresonanz werden durch die HF-Impulse freie Präzessionsbewegungen angeregt. Da die Spin-Spin-Relaxationszeit der Probe ($\sim$0,5 sec) sehr viel größer als die Pausenlänge ist, wird während den Pausen in der Schwingspule eine Spannung der Larmor-Frequenz induziert. Der Oszillator schwingt infolgedessen in Kernresonanz schneller an, und seine Impulsfolgefrequenz nimmt zu. Außerdem wird die Impulshöhe größer, da der Schwingeinsatz vom Pegel der induzierten Spannung anstatt vom Rauschpegel erfolgt. Es handelt sich also um eine Schaltung von der in Ziff. 14β besprochenen Art. Die zweite Stufe arbeitet als HF-Gittergleichrichter und als Spannungsverstärker. Die dritte Stufe ist als Brücke geschaltet.

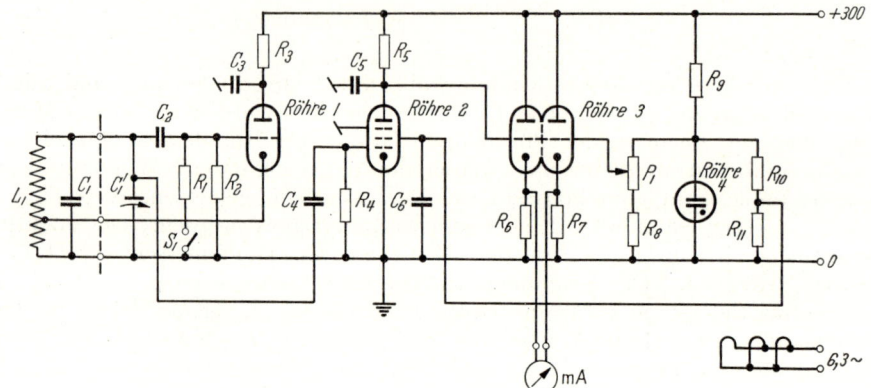

Fig. 89. Schaltung des Resonanzdetektors von Kirchner und Wilhelmy.

Die vier Brückenzweige sind: Triode, Triode, R_6 und R_7. Die Brücke wird so abgeglichen, daß außerhalb Kernresonanz das Amperemeter Stromlosigkeit anzeigt. In Kernresonanz wird das Brückengleichgewicht gestört und durch das Meßinstrument fließt ein Strom von etwa 1 mA. Die Einstellzeitkonstante des Instruments beträgt ungefähr 1 sec. Die Bandbreite der Beobachtungsanordnung wird dadurch auf 1 Hz eingeengt.

Als Frequenznormal dient der Sender Droitwich I (200 kHz $\pm 2 \cdot 10^{-8}$). Diesem Sender wurde im Empfänger die erste Oberwelle eines korrigierbaren 100 kHz-Quarzoszillators überlagert, der seinerseits durch Frequenzvergleich mittels Lissajousscher Figuren Eichpunkte auf der Skala eines variablen Meßsenders lieferte. Zur Messung der Resonanzfrequenz wurde dessen Ausgangsspannung mit der Oszillatorspannung gemischt und der Frequenzabstand ebenfalls auf dem Oszillographenschirm bsetimmt.

Die Basisbreite des Resonanzsignals, d.h. der Abstand der beiden Fußpunkte der Resonanzkurve betrug bei der zur Messung verwandten Probe 120 mGauß. Die Resonanzbreite, definiert als der Abstand der beiden Punkte größter Flankensteilheit, war etwa dreimal kleiner. Die Verfasser geben an, daß sich die Resonanzspitze noch etwa zehnfach feiner fixieren läßt und schätzen dementsprechend die Meßunsicherheit der Resonanzstelle zu $\pm 2 \cdot 10^{-5}$.

Es ist wenig wahrscheinlich, daß der Unterschied zwischen den beiden Meßergebnissen (21.1) und (21.6) physikalische Ursachen hat. Insbesondere ist keine

Beobachtung bekannt, die auf eine Abhängigkeit des gyromagnetischen Verhältnisses von der Feldstärke hinweisen würde. Man wird deshalb zunächst nach experimentellen Fehlerquellen suchen. Das Verfahren der Feldstärkemessung mit Hilfe einer COTTON-Waage ist zweifellos komplizierter als die Bestimmung der Feldstärke aus den Daten einer Zylinderspule. Andererseits ist zu beachten, daß bei der zuletzt beschriebenen Messung die Ungenauigkeit der Resonanzfrequenzbestimmung etwa ebenso groß ist, wie die der Feldstärkenmessung. Zur Aufklärung der obigen Differenz müßte also das bei der zweiten Methode zur Resonanzfrequenzmessung angewandte Verfahren ebenfalls überprüft werden.

Auf eine weitere Möglichkeit haben KIRCHNER und WILHELMY hingewiesen. Bei der Methode von THOMAS, DRISCOLL und HIPPLE wurde die Feldstärke auf Grund einer Kraftmessung gemäß $H = P/L \cdot J$ (P = Kraft, L = Länge des Leiters) bestimmt. Dementsprechend galt:

$$\gamma_{\text{T.D.H.}} = \frac{\omega}{H} = \frac{\omega \cdot L \cdot J_{\text{int}}}{P} = C' \cdot J_{\text{abs}}.$$

In dem Experiment von KIRCHNER und WILHELMY war die Feldstärke zu der Stromstärke proportional, es war also:

$$\gamma_{\text{K.W.}} = \frac{\omega}{H} = \frac{\omega}{K \cdot J_{\text{int}}} = \frac{C''}{J_{\text{abs}}}$$

Wären beide Ergebnisse fehlerfrei, so würde folgen, daß die zur Zeit gültige Relation 1 A_{int} = 0,99985 A_{abs} falsch ist und durch die Beziehung 1 A_{int} = 0,99990 A_{abs} ersetzt werden muß.

Aus den kernmagnetischen Resonanzexperimenten können nicht nur die magnetischen Dipolmomente der Kerne erschlossen werden, sondern auch die Vorzeichen ihrer gyromagnetischen Verhältnisse. Gemäß der üblichen Definition ist das Vorzeichen des gyromagnetischen Verhältnisses einer positiven rotierenden Ladung ebenfalls positiv. In diesem Fall haben der Spin-Vektor und der Kerndipolmoment-Vektor dieselbe Richtung. Entsprechend dieser Festlegung ist das Vorzeichen des gyromagnetischen Verhältnisses des Elektrons z. B. negativ. Wie in Ziff. 11 erwähnt, lassen sich in Brückenschaltungen mit gekreuzten Sender- und Empfänger-Spulen die Vorzeichen verschiedener Kernarten relativ zueinander bestimmen. Dazu beobachtet man nacheinander mit der gleichen Kernresonanz-Apparatur bei konstanter Frequenz die Präzessionsbewegung der zu vergleichenden Kernarten. Zur Einstellung der verschiedenen Resonanzstellen muß dementsprechend die Magnetfeldstärke geändert werden. Die Präzessionsbewegung wird durch ein in der Senderspule oszillierendes Hochfrequenzfeld angeregt. Genauer gesagt, wird diese Anregung nur durch eine der beiden gegensinnig rotierenden Komponenten, in welche man das oszillierende Feld zerlegen kann, bewirkt, und zwar durch diejenige, die mit dem resultierenden Magnetisierungsvektor des Kernspinsystems im gleichen Sinn um das konstante Feld präzessiert (vgl. auch Ziff. 4). In der Beobachtungsspule werden zwei Spannungen induziert. Die Phase und der Betrag der ersten, durch den Streufluß der Senderspule induzierten Spannung hängen von der Justierart der Brücke ab, sind also, allerdings nur in erster Näherung, von der Art der beobachteten Substanz unabhängig. Die Amplitude der zweiten, von dem jeweils untersuchten Kernspinsystem induzierten Spannung ändert sich mit der Resonanzfeldstärke, der Kernzahl, den Relaxationszeiten der Substanz und dem Betrag des gyromagnetischen Verhältnisses. Die Phase dieser Spannung hängt jedoch beim adiabatischen langsamen Resonanzdurchgang im wesentlichen nur vom Präzessionssinn der Kerne ab, und zwar ändert sie sich bei einem Vorzeichenwechsel des gyromagnetischen Verhältnisses um 180°. Solche Phasenwechsel sind bequem beobachtbar, dagegen ist es bisher nicht gelungen, auf diesem Wege den Präzessionssinn irgendeiner Kernart absolut zu ermitteln und sich damit einen Standardkern zu verschaffen. KELLOGG,

RABI, RAMSEY und ZACHARIAS[1] haben jedoch schon 1939 auf Grund von Molekularstrahluntersuchungen nachgewiesen, daß das Vorzeichen des gyromagnetischen Verhältnisses der Protonen positiv ist. 1949 haben dann ROGERS und STAUB[2,3] mit einer abgewandelten Kerninduktionsmethode dieses Ergebnis bestätigt.

Der Grundgedanke ihrer experimentellen Anordnung ist sehr einfach. Zur Anregung der Präzessionsbewegung benutzten sie statt einem aus zwei Komponenten bestehenden oszillierenden Hochfrequenzfeld ein rotierendes Feld. Stimmt dessen leicht meßbarer Umlaufsinn mit dem der Präzession überein, dann wird ein Kernsignal angeregt, im anderen Fall aber nicht.

Ein hochfrequent rotierendes Feld läßt sich, wie aus der Elektrotechnik bekannt, durch Überlagerung zweier orthogonaler, mit einer Phasendifferenz von 90° oszillierender Felder erzeugen. An Stelle eines koaxialen Senderspulenpaars benötigt man also zwei orthogonale Senderspulenpaare. Damit stößt man aber auf eine technische Schwierigkeit bei der Zusammenstellung der gesamten Spulenanordnung. In der ursprünglichen BLOCHschen Anordnung liegen die Achsen der Sender- und der Empfangsspulen zueinander senkrecht. Infolgedessen ist die durch das Streufeld der Senderspulen in der Empfängerspule induzierte Spannung nicht sehr groß und kann durch Feldtrimmer (vgl. S. 195) auf eine beliebige zur Beobachtung der Kernsignalspannung geeignete Größe reduziert werden. Im vorliegenden Fall durchdringt der größte Teil des Senderfeldflusses die Empfangsspule achsenparallel. Die in dieser Spule induzierte Senderspannung ist also zur Beobachtung der Kerninduktionsspannung viel zu groß. STAUB und ROGERS haben deshalb folgenden Kunstgriff angewandt. In der von ihnen entwickelten Anordnung (vgl. Fig. 90) liegt die Empfangsspule in der Mitte der vier Senderspulen und zwar koaxial zu einem der Senderspulenpaare. Auf der gleichen Achse (in der Fig. 90 als x-Richtung bezeichnet) befindet sich eine vierte Spule, deren Windungssinn dem der Empfangsspule entgegengesetzt ist. Der Abstand dieser Spulen vom Zentrum der Anordnung ist so groß, daß das präzessierende Kernspinsystem in ihr praktisch keine Spannung zu induzieren vermag. Dagegen umschließt sie ebenso wie die Empfangsspule den Fluß des Senderfelds in x-Richtung fast vollständig. Schaltet man somit beide Spulen in Serie, so läßt sich die von dem zur x-Achse parallelen Senderspulenpaar in der Empfangsspule induzierte Spannung auf einen sehr kleinen Betrag reduzieren, wenn man den Abstand der Kompensationsspule vom Zentrum entsprechend justiert. Da im allgemeinen auch durch den Fluß des in der y-Richtung liegenden Senderspulenpaars in der Beobachtungsspule eine kleine Spannung induziert wird, wurde noch eine dritte zur y-Richtung koaxiale Spule, allerdings sehr viel kleinerer Windungszahl, mit der Empfänger- und der Kompensationsspule in Serie verbunden. Damit kann auch diese Komponente der in der Empfangsspule induzierten Senderspannung neutralisiert werden. Zum endgültigen Abgleich des ganzen Systems wurden noch zwei um die x- und um die y-Richtung drehbare Feldtrimmer verwandt.

Zur Kontrolle der Phasen- und Betrags-Justierung der an den beiden Senderspulenpaaren liegenden HF-Spannungen, sowie zur Bestimmung des Drehsinns des zirkular polarisiert rotierenden Senderfelds, enthalten zwei dieser Spulen in ihrem Zentrum je eine weitere kleine Empfangsspule. Deren Zuleitungen wurden über einen Verstärker mit den Ablenkplatten eines Oszillographen verbunden.

---

[1] J. M. B. KELLOGG, I. I. RABI, N. F. RAMSEY u. J. R. ZACHARIAS: Phys. Rev. **56**, 728 (1939).

[2] E. H. ROGERS u. H. H. STAUB: Phys. Rev. **76**, 980 (1949).

[3] H. H. STAUB u. E. H. ROGERS: Helv. phys. Acta **23**, 63 (1950).

Der Drehsinn des Felds läßt sich aus einem Vergleich der Phasen der in den beiden Empfangsspulen induzierten Spannungen ermitteln.

STAUB und ROGERS haben mit der beschriebenen Anordnung bestätigt, daß das Vorzeichen des gyromagnetischen Verhältnisses der Protonen positiv ist, und außerdem sichergestellt, daß das Vorzeichen des gyromagnetischen Verhältnisses der Neutronen negativ ist. Bezüglich der technischen Einzelheiten des Neutronenexperiments (Atomstrahlresonanz-Methode) muß auf die Originalarbeiten der Verfasser verwiesen werden.

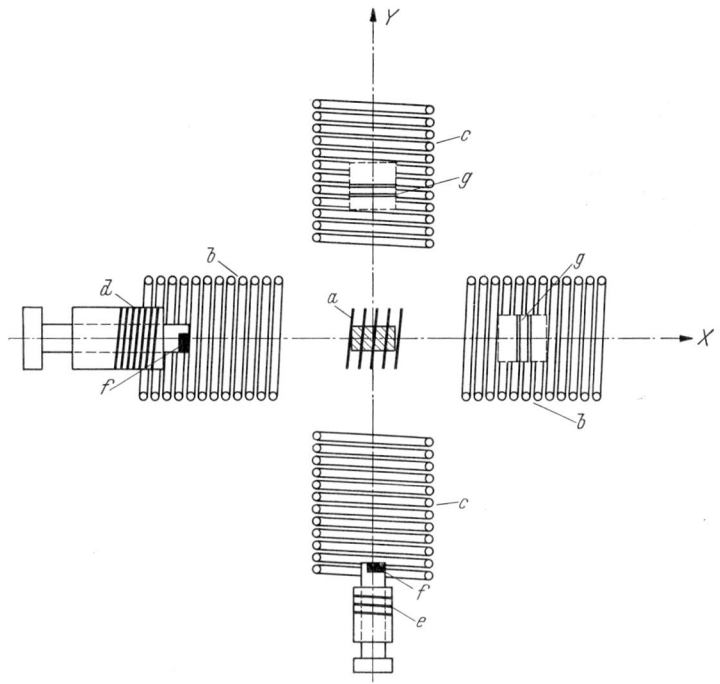

Fig. 90. Spulenanordnung nach STAUB und ROGERS. *a* Empfangsspule mit Probe, *bb* koaxiales Senderspulenpaar, *cc* orthogonales Senderspulenpaar, *d* zur Empfangsspule koaxiale Kompensationsspule, *e* zur Empfangsspule orthogonale Kompensationsspule, *ff* Feldtrimmer, *gg* Kontrollspulen zur Messung des Umlaufssinns des Senderfelds.

22. Vergleich der LARMOR-Präzessionsfrequenz der Protonen mit der Zyklotronfrequenz der Elektronen und der Protonen.

Frequenzvergleiche sind meßtechnisch besonders erfolgversprechend. So war es naheliegend, nach weiteren Versuchsanordnungen zu suchen, in denen ebenso wie in den LARMOR-Präzessions-Experimenten eine Frequenz mit der Feldstärke eines Magneten verknüpft ist. Vergleicht man die Frequenzen solcher physikalisch verschiedener Anordnungen im selben Magnetfeld, so erhält man stets Beziehungen, die entweder zur Bestimmung von Fundamentalkonstanten geeignet sind, oder aber die Absolutmessung einer der beteiligten Größen ermöglichen.

Es ist wohl bekannt, daß in einem homogenen Magnetfeld jedes sich orthogonal zu den Feldlinien fortbewegende geladene Elementarteilchen eine Kreisbahn beschreibt. Der Radius dieser Bahn folgt aus der Gleichsetzung der stets senkrecht zur Feld- und zur Bewegungsrichtung angreifenden magnetischen Kraft evH/c ($e =$ Ladung in elektrostatischen Einheiten, $v =$ Geschwindigkeit des Teilchens, $H =$ Feldstärke und $c =$ Lichtgeschwindigkeit) und der Zentrifugalkraft mv^2/r ($m =$ Masse des Teilchens und $r =$ Radius der Kreisbahn) zu $r = mvc/eH$. Die

Umlaufzeit τ_z des Teilchens auf einer solchen Bahn und damit auch die zur ihr reziproke Umlaufsfrequenz

$$\nu_z = \frac{1}{\tau_z} = \frac{v}{2\pi r} = \frac{v e H}{2\pi m v c} = \frac{e H}{2\pi m c} \qquad (22.1)$$

ist sowohl von der Geschwindigkeit als auch vom Bahnradius unabhängig. Dieser Sachverhalt wird im Zyklotron zur Beschleunigung von Elementarteilchen ausgenützt. Die Kreisbahnfrequenz (22.1) wird deshalb häufig als Zyklotronfrequenz bezeichnet. Für sie ist die obige Forderung erfüllt. Mehrere Autoren haben deshalb die LARMOR- und die Zyklotronfrequenzen verschiedener Elementarteilchen miteinander verglichen.

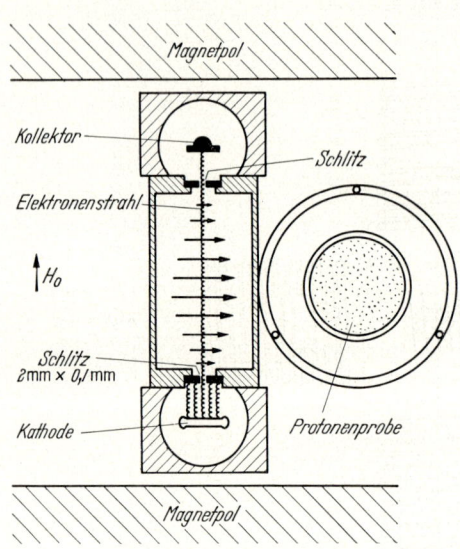

Fig. 91. Elektronenresonator nach GARDNER und PURCELL.

So haben z.B. GARDNER und PURCELL die Zyklotronfrequenz der Elektronen und die LARMOR-Frequenz der Protonen im gleichen Magnetfeld gemessen[1,2]. Aus den beiden Beziehungen

$$\nu_{Z,e} = \frac{e H}{2\pi m_e c}$$

und

$$\nu_{L,p} = \frac{\mu_p H}{2\pi \cdot I_p \cdot \hbar} = \frac{2\mu_p H}{h}$$

folgt die Gleichung

$$\frac{\nu_{L,p}}{\nu_{Z,e}} = \frac{\mu_p}{e\,\hbar/2 m_e c} = \frac{\mu_p}{\mu_B}, \qquad (22.2)$$

in der mit μ_B das BOHRsche Magneton bezeichnet wurde. Sie läßt sich zweifach interpretieren. Kann man voraussetzen, daß die Größen der in ihr enthaltenen Fundamentalkonstanten e/m_e, h und c, mit anderen Worten der Betrag des BOHRschen Magnetons, im absoluten Maßsystem ausreichend genau bekannt sind, so ermöglicht die Gl. (22.2) eine Absolutbestimmung des Dipolmoments der Protonen. Übertrifft andererseits die Genauigkeit mit der das Protonenmoment bereits bekannt ist (vgl. Ziff. 21) die auf dem vorliegenden Wege erzielbare, so wird die Beziehung (22.2) zu einer Bestimmungsgleichung für das BOHRsche Magneton, d.h. aber zu einer Bestimmungsgleichung für die Fundamentalkonstanten, aus denen es sich definitionsgemäß zusammensetzt.

Fig. 91 zeigt den im Magnetfeld befindlichen Teil der von GARDNER und PURCELL zur Messung der beiden Frequenzen $\nu_{L,p}$ und $\nu_{Z,e}$ entwickelten Apparatur und Fig. 92 deren gesamtes Schaltbild. Das BOHRsche Magneton ist ungefähr 657mal größer als das Protonenmoment. In einem Feld von einigen tausend Gauß liegt die Elektronenresonanzstelle also im Gebiet der cm-Wellen. Um kommerziell entwickelte Geräte benutzen zu können, wurde sie deshalb mit einem Sender der Frequenz 9370 MHz ($\lambda \approx 3{,}2$ cm) angeregt. Dementsprechend betrug die Resonanzfeldstärke des Magneten etwa 3300 Gauß und die Frequenz der Protonenresonanz ungefähr 14 MHz.

Die Protonenresonanzfrequenz wurde nach der Resonanzabsorptionsmethode gemessen. Die Einzelheiten dieses Verfahrens wurden bereits besprochen (vgl.

[1] J. H. GARDNER u. E. M. PURCELL: Phys. Rev. **76**, 1262 (1949).
[2] J. H GARDNER: Phys. Rev. **83**, 996 (1951).

Ziff. 9 und 10), es sollen deshalb hier nur Daten genannt werden. Der Sender enthielt einen 7,12 MHz-Steuerquarz. Er lieferte zur Anregung der Protonenresonanz eine Spannung der Frequenz 14,24 MHz und zum Vergleich mit der Elektronenresonanz eine zweite Ausgangsspannung der Frequenz $9 \cdot 14,24$ MHz. Die Protonenprobe bestand aus 0,13 cm³ Mineralöl. Das Magnetfeld wurde mit 60 Hz moduliert. Die Modulations-Spannungsquelle steuerte über einen Phasenschieber auch die Zeitablenkung eines Oszillographen. Mittels eines elektronischen Schalters konnten das Protonen- und das Elektronen-Signal gleichzeitig auf dessen Bildschirm beobachtet werden.

Das wesentlichste Schaltelement der Elektronenresonanz-Apparatur ist ein rechteckiger evakuierter Hohlleiter (vgl. Fig. 91), dessen beide Schmalseiten je einen 0,1 mm breiten und 2 mm langen Schlitz enthalten. Unterhalb des einen Schlitzes befindet sich eine indirekt geheizte Glühkathode. Ihre negative Vorspannung gegenüber dem Hohlleiter ist variabel. Die Kathode liefert einen Elektronenstrom, der durch den unteren Schlitz zum Strahl gebündelt wird. Werden die Elektronen auf ihrem Wege durch den Hohlleiter nicht

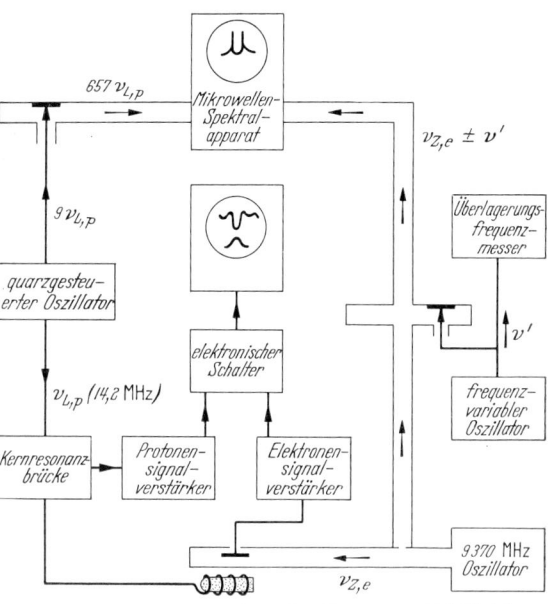

Fig. 92. Schematische Darstellung des Frequenzvergleich-Verfahrens.

abgelenkt, so treffen sie auf den zweiten Schlitz und werden oberhalb von diesem von einem Kollektor aufgefangen, der relativ zum Hohlleiter positiv mit etwa 20 V vorgespannt ist.

Der Elektronenstrahl verläuft parallel zu den Magnetfeldlinien. Für die einzelnen Elektronenbahnen gilt dies jedoch nicht. Nach dem Austritt aus dem unteren Schlitz beschreiben die Elektronen Spiralbahnen um die Feldlinien, deren Radien gemäß der Beziehung $r = mvc/eH$ von ihren individuellen anfänglichen Transversalgeschwindigkeiten abhängen. Dagegen stimmt die Kreisbahnfrequenz aller Elektronen überein und folgt aus Gl. (22.1). Das Magnetfeld verhindert also die Ausdehnung des Strahls, vorausgesetzt, daß Wechselwirkungen zwischen den Elektronen vernachlässigt werden können. Der Radius der Kreisbahnen beträgt für Elektronen thermischer Geschwindigkeit ($v \approx 2 \cdot 10^7$ cm/sec) in einem Feld von 3000 Gauß etwa 0,003 mm. Die Schlitze sind demnach ungefähr 30mal breiter, d.h. die meisten Elektronen können ungestört durch sie hindurchfliegen.

In dem Hohlleiter wird aber nun ein schwaches transversales elektrisches Mikrowellenfeld der Zyklotronfrequenz (22.1) angeregt (stehende Wellen im TE_{10}-Modus). Dadurch dehnt sich der Elektronenstrahl in Resonanz aus, denn das elektrische Wechselfeld vergrößert die transversale Geschwindigkeiten der Elektronen während ihrer vertikalen Bewegung immer mehr. Vernachlässigt man den anfänglichen thermischen Radius der Spiralbahnen, so findet man für den Bahnradius der Elektronen im Abstand a vom unteren Schlitz die Beziehung:

$$r_a = \left| \frac{a \cdot e \cdot E \cdot \cos\left(\frac{1}{2} a \frac{\Delta\omega}{v}\right)}{\pi \cdot m \cdot v \cdot \omega \cdot \left(1 - \left(\frac{a \Delta\omega}{\pi v}\right)^2\right)} \right|. \tag{22.3}$$

Darin ist v die vertikale Geschwindigkeit der Elektronen, E die maximale Amplitude des Mikrowellenfelds, ω dessen Frequenz und $\Delta\omega = \omega - \omega_z$ der Resonanzabstand. Wegen der Zunahme der Bahnradien trifft in der Nachbarschaft der Resonanzstelle nur noch ein Teil der Elektronen den oberen Schlitz und erreicht die Anode. Setzt man voraus, daß die Anodenstromstärke nur von den diesem Bild zugrunde liegenden geometrischen und kinematischen Daten des Experiments abhängt, so erhält man für sie die Gleichung

$$J(\alpha) = J_0(\alpha) \left(1 - \frac{2 \cdot A \cdot e \cdot E}{m \cdot \omega \cdot d} \left| \frac{\cos\left(\frac{1}{2}\pi \frac{\Delta\omega}{\alpha}\right)}{\alpha\left(1 - \left(\frac{\Delta\omega}{\alpha}\right)^2\right)} \right| \right), \tag{22.4}$$

wenn man mit $J_0(\alpha) = N(\alpha) \cdot e \cdot d \cdot l$ die Stromstärke weit außerhalb Resonanz bezeichnet und die Abkürzung $\alpha = \pi v/a$ benutzt. Ferner ist d die Schlitzbreite, l die Schlitzlänge und $N(\alpha)$ die Dichte der Elektronen, die pro Sekunde den unteren Schlitz passieren und eine Geschwindigkeit innerhalb des Intervalls v, $v + dv$ haben. Die Konstante A hat den Wert 1, wenn man annimmt, daß kein Elektron, das die Wände des oberen Schlitzes berührt hat, die Anode noch erreichen kann. Sie hat die Größe $1/\pi$, wenn alle Elektronen, die in den oberen Schlitz eintreten, auch vom Kollektor aufgefangen werden. Die Halbwertsbreite des Resonanzsignals folgt aus (22.4) zu:

$$\Delta\nu_{\frac{1}{2}} = 3,3 \frac{\alpha}{2\pi} = 1,65 \frac{v}{a}. \tag{22.5}$$

Für in vertikaler Richtung mit einheitlicher thermischer Geschwindigkeit fliegende Elektronen ergibt sich z.B. ($v = 2 \cdot 10^7$ cm/sec und $a = 2,3$ cm) $\Delta\nu_{\frac{1}{2}} = 14$ MHz $\hat{=} 4,3$ Gauß. Berücksichtigt man die MAXWELLsche Verteilung der Vertikalgeschwindigkeiten der Elektronen, dann erhält man einen etwas kleineren Wert. Allgemein ist die Signalbreite um so kleiner und die Meßgenauigkeit dementsprechend um so höher, je kleiner der Quotient v/a ist, je öfter also die Elektronen auf ihrem Weg durch den Hohlleiter die Feldlinien umkreisen.

Die Gl. (22.4) sagt in Übereinstimmung mit dem experimentellen Befund Seitenminima der Anodenstromkurve in Funktion des Resonanzabstands voraus. Man versteht deren Zustandekommen, wenn man bedenkt, daß das Mikrowellenfeld die transversale Geschwindigkeit der Elektronen mehr oder weniger beschleunigt oder verzögert, je nach der Größe der Phasendifferenz zwischen der Rotationsfrequenz der Elektronen und dem Hohlleiterwechselfeld. Dementsprechend ist die erste Rückkehr der Stromstärke auf ihren vollen Betrag beim Resonanzdurchgang genau dann zu erwarten, wenn die Elektronen in der ersten Hälfte ihrer Laufzeit durch den Hohlleiter transversale kinetische Energie gewinnen und diese in der zweiten Hälfte wieder verlieren. Die Fortsetzung dieses Gedankengangs liefert die Erklärung für die in Gl. (22.4) behauptete Existenz einer Folge von Seitenbandsignalen. Die im Experiment beobachteten Amplitudenunterschiede sind natürlich durch die MAXWELLsche thermische Geschwindigkeitsverteilung der Elektronen etwas verwischt.

Der Einfluß der Wechselwirkungen zwischen den Elektronen auf die Zyklotron-Resonanzfrequenz läßt sich wie folgt abschätzen. Jedes Elektron bewegt sich in dem resultierenden Feld aller anderen Elektronen. Setzt man voraus, daß für die Resonanzfrequenzänderung eines bestimmten Elektrons nur die Elektronen verantwortlich sind, die sich innerhalb des zylindrischen Strahls befinden, der durch die Spiralbahn des betrachteten Elektrons abgegrenzt wird, geht man also davon aus, daß sich der Einfluß aller übrigen Elektronen im Mittel über viele Umlaufperioden gegenseitig aufhebt, so erhält man für die auf das Elektron wirkenden Kräfte — ohne das Mikrowellenfeld — die Gleichgewichtsbeziehung:

$$m r \omega^2 = \frac{e v H}{c} - e E. \tag{22.6}$$

Darin ist $E = 2\pi r \varrho$ die Feldstärke der zylindrischen Ladungsverteilung, ϱ die Raumladungsdichte und r der Bahnradius des Elektrons. Es ist also

$$m \omega^2 = \frac{e H \omega}{c} - 2\pi e \varrho$$

und

$$\omega = \frac{eH}{2mc} \pm \frac{1}{2} \sqrt{\left(\frac{eH}{mc}\right)^2 - \frac{8\pi e \varrho}{m}}. \tag{22.7}$$

Kann man annehmen, daß der Raumladungseffekt nur eine kleine Korrektur der Zyklotronresonanzfrequenz bewirkt, so vereinfacht sich (22.7) zu:

$$\omega = \frac{eH}{mc}\left(1 - \frac{2\pi e \varrho/m}{(eH/mc)^2}\right) = \omega_0 \left(1 - \frac{2\pi e \varrho/m}{\omega_0^2}\right). \tag{22.8}$$

Der Effekt ist also um so kleiner, je niedriger die Raumladungsdichte ϱ ist. Deshalb wurde einmal — bestimmt durch die Schlitzform — ein bandförmiger Elektronenstrahl untersucht. Vor allem aber wurde in dem Hohlleiter die Kathodenstromstärke niedrig gehalten. Es ergab sich empirisch, daß die durch die Raumladung bedingte Resonanzverschiebung vernachlässigt werden kann, wenn die Vorspannung V_k des Hohlleiters relativ zur Kathode kleiner als etwa 1 V war.

Der Einfluß der Raumladungen auf den experimentellen Ablauf erwies sich im übrigen in einer anderen Hinsicht als meßtechnisch sehr günstig. Anfänglich wurde die Elektronenresonanz-Apparatur mit einer Vorspannung V_k von etwa 5 V betrieben, und der Anodenstrom mit einem empfindlichen Galvanometer gemessen. Beim langsamen monotonen Resonanzdurchgang ergab sich außerhalb Resonanz eine Kollektorstromstärke von ungefähr 0,4 μA und in Resonanz eine Stromstärke von etwa 0,3 μA. Unter diesen Bedingungen wurde also in Übereinstimmung mit Gl. (22.4) ein Resonanzminimum gefunden, dessen Breite durch (22.5) ungefähr richtig vorhergesagt wurde. Bei den eigentlichen Messungen wurde die Magnetfeldstärke mit 60 Hz moduliert, und der Verlauf des zuvor verstärkten Anodenstroms auf dem Bildschirm eines Oszillographen abgebildet. Außerdem wurde die Kathodenvorspannung zur Vermeidung der Raumladungsresonanzverschiebung unter ein Volt vermindert. Unter diesen Bedingungen fanden GARDNER und PURCELL in Resonanz ein Maximum der Anodenstromkurve, dessen Breite ungefähr 10mal kleiner war als die aus Gl. (22.5) folgende Signalbreite. Die Anodenstromstärke betrug dabei außerhalb Resonanz ungefähr 0,0005 μA und wurde in Resonanz einige hundertmal größer. Die Verfasser nehmen zur Deutung dieses Effekts an, daß außerhalb Resonanz der Elektronenstrom durch die Raumladung innerhalb des Hohlleiters begrenzt wird. In Resonanz wird die Elektronenwolke durch das Mikrowellenfeld ausgedehnt, dadurch vermindert sich die Tiefe des Potentialminimums im Hohlleiter und die Anodenstromstärke steigt an.

Durch die Raumladung werden die langsam fliegenden Elektronen besonders stark beeinflußt. Diese Elektronen umkreisen aber die Feldlinien öfter als die schnellen Elektronen [vgl. Gl. (22.5)]. Somit ist es verständlich, daß die Ausnutzung des Raumladungseffekts zur Feststellung der Resonanzstelle zugleich eine Geschwindigkeitsauswahl der Elektronen darstellt und damit eine Erhöhung des Auflösungsvermögens der Elektronenresonanz zur Folge hat.

Die Amplitude des Mikrowellenfelds konnte mittels eines Dämpfungsglieds innerhalb eines großen Bereichs geändert werden. Ist sie zu groß, dann wächst von einer gewissen Grenze an die Resonanzbreite des Signals mit ihr. Andererseits durfte sie auch nicht zu klein sein, da sonst die Signalhöhe zu niedrig wird.

Zur Bestimmung des Frequenzverhältnisses $v_{L,p}/v_{Z,e}$ wurde zunächst (vgl. Fig. 92), die Magnetfeldstärke H so justiert, daß die Protonenresonanzbedingung $\omega_{L,p} = \gamma H_0$ erfüllt war. Dann wurde die Frequenz des cm-Wellensenders gemäß der Elektronenresonanzbedingungen (22.1) der Feldstärke H_0 angepaßt. Nach diesen Justierungen befanden sich das Elektronensignal und das Kernsignal auf dem Bildschirm des Oszillographen genau untereinander. Der eigentliche Frequenzvergleich erfolgte auf dem Bildschirm eines Spektralapparats für Mikrowellen. Ein solches elektronisches Gerät[1] besteht im wesentlichen aus einem schmalbandigen Überlagerungsempfänger, dessen Oszillator durch eine Sägezahnspannung frequenzmoduliert wird, und dessen empfangenes Frequenzband sich damit also periodisch linear im Takte der Steuerspannung verschiebt. Dieselbe Steuerspannung liegt an den Horizontalplatten der Kathodenstrahlröhre, während den Vertikalplatten die Ausgangsspannung des Empfängers zugeführt wird. Auf dem Schirm der Kathodenstrahlröhre wird somit die Amplitudenverteilung eines beliebigen dem Eingang zugeleiteten Signals in Funktion der Frequenz abgebildet. Sollen insbesondere die Frequenzen zweier Spannungen miteinander verglichen werden, so legt man beide Spannungen zugleich an den Eingang des Frequenzanalysators und erhält auf dem Bildschirm eine durch zwei Impulse unterbrochene Grundlinie.

Im vorliegenden Fall war eine der beiden verglichenen Frequenzen die 657te Oberwelle der Protonenresonanzfrequenz $v_{L,p}$, und zwar wurde mit der bereits im Kernresonanzsender erzeugten 9ten Oberwelle dieser Frequenz ein Selenkristall in hohen Oberwellen angeregt. Dieser Kristall war in einem als Filter wirkenden Hohlleiter enthalten, der aus dem Frequenzgemisch die $(9 \times 73)\text{te} = 657\text{te}$ Oberwelle aussonderte. Die zweite zu vergleichende Spannung, also die Elektronenresonanzfrequenz $v_{Z,e}$ wurde in einem T-Glied mit einer variablen Frequenz v' amplitudenmoduliert. Am Eingang des Analysators lagen also die Frequenzen $657 \cdot v_{L,p}$ und $v_{Z,e} \pm v'$. Die Modulationsfrequenz v' ließ sich dann leicht so justieren, daß eines der Seitenbänder der Elektronenresonanzfrequenz mit der 657ten Oberwelle der Protonenresonanzfrequenz auf dem Bildschirm zur Deckung kam. Unter diesen Umständen gilt

$$\frac{\omega_{Z,e}}{\omega_{L,p}} = 657 \pm \frac{v'}{14,24}. \tag{22.9}$$

Die Frequenz v' wurde mit einem quarzgeeichten Überlagerungs-Frequenzmesser bestimmt.

---

[1] C. G. Montgomery: Technique of Microwave Measurements; Rad. Lab. Series. New York: McGraw-Hill Book Comp. Inc. 1947.

Die Feldstärken am Ort der Protonenprobe und längs des Elektronenstrahls stimmen nicht ganz überein. Um diesen Fehler zu eliminieren, wurden die Kernprobe und der Elektronenresonator nach jeder zweiten Messung vertauscht. Außerdem fanden die Verfasser, daß sich die Elektronenresonanzstelle auf dem Bildschirm verschob, wenn man die Richtung des Kathodenheizstroms änderte. Infolgedessen wurde auch die Richtung des Felds dieses Stroms bezüglich der Orientierung des großen Magnetfelds nach jeder Messung geändert, entweder durch Umpolen des Heizstroms oder durch Umpolen des Magnetstroms. Der wahre Wert des Frequenzverhältnisses $\omega_{Z,e}/\omega_{L,p}$ folgte somit aus einer Mittelwertbildung von vier Einzelmessungen. Die Abweichungen zwischen den Ergebnissen verschiedener solcher Versuchsreihen unterschieden sich untereinander um weniger als $1/_{150\,000}$. Der Mittelwert aller Messungen war ohne Berücksichtigung der diamagnetischen Korrektur der Protonen $\omega_{Z,e}/\omega_{L,p} = 657,4752$.

Die mittlere Streuung zwischen den Ergebnissen betrug 0,0037 und die maximale Abweichung 0,0056. Auf Grund einer sorgfältigen Abschätzung aller möglichen systematischen Fehlerquellen kamen die Verfasser zu dem abschließenden Ergebnis

$$\omega_{Z,e}/\omega_{L,p} = 657,475 \pm 0,008. \tag{22.10}$$

Mit dem diamagnetischen Korrekturfaktor $2,68 \cdot 10^{-5}$ folgt aus (22.2) und (22.10)

$$\mu_p = (1,521\,01 \pm 0,000\,02) \cdot 10^{-3} \frac{e\,\hbar}{2\,m_e\,c}. \tag{22.11}$$

Die Zyklotronfrequenz und die LARMOR-Frequenz der Protonen wurden von SOMMER, THOMAS und HIPPLE[1,2], sowie von JEFFRIES und BLOCH[3,4] im selben Magnetfeld miteinander verglichen. Aus den beiden Beziehungen

$$\nu_{L,p} = \frac{2\mu_p H}{h} \quad \text{und} \quad \nu_{Z,p} = \frac{e H}{2\pi m_p c}$$

folgt die (22.2) entsprechende Gleichung[5]

$$\frac{\nu_{L,p}}{\nu_{Z,p}} = \frac{\mu_p}{e\hbar/2 m_p c} = \frac{\mu_p}{\mu_{KM}} = \frac{\omega_{L,p}}{\omega_{Z,p}} \tag{22.12}$$

in der mit μ_{KM} die als Kernmagneton definierte Größe bezeichnet wurde.

SOMMER, THOMAS und HIPPLE haben zur Messung der Frequenz $\omega_{Z,p}$ einen von ihnen Omegatron genannten Kleinzyklotrontyp entwickelt. Sie wählten für das Gerät diesen Namen, da es sich um ein speziell zur Messung von Winkelgeschwindigkeiten (Kreisfrequenzen) konstruiertes Zyklotron handelt, zur Bezeichnung von Winkelgeschwindigkeiten wird aber in der Literatur üblicherweise der Buchstabe ω verwandt.

Sieht man von der vorliegenden Meßaufgabe ab, so handelt es sich bei einem derartigen Zyklotron um ein Instrument zur präzisen Messung von Massenverhältnissen. Für einfach geladene Ionenarten gilt dabei der Zusammenhang $m_1/m_2 = \omega_{Z,2}/\omega_{Z,1}$. Das experimentelle Problem war, das Auflösungsvermögen einer solchen, im Prinzip an sich bekannten Meßanordnung ausreichend zu steigern.

Um eine hohe Auflösung zu erzielen, müssen die Ionen die magnetischen Feldlinien möglichst oft umkreisen. Der Radienzuwachs pro Umlauf und damit

---

[1] J. A. HIPPLE, H. SOMMER u. H. A. THOMAS: Phys. Rev. **76**, 1877 (1949).
[2] H. SOMMER, H. A. THOMAS u. J. A. HIPPLE: Phys. Rev. **80**, 487 (1950); **82**, 697 (1951).
[3] F. BLOCH u. C. D. JEFFRIES: Phys. Rev. **80**, 305 (1950).
[4] C. D. JEFFRIES: Phys. Rev. **81**, 1040 (1951).
[5] L. W. ALVAREZ u. F. BLOCH: Phys. Rev. **57**, 111 (1940).

die Amplitude der Hochfrequenz-Spannung müssen dementsprechend möglichst klein sein. Die in den normalen Zyklotronen ausgenutzte Möglichkeit, die Bahn des Ionenstrahls mit Hilfe des beschleunigenden elektrischen Felds zu stabilisieren, entfällt somit. Magnetisch darf man ebenfalls nicht fokussieren, denn im inhomogenen Feld wächst die Resonanzbreite mit der Breite der Feldstreuung. Im Omegatron wird der Ionenstrahl deshalb elektrostatisch fokussiert. Fig. 93 zeigt schematisch die Anordnung und Schaltung seiner Elektroden. Das Elektrodensystem befindet sich in einer sorgfältig evakuierten ($1,9 \cdot 10^{-7}$ mm Hg) Glasröhre von 4,7 cm Durchmesser. Die Ionen werden in seinem Zentrum durch einen zur Magnetfeldrichtung parallelen Elektronenstrahl, also durch Stoßionisation des in der Röhre enthaltenen restlichen Wasserstoffgases erzeugt. Beschleunigt werden sie in dem elektrischen Wechselfeld zweier an einen Hochfrequenzgenerator angeschlossener paralleler, 3×5 cm großer Platten. Zwischen

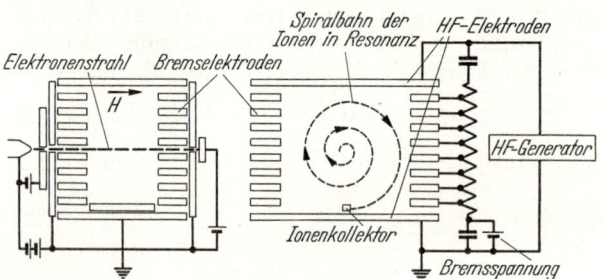

den beiden Hochfrequenz-Elektroden befindet sich ein Bremselektrodensystem. Es besteht aus acht parallelen rechteckigen Elektroden und ist positiv aufgeladen. Das statische elektrische Feld dieser Elektroden vermindert den Ionenverlust in axialer Richtung. Zur Homogenisierung des hochfrequenten

Fig. 93. Anordnung und Schaltung der Elektroden im Omegatron von SOMMER, THOMAS und HIPPLE.

elektrischen Felds wurden die Bremselektroden über zwei Kondensatoren und einen Spannungsteiler auch an den Hochfrequenz-Generator angeschlossen. Für ihre konstante Vorspannung ist der Spannungsteiler bedeutungslos. In einem Schlitz in einer der beiden Hochfrequenz-Elektroden befindet sich ein zur Beobachtung von Ionenströmen geeigneter Kollektor.

Eine Untersuchung der Bewegung geladener Teilchen im Omegatron zeigt, daß anfänglich ruhende Ionen — ihr Verhältnis Ladung zu Masse sei e/m — mit der Winkelgeschwindigkeit $(\omega + \omega_Z)/2$ eine Spiralbahn vom Radius

$$r = \frac{E_0 \cdot c}{H \cdot \varepsilon} \sin(\varepsilon t/2) \tag{22.13}$$

um die Richtung des konstanten Magnetfelds H beschreiben, wenn ein elektrisches Feld der Größe $E = E_0 \cdot \sin \omega t$ auf sie einwirkt, und zugleich der Frequenzabstand $\varepsilon = |\omega - \omega_Z|$ sehr klein gegen die Zyklotronkreisfrequenz $\omega_Z = eH/mc$ ist. Außerhalb Resonanz ($\varepsilon \neq 0$) ändert sich demnach der Radius der Spiralbahn periodisch mit der Kreisfrequenz $\frac{|\omega - \omega_Z|}{2}$. In Resonanz nimmt er monoton zu, und es ist

$$r = \frac{E_0 \cdot c \cdot t}{2H}. \tag{22.14}$$

Hat der Ionenkollektor vom Ursprung der Spirale den festen Abstand R_0, so erreichen ihn die geladenen Teilchen niemals, wenn R_0 größer als $E_0 c/H \varepsilon$ ist. Zu jedem bestimmten Tripel der experimentellen Daten H, E_0 und R_0 gehört also ein kritischer Resonanzabstand $\varepsilon' = E_0 c/H R_0$, bei dem die Ionen den Kollektor gerade noch erreichen. Sicher trennen lassen sich also zwei Elementarteilchen verschiedener Resonanzfrequenzen, wenn ihr relativer Frequenzunterschied $\nu/\Delta \nu = \omega_Z/2\varepsilon'$ ist. Aus $m = k/\nu$ ($k =$ Proportionalitätskonstante) und $\Delta m/\Delta \nu = -k/\nu^2 = -m/\nu$ folgt $m/|\Delta m| = \nu/|\Delta \nu|$. Für den als Auflösungsvermögen bezeichneten Quotienten $m/\Delta m$ gilt also (das Vorzeichen von Δm ist definitionsgemäß positiv):

$$\frac{m}{\Delta m} = \frac{\omega_Z \cdot H \cdot R_0}{2 \cdot E_0 \cdot c} = \frac{e\,H^2\,R_0}{2\,m\,c^2\,E_0}. \tag{22.15}$$

Aus Gl. (22.15) geht u. a. hervor, daß sich das Auflösungsvermögen umgekehrt proportional zur Masse der untersuchten Teilchen verhält. Das Omegatron eignet sich demnach vor allem zur vergleichenden Messung leichter Massen.

Ionen im kritischen Resonanzabstand erreichen den Kollektor nach der Zeit $t' = \pi/\varepsilon'$, es ist also auch

$$\frac{m}{\Delta m} = \frac{\omega_Z t'}{2\pi} = n', \tag{22.16}$$

wenn man mit n' die Zahl der Umläufe dieser Ionen um die Feldrichtung bezeichnet. In Resonanz ist gemäß (22.14) $t = 2R_0 H/E_0 c = 2/\varepsilon' = 2t'/\pi$, und es folgt

$$\frac{m}{\Delta m} = \frac{\pi}{2}\,n, \tag{22.17}$$

wenn man die Zahl der Umläufe der Ionen um die Feldrichtung in Resonanz vor dem Auftreffen auf den Kollektor n nennt.

Der maximale Radius r_m, den Ionen erreichen, deren Masse sich von derjenigen der Resonanzionen um den Betrag Δm unterscheidet, folgt aus (22.13) zu $r_m = \frac{E_0 \cdot c}{H \cdot \varepsilon} = \frac{E_0 \cdot c \cdot \omega_Z}{H(\omega - \omega_Z) \cdot \omega_Z} = \frac{E_0\,c\,m}{H \cdot \omega_Z\,\Delta m}$. Andererseits folgt aus (22.14) die Beziehung $R_0 = \frac{E_0\,c\,t}{2H} = \frac{E_0\,c\,2\pi\,n}{2H\,\omega_Z}$; es gilt somit

$$\frac{r_m}{R_0} = \frac{m}{\pi\,n\,\Delta m}. \tag{22.18}$$

Ist z.B. $\dfrac{\Delta m}{m} = \dfrac{1}{2200}$ und $n = 7000$ Umdrehungen, so folgt $r_m = \dfrac{R_0}{10}$. Außerhalb Resonanz bleiben die Ionen also in der unmittelbaren Nachbarschaft des Spiralursprungs.

Zur Berechnung der Ionenenergie in Resonanz unterteilt man das tatsächlich oszillierende Feld $E = E_0 \sin \omega t$ in zwei mit der Frequenz ω gegensinnig rotierende Felder derselben konstanten Amplitude $E_0/2$. Die Ionen erhalten ihre endgültige Energie V nur von jener Feldkomponente, die sie ständig in Richtung ihrer Bewegung beschleunigt. Es ist also $V = \frac{1}{2}E_0 L$, wenn man mit L die gesamte Länge der Ionenbahn ($L = n\pi R_0$) bezeichnet. Soll z.B. in einem Feld von etwa 5000 Gauß und bei 1 Kollektorabstand von 1 cm das Auflösungsvermögen $10\,000:1$ betragen, so folgt aus (22.15) für die Amplitude des elektrischen Feldes $E_0 = 0,1$ V/cm. Trotz dieser geringen Feldstärke ist die Endenergie der beschleunigten Teilchen $V = 1000$ Elektronenvolt. Die Laufzeit der Ionen beträgt in diesem Zahlenbeispiel ungefähr 1 msec, die Zyklotron-Resonanzfrequenz der Protonen etwa 7 MHz, demgemäß die Zahl der Umläufe der Ionen ungefähr 7000 und ihre Bahnlänge etwa 220 m.

Die im Omegatron beobachtete Resonanzfrequenz unterscheidet sich leider geringfügig von der wahren Zyklotron-Resonanzfrequenz. $\nu_Z = eH/2\pi\,m\,c$, da sich innerhalb der Röhre dem starken konstanten Magnetfeld noch schwache elektrostatische Felder überlagern. Diese Felder haben mindestens zwei Quellen, die positiv geladenen Bremselektroden und Raumladungen innerhalb der Röhre. Eine Berechnung des elektrischen Felds der Bremselektroden im Omegatron ergab, daß seine Feldstärke näherungsweise proportional mit dem Abstand r vom Feld-mittelpunkt wächst. Dasselbe galt für das Raumladungsfeld bei der Ableitung der Beziehung (22.7). Aus dieser Ableitung folgt somit für die im Omegatron beobachtete Frequenz die Beziehung

$$\omega = \omega_Z \left(1 - \frac{\bar{E} \, c^2 \, m}{e \, H^2}\right), \tag{22.19}$$

wenn man mit $\bar{E}$ die elektrische Feldstärke im Abstand $r = 1$ vom Mittelpunkt bezeichnet. Die Frequenzverschiebung ist demnach zwar vom Radius unabhängig, wächst aber mit der Teilchenmasse. Für die Resonanzfrequenzen v_1 und v_2 zweier im gleichen Feld untersuchter Teilchenarten der Masse m_1 und m_2 gelten also die Gleichungen

$$\left.\begin{aligned} v_1 &= v_{Z,1}(1 - K m_1), \\ v_2 &= v_{Z,2}(1 - K m_2), \end{aligned}\right\} \tag{22.20}$$

ferner ist

$$v_{Z,1}/v_{Z,2} = m_2/m_1. \tag{22.21}$$

Die Auflösung dieses Gleichungstripels liefert für den Korrekturfaktor K die Beziehung

$$K = \frac{m_2 v_2 - m_1 v_1}{m_1 m_2 (v_2 - v_1)} \tag{22.22}$$

und für die wahren Zyklotron-Resonanzfrequenzen die Bestimmungsgleichungen

$$v_{Z,1} = \frac{v_1}{1 - K m_1} \quad \text{und} \quad v_{Z,2} = \frac{v_2}{1 - K m_2}. \tag{22.23}$$

Zur Bestimmung der gesuchten Resonanzfrequenz $v_{Z,p}$ der Protonen mußte somit stets deren tatsächliche Resonanzfrequenz v_p und die Resonanzfrequenz v_x einer weiteren Ionenart gemessen werden. Zum Vergleich geeignete Ionen sind z.B. H_2^+, D_2^+ und H_2O^+. Die Elementarteilchenmassen m_p und m_x sind für den vor-liegenden Zweck genügend genau bekannt. Die experimentellen Untersuchungen zeigten, daß die Frequenzverschiebung etwa $^1/_{50\,000}$ pro 0,1 V Bremsspannung betrug.

Das Omegatron samt der an den Ionenkollektor angeschlossenen Elektro-meterröhre konnte leicht aus dem Luftspalt des Magneten herausgenommen und durch einen Kernresonanz-Probenkopf ersetzt werden. Die Konstruktion des kernmagnetisch stabilisierten Magneten und die zur Messung der LARMOR-Frequenz $v_{L,p}$ verwandte Anordnung wurden bereits in Ziff. 21 besprochen.

Das Ergebnis der Messung war

$$v_{L,p}/v_{Z,p} = 2{,}79268_5 \pm 0{,}00006. \tag{22.24}$$

Darin ist $v_{L,p}$ die kernmagnetische Resonanzfrequenz der Protonen in der Stan-dardölprobe, die auch bei der in Ziff. 21 beschriebenen Messung untersucht wurde. Der in (22.24) angegebene Fehler ist mehrfach größer als der geschätzte wahr-scheinliche Fehler. Die diamagnetische Korrektur der Standardölprobe beträgt[1] $28{,}1 \cdot 10^{-6}$. Berücksichtigt man dies, so erhält man aus (22.12) und (22.24) die Größe des magnetischen Protonenmoments in Kernmagnetonen

$$\mu_p = (2{,}79276 \pm 0{,}00006) \frac{e \, \hbar}{2 \, m_p \, c}. \tag{22.25}$$

JEFFRIES und BLOCH haben in ihrer etwa gleichzeitigen Arbeit zur Bestimmung der Zyklotronresonanzfrequenz der Protonen ein verzögerndes Zyklotron gebaut.

---

[1] H. A. THOMAS: Phys. Rev. **80**, 901 (1950).

Die Genauigkeit ihrer Messung (ungefähr 10^{-4}) war etwa dreimal niedriger als die obige. Ihr Ergebnis

$$\left.\begin{array}{l} v_{L,p}/v_{Z,p} \\ = 2{,}7924 \pm 0{,}0002 \end{array}\right\} (22.26)$$

bzw. das diamagnetisch korrigierte

$$\left.\begin{array}{l} \mu_p = (2{,}79250 \pm \\ \pm 0{,}00020)\, \dfrac{e\,\hbar}{2\,m_p\,c} \end{array}\right\} (22.27)$$

stimmt fast mit den entsprechenden Resultaten (22.24) und (22.25) überein, wenn man die beiderseitigen Fehlerintervalle berücksichtigt.

Fig. 94 zeigt Einzelheiten der Konstruktion des verzögernden Zyklotrons. In einem solchen „rückwärts laufenden" Zyklotron kann der Gasdruck beliebig niedrig sein, denn die Ionen werden außerhalb erzeugt und beschleunigt und dann durch ein Druckstufensystem in das Gerät eingeschossen. Sie treten tangential in eines der beiden Dee's ein und durchlaufen, praktisch ungestört durch Zusammenstöße mit Restgasatomen, eine normale Zyklotron-Spiralbahn in umgekehrter Richtung. In der Nähe des Zentrums der Anordnung befindet sich eine Sonde zur Messung der Ionenstromstärke. In Resonanz erreicht diese ihren Maximalbetrag.

Die relativistische Massenänderung der Protonen auf ihrer Bahn muß ver-

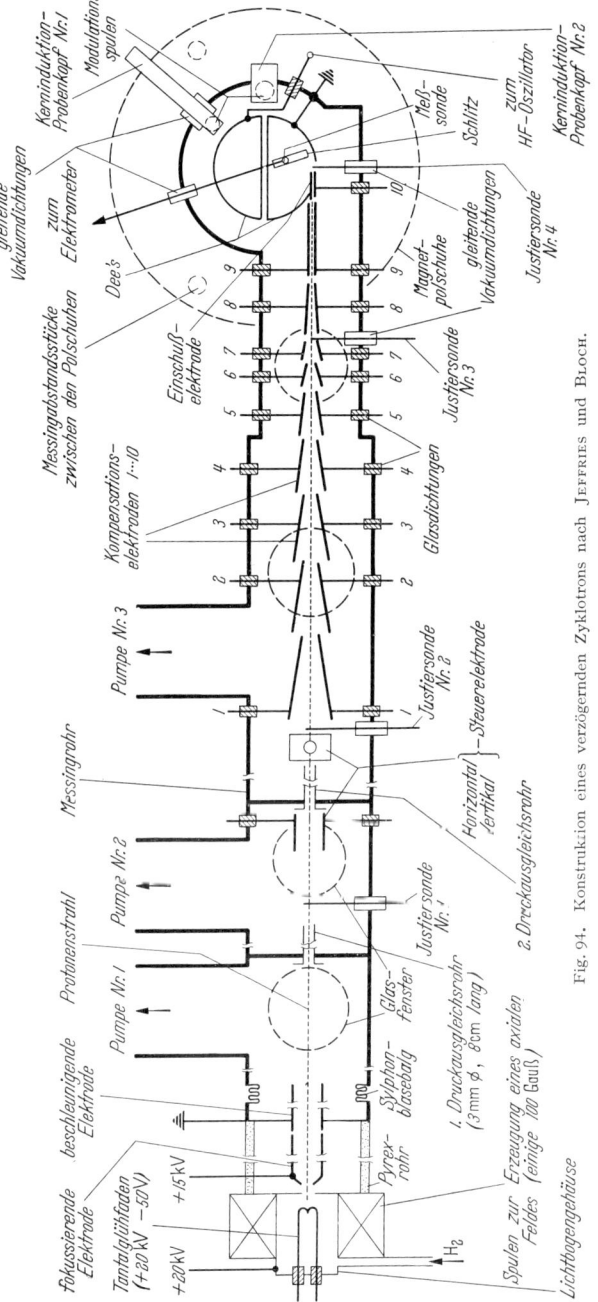

Fig. 94. Konstruktion eines verzögernden Zyklotrons nach JEFFRIES und BLOCH.

nachlässigbar sein. Dies begrenzt ihre maximale Energie auf einige 10^4 eV ($\Delta m/m = eV/mc^2 = 10^{-5}$ für 10^{-4} eV). Bei der vorliegenden Anordnung betrug die Anfangsenergie der Protonen $2 \cdot 10^4$ eV. Da ihre Bahn geradlinig in das Zyklotron einmünden soll, wurde die von dem Magnetfeld vor dem Eintritt in die Dee's auf die Protonen ausgeübte Kraft durch die elektrostatischen Anziehungs- bzw. Abstoßungs-Kräfte einer Folge vom Kompensationselektroden neutralisiert (in Fig. 94 sind die oberen Elektroden positiv und die unteren negativ aufgeladen). Die Amplitude des elektrischen Wechselfeldes zwischen den Dee's muß aus zwei Gründen eine Mindestgröße haben. Einmal konnten die Verfasser auf die elektrische axiale Fokussierung des Ionenstrahls nicht verzichten, vor allem aber folgt aus der Geometrie der Anordnung, daß der Bahnradius der phasenrichtig eingeschossenen Elektronen während ihres ersten Umlaufs mindestens um so viel verkleinert werden muß, daß die entsprechend verzögerten Protonen danach an der Einschußelektrode, also an der letzten der Kompensationselektroden vorbei fliegen können. Praktisch hat dies zur Folge, daß in einer mechanisch realisierbaren Konstruktion die Gesamtzahl der Umläufe der Protonen auf ungefähr 50 begrenzt ist. Das Auflösungsvermögen des verzögernden Zyklotrons ist also in der bis jetzt besprochenen Betriebsart sehr viel niedriger als das des Omegatrons. BLOCH und JEFFRIES haben jedoch bemerkt, daß man das verzögernde Zyklotron auch mit Oberwellen der Zyklotron-Resonanzfrequenz betreiben kann. Die Resonanzbreite des Signals wird aus zwei Gründen dadurch erheblich vermindert. Legt man zwischen die Dee's eine Spannung der n-ten ungeraden Oberfrequenz der Umlaufs-Resonanzfrequenz, so wird für eine feste vorgegebene Zahl von Umläufen der Protonen die Signalbreite um einen Faktor n kleiner, denn die Resonanzbedingung muß entsprechend präziser erfüllt werden, damit die Protonen die Empfangssonde erreichen können.

Außerdem erhöht sich aber auch die Zahl der Umläufe, die ein Proton benötigt, bis es auf die im Spiralinneren befindliche Sonde auftreffen kann. Mit der Abnahme des Spiralradius werden die Protonen immer langsamer und verweilen dementsprechend immer länger im Spalt zwischen den beiden Dee's. Die resultierende Verzögerungswirkung des elektrischen Feldes wird aber um so kleiner, je größer die Verweilzeit für ein festes $n > 1$ ist, und je größer n selbst ist. Dieser Effekt hatte für $n = 11$ in der Anordnung von BLOCH und JEFFRIES eine Zunahme der Umlaufzahl auf etwa 500 zur Folge.

Die Zahl n läßt sich jedoch nicht beliebig erhöhen. Höhere Oberwellen fokussieren den Ionenstrahl schlechter. Außerdem darf der Verweilzeit-Effekt nicht so groß werden, daß schon während des ersten Umlaufs der Bahnradius merklich weniger als für $n = 1$ verkleinert wird, da sonst die Protonen sämtlich auf die Einschußelektrode auftreffen würden. Immerhin gelang es den Verfassern auf diesem Weg die Resonanzbreite auf ungefähr 10^{-4} zu reduzieren. Weitere Einzelheiten zur Konstruktion des verzögernden Zyklotrons können der Fig. 94 entnommen werden.

23. Einordnung der kernmagnetischen Resonanzexperimente in das System der Meßmethoden physikalischer Elementarkonstanten. Die Genauigkeit der voranstehend beschriebenen Messungen ist so hoch, daß ihre Ergebnisse in das derzeitige System der Elementarkonstanten-Bestimmungen aufgenommen werden konnten. Einige Relationen zwischen den mitgeteilten Resultaten sind unmittelbar zu erkennen.

Bildet man das Verhältnis der Ergebnisse von THOMAS, DRISCOLL und HIPPLE[1] — Vergleich der in einem Magnetfeld auf einen stromführenden Leiter wirkenden Kraft mit der LARMOR-Präzessionsfrequenz der Protonen im selben Feld — und von GARDNER und PURCELL[2] — Vergleich der Elektronen-Zyklotronresonanzfrequenz mit der LARMOR-Präzessionsfrequenz der Protonen im selben

---

[1] H. A. THOMAS, R. L. DRISCOLL u. J. A. HIPPLE: J. Res. Nat. Bur. Stand. **44**, 569 (1950).
[2] J. H. GARDNER: Phys. Rev. **83**, 996 (1951).

Feld —, so ergibt sich aus (2.1) und (22.2) eine Bestimmungsgleichung für die spezifische Ladung des Elektrons[1]

$$\frac{\gamma_p}{\dfrac{\mu_p}{\mu_B}} = \frac{\dfrac{2\mu_p}{\hbar}}{\dfrac{2\mu_p m_e c}{e\hbar}} = \frac{e_{\text{e.s.E.}}}{m_e \cdot c} = \frac{e_{\text{e.m.E.}}}{m_e}, \tag{23.1}$$

und zwar folgt aus (21.1) und (22.10) der numerische Wert

$$\frac{e}{m_e} = (1{,}75890 \pm 0{,}00005) \cdot 10^7 \ [\text{e.m.E.}/\text{g}]. \tag{23.2}$$

Ebenso folgt aus der Kombination des Experiments von THOMAS, DRISCOLL und HIPPLE [Gl. (2.1)] mit dem von SOMMER, THOMAS und HIPPLE[2] — Vergleich der Zyklotronresonanzfrequenz der Protonen mit ihrer LARMOR-Präzessionsfrequenz [Gl. (22.12)] — eine Bestimmungsgleichung für die spezifische Ladung des Protons

$$\frac{\gamma_p}{\dfrac{\mu_p}{\mu_{\text{KM}}}} = \frac{\dfrac{2\mu_p}{\hbar}}{\dfrac{2\mu_p m_p c}{e\hbar}} = \frac{e_{\text{e.s.E.}}}{m_p \cdot c} = \frac{e_{\text{e.m.E.}}}{m_p}. \tag{23.3}$$

Deren numerischer Wert ergibt sich aus (21.1) und (22.24), und zwar ist

$$\frac{e}{m_p} = (0{,}95794_2 \pm 0{,}00003) \cdot 10^4 \ [\text{e.m.E.}/\text{g}] \tag{23.4}$$

Sowohl in Gl. (23.1) als auch in Gl. (23.3) hebt sich die nur auf vier Stellen bekannte PLANCKsche Konstante $\hbar$ heraus. Bemerkenswert ist ferner, daß in beiden Fällen die diamagnetisch unkorrigierten Ergebnisse miteinander verglichen werden konnten. Dies ist zulässig; denn die diamagnetischen Korrekturfaktoren der in den drei Experimenten zur Messung der Protonen-Präzessionsfrequenz untersuchten Substanzen unterscheiden sich nur unwesentlich; es gilt also z.B. auch $\gamma_{p,\text{unkorr.}} \Big/ \dfrac{\omega_{L,p,\text{unkorr.}}}{\omega_{Z,e}} = e_{\text{e.m.E.}}/m_e$.

Zur Messung der spezifischen Elektronen- und Protonenladungen hätte man natürlich im Prinzip auch die Zyklotronresonanzfrequenzen der beiden Elementarteilchen mit der auf einen stromführenden Leiter im selben Feld einwirkenden Kraft direkt vergleichen können. Die Messung der zu der jeweiligen Feldstärke gehörenden LARMOR-Resonanzfrequenz hat also nur eine vermittelnde Funktion zwischen den eigentlich zu vergleichenden Meßgrößen. Jedoch ist es aus mehreren Gründen lohnend, diesen scheinbaren meßtechnischen Umweg zu wählen. Einmal sind die kernresonanztechnischen Verfahren, verglichen mit den übrigen Magnetfeld-Meßmethoden außerordentlich bequem zu handhaben. Vor allem aber erlauben sie eine sehr genaue und fast punktweise Vermessung der Felder in denen experimentiert werden soll. Mit einigen Feld-Meßmethoden kann man zwar ebenfalls den Mittelwert der Feldverteilung eines größeren Gebiets präzis ermitteln, jedoch ist zur Zeit kein weiteres Verfahren zur genauen Messung der Feldstärke an einem bestimmten Ort bekannt.

Vergleicht man die experimentellen Ergebnisse von SOMMER, THOMAS und HIPPLE und von GARDNER und PURCELL, so erhält man aus (22.12) und (22.2)

---

[1] e.s.E. = elektrostatische Einheiten, e.m.E. = elektromagnetische Einheiten.
[2] H. SOMMER. H. A. THOMAS u. J. A. HIPPLE: Phys. Rev. **82**, 697 (1951).

eine Bestimmungsgleichung für das Verhältnis der Protonen- zur Elektronenmasse

$$\frac{\dfrac{\mu_p}{\mu_B}}{\dfrac{\mu_p}{\mu_{KM}}} = \frac{\dfrac{e\,\hbar}{2\,m_p\,c}}{\dfrac{e\,\hbar}{2\,m_e\,c}} = \frac{m_e}{m_p}. \tag{23.5}$$

Aus (22.24) und (22.10) folgt der numerische Wert

$$\frac{m_p}{m_e} = 1836{,}12 \pm 0{,}05. \tag{23.6}$$

Auch hier hätte man im Prinzip die Zyklotronresonanzfrequenzen der beiden
Elementarteilchen im selben Feld direkt miteinander vergleichen können.

Eine weitere wichtige Grundkonstante, die FARADAYsche Zahl (F) oder die
sog. Äquivalentladung erhält man durch Multiplikation der Gl. (23.4) mit dem
wohlbekannten Atomgewicht A_p des Protons ($A_p = 1{,}007\,593 \pm 0{,}000\,003$)

$$F = L\,e = A_p\,\frac{e}{m_p}. \tag{23.7}$$

In dieser Gleichung ist L die LOSCHMIDTsche Zahl[1]. In der physikalischen Skala
[1 atomphysikalische Masseneinheit (ME) = 1/16 der Masse des Sauerstoffisotops
O^{16}, $L = g/ME$, $A_p = m_p/ME$] gilt

$$F = (0{,}965\,20_3 \pm 0{,}00003) \cdot 10^4\,[\text{e.m.E./g-Mol}]. \tag{23.8}$$

CRAIG und HOFFMANN[2] haben elektrochemisch den Wert $F = (0{,}965\,19_3 \pm 0{,}000026)$
$\times 10^4$ [e. m. E./g-Mol] ermittelt. Berücksichtigt man die jeweiligen Fehlerintervalle, dann stimmen die beiden Ergebnisse überein. Zur Berechnung der Äquivalentladung in der chemischen Skala (1 chemische ME = 1/16 der Masse des
mittleren Sauerstoffatoms im natürlichen Isotopengemisch) muß man Gl. (23.8) mit
dem SMYTHschen Faktor ($k = 1{,}000\,279 \pm 0{,}000003$) dividieren ($F_{\text{phys}} = k \cdot F_{\text{chem}}$).

Für einen der jüngsten Zweige der theoretischen Physik, die Quantenelektro-
dynamik, wurde ein Vergleich der Ergebnisse des Experiments von GARDNER
und PURCELL mit dem Resultat einer von KOENIG, PRODELL und KUSCH[3] (vgl.
auch [4] und [5]) angestellten Untersuchung besonders wichtig.

Die letztgenannten Verfasser haben das Verhältnis der g-Faktoren der in
Wasserstoffatomen gebundenen Elektronen und der Protonen einer Mineralölprobe gemessen. Zur Bestimmung der Elektronen-Präzessionsfrequenz untersuchten sie die Hyperfeinstruktur des Wasserstoffs mit dem RABIschen Atomstrahl-
Resonanzverfahren. Bezüglich der experimentellen Einzelheiten kann auf den
in diesem Band enthaltenen Bericht über Atom- und Molekularstrahl-Resonanz-
untersuchungen verwiesen werden. Die Protonenresonanz wurde nach dem von
POUND und KNIGHT[6] angegebenen und auf S. 209 besprochenen Verfahren gemessen. Das außerordentlich genaue Ergebnis war

$$\frac{g_{e,j}}{g_{p,\text{kugelf. Ölprobe}}} = 658{,}2171 \pm 0{,}0006. \tag{23.9}$$

[1] In der amerikanischen Literatur wird die oben als LOSCHMIDTsche Zahl bezeichnete Konstante [$L = (6{,}024\,72 \pm 0{,}000\,36) \cdot 10^{23}$ g Mol^{-1}] zum Teil AVOGADROsche Konstante genannt. Als LOSCHMIDTsche Konstante wird dort das in der deutschen Literatur AVOGADROsche Zahl genannte Verhältnis $L_0 = L/V_0$ [$V_0 =$ Molvolumen, $L_0 = (2{,}687\,13 \pm 0{,}000\,16) \times 10^{19}$ cm^{-3}] bezeichnet.

[2] D. N. CRAIG u. J. I. HOFFMANN: Phys. Rev. **80**, 487 (1950).

[3] S. H. KOENIG, A. G. PRODELL u. P. KUSCH: Phys. Rev. **88**, 191 (1952).

[4] P. KUSCH u. H. TAUB: Phys. Rev. **75**, 1477 (1949).

[5] H. TAUB u. P. KUSCH: Phys. Rev. **75**, 1481 (1949).

[6] R. V. POUND u. W. D. KNIGHT: Rev. Sci. Instrum. **21**, 219 (1950).

Der in (23.9) angegebene Zahlenwert unterscheidet sich von den in verschiedenen Versuchsreihen direkt gewonnenen Meßergebnissen noch um kleine Korrekturbeträge. So war zu beachten, daß durch den Volum-Diamagnetismus[1] der jeweils untersuchten Protonenproben die Feldstärke am Ort der Kerne geringfügig verändert wird. Dieser Effekt verschwindet, wenn die Probe kugelförmig ist. In einem unendlich langen Zylinder wird das äußere Feld dagegen um den Betrag $\frac{2}{3}\pi\chi_d \cdot H$ vermindert ($\chi_d=$diamagnetische Suszeptibilität). Für die vorliegenden Messungen wurde eine zylinderförmige Probe benutzt, deren Länge 8,5mal so groß wie ihr Durchmesser war. Nach einer Arbeit von DICKINSON[2] ist es dann zulässig bei der Berechnung der volum-diamagnetischen Korrektur den Zylinder als unendlich lang zu betrachten. Mit den Zahlenwerten $\chi_{d,\mathrm{H_2O}} = -0,72 \cdot 10^{-6}$ und $\chi_{d,\mathrm{Öl}} = 1,01 \cdot \chi_{d,\mathrm{H_2O}}$ folgt somit $\frac{2}{3}\pi\chi_{d,\mathrm{Öl,H_2O}} \cdot H \approx 1,5 \cdot 10^{-6} \cdot H$. Einige Messungen wurden mit Wasserproben ausgeführt, in denen zur Verkürzung der Relaxationszeiten paramagnetische Ionen (Cu^{++}) gelöst waren. Durch das Feld dieser Ionen wird die Feldstärke an den Kernorten nach Messungen von DICKINSON[2] in einer 0,13-molaren Lösung um $0,7 \cdot 10^{-6}$ erhöht. Weiter unterscheiden sich nach Messungen von GUTOWSKY und McCLURE[3] die hüllen-diamagnetischen Korrekturen des Mineralöl- und des Wassermoleküls um den Betrag $3,4 \cdot 10^{-6}$. Zu den mit Wasserproben erhaltenen Meßwerten mußte also eine Gesamtkorrektur von $(1,5+3,4-0,7) \cdot 6,58 \cdot 10^{-4}$ addiert werden, um das Ergebnis (23.9) zu erhalten.

Eigentlich interessiert jedoch das Verhältnis der g-Faktoren des freien ruhenden Elektrons und des freien Protons. Man benötigt somit eine Beziehung zwischen den Faktoren $g_{e,j}$ des Wasserstoffatoms und $g_e = g_{e,s}$ des freien Elektrons. Besäße der Kern des Wasserstoffatoms keinen Drehimpuls, dann würde sich das Atom in einem verschwindend schwachen äußeren Magnetfeld in einem reinen S-Zustand befinden, sein g-Faktor würde also nur vom Elektronenspin herrühren, d.h. es wäre exakt

$$g_{e,j} = g_{e,s'}, \tag{23.10}$$

wenn man mit $g_{e,s'}$ den Spin-g-Faktor des im H-Atom gebundenen Elektrons bezeichnet. Sowohl durch die Existenz des Protonenkernmoments als auch durch ein äußeres Magnetfeld wird aber den zwei entarteten S-Grundniveaus des Wasserstoffatoms ($^2S_{\frac{1}{2}}$) ein kleiner Teil der beiden Niveaus $^2D_{\frac{3}{2}}$, $m_j = \pm\frac{1}{2}$ beigemischt. Dieser Beitrag läßt sich nach von PERL und HUGHES abgeleiteten Formeln berechnen. Aber auch eine grobe Abschätzung zeigt schon, daß die Gl. (23.10) bis auf einen Fehler von etwa $2 \cdot 10^{-12}$ richtig ist, der angeführte Mischungseffekt kann also vollständig vernachlässigt werden.

Relativistische Effekte, im wesentlichen verursacht durch die Bewegung des gebundenen Elektrons im Wasserstoffatom und die damit verknüpfte Zunahme der Elektronenmasse haben zur Folge, daß $g_{e,s'}$ kleiner ist als der g-Faktor des freien Elektrons $g_{e,s}$. Eine Berechnung dieses Effekts liefert die Beziehung[4]

$$\frac{g_{e,s'}}{g_{e,s}} = \frac{1}{3}\left[1 + 2(1-\alpha^2)^{\frac{1}{2}}\right] \approx 1 - \frac{\alpha^2}{3} = 1 - 17,8 \cdot 10^{-6}. \tag{23.11}$$

Damit folgt aus (23.9) bis (23.11) das Verhältnis der g-Faktoren der freien Elektronen und der in einer kugelförmigen Mineralölprobe enthaltenen Protonen, und zwar ist

$$\frac{g_{e,s}}{g_{p,\,\mathrm{kugelf.\,Ölprobe}}} = 658,2288 \pm 0,0006. \tag{23.12}$$

---

[1] Nicht zu verwechseln mit dem Hüllen-Diamagnetismus der Atome bzw. Moleküle!

[2] W. C. DICKINSON: Phys. Rev. 81, 717 (1951).

[3] H. S. GUTOWSKY u. R. E. McCLURE: Phys. Rev. 81, 276 (1951).

[4] N. F. MOTT u. H. S. W. MASSEY: Theory of Atomic Collisions, 2. Aufl., S. 72. Oxford: Clarendon Press 1949.

Gardner und Purcell haben die Zyklotron-Resonanzfrequenz der Elektronen ebenfalls mit der Larmor-Resonanzfrequenz der Protonen einer kugelförmigen Mineralölprobe verglichen. Ihr hüllen-diamagnetisch nicht korrigiertes Ergebnis (22.10) $\omega_{Z,e}/\omega_{L,p} = 2 g_{e,l}/g_{p,\,\text{kugelf. Ölprobe}} = 657{,}475 \pm 0{,}008$ liefert also kombiniert mit (23.12) den wichtigen Zusammenhang

$$\frac{g_{e,s}}{g_{e,l}} = 2\,(1{,}001\,146 \pm 0{,}000\,012)\,, \qquad (23.13)$$

bzw.

$$\mu_e = (1{,}001\,146 \pm 0{,}000\,012)\,\mu_B\,. \qquad (23.14)$$

Aus (23.12) ergibt sich das Verhältnis der g-Faktoren der beiden freien Elementarteilchen g_e/g_p, wenn man zu dem von Ramsey[1] berechneten diamagnetischen Abschirmungsfaktor der Elektronenhülle des H_2-Moleküls $(26{,}8 \cdot 10^{-6})$ die von Gutowsky und McClure[2] gemessene Differenz der Abschirmungsfaktoren der H_2-Moleküle und der Ölmoleküle $(3{,}7 \cdot 10^{-6})$ addiert. Demgemäß ist

$$\frac{g_e}{g_p} = 658{,}2087 \pm 0{,}0006\,. \qquad (23.15)$$

Diesen experimentellen Ergebnissen soll ein kurzer Überblick über die Entwicklung der theoretischen Berechnung des Elektroneneigenmoments gegenübergestellt werden[3]. Aus der Diracschen relativistischen Wellengleichung[4] ergibt sich der Elektronenspin zu $I = \frac{1}{2}$. Weiter folgt aus ihr für das magnetische Moment des Elektrons[5] die Beziehung $\mu_e = -g_e\, I \mu_B$, in welcher der g-Faktor $g_e = 2$ ist. Nach der klassischen Berechnung des magnetischen Moments einer mit einem gewissen Drehimpuls rotierenden elektrischen Ladung sollte $g_e = 1$ sein. Man bezeichnet deshalb häufig die von der Diracschen Theorie richtig beschriebene Erfahrung, daß g_e in Wirklichkeit doppelt so groß ist, als magnetomechanische Anomalie.

Das experimentelle Ergebnis (23.13) zeigt aber, daß auch dies nur näherungsweise gültig ist, und zwar ist der g_e-Faktor etwas größer als 2. Zur theoretischen Deutung dieser weiteren Anomalie kann man sog. Strahlungskorrekturen in der Form einer Reihe zunehmender Potenzen der Sommerfeldschen Feinstrukturkonstanten $\alpha = 2\pi\, e^2/hc$ berechnen. Schwinger[6] hat dies erstmals getan und gefunden, daß die Korrektur erster Ordnung die Größe $\alpha/2\pi$ hat. Die Strahlungskorrektur zweiter Ordnung wurde von Karplus und Kroll[7] zu $-2{,}973\,\alpha^2/\pi^2$ bestimmt. In dieser Näherung hat man also theoretisch zu erwarten:

$$g_e = 2\left(1 + \frac{\alpha}{2\pi} - 2{,}973\,\frac{\alpha^2}{\pi^2}\right). \qquad (23.16)$$

Benutzt man den von Dayhoff, Triebwasser und Lamb[8] aus Untersuchungen der Feinstruktur des H-Atoms ermittelten Wert der Sommerfeldschen Konstanten

$$\alpha = 1/137{,}036, \qquad (23.17)$$

so findet man für die Korrekturen erster und zweiter Ordnung die numerischen Werte

$$\frac{\alpha}{2\pi} = 0{,}001\,1614, \qquad -2{,}973\,\frac{\alpha^2}{\pi^2} = -0{,}0000160 \qquad (23.18)$$

---

[1] N. F. Ramsey: Phys. Rev. **78**, 699 (1950).
[2] Siehe Fußnote 2, S. 273.
[3] Für die ausführliche Darlegung der theoretischen Zusammenhänge in diesem Handbuch vgl. G. Källén in Bd. V, Teil 1, insbesondere S. 299—323.
[4] P. A. M. Dirac: Proc. Roy. Soc. Lond. **117**, 610 (1927).
[5] Vgl. auch F. Bloch: Physica, Haag **19**, 821 (1953).
[6] J. Schwinger: Phys. Rev. **73**, 416 (1948).
[7] R. Karplus u. N. M. Kroll: Phys. Rev. **77**, 536 (1950).
[8] E. S. Dayhoff, S. Triebwasser u. W. E. Lamb jr.: Phys. Rev. **89**, 106 (1953).

und entsprechend für g_e

$$g_e = 2 \times 1,001\,1454. \tag{23.19}$$

Die Übereinstimmung zwischen dem experimentellen Ergebnis (23.13) und dem theoretischen (23.19) ist ausgezeichnet. Jedoch muß im Hinblick auf die verhältnismäßig große Unsicherheit des experimentellen Resultats darauf verwiesen werden, daß entweder diese Übereinstimmung zufälligerweise so gut ist, oder daß der Betrag des experimentellen Meßfehlers zu skeptisch abgeschätzt wurde.

DuMond und Cohen[1] sowie Bearden und Watts[2] haben die in dieser Ziffer beschriebenen experimentellen Vergleiche zum Teil in das allgemeine System der Elementarkonstanten-Bestimmungen einbezogen und nach der Methode der kleinsten Quadrate die sog. derzeit besten Werte für die Konstanten ermittelt. Diese unterscheiden sich nur geringfügig von den hier wiedergegebenen speziellen Meßergebnissen.

II. Relativmessungen — Struktur der kernmagnetischen Spektren.

24. Aufgabenstellung, Wahl der Probe und der Meßmethode. Kernmagnetische Relativmessungen, also Vergleiche von Larmor-Resonanzfrequenzen lassen sich mit hoher Genauigkeit ausführen. So ist es nicht überraschend, daß sich an die ursprünglich einzige experimentelle Aufgabe, der Vermessung des Spektrums der Larmor-Frequenzen aller Kernarten, bald als zweites Problem die Erforschung der Spektren jeder einzelnen Kernart anschloß. Zwar hängen die Kernresonanzfrequenzen in erster Näherung nur von der äußeren Magnetfeldstärke und von der Größe der gyromagnetischen Verhältnisse, also von einer Kerneigenschaft und von einer experimentell beliebig wählbaren Größe ab. Präzisere Beobachtungen mit hoch auflösenden Spektrometern zeigten jedoch, daß nicht nur die Formen und Halbwertsbreiten der Resonanzlinien, sondern auch deren Frequenzen in mannigfaltiger Weise von der Art und dem Zustand der Substanzen abhängen, in denen die Kerne enthalten sind. In manchen chemisch einfachen Verbindungen sind die Resonanzfrequenzen nur mehr oder weniger verschoben. In den meisten Substanzen spaltet sich jedoch die Resonanzlinie jeder Kernart in ein ganzes, oft linienreiches Spektrum auf. Alle derartigen Frequenzänderungen lassen sich durch dem äußeren Feld überlagerte substanzinnere Magnetfelder bzw. durch magnetische und elektrische Wechselwirkungen zwischen den in der Substanz enthaltenen Elementarteilchen deuten. Es leuchtet unmittelbar ein, daß die Untersuchung der Struktur der kernmagnetischen Spektren somit ebenso wie die Beobachtung des kernmagnetischen Relaxationsverhaltens der Substanzen ein ausgezeichnetes neues Hilfsmittel zur Erforschung innerer Felder und Wechselwirkungen darstellt.

Vom Standpunkt der Kernphysik aus gesehen, interessieren jedoch nur die Resonanzfrequenzen der ungebundenen, sog. nackten Kerne. Der Themabegrenzung des vorliegenden Bandes entsprechend, sollen deshalb in den folgenden Ziffern die Spektren der zur Bestimmung ungestörter Kernresonanzfrequenzen geeigneten Substanzen genauer als die übrigen besprochen werden.

Will man die Resonanzfrequenzen der ungebundenen Kerne bestimmen, so ist das erste und vielleicht auch das wichtigste Problem, eine möglichst gute Probenwahl zu treffen. Drei Gesichtspunkte sind dabei zu beachten. Es hängt von

---

[1] J. W. M. DuMond u. E. R. Cohen: Rev. Mod. Phys. **20**, 82 (1948); **25**, 691 (1953). Vgl. auch dieses Handbuch, Bd. XXXV.

[2] J. A. Bearden u. H. M. Watts: Phys. Rev. **81**, 73 (1951).

der Substanz ab, welche Ansprüche an die *Empfindlichkeit* der Apparatur zu stellen sind, welcher Aufwand also zur Lösung einer bestimmten Aufgabenstellung erforderlich wird. Ebenso ist das *Auflösungsvermögen*, d.h. die obere Grenze der Frequenzmeßgenauigkeit eine durch die Substanz bestimmte Größe, es sei denn, daß es bereits durch Unvollkommenheiten der Apparatur begrenzt wird. Ebenso wichtig, wenn auch leider zum Teil unbeachtet geblieben, ist der letzte Gesichtspunkt. Man muß rechtzeitig berücksichtigen, daß die Größe der zu den gemessenen Resonanzfrequenzen hinzuzufügenden oder abzuziehenden *Korrekturen* ebenfalls von der Substanz abhängt. Alle an kompakten Proben gemessenen LARMOR-Frequenzen müssen zum mindesten diamagnetisch korrigiert werden, man muß also die inneren Felder der Elektronenhüllen der Moleküle bzw. Atome an den Kernorten berechnen können. Die gegenwärtig bekannten Rechenverfahren zur Bestimmung der Größe der Hüllenfelder liefern jedoch, außer in einigen Sonderfällen, nur unbefriedigende Resultate, wenn man ihre Genauigkeit mit derjenigen der experimentell erhaltenen Daten vergleicht. Somit empfiehlt es sich, die Größe dieser intramolekularen bzw. intraatomaren Feldkorrektur so klein als möglich zu halten, d.h. aber, daß die Kernresonanzfrequenzen nur in einfach gebauten und nur in rein diamagnetischen Molekülen gemessen werden sollten. Es darf erwartet werden, daß zu einem späteren Zeitpunkt präzisere Berechnungsunterlagen zur Verfügung stehen. Um deren nachträgliche Berücksichtigung zu ermöglichen, sollten alle Veröffentlichungen kernmagnetischer Meßergebnisse eine genaue Beschreibung der verwandten Probensubstanz enthalten. Dem Magnetfeld am Kernort überlagern sich aber außer dem Feld der Elektronenhülle des eigenen Moleküls auch noch die Felder der übrigen Moleküle der Probensubstanz. Dementsprechend muß im allgemeinen zu dem gemessenen Ergebnis auch noch eine intermolekulare Korrektur hinzugefügt werden. In Flüssigkeiten befinden sich die Moleküle in BROWNscher Bewegung, ihre Dipolfelder mitteln sich also, außer einem als makroskopische Volumsuszeptibilität behandelbaren und u. a. auch von der Probenform abhängenden kleinen Anteil (vgl. Ziff. 25) zeitlich aus. In Festkörpern sind dagegen die Moleküle und mit diesen deren Felder mehr oder weniger stark fixiert, dem äußeren Feld überlagert sich also ein inneres meist sehr inhomogenes Feld von möglicherweise beträchtlicher Größe. Damit wurde bereits ein Argument genannt, das gegen die Verwendung fester Substanzen spricht, wenn reine Kernresonanzfrequenzen gemessen werden sollen.

Im Hinblick auf die stets begrenzte apparative Empfindlichkeit muß die Probe die zu untersuchenden Kerne in genügender Anzahl pro Volumeinheit enthalten. Die Signalspannung wächst, wie früher auseinandergesetzt, proportional mit dem statischen resultierenden Moment, also mit der Kernzahl. Der im Luftspalt des Magneten zur Verfügung stehende Raum ist aber fest vorgegeben. Es ist keineswegs trivial, diese Bedingung auszusprechen. Da sie für Gase schlecht erfüllt ist, wurden bisher Kernresonanzfrequenzen selten an gasförmigen Substanzen bestimmt, obwohl diese alle anderen Forderungen gut erfüllen. In den letzten Jahren konnte die apparative Empfindlichkeit jedoch erheblich gesteigert werden. Es darf daher angenommen werden, daß in Zukunft häufiger als bisher die Resonanzfrequenzen einfach gebauter gasförmiger Verbindungen gemessen werden. Zur Erhöhung der Kernkonzentration kann man Gase natürlich komprimieren. Auf diesem Wege konnten z.B. die LARMOR-Frequenzen der Kerne der chemisch nicht reagierenden Edelgase bestimmt werden[1-3]. Außerdem haben

---

1 H. L. ANDERSON: Phys. Rev. **76**, 1460 (1949).

2 W. G. PROCTOR u. F. C. YU: Phys. Rev. **81**, 20 (1951).

3 E. BRUN, J. OESER, H. H. STAUB u. C. G. TELCHOW: Phys. Rev. **93**, 904 (1954).

mehrere Autoren[1-3] die Resonanzfrequenz der Protonen im Wasserstoffgas gemessen, da sich die hüllendiamagnetische Korrektur dieses Moleküls genau berechnen läßt (vgl. Ziff. 26).

Die Mehrzahl der veröffentlichten Messungen wurden in Lösungen ausgeführt. Für die meisten Kerne lassen sich leicht Verbindungen ausfindig machen, die sich in geeigneten Lösungsmitteln in ausreichend hoher Konzentration lösen.

Ebenfalls mit Rücksicht auf die begrenzte apparative Empfindlichkeit sollte sich in einer geeigneten Substanz die reziproke, in Hz gemessene Linienbreite, d.h. die effektive transversale Relaxationszeit ($1/T_{2,\,\mathrm{eff}} = 1/T_2 + 1/T_2' + 1/T_2'' + \cdots$, vgl. Ziff. 3) nicht zu sehr von der Spin-Gitter-Relaxationszeit unterscheiden. Bei der fast immer zur Beobachtung der Resonanzlinien angewandten Methode des langsamen adiabatischen Resonanzdurchgangs wächst z.B. die optimale Absorptions-Signalspannung proportional mit $\sqrt{T_{2,\,\mathrm{eff}}/T_1}$ [vgl. Gl. (6.17)].

In typischen Festkörpern — gemischte Aggregatzustände sollen in diesem Zusammenhang nicht besprochen werden — ist die Zeit T_1 um viele Größenordnungen länger als $T_{2,\,\mathrm{eff}}$. Einmal ist infolge der starren räumlichen Verteilung der Moleküle die für die Größe von T_1 verantwortliche Intensität des magnetischen Rauschens sehr viel kleiner als etwa in Flüssigkeiten und T_1 dementsprechend sehr lang. Zum anderen ist aber aus dem gleichen Grund die Linienbreite ($1/T_{2,\,\mathrm{eff}}$) sehr groß, da sie im wesentlichen durch die beträchtliche Inhomogenität des inneren Felds bestimmt wird. Daraus folgt, daß zur Beobachtung der Kernsignale von Festkörpern erheblich empfindlichere Apparaturen benötigt werden, als zur Beobachtung der Kernsignale von Flüssigkeiten und Gasen, deren wirkliche Relaxationszeiten stets der gleichen Größenordnung angehören. Jedoch kann auch in diesem Fall $T_{2,\,\mathrm{eff}}$ auf Grund der Inhomogenität des äußeren Magnetfelds sehr viel kleiner als T_2 und damit erst recht sehr viel kleiner als T_1 sein. Man kann dann in Flüssigkeiten, wie schon früher erwähnt, durch einen Zusatz paramagnetischer Ionen T_1 und T_2 verkürzen und damit den Quotienten $T_{2,\,\mathrm{eff}}/T_1$ vergrößern, denn $T_{2,\,\mathrm{eff}}$ ändert sich infolge der Beimischung der Ionen nur wenig. Ebenso kann man die Relaxationszeiten gasförmiger Proben verkleinern und damit eventuell einander annähern, wenn man sie etwa mit dem hüllenparamagnetischen Sauerstoffgas vermischt[4], oder das zu untersuchende Gas in ein Gefäß füllt, das gepulvertes paramagnetisches Eisenoxyd (Fe_2O_3) enthält[5,6].

Es muß jedoch darauf hingewiesen werden, daß die Ausnützung dieses katalytischen Effekts zur Herabsetzung der Forderung an die Apparatur-Empfindlichkeit aus anderen Gründen sehr bedenklich ist. Die Beifügung paramagnetischer Ionen hat eine Verschiebung der Kernresonanzfrequenz zur Folge, deren Größe sich zur Zeit nicht exakt berechnen läßt (vgl. Ziff. 25). Außerdem wird ein äußeres homogenes Feld durch sie innerhalb des Probevolumens inhomogen verzerrt, d.h. aber daß sie auch die apparative Linienbreite etwas vergrößern. Für kugelförmige Proben gilt das zuletzt gesagte allerdings nicht, wie in der Magnetostatik gezeigt wird.

Der bessere Weg zur Vergrößerung der Signalspannung ist zweifellos, die apparative Linienbreite so weit als möglich zu vermindern. In den letzten Jahren wurden sehr erfolgreich Erfahrungen hierzu gesammelt (vgl. die Ziff. 17 und 18).

[1] E. M. PURCELL, R. V. POUND u. N. BLOEMBERGEN: Phys. Rev. **70**, 986 (1949).
[2] H. S. GUTOWSKY u. R. E. McCLURE: Phys. Rev. **81**, 276 (1951).
[3] B. SMALLER, E. YASAITIS u. H. L. ANDERSON: Phys. Rev. **81**, 896 (1951).
[4] H. L. ANDERSON: Phys. Rev. **76**, 1460 (1949).
[5] W. G. PROCTOR u. F. C. YU: Phys. Rev. **81**, 20 (1951).
[6] F. BLOCH: Phys. Rev. **83**, 1062 (1951).

Zwar zwingt dies dazu, die Kernsignale nur mit sehr schwachen Hochfrequenzfeldern anzuregen [vgl. Gl. (6.16)], aber dieses Problem läßt sich technisch leicht lösen. Ein solches Vorgehen erhöht vor allem zugleich das Auflösungsvermögen bis zu der durch die Substanz vorgegebenen Grenze.

Allgemein lassen sich zwei Resonanzlinien um so leichter trennen, und ebenso läßt sich ihr Frequenzabstand um so leichter messen, je schmaler die beiden Linien sind. Die Form und Breite einer kernmagnetischen Resonanzlinie wird im wesentlichen durch die Dauer der Phasengedächtniszeit T_2 der Kernpräzession, durch die Größe des substanzinneren statischen Magnetfelds und durch die Breite der äußeren Feldverteilung (Inhomogenität) bestimmt. Der zuletzt genannte Parameter ist meistens allein verantwortlich für die oben erwähnte apparative Linienbreite. Ist deren Größe vernachlässigbar, so ist es bezüglich des Auflösungsvermögens sinnvoll, bei der Probenwahl nach einer Substanz zu suchen, in der die Linienbreiten der Kernresonanzfrequenzen möglichst klein sind, in der die Zeit T_2 also möglichst lang und das innere Magnetfeld möglichst homogen ist. Zahlreiche Flüssigkeiten erfüllen diese Bedingungen ausgezeichnet (vgl. Ziff. 3), Festkörper dagegen nur in Sonderfällen. Gase verhalten sich ähnlich wie Flüssigkeiten niedriger Viskosität.

Ein weiterer Gesichtspunkt der Probenwahl soll ebenfalls noch erwähnt werden. Bei allen Relativmessungen innerhalb des kernmagnetischen Spektrums werden die Resonanzfrequenzen von zwei Kernarten im selben Feld miteinander verglichen. Praktisch kann man z.B. die Messungen mit verschiedenen Proben, meist auch mit verschiedenen Probenköpfen, aber mit der gleichen elektronischen Apparatur am selben Feldort nacheinander ausführen. Man muß sich dann allerdings darauf verlassen, daß die zeitliche Stabilität des Magnetfeldes ausreichend gut ist. Andererseits kann man die Resonanzfrequenzen der zwei Proben in nebeneinanderliegenden Spulen mit zwei getrennten elektronischen Apparaturen gleichzeitig anregen und beobachten. In diesem Fall muß man die Differenz der Feldstärken an den beiden Meßorten genügend genau kennen. Besser geschützt ist man jedoch gegen systematische Meßfehler, wenn man Proben verwendet, die sowohl die zu untersuchenden Kerne als auch die Vergleichskerne enthalten. Ein gutes Beispiel hierfür sind wäßrige Lösungen von Verbindungen der interessierenden Atomarten. Die Standardkerne — Protonen — befinden sich in diesem Fall im Lösungsmittel. Umschließt man die Probe mit zwei Spulen verschiedener Induktivität, die in zwei zu den Resonanzfrequenzen der verglichenen Kernarten passenden Schwingkreisen enthalten sind, so kann man die beiden Resonanzfrequenzen mit zwei elektronischen Apparaturen tatsächlich gleichzeitig am selben Feldort messen. Zur Verminderung der gegenseitigen Einstreuung ordnet man die beiden Spulen zweckmäßigerweise orthogonal an. Einen besonderen Vorteil haben ferner gemischte Proben, deren Moleküle rein diamagnetisch sind. In ihnen stimmen die volumdiamagnetischen Korrekturen der Resonanzfrequenzen aller verglichenen Kernarten stets überein und können somit das Ergebnis von Relativmessungen nicht verfälschen. In gemischten Proben, die auch hüllenparamagnetische Dipole enthalten, sind die Verhältnisse dagegen komplizierter (vgl. Ziff. 25).

Zur Messung der Kernresonanzfrequenzen verwendet man meist automatisch registrierende Hochfrequenz-Spektrographen. Als solche, auch Kernspektrometer genannte Geräte, kann man alle Kombinationen eines geeigneten Magneten mit einem der im Abschnitt B dieses Beitrags beschriebenen elektronischen Apparaturtypen bezeichnen, sofern sie vorwiegend zur Messung von Resonanzfrequenzen konstruiert wurden und deren Bestimmung in einem genügend großen Frequenz- bzw. Feldstärke-Intervall erlauben. Soll ein solches Spektrometer die Registrierung der Resonanzlinien von möglichst vielen Kernarten erlauben, so muß sein Meßbereich sehr groß sein. In einem Feld von 10000 Gauss unterscheiden sich z.B. die Resonanzfrequenzen der Protonen ($\sim$40 MHz) und der F^{19}-Kerne von den Resonanzfrequenzen der Isotope Pb^{204} ($\sim$400 kHz) und Sm^{149} um einen Faktor der Größenordnung 100. Erinnert man weiter daran, daß die Halbwertsbreiten der kernmagnetischen Resonanzlinien häufig kleiner als 1 Hz sind, so wird verständlich, daß es ein schwieriges experimentelles Problem sein kann, eine unbekannte Resonanzlinie aufzufinden. Die Geschwindigkeit mit welcher das Frequenz- bzw. Feldstärken-Intervall, innerhalb dessen die Resonanzlinie vermutet wird, abgesucht werden darf, ist stets aus einem der zwei folgenden Gründe begrenzt. Soll der Resonanzdurchgang adiabatisch erfolgen, und im Hinblick auf das Auflösungsvermögen ist dies zu empfehlen, so darf gemäß den Ungleichungen (6.1)

$$\frac{d\omega}{dt} \ll (\Delta\,\omega_h)^2, \qquad \frac{dH}{dt} \ll |\gamma|\,(\Delta H_h)^2$$

die Änderung pro Zeiteinheit des bei der Messung variierten Parameters der Gleichung $\omega = \gamma H$ nur klein gegen eine durch die Linienbreite bestimmte Grenze sein. Weiter wird meistens zur Erhöhung der apparativen Empfindlichkeit die Resonanzlinie differentiell abgetastet, und die Kernsignalspannung äußerst schmalbandig und phasenempfindlich gleichgerichtet (vgl. Ziff. 12). Damit unterliegt aber die Registriergeschwindigkeit einer zweiten, im allgemeinen noch wesentlich einschneidenderen Begrenzung. Die Zeitkonstante, mit welcher der Verstärker- und Registrierteil einer solchen Apparatur einer Signalspannungsänderung folgt, ist gleich dem Reziproken der Bandbreite der gesamten Empfangsanordnung. Soll diese die Spannungsänderungen bei einem Resonanzdurchgang korrekt wiedergeben können, so muß die Registriergeschwindigkeit aber so klein sein, daß die relative Spannungsänderung am Eingang des Verstärkers in jedem Zeitabschnitt von der Größenordnung der Empfangs-Zeitkonstanten klein gegen eins ist. Praktisch erreicht werden Bandbreiten zwischen $^1/_{100}$ und $^1/_{1000}$ Hz, die entsprechenden Zeitkonstanten, 100 bis 1000 sec, sind also sehr groß.

Zum Glück kennt man die magnetischen Momente und mit diesen die gyromagnetischen Verhältnisse der meisten stabilen Kerne bereits aus den Untersuchungen der Hyperfeinstruktur der optischen Linienspektren und zwar bis auf einen Fehler von 1 bis 10%. Dieser Umstand hat wesentlich dazu beigetragen, daß die kernmagnetischen Resonanzlinien der verschiedenen Kernarten in den letzten Jahren in rascher Folge aufgefunden und vermessen werden konnten. Ein wesentliches Stück Vorarbeit hierzu wurde außerdem schon in der Molekularstrahl-Spektroskopie geleistet.

Ebenso wie die Probenwahl hängt auch die Entscheidung, welcher Apparatur-Typ zu wählen ist, besser gesagt, welche experimentellen Probleme bei der Konstruktion der Apparatur besonders sorgfältig zu lösen sind, zum Teil von der Aufgabenstellung ab. Unterschieden werden müssen zunächst die Fragestellungen, welche Anregungs- und Beobachtungsmethoden und welche elektronischen Schaltungsarten sich allgemein oder für bestimmte Aufgabenstellungen am besten bewährt haben oder sich voraussichtlich am weitesten entwickeln lassen. In dieser umfassenden Form können die angeschnittenen Fragen gegenwärtig kaum zweifelsfrei beantwortet werden. Es sollen daher im folgenden nur einige Teilantworten genannt werden.

Sowohl das Auflösungsvermögen als auch die Empfindlichkeit eines Kernspektrometers werden vor allem durch die Größen der räumlichen bzw. zeitlichen Änderungen des Magnetfelds bestimmt, in dem die Kernresonanzfrequenzen gemessen werden. Es ist trivial, daß die Auflösung nahe benachbarter

Resonanzlinien voraussetzt, daß die Breite der räumlichen Feldstärkestreuung innerhalb des Probenbereichs (Homogenität) kleiner als der zu messende Linienabstand ist. Dagegen hängt die Empfindlichkeit eines Spektrometers in erster Linie von der maximal zulässigen Beobachtungsdauer, also von der zeitlichen Konstanz (Stabilität) des Magnetfeldes ab. Dies entspricht der allgemeinen Erfahrung, daß sich aus statistisch schwankenden Meßwerten die Meßgröße um so genauer ermitteln läßt, je zahlreicher die Meßwerte sind. Im Fall der kernmagnetischen Resonanz ist die Empfindlichkeit (Signal-Rausch-Verhältnis) eine Funktion der Beobachtungsbandbreite. Mit deren Einengung wird aber nicht nur die Rauschspannung vermindert, sondern auch wie oben diskutiert die erforderliche Beobachtungsdauer vergrößert. Bei der voranstehenden Diskussion darf jedoch nicht übersehen werden, daß das Auflösungsvermögen bzw. die Feldhomogenität und die Empfindlichkeit bzw. die Feldstabilität auch nicht ganz unabhängig voneinander sind. Langsame zeitliche Schwankungen des Felds innerhalb der Meßdauer begrenzen das Auflösungsvermögen ebenso wie die räumliche Inhomogenität des Felds. Andererseits kann man ein Magnetfeld um so präziser zeitlich stabilisieren je definierter, also je räumlich-homogener das Feld am Ort der Meßsonde ist.

Bezüglich des Auflösungsvermögens nahe benachbarter Resonanzlinien und der Frequenzmeßgenauigkeit beliebig weit voneinander entfernter Resonanzlinien besteht nach Ansicht des Verfassers zwischen den verschiedenen in Abschnitt B besprochenen elektronischen Apparatur-Typen kein grundsätzlicher Unterschied. In Ziff. 16 wurde dies zum Teil bereits im Rahmen eines Vergleichs der kontinuierlichen und der impulstechnischen Verfahren diskutiert. Es bleibt nur noch nachzutragen, daß das Auflösungsvermögen bei den kontinuierlichen Methoden apparativ begrenzt wird, wenn das anregende Feld H_1 zu stark ist. Gemäß Gl. (6.12) hat z.B. beim langsamen adiabatischen Resonanzdurchgang das Absorptionssignal die mit der Hochfrequenzfeldstärke H_1 zunehmende Halbwertsbreite

$$\Delta \omega_h = \frac{2}{T_2} \sqrt{1 + \gamma^2 H_1^2 T_1 T_2}\,.$$

Bezüglich des Signal-Rausch-Verhältnisses folgt der optimale Wert von H_1 in diesem Fall aus der Beziehung $\gamma^2 H_1^2 T_1 T_2 = 1$ [vgl. Gl. (6.16)]. Die Vergrößerung der natürlichen Signalbreite bei der optimalen Anregung um einen Faktor 1,4 dürfte aber im allgemeinen unwesentlich sein.

Auf einen weiteren Grund, die Kernsignale nur mit schwachen Hochfrequenzfeldern anzuregen, haben Bloch und Siegert[1] hingewiesen. Wie früher ausgeführt, erfolgt die Anregung der Kernpräzessionsbewegung nur durch eine der beiden gegensinnig rotierenden Komponenten des in Wirklichkeit oszillierenden Hochfrequenzfelds und zwar durch die im Präzessionssinn rotierende. Die genannten Verfasser haben aber gezeigt, daß die gegensinnig rotierende Komponente Ursache einer apparativen Resonanzverschiebung ist, deren Größe ungefähr gleich dem Quadrat des Feldstärkenverhältnisses $(H_1/H_0)^2$ ist, wenn man das konstante Feld mit H_0 bezeichnet. Praktisch ist dies nicht sehr wichtig, selbst wenn H_1/H_0 den ungewöhnlich großen Wert 10^{-3} hat, ändern sich die Resonanzfrequenzen relativ nur um einen Faktor der Größenordnung 10^{-6}.

Wie in Ziff. 16 (S. 230) bereits ausgeführt, läßt sich bei den kontinuierlichen Verfahren die apparative Empfindlichkeit weiter steigern als bei den impulstechnischen Resonanzmethoden. Mehrere Autoren[2] haben ferner mitgeteilt, daß

---

[1] F. Bloch u. A. Siegert: Phys. Rev. 57, 522 (1940).
[2] Zum Beispiel: A. Lösche: Exp. Techn. Physik 1, 19, 69, 128 (1953).

auch zwischen den verschiedenen kontinuierlichen elektronischen Nachweis-
prinzipien der Kernresonanz Unterschiede bezüglich des optimal erzielbaren
Signal-Rausch-Verhältnisses bestehen. Sicher ist, daß dieses die Empfindlichkeit charakterisierende Verhältnis außerdem auch noch von der Anregungs-
und Beobachtungsart abhängt, ob also etwa stationäre oder instationäre Kern-
präzessionsbewegungen angeregt werden, ob ferner bei der Abtastung der Resonanzlinie das Feld niederfrequent moduliert oder nur monoton[1] geändert wird.
Im einzelnen kann hierauf nicht eingegangen werden. Häufig zitierte Kernspektrometer wurden z.B. von POUND, KNIGHT[2] und WATKINS[3], von PROCTOR[4]
und von WEAVER[5] u.a. beschrieben.

**25. Abhängigkeit der Kernresonanzfrequenz von der makroskopischen Sus-
zeptibilität der Substanz.** In allen Proben weichen die Feldstärken an den Kernorten von der äußeren Feldstärke mehr oder weniger ab. Es ist sinnvoll, bei der
Untersuchung der dem äußeren Feld (H) überlagerten substanzinneren Felder,
innere makroskopische (H') und innere mikroskopische Felder (H'') zu unterscheiden. Einer solchen Gliederung liegt die Vorstellung zugrunde, daß sich die
ganze Probe in dem äußeren Feld der räumlich mittleren Stärke H, jedes Molekül,
Atom oder Ion der Probe im räumlichen Mittel in einem Feld der Stärke $H+H'$
und jeder Kern im räumlichen Mittel in einem Feld der Stärke $H+H'+H''$
befindet. Zwar haben die Felder H' — Zusatzfeld im Inneren der Probe und H'' —
Zusatzfeld im Inneren der Elektronenhüllen — zum Teil die gleichen Quellen,
z.B. den Paramagnetismus der untersuchten oder anderer in der Substanz enthaltenen Atomkerne, oder den Diamagnetismus bzw. Paramagnetismus der
Elektronenhüllen; trotzdem ist es zweckmäßig, bei der theoretischen Behandlung
des Problems der Resonanzfrequenz-Verschiebungen die sog. volummagnetischen
und die sog. hüllenmagnetischen Korrekturen getrennt zu berechnen.

In der vorliegenden Ziffer soll zunächst die Berechnung der volummagnetischen
Felder besprochen werden. Diese werden in Flüssigkeiten und Gasen ausschließlich durch den Diamagnetismus und durch den Paramagnetismus der Kerne und
Elektronenhüllen erzeugt. In Festkörpern müssen darüber hinaus auch jene
inneren Felder berücksichtigt werden, welche ihre Entstehung dem Zusammenwirken von größeren Molekül- bzw. Atomgruppen verdanken, also z.B. ferro-
und ferrimagnetische Felder. Speziell in Metallen werden weiter beträchtliche
Resonanzverschiebungen durch den Paramagnetismus der Leitungselektronen
verursacht.

Zur Bestimmung der zusätzlichen magnetischen Feldstärke H' im Inneren
einer aus irgendeiner Art von Elementarmagneten bestehenden Substanz kann
man Rechenverfahren anwenden, die von DEBYE[6] und LORENTZ[7] angegeben
worden sind, vgl. hierzu auch DICKINSON[8].

Hat die Probe die Form eines Ellipsoids und befindet sie sich in einem völlig
homogenen äußeren Feld H, dessen Richtung zu einer der Achsenrichtungen des
Ellipsoids parallel ist, so ist auch das Feld H_i im Probeninnern völlig homogen.
Durch das äußere Feld wird die Probe magnetisiert und zwar gilt unter diesen
Bedingungen in der ganzen Probe $M = \chi H_i$. Die bekannten paramagnetischen

[1] J. T. ARNOLD: Phys. Rev. **102**, 136 (1956).
[2] R. V. POUND u. W. D. KNIGHT: Rev. Sci. Instrum. **21**, 219 (1950).
[3] G. D. WATKINS u. R. V. POUND: Phys. Rev. **82**, 343 (1951).
[4] W. G. PROCTOR: Phys. Rev. **79**, 35 (1950).
[5] H. E. WEAVER: Phys. Rev. **89**, 923 (1953).
[6] P. DEBYE: Phys. Z. **13**, 97 (1912). — Siehe auch P. DEBYE: Polare Molekeln. Leipzig:
S. Hirzel 1929. — Polar Molecules. New York: Dover 1945.
[7] H. A. LORENTZ: The Theory of Electrons. 1909. Neuauflage New York: Dover 1952.
[8] W. C. DICKINSON: Phys. Rev. **81**, 717 (1951).

Permeabilitäten ($\mu = 1 + 4\pi\chi$) und erst recht die diamagnetischen sind jedoch nur so wenig von eins verschieden, daß man in der voranstehenden Beziehung unbedenklich H_i durch H ersetzen kann. Die zusätzliche Feldstärke H' am Ort der Substanzpartikel (Moleküle, Atome oder Ionen) kann man sich aus drei Komponenten zusammengesetzt vorstellen

$$H' = H_a' + H_b' + H_c'. \tag{25.1}$$

Dazu zeichnet man um jedes speziell betrachtete Partikel in Gedanken eine Kugel, deren Mittelpunkt mit demjenigen des betrachteten Teilchens übereinstimmt und deren Radius so klein ist, daß ihre ganze Oberfläche in der Probe enthalten ist. Formal kann man dann mit H_a' das Entmagnetisierungsfeld in der Probe bezeichnen. Besteht diese aus paramagnetischen Dipolen, so ist die Richtung dieses Felds antiparallel zur Richtung des äußeren Felds, besteht sie dagegen aus diamagnetischen Dipolen, so stimmt die Richtung dieses Felds im Probeninnern mit der äußeren Feldrichtung überein. Weiter kann man das Feld der an der Kugeloberfläche befindlichen Dipole im Inneren der hohl gedachten Kugel H_b' nennen. Seine Richtung ist offenbar stets der Richtung des Felds H_a' entgegengesetzt, also im Fall einer paramagnetischen Dipolbelegung der Kugeloberfläche parallel zum äußeren Feld und im Fall einer Belegung der Fläche mit diamagnetischen Dipolen antiparallel zum äußeren Feld. Endlich kann man unter H_c' das Feld aller Dipole innerhalb der Kugeloberfläche am Ort des betrachteten Partikels, also im Kugelmittelpunkt verstehen.

Die Größe des Teilfelds H_b', des sog. LORENTZ- oder Hohlkugelfelds läßt sich unter den obigen Voraussetzungen leicht berechnen; bekanntlich ergibt sich

$$H_b' = \frac{4\pi}{3} M = \frac{4\pi}{3} \chi H. \tag{25.2}$$

Hat die Probe selbst ebenfalls Kugelform, so ist die Dichte der Dipole an ihrer Oberfläche offenbar gleich derjenigen auf der inneren LORENTZ-Kugel. Jedoch tragen die jeweils zueinander gehörenden Halbkugelschalen stets freie Pole entgegengesetzten Vorzeichens. Daraus folgt, daß im speziellen Fall einer kugelförmigen Probe die Summe der Felder $H_a' + H_b'$ stets verschwindet. Hat die Probe dagegen eine beliebige Form, so kann man für das Entmagnetisierungsfeld H_a' ansetzen:

$$H_a' = -\alpha M = -\alpha \chi H. \tag{25.3}$$

Die in (25.3) enthaltene Konstante α wird Entmagnetisierungsfaktor genannt. Der Entmagnetisierungsfaktor der Kugel hat demnach die Größe $4\pi/3$, während der Entmagnetisierungsfaktor eines unendlich langen Zylinders, dessen Achse senkrecht zu den Feldlinien eines unendlich ausgedehnten homogenen Felds liegt, z.B. die Größe 2π hat. Zwischen diesen beiden theoretisch behandelbaren Grenzfällen liegen die experimentell häufig benutzten zylindrischen Proben endlicher Länge. Den im Anschluß besprochenen Messungen von DICKINSON[1] (vgl. hierzu Fig. 95) kann man ungefähr folgende Interpolationswerte entnehmen: Hat das Verhältnis Zylinderlänge/Zylinderdurchmesser nacheinander die Größen 20; 3; 1,5 und 1, so haben die zu diesen Probenformen gehörenden Entmagnetisierungsfaktoren die Werte $^{12}/_6\pi$, $^{11}/_6\pi$, $^{10}/_6\pi$ und $^9/_6\pi$. Der Entmagnetisierungsfaktor der Kugel $^8/_6\pi$ schließt stetig an diese Reihe an. Der Entmagnetisierungsfaktor eines zur Feldrichtung parallelen unendlich langen Zylinders ist Null.

Zu beachten ist jedoch, daß die obige Definition von α streng nur zulässig ist, wenn das Feld im Probeninnern homogen ist. Diese Voraussetzung ist aber nur

---

[1] W. C. DICKINSON: Phys. Rev. 81, 717 (1951).

erfüllt, wenn sich eine ellipsoidförmige Probe in einem völlig homogenen Feld befindet, dessen Richtung zu einer der drei Achsen des Ellipsoids parallel ist.

Zur Berechnung des Felds H_c' im Mittelpunkt der LORENTZ-Kugel muß man die Summe der Felder aller in dieser Kugel enthaltenen magnetischen Dipole bilden. Die zur äußeren Feldrichtung H transversalen Komponenten dieser Felder heben sich im zeitlichen Mittel in Flüssigkeiten und Gasen sicher gegenseitig auf. Die Komponente des Felds eines einzelnen Dipols parallel zur Richtung des äußeren Felds hat die mittlere Größe

$$H_{\parallel} = \bar{\mu} \cdot r^{-3} (1 - 3 \cos^2 \Theta), \tag{25.4}$$

wenn man den Abstand zwischen dem Kugelmittelpunkt und dem Mittelpunkt des betrachteten Dipols r nennt, den Winkel zwischen der Magnetfeldrichtung und der Verbindungslinie der beiden Mittelpunkte mit Θ bezeichnet und unter $\bar{\mu}$ den Betrag des über alle möglichen Spinorientierungen nach BOLTZMANN gemittelten und somit zur äußeren Feldrichtung parallelen Dipolmoments versteht. Ist $\bar{\mu}$ innerhalb der ganzen Probe im zeitlichen Mittel konstant und haben alle vom Mittelpunkt der LORENTZ-Kugel ausgehenden Halbstrahlen im gleichen Abstand von diesem dieselbe Besetzungswahrscheinlichkeit mit Dipolen, ist also die Probe völlig isotrop, d.h. im zeitlichen Mittel frei von Nahordnungen, können die Wechselwirkungen zwischen den Dipolen mit anderen Worten also vernachlässigt werden, so heben sich auch die zur Feldrichtung parallelen Komponenten der Dipole innerhalb der LORENTZ-Kugel gegenseitig auf, denn es ist:

$$\int_0^{\pi} (1 - 3 \cos^2 \Theta) \sin \Theta \, d\Theta = 0. \tag{25.5}$$

Experimentell wurde jedoch gefunden, daß die voranstehenden Voraussetzungen häufig nicht erfüllt sind. Man kann dann zur Charakterisierung des Felds H_c' einen Wechselwirkungsfaktor q nach der Beziehung

$$H_c' = q M = q \chi H \tag{25.6}$$

einführen. Dieser Definition liegt die Vorstellung zugrunde, daß die Stärke des inneren Wechselwirkungsfelds proportional mit der Substanzsuszeptibilität und mit der äußeren Feldstärke wächst. Faßt man (25.2), (25.3) und (25.6) nach (25.1) zusammen, so erhält man für das gesamte makroskopische innere Feld H' den Ansatz

$$H' = \left(\frac{4\pi}{3} - \alpha + q \right) \chi H, \tag{25.7}$$

wobei die Konstante α nur von der Probenform und die Konstante q nur von Substanzeigenschaften abhängen soll.

Die Größenordnung der Feldstärke H' wird im wesentlichen durch diejenige der Suszeptibilität χ bestimmt. Diskutiert man nacheinander die verschiedenen möglichen Quellen des inneren Felds, so kann man sofort zeigen, daß beim gegenwärtigen Stand der Meßtechnik das kernparamagnetische innere Feld sicher vernachlässigt werden kann. Bei Zimmertemperatur beträgt z.B. die statische Suszeptibilität der Protonen des Wassers trotz deren hoher Dichte nur $3{,}2 \cdot 10^{-10}$. Weiter kann auch der Anteil des inneren Felds vernachlässigt werden, dessen Quellen die durch die Molekülrotation erzeugten magnetischen Momente sind, denn es ist wohlbekannt, daß diese von derselben Größenordnung wie die Kernmomente sind.

Sorgfältiger muß bei Präzisionsmessungen die Größe des volumdiamagnetischen Felds der Elektronenhüllen abgeschätzt werden. In Tabelle 2 wurden die Volumsuszeptibilitäten einiger diamagnetischer Flüssigkeiten zusammengestellt. Die Zahlenwerte sind sehr klein und unterscheiden sich untereinander maximal um $5 \cdot 10^{-7}$. Dem würde gemäß Gl. (25.7) für $q = 0$ eine relative Differenz $\Delta H'/H$ der inneren Feldstärken in den beiden verglichenen Substanzen von $1 \cdot 10^{-6}$ entsprechen, wenn die Probe zylindrisch ist, während in einer kugelförmigen Probe die volumdiamagnetische Feldstärke Null wäre. Es darf angenommen werden, daß die Voraussetzung $q = 0$ in rein diamagnetischen Substanzen erfüllt ist, da die Dipolmomente in diesem Fall wohl zu klein sind, um merklich untereinander wechselwirken zu können.

Der als nächstes zu besprechende Fall, einer aus lauter gleichen paramagnetischen Partikeln bestehenden Substanz spielt praktisch keine Rolle, da die von den paramagnetischen Elektronenhüllen in ihrem Innern, also an den Kernorten erzeugten Felder H'' im allgemeinen sehr viel stärker als das äußere Feld sind. Damit wird aber mindestens die Messung ungestörter Kernresonanzfrequenzen unmöglich.

Tabelle 2. *Volumdiamagnetische Suszeptibilitäten einiger Flüssigkeiten*[1].

Flüssigkeit		χ_{Vol}
HNO_3	Salpetersäure	$-0,700 \cdot 10^{-6}$
H_2O	Wasser	$-0,721 \cdot 10^{-6}$
$(C_2H_5)_2O$	Äthyläther	$-0,546 \cdot 10^{-6}$
$C_2H_4O_2$	Essigsäure	$-0,552 \cdot 10^{-6}$
C_6H_6	Benzol	$-0,636 \cdot 10^{-6}$
C_2H_6O	Äthylalkohol	$-0,587 \cdot 10^{-6}$
CCl_4	Tetrachlorkohlenstoff	$-0,685 \cdot 10^{-6}$
Cl_2	(flüssig)	$-0.887 \cdot 10^{-6}$

Sehr häufig wurden dagegen gemischte Substanzen untersucht, vor allem Lösungen paramagnetischer Ionen in einem aus diamagnetischen Molekülen bestehenden Lösungsmittel. Die Suszeptibilitäten solcher Lösungen wachsen proportional mit der Ionenkonzentration und liegen meistens zwischen 10^{-4} bis 10^{-6}. Sie dürfen also sicher nicht mehr vernachlässigt werden. Ferner ist der in (25.6) definierte Wechselwirkungsfaktor q solcher gemischter Lösungen im allgemeinen von Null verschieden. Zur experimentellen Prüfung des Ansatzes (25.7) hat Dickinson[1] die Resonanzfrequenzverschiebungen wäßriger paramagnetischer Lösungen in Abhängigkeit von der Suszeptibilität (Ionenkonzentration), in Funktion der Magnetfeldstärke und in Abhängigkeit von der Probenform gemessen. Er fand die in (25.7) vorausgesetzte Proportionalität zwischen der inneren Feldstärke, der äußeren Feldstärke und der Suszeptibilität bestätigt. Fig. 95 gibt mit verschiedenen Probenformen erhaltene Meßergebnisse der inneren Feldstärke H' in verschieden konzentrierten wäßrigen Eisen(II)-Chlorid-Lösungen wieder. Experimentell bestimmt wurden selbstverständlich bei all diesen Messungen die Differenzen der Resonanzfrequenzen der Protonen in den verschiedenen Proben.

Aus dem Verlauf der Geraden a in Fig. 95 kann man die Größe des Wechselwirkungsfaktors q ermitteln, da bei einer kugelförmigen Probe $H' = H_c'$ ist. In der kleinen Zeichnung innerhalb der Fig. 95 wurden die bei einer Konzentration von $2 \cdot 10^{21}$ Fe^{++}-Ionen/cm^3 gemessenen Feldstärken H' verschieden langer zylindrischer Proben über deren Zylinderlänge aufgetragen. Es zeigt sich, daß die so erhaltene Kurve mit zunehmender Zylinderlänge bei konstantem Zylinderradius schnell gegen einen Grenzwert, nämlich den des unendlich langen Zylinders konvergiert. Somit ist es berechtigt, die Probe von 5,4 mm Durchmesser und 100 mm Länge als ausreichende Approximation eines unendlich langen Zylinders

---

[1] W. C. Dickinson: Phys. Rev. **81**, 717 (1951).

anzusehen. Dessen Entmagnetisierungsfaktor $(\alpha = 2\pi)$ kennt man aber und kann somit auch aus dem Verlauf der Geraden e in Fig. 95 den Wechselwirkungsfaktor q errechnen. Der so erhaltene Wert $(q = 1,1)$ stimmt mit dem der kugelförmigen Probe in diesem Fall besonders gut überein. Dieses Ergebnis rechtfertigt die in (25.2) und (25.6) vorgenommene Einführung zweier Konstanten, deren eine substanzunabhängig sein soll, während die andere formunabhängig sein soll. In Tabelle 3 sind weitere Meßergebnisse von q-Werten zusammengestellt. Nach Ansicht von DICKINSON sind die dort zum Teil zu bemerkenden Unterschiede zwischen den q-Werten kugelförmiger und unendlich langer zylindrischer Proben als experimentelle Meßfehler zu erklären. Dabei dürften die mit kugelförmigen Proben gemessenen q-Werte auf Grund des Meniskuseffekts unzuverlässiger sein.

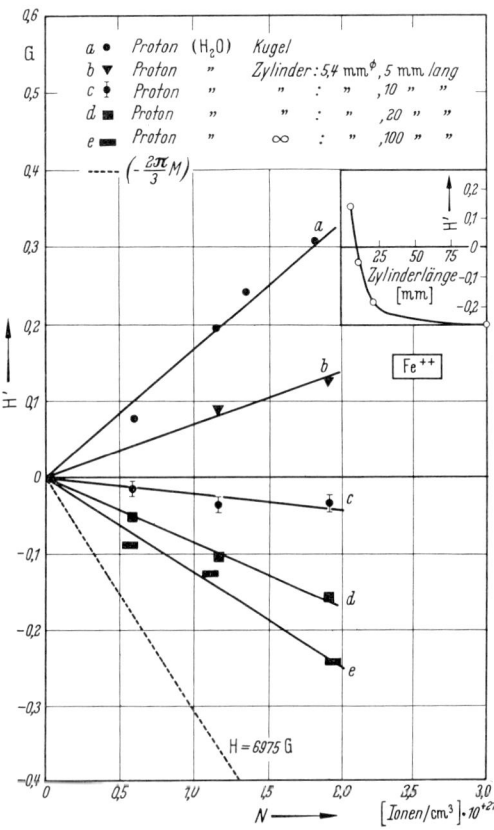

Die experimentelle Erfahrung zeigt (vgl. Tabelle 3), daß die Größe des Wechselwirkungseffekts nicht nur von der paramagnetischen Ionenart, sondern auch von der im Resonanzexperiment beobachteten Kernart, sowie von deren chemischer Bindungsart abhängt. Eine Sonderstellung nimmt der Resonanzfrequenzvergleich von Isotopen in der gleichen Molekülbindung ein. Die q-Werte von Isotopen sind, wie experimentell nachgewiesen wurde, untereinander gleich. Da in diesem Fall außerdem auch die hüllendiamagnetischen Korrekturen, bzw. allgemeiner die inneren Felder H'' übereinstimmen, wird in diesem Fall die experimentell erreichbare hohe Meßgenauigkeit voll ausnutzbar.

Fig. 95. Zur Abhängigkeit der inneren Feldstärke H' von der Probenform. Gemessen wurden die Differenzen der Protonen-Resonanzfrequenzen in verschieden konzentrierten wäßrigen Lösungen von Eisen(II)Chlorid in Proben verschiedener Formen und Abmessungen. In der eingefügten kleinen Zeichnung wurden die durch Extrapolation aus den Kurven b, c, d und e folgenden Größen des inneren Felds H' für zylindrische Proben verschiedener Längen aber gleicher Ionen-Konzentration ($2 \cdot 10^{21}$ Fe++/cm³) aufgetragen.

Bemerkenswert ist, daß der q-Faktor auch negative Werte haben kann (vgl. Tabelle 3; Zeile 1, 10 und 16). Jeder Deutungsversuch des Phänomens der Wechselwirkungsfelder muß dies erklären können. Solche Versuche sind von DICKINSON und BLOEMBERGEN[1], von PURCELL sowie von GRIVET und AYANT[2,3] unternommen worden. Es liegt nahe, die Abweichungen von der ursprünglichen einfachen DEBYE-Theorie $(q = 0)$ ebenso zu erklären, wie man es im Fall elektrischer Dipole mit Erfolg tut, nämlich durch die Annahme einer zeitweisen Bildung von Nahordnungen, deren gemeinsames Moment sich wesentlich von dem der

[1] N. BLOEMBERGEN u. W. C. DICKINSON: Phys. Rev. 79, 179 (1950).
[2] P. GRIVET u. Y. AYANT: C. R. Acad. Sci., Paris 232, 1094 (1951).
[3] Y. AYANT: C. R. Acad. Sci., Paris 232, 1203, 1298 (1951).

Summe der Partner solcher Gebilde unterscheidet. Jedoch befriedigt eine analoge Erklärung im magnetischen Fall weniger, da ein solcher Assoziationseffekt praktisch eine Vorstufe des Ferromagnetismus darstellen würde, dies ist aber auf Grund der großen magnetischen Verdünnung der Lösungen unwahrscheinlich. PURCELL hat zur Deutung des Wechselwirkungsfelds vorgeschlagen, einen Elektronenaustauschprozeß zwischen den paramagnetischen Ionen und den übrigen normalerweise diamagnetischen Ionen oder Molekülen, innerhalb deren Hüllen sich die zu beobachtenden Kerne befinden, anzunehmen. Auf Grund dieses Prozesses sollen die diamagnetischen Partikel zeitweise paramagnetisch

Tabelle 3. *Experimentelle Werte des Wechselwirkungsfaktors q.* w. L. = wäßrige Lösung[1].

Paramagnetisches Ion	Beobachtete Kernart	Chemische Verbindung	q-Wert	
			Zylinderprobe (∞-lang)	Kugelprobe
$Ni^{++}(Cl_2)$	H^1	H_2O	$-1,0$	$-0,1$
	F^{19}	SbF_3 w. L.	$2,3$	$2,2$
	Li^7	$LiCl$ w. L.	$0,9$	
$Ni^{++}(SO_4)$	H^1	H_2O	$0,0$	$0,0$
	F^{19}	SbF_3 w. L.	$3,5$	$2,5$
	Li^7	$LiCl$ w. L.	$0,3$	
$Co^{++}(Cl_2)$	H^1	H_2O	$0,7$	$0,9$
	F^{19}	SbF_3 w. L.	$3,1$	$2,3$
	F^{19}	HF w. L.	$19,0$	
$Cu^{++}(Cl_2)$	H^1	H_2O	$-1,8$	$-1,5$
$Fe^{++}(Cl_2)$	H^1	H_2O	$1,1$	$1,1$
	D^2	D_2O	$1,2$	
	F^{19}	SbF_3 w. L.	$5,1$	$4,0$
	Li^7	$LiCl$ w. L.	$0,0$	
$Mn^{++}(Cl_2)$	Li^7	$LiCl$ w. L.	$0,5$	
$Cr^{+++}(Cl_3)$	H^1	H_2O	$-3,0$	
	F^{19}	HF w. L.	$6,0$	
$Er^{+++}(Cl_3)$	H^1	H_2O	$1,1$	

sein. Austauschvorgänge dieser Art müßten mit einer sehr viel höheren Frequenz als die LARMOR-Frequenz der Kerne erfolgen, da sie sich sonst auch durch eine Resonanzlinienverbreiterung bemerkbar gemacht hätten. Zwar erklärt diese Deutung die Erfahrung (vgl. Tabelle 3), daß die q-Werte negativer Ionen, z.B. der Fluor-Ionen, besonders groß sind — die elektrostatische Anziehung zwischen den positiven paramagnetischen Ionen und den negativen Fluor-Ionen könnte den Elektronenaustauschprozeß erleichtern —, jedoch gibt sie keinen Hinweis zur Erklärung der negativen q-Werte. Die übrigen genannten Autoren haben den Einfluß der elektrischen Felder innerhalb der Substanz, sowie das anisotrope Verhalten der paramagnetischen Ionen in Gegenwart stark schwankender elektrischer Felder diskutiert. Vermutlich ist die endgültige Lösung des Problems der Wechselwirkungsfelder in dieser Richtung zu suchen.

26. Abhängigkeit der Kernresonanzfrequenz von der mikroskopischen Abschirmung der Elektronenhülle. Innerhalb jedes Atoms oder Moleküls überlagert sich dem äußeren Feld H und der Summe H' der Felder aller übrigen kern- oder hüllenmagnetischen Partikel der Substanz als drittes Feld H'' das innere Feld der eigenen Elektronenhülle. Die innere Feldstärke H'' am Kernort läßt sich leicht berechnen, wenn man voraussetzt, daß die betrachtete Elektronenhülle in einem äußeren Feld rein diamagnetisch reagiert, und wenn man ferner

---

[1] Nach W. C. DICKINSON: Phys. Rev. **81**, 717 (1951).

annimmt, daß die Ladungsverteilung um den Kernmittelpunkt kugelsymmetrisch ist. Praktisch bedeutet dies, daß der Kern in einem freien Atom oder in einem einatomigen Ion enthalten sein soll, dessen Elektronenhülle sich in einem S-Zustand befindet. Dieser Fall ist von LAMB[1] erstmals behandelt worden.

Bringt man ein solches Atom in ein Magnetfeld der Stärke H oder schaltet man an seinem Ort ein Feld der Stärke H ein (H kann innerhalb des vorliegenden Problemkreises unbedenklich mit $H_i = H + H'$ gleichgesetzt werden), so werden in der den Kern umgebenden Ladungshülle diamagnetische Ströme induziert, deren Magnetfeld das äußere Feld schwächt. Die gyromagnetischen Verhältnisse der freien Kerne sind aus diesem Grunde größer als die aus der Beziehung $\gamma = \omega/H$ experimentell erhaltenen Werte. Zur Berechnung der diamagnetischen Stromdichte in der Elektronenhülle betrachtet man zweckmäßigerweise ein ringförmiges Volumelement, dessen Ebene orthogonal zur Richtung des äußeren Feldes ist und dessen Mittelpunkt auf der durch den Kernmittelpunkt gehenden Feldlinie liegt. An einer in diesem Volumelement enthaltenen Leiterschleife läge die Spannung

$$U = -\frac{d\Phi}{dt} = -\pi a^2 \frac{dH}{dt}, \tag{26.1}$$

wenn man den magnetischen Fluß durch den Ring Φ nennt und den Ringradius mit a bezeichnet. Entlang dieser Ringbahn wirkt auf geladene Partikel die Feldstärke

$$E = \frac{U}{2\pi a} = -\frac{1}{2} a \frac{dH}{dt}. \tag{26.2}$$

Da wir annehmen können, daß sich die Bahnen der Hüllenelektronen durch das Einschalten des Feldes H nicht ändern, werden die Elektronen durch das elektrische Feld E nur beschleunigt oder verzögert, je nach ihrem Umlaufssinn bezüglich der Feldrichtung und zwar gilt

$$-\frac{e}{c} \cdot E = \pm m \frac{dv}{dt}. \tag{26.3}$$

Faßt man die Ansätze (26.2) und (26.3) zusammen, so folgt für die positive bzw. negative zusätzliche Geschwindigkeit aus $\frac{dv}{dt} = \pm \frac{1}{2} \frac{e \cdot a}{m \cdot c} \frac{dH}{dt}$ die Beziehung

$$v = \pm \frac{1}{2} \cdot \frac{e \cdot a}{m \cdot c} \cdot H. \tag{26.4}$$

Die diamagnetisch induzierte Stromdichte s ist an jedem Ort gleich dem Produkt dieser Geschwindigkeit mit der laut Voraussetzung kugelsymmetrischen, also nur vom Radius abhängigen Ladungsdichte $\varrho(r)$; es ist also

$$s = \frac{1}{2} \frac{e\,\varrho(r)}{m \cdot c} \cdot a\,H. \tag{26.5}$$

Zur Berechnung des Feldes dieser Stromverteilung am Kernort geht man am besten davon aus, daß für die Feldstärke eines vom Strom J (gemessen in elektrostatischen Einheiten) durchflossenen Leiterelementes der Länge dl an einem Aufpunkt im Abstand r, dessen Verbindungslinie orthogonal zu dem Leiterelement ist, die BIOT-SAVARTsche Beziehung

$$dH = \frac{J}{c \cdot r^2}\,dl \tag{26.6}$$

gilt[2]. Bezeichnet man den Winkel, unter dem man vom Kernmittelpunkt aus den Halbmesser a eines Stromringes sieht, mit ψ, so ist die Entfernung eines

---

[1] W. E. LAMB jr.: Phys. Rev. 60, 817 (1941).
[2] P. HERTZ: Magnetische Felder von Strömen. In GEIGER-SCHEEL, Handbuch der Physik, Bd. XV. Berlin: Springer 1927.

Ringelementes dl vom Kern $r = a/\sin\psi$. Für die Feldstärke dieses Elementes am Kernort gilt also

$$dH = \frac{J}{c} \cdot \frac{\sin^2\psi}{a^2} \, dl \qquad (26.7)$$

und für deren zur Richtung des äußeren Feldes parallele Komponente

$$dH_\| = \frac{J}{c} \cdot \frac{\sin^3\psi}{a^2} \, dl. \qquad (26.8)$$

Somit ist die Feldstärke eines ganzen Ringes

$$dH = \frac{2\pi}{c\,a} \sin^3\psi \cdot s \cdot dF, \qquad (26.9)$$

wenn man den Ringquerschnitt dF nennt. Durch Integration über alle Ringelemente $dF = r\,d\psi\,dr$ erhält man für das diamagnetische Feld am Kernort die Beziehung

$$H'' = \frac{\pi \cdot e}{m \cdot c^2} \cdot H \int\limits_0^\pi \sin^3\psi \, d\psi \int\limits_0^\infty \varrho(r)\, r\, dr = \frac{e}{3\, m c^2} H \int\limits_0^\infty \frac{\varrho(r)}{r} 4\pi\, r^2\, dr. \qquad (26.10)$$

Das letzte Integral über r ist gleich dem elektrostatischen Potential der Ladungsverteilung um den Kern in dessen Mittelpunkt. Bezeichnet man dieses Potential mit $\varphi(0)$, so kann man also statt (26.10) auch schreiben

$$H'' = \frac{e}{3\,m c^2} \cdot \varphi(0)\, H. \qquad (26.11)$$

Um diese Gleichung zur Bestimmung von Korrekturen verwenden zu können, muß man das Potential $\varphi(0)$, d.h. aber die Ladungsverteilung $\varrho(r)$ kennen, man muß somit im weiteren spezielle Atommodelle betrachten.

Legt man der Betrachtung das Modell von FERMI-THOMAS zugrunde, wie es LAMB zunächst getan hat, so kann man formal stets schreiben $\varphi(0) = -Z e\left(\dfrac{1}{r}\right)$, wobei unter $\left(\dfrac{1}{r}\right) = \dfrac{1}{r_{\text{eff}}}$ ein gemittelter Wert der reziproken Ladungsabstände vom Kern und unter Z die Ordnungszahl des Atoms verstanden werden soll. Beim Wasserstoffatom ist die Annahme naheliegend, daß $r_{\text{eff}} \approx a_0$ dem Radius der ersten Bohrschen Bahn sein soll.

Vergleicht man weiter das mit negativer Ladung gefüllte Volumen eines Atoms der Ordnungszahl Z mit dem entsprechenden Volumen des Wasserstoffatoms, so kommt man zu dem Schluß, daß der effektive Radius eines solchen Atoms von der Größenordnung $a_0/Z^{\frac{1}{3}}$ sein sollte. Damit folgt aus (26.11) die Beziehung

$$H'' = -\gamma\, \frac{e^2}{m c^2 a_0} Z^{\frac{4}{3}} H = -\gamma\, \alpha^2 Z^{\frac{4}{3}} H, \qquad (26.12)$$

in der $\alpha = \dfrac{e^2}{\hbar \cdot c}$ die Sommerfeldsche Feinstrukturkonstante und γ eine noch genauer zu berechnende Konstante von der Größenordnung eins ist. Auf Grund des Fermi-Thomas-Atommodells findet man für γ den Zahlenwert 0,598, vgl. hierzu[1-3]. Somit ist in der Näherung dieses Modells die diamagnetische Feldstärke am Kernort

$$H'' = -0{,}319 \cdot 10^{-4} \cdot Z^{\frac{4}{3}} \cdot H = -\sigma H. \qquad (26.13)$$

[1] W. E. LAMB jr.: Phys. Rev. **60**, 817 (1941).
[2] E. CONDON u. G. SHORTHLEY: Theory of Atomic Spectra, S. 336. Cambridge 1935.
[3] L. BRILLOUIN: L'Atome de Thomas-Fermi. Paris: Herrmann 1934.

Die mit σ bezeichnete Größe wird diamagnetische Abschirmungskonstante genannt.

Einen verbesserten Ansatz erhält man, wenn man der Betrachtung das Hartree- bzw. das Hartree-Fock-Atommodell zugrunde legt. Für eine ganze Anzahl von Atomen wurden die aus diesem Modell folgenden elektrostatischen Potentiale am Kernort berechnet. DICKINSON [1] hat ein in Tabelle 4 wieder-

Tabelle 4. *Werte des Potentials* $-\varphi'(0)$ *berechnet aus Hartree- und Hartree-Fock-Wellengleichungen.*

Ord-nungs-zahl	$-\varphi'(0)$ in atomaren Einheiten		Zitat	Ord-nungs-zahl	$-\varphi'(0)$ in atomaren Einheiten		Zitat
	a) berechnet ohne Austauschterme (Hartree-Näherung)	b) berechnet mit Austauschtermen (Hartree-Fock-Näherung)			a) berechnet ohne Austauschterme (Hartree-Näherung)	b) berechnet mit Austauschtermen (Hartree-Fock-Näherung)	
1	1,000	1,000	a a	18	68,89	69,67	q q
2^+	2,000	2,000	a a	19^+	73,91	74,67	q q
2	3,37		b	19		74,90	o
3^+		5,369	c	19^-		75,02	o
3	5,463	5,714	d c	20^+		79,95	r
4	8,365	8,410	e f	20	79,3	80,20	e r
5		11,23	g	26	114,0		s
6	14,51	14,69	h i	29^+	133,5	135,1	t t
7		18,32	j	30	139,3		u
7^-		18,70	j	31^+	145,2		u
8^+	21,373	21,61	k l	31	146,5		u
8	21,944	22,26	k l	32	153,7		v
8^-		22,72	l	33^+	160,0		u
9		26,12	m	33	160,3		u
9^-	26,5	26,56	n m	37^+	187,5		n
10		30,811	m	47^+	262		w
11^+	34,8	35,13	b j	55^+	323		x
11		35,43	j	74	490		y
11^-		35,57	o	80	543		e
17^-	63,88	64,67	p p				

a) Exakte Wellengleichung.
b) D. R. HARTREE: Proc. Cambridge Phil. Soc. **24**, 111 (1927/28).
c) V. FOCK u. M. J. PETRASHEN: Phys. Z. Sowjet. **8**, 547 (1935).
d) J. HARGREAVES: Proc. Cambridge Phil. Soc. **25**, 75 (1928/29).
e) D. R. u. W. HARTREE: Proc. Roy. Soc. Lond., Ser. A **149**, 210 (1935)
f) D. R. u. W. HARTREE: Proc. Roy. Soc. Lond., Ser. A **150**, 9 (1935).
g) BROWN, BARTLETT u. DUNN: Phys. Rev. **44**, 296 (1933).
h) C. C. TORRANCE: Phys. Rev. **46**, 388 (1934).
i) A. JUCYS: Proc. Roy. Soc. Lond., Ser. A **173**, 59 (1939).
j) D. R. u. W. HARTREE: Proc. Roy. Soc. Lond., Ser. A **193**, 299 (1948).
k) D. R. HARTREE u. M. M. BLACK: Proc. Roy. Soc. Lond., Ser. A **139**, 311 (1933).
l) D. R. HARTREE u. B. SWIRLES: Phil. Trans. Roy. Soc. Lond., Ser. A **238**, 229 (1939/40)
m) F. W. BROWN: Phys. Rev. **44**, 214 (1933).
n) D. R. HARTREE: Proc. Roy. Soc. Lond., Ser. A **151**, 96 (1935).
o) D. R. u. W. HARTREE: Proc. Cambridge Phil. Soc. **34**, 550 (1938).
p) D. R. u. W. HARTREE: Proc. Roy. Soc. Lond., Ser. A **156**, 45 (1936).
q) D. R. u. W. HARTREE: Proc. Roy. Soc. Lond., Ser. A **166**, 450 (1938).
r) D. R. u. W. HARTREE: Proc. Roy. Soc. Lond., Ser. A **164**, 167 (1938).
s) M. F. MANNING u. L. GOLDBERG: Phys. Rev. **53**, 662 (1938).
t) D. R. u. W. HARTREE: Proc. Roy. Soc. Lond., Ser. A **157**, 490 (1936).
u) W. u. D. R. HARTREE u. M. F. MANNING: Phys. Rev. **59**, 299 (1941).
v) W. u. D. R. HARTREE u. M. F. MANNING: Phys. Rev. **59**, 306 (1941).
w) M. M. BLACK: Mem. Manchester Lit. Phil. Soc. **79**, 29 (1935).
x) D. R. HARTREE: Proc. Roy. Soc. Lond., Ser. A **143**, 506 (1933/34).
y) M. F. MANNING u. J. MILLMAN: Phys. Rev. **49**, 848 (1936).

[1] W. C. DICKINSON: Phys. Rev. **80**, 563 (1950).

gegebenes Verzeichnis der Ergebnisse dieser Rechnungen veröffentlicht. Die Potentiale φ werden in dieser Zusammenstellung in atomaren Einheiten angegeben (angedeutet durch einen Indexstrich). Die in der Tabelle enthaltenen Werte müssen demnach zur Berechnung der Korrekturen in die aus (26.11) folgende Gleichung

$$\frac{H''}{H} = \frac{1}{3} \cdot \frac{e}{m \cdot c^2} \, \varphi(0) = \frac{1}{3}\, \alpha^2\, \varphi'(0) \tag{26.14}$$

eingesetzt werden. Da genügend Potentiale direkt berechnet worden sind, kann man die Potentiale der übrigen Atome einfach durch Interpolieren zwischen den

Tabelle 5. *Diamagnetische Korrekturen neutraler Atome nach Dickinson.*

z	H''/H (%)	z	H''/H (%)	z	H''/H (%)	z	H''/H (%)	z	H''/H (%)	z	H''/H (%)
1	0,0018	17	0,115	33	0,285	49	0,491	65	0,724	81	0,982
2	0,0060	18	0,124	34	0,296	50	0,504	66	0,740	82	0,998
3	0,0101	19	0,133	35	0,308	51	0,517	67	0,756	83	1,01
4	0,0149	20	0,142	36	0,321	52	0,531	68	0,772	84	1,03
5	0,0199	21	0,151	37	0,333	53	0,545	69	0,788	85	1,05
6	0,0261	22	0,161	38	0,345	54	0,559	70	0,804	86	1,06
7	0,0325	23	0,171	39	0,358	55	0,573	71	0,820	87	1,08
8	0,0395	24	0,181	40	0,371	56	0,587	72	0,837	88	1,10
9	0,0464	25	0,191	41	0,384	57	0,602	73	0,853	89	1,11
10	0,0547	26	0,202	42	0,397	58	0,616	74	0,869	90	1,13
11	0,0629	27	0,214	43	0,411	59	0,631	75	0,885	91	1,15
12	0,0710	28	0,226	44	0,425	60	0,647	76	0,901	92	1,16
13	0,0795	29	0,238	45	0,438	61	0,662	77	0,917		
14	0,0881	30	0,249	46	0,452	62	0,678	78	0,933		
15	0,0970	31	0,261	47	0,465	63	0,693	79	0,949		
16	0,106	32	0,273	48	0,478	64	0,709	80	0,965		

Werten der Tabelle 4 ermitteln. Dies ist bei der Zusammenstellung der Tabelle 5 geschehen, in der die diamagnetischen Korrekturwerte für alle neutralen Atome enthalten sind. Zur Prüfung der Zuverlässigkeit dieser Werte hat Dickinson andere aus den Hartree-Fock-Näherungen folgende Aussagen mit den entsprechenden experimentellen Resultaten verglichen und kommt zu dem Schluß, daß die Korrekturen der Tabelle 5 bis auf etwa 5% richtig sein dürften, wenn man von den schwersten Atomen absieht.

Zur experimentellen Bestimmung von Korrekturdaten müßte man Kernresonanzexperimente mit der erforderlichen Präzision an total ionisierten Atomen ausführen können. Dies ist bisher in keinem Fall geschehen, eine direkte Bestätigung der Werte von Tabelle 5 liegt also nicht vor. Dagegen weisen zahlreiche Meßergebnisse[1-5] darauf hin, daß die voranstehende Behandlungsart des Problems der hülleninneren Felder zu einfach ist. Sieht man von den wenigen Fällen ab, in denen sich Kerne tatsächlich in nicht weiter gebundenen Atomhüllen befinden (z.B. in Ionen, in Flüssigkeiten oder in Edelgasatomen), so ist die Voraussetzung der kugelsymmetrischen Ladungsverteilung um den betrachteten Kern bei der Mehrzahl der üblicherweise als Probesubstanz zur Auswahl stehenden Molekülbindungsarten sicher nicht erfüllt. Die Meßergebnisse zeigen, daß sich die Resonanzfrequenzen derselben Kernart in verschiedenen Molekülbindungen

[1] W. D. Knight: Phys. Rev. 76, 1259 (1949).
[2] W. G. Proctor u. F. C. Yu: Phys. Rev. 77, 717 (1950).
[3] W. G. Proctor u. F. C. Yu: Phys. Rev. 81, 20 (1951).
[4] W. C. Dickinson: Phys. Rev. 77, 736 (1950).
[5] W. C. Dickinson: Phys. Rev. 81, 717 (1951).

und sogar innerhalb der gleichen Moleküle an verschiedenen Bindungsorten zum Teil beträchtlich voneinander unterscheiden. Dieser als chemische Resonanzfrequenzverschiebung (chemical shift) bezeichnete Effekt läßt sich bis jetzt nur in einigen Sonderfällen berechnen. Zwar hat RAMSEY[1-6] Verfahren angegeben, mit deren Hilfe sich prinzipiell alle chemischen Frequenzverschiebungen berechnen ließen, explizite Rechenergebnisse liegen jedoch nur für das allerdings praktisch auch besonders wichtige Wasserstoffmolekül H_2 vor[3], vgl. hierzu auch [7,8], sowie die Publikationen von HYLLERAAS und SKAVLEM[9] und von NEWELL[10].

RAMSEY fand, daß sich die Abschirmkorrektur allgemein aus zwei Termen zusammensetzt, deren erster, die sog. diamagnetiche Korrektur mit dem Ausdruck (26.10) der Lambschen Theorie übereinstimmt (abgesehen davon, daß das dortige Integral über das ganze, im allgemeinen nicht kugelsymmetrische Molekül erstreckt werden muß), während der zweite Term von einem induzierten Paramagnetismus zweiter Ordnung herrührt. Seine Berechnung setzt die Kenntnis der Wellenfunktionen der angeregten Zustände des Moleküls voraus; diese sind aber in der Regel nicht bekannt.

Das Problem der Resonanzverschiebungen ist mit der in den letzten Jahren erfolgten Genauigkeitssteigerung der kernmagnetischen Messungen sehr aktuell geworden. Die Genauigkeit, mit der die magnetischen Dipolmomente der Kerne angegeben werden können, ist gegenwärtig im wesentlichen durch die noch bestehenden Unklarheiten bezüglich der an den Meßergebnissen anzubringenden Korrekturen begrenzt. Beim Studium der in Abschnitt III zusammengestellten Ergebnisse der kernmagnetischen Messungen muß deshalb stets bedacht werden, daß die von den Autoren angegebenen sog. korrigierten μ-Werte noch systematische Korrekturfehler enthalten können. Die nach der Lambschen Theorie berechneten Korrekturen steigen von dem ungefähren Wert $2 \cdot 10^{-5}$ bei dem leichtesten, dem H-Atom, bis auf etwa $1 \cdot 10^{-2}$ bei den schwersten Elementen an. Glücklicherweise unterscheiden sich andererseits aber auch gerade bei den schwersten Elementen die atomaren Lambschen Korrekturen am wenigsten von den molekularen, da die bindenden Elektronen bei ihnen vom Kern weit entfernt sind und somit relativ wenig zum Wert des Integrals in (26.10) beitragen. Besonders groß sind die Unterschiede zwischen den molekularen und atomaren Korrekturen einerseits bei leichten Kernen — verstandlicherweise, denn durch die Bindung werden die räumlichen Verteilungen der Elektronenwolken um sie und damit die möglichen Bahnen der diamagnetisch induzierten Ströme besonders gründlich geändert —, andererseits aber auch bei Molekülen, in denen der Abstand zwischen ihrem Grundzustand und dem ersten angeregten Zustand verhältnismäßig klein ist. So fanden z.B. PROCTOR und YU[11], daß sich die Resonanzfrequenzen des Co^{59} in wäßrigen Lösungen von $K_3Co(CN)_6$ und $K_3Co(C_2O_4)_3$ um 1,3% unterscheiden. Trifft die Ramseysche Deutung zu, daß die anomal große

[1] N. F. RAMSEY: Phys. Rev. 77, 567 (1950).
[2] N. F. RAMSEY: Phys. Rev. 78, 339 (1950).
[3] N. F. RAMSEY: Phys. Rev. 78, 699 (1950).
[4] N. F. RAMSEY: Physica, Haag 17, 303 (1951).
[5] N. F. RAMSEY: Phys. Rev. 86, 243 (1952).
[6] N. F. RAMSEY: Nuclear Moments, S. 71. New York: John Wiley & Sons, Inc. 1953; London: Chapman & Hall, Limited.
[7] H. G. KOLSKY, T. E. PHIPPS jr., N. F. RAMSEY u. H. B. SILBSEE: Phys. Rev. 79, 883 (1950).
[8] H. G. KOLSKY, T. E. PHIPPS jr., N. F. RAMSEY u. H. B. SILBSEE: Phys. Rev. 87, 395 (1952).
[9] E. HYLLERAAS u. S. SKAVLEM: Phys. Rev. 79, 117 (1950).
[10] G. F. NEWELL: Phys. Rev. 80, 476 (1950).
[11] W. G. PROCTOR u. F. C. YU: Phys. Rev. 77, 717 (1950).

chemische Resonanzverschiebung zwischen diesen beiden Molekülen durch die Existenz von dem Grundzustand nahe benachbarten angeregten Energieniveaus der Elektronen bewirkt wird, so sollte diese Resonanzverschiebung eine leichte Temperaturabhängigkeit aufweisen und eine solche ist von den genannten Verfassern tatsächlich auch beobachtet worden. Besonders wichtig sind für die Deutung natürlich alle experimentellen Beweise, vgl. z.B.[1-3], daß der Betrag der chemischen Verschiebungen proportional mit der Feldstärke wächst, daß es sich also tatsächlich um eine durch das äußere Feld verursachte Änderung des Zustands der Elektronenhülle handelt.

Die von RAMSEY angegebene allgemeine Gleichung des auf dem induzierten Molekülparamagnetismus zweiter Ordnung beruhenden Teiles des Abschirmfeldes am Kernort ist recht kompliziert. Ihre Ableitung schließt an die unter anderem von VAN VLECK[4] dargestellte Theorie des Diamagnetismus und des Paramagnetismus zweiter Ordnung von Molekülen an. Für lineare Moleküle fand RAMSEY eine einfachere Berechnungsart dieses Feldes, die einen Vergleich mit den von BROOKS[5] und WICK[6] angegebenen Theorien der Spin-Rotations-magnetischen Wechselwirkungen erlaubt. Damit wird es möglich, experimentelle Meßergebnisse auf diesem Gebiet bei der Berechnung des zweiten Terms des Abschirmfeldes zu verwenden. Speziell für das H_2-Molekül erhält man auf diesem Wege die Beziehung

$$\sigma^{H_2} = \sigma_d + \sigma_p = \frac{e^2}{3\,m\,c^2} \int\limits_{(\infty)} \frac{\varrho}{r}\,dv - \frac{\alpha^2\,a_0\,a^2}{6\,\mu_{KM}}\left(\frac{2\,\mu_{KM}}{a^3} - \frac{A\,H_R}{J\,M}\right). \tag{26.15}$$

Darin ist das erste Glied das verallgemeinerte Volumintegral von (26.10). In dem zweiten Glied ist α wiederum die Sommerfeldsche Feinstrukturkonstante und a_0 der Radius der ersten Bohrschen Bahn; a ist der Abstand der beiden Protonen im H_2-Molekül, $\mu_{KM} = e\hbar/2\,m\,c$ das Kernmagneton, M die Protonenmasse, A die reduzierte Molekülmasse ($A\,a^2$ ist also gleich dem Trägheitsmoment des Moleküls), J die Rotationsquantenzahl des Moleküls und H_R das durch die mit der Kreisfrequenz $J\hbar/A\,a^2$ erfolgende Rotation des Moleküls am Kernort erzeugte Magnetfeld. Der Quotient $A\,H_R/J\,M$ wurde von KELLOGG, RABI, RAMSEY und ZACHARIAS[7] in einem Molekularstrahlexperiment gemessen, und zwar fanden sie für das H_2-Molekül den numerischen Wert (in der Originalarbeit wird H_R mit H' bezeichnet):

$$A\,H_R/J\,M = (13{,}66 \pm 0{,}20)\ \text{Gauß}. \tag{26.16}$$

Damit folgt für σ_p

$$\left.\begin{aligned} \sigma_p &= -\frac{0{,}5291 \cdot 0{,}7414^2}{6 \cdot 137^2 \cdot 5{,}05}\,(24{,}62 - 13{,}66) \\ &= -(0{,}56 \pm 0{,}01) \cdot 10^{-5}. \end{aligned}\right\} \tag{26.17}$$

Für das erste Glied von (26.15) hat ANDERSON[8] unter Verwendung der von NORDSIECK[9] angegebenen Wellengleichung des H_2-Moleküls den Wert $\sigma_d = 3{,}2 \cdot 10^{-5}$ berechnet. Auf Grund anderer etwas abweichender Wellengleichungen

[1] W. C. PROCTOR u. F. C. YU: Phys. Rev. 77, 717 (1950).
[2] W. G. PROCTOR u. F. C. YU: Phys. Rev. 81, 20 (1951).
[3] W. C. DICKINSON: Phys. Rev. 81, 717 (1951).
[4] J. H. VAN VLECK: Electric and Magnetic Susceptibilities. London: Oxford University Press 1932.
[5] H. BROOKS: Phys. Rev. 59, 925 (1941); 60, 168 (1941).
[6] G. C. WICK: Phys. Rev. 73, 51 (1948).
[7] J. M. B. KELLOGG, I. I. RABI, N. F. RAMSEY u. J. R. ZACHARIAS: Phys. Rev. 56, 728 (1939).
[8] H. L. ANDERSON: Phys. Rev. 76, 1460 (1949).
[9] A. NORDSIECK: Phys. Rev. 58, 310 (1940).

fanden HYLLERAAS und SKAVLEM[1] den wenig abweichenden Zahlenwert $\sigma_d = 3,16 \cdot 10^{-5}$. Benutzt man den erstgenannten Wert, so erhält man zusammenfassend für die Abschirmkonstante des H_2-Moleküls das Ergebnis

$$\sigma^{H_2} = 2,6_8 \cdot 10^{-5}. \tag{26.18}$$

Auf Grund dieses theoretischen Ergebnisses kann man die Abschirmkonstanten der Protonen in allen Verbindungen und an allen Verbindungsorten experimentell bestimmen, wenn man die Resonanzfrequenzen der Protonen in diesen Verbindungen mit der Resonanzfrequenz der Protonen im H_2-Molekül vergleicht. Für die häufig in Standardproben verwandten Substanzen Wasser und Mineralöl haben dies z.B. THOMAS[2] sowie GUTOWSKY und McCLURE[3] getan.

27. Struktur der Spektren von Flüssigkeiten und Gasen. Während in den letzten beiden Ziffern die Resonanzfrequenz-Änderungen nach ihren Ursachen gegliedert behandelt wurden, soll die Diskussion der kernmagnetischen Spektren in dieser und der folgenden Ziffer von dem Aggregatzustand der untersuchten Substanz ausgehen. Im Prinzip sind beide Arten der Gliederung des Stoffgebiets gleichwertig. Bei der ersten Einteilungsart geht man von der Deutung der experimentellen Ergebnisse aus und ordnet demgemäß die Effekte, welche den Resonanzfrequenz-Änderungen zugrunde liegen. Die zweite Einteilungsart folgt der unmittelbaren Erfahrung des Experimentalphysikers, daß die Struktur aller kernmagnetischen Spektren charakteristisch vom Aggregatzustand der Probe abhängt. In Gasen und Flüssigkeiten sind die Resonanzlinien scharf, d.h. schmal und hoch. Die Atome oder Moleküle der Proben verfügen in diesen Zuständen in der Regel über eine so große Translations- und Rotations-Bewegungsfreiheit, daß sich die Felder ihrer magnetischen Dipole, unter anderen auch diejenigen der beobachteten Kernmagnete selbst, ausreichend schnell zeitlich ausmitteln. In Festkörpern stehen der BROWNschen Temperaturbewegung der Substanzpartikel praktisch nur die Schwingungs-Freiheitsgrade zur Verfügung. Infolge dieser stark verminderten Beweglichkeit können die Dipole fester Proben untereinander auch statisch wechselwirken. Ihre dem äußeren Feld überlagerten zeitlich mittleren Felder sind aber recht inhomogen und verursachen dementsprechend eine starke Verbreiterung der Resonanzlinien. Feinere Resonanzverschiebungs- und Aufspaltungseffekte, wie sie in der letzten und in der vorliegenden Ziffer behandelt werden, können infolgedessen an festen Proben nicht beobachtet werden, obwohl sie in diesen ebenso wie in flüssigen und gasförmigen Proben vorhanden sind.

Sieht man von makroskopischen Effekten (vgl. Ziff. 25) ab, so sind für die Aufspaltung der Resonanzfrequenzen fast aller Kernarten (abgesehen von denen der Edelgase) in linienreiche Spektren zwei Effekte verantwortlich, die in der letzten Ziffer beschriebene Abschirmungs-Resonanzverschiebung (chemical shift) und die sog. Spin-Spin-Wechselwirkung. Gemäß dem ersten bereits besprochenen Effekt (Ziff. 26) sind die Resonanzfrequenzen an verschiedenen, nicht äquivalenten Bindungsorten voneinander verschieden, und zwar hängt der Betrag und das Vorzeichen dieser Verschiebung bezüglich der reinen Kernresonanzfrequenz von der Beschaffenheit der Elektronenumgebung des Bindungsorts ab, außerdem wächst ihr Betrag mit der äußeren Feldstärke. Hätten alle nicht makroskopisch erklärbaren Kernresonanzfrequenz-Änderungen nur diese eine Ursache, so dürften z.B. in Ammoniak (NH_3) und in Wasser (H_2O) die Protonen

[1] E. HYLLERAAS u. S. SKAVLEM: Phys. Rev. **79**, 117 (1950); vgl. jedoch die Berechnung einer Korrektur 2. Ordnung in dieser Arbeit.
[2] H. A. THOMAS: Phys. Rev. **80**, 901 (1950).
[3] H. S. GUTOWSKY u. R. E. McCLURE: Phys. Rev. **81**, 276 (1951).

jeweils nur *eine* Resonanzfrequenz besitzen, denn aus naheliegenden Symmetrie-vorstellungen folgt, daß die verschiedenen Bindungsorte der Protonen in diesen Molekülen äquivalent sein müssen. Erst recht sollte das Kernresonanz-Spektrum aller Moleküle, in denen überhaupt nur ein Kern der untersuchten Art enthalten ist, aus genau einer Linie bestehen. Experimentell hat sich dies jedoch nicht bestätigt. Fig. 96 zeigt als typisches Gegenbeispiel die Dublettstruktur der Fluorresonanz des Moleküls Fluorphosphoroxydichlorid ($POCl_2F$). Obwohl nur ein Fluorkern in dieser Verbindung enthalten ist, fanden Gutowsky, McCall und Slichter[1] zwei Fluor-Resonanzlinien gleicher Intensität in einem Abstand von ungefähr 0,3 Gauß. Außer dem F^{19}-Kern hat auch der in der Verbindung in der Regel enthaltene Phosphorkern P^{31} einen endlichen Spin ($I = \frac{1}{2}$), während der Drehimpuls der beiden anderen Kerne im allgemeinen Null ist. Experimentell erweist sich, daß das Kernresonanz-Spektrum des Phosphors in

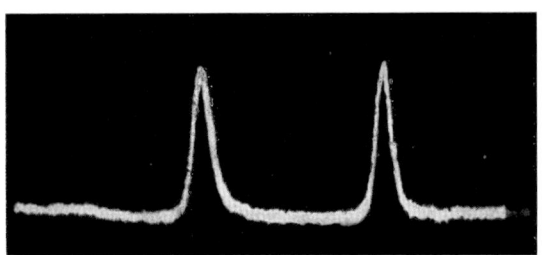

Fig. 96. Oszillographische Registrierung der Dublettstruktur der Fluor-resonanz des Moleküls $POCl_2F$. Die Resonanzkurve wurde in einem Magnetfeld von 6365 Gauß beobachtet. Der von der Gleichfeldstärke unabhängige Linienabstand des Dubletts beträgt 0,294 Gauß.

der genannten Verbindung ebenfalls zwei Linien hat. Dagegen besteht das Phosphorspektrum der ähnlichen Verbindung Difluorphosphoroxyfluorid ($POClF_2$) aus drei Resonanzkurven im gleichen Abstand, und zwar umschließt deren mittlere eine doppelt so große Fläche wie die beiden äußeren. Das Fluorspektrum dieses Moleküls hat wiederum zwei Linien gleicher Intensität.

Die Existenz des Struktureffekts, dessen Merkmale soeben beschrieben werden, wurde zuerst von Proctor und Yu entdeckt[2,3]. Ordnet man das experimentelle Material der genannten und späterer Autoren[1, 4-10], so findet man empirisch eine Anzahl allgemein gültiger Regeln bezüglich des Vorkommens der Linienaufspaltung, über die Abstände und die Zahl der Linien und bezüglich der Intensitätsverteilung in den Liniengruppen.

Beobachtet wird der Effekt nur, wenn in der als Probesubstanz verwendeten Molekülsorte außer der beobachteten Kernart noch mindestens eine zweite Kernart enthalten ist, deren Spin und deren magnetisches Moment von Null verschieden ist, oder wenn die beobachteten Kerne in der als Probesubstanz verwendeten Molekülart mindestens zwei nicht äquivalente Bindungsorte besitzen. In den Wasserarten H_2O^{16} oder D_2O^{16} gibt es also z.B. keine Linienaufspaltung, wohl aber in HDO^{16} und in H_2O^{17}.

Die Größe der Linienabstände erweist sich experimentell — ganz im Gegensatz zu den Erfahrungen über die Abschirmungs-Resonanzverschiebungen — als unabhängig vom Betrag der äußeren Feldstärke. Besonders dieses Ergebnis

---

[1] H. S. Gutowsky, D. W. McCall u. C. P. Slichter: J. Chem. Phys. **21**, 279 (1953).
[2] W. G. Proctor u. F. C. Yu: Phys. Rev. **78**, 471 (1950).
[3] W. G. Proctor u. F. C. Yu: Phys. Rev. **81**, 20 (1951).
[4] S. S. Dharmatti u. H. E. Weaver: Phys. Rev. **87**, 675 (1952).
[5] H. S. Gutowsky u. D. W. McCall: Phys. Rev. **82**, 748 (1951).
[6] H. S. Gutowsky, D. W. McCall u. C. P. Slichter: Phys. Rev. **84**, 589 (1951).
[7] E. L. Hahn u. D. E. Maxwell: Phys. Rev. **84**, 1246 (1951).
[8] E. L. Hahn u. D. E. Maxwell: Phys. Rev. **88**, 1070 (1952).
[9] E. B. McNeil, C. P. Slichter u. H. S. Gutowsky: Phys. Rev. **84**, 1245 (1951).
[10] W. E. Quinn u. R. M. Brown: J. Chem. Phys. **21**, 1605 (1953).

ermöglicht es, die beiden zur Diskussion stehenden Effekte meßtechnisch leicht und zuverlässig zu unterscheiden. Untersucht man die Größe der Linienabstände in Gruppen chemisch sehr ähnlicher Moleküle, in denen also etwa die Atome mit den beobachteten Kernen oder eines von deren Nachbaratomen gegen Atome derselben Wertigkeit (z.B. aus derselben Spalte des periodischen Systems der Elemente) ausgetauscht wurden, oder mißt die Resonanzaufspaltung gar in chemisch identischen Molekülen, in denen nur verschiedene Isotope der beobachteten Kernart oder derer Nachbarn enthalten sind, so findet man, daß die den Linienabständen entsprechenden Energieniveau-Differenzen proportional mit den Beträgen der magnetischen Momente der beobachteten Kerne und der nichtidentischen Nachbarkerne wachsen. Sind speziell die beobachteten Kerne und die Nachbarkerne identisch, jedoch nicht äquivalent gebunden, so wächst der Linienabstand, übereinstimmend mit dem voranstehenden Ergebnis proportional mit dem Quadrat der Kerndipolmomente. Andere Versuche zeigen, daß die Linienabstände um so kleiner werden, je weiter die Kerne, welche ihre Resonanzfrequenzen gegenseitig beeinflussen, in den Molekülen voneinander entfernt sind.

Für die Zahl der Linien findet man folgende Regeln: Haben die beobachteten Kerne nur einen Nachbarkern, dessen magnetomechanische Momente von Null verschieden sind, so stimmt die Linienzahl mit der Zahl der möglichen Einstellungen des Spins I dieses Kerns im äußeren Feld überein, beträgt also $2I+1$, unabhängig davon, ob der Nachbarkern nicht identisch oder identisch, aber nicht äquivalent gebunden ist. Im HD-Molekül, dessen Kernresonanzspektrum zur quantitativen Prüfung der theoretischen Ansätze besonders ausführlich untersucht wurde [1-3], hat z.B. die Protonenresonanz Triplettstruktur ($I_D = 1$) und die Deuteronenresonanz ($I_H = \frac{1}{2}$) Dublettstruktur. Haben mehrere Nachbarkerne von Null verschiedene magnetomechanische Momente, dann muß man zur richtigen Vorhersage der experimentellen Resultate den Fall der untereinander identischen und äquivalent gebundenen Nachbarn von allen übrigen Möglichkeiten unterscheiden. Für Nachbarkerne, welche beide Bedingungen erfüllen, gilt die empirische Regel, daß ihr resultierender Spin für die Linienzahl maßgebend ist, n solcher Nachbarn vom Spin I erzeugen also eine $(2nI+1)$-fache Resonanzaufspaltung. In allen anderen Fällen überlagern sich dagegen die Resonanzaufspaltungen, m nicht identische oder nicht äquivalent gebundene Nachbarkerne bzw. Nachbarkern-Gruppen vom resultierenden Spin I_m verwandeln dementsprechend eine ursprünglich einfache Linie in ein $\prod_m (2I_m+1)$-Linienspektrum.

In sehr hoher Auflösung werden die Spektren übrigens, wie später gezeigt wird, noch wesentlich komplizierter.

Ebenso wie die Zahl der Linien hängt auch deren Intensität d.h. die Größe der von den Resonanzkurven umschlossenen Flächen, von den möglichen Spineinstellungen der Nachbarkerne ab und zwar ist sie proportional zu den statistischen Gewichten der Einstellungsmöglichkeiten.

Sieht man von dem Begriff der äquivalenten Bindung ab, so ist das wohlbekannte Erfahrungsmaterial der optischen Spektroskopie dem geschilderten kernmagnetischen in mancher Hinsicht ähnlich. Es lag daher nahe, in Analogie zur optischen Hyperfeinstruktur, auf die Existenz von Wechselwirkungen zwischen den Kernen der Moleküle zu schließen, eine Folgerung, die zuerst von GUTOWSKY, McCALL und SLICHTER[4] sowie von HAHN und MAXWELL[5] gezogen wurde.

[1] B. SMALLER, E. L. YASAITIS, E. C. AVERY u. D. A. HUTCHISON: Phys. Rev. **88**, 414 (1952).
[2] H. Y. CARR u. E. M. PURCELL: Phys. Rev. **88**, 415 (1952).
[3] T. F. WIMETT: Phys. Rev. **91**, 476 (1953).
[4] H. S. GUTOWSKY, D. W. McCALL u. C. P. SLICHTER: Phys. Rev. **84**, 589 (1951).
[5] E. L. HAHN u. D. E. MAXWELL: Phys. Rev. **84**, 1246 (1951).

Übrigens ist dieser Analogie wegen vorgeschlagen worden, vgl. z.B.[1], entsprechend der Nomenklatur der Atomspektren, die auf dem Einfluß der Elektronenhülle beruhende (chemische) Abschirmungsverschiebung der Resonanzfrequenzen als „Feinstruktur" und die durch Kopplungen zwischen den magnetomechanischen Momenten der Atomkerne bewirkte Aufspaltung als „Hyperfeinstruktur" der kernmagnetischen Spektren zu bezeichnen.

Die dem Anscheine nach so selbstverständliche Wechselwirkungs-Hypothese führt bei einer genaueren Betrachtung sofort auf eine wesentliche Schwierigkeit. Eine direkte Dipol-Dipol-Wechselwirkung zwischen zwei Kernmomenten μ_1 und μ_2 hat die Form:

$$H_{DD} = r^{-3} \left[\mu_1 \cdot \mu_2 - 3 (\mu_1 \cdot r)(\mu_2 \cdot r) \, r^{-2} \right] \tag{27.1}$$

kann also in Flüssigkeiten und Gasen, außer im Bereich sehr tiefer Temperaturen, nicht existieren, denn die Verbindungslinien zwischen den Dipolen (r) ändern auf Grund der Temperaturbewegung ihre Orientierung bezüglich der äußeren Feldrichtung so schnell und unregelmäßig, daß alle Richtungen gleich wahrscheinlich sind. Das Integral von (27.1) über alle Richtungen und damit sowohl das räumliche als auch das zeitliche Mittel von (27.1) verschwinden aber. Gerade dieser Sachverhalt war ja, wie eingangs auseinandergesetzt, das charakteristische Kennzeichen der Proben mit beweglichen, unbehindert rotationsfähigen Partikeln. Angesichts der eindeutigen experimentellen Resultate mußte man daher nach Mechanismen suchen, die eine von der Orientierung der Dipol-Dipol-Verbindungslinien unabhängige oder wenigstens im räumlichen Mittel endliche Kopplung übertragen. Der erste Versuch, das Zustandekommen indirekter statischer Dipol-Dipol-Wechselwirkungen zu verstehen wurde von GUTOWSKY, McCALL und SLICHTER[2] unternommen. Berücksichtigt man nämlich die magnetische Abschirmung der Kernwechselwirkung durch die Bahnbewegung der Elektronen, dann wird der Ausdruck (27.1) komplizierter und sein Integral über alle Richtungen verschwindet nicht mehr. Berechnungen dieses Effekts zeigten jedoch, daß die durch ihn erklärbaren Linienaufspaltungen wesentlich kleiner als die experimentell beobachteten sind.

In der Folge wiesen RAMSEY und PURCELL[3,4] darauf hin, daß die Vermittlung der Wechselwirkung nicht nur über die magnetischen Bahnmomente der Moleküle, wie soeben erwähnt, sondern auch über die magnetischen Eigenmomente der Hüllenelektronen erfolgen kann. Zwischen jedem Kern und dem Elektronenspin seiner eigenen Atomhülle besteht bekanntlich eine magnetische Wechselwirkung. Andererseits sind aber auch die Elektronen-Eigenmomente der Atome in der Molekülbindung mittels der Austauschkräfte untereinander verkoppelt. Das Zusammenwirken dieser Effekte liefert den Mechanismus der sog. elektronengekoppelten Spin-Spin- oder Dipol-Dipol-Wechselwirkungen.

Sein wesentliches Kennzeichen, die Übertragung der Kopplung über bindende Elektronenpaare, läßt sich am leichtesten verstehen, wenn man zweiatomige im $^1\Sigma$-Zustand befindliche Moleküle betrachtet, deren zwei Kerne speziell den Spin $I = \frac{1}{2}$ besitzen. Eines ihrer beiden bindenden Elektronen E_1 möge sich näher bei dem ersten Kern K_1 und das andere Elektron E_2 näher bei dem zweiten Kern K_2 befinden. Die Wahrscheinlichkeit antiparalleler Einstellung der Drehimpulsvektoren des Elektrons E_1 und des Kerns K_1 einerseits und des Elektrons E_2 und des Kerns K_2 andererseits ist dann größer als die paralleler, da das Vorzeichen

[1] K. H. HAUSSER: Angew. Chem. **68**, 729 (1956).
[2] H. S. GUTOWSKY, D. W. McCALL u. C. P. SLICHTER: Phys. Rev. **84**, 589 (1951).
[3] N. F. RAMSEY u. E. M. PURCELL: Phys. Rev. **85**, 143 (1952).
[4] N. F. RAMSEY: Phys. Rev. **91**, 303 (1953).

des gyromagnetischen Verhältnisses der Elektronen negativ ist, ihr Drehimpuls und ihr magnetisches Moment also entgegengesetzt gerichtet sind. Nach dem PAULI-Prinzip sind aber im Singulett-Grundzustand des Moleküls auch die Elektronenspins von E_1 und E_2 antiparallel. Die Kombination dieser Wechselwirkungen liefert das Ergebnis, daß die antiparallele Einstellung der beiden Kerne, unabhängig von der Orientierung ihrer Verbindungslinie, energetisch günstiger ist als ihre parallele. Jedoch sind die Energieunterschiede zwischen den Molekülen mit parallelen und den Molekülen mit antiparallelen Kernen so klein, daß die beiden Molekülsorten in den Proben ungefähr gleich häufig vorkommen. Jede Absorption eines Quants der LARMOR-Frequenz durch einen Kern führt stets ein Molekül der einen Sorte in ein Molekül der anderen Sorte über. Aus dieser Auswahlregel folgt, daß zwischen den zwei Energieniveau-Dubletts, in welche sich die zwei ursprünglich einfachen ZEEMAN-Niveaus in dem hier diskutierten Sonderfall $I_{K_1} = \frac{1}{2}$ und $I_{K_2} = \frac{1}{2}$ auf Grund der Spin-Spin-Wechselwirkung verwandeln, nur zwei Übergänge möglich sind, und zwar haben diese Übergänge nach dem vorletzten Satz die gleiche Wahrscheinlichkeit und die entsprechenden Resonanzlinien somit die gleiche Intensität. Außerdem folgt aus der beschriebenen Vorstellung sofort, daß die Linienaufspaltung symmetrisch zu der durch Spin-Spin-Wechselwirkung ungestörten Resonanzfrequenz erfolgt.

In der Terminologie der Störungstheorie entspricht der von RAMSEY und PURCELL vorgeschlagene Mechanismus einer Störung zweiter Ordnung durch die energetisch höher liegenden Triplettzustände der Moleküle, und zwar ist die störende Wechselwirkung die magnetische Kopplung der Kerne mit den Elektronenspins. Eine quantitative Behandlung des vorliegenden Problemkreises, in welcher sowohl der Betrag der über die Elektronenspins vermittelten Kernspin-Wechselwirkung als auch der Anteil der über die Elektronenbahnbewegung erfolgenden Kernspin-Wechselwirkung berechnet wurden, ist von RAMSEY[1] veröffentlicht worden. Die Arbeit zeigt am Beispiel des HD-Moleküls, daß unter vernünftigen Annahmen über die Größe der in die Rechnung eingehenden mittleren Energie der angeregten Elektronenzustände, die errechnete Größe der Spin-Spin-Aufspaltung mit dem gemessenen Wert[2-4] $(43 \pm 0,5)$ Hz befriedigend übereinstimmt. Außerdem bestätigt sie, daß der weitaus überwiegende Anteil der Wechselwirkung durch die Elektronenspins übertragen wird.

Die experimentellen Untersuchungen der kernmagnetischen Spektren flüssiger Proben führten bald auf eine grundsätzliche Schwierigkeit. Manche Molekülspektren, welche nach den zuvor gesammelten Erfahrungen unbedingt eine durch Spin-Spin-Wechselwirkung erzeugte Multiplettstruktur zeigen sollten, erwiesen sich im Experiment überraschenderweise als einfache Resonanzlinie. Ebenso fand man, daß die Existenz der chemischen Resonanzverschiebung, oder deren Größe in einigen flüssigen Substanzen, vor allem in Mischungen und Lösungen von der Probentemperatur, von der Konzentration der Anteile und sogar gelegentlich von der Konzentration geringfügiger Verunreinigungen abhängt. Solche und ähnliche Beobachtungen wurden von ARNOLD, DHARMATTI und PACKARD[5-7], von GUTOWSKY, SAIKA und FUJIWARA[8,9], von ZIMMERMAN und

[1] N. F. RAMSEY u. E. M. PURCELL: Phys. Rev. **85**, 143 (1952).
[2] B. SMALLER, E. YASAITIS, E. C. AVERY u. D. A. HUTCHISON: Phys. Rev. **88**, 414 (1952).
[3] H. Y. CARR u. E. M. PURCELL: Phys. Rev. **88**, 415 (1952).
[4] T. F. WIMETT: Phys. Rev. **91**, 476 (1953).
[5] J. T. ARNOLD, S. S. DHARMATTI u. M. E. PACKARD: J. Chem. Phys. **19**, 507 (1951).
[6] J. T. ARNOLD u. M. E. PACKARD: J. Chem. Phys. **19**, 1608 (1951).
[7] J. T. ARNOLD: Phys. Rev. **102**, 136 (1956).
[8] H. S. GUTOWSKY u. A. SAIKA: J. Chem. Phys. **21**, 1688 (1953).
[9] H. S. GUTOWSKY u. S. J. FUJIWARA: J. Chem. Phys. **22**, 1782 (1954).

WEINBERG [1-3], von OGG [4], von SHOOLERY und ALDER [5] und von anderen veröffentlicht. Die Art der Effekte und die Umstände, unter welchen sie in den Experimenten besonders ausgeprägt beobachtet werden, weisen deutlich darauf hin, daß man ihren Ursprung in schnell ablaufenden chemischen Austauschprozessen zu suchen hat, eine Schlußfolgerung, die von den genannten Autoren auch sofort gezogen worden ist.

Man versteht dies, wenn man bedenkt, daß die Wechselwirkungen der Kerne einer bestimmten Art mit ihrer Umgebung in irgendeiner speziellen Proben-Zusammensetzung nur dann zu einem mehrlinigen Spektrum führen, wenn diese Wechselwirkungen konstante Anteile haben. Dies gilt für Abschirmungs-Resonanzverschiebungen ebenso wie für Spin-Spin-Wechselwirkungen. Der bisherigen Betrachtung lag zwar die Anschauung zugrunde, daß zwischen den Substanzpartikeln flüssiger Proben im zeitlichen Mittel keine magnetischen Kopplungen existieren (abgesehen von den in diesem Zusammenhang uninteressanten makroskopischen Suszeptibilitäts-Effekten), jedoch wurde stillschweigend angenommen, daß die Struktur der einzelnen Probenpartikel, also der Moleküle, und die Identität ihrer Bestandteile während der Beobachtungszeit erhalten bleiben. Dementsprechend konnten sich die trotz der Temperaturbewegung zwischen diesen Bestandteilen wirkenden magnetischen Kopplungen zeitlich nicht ausmitteln. Vereinfacht kann man sagen, daß die Probe als dynamisches Gebilde und ihre Partikel als statische Gebilde angesehen wurden, ganz entsprechend der elementaren Vorstellung über das Wesen der Molekülbindung. Im flüssigen Zustand sind jedoch die Moleküle wie wohl bekannt, keineswegs immer konstante Atomzusammensetzungen, man denke nur an den Vorgang der Gleichgewichtseinstellung bei der Ionenbildung in elektrolytischen Lösungen. Aber auch in weniger krassen Fällen kann die Molekülbindung typisch dynamische Merkmale erhalten. Betrachtet sei etwa eine aus zwei Partnern gemischte Flüssigkeit, deren beide Molekülsorten eine bestimmte Kernart neben anderen verschiedenen enthalten. Die Abschirmungs-Resonanzverschiebung der in den beiden Molekülsorten gebundenen Kernart unterscheiden sich dann mehr oder weniger, und man erhält im Experiment ein aus mindestens zwei Linien bestehendes Spektrum. Vertauschen die Kerne aber während der resonanzspektroskopischen Beobachtung ihre Bindungsorte in den beiden Molekülsorten (betrachtet seien zunächst nur Übergänge zwischen verschiedenen Molekülen), so erfahren sie alle dieselbe zeitlich mittlere Abschirmung, wenn der Platzwechsel nur genügend häufig erfolgt, seine Frequenz also ausreichend groß ist. Statt einer Dublett-Feinstruktur erhält man dementsprechend im Experiment eine einfache Linie, deren Abschirmungskonstante eine mittlere Größe hat. Allgemein findet man, daß mehrlinige Spektren nur beobachtet werden, wenn die Verweilzeit τ (mittlere Lebensdauer) der Kerne in ihren Bindungsorten sehr viel größer als die Schwingungsdauer $1/\Delta\nu$ der Differenzfrequenz der Linien (reziproke Aufspaltungsbreite) ist. Erfolgt der Austausch sehr schnell, ist also $\tau \ll 1/\Delta\nu$, so erhält man nur eine scharfe Resonanzlinie. Im Übergangsgebiet, d.h. in dem Bereich in dem die beiden Zeiten von der gleichen Größenordnung sind, entsteht eine breite noch mehr oder weniger differenzierte Resonanzkurve, deren Struktur und Linienbreite von der Verweilzeit τ stark abhängen. In diesem Bereich kann man die Dauer von τ recht genau abschätzen, vgl. hierzu z.B. [6]. Man sieht leicht ein,

1 J. R. ZIMMERMAN: J. Chem. Phys. **21**, 1605 (1953).
2 J. R. ZIMMERMAN: J. Chem. Phys. **22**, 950 (1954).
3 I. WEINBERG u. J. R. ZIMMERMAN: J. Chem. Phys. **23**, 748 (1955).
4 R. A. OGG: J. Chem. Phys. **22**, 560 (1954).
5 J. N. SHOOLERY u. B. J. ALDER: J. Chem. Phys. **23**, 805 (1955).
6 J. T. ARNOLD: Phys. Rev. **102**, 136 (1956).

daß die Verweilzeit von der Partnerkonzentration, von der Probentemperatur und von anderen mit den genannten Größen verknüpften makroskopischen Substanzdaten abhängt. Damit sind die geschilderten kernmagnetischen Resultate qualitativ erklärt. Zwei Beispiele sollen die dargelegten Zusammenhänge noch verdeutlichen. WEINBERG und ZIMMERMAN[1] haben das Protonenspektrum von Wasser-Alkoholgemischen verschiedener Konzentration gemessen. In erster Näherung, d.h. in grober Auflösung besteht das Spektrum aus vier Linien, wenn der Wassergehalt kleiner als etwa 20 Gewichtsprozent ist. Drei dieser Linien sind den drei äquivalenten Bindungsmöglichkeiten der Protonen im Alkoholmolekül, vgl. hierzu Fig. 97, und die vierte der Bindungsmöglichkeit im Wassermolekül zuzuordnen. Bei höheren Wasserkonzentrationen nähern sich die zur Hydroxylgruppe (OH) des Alkohols und die zum Wasser gehörenden Linien

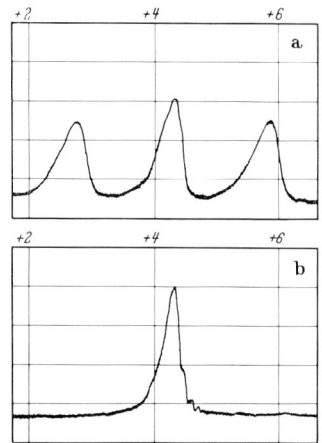

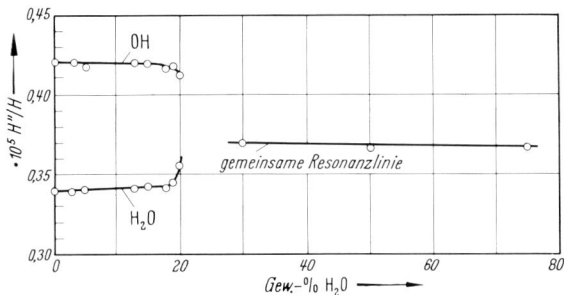

Fig. 97. Verlauf der chemischen Verschiebungen der Protonen-Resonanzlinien des Wassers und der Hydroxylgruppe von Äthanol in einem Gemisch beider Flüssigkeiten. Parameter der Meßreihe ist das Mischungsverhältnis. Die Resonanzverschiebungen wurden bezüglich der mittleren Resonanzlinie der Methylgruppe (CH$_3$) des Äthanols gemessen. 27 Gewichtsprozent H$_2$O entsprechen ungefähr einem Verhältnis der Molekülzahlen pro Volumeinheit von 1:1.

Fig. 98 a u. b. Protonenresonanzspektrum von flüssigem Ammoniak (N^{14}H$_3$). Meßfrequenz: $f = 30$ MHz, Abszisseneinheit: $10^{-6} \times$ $f = 30$ Hz. Abszissennullpunkt: Resonanzfrequenz des reinen Wassers: a „Völlig trockenes" Ammoniak. b „Nasses" Ammoniak (10^{-7} Beimischung H$_2$O).

einander. Die Protonen-Austauschfrequenz zwischen diesen beiden Gruppen wird also ungefähr gleich ihrer Differenzfrequenz. Die Linien vereinigen sich schließlich bei einer Gewichtskonzentration des Wassers von etwa 30% (27% Gewichtskonzentration H$_2$O bedeutet Wassermolekül- = Alkoholmolekülzahl), vgl. Fig. 97. Aus den Messungen folgt z.B., daß die Protonenverweilzeit in der Hydroxylgruppe im kritischen Übergangsgebiet (20 bis 25% H$_2$O) ungefähr 0,05 sec beträgt. Über die Verweilzeiten der Protonen in den beiden anderen Bindungsorten des Alkoholmoleküls kann man nur aussagen, daß sie sicher sehr viel länger sind. Zur Erklärung des schnellen Protonen-Platzwechsels zwischen den Hydroxylgruppen der Alkoholmoleküle und den Wassermolekülen nimmt man an, daß sich zwischen diesen Molekülen bzw. Molekülteilen sog. Wasserstoffbrücken ausbilden, über welche Protonen leicht ausgetauscht werden.

Beobachtungen der Protonenresonanz von OGG[2] an gasförmigem und flüssigem Ammoniak (NH$_3$) der beiden Isotope N^{14} ($I = 1$) und N^{15} ($I = \frac{1}{2}$) zeigen, vgl. Fig. 98, daß die Dipol-Dipol-Aufspaltung bei schnellem Kernaustausch ebenso wie die Abschirmungs-Resonanzverschiebung verschwindet. Die Arbeit von OGG ist zugleich ein schönes Beispiel dafür, welch geringe Verunreinigungen zur Hervorrufung dieses Effekts ausreichen können. Im gasförmigen Zustand

[1] I. WEINBERG u. J. R. ZIMMERMAN: J. Chem. Phys. 23, 748 (1955).
[2] R. A. OGG: J. Chem. Phys. 22, 560 (1954).

besteht das Protonenspektrum des $N^{14}H_3$-Moleküls aus drei Linien. Dies ist gut zu verstehen, da der mit den drei äquivalent gebundenen Protonen wechselwirkende Stickstoffkern den Spin $I = 1$ hat. In flüssigen und zwar auch in gut getrockneten Ammoniakproben fand man dagegen immer nur eine mittlere scharfe Resonanzlinie. Erst als es gelang, durch extrem sorgfältige Trocknungsmaßnahmen (unter anderem Verwendung von destilliertem metallischem Natrium als Trockenmittel) den Wassergehalt der Proben bis auf einen Bruchteil von etwa 10^{-7} zu senken, wurde der Austauscheffekt unwirksam, und die Protonenspektren der flüssigen $N^{14}H_3$- und $N^{15}H_3$-Moleküle erhielten die erwartete Triplett- bzw. Dublettstruktur. Die Wirksamkeit solch geringer Wasserspuren muß man der Reaktion:

$$H_2O + NH_3 \rightleftharpoons OH^- + NH_4^+$$

zuordnen. Sie ermöglicht es, daß die Protonen zwischen den Ammoniakmolekülen schnell ausgetauscht werden. Auf ihrem Wege durch verschiedene Moleküle erfahren die Protonen nacheinander die drei möglichen Wechselwirkungen, entsprechend der Zahl der möglichen Einstellungen des Stickstoffkerns im äußeren Feld. Beobachtet wird aber nur deren Mittelwert, also eine scharfe Resonanzlinie. Aus anderen Daten kann man übrigens schließen, daß zu der oben angeschriebenen endothermen Reaktion nur wenige Kilocalorien Aktivierungsenergie pro Mol benötigt werden, ihr schneller Verlauf ist demnach plausibel. Weiter ist sie, da es sich um eine Ionenreaktion handelt, in der Gasphase unmöglich. Auch dies stimmt mit der experimentellen Erfahrung überein.

Aus den geschilderten Ergebnissen geht hervor, daß das Studium chemischer Austauschprozesse mit Hilfe der kernmagnetischen Methode Aufschlüsse über Molekülreaktionen und über deren zeitlichen Ablauf liefern kann, die schwerlich auf eine andere Weise gewonnen werden können. Wie so oft in der Physik, wurde also auch hier ein zunächst als Störeffekt bemerktes Phänomen nach seiner Deutung zu einer neuen Informationsquelle.

Die qualitative Einzelbesprechung der für die Struktur der Flüssigkeits- und Gas-Spektren verantwortlichen Effekte ist damit abgeschlossen. Im Verlauf der in den letzten Jahren erfolgten Entwicklung der Resonanzapparaturen, insbesondere der außerordentlichen Erhöhung ihres Auflösungsvermögens, ist jedoch das experimentelle Erfahrungsmaterial so umfangreich und differenziert geworden, daß die feineren Einzelheiten der Spektren nur noch auf Grund einer quantitativen Behandlung verstanden werden können. Ein Musterbeispiel für das Anwachsen der beobachteten Linienzahl und zugleich für die Verfeinerung der experimentellen Technik ist der im Laufe von 7 Jahren erzielte Fortschritt bei der Erforschung der Resonanzlinien des Äthanols. Deshalb soll im folgenden das Protonenspektrum dieser Verbindung und dessen quantitative Deutung in den verschiedenen Entwicklungsstufen beschrieben werden.

In sehr grober Auflösung fand man ursprünglich nur eine Resonanzlinie. Dabei bedeutet „sehr grob" im heutigen Sprachgebrauch, daß bei einer Arbeitsfrequenz von etwa 30 MHz im günstigsten Fall zwei Linien getrennt werden können, deren Frequenzdifferenz ungefähr 100 Hz beträgt. Das maximale Auflösungsvermögen kann also immerhin schon den Wert $\nu/\Delta\nu = 3 \cdot 10^5$ haben. Vermindert man den auflösbaren Frequenzabstand auf etwa 15 Hz ($\nu/\Delta\nu = 2 \cdot 10^6$), so erhält man das in Fig. 99 wiedergegebene Spektrum. Es besteht aus drei Resonanzlinien, deren umschlossene Flächen (Intensitäten) sich wie $3:2:1$ verhalten. Dieser Befund erleichtert die Zuordnung der Linien sehr, denn das Äthanolmolekül CH_3CH_2OH besteht bekanntlich aus drei Bindungsgruppen und zwar der Methyl-, der Methylen- und der Hydroxylgruppe, und diese Gruppen enthalten gerade drei, zwei bzw. ein chemisch äquivalent gebundenes Proton. Die beobachtete Linienaufspaltung ist also eine Folge der in den drei Molekülteilen verschiedenen Abschirmungs-Verschiebungen der Resonanzfrequenzen.

Resonanzverschiebungen dieses Typs lassen sich in der HAMILTON-Funktion des Moleküls, d.h. in der Darstellung seiner Spin-Energie, dadurch beschreiben, daß man zu dem äußeren Feld H_0 die molekülinneren Felder h_k, welche in den verschiedenen Teilen des Moleküls in verschiedener Stärke induziert werden, addiert. Unter Vernachlässigung der möglichen Spin-Spin-Wechselwirkungen zwischen den Protonen der verschiedenen chemischen Gruppen erhält der HAMILTON-Operator des Äthylalkohol-Moleküls in der vorliegenden sog. nullten Näherung die einfache Form.

$$\mathfrak{H}^{(0)} = -\hbar|\gamma|\left\{ \begin{matrix}(H_0+h_A)\cdot I_A + \\ + (H_0+h_B)\cdot I_B + \\ + (H_0+h_C)\cdot I_C.\end{matrix}\right\} \quad (27.2)$$

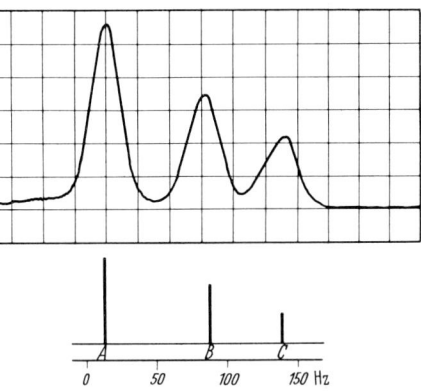

Darin ist $\hbar = h/2\pi$ das Wirkungsquantum, γ das gyromagnetische Verhältnis der Protonen und I_K der Gesamtspin der K-ten Protonengruppe. Das Quadrat des Spinoperators I_K kann alle positiven Werte der Reihe $\left(\frac{1}{2n}+1\right)\left(\frac{1}{2n}\right), \left(\frac{1}{2n}\right)\left(\frac{1}{2n}-1\right), \ldots$ haben, in der n gleich der Protonenzahl der K-ten Gruppe ist. Für die Energieniveaus von (27.2), charakterisiert durch die Eigenwerte m_A, m_B und m_C der zur äußeren Feldrichtung H_0 parallelen z-Komponenten I_{zA}, I_{zB}, und I_{zC} der Spinoperatoren gilt somit in nullter Näherung:

Fig. 99. Spektrum der Protonen des Äthylalkohols in grober Auflösung. A Methylgruppe (CH₃). B Methylengruppe (CH₂). C Hydroxylgruppe (OH), $H_0 \approx 7000$ Gauß, $f \approx 30$ MHz. Einzige Ursache der zur Feldstärke proportionalen Resonanzfrequenz-Unterschiede ist die in den einzelnen Gruppen des Äthanols verschieden große Abschirmung der Protonen durch die Elektronen der Molekülhülle (chemical shift). Die Abschirmung ist in der Methylgruppe am größten, deren Protonen haben also in einem Magnetfeld vorgegebener Feldstärke die niederste Resonanzfrequenz, bzw. in einem Spektrometer mit fester Meßfrequenz die höchste Resonanzfeldstärke. Die Wahl des Nullpunkts der Frequenzskala erfolgte willkürlich.

$$E^{(0)}_{m_A m_B m_C} = -\hbar\left\{(\omega_0+\omega_A)\,m_A + (\omega_0+\omega_B)\,m_B + (\omega_0+\omega_C)\,m_C\right\}. \quad (27.3)$$

Darin ist $\omega_0 = |\gamma|H_0$ und $\omega_K = |\gamma|h_K$. Experimentell bestimmbar sind nur die Differenzen der Resonanzverschiebungen $\delta_{KL} = \omega_K - \omega_L$. Setzt man die für das Äthanolmolekül bei 30,5 MHz gemessenen Werte (Probentemperatur 27° C)

$$\delta_{BA}/2\pi = (75,0 \pm 1)\ \text{Hz} \quad \text{und} \quad \delta_{CB}/2\pi = (50.5 \pm 1)\ \text{Hz}$$

in das Energieniveau-Schema (27.3) ein, so liefert dieses genau die drei in Fig. 99 unter der experimentellen Registrierung wiedergegebenen Linien, wenn nur Übergänge erlaubt sind, bei denen sich der Eigenwert m einer Gruppe um ± 1 ändert. Die Wahl des Nullpunkts auf der Energie- bzw. Frequenzskala ist natürlich willkürlich. Das durch die Linienhöhe charakterisierte Intensitätsverhältnis folgt wie bereits ausgeführt, in diesem Fall direkt aus der Protonenverteilung auf die Gruppen des Äthanols.

Fig. 100 zeigt das Protonenspektrum des Äthylalkohols, das man mit einer Apparatur erhält, deren Auflösungsvermögen etwa $1 \cdot 10^7$ beträgt, dies entspricht einem absoluten Frequenz-Unterscheidungsvermögen von etwa 3 Hz. Es zeigt sich, daß die Struktur des Spektrums vom Reinheitsgrad der Probe abhängt. In der Figur sind die beiden Grenzfälle einer „unreinen" und einer „reinen" Probe wiedergegeben. Präziser gefaßt, besagen diese Bezeichnungen, daß die Proben entweder ausreichend viel oder fast gar keine störenden H$^+$ oder OH$^-$-Ionen enthalten. Die Linienzahlen und die Intensitätsverteilung innerhalb der drei Liniengruppen, die in der Auflösung der Fig. 99 noch als drei einfache Linien erschienen, weisen darauf hin, daß die Aufspaltungen als Folge von Spin-Spin-Wechselwirkungen zwischen benachbarten Protonengruppen zu erklären sind.

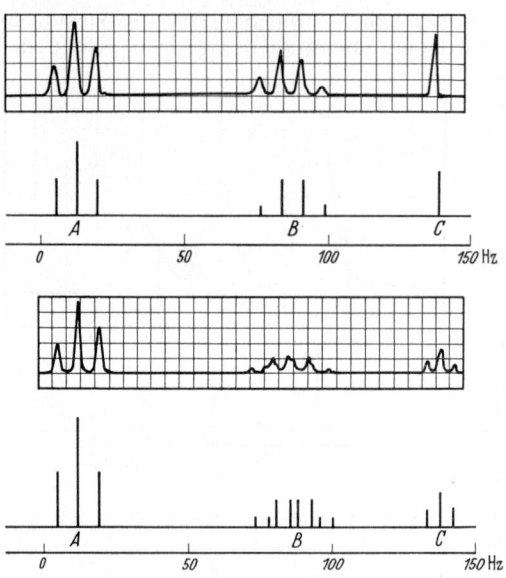

Am Schluß dieser Ziffer wird übrigens ein Verfahren beschrieben, das in weniger eindeutigen Fällen eine experimentelle Entscheidung erlaubt, ob Spin-Spin-Wechselwirkungen für eine bestimmte Aufspaltung verantwortlich sind, und welche Nachbarn an dieser beteiligt sind.

Aus dem in Fig. 100 wiedergegebenen Spektrum des unreinen Äthanols folgt, daß in dem Alkoholmolekül in diesem Fall nur eine wechselseitige Spin-Spin-Kopplung, und zwar zwischen den Protonen der Methyl- und den Protonen der Methylengruppe im zeitlichen Mittel existiert (natürlicher Kohlenstoff besteht zu fast 99% aus C^{12}, der Spin dieses Isotops ist Null). Die Resonanzkurve der Methylgruppe hat Triplettstruktur ($I_{\text{Methylen}} = 1$) und diejenige der Methylengruppe hat Quartettstruktur ($I_{\text{Methyl}} = \frac{3}{2}$). Die Hydroxylprotonen besitzen nur eine Resonanzlinie und haben außerdem keinen Einfluß auf die Resonanzstruktur der Methylenprotonen. Daraus folgt, daß sich ihre Wechselwirkungen mit Nachbarprotonen zeitlich ausmitteln, diese Protonen vertauschen also ständig ihre

Fig. 100. Spektrum der Protonen des Äthylalkohols in mittlerer Auflösung. *A* Methylgruppe (CH$_3$). *B* Methylengruppe (CH$_2$). *C* Hydroxylgruppe (OH), $H_0 \approx 7000$ Gauß, $f \approx 30$ MHz. Oben: „unreiner" Alkohol. Unten: „reiner" Alkohol, d.h. H$^+$- bzw. OH$^-$-Ionen-Konzentration kleiner als 10^{-6} normal. Die Diagramme unter den experimentellen Registrierungen zeigen die Anordnung der theoretisch zu erwartenden Spektren, wenn man außer den in den einzelnen Molekülteilen verschiedenen Frequenzverschiebungen die Spin-Spin-Wechselwirkung zwischen den Protonengruppen des Äthanols in erster Ordnung berücksichtigt. Die Höhe der Linien in diesen Diagrammen ist ein Maß für die in der angegebenen Näherung zu erwartenden Intensitäten der einzelnen Spektrallinien. Zur Prüfung der Gültigkeit dieser Näherung muß man dementsprechend die Höhe dieser Linien mit der Größe der von der zugehörigen Resonanzlinie der experimentellen Registrierung umschlossenen Fläche vergleichen. Kleine Abweichungen sind erkennbar.

Bindungsorte in den verschiedenen Nachbarmolekülen. Die Linienabstände sind übrigens in der Methyl- (*A*) und in der Methylengruppe (*B*) gleich groß. Ihr experimentell ermittelbarer Kreisfrequenzabstand wird üblicherweise mit dem Symbol J_{AB} bezeichnet. In dem vorliegenden Beispiel beträgt der Frequenzabstand $J_{AB}/2\pi$ zwischen den Hyperfeinstrukturlinien (7.15 $\pm$ 0.25) Hz, also ungefähr 10% der Feinstruktur-Frequenzdifferenz $\delta_{BA}/2\pi$. In Alkohol, der keinen H$^+$- oder OH$^-$-Ionen-Überschuß größer als ungefähr 10^{-5} normal enthält, verläuft der Austausch der Hydroxylprotonen sehr langsam. Dementsprechend spiegelt sein kernmagnetisches Molekülspektrum, vgl. Fig. 100 unten auch die Spin-Spin-Wechselwirkung zwischen den Protonen der Hydroxyl- und der Methylengruppe wider. Die Methylenresonanzkurve hat in diesem Fall die Form eines Oktetts, da $(2I_{\text{OH}} + 1)(2I_{\text{CH}_3} + 1) = 8$ ist. Die Resonanzkurve der Hydroxylprotonen ist ebenso wie diejenige der Methylprotonen triplettförmig, denn die beiden Protonengruppen wechselwirken mit denselben Partnern, den

Methylenprotonen. Die Größen der beiden Spin-Spin-Kopplungen sind natürlich verschieden, und zwar ist diejenige, an der die Hydroxylprotonen beteiligt sind, etwas schwächer. Der zu ihr gehörende Frequenzabstand beträgt $J_{BC}/2\pi =$ (4,8 ± 0,2) Hz.

Zur quantitativen Beschreibung der Spektrenstruktur muß man zunächst zu dem HAMILTON-Operator (27.2) Spin-Spin-Wechselwirkungsglieder hinzufügen. Die experimentellen Ergebnisse zeigen, daß dies in der Form

$$\mathfrak{H}^{(1)} = \mathfrak{H}^{(0)} - \hbar \{ J_{AB} I_A \cdot I_B + J_{BC} I_B \cdot I_C \} \tag{27.4}$$

erfolgen kann. Eine Wechselwirkung zwischen den Hydroxyl- und den Methylprotonen ist experimentell nicht beobachtbar, J_{AC} ist also, verglichen mit den beiden anderen Kopplungskonstanten, wegen des größeren Abstandes der beiden äußeren Protonengruppen verschwindend klein. Außerdem wird auf Grund des chemischen Austauschs im unreinen Alkohol auch die Konstante J_{BC} Null. Die kernmagnetischen Energieniveaus des Äthanolmoleküls können in erster Ordnung durch eine Berechnung der Diagonalelemente von $\mathfrak{H}^{(1)}$ bestimmt werden. Diese folgen aus (27.4) in dieser Näherung einfach durch den Ersatz der Operatoren I_A, I_B und I_C durch die entsprechenden Eigenwerte m_A, m_B und m_C ihrer z-Komponenten, es gilt also:

$$E^{(1)}_{m_A m_B m_C} = -\hbar \left\{ (\omega_0 + \omega_A) m_A + (\omega_0 + \omega_B) m_B + \atop + (\omega_0 + \omega_C) m_C + J_{AB} m_A m_B + J_{BC} m_B m_C \right\}. \tag{27.5}$$

Aus (27.5) folgen die in Fig. 100 unter den experimentellen Registrierungen wiedergegebenen Linienverteilungen, wenn sich immer nur ein Eigenwert m_K und dieser nur um ±1 ändern kann. Die Absorptionsfrequenzen der Hydroxylprotonen (Gruppe A) erhält man z.B. aus der Beziehung:

$$\omega_{m_A \to m_A - 1} = \frac{1}{\hbar} (E^{(1)}_{(m_A-1) m_B m_C} - E^{(1)}_{m_A m_B m_C}) = \omega_0 + \omega_A + J_{AB} m_B. \tag{27.6}$$

Darin kann m_B die Eigenwerte 1, 0 und −1 haben, und dementsprechend besteht das Spektrum der Hydroxylgruppe aus drei äquidistanten Linien. Sieht man von der Unbestimmtheit des Vorzeichens von J_{AB} ab, so kann man zu jeder Linie die zugehörige magnetische Quantenzahl der Nachbargruppe nennen. Entsprechend gilt für das Methylenspektrum des reinen Alkohols:

$$\omega_{m_B \to m_B - 1} = \frac{1}{\hbar} (E^{(1)}_{m_A (m_B-1) m_C} - E^{(1)}_{m_A m_B m_C}) = \omega_0 + \omega_B + J_{AB} m_A + J_{BC} m_C. \tag{27.7}$$

Darin können m_A bzw. m_C die Eigenwerte $\frac{3}{2}$, $\frac{1}{2}$, $-\frac{1}{2}$, $-\frac{3}{2}$ bzw. $\frac{1}{2}$, $-\frac{1}{2}$ haben.

Für die theoretische Intensitätsverteilung der Linien gilt gemäß der Protonenverteilung im Äthanolmolekül nach wie vor, daß sich die Intensitäten der Gruppen A, B und C wie 3:2:1 verhalten, d.h. aber, daß auch für die Summen der Linienlängen der drei Gruppen in den theoretischen Linienverteilungen der Fig. 100 dieselbe Verhältnisbeziehung gültig ist. Innerhalb jeder Gruppe ist das Intensitätsverhältnis der Linien gleich dem Verhältnis der statistischen Gewichte der Eigenwerte m_K der Spin-z-Komponenten I_{zK} der Nachbargruppen zu denen eine Wechselwirkungs-Beziehung besteht. Das statistische Gewicht einer magnetischen Quantenzahl wächst proportional mit der Zahl der Einstellmöglichkeiten der Einzelspins, welche gerade diese Quantenzahl ergeben. Im Äthanolmolekül gilt dementsprechend für die statistischen Gewichte der Eigenwerte:

$$I_{zA} = \tfrac{3}{2}:$$

$$m_A: \quad \tfrac{3}{2} \qquad\qquad \tfrac{1}{2} \qquad\qquad -\tfrac{1}{2} \qquad\qquad -\tfrac{3}{2}$$

$\tfrac{1}{2}, \tfrac{1}{2}, \tfrac{1}{2}$	$\tfrac{1}{2}, \quad \tfrac{1}{2}, -\tfrac{1}{2}$	$\tfrac{1}{2}, -\tfrac{1}{2}, -\tfrac{1}{2}$	$-\tfrac{1}{2}, -\tfrac{1}{2}, -\tfrac{1}{2}$
	$\tfrac{1}{2}, -\tfrac{1}{2}, \quad \tfrac{1}{2}$	$-\tfrac{1}{2}, \quad \tfrac{1}{2}, -\tfrac{1}{2}$	
	$-\tfrac{1}{2}, \quad \tfrac{1}{2}, \quad \tfrac{1}{2}$	$-\tfrac{1}{2}, -\tfrac{1}{2}, \quad \tfrac{1}{2}$	

$$\text{Gewicht: } 1 \quad : \quad 3 \quad : \quad 3 \quad : \quad 1$$

$$I_{zB} = 1: \qquad\qquad\qquad\qquad\qquad I_{zC} = \tfrac{1}{2}:$$

$$m_A: \quad 1 \qquad\quad 0 \qquad\quad 1 \qquad\qquad m_A: \quad \tfrac{1}{2} \quad -\tfrac{1}{2}$$

$\tfrac{1}{2}, \tfrac{1}{2}$	$\tfrac{1}{2}, -\tfrac{1}{2}$	$-\tfrac{1}{2}, -\tfrac{1}{2}$		$\tfrac{1}{2}$	$-\tfrac{1}{2}$
	$-\tfrac{1}{2}, \quad \tfrac{1}{2}$				

$$\text{Gewicht: } 1 \quad : \quad 2 \quad : \quad 1 \qquad\qquad \text{Gewicht: } 1 \quad : \quad 1$$

Die Intensitäten aller Linien des reinen Äthanols, vgl. Fig. 100 unten stehen damit in erster Näherung zueinander im Verhältnis:

$$\tfrac{3}{4} \cdot \tfrac{3\cdot 2}{4} \cdot \tfrac{3}{4} \cdot \tfrac{2}{16} \cdot \tfrac{2}{16} \cdot \tfrac{2\cdot 3}{16} \cdot \tfrac{2\cdot 3}{16} \cdot \tfrac{2\cdot 3}{16} \cdot \tfrac{2}{16} \cdot \tfrac{2}{16} \cdot \tfrac{1}{4} \cdot \tfrac{1\cdot 2}{4} \cdot \tfrac{1}{4}.$$

Ein Vergleich der experimentell registrierten Spektren mit den theoretischen Diagrammen zeigt, daß diese in der Auflösung der Fig. 100 und in der Näherung der soeben besprochenen Theorie zweiter Ordnung recht gut, wenn auch nicht vollständig übereinstimmen. Steigert man das Auflösungsvermögen jedoch weiter, in Fig. 101 beträgt es etwa $6 \cdot 10^7$, entsprechend einem absoluten Unterscheidungsvermögen von etwa 0.5 Hz, so wird die Struktur des Äthanolspektrums wesentlich komplizierter. Alle in der voranstehenden Figur beobachteten einfachen Linien erweisen sich nunmehr als aus zwei oder mehreren Linienkomponenten bestehend. So besteht z.B. in der höheren Auflösung jede der beiden äußeren zu den Quantenzahlen $m_B = \pm 1$ gehörenden Linien der Methylgruppe aus drei und die mittlere zur Quantenzahl $m_B = 0$ gehörende Linie dieser Gruppe aus zwei Komponenten.

Die Erklärung dieser Strukturkomplikation findet man, wenn man die Eigenwerte von $\mathfrak{H}^{(1)}$ in höherer Ordnung mit den Größen $J_{AB}/\omega_A - \omega_B$ bzw. $J_{BC}/\omega_B - \omega_C$ als Störungsparameter berechnet. In zweiter Ordnung fanden Bloch[1] und Anderson[2] für die Energieniveaus von (27.4) die Beziehung:

$$
\begin{aligned}
E^{(2)}_{m_A m_B m_C I_A I_B I_C} = &-\hbar \Big\{ (\omega_0 + \omega_A) m_A + (\omega_0 + \omega_B) m_B + (\omega_0 + \omega_C) m_C + \\
&+ J_{AB} m_A m_B + \frac{J^2_{AB}}{2(\omega_A - \omega_B)} \left[m_A (I_B^2 + I_B - m_B^2) - m_B (I_A^2 + I_A - m_A^2) \right] + \\
&+ J_{BC} m_B m_C + \frac{J^2_{BC}}{2(\omega_B - \omega_C)} \left[m_B (I_C^2 + I_C - m_C^2) - m_C (I_B^2 + I_B - m_B^2) \right] \Big\}.
\end{aligned}
\tag{27.8}
$$

Für die Frequenzen der Übergänge $m_A \to m_A - 1$ gilt also in dieser Näherung:

$$
\begin{aligned}
\omega^{(2)}_{m_A \to m_A - 1} = &\omega_0 + \omega_A + J_{AB} m_B + \frac{J^2_{AB}}{2(\omega_A - \omega_B)} \times \\
&\times \left[I_B (I_B + 1) - m_B (m_B + 1) + 2 m_A m_B \right].
\end{aligned}
\tag{27.9}
$$

[1] F. Bloch: Phys. Rev. **102**, 104 (1956).
[2] W. A. Anderson: Phys. Rev. **102**, 151 (1956).

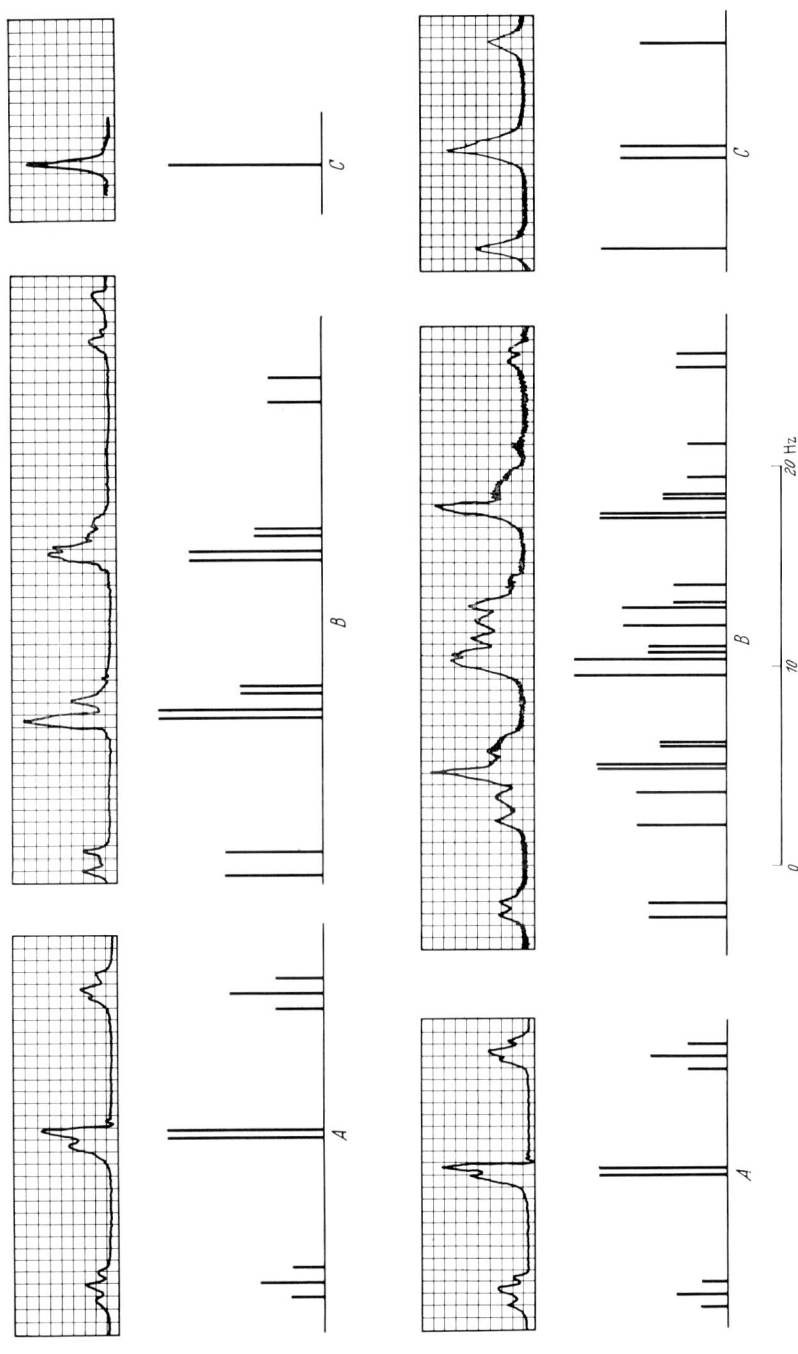

Fig. 101. Spektrum der Protonen des Äthylalkohols in hoher Auflösung. A Methylgruppe (CH_3). B Methylengruppe (CH_2). C Hydroxylgruppe (OH). $H_0 \approx 7000$ Gauß, $f \approx 30$ MHz. Oben: „unreiner" Alkohol. Unten: „reiner" Alkohol, d.h. H^+- bzw. O^--Ionen-Konzentration kleiner als 10^{-5} normal. Die Diagramme unter den experimentellen Registrierungen zeigen die theoretisch zu erwartende Anordnung und Intensitätsverteilung der Spektren nach einer von ANDERSON ausgeführten Störungsrechnung, in welcher die Spin-Spin-Wechselwirkungen zwischen den Protonengruppen in zweiter Ordnung berücksichtigt wurde.

Die Gl. (27.8) enthält Glieder, die proportional zu J_{AB}^2 und quadratisch in m_A sind. Infolgedessen hängen die Übergangsfrequenzen (27.9) für $\Delta m_A = \pm 1$ von der Quantenzahl m_A vor bzw. nach dem Übergang ab, sie haben also verschiedene Größen für die drei Übergänge von m_A: $\frac{3}{2} \leftrightarrow \frac{1}{2}, \frac{1}{2} \leftrightarrow -\frac{1}{2}, -\frac{1}{2} \leftrightarrow -\frac{3}{2}$. Die Beziehung (27.9) gibt somit die Triplettstruktur der beiden äußeren zu $m_B = \pm 1$ gehörenden Linien der Gruppe A richtig wieder. Für $m_B = 0$ verschwindet der in m_A quadratische Term, die mittlere Linie der Gruppe A sollte also einfach sein. Die Erklärung für ihre im Experiment beobachtete Dublettstruktur (vgl. Fig. 101) ist nach (27.8), daß die Höhen der Energieniveaus nicht nur von den Größen der magnetischen Quantenzahlen m_R, sondern auch von den Größen der Gesamtspins I_R der einzelnen Gruppen abhängen. So kann z. B. $I_B = 1$ und $I_B = 0$ sein. Im ersten Fall kann auch $m_B = 0$ sein, im zweiten Fall ist sicher $m_B = 0$.

Mit Hilfe dieser Vorstellung läßt sich übrigens noch ein weiterer experimenteller Befund qualitativ erklären. Die mittlere Linie der Gruppe A in Fig. 101, welche zu dem Singulettzustand $I_B = 0$ der Methylengruppe gehört, ist beträchtlich schmaler als die Linien, welche zu den Triplettzuständen $I_B = 1$ gehören. Die Breite einer Linie hängt aber auch von der Lebensdauer der Zustände der Nachbargruppen ab. Im vorliegenden Fall ist ein Übergang der Quantenzahl m_B von 0 zu $+1$ oder -1 ein relativ unwahrscheinliches Ereignis, wenn der Anfangswert von I_B Null ist, da es sich dann um einen Übergang von einem Singulett- zu einem Triplettzustand handelt. Für $I_B = 1$ sind dagegen Änderungen von m_B um ± 1 nur Übergänge innerhalb des Triplettzustands, diese haben aber eine wesentlich höhere Wahrscheinlichkeit.

Die Resonanzlinienstruktur der Hydroxylgruppe des reinen Alkohols sollte eigentlich derjenigen der Methylgruppe ähnlich sein. Die experimentell beobachteten Abweichungen dürften dadurch bedingt sein, daß selbst in den reinsten Proben die Austauschverbreiterung der Linien die Beobachtung der feineren Effekte verhindert. Die Verteilungen der Resonanzlinien in den unter den experimentellen Registrierungen der Fig. 101 wiedergegebenen Diagrammen wurden aus Gl. (27.8) berechnet. Ein Vergleich der experimentellen und der theoretischen Frequenzverteilungen zweiter Ordnung zeigt, daß sie nicht exakt übereinstimmen. Führt man die Störungsrechnung bis zur dritten Ordnung weiter, so werden die Abweichungen wesentlich geringer. Anderson[1] hat dies getan und gefunden, daß für die Energieniveaus eines Moleküls mit beliebig vielen Gruppen äquivalenter Kerne in dritter Näherung folgende Beziehung gilt:

$$
\begin{aligned}
E^{(3)}_{m_A m_B \ldots I_A I_B \ldots} = -\hbar \Bigg\{ & \sum_{R=A,B,\ldots} (\omega_0 + \omega_R) m_R + \sum_{R=A,B,\ldots} \sum_{S=A,B,\ldots,\neq R} \frac{1}{2} J_{RS} m_R m_S + \\
& + \sum_{R=A,B,\ldots} \sum_{S=A,B,\ldots,\neq R} \frac{J_{RS}^2}{4(\omega_R - \omega_S)} \left[m_R(I_S^2 + I_S - m_S^2) - m_S(I_R^2 + I_R - m_R^2) \right] - \\
& - \sum_{R=A,B,\ldots} \sum_{S=A,B,\ldots,\neq R} \frac{J_{RS}^3}{2(\omega_R - \omega_S)^2} (I_R - m_R)(I_R + m_R + 1) \times \\
& \qquad \times \left[\frac{1}{2}(I_S - m_S)(I_S + m_S + 1) + m_S(1 - m_S + m_R) \right] - \\
& - \sum_{R=A,B,\ldots} \sum_{S=A,B,\ldots,\neq R} \sum_{T=A,B,\ldots,\neq R,S} \frac{J_{RS}^2 J_{RT}}{2(\omega_R - \omega_S)^2} \left[m_T m_R (I_S - m_S) \times \right. \\
& \qquad \left. \times (I_S + m_S + 1) - m_T m_S(I_R - m_R)(I_R + m_R + 1) \right] - \\
& - \sum_{R=A,B,\ldots} \sum_{S=A,B,\ldots,\neq R} \sum_{T=A,B,\ldots,\neq R,S} \frac{J_{RS} J_{ST} J_{RT}}{2(\omega_S - \omega_T)(\omega_T - \omega_R)} \times \\
& \qquad \times m_R m_S(I_T - m_T)(I_T + m_T + 1) \Bigg\}.
\end{aligned} \tag{27.10}
$$

[1] W. A. Anderson: Phys. Rev. 102, 151 (1956).

Die Berechnung der Verhältnisse der Intensitätsmaxima ist in höherer Ordnung sehr kompliziert. Streng ist sie überhaupt nur ausführbar, wenn man zugleich die theoretische Behandlung des Problems der Linienbreiten beherrscht, da die Breiten der verschiedenen Linien eines Spektrums im allgemeinen ungleich groß sind. Man kann indessen zeigen, vgl. hierzu[1], daß die integrierte Intensität jeder Resonanzlinie, d.h. die von ihr umschlossene Fläche proportional mit dem Quadrat des Matrixelements von $I_x + iI_y$ wächst, wenn man mit I_x bzw. I_y die x- bzw. y-Komponente des zu den fraglichen Kernen gehörenden Spin-Operators bezeichnet. Im selben Grad, in dem die Linienbreiten eines Spektrums als gleich vorausgesetzt werden können, sind die integrierten Intensitäten der Resonanzlinien ein Maß für deren Intensitätsmaxima. Explizite Berechnungen von Matrixelementen wurden von ANDERSON[2] ausgeführt, der die Übergangswahrscheinlichkeiten zwischen den benachbarten Zuständen eines aus mehreren Gruppen äquivalenter Kerne bestehenden Kernspinsystems für den Fall einer schwachen Störung dieses Systems durch ein Hochfrequenzfeld der LARMOR-Frequenz berechnete. Dabei benutzte er für die Wellenfunktionen vor und nach der Zustandsänderung die in erster Näherung erhaltenen Ausdrücke. Die Linienhöhen in den theoretischen Diagrammen der Fig. 101 geben die Ergebnisse von ANDERSON wieder. Allgemein fand er, daß die Wahrscheinlichkeit W bzw. die zu ihr proportionale Intensität eines Übergangs zwischen einem Zustand mit der Energie $E_{m_A m_B \dots}$ und einem anderen mit der Energie $E_{(m_A-1)m_B \dots}$ proportional ist zu dem Ausdruck:

$$W_{m_A \to m_A - 1} \sim (I_A - m_A + 1)(I_A + m_A)\left\{1 - \sum_{R \neq A} \frac{2 J_{AR} m_R}{\omega_A - \omega_R}\right\}. \qquad (27.11)$$

Aus Gl. (27.11) folgt, daß in jedem Multiplett (Liniengruppe für verschiedene m_R) die Intensität der Linien um so kleiner ist, je größer der Frequenzabstand von den die Aufspaltung verursachenden Gruppen ist, denn sowohl die Abstände als auch die Intensitäten der Linien sind in der beschriebenen Weise von den Größen der Produkte $J_{AR} m_R$ abhängig. Dieses Ergebnis stimmt mit dem experimentellen Befund überein, (vgl. Fig. 101) und kann als Regel zur Identifizierung der Linien komplizierterer Spektren ein nützliches Hilfsmittel sein. In diesem Zusammenhang ist noch zu bemerken, daß sich zwar aus der Struktur eines kernmagnetischen Spektrums die Vorzeichen der Spin-Spin-Kopplungskonstanten J_{RS} der untersuchten Verbindung grundsätzlich nicht bestimmen lassen. Sind indessen die Kernspins einer Gruppe äquivalenter Kerne dieser Verbindung mit den Kernspins von zwei oder mehreren anderen Gruppen gekoppelt, so kann man die Vorzeichen der zugehörigen Kopplungskonstanten relativ zueinander ermitteln, wenn man die nach der Störungstheorie zweiter Ordnung vorhergesagten Effekte mit der experimentell ausreichend hoch aufgelösten Spektrumsstruktur vergleicht.

Die Berechnung der Übergangsfrequenzen und der Linienintensitäten wird besonders einfach, wenn nur die Kerne zweier Gruppen miteinander wechselwirken, und wenn außerdem eine dieser Gruppen den maximalen Spin $\frac{1}{2}$ hat, also nur aus einem Kern mit dem Spin $\frac{1}{2}$ besteht. BANERJEE, DAS und SAHA[3] haben gezeigt, daß man in diesem Fall zur Bestimmung der Energieniveaus nur quadratische Gleichungen lösen muß. Am einfachsten sind die Verhältnisse natürlich, wenn die zwei wechselwirkenden Kerngruppen beide nur aus einem Kern mit dem Spin $I = \frac{1}{2}$ bestehen. HAHN und MAXWELL[4] haben diesen Fall

[1] F. BLOCH: Phys. Rev. **102**, 104 (1956).
[2] W. A. ANDERSON: Phys. Rev. **102**, 151 (1956).
[3] M. K. BANERJEE, T. P. DAS u. A. K. SAHA: Proc. Roy. Soc. Lond., Ser. A **226**, 490 (1954).
[4] E. L. HAHN u. D. E. MAXWELL: Phys. Rev. **88**, 1070 (1952).

zuerst theoretisch und experimentell untersucht. Die Ergebnisse ihrer Berechnungen sind in Tabelle 6 wiedergegeben. Fig. 102 zeigt ihre experimentellen Registrierungen der Protonensignale einer Verbindung dieses Typs, und zwar des 2-Brom-5-Chlorthiophens.

Tabelle 6.

Linie	Übergang $(J \to 0)$				Relative Frequenz der Linien	Relative Intensität der Linien
	m_A	$m_B \leftrightarrow m_A$		m_B		
ω_{13}	$\frac{1}{2}$	$\frac{1}{2}$	$-\frac{1}{2}$	$\frac{1}{2}$	$\frac{1}{2}J + \frac{1}{2}(\delta^2 + J^2)^{\frac{1}{2}}$	$1 - J/(\delta^2 + J^2)^{\frac{1}{2}}$
ω_{24}	$\frac{1}{2}$	$-\frac{1}{2}$	$-\frac{1}{2}$	$-\frac{1}{2}$	$-\frac{1}{2}J + \frac{1}{2}(\delta^2 + J^2)^{\frac{1}{2}}$	$1 + J/(\delta^2 + J^2)^{\frac{1}{2}}$
ω_{12}	$\frac{1}{2}$	$\frac{1}{2}$	$\frac{1}{2}$	$-\frac{1}{2}$	$\frac{1}{2}J - \frac{1}{2}(\delta^2 + J^2)^{\frac{1}{2}}$	$1 + J/(\delta^2 + J^2)^{\frac{1}{2}}$
ω_{34}	$-\frac{1}{2}$	$\frac{1}{2}$	$-\frac{1}{2}$	$-\frac{1}{2}$	$-\frac{1}{2}J - \frac{1}{2}(\delta^2 + J^2)^{\frac{1}{2}}$	$1 - J/(\delta^2 + J^2)^{\frac{1}{2}}$

Bei der von Hahn und Maxwell angewandten Methode der Beobachtung impulstechnisch angeregter freier Präzessionssignale (vgl. Ziff. 4 und 15) kann die Struktur des Spektrums nicht in der vertrauten Weise unmittelbar aus der experimentellen Registrierung abgelesen werden. Bei den kontinuierlichen Verfahren der Beobachtung erzwungen angeregter Präzessionsbewegungen im langsamen adiabatischen Resonanzdurchgang ist dies möglich, und die Abszissenachse der Registrierung kann direkt in Frequenzeinheiten geeicht werden. Die Amplitude der freien Präzessionssignale wird in den impulstechnischen Experimenten als Funktion der Zeit beobachtet. Sie ist mehr oder weniger kompliziert moduliert, wenn die beobachteten Kerne verschiedene nahe benachbarte Resonanzfrequenzen haben, so wie es bei Molekülspektren allgemein der Fall ist. Damit wird die Bestimmung der Frequenzabstände des

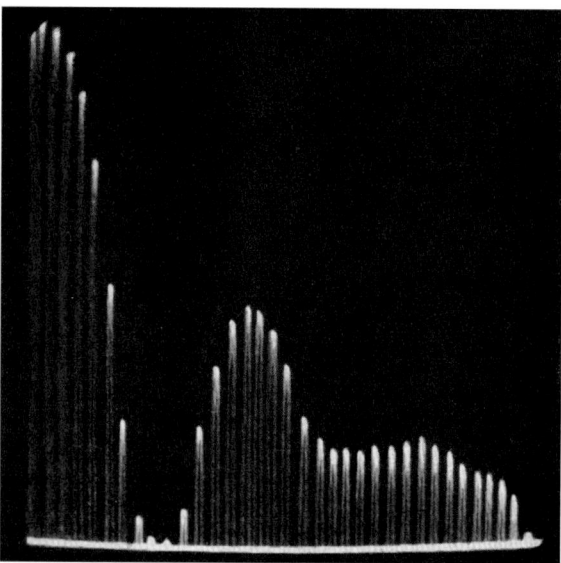

Fig. 102. Oszillographische Registrierung impulstechnisch angeregter Protonen-Echos der Verbindung 2-Brom-5-Chlorthiophen. Die Registrierung zeigt etwa 35 nacheinander mit zunehmendem Impulsabstand t' aufgenommene Echos. Deren Entfernung vom ersten Impuls $(2t')$ wuchs dementsprechend ebenfalls im Verlauf der Meßreihe in ungefähr äquidistanten Schritten. Der erste Impuls löste stets auch die Zeitablenkung des Oszillographen aus. Deren gesamte Dauer betrug 0,66 sec. Die Impulse sind auf der Aufnahme nicht zu erkennen. Gemessen wurde bei der Larmor-Frequenz 31 MHz.

untersuchten Spektrums zu einer Analyse der Schwebungsfrequenzen der freien Präzessionssignale. Zur Demonstration dieses Verfahrensunterschieds zeigt Fig. 103 das von Anderson mit der üblicheren Methode des kontinuierlichen Resonanzdurchgangs gemessene Spektrum der gleichen Substanz wie in Fig. 102. Das Ergebnis der Messungen ist, daß im 2-Brom-5-Chlorthiophen bei der Meßfrequenz von etwa 30 MHz die durch Spin-Spin-Wechselwirkung verursachte Linientrennung fast ebenso groß ist, wie die chemische Abschirmungs-Resonanzverschiebung, und zwar ist nach der Messung von ANDERSON

$$\delta/2\pi = \omega_A - \omega_B/2\pi = (4,7 \pm 0,2)\,\text{Hz} \quad \text{und} \quad J/2\pi = (3,9 \pm 0,2)\,\text{Hz}.$$

In ausreichend schwachen äußeren Feldern H_0 sind darüber hinaus wegen der Proportionalität von δ und H_0 die chemischen Verschiebungen δ stets kleiner als die konstanten Spin-Spin-Auf-spaltungen J. Verwendet man also die Bezeichnungen Feinstruktur und Hyperfeinstruktur für die beiden Effekte, so muß man stets bedenken, daß diese Bezeichnungen in der kernmagnetischen Spektro-skopie im Gegensatz zu ihrer Bedeutung in der optischen Spek-troskopie keinen Hinweis auf die relativen Größen der zwei Auf-spaltungsarten enthalten.

Als letztes Beispiel zeigt Fig. 104 das Protonenspektrum von β-Pro-piolacton $(\underline{CH_2CH_2COO})$. Man er-hält eine mit der experimentellen Registrierung quantitativ gut über-einstimmende theoretische Linien- und Intensitätsverteilung (vgl. Fig.104 unten), wenn man annimmt, daß das Molekül zwei Gruppen von

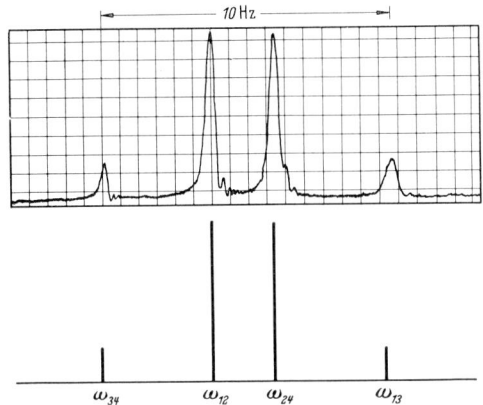

Fig. 103. Protonenspektrum von 2-Brom-5-Chlorthiophen. Das theoretische Spektrum unter der experimentellen Registrierung wurde für $J/\delta = 0,835$ nach Tabelle 6 berechnet. Die Bezeichnungs-weise der Frequenzen stimmt mit der dort benutzten überein.

je zwei äquivalent gebundenen Protonen enthält. Studiert man die Struktur des Moleküls, so erkennt man, daß dieses Ergebnis nicht trivial ist. Es bestätigt

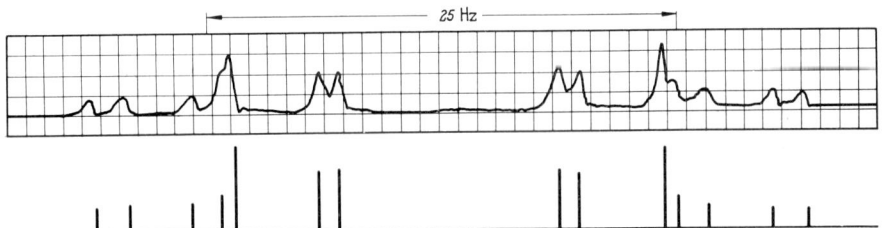

Fig. 104. Protonenspektrum von β-Propiolacton. Das theoretische Spektrum unter der experimentellen Registrierung wurde für $J/\delta = 0,265$ berechnet. Die chemische Resonanzfrequenz-Verschiebung und die Spin-Spin-Wechselwirkungskonstante des Moleküls haben die Größen: $\delta/2\pi = (23,0 \pm 1,0)\,\text{Hz}$ (bei 30 MHz), $J/2\pi = (6,1 \pm 0,3)\,\text{Hz}$.

im übrigen die schon früher ausgesprochene Vermutung, daß das Molekül eine Symmetrieebene besitzt, in welcher sich die Kohlenstoff- und die Sauerstoffatome befinden, während die Bindungsorte der Protonen symmetrisch ober- und unter-halb dieser Ebene liegen.

Die beschriebenen Beispiele zeigten bereits, daß die kernmagnetische Spektro-skopie flüssiger und gasförmiger Proben neben ihrer Anwendbarkeit zur Erfor-schung von chemischen Austauschprozessen auch ein ausgezeichnetes Hilfsmittel zur Klärung von chemischen Strukturproblemen, zur Ermittlung von räumlichen Atomanordnungen und zur Ausführung von Analysen darstellt[1]. Die zuletzt genannte Anwendungsmöglichkeit ist im wesentlichen zur Lösung industrieller Probleme geeignet, denn sie erfolgt einerseits ohne chemische Umwandlungen, setzt aber andererseits voraus, daß die Spektrenstrukturen der Verbindungen, die in der Analyse identifiziert werden sollen, bereits bekannt sind. Entsprechende Tabellen wurden von Wertz[2], von Gutowsky[3] und von anderen veröffentlicht. Zur Lösung der meisten chemischen Konstitutionsprobleme ist es übrigens ausreichend, wenn das Auflösungsvermögen der Apparatur etwa $1 \cdot 10^7$ beträgt (wie z.B. in Fig. 100), extreme Homogenisierungs-Maßnahmen sind also nicht erforderlich. Ein typisches Beispiel für diese Anwendungsart ist die von Kende[4] mitgeteilte Strukturaufklärung der Feist-Säure. Mit chemischen Methoden konnte nicht eindeutig entschieden werden, welche der folgenden drei Strukturformeln

$$
\text{a)} \quad H_2C=C \begin{subarray}{l} \\ \\ \end{subarray} \begin{array}{c} H \\ | \\ C-COOH \\ C-COOH \\ | \\ H \end{array} \qquad \text{b)} \quad C_3H-C \begin{array}{c} H \\ | \\ C-COOH \\ \| \\ C-COOH \end{array} \qquad \text{c)} \quad H_3C-C \begin{array}{c} H \\ | \\ C-COOH \\ | \\ C-COOH \end{array}
$$

richtig ist. Aus der Anzahl, den Lagen und dem Intensitätsverhältnis der kernmagnetischen Resonanzlinien ergab sich sofort die Strukturformel a, und zwar genügte es zur Entscheidung dieser Frage — analog Fig. 99 —, das Spektrum der Abschirmungs-Resonanzverschiebungen (Feinstruktur) zu studieren. Ein typisches Beispiel für die Ausnützung der Spin-Spin-Aufspaltung zur Strukturuntersuchung ist der Vergleich der kernmagnetischen Spektren der verschiedenen Phosphorsäuren[5]:

$$
O=P \begin{array}{c} OH \\ OH \\ OH \end{array} \qquad O=P \begin{array}{c} OH \\ OH \\ H \end{array} \qquad O=P \begin{array}{c} OH \\ H \\ H \end{array}
$$

a) Phosphorsäure   b) Phosphorige Säure   c) Unterphosphorige Säure

In mittlerer Auflösung beobachtet man nur Linienaufspaltungen, welche durch Spin-Spin-Wechselwirkungen zwischen unmittelbaren Nachbarn hervorgerufen werden. Infolgedessen besteht das Phosphorresonanz-Spektrum der Moleküle a, b bzw. c in erster Näherung aus 1, 2 bzw. 3 Linien; eine Unterscheidung der Säuren ist dementsprechend leicht möglich.

Gelegentlich kann es bei der Untersuchung der Spektren komplizierter Moleküle vorkommen, daß nicht eindeutig entschieden werden kann, zwischen welchen Nachbarkernen oder Nachbarkerngruppen die für die beobachteten Linienaufspaltungen verantwortlichen Spin-Spin-Wechselwirkungen stattfinden. Einer Anregung von Bloch[6] folgend, haben deshalb Bloom und Shoolery[7] sowie Anderson[8] eine Methode entwickelt, die es gestattet, die an einer Spin-Spin-Kopplung beteiligten Partner experimentell zu identifizieren. Das Verfahren ist

[1] K. H. Hausser: Angew. Chem. 68, 729 (1956).
[2] J. E. Wertz: Chem. Reviews 55, 849 (1953).
[3] H. S. Gutowsky: Prospekt der Firma Varian Associates, Palo Alto, Calif. USA (Hersteller von Kernresonanz-Spektrometern).
[4] A. S. Kende: Chem. a. Ind. 1956, 437.
[5] H. S. Gutowsky, D. W. McCall u. C. P. Slichter: J. Chem. Phys. 21, 279 (1953).
[6] F. Bloch: Phys. Rev. 93, 944 (1954).
[7] A. L. Bloom u. J. N. Shoolery: Phys. Rev. 97, 1261 (1955).
[8] W. A. Anderson: Phys. Rev. 102, 151 (1956).

nur anwendbar, wenn die Abschirmungs-Verschiebung groß ist, verglichen mit der untersuchten Spin-Spin-Kopplung. Es besteht darin, eine der beiden vermutlich wechselwirkenden Kerngruppen mit einem relativ starken Hochfrequenz-Feld der LARMOR-Frequenz ihrer Kerne anzuregen, während zur gleichzeitig erfolgenden Beobachtung der Resonanzlinien der zweiten Gruppe nur ein relativ schwaches Hochfrequenzfeld der LARMOR-Frequenz der zu dieser Gruppe gehörenden Kerne verwendet wird. Dies hat zur Folge, daß die Abstände der Linien des Multipletts der zweiten Gruppe stark reduziert werden, wenn für dessen Struktur tatsächlich Spin-Spin-Kopplungen mit der stark angeregten ersten Kerngruppe verantwortlich sind. Technische Einzelheiten über die Anwendung des Verfahrens sowie eine Erklärung des ausgenutzten Effekts findet man in der Arbeit von ANDERSON.

28. Struktur der Spektren von Festkörpern. Das Studium der kernmagnetischen Spektren fester Körper hat zur Entdeckung zahlreicher Kernresonanz-Phänomene geführt, die in ihrer Mannigfaltigkeit getreulich das Bild wiederspiegeln, das die gesamte Festkörperphysik dem in anderen Disziplinen tätigen Physiker bietet. Um eine komprimierte Darstellung dieses neuen, Kern- und Festkörperphysik verbindenden Forschungszweigs zu ermöglichen, soll zunächst seine Gliederung besprochen werden. In der Kernresonanz-Spektroskopie wird geprüft, welche magnetischen und welche elektrischen Wechselwirkungen zwischen den Partikeln der Substanzen und zwischen diesen und äußeren Feldern existieren. Somit ist zur Charakterisierung der Eigenschaften eines Festkörpers vom Standpunkt des hier behandelten Fachgebiets aus nacheinander zu fragen, in welcher Form die Partikel in dem Körper angeordnet sind, welche magnetischen oder elektrischen Momente die in ihm enthaltenen Kerne und Elektronenhüllen besitzen, und über welche Bewegungsmöglichkeiten sie innerhalb des Körpers verfügen.

Komplizierte Vorgänge lassen sich schneller verstehen, wenn es gelingt einfache Sonderfälle ausfindig zu machen, zu deren Verständnis nur ein Teil der in Frage kommenden Parameter berücksichtigt werden muß. Dementsprechend werden in dem vorliegenden Bericht in der Reihenfolge der obigen Aufzählung zunächst nur Festkorper behandelt, die tatsächlich als starr betrachtet werden dürfen, und welche darüber hinaus nur Kernmomente und unter diesen nur magnetische Dipolmomente enthalten. In der damit abgegrenzten Festkörperklasse werden dann die Formen der kernmagnetischen Resonanzkurven für zunehmend verwickeltere Partikelanordnungen besprochen. In dem zweiten Teil der Darstellung wird der Einfluß von Kernquadrupol-Wechselwirkungen mit dem Gradienten innerer elektrischer Felder auf die Linienform der kernmagnetischen Resonanz in ebenfalls zunächst als völlig starr vorausgesetzten Festkörpern beschrieben. Im Anschluß daran sollen jene Resonanzlinien-Effekte besprochen werden, die als Beweglichkeitseffekte erklärt werden. Eine Aufzählung solcher Bewegungsmöglichkeiten umschließt sowohl einfache Umorientierungen im Gefüge des Festkörpers, als auch typische Diffusionserscheinungen. Insbesondere ist in diesem Zusammenhang das kernmagnetische Verhalten einer bekanntlich sehr wichtigen Festkörperklasse, nämlich derjenigen der Metalle, zu besprechen, deren besonderes Kennzeichen ja gerade die weitgehende Bewegungsfreiheit eines Teils ihrer Partikel, der Leitungselektronen ist. Das Relaxationsverhalten aller Substanzen spiegelt, wie in Ziff. 3 bereits besprochen, gerade deren Bewegungszustand wieder; somit wäre es sinnvoll, die Relaxations-Erscheinungen sowie deren indirekten Einfluß auf die Resonanzlinienform im Anschluß an die Besprechung der direkten Beweglichkeitseffekte zu diskutieren. Um die Einheitlichkeit der

Darstellung zu wahren, soll jedoch bezüglich der in Festkörpern wirksamen Relaxations-Mechanismen an die Ausführungen in Ziff. 3 erinnert werden.

Dank der Beugungsversuche mit Röntgen-, Elektronen- und Neutronenstrahlen kennt man die Struktur der meisten Kristalle recht genau. Es lag daher nahe, zunächst einen bekannten Einkristall möglichst einfacher Struktur kernmagnetisch zu untersuchen. Bei der Auswahl der einfachsten Kristallstruktur kommt es allerdings in der magnetischen Kernresonanz-Spektroskopie weniger darauf an, daß der Kristall eine möglichst hohe Symmetrie besitzt, wie man es in der eigentlichen Festkörperphysik fordern würde, sondern man muß danach fragen, wieviel Kerne in ihm gruppenweise magnetisch wechselwirken, und wieviele verschiedene Typen solcher Wechselwirkungs-Gruppen in dem Kristall existieren. Diese Fragestellung wird verständlich, wenn man bedenkt, daß die Wechselwirkung zweier Dipole μ_1 und μ_2

$$\mathfrak{H}_{00} = r^{-3} \left[\mu_1 \cdot \mu_2 - 3 (\mu_1 \cdot r) (\mu_2 \cdot r) r^{-2} \right] \tag{28.1}$$

umgekehrt proportional zu der dritten Potenz ihres Abstands ist. Unterscheiden sich demnach die Beträge zweier Radiusvektoren zwischen magnetisch gekoppelten Nachbarkernen nur um etwa einen Faktor 2, so kann die Wechselwirkung des entfernteren Partners praktisch schon vernachlässigt werden. Im allgemeinen gehören die zu berücksichtigenden Kopplungspartner übrigens zu demselben Molekül.

Der einfachste Fall einer direkten magnetischen Kopplung in einem festen Körper liegt demnach vor, wenn die Kernmagnete in einem Einkristall paarweise enthalten sind. Unter den aus der Kristallographie bekannten Substanzen ist diese Kristallart glücklicherweise gerade besonders häufig vertreten, z.B. gehören ihr fast alle Salzkristalle an, welche Hydratmoleküle enthalten. In einer musterhaft ausgeführten Arbeit hat zuerst PAKE[1] das kernmagnetische Spektrum eines solchen Salzes gemessen und gedeutet. Als Probensubstanz verwandte er Gips ($CaSO_4 \cdot 2 H_2O$) in mono- und polykristalliner Form. Gips enthält keine paramagnetischen Ionen und außer den Protonen der Wassermoleküle, sowie einigen natürlich selten vorkommenden Isotopen von Calcium, Schwefel und Sauerstoff auch keine Kerne mit nicht verschwindendem Spin. Seine Struktur wurde von WOOSTER[2] ausführlich untersucht und war bekannt, wenn man von der genauen Lage der Protonen absieht, welche sich wie die aller leichten Kerne röntgenographisch kaum ermitteln läßt. Aus der elementaren klassischen Physik ist bekannt, daß die Feldstärke eines kleinen Magneten mit dem Dipolmoment μ an einem Ort im Abstand r, dessen Verbindungslinie zu dem Dipol mit der Dipolachse den Winkel Θ einschließt, den Wert $\mu/r^3 (3 \cos^2 \Theta - 1)$ hat. Befindet sich demnach ein einzelnes Protonenpaar unbeweglich in einem äußeren Feld $\boldsymbol{H}^*$, so beträgt die Gesamtfeldstärke an den Orten der beiden Protonen [vgl. (28.1)]

$$\boldsymbol{H} = \boldsymbol{H}^* \pm \mu_p \, r^{-3} (3 \cos^2 \Theta - 1), \tag{28.2}$$

je nach dem ob das wechselwirkende Nachbarproton parallel oder antiparallel zu dem äußeren Feld liegt. Zwar sind nach der Quantentheorie die beiden genannten Orientierungen nicht möglich, da die Kerne aber stets mit der LARMOR-Frequenz um die Feldrichtung präzessieren, liegt ihr zeitlich mittleres magnetisches Moment tatsächlich zur äußeren Feldrichtung parallel oder antiparallel. Übrigens ist μ, das sog. magnetische Moment, in Wirklichkeit ja nur die maximale

[1] G. E. PAKE: J. Chem. Phys. 16, 327 (1948). Vgl. auch G. E. PAKE: Amer. J. Phys. 18, 438, 473 (1950).
[2] W. A. WOOSTER: Z. Kristallogr. 94, 375 (1936).

zur äußeren Feldrichtung parallele Komponente des eigentlichen magnetischen Moments (vgl. Ziff. 1) und deren Größe ist gleich dem zeitlichen Mittelwert des Moments. Der genannte Einwand beeinträchtigt also auch quantitativ die Gültigkeit der halbklassisch abgeleiteten Beziehung (28.2) nicht. Dagegen zwingt ein zweiter quantentheoretischer Einwand zu einer genaueren Behandlung des Problems. Sind die wechselwirkenden Kerne identisch, wie in dem hier besprochenen Fall eines Protonenpaars, so stimmen auch ihre LARMOR-Frequenzen überein. Es existiert demnach außer der durch (28.2) beschriebenen statischen Kopplung noch ein zweiter dynamischer Wechselwirkungs-Prozeß und zwar das gleichzeitige Umklappen zweier Kerne, wechselseitig induziert durch ihre präzessierenden Felder, energetisch also der Übergang eines Quants $h\nu$ der LARMOR-Frequenz von einem der beiden Kerne zu dem andern (vgl. Ziff. 3). Zur Lösung des Problems hat PAKE die magnetische Wechselwirkung zwischen den Kernen eines Paars als Störung der ZEEMAN-Energieniveaus der Kerne im äußeren Feld behandelt und in erster Ordnung berechnet, daß die Resonanzfeldstärke eines Paars identischer Kerne die Werte

$$H = H^* \pm \tfrac{3}{2}\,\mu\,r^{-3}\,(3\cos^2\Theta - 1) \qquad (28.3)$$

hat, wenn man unter H^* die Resonanzfeldstärke der ungestörten Kerne versteht. Der Ausdruck (28.3) unterscheidet sich von (28.2) um einen Faktor $\tfrac{3}{2}$, der nach der PAKEschen Rechnung und nach der obigen anschaulichen Erklärung erwartungsgemäß für ein Paar nicht identischer Kerne den Wert 1 erhält, in diesem Fall ist also die Beziehung (28.2) richtig.

Enthält die Elementarzelle des untersuchten Einkristalls n Typen von Protonenpaaren, welche sich in der Regel nur in der Orientierung Θ ihrer Verbindungslinien bezüglich des äußeren Felds, nicht aber in deren Größe r unterscheiden, so ist gemäß (28.3) zu erwarten, daß das kernmagnetische Spektrum des Kristalls aus $2n$ mehr oder weniger deutlich unterscheidbaren und symmetrisch um die Resonanzfrequenz der ungestörten Kerne verteilten Resonanzlinien besteht. Die Linienzahl kann kleiner als $2n$ sein, wenn sich eines oder mehrere der Linienpaare in der Spektrumsmitte vereinigen, wenn also für die Orientierungswinkel Θ_ν der zugehörigen Protonenpaare die Beziehung $3\cos^2\Theta = 1$ gültig ist, oder wenn die Orientierungswinkel Θ_ν zweier oder mehrerer Protonenpaare gleich groß sind, wenn also die Maxima der zugehörigen Linienpaare von der Spektrumsmitte den gleichen Abstand haben und sich somit überdecken. Darüber hinaus kann die beobachtbare Linienzahl auch noch kleiner als die Zahl der in Wirklichkeit existierenden Linien sein. Sind die Linienabstände kleiner als die Linienbreiten, dann ist wegen des praktisch immer begrenzten Auflösungsvermögens der Apparaturen eine Trennung der Linien im allgemeinen nicht mehr möglich.

Während die Größen der Kernverbindungslinien und deren Orientierungen bezüglich der Kristallachsen Konstanten der Kristallstruktur sind, sind die Orientierungswinkel Θ_ν dieser Verbindungslinien bezüglich des äußeren Felds H^* frei wählbar, wenn man davon absieht, daß die zwischen ihnen bestehenden Winkelbeziehungen natürlich stets erhalten bleiben. Zur Bestimmung der Größen und der Winkelkoordinaten der Radiusvektoren im Kristallgefüge kann man den Kristall etwa zunächst im Magnetfeld so orientieren, daß die Feldlinien parallel zu einer Ebene einer Kristallebenen-Schar verlaufen, deren Normalen eine von den Feldlinien senkrecht durchstoßene Kreisfläche bilden. Dann kann man nacheinander die zu diesen verschiedenen Orientierungen des Kristalls gehörenden kernmagnetischen Spektren in Funktion der Lage der Feldlinien innerhalb jeder der zum äußeren Feld parallelen Ebenen beobachten. Dreht man auf diese Weise den Kristall einmal um die Feldrichtung um 360° und zum anderen um jede der Normalen der zur Feldrichtung parallelen Ebenen-Schar nacheinander ebenfalls um $360°$, so findet man sicher alle Kristallorientierungen, in denen die Abstände der verschiedenen Linienpaare maximal oder minimal sind oder untereinander übereinstimmen. Aus diesen Beobachtungsergebnissen kann man nach (28.3) die interessierenden Daten der Kristallstruktur auch in recht komplizierten Fällen ermitteln.

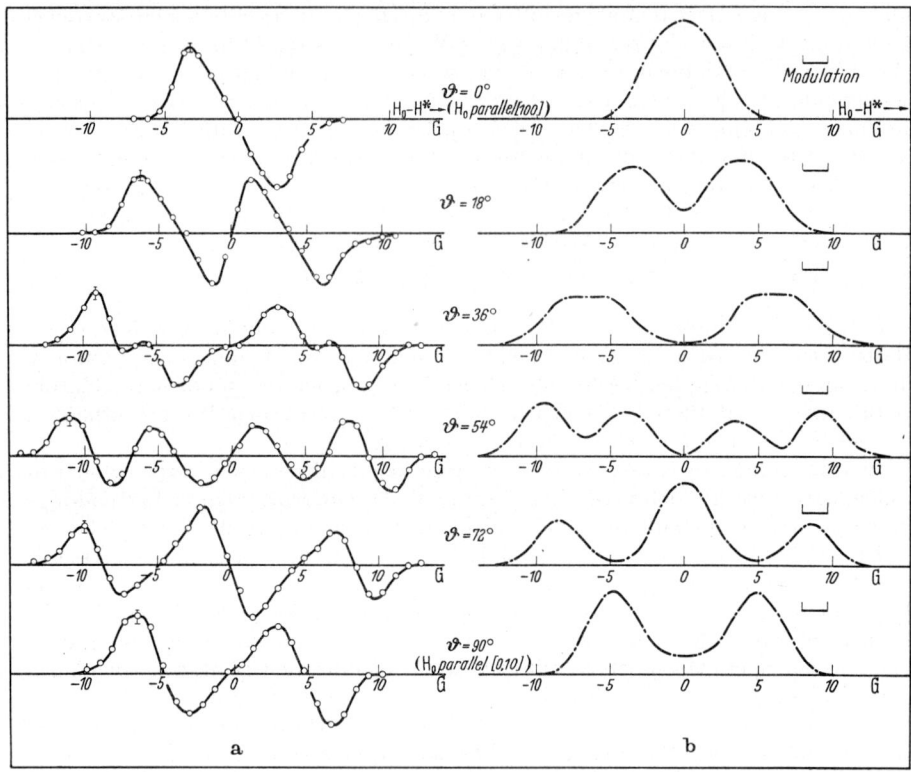

Fig. 105. Verlauf der Protonenresonanzkurve in einem Gips-Einkristall (CaSO$_4$ · 2H$_2$O). Die Figur zeigt experimentelle Registrierungen des Verlaufs der Absorptionskurven-Ableitungen (a) sowie der Integrale dieser Kurven (b). Parameter der Messungen ist der Winkel ϑ zwischen der Richtung des äußeren, stets zur (001)-Ebene parallelen Felds H_0 und der Achsenrichtung (100) des Gipskristalls. Die Ordinatenmaßstäbe wurden willkürlich gewählt. Die Größe der Rausch-spannungs-Schwankungen wurde jeweils auf dem ersten Maximum der Ableitungskurven angedeutet. Die Amplitude der Modulation des Gleichfelds H_0, gemessen von Maximum zu Maximum, betrug 1,5 Gauß. Die Striche über den Absorptions-kurven sollen eine anschauliche Vorstellung über die relative Größe dieser Amplitude vermitteln. Die Messungen wurden in einem Feld von 6823 Gauß bei der Frequenz 29,01 MHz ausgeführt.

In den meisten praktischen Fällen ist die Zahl der verschiedenen Kernpaar-Typen verhältnismäßig klein, außerdem weiß man in der Regel von den Röntgen-struktur-Untersuchungen bereits, welche Orientierungen der Kernpaare bezüg-lich der Kristallachse möglich sind, so daß die kernmagnetische Strukturunter-suchung des Kristalls oft nur zur Auswahl unter verschiedenen oder zur Prüfung bestimmter Orientierungs-Hypothesen erforderlich ist. In diesen Fällen genügt es in der Regel, wenn man den Kristall parallel zu einer entsprechend der Struktur-vermutung passend gewählten Hauptebene schneidet, und dann die Spektrums-formen in Funktion des Drehwinkels des Kristalls um die Normale dieser Ebene bestimmt. Fig. 105 zeigt als Beispiel einer solchen Meßreihe die von Pake ge-messenen Protonen-Resonanzkurven eines Gipseinkristalls, dessen (001)-Ebene parallel zu den äußeren Feldlinien orientiert war und dann um die Normale dieser Ebene gedreht wurde. Es zeigt sich, daß die Elementarzelle des Gipskristalls zwei verschieden orientierte Protonenpaar-Typen enthält (übereinstimmend mit der Anzahl der Hydratmoleküle des Calciumsulfats), denn die Resonanzkurve zeigt im allgemeinen eine aus zwei symmetrischen Linienpaaren bestehende Feinstruktur. Liegt das äußere Feld H_0 in der (001)-Ebene speziell parallel zu der [100]-Achse des Kristalls (Drehwinkel $\vartheta = 0$), so fallen alle vier Linien nahezu zusammen (vgl. Fig. 105). Dagegen haben in den beiden Stellungen $\vartheta = 18°$ und $\vartheta = 90°$ des Kristalls die beiden Linienpaare fast den gleichen Abstand. Die vierte Registrierung ($\vartheta = 54°$) zeigt den allgemeinen Fall der aus vier symmetrischen Kernsignalen bestehenden Resonanzkurve, und die zu dem Drehwinkel $\vartheta = 72°$ gehörende Registrierung als weiteren Sonderfall die Vereinigung der beiden Linien einer Protonenpaar-Gruppe im Zentrum. Ein Ergebnis der Messungen war, daß die Protonenabstände in den Wassermolekülen des Kristalls 1,58 Å betragen. Verbindet man dieses Resultat mit der Annahme, daß der Öffnungswinkel des Wassermoleküls ($\not< HOH$) 108° beträgt, so folgt für seinen OH-Abstand 0,98 Å.

Das geschilderte Verfahren läßt sich in der gleichen Form auch zur Untersuchung komplizierter Salzkristalle mit zahlreichen verschieden orientierten Protonenpaar-Typen anwenden, nur müssen dann natürlich auch die Spektren in ausreichend vielen verschiedenen Ebenen untersucht werden. Ein schönes Beispiel hierzu hat LÖSCHE[1] veröffentlicht, der die Struktur des Seignettesalzes, eines Hydratkristalls mit 16 verschieden orientierten Wassermolekülen auf diese Weise klären konnte.

Auch aus den kernmagnetischen Spektren polykristalliner Substanzen (also gewissermaßen aus den kernresonanz-spektroskopischen DEBYE-SCHERRER-Diagrammen der Kristalle) lassen sich in vielen Fällen noch Informationen über die Struktur des Kristallgefüges gewinnen. Allerdings ist sofort einzusehen, daß solche Aussagen nur die Abstände der wechselwirkenden Kerne betreffen können, nicht aber deren Orientierungen. Um eine quantitative Behandlung des Problems der Resonanzkurven polykristalliner Festkörper überhaupt zu ermöglichen, muß man voraussetzen können, daß alle Orientierungen der Kristallkörner der Probesubstanz gleich wahrscheinlich sind. Damit wird es aber unwesentlich, welche Richtung die verschiedenen Kernpaar-Typen in den einzelnen Kristalliten haben. Aus der genannten Voraussetzung folgt für die Wahrscheinlichkeit $w(\Theta)$, daß die Verbindungslinie irgendeines Kernpaars mit der äußeren Feldrichtung den Winkel Θ einschließt, die Beziehung $w(\Theta) = 1/2 \cdot \sin \Theta$. Nach Gl. (28.3) sind der Winkel Θ und der Resonanzfeldabstand $h = H_0 - H^*$ der Zentren H_0 der Linienpaare vom Zentrum H^* der gesamten Resonanzkurve miteinander verknüpft:

$$h = \pm \frac{3}{2} \mu r^{-3} (3 \cos^2 \Theta - 1) \quad \text{bzw.} \quad \cos \Theta = \pm \frac{1}{\sqrt{3}} \left(\frac{\pm 2h}{3 \mu r^{-3}} + 1 \right)^{\frac{1}{2}}. \quad (28.4\text{a, b})$$

Diese Beziehungen sind zweideutig, denn einerseits gehören zu jedem Wert von Θ zwei h-Werte ($h^+ = +\frac{3}{2} \cdots$ und $h^- = -\frac{3}{2} \cdots$), andererseits erhält man aber auch einen Teil der h-Werte, nämlich alle die in dem Intervall $(-\frac{3}{2} \cdot \mu \cdot r^{-3}, +\frac{3}{2} \cdot \mu \cdot r^{-3})$ enthaltenen für zwei Werte von Θ, denn $(3 \cos^2 \Theta - 1)$ kann für geeignet gewählte Winkel Θ sowohl alle positiven als auch alle negativen Werte zwischen 0 und 1 haben. Diese kleine Komplikation ist bei der Ausführung des nächsten Schritts der Überlegung, dem Übergang von der Wahrscheinlichkeit $w(\Theta)$ der Kernpaar-Orientierung Θ zu der Wahrscheinlichkeit $p(h)$ des Resonanzabstands h der

---

[1] A. LÖSCHE: Exp. Techn. Physik 3, 18 (1956).

Linienzentren zu beachten. Zwischen den beiden Wahrscheinlichkeiten würde offenbar ohne diese Vorzeichen-Zweideutigkeit in (28.4) der einfache Zusammenhang

$$p(h)\, dh = 2w(\Theta)\, d\Theta \qquad (28.5)$$

bestehen, da dann der Bruchteil der Linienzentren, welche sich in dem Intervall $(h, h + dh)$ befänden, gleich dem Bruchteil der Kernpaar-Orientierungen wäre, deren Winkel bezüglich der äußeren Feldrichtung nach (28.4) entweder Θ oder $180°{-}\Theta$ betragen [h hängt in (28.4) quadratisch von $\cos\Theta$ ab]. Enthielte somit die Beziehung (28.4) nur das Plus- bzw. nur das Minuszeichen, so hätten die entsprechenden Wahrscheinlichkeiten $p^+(h)$ und $p^-(h)$ der Linienzentren-Abstände h die Größen

$$\begin{matrix} p^+(h) \\ p^-(h) \end{matrix} = \frac{2\cdot w(\Theta)}{dh/d\Theta} = 3^{-\frac{3}{2}}\cdot \mu^{-1}\cdot r^{+3}\left[1 \pm h\Big/\left(\frac{3}{2}\,\mu\, r^{-3}\right)\right]^{-\frac{1}{2}} \qquad (28.6)$$

und zwar die Wahrscheinlichkeit $p^+(h)$ in dem Intervall $-\frac{3}{2}\mu r^{-3} < h < 3\mu r^{-3}$ und die Wahrscheinlichkeit $p^-(h)$ in dem Intervall $-3\mu r^{-3} < h < \frac{3}{2}\mu r^{-3}$. (Außerhalb der genannten Intervalle haben beide Wahrscheinlichkeiten den Wert Null.) Tatsächlich sind die Wahrscheinlichkeiten aber wegen der Existenz des jeweiligen anderen Vorzeichens nur halb so groß. Die Wahrscheinlichkeiten der zu den beiden Vorzeichen gehörenden Übergänge sind praktisch gleich groß, haben also den Wert $\frac{1}{2}$, da die Boltzmann Faktoren der zugehörigen Energieniveaus bei Zimmertemperatur um weniger als 10^{-5} von 1 abweichen. Die Besetzungszahlen der Niveaus unterscheiden sich dementsprechend praktisch nicht. Die Gesamtwahrscheinlichkeit des Linienzentrums-Abstands $h = H_0 - H^*$ ist somit gleich der halben Summe der Teilwahrscheinlichkeiten (28.6):

$$p(h) = 3^{-\frac{1}{2}}\cdot 2^{-2}\cdot \left(\tfrac{3}{2}\mu r^{-3}\right)^{-1}\left[\left(1 + h/\tfrac{3}{2}\mu r^{-3}\right)^{-\frac{1}{2}} + \left(1 - h/\tfrac{3}{2}\mu r^{-3}\right)^{-\frac{1}{2}}\right], \qquad (28.7)$$

wobei die obigen Intervallvorschriften zu beachten sind.

Berechnet man den Verlauf dieser Funktion für den am Gips-Einkristall ermittelten Wert $\frac{3}{2}\cdot \mu\cdot r^{-3} = 5{,}4$ Gauß, so erhält man die in Fig. 106 im oberen Diagramm gestrichelt wiedergegebene Kurve. Die resultierende im Experiment beobachtete Absorptionskurve einer Gipspulver-Probe ist jedoch keine Aneinanderreihung von Linienzentren, sondern eine Überlagerung infinitesimaler Absorptionskurven endlicher Breite. Bezeichnet man die normierte Funktion, welche die Form dieser einzelnen Resonanzkurven beschreibt, mit $s(H - H_0)$, so gilt (normiert heißt, daß $\int\limits_{-\infty}^{\infty} s\cdot dH = 1$ ist) für die resultierende Absorptionskurve einer polykristallinen Probe offenbar allgemein:

$$f(h) = \int\limits_{-\infty}^{+\infty} p(H_0 - H^*)\cdot s(H - H_0)\, dH_0. \qquad (28.8)$$

Die obere ausgezeichnete Kurve in Fig. 106 wurde nach (28.8) und (28.7) unter der Annahme berechnet, daß $s(H - H_0)$ eine Gaußsche Funktion ist, also die Form $e^{-(H-H_0)^2/2\beta^2}$ hat. Dazu wurde die Linienbreite β der Komponenten aus den Breiten der äußeren Linien der Registrierung $\vartheta = 72°$ der Fig. 105 zu 1,54 Gauß bestimmt. Verantwortlich für die endliche Linienbreite der Komponenten sind die in dem voranstehenden Text vernachlässigten Wechselwirkungen zwischen weiter entfernten Protonen, deren zweitnächstes in Übereinstimmung mit dem voranstehenden β-Wert jeweils etwa den Abstand 2,8 Å hat, während der intramolekulare Protonenabstand des Wassers, wie bereits erwähnt, zu 1,58 Å ermittelt wurde.

Aus der Ableitung geht hervor, daß alle polykristallinen Substanzen, deren Kerndipole genügend ausgeprägt paarweise angeordnet sind, im Resonanzexperiment die für sie typische, in Fig. 106 wiedergegebene Zweimaxima-Resonanzkurve zeigen sollten. Diese Folgerung konnte in zahlreichen Fällen experimentell bestätigt werden. Insbesondere ist die Anwendbarkeit der besprochenen Theorie natürlich nicht auf Hydratkristalle beschränkt, so konnten z.B. GUTOWSKY, KISTIAKOWSKY, PAKE und PURCELL[1] den Protonenabstand im 1,2-Dichloräthan-Molekül (CH_2Cl-CH_2Cl) aus dessen Polykristall-Spektrum zu 1,7 Å ermitteln.

In manchen Substanzen sind die Kerne, deren Spektrum untersucht werden soll, in mehr oder weniger deutlich getrennten Gruppen von drei oder vier Kernen angeordnet. Unterscheiden sich die Abstände der identischen oder verschiedenen Kerne solcher Gruppen ausreichend von den Beträgen ihrer Radiusvektoren zu allen Kernen der Nachbargruppen, so sind zum Verständnis der Spektren dieser nacheinander immer komplizierteren Festkörperarten keine neuen oder bei der voranstehenden Besprechung der Paar-Kristalle nicht erwähnten physikalischen Gedankengänge erforderlich. Dagegen spiegelt der Umfang der zur quantitativen Deutung der Spektren notwendigen Berechnungen die zunehmende Komplikation der Festkörperstruktur wieder. Typische Beispiele für Festkörper

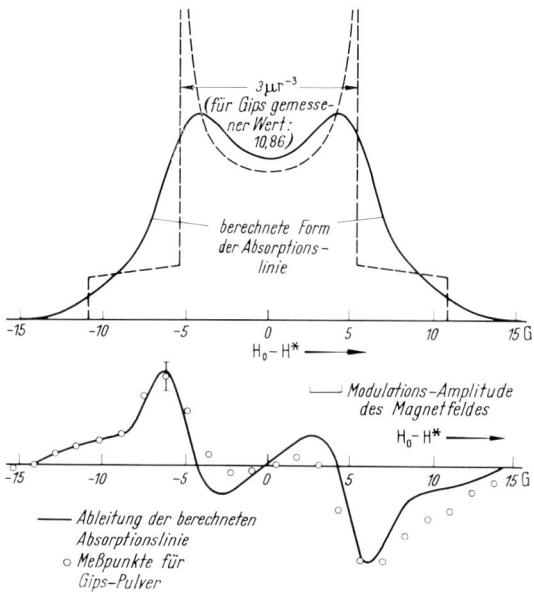

Fig. 106. Verlauf der Protonenresonanzkurve in polykristallinem Gips ($CaSO_4 \cdot 2H_2O$). Die gestrichelte Kurve des oberen Diagramms zeigt den Verlauf der berechneten Verteilungsfunktion aller Resonanzlinien-Mittelpunkte für den Fall, daß nur die Wechselwirkungen zwischen den beiden nächsten Nachbarn, also zwischen den zwei Protonen jedes H_2O-Moleküls berücksichtigt werden. Beachtet man, daß jede einzelne Resonanzlinie eine endliche Breite hat, womit auch die Wechselwirkungen zwischen weiter voneinander entfernten Protonen in die Betrachtung einbezogen werden, so erhält man für die Form des Absorptionssignals von polykristallinem Gips die ausgezogene Kurve des oberen Diagramms. In der Zeichnung wurde angenommen, daß alle einzelnen Resonanzlinien eine Gaußsche Form der Breite 1,54 Gauß haben. Deren Überlagerung gemäß der gestrichelt gezeichneten Verteilungsfunktion liefert dann die berechnete Absorptionskurve. Experimentell beobachtet wurde die Ableitung dieser Kurve. Das untere Diagramm dient also zur Erleichterung des Vergleichs der berechneten Kurve mit den Meßwerten.

mit in Dreiergruppen geordneten Kernen sind u. a. zahlreiche feste organische Verbindungen, welche Methylgruppen (CH_3) enthalten. So zeigt Fig. 107, daß ebenfalls von den zuletztgenannten Autoren[1] gemessene Protonenspektrum von polykristallinem 1,1,1-Trichloräthan (CH_3-CCl_3). Obwohl in diesem speziellen Fall alle drei Kerne identisch sind, den Spin $\frac{1}{2}$ besitzen und untereinander sogar gleiche Abstände haben, also im Körper gleichseitige Dreiecke bilden, liefert die zum Verständnis des Spektrums der polykristallinen Substanz zunächst wiederum erforderliche Berechnung der Linienzentren-Verteilung — analog (28.7) — einen verglichen mit der gestrichelten Kurve der Fig. 106 überraschend komplizierten Verlauf dieser Wahrscheinlichkeitsverteilung. Allerdings vermischen sich die Details der Kurve — zehn verschieden große Maxima in unregelmäßig großen Abständen —

[1] H. S. GUTOWSKY, G. B. KISTIAKOWSKY, G. E. PAKE u. E. M. PURCELL: J. Chem. Phys. 17, 972 (1949); 18, 162 (1950).

im allgemeinen bei der Ausführung des nächsten Rechenschritts, der Integration nach (28.8) wegen der endlichen Linienbreite der Komponenten, d.h. wegen der Wechselwirkungen zwischen den Dreiergruppen. Man erhält als für diese Unterklasse der Dreiergruppen-Polykristalle typische Spektrumsform eine Dreimaxima-Resonanzkurve, wie sie in Fig. 107 wiedergegeben ist. Weitere Beispiele für diesen einfachsten Fall der Anordnung der Dreiergruppe sind eine Anzahl saurer Hydrate, welche Oxoniumionen (H_3O^+) enthalten. Festkörper dieses Typs wurden untersucht von RICHARDS und SMITH[1], sowie von KAKIUCHI, SHONO, KOMATSU und KIGOSHI[2]. Der allgemeinere Fall der Wechselwirkungen dreier identischer Kerne vom Spin $\frac{1}{2}$, deren Dreiecke beliebige Formen haben, wurde von ANDREW und BERSOHN[3] berechnet. WAUGH, HUMPHREY und YOST[4] untersuchten analytisch und experimentell am Beispiel der Bifluorid-Ionen der Verbindungen KHF_2 und $NaHF_2$ die Spektren von geradlinig angeordneten Dreiergruppen, welche aus einem verschiedenen und zwei gleichen Kernen bestehen.

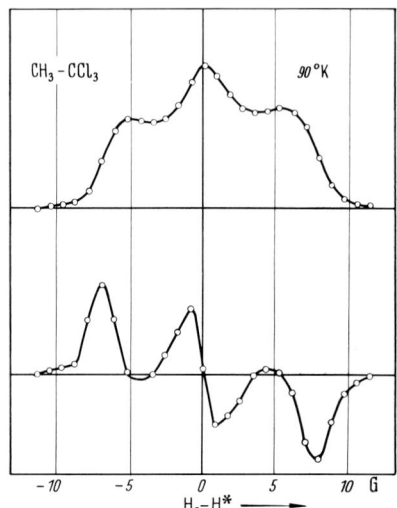

Will man die beschriebenen Untersuchungsverfahren auf feste Substanzen übertragen, die aus vier Kernen bestehende Gruppen enthalten (z.B. CH_4, NH_4^+, $2H_2O$), so liefert in der Regel nur noch der Vergleich der experimentellen und der theoretischen Einkristallspektren Aussagen über die Festkörperstruktur, deren Wert dem Umfang der analytischen Behandlung angemessen ist. Die Linienzentren-Verteilung polykristallin angeordneter Vierergruppen [vgl. Gl. (28.7)] ist so kompliziert, daß die durch eine Integration nach (28.8) über alle Linienkomponenten erhaltenen theoretischen Spektren, auch von Substanzen mit kleinen Linienkomponenten-Breiten, nur noch eine verwaschene und breite Struktur, praktisch ohne irgendwelche charakteristischen Extrema be-

Fig. 107. Verlauf der Protonenresonanzkurve in polykristallinem 1,1,1-Trichloräthan bei 90° K. Der Verlauf des Absorptionssignals ist typisch für Kernspinsysteme in denen zufällig orientierte Gruppen von drei in gleichen Abständen angeordneten identischen Kernen enthalten sind, deren Spin $I = \frac{1}{2}$ ist. Experimentell beobachtet wurde die in dem unteren Diagramm wiedergegebene Ableitung der Absorptionskurve. Die Meßfrequenz betrug 29 MHz.

sitzen. Experimentelle und theoretische Untersuchungen über Vierergruppen-Kristalle wurden u. a. von GUTOWSKY, KISTIAKOWSKY, PAKE und PURCELL[5], von TOMITA[6], von ITOH, KUSAKA, YAMAGATA, KIRIYAMA und IBAMOTO[7], sowie von BERSOHN und GUTOWSKY[8] veröffentlicht.

Es ist leicht einzusehen, daß die beschriebene analytische Behandlungsart in noch komplizierteren Fällen der Kernanordnung, wenn also im Kristallgefüge

[1] R. E. RICHARDS u. J. A. S. SMITH: Trans. Faraday Soc. **47**, 1261 (1951).

[2] Y. KAKIUCHI, H. SHONO, H. KOMATSU u. K. KIGOSHI: J. Chem. Phys. **19**, 1069 (1951). — J. Phys. Soc. Japan **7**, 102 (1952).

[3] E. R. ANDREW u. R. BERSOHN: J. Chem. Phys. **18**, 159 (1950); **20**, 924 (1952).

[4] J. S. WAUGH, F. B. HUMPHREY u. D. M. YOST: J. Phys. Chem. **57**, 486 (1953).

[5] H. S. GUTOWSKY, G. B. KISTIAKOWSKY, G. E. PAKE u. E. M. PURCELL: J. Chem. Phys. **17**, 972 (1949); **18**, 162 (1950).

[6] K. TOMITA: Progr. Theor. Phys. **8**, 138 (1952). — Phys. Rev. **89**, 429 (1953).

[7] J. ITOH, R. KUSAKA, Y. YAMAGATA, R. KIRIYAMA u. H. IBAMOTO: J. Chem. Phys. **20**, 1503 (1952); **21**, 190 (1953). — J. Phys. Soc. Japan **8**, 293 (1953).

[8] R. BERSOHN u. H. S. GUTOWSKY: J. Chem. Phys. **22**, 651 (1954).

nur sehr verwickelte, nur wenig ausgeprägte, oder gar überhaupt keine Gruppen-bildungen von Kernen zu erkennen sind, vollständig versagt. Auf Grund von Überlegungen, welche zuerst von VAN VLECK[1] mitgeteilt wurden, können auch in diesem allgemeinsten Fall der durch magnetische Wechselwirkungen geformten Resonanzkurven als starr betrachtbarer Festkörper die experimentellen Re-sonanzkurven mit den Ergebnissen theoretischer Vorstellungen verglichen werden.

Allgemein kann man zur Auswertung experimentell erhaltener Absorptions-kurven resonanzfähiger Gebilde zunächst durch eine passende Ordinatentrans-formation die normierte Funktion des zu untersuchenden Kurvenverlaufs be-stimmen. Je nach dem, ob man den Verlauf dieser Funktion in Abhängigkeit von der Frequenz $[g(\nu)$ bzw. $g(\nu-\nu^*)]$, oder im vorliegenden Fall auch in Ab-hängigkeit von der Feldstärke $[g(H)$ bzw. $g(H-H^*)]$ angibt, gilt definitions-gemäß für die normierten Funktionen ($\nu^* =$ Resonanzfrequenz und $H^* =$ Reso-nanzfeldstärke des Kurvenzentrums):

$$\left.\begin{aligned}
\text{oder}\quad &\int_0^\infty g(\nu)\,d\nu = 1 \quad \text{bzw.}\quad \int_{-\infty}^\infty g(\nu-\nu^*)\,d(\nu-\nu^*)=\int_{-\infty}^\infty g(\Delta\nu)\,d(\Delta\nu)=1\,,\\
&\int_0^\infty g(H)\,dH = 1 \quad \text{bzw.}\quad \int_{-\infty}^\infty g(H-H^*)\,d(H-H^*)=\int_{-\infty}^\infty g(\Delta H)\,d(\Delta H)=1\,.
\end{aligned}\right\} \quad (28.9)$$

Zur zahlenmäßigen Charakterisierung des beobachteten Kurvenverlaufs mißt man zweckmäßigerweise die sog. Momente der Kurve aus. Man versteht unter dem n-ten Moment einer Resonanzkurve definitionsgemäß den Mittelwert der n-ten Potenz des mit der normierten Amplitudenfunktion des Kurvenverlaufs gewogenen Frequenz- bzw. Feldstärkenabstands vom Zentrum ν^* bzw. H^* der Resonanzkurve:

$$\left.\begin{aligned}
\langle(\nu-\nu^*)^n\rangle &= \int_0^\infty g(\nu)\,(\nu-\nu^*)^n d\nu \quad \text{bzw.}\quad \langle(H-H^*)^n\rangle = \int_{-\infty}^\infty g(H)\,(H-H^*)^n dH\\
&= \int_{-\infty}^\infty g(\Delta\nu)\,(\Delta\nu)^n\,d(\Delta\nu) \qquad\qquad\quad = \int_{-\infty}^\infty g(\Delta H)\,(\Delta H)^n d(\Delta H)\,.
\end{aligned}\right\} (28.10\text{a,b})$$

Für ungerade n verschwinden in dem vorliegend diskutierten Fall der durch magnetische Dipol-Dipol-Wechselwirkungen bestimmten Absorptionskurven die Integrale (28.10), da unter dieser Voraussetzung die Funktionen $g(\Delta\nu)$ und $g(\Delta H)$ gerade sind, bzw. die Funktionen $g(\nu)$ und $g(H)$ spiegelsymmetrisch ver-laufen. Die Absorptionskurven haben also in diesem Fall nur gerade endliche Momente.

Die Beschreibung der Form einer Kurve durch deren Momente erinnert an das FOURIER-Verfahren zur Analyse eines vorgegebenen Kurvenverlaufs. In der Tat konnte YOKOTA[2] zeigen, daß die beiden Beschreibungsweisen verknüpft sind und zwar gilt nach dem Fourierschen Integraltheorem z.B. für die Trans-formierte der Funktion $g(\Delta\nu)$, sie sei mit $f(\tau)$ bezeichnet, die Beziehung:

$$\left.\begin{aligned}
f(\tau) &= \frac{1}{\sqrt{2\pi}} \int_{-\infty}^\infty g(\Delta\nu)\,e^{i\tau\Delta\nu}\,d(\Delta\nu)\\
&= \frac{1}{\sqrt{2\pi}} \sum_{n=0}^\infty \frac{(i\tau)^n}{n!} \int_{-\infty}^\infty g(\Delta\nu)\,(\Delta\nu)^n d(\Delta\nu) = \frac{1}{\sqrt{2\pi}} \sum_{n=0}^\infty \frac{(i\tau)^n}{n!}\,\langle(\Delta\nu)^n\rangle\,.
\end{aligned}\right\} (28.11)$$

Die Fourier-Transformierte einer Resonanzkurve ist also gleich einer unendlichen Reihe derer Momente.

Im Anschluß an frühere Arbeiten von WALLER[1] und BROER[2] gelang es VAN VLECK[3], quantentheoretisch die Größe des zweiten und vierten Moments kernmagnetischer Absorptionskurven zu berechnen. Arbeiten über das gleiche Thema haben außerdem PRYCE und STEVENS[4], sowie KAMBE und USUI[5] veröffentlicht. VAN VLECK's häufig zitierte Gleichung des zweiten Moments hat für einen monokristallinen Festkörper, der nur eine Kernart enthält, also im Fall der Wechselwirkung identischer Kerne die Formen

bzw.

$$\left. \begin{aligned} \langle(\Delta\nu)^2\rangle &= \tfrac{3}{4} N^{-1} g^4 \mu_B^4 h^{-2} I(I+1) \sum_{j\neq k} (3\cos^2\Theta_{jk}-1)^2 r_{jk}^{-6}, \\ \langle(\Delta H^2)\rangle &= \tfrac{3}{4} N^{-1} g^2 \mu_B^2 I(I+1) \sum_{j\neq k} (3\cos^2\Theta_{jk}-1)^2 r_{jk}^{-6}, \end{aligned} \right\} \qquad (28.12\text{a, b})$$

je nach dem, ob die Resonanzkurve in Funktion des Frequenz- oder des Feldabstands vom Kurvenzentrum aufgenommen wurde. Darin ist r_{jk} der Betrag des die beiden Kerne j und k verbindenden Radiusvektors, Θ_{jk} der Winkel zwischen diesem Vektor und der äußeren Feldrichtung, g der LANDÉ-Faktor der Kerne, I ihr Spin und μ_B das BOHRsche Magneton.

An sich sollte in (28.12) die Summierung $j\neq k$ und damit die entsprechende Mittelwertsbildung (Division mit N) über alle $N(N-1)$ Verbindungslinien der N-Kerne mit den jeweils $(N-1)$ übrigen Kernen der Substanz erfolgen. Je zwei dieser Vektoren stimmen natürlich überein, der Zahlenfaktor $\tfrac{3}{4}$ in (28.12) ist also eigentlich als $\tfrac{3}{2}\cdot\tfrac{1}{2}$ zu lesen. Da aber die Kernanordnungen in allen Elementarzellen eines Kristalls übereinstimmen, ist es für die Mittelwertsbildung in (28.12) auch ausreichend, wenn man über die $n(N-1)$ Verbindungslinien der n Kerne einer Elementarzelle mit allen $(N-1)$ jeweils übrigen Kernen der Substanz, in und außerhalb der betrachteten Elementarzellen summiert, und diese Summe durch n statt N dividiert. Sind in einem speziellen Kristall alle Atome, welche die untersuchten Kerne enthalten, gleich orientiert, so kann z.B. n einfach den Wert 1 haben. Überdies kann der größte Teil der Glieder der obigen Summation auf Grund der r^{-6}-Abhängigkeit des zweiten Moments immer vernachlässigt werden. Praktisch genügt es, die Wechselwirkungen zwischen den n Kernen einer Elementarzelle und zwischen diesen und den Kernen der unmittelbaren Nachbarzelle zu berücksichtigen.

Enthält der als monokristallin vorausgesetzte Festkörper außer den resonanz-spektroskopisch beobachteten Kernen (Anzahl N, Spin I, LANDÉ-Faktor g) noch eine weitere Kernart (Anzahl N', Spin I', LANDÉ-Faktor g'), deren Spin ebenfalls endlich ist, so ist zur Berechnung des zweiten Moments der Resonanzkurve zu dem Ausdruck (28.12) noch eine weitere Summe und zwar diejenige der Wechselwirkungen zwischen jedem der N beobachteten Kerne mit jedem der N' Kerne der anderen Art hinzuzufügen. VAN VLECK findet — analog zu dem Unterschied der beiden Gln. (28.2) und (28.3) —, daß sich das zweite Glied von dem ersten Glied (28.12) nur um einen Faktor $(\tfrac{2}{3})^2$ unterscheidet, um den die Kopplungen zwischen verschiedenen Kernen stets schwächer sind, als die zwischen identischen Kernen. Unter der ausdrücklichen Voraussetzung, daß der

---

[1] I. WALLER: Z. Physik **79**, 381 (1932).
[2] L. J. F. BROER: Physica, Haag **10**, 801 (1943).
[3] J. H. VAN VLECK: Phys. Rev. **74**, 1168 (1948).
[4] M. H. L. PRYCE u. K. W. H. STEVENS: Proc. Phys. Soc. Lond. A **63**, 36 (1950).
[5] K. KAMBE u. T. USUI: Progr. Theor. Phys. **8**, 302 (1952).

Körper keine elektronen-paramagnetischen Partikel enthält und damit Austausch-Wechselwirkungen unberücksichtigt bleiben können, gilt für das zweite Moment der Resonanzkurve der mit ungestrichenen Buchstaben bezeichneten Kernart die Gleichung

$$\langle (\varDelta v)^2 \rangle = g^2 \mu_B^2 h^{-2} \langle (\varDelta H)^2 \rangle = \tfrac{3}{4} N^{-1} g^2 \mu_B^4 h^{-2} \times$$
$$\times \left[g^2 I(I+1) \sum_{j \neq k} (3 \cos^2 \Theta_{jk} - 1) r_{jk}^{-6} + \tfrac{4}{9} g'^2 I'(I'+1) \sum_{j \neq k} (3 \cos^2 \Theta_{jk} - 1) r_{jk}^{-6} \right]. \quad (28.13)$$

Bezüglich der zweiten Summenbildung gelten natürlich die voranstehend genannten Erleichterungen sinngemäß.

Zur Berechnung der Resonanzkurve polykristalliner Kristalle kann man annehmen, daß alle Orientierungen Θ_{jk} gleich wahrscheinlich sind. Der räumliche Mittelwert von $(3 \cos^2 \Theta_{jk} - 1)^2$ hat die Größe $\tfrac{4}{5}$, folglich gilt für das zweite Moment des Polykristalls

$$\langle (\varDelta v)^2 \rangle = g^2 \mu_B^2 h^{-2} \langle (\varDelta H)^2 \rangle = \tfrac{3}{5} N^{-1} g^2 \mu_B^4 h^{-2} \times$$
$$\times \left[g^2 I(I+1) \sum_{j \neq k} r_{jk}^{-6} + \tfrac{4}{9} g'^2 I'(I'+1) \sum_{j \neq k} r_{jk}^{-6} \right]. \quad (28.14)$$

VAN VLECK hat auch das vierte Moment kernmagnetischer Resonanzkurven berechnet, wegen der Kompliziertheit des Ergebnisses soll jedoch auf die Originalarbeit verwiesen werden. Damit wird es u. a. möglich, Voraussetzungen über die Form der Resonanzkurve zu prüfen, z.B. ob diese einen durch eine GAUSSsche Fehlerfunktion beschreibbaren Verlauf hat [vgl. (4.5)], wie häufig vorausgesetzt wird.

Die Signalspannungen sind bei der Beobachtung von Festkörperspektren verhältnismäßig klein (vgl. S. 230). Zur Verbesserung des Signal-Rausch-Verhältnisses muß man daher im allgemeinen die Resonanzkurve differentiell abtasten, und die Signalspannung extrem schmalbandig verstärken (vgl. Ziff. 12). Die Signale von Festkörpern werden dementsprechend fast immer in der Form der ersten Ableitungen der eigentlichen Absorptionskurven beobachtet. PAKE und PURCELL[1] haben deshalb darauf hingewiesen, daß man sich bei der experimentellen Bestimmung des zweiten und vierten Moments der Resonanz-Kurven eine Integration ersparen kann, wenn man die Beziehung

$$\langle (\varDelta v)^{2n} \rangle = -\frac{1}{2n+1} \int_0^\infty (\varDelta v)^{2n+1} \frac{dg(v)}{dv} dv \quad (28.15)$$

beachtet, welche durch partielle Integration aus (28.10a) folgt, da die Funktion $g(v)$ außerhalb eines kleinen Frequenzintervalls um die Resonanzfrequenz v^* stets verschwindet.

Die Gleichungen des zweiten Moments (28.13) und (28.14) sind inzwischen von mehreren Autoren an bekannten Kristallstrukturen geprüft worden und dürfen als bestätigt bezeichnet werden. Man kann sie dementsprechend ebenso wie die Beziehungen (28.2), (28.3), (28.7) und (28.8) in zahlreichen Fällen als Hilfsmittel zur kristallographischen Strukturuntersuchung benutzen. Dazu kann man z.B. unter verschiedenen Annahmen über einige Strukturdetails die zweiten Momente der zugehörigen Absorptionskurven berechnen, und die Ergebnisse mit dem zweiten Moment der experimentellen Resonanzkurve vergleichen. Von dieser Möglichkeit haben u. a. BERSOHN und GUTOWSKY[2], sowie GUTOWSKY, PAKE und

[1] G. E. PAKE u. E. M. PURCELL: Phys. Rev. **74**, 1184 (1948); **75**, 534 (1949).
[2] R. BERSOHN u. H. S. GUTOWSKY: J. Chem. Phys. **22**, 651 (1954).

BERSOHN[1], ANDREW und HYNDMAN[2], sowie ANDREW und EADES[3] Gebrauch gemacht. Es ist klar, daß auch in dem allgemeinen Fall der beliebigen Kernanordnung die Untersuchung der Einkristallspektren mehr Informationen über die Kristallstruktur als die Untersuchung der Polykristallspektren liefert.

Alle bisher beschriebenen Kernresonanz-Phänomene sind in Experimenten beobachtet worden, in denen das kernmagnetische Spektrum irgendeiner Substanz in einem äußeren Magnetfeld beobachtet wurde. Die Stärke dieses Felds war immer so bemessen, daß die Resonanzfrequenzen der untersuchten Kerne sehr viel größer als die Resonanzfrequenz-Differenzen innerhalb ihrer mehr oder weniger komplizierten Spektren waren. Es konnte somit stets vorausgesetzt werden, daß $h \ll H$ ist, vgl. (28.4a), daß also die Dipol-Dipol-Wechselwirkung analytisch als kleine Störung der eigentlichen, durch den Betrag des äußeren Felds bestimmten ZEEMAN-Energieniveaus der Kerne behandelt werden darf. Während im Fall der indirekten elektronen-gekoppelten Dipol-Dipol-Wechselwirkung in Flüssigkeiten und Gasen (Ziff. 27) die Frequenzaufspaltungen der Linien im allgemeinen einige Hertz betrugen ($h/H \approx 10^{-6}$), ergaben die in dieser Ziffer diskutierten Untersuchungen von Festkörpern, daß deren Spektren auf Grund der direkten Dipol-Dipol-Wechselwirkung immerhin in der Regel etwa 10 bis 100 kHz breit sind ($h/H \approx 10^{-3}$). Das genannte Frequenzintervall liegt in einem hochfrequenztechnisch gut beherrschten Bereich. Damit lag die Frage nahe, ob sich nicht die Resonanzabsorption der Kerndipole in den in starren Körpern konstanten Feldern ihrer Nachbarn unmittelbar, ohne ein äußeres Feld H nachweisen läßt. Dieser Effekt ist von REIF und PURCELL[4] tatsächlich gefunden worden. Um Verwechslungen auszuschließen, bezeichnet man zweckmäßigerweise die Beobachtung der direkten Dipol-Dipol-Wechselwirkung im Nullfeld entweder in Analogie zur „Reinen Kern-Quadrupolresonanz" (vgl. S. 329) als „Reine Kerndipol-Resonanz", oder in Erinnerung an den zugrunde liegenden Effekt als „Dipolwechselwirkungs-Resonanz".

Im Prinzip sollten sich an allen in dieser Ziffer bisher besprochenen kristallinen Körpern reine Kerndipol-Resonanzphänomene nachweisen lassen. Praktisch steht dem jedoch in der Regel entgegen, daß die im Experiment beobachtbare Signalspannung und damit das Signal-Rausch-Verhältnis bei dieser Resonanzart noch sehr viel kleiner ist, als bereits bei Festkörpern allgemein. REIF und PURCELL haben deshalb als Mustersubstanz festen Wasserstoff gewählt. Sie untersuchten den bei 14° K schmelzenden Wasserstoff in der üblichen Weise kernmagnetisch bei etwa 1,15° K in einem normalen äußeren Feld (dies haben HATTON und ROLLIN[5] schon früher ebenfalls getan), in einem sehr schwachen äußeren Feld und ganz ohne ein äußeres Feld mit dem in Ziff. 13 beschriebenen technischen Verfahren von POUND, KNIGHT und WATKINS. In dem bekanntlich diamagnetischen Wasserstoffmolekül ist der intramolekulare Kernabstand mit 0,75 Å ungewöhnlich klein und damit die Dipol-Dipol-Wechselwirkung besonders groß. Dagegen sind alle intermolekularen Kernabstände ungewöhnlich groß. Übereinstimmend damit kann man u. a. anführen, daß der Wasserstoff unter den Festkörpern die niedrigste Dichte, eine besonders große Nullpunktsenergie und eine besonders kleine Schmelzwärme besitzt. Die Moleküle sind somit im festen Wasserstoff sehr locker gebunden, und dementsprechend ist auch seine Spin-

[1] H. S. GUTOWSKY, G. E. PAKE u. R. BERSOHN: J. Chem. Phys. 22, 643 (1954).
[2] E. R. ANDREW u. D. HYNDMAN: Proc. Phys. Soc. Lond. A 66, 1187 (1953).
[3] E. R. ANDREW u. R. G. EADES: Proc. Roy. Soc. Lond., Ser. A 216, 398 (1953); 218, 537 (1953). — Proc. Phys. Soc. Lond. A 66, 415 (1953).
[4] F. REIF u. E. M. PURCELL: Phys. Rev. 91, 631 (1953).
[5] J. HATTON u. B. V. ROLLIN: Proc. Roy. Soc. Lond., Ser. A 199, 222 (1949).

Gitter-Relaxationszeit verhältnismäßig kurz. Alle diese Umstände tragen dazu bei, das Problem eines ausreichenden Signal-Rausch-Verhältnisses zu verringern. Fester Wasserstoff besteht in dem untersuchten Temperaturbereich zu etwa 75% aus Orthomolekülen, deren resultierender Kernspin $I = 1$ ist und zu etwa 25% aus Paramolekülen, deren resultierender Kernspin $I = 0$ ist. Beobachtbar sind im Kernresonanz-Experiment nur die Kerne der Orthomoleküle. REIF und PURCELL fanden die Dipolwechselwirkungs-Resonanzlinie in guter Übereinstimmung mit den von ihnen abgeleiteten theoretischen Beziehungen bei $(165{,}7 \pm 0{,}1)$ kHz, und zwar erweist sich diese Frequenz erwartungsgemäß als ebenso groß wie die Frequenzdifferenz der beiden Hauptmaxima der Resonanzkurve des festen Wasserstoffs in einem starken äußeren Feld von ungefähr 3500 Gauß. Die gemessene Halbwertsbreite der Nullfeldlinie betrug 23 kHz. Auf Grund dieses Werts läßt sich die intermolekulare Kopplung abschätzen.

Kerne, deren Spin $I > \frac{1}{2}$ ist, können bekanntlich ein elektrisches Quadrupolmoment besitzen und haben auch erfahrungsgemäß neben ihrem magnetischen Dipolmoment ein solches elektrisches Moment. Die Definitionsgleichung des skalaren Quadrupolmoments Q lautet, je nach dem, ob man sich die Kernladung kontinuierlich oder punktförmig verteilt vorstellt:

$$e\,Q = \int\limits_{(\text{Kern})} (3z' - r^2)\,\varrho\,dv = \int\limits_{(\text{Kern})} (3\cos^2\Theta_{r,\,z'} - 1)\,r^2\,\varrho\,dv \qquad (28.16\text{a})$$

oder

$$e\,Q = \sum_i (3\cos^2\Theta_{r_i z'} - 1)\,r_i^2\,e_i. \qquad (28.16\text{b})$$

Darin sind e die Elektronenladung, ϱ die elektrische Ladungsdichte im Kern, r der Abstand des Kern-Volumelements dV vom Kernmittelpunkt, z' seine vom selben Mittelpunkt aus gemessene Koordinate bezüglich der Kern-Symmetrieachse (Drehimpulsachse) und $\Theta_{r,\,z'}$, der von r und I eingeschlossene Winkel.

Das Moment Q hat demnach die Dimension Länge². Das Integral (28.16a) verschwindet, wenn die Ladungsverteilung kugelsymmetrisch ist, d.h. nur von r abhängt. Q kann als Maß für die Abweichung der Kernladungsverteilung von der Kugelsymmetrie bezeichnet werden. Auf ein elektrisches Kernquadrupolmoment wirkt in einem elektrostatischen Feld ein Drehmoment ein, dessen Betrag proportional mit den Beträgen des Quadrupolmoments und der ersten Ableitung des Feldes wächst. Analog galt, daß in einem magnetostatischen Feld auf ein magnetisches Kern-Dipolmoment ein zu den Beträgen des Moments und des Felds proportionales Drehmoment einwirkt. Daraus folgt, daß die Wechselwirkungsenergie eines Kernquadrupols in einem ausreichend inhomogenen elektrostatischen Feld z.B. im Innern eines Kristalls, dessen Symmetrie niedriger als kubisch ist, von der Orientierung des Kerns in dem Kristall abhängt. Befindet sich der Kristall in einem äußeren Magnetfeld H_0, so kann der in ihm enthaltene betrachtete Kern mit den Momenten I, μ und Q eine der durch die magnetischen Quantenzahlen m beschriebenen $2I + 1$ möglichen Orientierungen haben. Die zu diesen Orientierungen gehörenden ZEEMAN-Energieniveaus $-m\mu H/I$ (vgl. Ziff. 1) unterscheiden sich um äquidistante Beträge, wenn weder störende Wechselwirkungen des Kerndipols μ mit dem äußeren Feld überlagerten kristallinneren magnetischen Feldern, noch störende Wechselwirkungen des Kernquadrupolmoments Q mit kristallinneren elektrischen Feldern existieren. Bei der vorangegangenen Besprechung des zuerst genannten Falls der magnetischen Dipol-Dipol-Kopplungen der Kerne in Festkörpern konnte stets vorausgesetzt werden, daß die Änderung der ZEEMAN-Energieniveaus sehr klein ($\sim 1/1000$) gegen deren Höhe ist. Damit war es zulässig, die Niveau-Änderungen im Rahmen von Störungsrechnungen in erster Ordnung zu bestimmen. Quadrupol-Wechselwirkungen konnten auf Grund der Voraussetzung, daß die untersuchten Kerne den Spin $\frac{1}{2}$ haben sollten, exakt vernachlässigt werden. In dem jetzt zu diskutierenden Fall der Quadrupol-Wechselwirkung mit dem kristallelektrischen Feld sind die Verhältnisse zwar etwas weniger günstig, aber doch wesentlich leichter analytisch behandelbar, als man zunächst auf Grund der Erfahrung, daß Kerne mit einem Spin $I \geq 1$ in der Regel sowohl ein magnetisches Dipolmoment als auch ein elektrisches Quadrupolmoment besitzen, annehmen würde. Aus den experimentellen Beobachtungen folgt nämlich der glückliche Umstand, daß die Energiebeträge E_{QF} der Kernquadrupol-Kristallfeld-Kopplungen in Magnetfeldern des in Resonanzexperimenten üblicherweise untersuchten Feldstärkenbereichs 1000 bis 10000 Gauß meist viel kleiner als die Energiebeträge E_{DH} der Wechselwirkung Kerndipol-Außenfeld, andererseits aber auch viel größer als die Energiebeträge E_{DD} der Wechselwirkungen der Kerndipole untereinander sind. Somit kann man in allen diesen Fällen die Quadrupol-Wechselwirkung als Störung erster Ordnung der ZEEMAN-Niveaus und die Dipol-Dipol-Wechselwirkung in erster Näherung als nicht existent betrachten.

Bei der Besprechung der Dipol-Dipol-Wechselwirkungen erwiesen sich die intramolekularen Kopplungen (genauer gesagt die Kopplungen innerhalb der Kerngruppen) als für die Linienzahl und die wesentlich kleineren intermolekularen Kopplungen als für die Struktur der einzelnen Linien verantwortlich. Durchaus analog spiegeln in den Festkörperspektren der Kerne $I \geq 1$ die Anzahl und Verteilung der Linienkomponenten die Quadrupol-Wechselwirkung und die Form der Linien die gleichzeitig existierende Dipol-Dipol-Wechselwirkung wieder.

Gelegentlich ist die linke Hälfte der Ungleichung

$$E_{DH} \gg E_{QF} \gg E_{DD} \tag{28.17}$$

mangelhaft erfüllt, z.B. für Al^{27} in Al_2O_3. Man muß dann die Störungsrechnung in höherer Ordnung, in dem genannten Beispiel bis zur dritten Ordnung weiterführen, um Übereinstimmung zwischen den experimentellen und den theoretischen Resultaten zu erhalten. Die rechte Hälfte von (28.17) ist um so besser erfüllt, je größer der Quotient $e^2 Q/\mu^2$ der untersuchten Kerne ist, denn μ^2/r^3 ist ein Maß für die Wechselwirkung zweier identischer Dipole im Abstand r und $e/r^3 \cdot eQ$ in grober Abschätzung ein Maß für die Wechselwirkung eines Quadrupols mit dem kristallelektrischen Feld. Dazu kann man sich etwa vorstellen, daß der Nachbarkern in einer einfach ionisierten Elektronenhülle enthalten ist, welche am Ort des betrachteten Kerns einen Feldgradienten der Größenordnung e/r^3 erzeugt. Hat die Bindung partiell kovalenten Charakter, so sind der Feldgradient und damit die Quadrupol-Wechselwirkung übrigens noch größer.

Die Entdeckung der Quadrupolaufspaltung kernmagnetischer Resonanzlinien von Festkörpern verdanken wir POUND[1], der die Spektren der Ein- und Polykristalle von $Na^{23}NO_3$, $Li_2^7CO_3$, $Li_2^7SO_4 \cdot H_2O$ und $Al_2^{27}O_3$ experimentell untersucht und theoretisch korrekt gedeutet hat. Zwei seiner Registrierungen sind in den Fig. 108 und 109 wiedergegeben. POUND berechnete für die Energie E_m des durch Quadrupol-Wechselwirkung gestörten m-ten ZEEMAN-Niveaus eines einzelnen Kerns und für die Kreisfrequenz seines Übergangs $m \leftrightarrow m-1$ die in erster Ordnung gültigen Gleichungen

$$E_m = -m\,\mu\,H/I + [e\,Q/4I(2I-1)]\,[3\,m^2 - I(I+1)]\,\frac{\partial F z}{\partial z}, \tag{28.18}$$

$$\omega_{m \leftrightarrow m-1} = \mu\,H/I\,\hbar - [3\,e\,Q/4I(2I-1)]\,[1 - 2m]\,\frac{\partial F z}{\partial z}. \tag{28.19}$$

¹ R. V. POUND: Phys. Rev. **79**, 685 (1950).

In diesen Beziehungen wurde die erste Ableitung der elektrischen Feldstärke in Richtung der zum äußeren Feld H parallelen z-Achse mit $\partial F/\partial z$ bezeichnet. Ist die elektrische Umgebung der untersuchten Kerne axialsymmetrisch, wie z.B. in zweiatomigen Molekülen, so kann man ihren Verlauf ebenso wie das Quadrupolmoment Q [vgl. Gl. (28.16)] durch eine skalare Größe q beschreiben, deren Definitionsgleichung (parallel zu der Symmetrieachse S liege eine z''-Achse)

$$e\,q = \int\limits_{(\text{Kernumgebung})} (3\,z''-r^2)\,\varrho\,r^{-5}\,dv = \int\limits_{(\text{Kernumgebung})} (3\cos^2\Theta_{r,\,z''}-1)\,\varrho\,r^{-3}\,dv, \quad (28.20\text{a})$$

oder

$$e\,q = \sum_j (3\cos^2\Theta_{r_j,\,z''}-1)\,r_j^{-3}\,e_j \qquad (28.20\text{b})$$

lautet, je nach dem, ob man sich die den Kern umgebende Ladung kontinuierlich oder punktförmig verteilt vorstellt. Für die Komponente des Kristallfeld-Gradienten in Richtung des äußeren Felds findet man mit Hilfe der Definition (28.20) den einfachen Ausdruck

$$\frac{\partial f}{\partial z} = \frac{1}{2}\,(3\cos^2\Theta_{z,\,z''}-1)\,e\,q. \qquad (28.21)$$

Um Verwechslungen auszuschließen sei an die geometrischen Festlegungen

$$z \uparrow\uparrow \vec{H}, \qquad z' \uparrow\uparrow \vec{I}, \qquad z'' \uparrow\uparrow S$$

erinnert.

Aus den Gln. (28.19) und (28.21) kann man die Struktur der Einkristall-Spektren ($I=1$) sofort ablesen. Alle $2I+1$ ZEEMAN-Niveaus werden durch die Quadrupolkopplung um verschiedene Beträge verschoben. Damit werden aber auch ihre $2I$ Abstände ungleich groß, und das Spektrum besteht dementsprechend aus $n \cdot 2I$ nach (28.19) symmetrisch um die Resonanzfrequenz ν^* der ungestörten Kerne verteilten Linien, wenn jede Elementarzelle des Kristalls n ungleich gebundene Kerne der untersuchten Art enthält. Ist I speziell ungerade, so hat der Übergang $m=\frac{1}{2} \leftrightarrow m=-\frac{1}{2}$ übrigens gerade die mittlere Resonanzfrequenz ν^*, allerdings gilt dies nur in erster Ordnung. Der Abstand der Linien hängt gemäß Gl. (28.21) von der Orientierung des Kristalls im Magnetfeld ab. Ein typisches Spektrum dieser Art zeigen die in Fig. 108 wiedergegebenen Registrierungen der Na^{23}-Resonanzlinien ($I_{\text{Na}^{23}} = \frac{3}{2}$) von hexagonal-einkristallinem Natriumnitrat (NaNO$_3$). Die drei Linien haben eine ungefähre Breite von 2 bis 3 kHz, die Spin-Spin-Wechselwirkung ist also in diesem Fall verglichen mit der Quadrupol-Wechselwirkung sicher vernachlässigbar. Der maximale Frequenzabstand der äußeren Linien von der mittleren Linie beträgt (167 ± 10) kHz, der Quotient $e^2 q\,Q/h$ hat demnach den experimentellen Wert (334 ± 2) kHz. Leider kann man das skalare in (28.20) definierte Maß $e\,q$ der elektrischen Feldverteilung zur Zeit noch nicht zuverlässig berechnen, da die Ladungsverteilungen in den Kristallen zu wenig bekannt sind. Wegen dieser Schwierigkeit ist es bisher noch in keinem Fall und auch mit keinem anderen resonanz-spektroskopischen Verfahren gelungen. die Absolutwerte von Kernquadrupolmomenten genau zu bestimmen. Dagegen ist es möglich, die Quadrupolmomentverhältnisse Q_1/Q_2 von Isotopen sehr genau zu messen, wenn man etwa deren Quadrupolaufspaltungen im selben Kristallgefüge und damit auch in der gleichen elektrischen Umgebung vergleicht.

In pulverförmigen Kristallproben sind alle Orientierungen der Symmetrieachsen der kristallelektrischen Felder bezüglich des äußeren Magnetfelds gleich wahrscheinlich. Zur Berechnung der Linienzentren-Verteilung des Polykristall-Spektrums muß man deshalb, ebenso wie in dem auf S. 315 besprochenen Fall

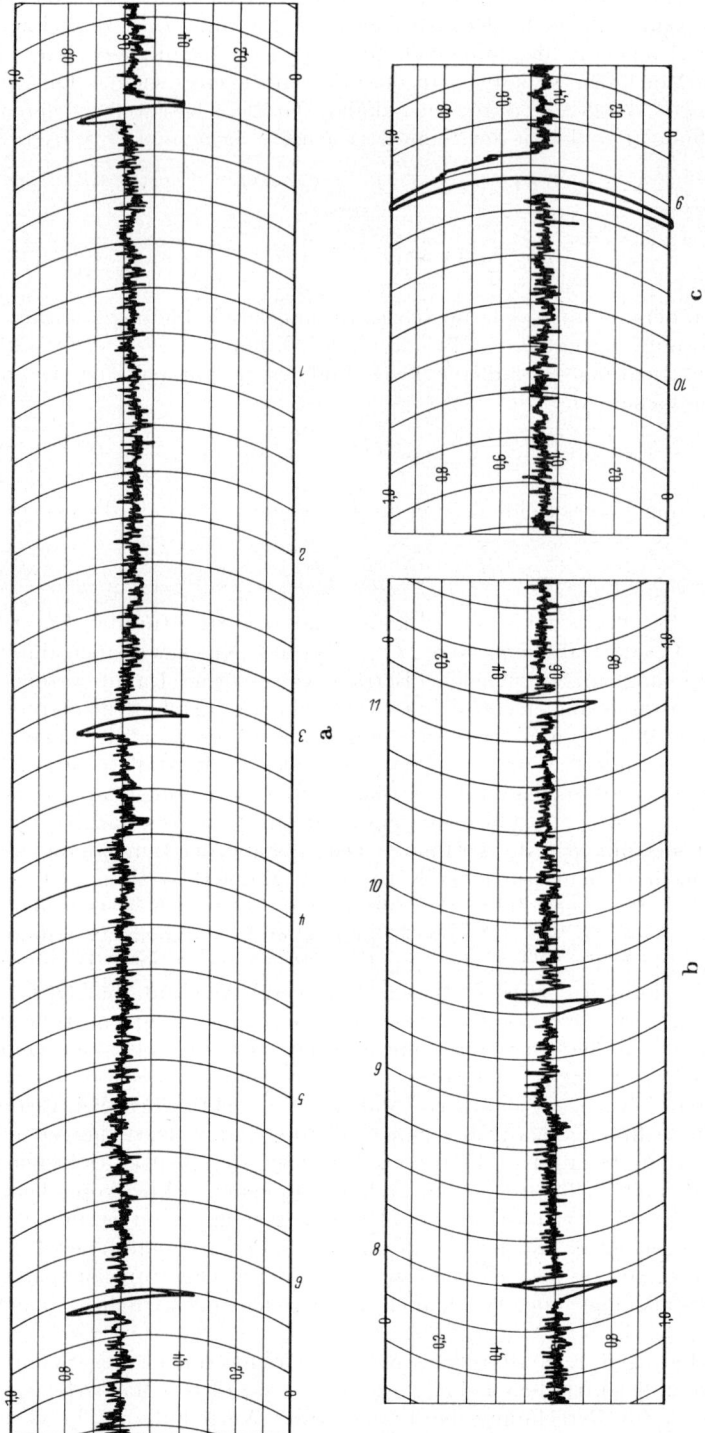

Fig. 108 a—c. Na²³-Resonanzspektrum beobachtet in einem Natriumnitrat-Einkristall (NaNO₃). Der zylindrisch geformte hexagonale Kristall war so geschnitten, daß die Zylinderachse senkrecht zu der dreizähligen Kristallachse lag. Der Winkel zwischen dieser Kristallachse und der Richtung des äußeren Felds H_0 betrug in den Registrierungen nacheinander 0° (a), 54° (b) und 90° (c). Der Frequenzmaßstab der horizontalen Achsen hat die ungefähre Größe 14 kHz pro Teilstrich. Die Resonanzfrequenz der mittleren Linie beträgt 7,18 MHz.

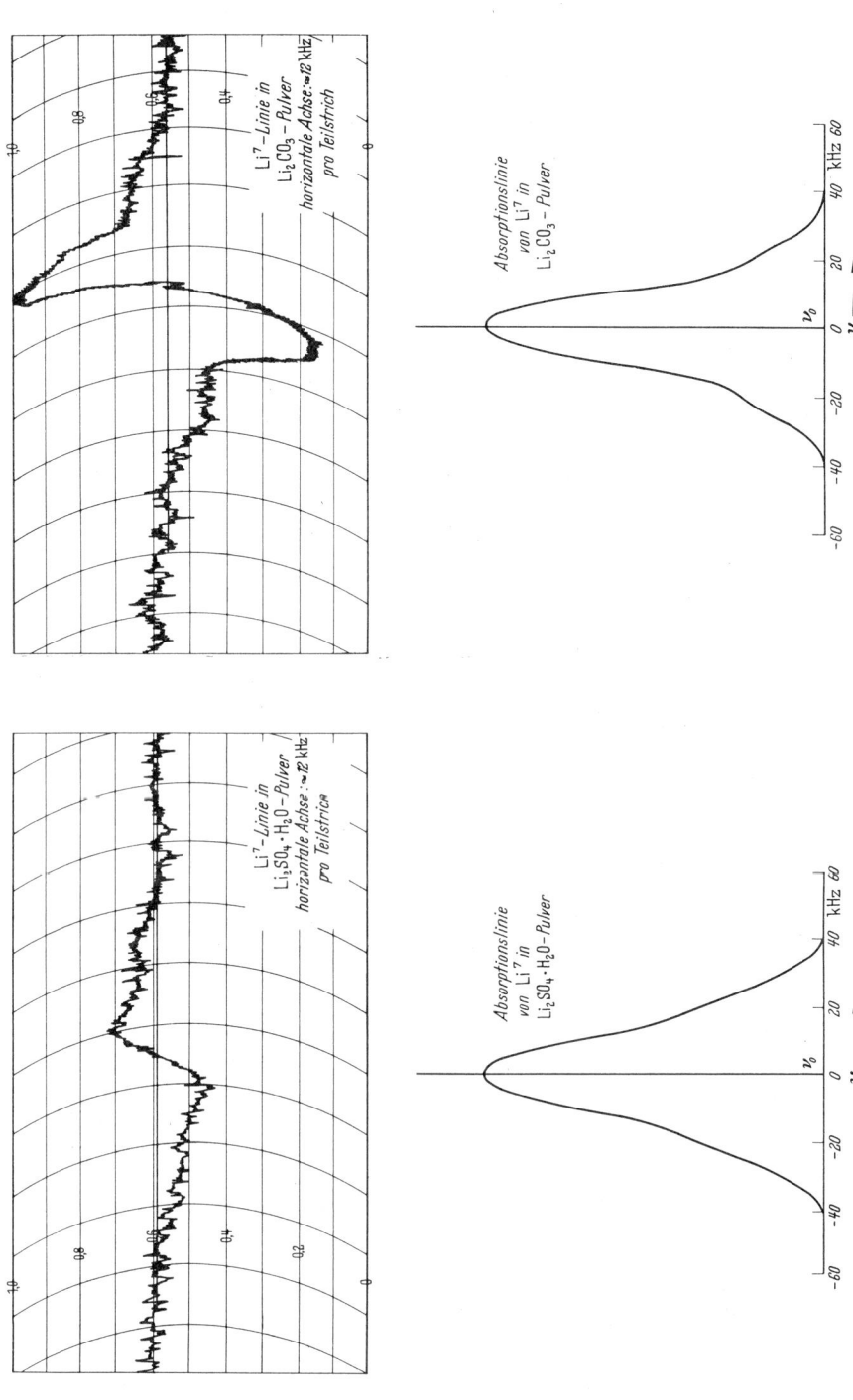

Fig. 109. Registrierungen der Li^7-Resonanzlinien in pulverförmigem Li$_2$SO$_4 \cdot$ H$_2$O und in pulverförmigem Li$_2$CO$_3$. Die unten wiedergegebenen Absorptionskurven wurden aus den experimentell gewonnenen Diagrammen ihrer Ableitungen bestimmt. Die Resonanzfrequenz betrug etwa 3 MHz.

der Dipol-Dipol-Wechselwirkung in Polykristallen den räumlichen Mittelwert von (28.21) bilden. Die resultierenden Spektrenstrukturen lassen sich auf diese Weise leicht für alle Spingrößen berechnen und werden um so komplizierter, je größer I ist. Ebenso wie bei der Dipol-Dipol-Kopplung kann man jedoch wegen der endlichen Linienbreite der Komponenten feinere Einzelheiten der Spektrums-form im allgemeinen experimentell nicht unterscheiden und muß sich dement-sprechend mit einer groben Abschätzung der Spektrumsbreite und des daraus folgenden Werts des Quotienten $e^2 q Q/h$ begnügen.

Polykristall-Spektren von Kernen, deren Spin ungeradzahlig ist $(I \geq \frac{3}{2})$, haben ein besonderes Merkmal. Nach (28.19) ist der Übergang $m = \frac{1}{2} \leftrightarrow m = -\frac{1}{2}$ von der Orientierung Θ der Kristallite in erster Näherung unabhängig. Folglich ist in den Polykristall-Spektren dieser Kerne die Resonanzabsorption im Zentrum sehr viel stärker als in den übrigen breiten Spektrumsbereichen beiderseits der mittleren Resonanzlinie, auf welche die zu den anderen Übergängen gehörenden Absorptionsfrequenzen mehr oder weniger gleichmäßig verteilt sind. Beispiele solcher Spektren sind die in Fig. 109 wiedergegebenen Resonanzkurven der Li⁷-Kerne $(I_{Li^7} = \frac{3}{2})$ in pulverförmigen Proben. Der Effekt wird weniger deutlich erkennbar, wenn die Quadrupol-Wechselwirkung ihrer Größe wegen als Störung höherer Ordnung der Zeeman-Terme behandelt werden muß, denn dann hängt auch die Frequenz des Übergangs $m = \frac{1}{2} \leftrightarrow m = -\frac{1}{2}$ von der Orientierung der Kri-stallite im äußeren Magnetfeld ab.

Zahlreiche Autoren haben sich in den letzten Jahren darum bemüht, den Anwendungsbereich der Poundschen Theorie durch Störungsrechnungen höherer Ordnung und durch die Berücksichtigung nicht-axialsymmetrischer Verteilungen des kristallelektrischen Felds zu erweitern. Zu nennen sind unter anderem Ar-beiten von Bersohn[1], von Petch, Smellie und Volkoff[2], von Volkoff[3] sowie von Petch, Volkoff und Cranna[4].

In einem Einkristall kubischer Symmetrie sollte es an sich, unabhängig von der Größe der Kernquadrupole, keine Quadrupol-Aufspaltung der Resonanz-linien geben. Häufig ist die Kristallstruktur jedoch mehr oder weniger verzerrt oder verspannt. Dadurch wird die Symmetrie der elektrischen Kernumgebungen niedriger als kubisch, und die Gradienten des kristallelektrischen Felds an den Orten der untersuchten Kerne endlich und voneinander verschieden. Infolge dieses Effekts kann das Kernresonanz-Spektrum von Festkörpern auf einen breiten Frequenzbereich ausgeweitet werden, vgl. z.B. eine Veröffentlichung von Watkins und Pound[5].

In einer Anzahl fester Körper, vor allem in solchen, die aus kovalent gebunde-nen Molekülen bestehen, ist das am Kernort wirkende elektrische Feld intra-molekularen Ursprungs. Die experimentelle Erfahrung zeigt, daß der Gradient des elektrischen Felds, und damit die Quadrupol-Wechselwirkung in diesen Substanzen wesentlich stärker als in den bisher besprochenen Fällen ist. Unter diesen Umständen kann man die Quadrupol-Aufspaltung natürlich nicht mehr als Störung der Zeeman-Niveaus betrachten. Man muß im Gegenteil gelegentlich die in praktisch herstellbaren Magnetfeldern vorliegende Wechselwirkung Kern-

---

[1] R. Bersohn: J. Chem. Phys. **20**, 1505 (1952).
[2] H. E. Petch, D. W. Smellie u. G. M. Volkoff: Phys. Rev. **84**, 602 (1951). — G. M. Volkoff, H. E. Petch u. D. W. Smellie: Canad. J. Phys. **30**, 270 (1952).
[3] G. M. Volkoff: Canad. J. Phys. **31**, 820 (1953).
[4] H. E. Petch, G. M. Volkoff u. N. G. Cranna: Phys. Rev. **88**, 1201 (1952). — H. E. Petch, N. G. Cranna u. G. M. Volkoff: Canad. J. Phys. **31**, 837 (1953).
[5] G. D. Watkins u. R. V. Pound: Phys. Rev. **89**, 658 (1953).

dipol-Magnetfeld als Störung der Quadrupol-Aufspaltung behandeln. Es lag daher der Gedanke nahe, die Übergangsfrequenzen zwischen den Quadrupol-Wechselwirkungstermen direkt, ohne Verwendung eines äußeren magnetischen Felds, zu messen. Das erste gelungene Experiment dieser Art ist von DEHMELT und KRÜGER[1] veröffentlicht worden, welche für die Quadrupol-Resonanzfrequenzen von Cl^{35} und Cl^{37} (beide Isotope haben den Spin $\frac{3}{2}$) in festem Transdichloräthylen die Werte 35,4 MHz und 27,96 MHz fanden. Diese heute als „Reine Kernquadrupol-Resonanz" bezeichnete hochfrequenz-spektroskopische Resonanzart unterscheidet sich in ihrer Ausführung, ebenso wie die ihr analoge „Reine Kerndipol-Resonanz" (vgl. S. 322), von den im Teil II dieser Abhandlung beschriebenen Methoden nur dadurch, daß man keinen Magneten benötigt. Damit entfällt die Möglichkeit, zur Resonanzbestimmung das Magnetfeld zu variieren, und man benötigt dementsprechend ein frequenzvariables Spektrometer, dessen Signal-Rausch-Verhältnis, wie bei der Untersuchung von Festkörper-Spektren allgemein, möglichst gut sein soll. Übrigens ist es auch gelungen, Quadrupol-Resonanzen in Form impulstechnisch angeregter freier Präzessionsbewegungen (vgl. Ziff. 15) zu beobachten[2].

Die reine elektrische Kernquadrupol-Resonanz wird heute üblicherweise als abgetrenntes neues Gebiet neben der kernmagnetischen Resonanz behandelt. So enthält der vorliegende Band des Handbuches einen besonderen Beitrag über dieses Gebiet, seine Einzelheiten sollen deshalb hier nicht weiter diskutiert werden. Zur Verdeutlichung des Verflochtenseins der magnetischen und der elektrischen Kernresonanz-Phänomene soll nur noch erwähnt werden, daß bei der reinen Quadrupol-Resonanz zwar die Wechselwirkungen, welche die Niveauaufspaltungen bewirken, elektrischer Natur sind. Die Wechselwirkungen des Kernspin-Systems mit den elektromagnetischen Schwingungen in der die Probe umschließenden Spule (z.B. deren Absorption) werden jedoch über das in der Spule oszillierende Magnetfeld und über die magnetischen Dipole der untersuchten Kerne übertragen.

Entsprechend der Sonderstellung, welche der Metallphysik in der übrigen Festkörperphysik zukommt, unterscheiden sich auch die an Metallen beobachteten Kernresonanz-Phänomene wesentlich von denen anderer Festkörperarten. Die für metallische Festkörper charkteristische Eigenschaft ist, daß in ihnen frei bewegliche Leitungselektronen existieren, welche quantentheoretisch (als Teilchen mit dem Spin $\frac{1}{2}$) als Fermionen zu behandeln sind. Dementsprechend gehorchen die Elektronen dem PAULI-Prinzip und können nach der bekannten derzeitigen Vorstellung als im Metallgefüge enthaltenes entartetes Elektronengas betrachtet werden.

Die Existenz der beweglichen Elektronen hat in der Resonanz-Spektroskopie der Metalle zwei Konsequenzen. Einmal wird die Präparation der Proben zu einem besonderen technischen Problem, zum anderen beeinflussen die Leitungselektronen, wie KNIGHT[3] entdeckte, die Lage und die Form der Kernresonanzlinien von Metallen (zum Teil weil sie auch deren Relaxationsverhalten bestimmen).

Das zuerst genannte Problem ist rein praktischer Natur und einfach eine Folge des Skin-Effekts. Hochfrequente Magnetfelder können bekanntlich in Metallen nur bis zu einer Tiefe der Größenordnung einiger 10^{-3} cm eindringen. Infolgedessen lassen sich in Resonanzexperimenten an metallischen Proben nur

---

[1] H. G. DEHMELT u. H. KRÜGER: Naturwiss. **37**, 111 (1950). Vgl. auch H. G. DEHMELT: Z. Physik **130**, 356 (1951) und H. KRÜGER: Z. Physik **130**, 371 (1951).
[2] M. BLOOM, E. L. HAHN u. B. HERZOG: Phys. Rev. **97**, 1699 (1955).
[3] W. D. KNIGHT: Phys. Rev. **76**, 1259 (1949).

jene Kerne beobachten, welche in einer dünnen Oberflächenschicht der Probe enthalten sind. Trotz der auf Grund dieses Umstands extrem kleinen Signal-spannungen ist übrigens die erste Beobachtung einer Metall-Resonanzlinie sogar ohne Verwendung einer eigentlichen Probe gelungen. POUND[1] fand nämlich die Resonanzfrequenzen der in dem Kupferdraht seiner Probenspule enthaltenen Cu^{63}- und Cu^{65}-Kerne. Zum genauen Studium von Kernresonanz-Spektren wird man sich jedoch stets bemühen, die Signalspannung, d.h. aber im Fall der Metalle die Probenoberfläche nach Möglichkeit zu vergrößern. Dazu unterteilt man die Probe zweckmäßigerweise, etwa mechanisch, möglichst fein. Sind alle Abmessungen der Partikel metallischer Proben (z.B. von Metall-Pulvern

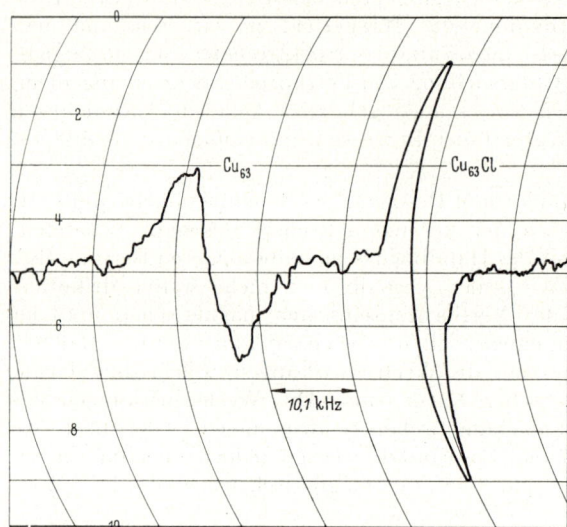

Fig. 110. Resonanzlinien von Cu^{63}. Die untersuchte Probe bestand aus einem Gemisch von pulverförmigem Kupferchlorid und von pulverförmigem Kupfer-metall im Gewichtsverhältnis 100:64. Die breitere und zugleich schwächere Resonanzlinie gehört zu dem Metallpulver. Ihre Frequenz ist bei einer Re-sonanzfrequenz von ungefähr 10 MHz etwa 25 kHz höher als die Frequenz der Kupferchlorid-Linie.

oder von Metall-Kolloiden) klein gegen die Eindring-tiefe des hochfrequenten magnetischen Felds, so durchdringt das Feld die gesamte Probe ohne wesent-lich geschwächt zu werden. Eine Untersuchung von BLOEMBERGEN[2] zeigt dar-über hinaus, daß eine solche feine Unterteilung der Pro-ben auch noch aus einem weiteren Grund empfehlens-wert ist. In Kernresonanz ändern sich bekanntlich die magnetischen Suszeptibili-täten aller Proben, in Me-tallen damit aber auch die Eindringtiefen der elektro-magnetischen Wellen und damit deren Leitungsver-luste. Der eigentliche Kern-absorptions-Verlauf wird durch diesen indirekten Resonanzeffekt zwar ver-stärkt, denn die Gesamtabsorption ist größer, aber auch verzerrt und somit schwieriger deutbar. Der Schwierigkeitsgrad der bei der Herstellung fein unter-teilter Metallproben zu lösenden technischen Probleme hängt außer von der verlangten Größe der Partikel, auch von den Forderungen an deren Reinheit und Verspannungs-Freiheit ab.

Vergleicht man die kernmagnetischen Resonanzfrequenzen von Metallatom-Kernen in gemischten Proben, welche etwa zum Teil aus Metallpulver und zum Teil aus einer pulverförmigen Salzverbindung der untersuchten Kernart be-stehen, so findet man stets, daß die Resonanzfrequenzen der in den Metall-Körnern enthaltenen Kerne einige Promille größer sind, als die Resonanzfrequenzen der nicht-metallisch gebundenen Kerne. In Fig. 110 sind als Beispiel hierzu die beiden Cu^{63}-Resonanzlinien einer aus Kupferchlorid und aus Kupfermetall be-stehenden Probe wiedergegeben. Der Betrag der, nach ihrem experimentellen Entdecker häufig KNIGHT-Resonanzverschiebung (KNIGHT resonance shift) ge-nannten Frequenzverschiebung wächst proportional mit dem Betrag des äußeren

[1] R. V. POUND: Phys. Rev. 73, 1112 (1948).
[2] N. BLOEMBERGEN: J. Appl. Phys. 23, 1383 (1952).

Magnetfelds und ist außerdem um so größer, je höher die Ordnungszahl der untersuchten Metall-Atomart ist, vgl. hierzu[1,2].

Die Resonanzverschiebung in Metallen verhält sich auf Grund ihrer Feldproportionalität qualitativ ebenso wie die durch die makroskopische Proben-Suszeptibilität (Ziff. 25) und ebenso wie die durch die diamagnetische Abschirmung der Kerne durch ihre Elektronenhülle (Ziff. 26) bewirkten Resonanzverschiebungen. Ihrer Größe wegen kann sie jedoch nicht als einer der genannten Effekte gedeutet werden. TOWNES[1,2] hat deshalb als Erklärung vorgeschlagen, daß für die metallische Resonanzverschiebung die Leitungselektronen verantwortlich sind, welche die Energieniveaus nahe der Nullpunkts-Grenzenergie der FERMI-Verteilung besetzen. Bekanntlich stimmt in Metallen die Verteilung der Elektronen auf ihre quantentheoretisch möglichen Energieniveaus wegen der Gültigkeit des PAULI-Prinzips für Fermionen fast mit der Elektronenverteilung am absoluten Nullpunkt überein. Nur etwa 1% der Elektronen besitzen bei Zimmertemperatur Energien, welche um Beträge der Größenordnung kT größer oder kleiner als die FERMI-Grenzenergie sind. Die restlichen 99% der Elektronen besetzen mit paarweise antiparallelen Spins den unteren Teil ihres Energieniveau-Schemas lückenlos. Wegen dieses Umstands ist die makroskopische mittlere Suszeptibilität der Metalle sehr klein. Dies schließt jedoch nicht aus, daß die lokale magnetische Suszeptibilität sehr viel größer sein kann, und sie ist es zweifellos in der Nachbarschaft der Kernorte, da dort die Wahrscheinlichkeitsdichte der Elektronen sicher größer ist, als in den Zwischenbereichen des Metallgefüges. Die quantitative Ausführung dieses Gedankenganges am Beispiel einiger Metalle, unter anderen durch TOWNES, HERRING und KNIGHT[2], durch KOHN und BLOEMBERGEN[3], sowie durch JONES und SCHIFF[4] lieferte theoretische Werte für die Größe der Resonanzverschiebung, deren Übereinstimmung mit den experimentell gefundenen Werten als gut bezeichnet werden darf, wenn man berücksichtigt, daß ein Teil der in die Rechnung eingehenden Parameter, z.B. die Wellenfunktionen der Elektronen nicht genügend genau bekannt sind.

In Metallkristallen, deren Symmetrie niedriger als kubisch ist, sind die Elektronen räumlich anisotrop verteilt. Die Experimente zeigen, daß dann auch die durch sie bewirkten Verschiebungen der Kernresonanzfrequenz anisotrop sind. Allerdings kann man dies kaum direkt feststellen, da die aus den genannten Grunden fein unterteilten Metallproben natürlich stets polykristallin sind. Eine Anisotropie der Resonanzverschiebung hat somit einfach eine Verbreiterung der Resonanzlinie zur Folge. So konnten z.B. BLOEMBERGEN und ROWLAND[5] die an Zinn tetragonaler Symmetrie beobachtete anomale Linienbreite als Anisotropie-Effekt deuten.

Während die Lagen der Kernresonanz-Spektren oder Linien von Metallen, von denen der übrigen Festkörper stets erheblich abweichen, unterscheiden sich die Formen ihrer Spektren (bzw. ihrer mehr oder weniger breiten Resonanzlinien) weniger von denen anderer Polykristalle. In manchen Metallproben wird die Spektrums-Struktur, ebenso wie in den zuvor in dieser Ziffer besprochenen nichtmetallischen Festkörpern durch Spin-Spin-Kopplungen, durch Quadrupol-Wechselwirkungen und durch die Relaxationsmechanismen bestimmt. Andere Proben zeigen dagegen den soeben erwähnten nur nicht-kubischen Metallen eigentümlichen Linienverbreiterungs-Effekt auf Grund der anisotropen Leitungselektronen-Verteilung. RUDERMAN und KITTEL[6] haben ferner theoretisch

---

[1] Siehe W. D. KNIGHT: Phys. Rev. **76**, 1259 (1949).
[2] C. H. TOWNES, C. HERRING u. W. D. KNIGHT: Phys. Rev. **77**, 852 (1950).
[3] W. KOHN u. N. BLOEMBERGEN: Phys. Rev. **80**, 913 (1950); **82**, 283 (1951).
[4] H. JONES u. B. SCHIFF: Proc. Phys. Soc. Lond. A **67**, 217 (1954).
[5] N. BLOEMBERGEN u. T. J. ROWLAND: Acta metallurg. **1**, 731 (1953).
[6] M. A. RUDERMAN u. C. KITTEL: Phys. Rev. **96**, 99 (1954).

untersucht, welchen Einfluß eine indirekte Austausch-Kopplung der Kernspins durch die Leitungselektronen auf die Linienform von Metall-Resonanzlinien haben kann. Sie finden, daß dieser ebenfalls nur in Metallen mögliche Effekt, in Proben reiner Isotope die Resonanzlinie verschärft, in Isotopen-Gemischen dagegen verbreitert.

Zahlreiche Ergebnisse der Kernresonanz-Spektroskopie an festen Substanzen lassen sich nur verstehen, wenn man annimmt, daß viele Festkörper in Wirklichkeit gar nicht so starr sind, wie man es sich üblicherweise vorstellt. In den kernmagnetischen Experimenten ist das eindeutige Kennzeichen des Beweglichseins oder des Beweglichwerdens (etwa infolge einer Erwärmung) der Partikel oder eines Teils der Partikel einer Substanz die wesentlich größere Resonanzlinien-Schärfe der in den beweglichen Partikeln enthaltenen Kerne. Phänomene dieser Art hat man in den meisten Festkörperklassen unter geeigneten Bedingungen gefunden. Die Wirkungsweise der allen derartigen Beobachtungen zugrunde liegenden sog. Beweglichkeits-Effekte wurde in Ziff. 27 bei der Besprechung der Spektren der Flüssigkeiten und Gase, deren charakteristisches Kennzeichen ja gerade die weitgehende Beweglichkeit ihrer Teilchen ist, bereits beschrieben. Wie dort auseinandergesetzt, beobachtet man immer dann ein Anwachsen der Signalamplitude, verbunden mit einer Abnahme der Halbwertsbreite des Signals, wenn sich durch genügend schnell verlaufende Bewegungen der ganzen Probe (vgl. S. 234) oder der Partikel der Probe (vgl. S. 296) die ortsabhängigen inneren Feldstärke-Differenzen bzw. inneren Wechselwirkungen ganz oder teilweise zeitlich ausmitteln. An Stelle einer mehr oder weniger breiten spektralen Verteilung der Resonanzfrequenzen erhält man auf Grund dieses Effekts nur noch schmale Resonanzlinien, deren Frequenz gleich der mittleren Frequenz der ursprünglichen Feldverteilung ist und deren Linienbreite nur noch durch Relaxationsprozesse bestimmt wird, wenn das äußere Magnetfeld genügend homogen ist.

Diese Hinweise zeigen bereits ausreichend deutlich, daß die resonanz-spektroskopische Verfolgung von inneren Umwandlungen und Bewegungen in Festkörpern (Umorientierungs-, Rotations-, und Diffusions-Vorgänge u. a.) ein wertvolles Hilfsmittel zur Beobachtung und zur Analysierung solcher Vorgänge darstellt. Einige verhältnismäßig willkürlich aus den zahlreichen Veröffentlichungen zu diesem Thema herausgegriffene Arbeiten sollen noch erwähnt werden. Die ersten Beobachtungen und Deutungen von Beweglichkeiten in Festkörpern wurden von Alpert[1], von Hatton und Rollin[2] sowie von Gutowsky und Pake[3] mitgeteilt. Interessante Beiträge haben Purcell[4], Andrew und Eades[5], Thomas, Alpert und Torrey[6], Powles und Gutowsky[7], Roshworth[8], Gutowsky[9], Seymour[10], Gutowsky, Pake und Bersohn[11] sowie Odajima, Sohma und Koike[12] u.a. veröffentlicht.

Zum Abschluß der Besprechung der kernmagnetischen Resonanzphänomene an kompakten Substanzen soll noch ein Effekt besprochen werden, der von

---

[1] N. L. Alpert: Phys. Rev. **72**, 637 (1947); **75**, 398 (1949).

[2] J. Hatton u. B. V. Rollin: Proc. Roy. Soc. Lond., Ser. A **199**, 222 (1949).

[3] H. S. Gutowsky u. G. E. Pake: J. Chem. Phys. **16**, 1164 (1948); **18**, 162 (1950).

[4] E. M. Purcell: Physica, Haag **17**, 282 (1951).

[5] E. R. Andrew u. R. G. Eades: Proc. Phys. Soc. Lond. A **65**, 371 (1952).

[6] J. T. Thomas, N. L. Alpert u. H. C. Torrey: J. Chem. Phys. **18**, 1511 (1950).

[7] J. G. Powles u. H. S. Gutowsky: J. Chem. Phys. **21**, 1695 (1953).

[8] F. A. Roshworth: Proc. Roy. Soc. Lond., Ser. A **222**, 526 (1954).

[9] H. S. Gutowsky: Phys. Rev. **83**, 1073 (1951).

[10] E. F. W. Seymour: Proc. Phys. Soc. Lond. A **66**, 85 (1953).

[11] H. S. Gutowsky, G. E. Pake u. R. Bersohn: J. Chem. Phys. **22**, 643, 651 (1954).

[12] A. Odajima, J. Sohma u. M. Koike: J. Chem. Phys. **23**, 1959 (1955).

OVERHAUSER[1] als Folge der Wechselwirkungen von Kernen und Leitungselektronen in Metallen vorhergesagt wurde und von CARVER und SLICHTER[2] erstmals experimentell an Alkalimetallen nachgewiesen wurde. In der Zwischenzeit konnte theoretisch und experimentell gezeigt werden[3-8], daß der nach OVERHAUSER benannte Wechselwirkungs-Effekt auch in nichtmetallischen Festkörpern und sogar in Flüssigkeiten existiert. Darüber hinaus sind Wechselwirkungen nach der Art des OVERHAUSER-Effekts nicht nur zwischen Kernen und Elektronen, sondern auch zwischen Kernen möglich (Kern-OVERHAUSER-Effekt), vgl. z.B.[9, 10]. Die Behandlung des OVERHAUSER-Effekts am Ende der den Festkörperspektren gewidmeten vorliegenden Ziff. 28 ist also eigentlich nicht korrekt, sondern nur historisch und didaktisch begründbar.

Der experimentelle Sachverhalt des Effekts läßt sich leicht beschreiben. Beobachtet man in einem Doppelresonanz-Experiment die Kernresonanz — etwa eines Metalls — und regt zugleich die paramagnetische Resonanz der Leitungselektronen (deren Resonanzfrequenz etwa 1000mal größer als die der Kerne ist) in einem Absorptionsexperiment bis zur Sättigung an, so findet man, daß die kernmagnetische Signalspannung sehr viel größer ist, als in demselben, aber ohne gleichzeitige Elektronenresonanz ausgeführten Kernresonanz-Experiment.

Zum Verständnis dieses Phänomens sei, einer Beschreibung von KLEIN[11] folgend, eine Substanz in einem Magnetfeld H betrachtet, welche N Kerne mit dem magnetischen Moment μ_k und dem Spin $I = \frac{1}{2}$ und N Elektronen mit dem magnetischen Moment μ_e und dem Spin $S = \frac{1}{2}$ enthält. Für die Aufenthaltswahrscheinlichkeiten der Kerne und der Elektronen in den zu den Quantenzahlen $m_{I,S} = \pm \frac{1}{2}$ gehörenden ZEEMAN-Energieniveaus $\mp \mu_k H$ bzw. $\pm \mu_e H$ gilt dann

und
$$w\left(I_1 + \tfrac{1}{2}\right) + w\left(I_1 - \tfrac{1}{2}\right) = 1 \quad \left.\begin{array}{l} \\ \\ \end{array}\right\}$$
$$w\left(S_1 + \tfrac{1}{2}\right) + w\left(S_1 - \tfrac{1}{2}\right) = 1. \quad \right. \tag{28.22}$$

Es sei vorausgesetzt, daß Übergänge zwischen den beiden kernmagnetischen Energieniveaus ausschließlich durch Wechselwirkungen zwischen den Kernen und den Elektronen in der Form der direkten Dipol-Dipol-Kopplung hervorgerufen werden. Dagegen sollen Übergänge zwischen den beiden elektronenmagnetischen Energieniveaus zum Teil durch die Wechselwirkung der Elektronenspins mit dem Gitter G der Substanz, zum Teil durch die Wechselwirkung der Elektronenspins mit einem hochfrequenten Magnetfeld H_1 ihrer LARMOR-Frequenz (Elektronenresonanz) und zum Teil auch durch die Wechselwirkung der Elektronenspins mit den Kernspins verursacht werden.

Die direkte Dipol-Dipol-Wechselwirkung koppelt nur die Zustandspaare $(m_I = +\frac{1}{2}, m_S = -\frac{1}{2})$ und $(m_I = -\frac{1}{2}, m_S = +\frac{1}{2})$ miteinander, denn gekoppelte Übergänge von Kernen und Elektronen sind nur möglich, wenn der Drehimpulserhaltungssatz erfüllt ist, wenn also etwa die Übergänge $m_I = +\frac{1}{2} \to m_I = -\frac{1}{2}$ und $m_S = -\frac{1}{2} \to m_S = +\frac{1}{2}$ eines Kerns und eines Elektrons gleichzeitig erfolgen. Die Übergangswahrscheinlichkeiten zwischen Energieniveaus, d.h. die zeitlichen

[1] A. W. OVERHAUSER: Phys. Rev. 91, 476 (1953); 92, 411 (1953).
[2] T. R. CARVER u. C. P. SLICHTER: Phys. Rev. 92, 212 (1953); 102, 975 (1956).
[3] F. BLOCH: Phys. Rev. 93, 944 (1954).
[4] A. W. OVERHAUSER: Phys. Rev. 94, 768 (1954).
[5] J. KORRINGA: Phys. Rev. 94, 1388 (1954).
[6] C. KITTEL: Phys. Rev. 95, 589 (1954).
[7] H. G. BELJERS, L. VAN DER KINT u. J. S. VAN WIERINGEN: Phys. Rev. 95, 1683 (1954).
[8] A. ABRAGAM: Phys. Rev. 98, 1729 (1955).
[9] I. SOLOMON: Phys. Rev. 99, 559 (1955).
[10] I. SOLOMON u. N. BLOEMBERGEN: J. Chem. Phys. 25, 261 (1956).
[11] M. J. KLEIN: Phys. Rev. 98, 1736 (1955).

Änderungen der Aufenthaltswahrscheinlichkeiten sind proportional zu den Besetzungszahlen der Niveaus, also proportional zu den Aufenthaltswahrscheinlichkeiten selbst. Infolgedessen gelten z.B. für die zeitlichen Änderungen der Aufenthaltswahrscheinlichkeiten $w(I_1 + \frac{1}{2})$ und $w(S_1 + \frac{1}{2})$ die Beziehungen

$$\frac{dw(I_1+\frac{1}{2})}{dt} = p(m_I:-\tfrac{1}{2}\to+\tfrac{1}{2}; m_S:+\tfrac{1}{2}\to-\tfrac{1}{2})\cdot w(I_1-\tfrac{1}{2})\cdot w(S_1+\tfrac{1}{2}) -$$
$$- p(m_I:+\tfrac{1}{2}\to-\tfrac{1}{2}; m_S:-\tfrac{1}{2}\to+\tfrac{1}{2})\cdot w(I_1+\tfrac{1}{2})\cdot w(S_1-\tfrac{1}{2}),$$
$$\frac{dw(S_1+\frac{1}{2})}{dt} = [p(m_S:-\tfrac{1}{2}\to+\tfrac{1}{2}; G)+p(m_S:-\tfrac{1}{2}\to+\tfrac{1}{2}; H_1)]\cdot w(S_1-\tfrac{1}{2}) -$$
$$- [p(m_S:+\tfrac{1}{2}\to-\tfrac{1}{2}; G)+p(m_S:+\tfrac{1}{2}\to-\tfrac{1}{2}; H_1)]\cdot w(S_1+\tfrac{1}{2}) +$$
$$+ p(m_I:+\tfrac{1}{2}\to-\tfrac{1}{2}; m_S:-\tfrac{1}{2}\to+\tfrac{1}{2})\cdot w(I_1+\tfrac{1}{2})\cdot w(S_1-\tfrac{1}{2}) -$$
$$- p(m_I:-\tfrac{1}{2}\to+\tfrac{1}{2}; m_S:+\tfrac{1}{2}\to-\tfrac{1}{2})\cdot w(I_1-\tfrac{1}{2})\cdot w(S_1+\tfrac{1}{2}),$$

(28.23 a,b)

wenn man mit den Größen p die Übergangswahrscheinlichkeiten pro Zeiteinheit der in Klammern angegebenen Übergänge bezeichnet.

Im thermodynamischen Gleichgewicht verhalten sich die inversen Übergangswahrscheinlichkeiten wie die BOLTZMANN-Faktoren der zugehörigen Niveaus, es ist also[1]

$$p(m_S:-\tfrac{1}{2}\to+\tfrac{1}{2}; G)\, e^{-\mu_e H/kT} = p(m_S:+\tfrac{1}{2}\to-\tfrac{1}{2}; G)\, e^{+\mu_e H/kT}$$
und
$$p(m_I:-\tfrac{1}{2}\to+\tfrac{1}{2}; m_S:+\tfrac{1}{2}\to-\tfrac{1}{2})\, e^{-(\mu_k+\mu_e)H/kT}$$
$$= p(m_I:+\tfrac{1}{2}\to-\tfrac{1}{2}; m_S:-\tfrac{1}{2}\to+\tfrac{1}{2})\, e^{+(\mu_k+\mu_e)H/kT}.$$

(28.24a, b)

(Im Fall der Kern-Elektron-Wechselwirkung sind die Dipolmomente der Kerne und Elektronen parallel, wenn ihre Spinvektoren antiparallel sind.)

Im stationären Zustand verschwinden definitionsgemäß die zeitlichen Ableitungen dp/dt, damit folgt aus (28.23) und (28.24):

$$\frac{w(S_1-\tfrac{1}{2})}{w(S_1\pm\tfrac{1}{2})} = \frac{p(m_S:-\tfrac{1}{2}\to+\tfrac{1}{2}; G)\, e^{+2\mu_e H/kT}+p(m_S:\pm\tfrac{1}{2}\to\mp\tfrac{1}{2}; H_1)}{p(m_S:-\tfrac{1}{2}\to+\tfrac{1}{2}; G)+p(m_S:\pm\tfrac{1}{2}\to\mp\tfrac{1}{2}; H_1)}$$
und
$$\frac{w(I_1-\tfrac{1}{2})}{w(I_1+\tfrac{1}{2})} = \frac{w(S_1-\tfrac{1}{2})}{w(S_1+\tfrac{1}{2})}\cdot e^{-2\mu_k H/kT}\cdot e^{-2\mu_e H/kT}.$$

(28.25 a, b)

In zwei Sonderfällen haben die Gleichungen eine besonders einfache Form. Wird kein äußeres Mikrowellenfeld in die Substanz eingestrahlt [$H_1=0$, $p(m_S:\pm\tfrac{1}{2}\to\mp\tfrac{1}{2}; H_1)=0$], so wird

$$\frac{w(S_1-\tfrac{1}{2})}{w(S_1+\tfrac{1}{2})} = e^{+2\mu_e H/kT} \quad\text{und}\quad \frac{w(I_1-\tfrac{1}{2})}{w(I_1+\tfrac{1}{2})} = e^{-2\mu_k H/kT}, \qquad (28.26a, b)$$

d.h. die Gln. (28.25) gehen wieder in die Voraussetzung (28.24) über, daß im ungestörten thermodynamischen Gleichgewicht die Verteilung der Kerne und Elektronen auf ihre ZEEMAN-Niveaus der BOLTZMANN-Beziehung gehorcht. Wird dagegen ein Mikrowellenfeld der Elektronen-LARMOR-Frequenz in die Substanz eingestrahlt, so entfernt sich das Elektronenspin-System nach (28.25a) vom Zustand der thermodynamischen Gleichgewichtsverteilung und nähert sich dem Gleichverteilungszustand (Sättigung). Es erreicht den Sättigungszustand, wenn die mit der Amplitude von H_1 wachsende Übergangswahrscheinlichkeit $p(m_S:\pm\tfrac{1}{2}\to\mp\tfrac{1}{2}; H_1)$ sehr viel größer als alle anderen Übergangswahrscheinlichkeiten ist. Damit folgt der OVERHAUSER-Effekt unmittelbar aus (28.25b). Für

---

[1] Die durch die Wechselwirkung der Elektronenspins mit dem magnetischen Wechselfeld H_1 ihrer LARMOR-Frequenz bewirkten Übergangswahrscheinlichkeiten $p(m_S:+\tfrac{1}{2}\to-\tfrac{1}{2}; H_1)$ und $p(m_S:-\tfrac{1}{2}\to+\tfrac{1}{2}; H_1)$ sind gleich groß (erzwungene Emission und Absorption).

$w(S_1 - \tfrac{1}{2}) = w(S_1 + \tfrac{1}{2})$ gilt:

$$\frac{w(I_1 - \tfrac{1}{2})}{w(I_1 + \tfrac{1}{2})} = e^{-2\mu_k H/kT} \cdot e^{-2\mu_e H/kT}. \tag{28.27}$$

Die Verteilung der Kerne auf ihre beiden Niveaus gehorcht nach (28.27) primär dem Elektronen-BOLTZMANN-Faktor $e^{-2\mu_e H/kT}$, denn der Kern-BOLTZMANN-Faktor $e^{-2\mu_k H/kT}$ unterscheidet sich sehr viel weniger von 1 als dieser ($\mu_k \ll \mu_e$). Besonders deutlich tritt der Sachverhalt des OVERHAUSER-Effekts zutage, wenn man beachtet, daß $2\mu_k H/kT \ll 1$ und $2\mu_e H/kT \ll 1$ ist und statt des Verhältnisses der Aufenthaltswahrscheinlichkeiten w die Differenz ΔN der Besetzungszahlen $[N(I_1 \pm \tfrac{1}{2}) = N \cdot w(I_1 \pm \tfrac{1}{2})]$ der ZEEMAN-Niveaus ausrechnet. Man findet so:

und damit

$$\left. \begin{aligned} \frac{w(I_1 - \tfrac{1}{2})}{w(I_1 + \tfrac{1}{2})} &= \left(1 - \frac{2\mu_k H}{kT} + \cdots\right)\left(1 - \frac{2\mu_e H}{kT} + \cdots\right) \\ \Delta N = N(I_1 + \tfrac{1}{2}) - N(I_1 - \tfrac{1}{2}) &= \frac{2(\mu_k + \mu_e)H}{kT}, \end{aligned} \right\} \tag{28.28}$$

d.h. aber, daß die Differenz der Besetzungszahlen der Kernspin-Niveaus ungefähr ebenso groß wird, wie die Differenz der Besetzungszahlen der Elektronenspin-Niveaus. Die resultierende Kernmagnetisierung ist dementsprechend bei gleichzeitiger Elektronenresonanz-Sättigung ungefähr 1000mal größer als im thermodynamischen Gleichgewicht.

In wirklichen Substanzen ist jedoch weder die Sättigung des Elektronenspin-Systems vollständig noch erfolgt die Kernrelaxation ausschließlich wie in Gl. (28.23 a) vorausgesetzt wurde, über die Kern-Elektron-Wechselwirkung. Analog zum Sättigungsfaktor der Kernresonanz (vgl. S.175) kann man zur Charakterisierung der Sättigung des Elektronen-Spin-Systems einen Sättigungsfaktor

$$\sigma = \frac{p(m_S : -\tfrac{1}{2} \to +\tfrac{1}{2}; H_1)/p(m_S : +\tfrac{1}{2} \to -\tfrac{1}{2}; G)}{1 + p(m_S : -\tfrac{1}{2} \to +\tfrac{1}{2}; H_1)/p(m_S : +\tfrac{1}{2} \to -\tfrac{1}{2}; G)} \tag{28.29}$$

einführen. Bezeichnet man die Summe aller nicht durch die Kern-Elektron-Kopplung verursachten Beiträge zur Spin-Gitter-Übergangswahrscheinlichkeit des Kernspinsystems mit $p''_{I,1} = 1/T''_{I,1}$ und den Beitrag der Kern-Elektron-Kopplung zur Spin-Gitter-Übergangswahrscheinlichkeit des Kernspinsystems mit $p'_{I,1} = 1/T'_{I,1}$, so ist die gesamte Spin-Gitter-Übergangswahrscheinlichkeit $p_{I,1} = \dfrac{1}{T_{I,1}} = \dfrac{1}{T'_{I,1}} + \dfrac{1}{T''_{I,1}}$, und man findet für die Differenz der Besetzungszahlen die Gleichung:

$$\Delta N = N(I_1 + \tfrac{1}{2}) - N(I_1 - \tfrac{1}{2}) = \frac{2\left(\mu_k + \sigma \cdot \dfrac{p'_{I,1}}{p_{I,1}}\,\mu_e\right)H}{kT}, \tag{28.30}$$

welche für $\sigma \approx 1$ und $T'_{I,1} \ll T''_{I,1}$ in (28.28) übergeht.

Die obige Ableitung stützt sich auf die Voraussetzung, daß die Wechselwirkungs-Partner der Kerne Elektronen sind, und daß die Wechselwirkungen überwiegend die Form der direkten Dipol-Dipol-Kopplung hat. Bei den meisten Kernen sind die Spinvektoren und die Dipolmomente parallel. Dementsprechend ändern sich in der voranstehenden Ableitung einige Vorzeichen, wenn der OVERHAUSER-Effekt auf Grund einer Kern-Kern-Wechselwirkung entsteht (sogen. Kern-OVERHAUSER-Effekt). Außerdem ist die Polarisationswirkung der Kern-Kern-Kopplung dem Betrage nach sehr viel geringer. Der Betrag und das Vorzeichen des durch die Sättigung der Resonanz einer Gruppe von Wechselwirkungs-Partnern verursachten Polarisationseffekts hängt aber auch, wie z.B. ABRAGAM[1] gezeigt hat, von der Art der Wechselwirkung ab. SOLOMON und BLOEMBERGEN[2] haben deshalb in einer Arbeit die Beobachtung des OVERHAUSER-Effekts als Hilfsmittel zur

---

[1] A. ABRAGAM: Phys. Rev. 98, 1729 (1955).
[2] I. SOLOMON u. N. BLOEMBERGEN: J. Chem. Phys. 25, 261 (1956).

Identifizierung des Wechselwirkungs-Typs benutzt und konnten so klären, unter welchen Umständen eine Kerndipol-Kopplung in einer bestimmten Substanz überwiegend direkt bzw. überwiegend elektronengekoppelt erfolgt.

Eine weitere naheliegende Anwendung des Kernpolarisations-Effekts durch Elektronenresonanz-Sättigung ist seine Ausnutzung in Experimenten zum Nachweis einer eventuellen räumlichen Anisotropie der Strahlenemission radioaktiver Kerne bezüglich ihrer Spinrichtung, vgl. hierzu eine Arbeit von ABRAHAM, KEDZIE und JEFFRIES[1].

In zahlreichen experimentellen und theoretischen Arbeiten werden zur Zeit Beiträge zu dem mit der Entdeckung des OVERHAUSER-Effekts neu erschlossenen Gebiet der wechselseitigen Beziehungen der Kern- und der Elektronenresonanz veröffentlicht. Ohne Anspruch auf Vollständigkeit soll, neben den bereits zitierten Arbeiten, noch auf Publikationen von BARKER und MENCHER[2], von KAPLAN[3], von BLOCH[4], von AZBEL, GERASIMENKO und LIFSCHITZ[5], von ABRAGAM, COMBRISSON und SOLOMON[6], sowie von JEFFRIES[7] hingewiesen werden.

III. Kernmomenten-Bericht.

Der vorliegende Bericht[8] enthält vor dem 31. Januar 1957 veröffentlichte Ergebnisse der Kernmomenten-Forschung. Es wurde versucht, alle bis zu diesem Datum bekanntgewordenen Messungen magnetischer Kernmomente zu zitieren, unabhängig von der Art des angewandten Meßverfahrens. Dementsprechend enthält der Bericht außer den Ergebnissen der Kerndipolmoment-Messungen auch die von drei Arbeitsgruppen über magnetische Kernoktopolmomente (μ_3) mitgeteilten experimentellen Daten (vgl. Ga^{69}, Ga^{71}, In ^{115} und J^{127}). Kern-Spin-Bestimmungen wurden in die Tabelle aufgenommen, wenn kein sicheres Ergebnis vorlag, wenn bisher kein magnetisches Moment gemessen wurde, oder wenn die Größe eines Spins mit der kernmagnetischen Resonanzmethode ermittelt oder bestätigt wurde. Die Tabelle enthält keine Daten über elektrische Kernmomente.

Im einzelnen wurden bei der Zusammenstellung der Daten folgende Regeln beachtet.

In der *ersten* Spalte (Kern) wurden alle stabilen Kernarten aufgeführt, unabhängig davon, ob Kernmomenten-Messungen vorlagen oder nicht. Instabile Elemente wurden ebenfalls stets in die Tabelle aufgenommen, auch wenn bisher keine Bestimmungen ihrer Momente veröffentlicht worden sind. Instabile Isotope wurden nur zitiert, wenn Messungen ihres Spins oder ihres magnetischen Moments vorlagen. Angeregte Kernzustände wurden nur dann aufgeführt — gekennzeichnet durch einen Stern (z. B *F$_9^{19}$) — wenn Messungen ihrer magnetischen Momente veröffentlicht worden sind. Die Elemente sind nach steigender Ladungszahl, die Isotope nach steigender Massenzahl angeordnet.

In der *zweiten* Spalte (Vork.) sind die von MATTAUCH und FLAMMERSFELD[9] sowie von COLLINS, ROURKE und WHITE[10] angegebenen relativen Häufigkeiten der Isotope im natürlichen Vorkommen eingetragen worden. Von anderen Autoren

---

[1] M. ABRAHAM, R. W. KEDZIE u. C. D. JEFFRIES: Phys. Rev. **106**, 165 (1957).
[2] W. A. BARKER u. A. MENCHER: Phys. Rev. **98**, 1868 (1955).
[3] J. I. KAPLAN: Phys. Rev. **99**, 1322 (1955).
[4] F. BLOCH: Phys. Rev. **120**, 104 (1956).
[5] M. Y. AZBEL, V. I. GERASIMENKO u. I. M. LIFSCHITZ: Z. exp. theor. Phys. USSR. **32**, 1212 (1957).
[6] A. ABRAGAM, J. COMBRISSON u. I. SOLOMON: Arch. d. Sci. **10**, 240 (1957).
[7] C. D. JEFFRIES: Phys. Rev. **106**, 164 (1957).
[8] Meinen Mitarbeitern, den Herren R. HAUSSER, G. J. KRÜGER und F. NOACK habe ich für ihre Hilfe bei der Zusammenstellung der Zitate ganz besonders zu danken.
[9] Isotopenbericht, Verlag der Z. Naturforsch. Tübingen 1949.
[10] F. A. WHITE, T. L. COLLINS u. F. M. ROURKE: Phys. Rev. **101**, 1786 (1956). — T. L. COLLINS, F. M. ROURKE u. F. A. WHITE: Phys. Rev. **105**, 196 (1957).

mitgeteilte Häufigkeitswerte unterscheiden sich zum Teil etwas von diesen. Da Häufigkeitsangaben im Rahmen der Kernmomenten-Forschung nur als Hinweis auf den ungefähren experimentellen Schwierigkeitsgrad benötigt werden, konnte darauf verzichtet werden, solche Abweichungen in der Tabelle wiederzugeben.

In der *dritten* Spalte (Spin) wurden die als richtig angenommenen Werte des Drehimpulses in Einheiten des PLANCKschen Wirkungsquantums $\hbar = h/2\pi$ eingetragen. Zweifelhafte Daten wurden durch Klammern gekennzeichnet. Fehlen spezielle Literaturhinweise bezüglich der Messung des Kernspins, so kann im allgemeinen auf die in den angeführten Publikationen über die kernmagnetischen Momente enthaltenen Literaturangaben verwiesen werden. Für gg-Kerne wurde immer $I = 0$ angegeben. Experimentelle Nachprüfungen wurden durch $\mu \approx 0$ gekennzeichnet.

In der *vierten* Spalte (Zitat) wurden alle aufgefundenen Veröffentlichungen zusammengestellt. Geordnet wurden diese zunächst nach der bearbeiteten Kernart (ein Teil der Publikationen wird somit mehrfach zitiert), innerhalb jeder Kernart nach dem Publikationsjahr und innerhalb jedes Jahrgangs alphabetisch nach dem ersten Verfassernamen.

In der *fünften* Spalte (Methode) wurden zur Vereinfachung der Darstellung Verfahren, die in dem vorliegendem Beitrag zum Handbuch nicht behandelt worden sind, nur summarisch unterschieden. Dementsprechend wurden unter der Bezeichnung Hfs = Hyperfeinstruktur alle optischen Methoden zusammengefaßt. Die Bezeichnung Atomstrahl umschließt sowohl einfache Strahlablenkexperimente als auch hochfrequenzspektroskopische Resonanzexperimente an Atom- und Molekülstrahlen. Ebenso sind die Bezeichnungen „Paramagnetische Resonanz", „Quadrupolresonanz", „Mikrowellenspektroskopie", „Winkelverteilung der γ-Strahlung" und „Daten der Zerfallsreihe" als Namen von Verfahrensgruppen zu verstehen. Dagegen wurden innerhalb der in dem vorliegenden Beitrag beschriebenen Methodengruppe der kernmagnetischen Resonanzspektroskopie an kompakten Substanzen im einzelnen Anordnungen unterschieden, in denen „Gekreuzte Spulen", eine normale „Brücke", ein „Empfindlicher Oszillator" (bzw. ein „Empfindlicher Verstärker") oder eine „Pendelrückkopplungsschaltung" benutzt worden sind. Arbeiten von Autoren, welche Kernmomente auf Grund früher veröffentlichter Meßdaten berechneten, wurden als solche gekennzeichnet. Arbeiten, welche sich mit Absolutmessungen der kernmagnetischen Momente bzw. deren Vorzeichen befassen, sind daran erkennbar, daß neben einem Verfahren zur Präzessionsfrequenzmessung noch eine zweite Meßvorrichtung (z. B. ein verzögerndes Zyklotron, ein Omegatron, ein β-Spektrograph, eine Cottonwaage, ein Solenoid oder ein Rotierendes Hochfrequenzfeld) benutzt worden ist.

In der *sechsten* Spalte (Substanz) wurden die zur Messung benutzten Substanzen eingetragen, soweit Angaben der Autoren vorlagen. Wurden die Meßbzw. Vergleichssubstanzen in verschiedenen Proben untersucht, so sind sie durch die Buchstaben a) und b) unterschieden. Dabei bedeuten die Abkürzungen: w.L. = wäßrige Lösung; ges. = gesättigt; m = molar. Bei Atomstrahlexperimenten wurden die im Strahl enthaltenen Moleküle oder Atome angegeben.

In der *siebenten* Spalte (Ergebnisse) wurden, soweit es sich um Resonanzverfahren handelt, in der Regel die direkt gemessenen Daten eingetragen. In einigen Fällen haben die Autoren nur korrigierte Werte angegeben; diese sind also solche gekennzeichnet.

In der *achten* und *neunten* Spalte (μ_{korr} und μ_{unkorr}) sind die von den Autoren angegebenen unkorrigierten bzw. korrigierten Werte der magnetischen Kerndipolmomente im allgemeinen in Kernmagnetonen μ_K angegeben. Später veröffentlichte genauere Umrechnungsfaktoren und Korrekturen wurden nicht berücksichtigt.

Kern	Vork. %	Spin	Zitat	Methode	Substanz	Ergebnisse	μ_{unkorr}	μ_{korr}
n_0^1	—	$\frac{1}{2}$	P. N. Powers Phys. Rev. 54, 827 (1938)	Atomstrahl	Neutronenstrahl		-2 ± 1	
			L. W. Alvarez, F. Bloch Phys. Rev. 57, 111 (1940) Phys. Rev. 57, 352 (1940)	a) Atomstrahl b) Zyklotron	a) Neutronen- strahl b) Protonenstrahl		$-1,935$ $\pm 0,02$	
			W.R.Arnold, A.Roberts Phys. Rev. 70, 766 (1946) Phys. Rev. 71, 878 (1947)	a) Atomstrahl b) Brücke	a) Neutronen- strahl b) H_2O, Paraffin u. a.	$\nu_{n1}/\nu_{H1}=0,68479$ $\pm 0,0004$	$-1,9103$ $\pm 0,0012$	$-1,9103$ $\pm 0,0012$
			F. Bloch et al. Phys. Rev. 74, 1025 (1948)	a) Atomstrahl b) Gekreuzte Spulen	a) Neutronen- strahl b) 0,1 m $MnSO_4$, w. L.	$\nu_{n1}/\nu_{H1}=0,685001$ $\pm 0,00003$ $\mu < 0$	$-1,91307$ $\pm 0,0006$	
			E. H. Rogers, H.H. Staub Phys. Rev. 76, 980 (1949) H. H. Staub, E.H. Rogers Helv. phys. Acta 23, 63 (1950)	a) Rotierendes HF-Feld b) Atomstrahl	b) Neutronen- strahl	$\mu < 0$		
			V. W. Cohen, N. R. Corn- gold, N. F. Ramsey Phys. Rev. 104, 283 (1956)	a) Atomstrahl b) Brücke	a) Neutronen- strahl b) 0,25 m Fe^{++}, w. L.	$\nu_{n1}/\nu_{H1}=0,685057$ $\pm 0,000017$		$-1,913148$ $\pm 0,000066$
H_1^1	$99,98_5$	$\frac{1}{2}$	I. Estermann, O. Stern Z. Physik 85, 17 (1933)	Atomstrahl	H_2		$2,5$ $\pm 0,25$	
			R. Frisch, O. Stern Z. Physik 85, 4 (1933)	Atomstrahl	H_2		$2-3$	
			I. I. Rabi et al. Phys. Rev. 46, 157 (1934)	Atomstrahl	H		$3,25$ $\pm 0,32$	
			J. M. B. Kellogg et al. Phys. Rev. 49, 421 (1936) Phys. Rev. 50, 472 (1936)	Atomstrahl	H		$+2,85$ $\pm 0,15$	
			B. G. Lasarew et al. Phys. Z. Sowjet. 10, 117(1936) Phys. Z. Sowjet. 11, 445(1937)	Statische Suszept.- Messung	H_2, fest		$2,7$ $\pm 0,3$	
			I. Estermann et al. Phys. Rev. 52, 535 (1937)	Atomstrahl	H_2, HD		$2,46$ $\pm 0,07$	
			J. M. B. Kellogg et al. Phys. Rev. 56, 728 (1939)	Atomstrahl	H_2, HD		$+2,785$ $\pm 0,02$	
			S. Millman, P. Kusch Phys. Rev. 60, 91 (1941)	Atomstrahl	KOH, NaOH	$g = 5,5791$ $\pm 0,0016$	$2,7896$ $\pm 0,0008$	$2,7928$ $\pm 0,0008$
			S. Millman, P. Kusch Phys. Rev. 60, 91 (1941)	Berechnung aus früheren Daten			$+2,7876$	
			J. R. Zimmerman et al. Phys. Rev. 73, 94 (1948)	Pendelrückk.			$2,788$ $\pm 0,028$	
			J.H. Gardner et al. Phys. Rev. 76, 1262 (1949)	a) Hohlleiter- zyklotron b) Brücke	a) Elektronen- strahl b) Mineralöl	$\nu_{El., Zykl.}/$ $\nu_{H1, Larm.}$ $= 657,475$ $\pm 0,008$		$(1,52100$ $\pm 0,00002) \times$ $10^{-3} \mu_B$
			J. A. Hipple et al. Phys. Rev. 76, 1877 (1949)	a) Omegatron b) Brücke	a) Protonenstrahl b) —	$\nu_{H1, Zykl.}/$ $\nu_{H1, Larm.}$ $= 0,358106$ $\pm 0,000010$		
			E.H.Rogers, H.H.Staub Phys. Rev. 76, 980 (1949) H. H. Staub, E.H. Rogers Helv. phys. Acta 23, 63 (1950)	a) Rotierendes HF-Feld b) Brücke	b) 0,1 m $MnSO_4$, w.L.	$\mu > 0$		
			H. Taub, P. Kusch Phys. Rev. 75, 1326 (1949) Phys. Rev. 75, 1481 (1949)	Atomstrahl	NaOH, Cs^{133}, In^{115}	$\Delta\nu_{el}$ in $Cs,In/$ $\nu_{H1, Larm.}$		$2,7935$ $\pm 0,00014$
			H. A. Thomas et al. Phys. Rev. 75, 902 (1949)	a) Cotton-Waage b) Brücke	Eisen-(III)- oxalat, w. L.	$\gamma_{H1} = (2,6752$ $\pm 0,0002) \times$ 10^4 sec^{-1} Gauß$^{-1}$	$(1,4100 \pm 0,0003) \times$ 10^{-23} dyn $\cdot$ cm $\times$ Gauß$^{-1}$	
			F. Bloch, G. D. Jeffries Phys. Rev. 80, 305 (1950)	a) Verzögerndes Zyklotron b) Gekreuzte Spulen	a) Protonenstrahl b) H_2O	$\nu_{H1, Larm.}/\nu_{Zykl.}$		$2,79245$ $\pm 0,0002$
			H. Sommer et al. Phys. Rev. 80, 487 (1950)	a) Omegatron b) Brücke	a) Protonenstrahl b) H_2O	$\nu_{H1, Larm.}/$ $\nu_{H1, Zykl.}$ $= 2,79268$ $\pm 0,00006$		$2,79268$ $\pm 0,00006$
			H. A. Thomas et al. J. Res. Nat. Bur. Stand. 44, 569 (1950) Phys. Rev. 78, 787 (1950)	a) Cotton-Waage b) Brücke	Fe $(NO_3)_3$, w.L. (Stand.-Ölprobe)	$\gamma_{H1} = (2,67528$ $\pm 0,00006) \times$ 10^4 sec^{-1} Gauß$^{-1}$	$1,4100 \cdot 10^{-23}$ $\pm 0,0002$ dyn $\cdot$ cm $\cdot$ Gauß$^{-1}$	

Kern	Vork. %	Spin	Zitat	Methode	Substanz	Ergebnisse	μ_{unkorr}	μ_{korr}
			J. H. Gardner Phys. Rev. 83, 996 (1951)	a) Hohlleiter- zyklotron b) Brücke	a) Elektronen- strahl b) Mineralöl	$\nu_{El., Zykl.}/$ $\nu_{H^1, Larm.}$ $= 657,475$ $\pm\quad 0,008$		$(1,52101$ $\pm 0,00002) \times$ $10^{-3}\,\mu_B$
			C. D. Jeffries Phys. Rev. 81, 1040 (1951)	a) Verzögerndes Zyklotron b) Gekreuzte Spulen	a) Protonenstrahl b) 0,02 m MnSO$_4$, w. L.	$\nu_{H^1, Larm.}/$ $\nu_{H^1, Zykl.}$	2,7924 $\pm 0,0002$	
			G. Lindström Phys. Rev. 83, 465 (1951)	a) β-Spektrograph b) Brücke		$\gamma_{H^1} = (2,67517$ $\pm 0,00020) \times$ $10^4\, sec^{-1}\, Gauß^{-1}$		
			H. Sommer et al. Phys. Rev. 82, 697 (1951)	a) Omegatron b) Brücke	a) Protonenstrahl b) H$_2$O + Katalys. (verglichen mit Öl-Stand.)	$\nu_{H^1, Larm.}/$ $\nu_{H^1, Zykl.}$ $= 2,792685$ $\pm 0,000006$		2,79276 $\pm 0,00006$
			D. J. Collington et al. Phys. Rev. 99, 1622 (1955)	a) Verzögerndes Zyklotron b) Gekreuzte Spulen	a) Protonenstrahl b) —	$\nu_{H^1, Larm.}/$ $\nu_{H^1, Zykl.}$		2,79281 $\pm 0,00004$
			F. Kirchner, W. Wilhelmy Z. Naturforsch. 10a, 657 (1955)	a) Solenoid b) Empfindlicher Oszillator	FeCl$_3$, 10^{17} Ionen/cm^3, w. L.	$\gamma_{H^1} = (2,67562$ $\pm 0,00016) \times$ $10^4\, sec^{-1}\, Gauß^{-1}$		
			W. Wilhelmy Private Mitteilung	a) Solenoid b) Empfindlicher Oszillator		$\gamma_{H^1, korr.}$ $= (2,67556$ $\pm 0,00017) \times$ $10^4\, sec^{-1}\, Gauß^{-1}$		
			P. Franken, S. Liebes jr. Phys. Rev. 104, 1197 (1956)	a) Zyklotron b) Kernresonanz	a) Elektronen- strahl b) Mineralöl	$\nu_{El., Zykl.}/$ $\nu_{H^1, Larm.}$ $= 657,462$ $\pm\quad 0,006$		$(1,521042$ $\pm 0,000016)$ $\times 10^{-3}\,\mu_B$
			K. R. Trigger Bull. Amer. Phys. Soc. II 1, 220 (1956)	a) Verzögerndes Zyklotron b) Gekreuzte Spulen	a) Protonenstrahl b) H$_2$O	$\nu_{H^1, Larm.}/$ $\nu_{H^1, Zykl.}$		2,79275 $\pm 0,00010$
H_1^2	0,015	1	I. Estermann, O. Stern Phys. Rev. 45, 761 (1934)	Atomstrahl	D$_2$		0,75 $\pm 0,25$	
			I. I. Rabi et al. Phys. Rev. 46, 163 (1934)	Atomstrahl	D$_2$		0,75 $\pm 0,2$	
			A. Farkas, L. Farkas Proc. Roy. Soc. Lond., Ser. A 152, 152 (1935)	Ortho-Para- Konversion	D$_2$, H$_2$, O$_2$	$\mu_{H^1}/\mu_{D^2} = 3,96$ $\pm 0,11$		
			J. M. B. Kellogg et al. Phys. Rev. 50, 472 (1936)	Atomstrahl	D		$+0,85$ $\pm 0,03$	
			J. M. B. Kellogg et al. Phys. Rev. 56, 728 (1939)	Atomstrahl	D$_2$, HD	$\mu_{H^1}/\mu_{D^2} = 3,2570$ $\pm 0,001$	$+0,855$ $\pm 0,006$	
			S. Millman, P. Kusch Phys. Rev. 60, 91 (1941)	Berechnung aus früheren Daten				0,8565
			W. R. Arnold, A. Roberts Phys. Rev. 70, 766 (1946) Phys. Rev. 71, 878 (1947)	Brücke	a) D$_2$O b) H$_2$O, Paraffin, Benzol, H$_2$O + 1,5% FeCl$_3$	ν_{D^2}/ν_{H^1} $= 0,153512$ $\pm 0,00001$	0,85647 $\pm 0,0004$	
			F. Bitter et al. Phys. Rev. 72, 1271 (1947)	Brücke	H$_2$ + D$_2$ flüssig	μ_{D^2}/μ_{H^1} $= 0,307021$ $\pm 0,000005$		
			F. Bloch et al. Phys. Rev. 72, 1125 (1947)	Gekreuzte Spulen	0,1 m MnSO$_4$ in 50% D$_2$O	μ_{H^1}/μ_{D^2} $= 3,257195$ $\pm 0,00002$		
			A. Roberts Phys. Rev. 72, 979 (1947)	Brücke und Pendelrückk.	a) H$_2$O, D$_2$O + H$_2$O, auch mit FeCl$_3$ b) D$_2$O	μ_{H^1}/μ_{D^2} $= 3,25731$ $\pm 0,00015$		
			K. Siegbahn et al. Ark. Fysik 1, 193 (1949) Nature, Lond. 163, 211 (1949)	Brücke	a) D$_2$O b) H$_2$O	μ_{D^2}/μ_{H^1} $= 0,3070183$ $\pm 0,0000015$		
			J. R. Zimmerman et al. Phys. Rev. 76, 163 (1949) Phys. Rev. 76, 350 (1949)	Pendelrückk.	a) 95% D$_2$O b) H$_2$O	ν_{D^2}/ν_{H^1} $= 0,15355$ $\pm 0,00005$		
			H. L. Anderson et al. AECU-1019 (1950) UAC-320 (1950)	Brücke	H$_2$O, D$_2$O	μ_{H^1}/μ_{D^2} $= 3,25720013$ $\pm 0,00000037$		
			E. C. Levinthal Phys. Rev. 78, 204 (1950)	Gekreuzte Spulen	0,1 m MnSO$_4$ in 50% D$_2$O	μ_{D^2}/μ_{H^1} $= 0,3070117$ $\pm 0,0000015$, I	> 0	

22*

Kern	Vork. %	Spin	Zitat	Methode	Substanz	Ergebnisse	μ_{unkorr}	μ_{korr}
			G. Lindström Phys. Rev. **78**, 817 (1950) Ark. Fysik **4**, 1 (1951) Physica, Haag **17**, 412 (1951)	Brücke	a) D_2O b) H_2O	μ_{D^2}/μ_{H^1} $=0{,}30701337$ $\pm 0{,}00000050$		
					a) D_2O b) Paraffinöl	μ_{D^2}/μ_{H^1} $=0{,}30701466$ $\pm 0{,}00000050$		
			B. Smaller et al. Phys. Rev. **80**, 137 (1950)	Brücke	Mn^{++} in H_2O + D_2O(verschied. Konz.)	μ_{H^1}/μ_{D^2} $=3{,}25720037$ $\pm 0{,}00000056$		
			B. Smaller Phys. Rev. **83**, 812 (1951)	Brücke	H_2+D_2, Gas	μ_{H^1}/μ_{D^2} $=3{,}2571990$ Korr $\pm 0{,}0000010$		
					$MnSO_4$ in H_2O $+D_2O$ (ver- schied. Konz.)	μ_{H^1}/μ_{D^2} $=3{,}2571999$ Korr $\pm 0{,}0000012$		
			B. Smaller et al. Phys. Rev. **81**, 896 (1951) Supplement 1 to Nuclear data NBS Circular No. 449 (1951), S. 1	Brücke	H_2+D_2, Gas	μ_{H^1}/μ_{D^2} $=3{,}25719902$ $\pm 0{,}00000060$	0,857606	
			P. J. Gray et al. Phys. Rev. **87**, 229 (1952)	Atomstrahl	a) HD b) D_2	μ_{H^1}/μ_{D^2} $=3{,}257207$ $\pm 0{,}000025$		
			T. F. Wimett Phys. Rev. **91**, 499 (1953)	Brücke	HD-Gas	μ_{D^2}/μ_{H^1} $=0{,}307012192$ $\pm 0{,}000000015$		
H_1^3	$<10^{-10}$	$\frac{1}{2}$	H. L. Anderson et al. Phys. Rev. **71**, 372 (1947)	Brücke	a) H_2^3O b) $Fe(NO_3)_3$ w.L.	ν_{H^3}/ν_{H^1} $=1{,}06666$ $\pm 0{,}00010$		
			F. Bloch et al. Phys. Rev. **71**, 373 (1947)	Gekreuzte Spulen	0,3 m $MnSO_4$ in 80% H_2^3O	μ_{H^3}/μ_{H^1} $=1{,}067 \quad \mu>0$ $\pm 0{,}001, \quad I$		
			F. Bloch et al. Phys. Rev. **71**, 551 (1947)	Gekreuzte Spulen	0,1 m $MnSO_4$ in 50% H_2^3O	ν_{H^3}/ν_{H^1} $=1{,}066636, \mu>0$ $\pm 0{,}00001, \quad I$		
He_2^3	$1{,}3\cdot10^{-4}$	$\frac{1}{2}$	H. L. Anderson et al. Phys. Rev. **73**, 919 (1948)	Brücke	a) He^3+O_2, Gas b) Polystyren	g_{He^3}/g_{H^1} $=0{,}763$ $\pm 0{,}007$		
			H. L. Anderson Phys. Rev. **76**, 1460 (1949) AECU-538 (UAC-122), 1949	Brücke	He^3, O_2, H_2 im Volumverh. $1:1:0{,}37$, 24 at	ν_{He^3}/ν_{H^1} $=0{,}7617866$ $\pm 0{,}0000012, I$		2,12815
					He^3+O_2, 24 at $V:6{,}7\cdot10^{17}$ Ionen/cm^3 Mn^{++} w.L.	ν_{He^3}/ν_{H^1} $=0{,}761779$ $\pm 0{,}000014$		
He_2^4	100	0	H.A.Bethe, R.F.Bacher Rev. Mod. Phys. **8**, 82 (1936)					
Li_3^6	7,30	1	M. Fox, I. I. Rabi Phys. Rev. **48**, 746 (1955)	Atomstrahl	Li	$\mu_{Li^6}/\mu_{Li^7}=0{,}20$ $\pm 0{,}05$	$\approx 0{,}7$	
			J. H. Manley, S. Millman Phys. Rev. **50**, 380 (1936) Phys. Rev. **51**, 19 (1937)	Atomstrahl	Li	$\mu_{Li^6}/\mu_{Li^7}=0{,}258$ $\pm 0{,}002$	0,85	
			H. Schüler, T. Schmidt Z. Physik **99**, 285 (1936)	Hfs.			$+0{,}6$ $\pm 0{,}2$	
			J. E. Gorham Phys. Rev. **53**, 563 (1938)	Atomstrahl		$\mu>0$	>0	
			I. I. Rabi et al. Phys. Rev. **53**, 495 (1938) Phys. Rev. **55**, 526 (1939)	Atomstrahl	LiCl, Li_2, LiF	$g=0{,}820$ $\pm 0{,}005$		0,820 $\pm 0{,}005$
			S. Millman Phys. Rev. **55**, 628 (1939)	Atomstrahl	LiCl	$\mu>0$	>0	
			S. Millman, P. Kusch Phys. Rev. **60**, 91 (1941)	Berechnung aus früheren Daten			$+0{,}8213$	
			H. E. Walchli ORNL-1775 (1954)			ν_{Li^6}/ν_{H^2} $=0{,}958638$ $\pm 0{,}000038$	$+0{,}821921$	
Li_3^7	92,70	$\frac{3}{2}$	G. Breit, F.W.Doermann Phys. Rev. **36**, 1732 (1930)	Berechnung aus früheren Daten			2,13 $\pm 0{,}04$	
			L. P. Granath Phys. Rev. **42**, 44 (1932)	Hfs.			3,29	

Kern	Vork. %	Spin	Zitat	Methode	Substanz	Ergebnisse	μ_{unkorr}	μ_{korr}
			E. Fermi, E. Segrè Z. Physik **82**, 729 (1933)	Berechnung aus früheren Daten			$\mu_B/\mu = 575$	
			S. Goudsmit Phys. Rev. **43**, 636 (1933)	Berechnung aus früheren Daten		$g = 2,19$	3,29	
			M. Fox, I. I. Rabi Phys. Rev. **48**, 746 (1935)	Atomstrahl	Li		3,20	
			J. H. Bartlett et al. Phys. Rev. **50**, 315 (1936)	Berechnung aus früheren Daten			3,33	
			S. Millman, J. R. Zacharias Phys. Rev. **51**, 1049 (1937)	Atomstrahl			> 0	
			I. I. Rabi et al. Phys. Rev. **53**, 495 (1938) Phys. Rev. **55**, 526 (1939)	Atomstrahl	LiCl, LiF	$g = 2,167 \pm 0,010$ $\mu_{\text{Li}^7}/\mu_{\text{Li}^6} = 3,963$ $\pm 0,004$	3,250 $\pm 0,016$	
			D. A. Jackson, H. Kuhn Proc. Roy. Soc. Lond., Ser. A **173**, 278 (1939)	Hfs.	Atomstrahl		3,25	
			S. Millman Phys. Rev. **55**, 628 (1939)	Atomstrahl	Li₂	$\mu > 0$	> 0	
			P. Kusch et al. Phys. Rev. **57**, 765 (1940)	Atomstrahl		$g = -2,18$ $\mu_{\text{Li}^7}/\mu_{\text{Li}^6} = 3,9610$ $\pm 0,0004$		
			S. Millman, P. Kusch Phys. Rev. **60**, 91 (1941)	Atomstrahl	Li₂, LiCl	$g = 2,1696$ $\pm 0,0020$		
			S. Millman, P. Kusch Phys. Rev. **60**, 91 (1941)	Berechnung aus früheren Daten			+ 3,2532	
			A. Bolle et al. Nuovo Cim. **9**, 3, 412 (1946)	Gekreuzte Spulen	Fe(NO₃)₃, w.L.	$\gamma_{\text{H}^1}/\gamma_{\text{Li}^7} = 2,60$ $\pm 0,02$		
			F. Bitter Phys. Rev. **75**, 1326 (1949)	Brücke		$\nu_{\text{Li}^7}/\nu_{\text{H}^1}$ $= 0,388625$ $\pm 0,00004$		
			P. Kusch, A. K. Mann Phys. Rev. **76**, 707 (1949)	Atomstrahl	LiBr, LiJ	$g_{\text{Li}^7}/g_{\text{Li}^6}$ $= 2,64094$ $\pm 0,00005$		
			K. Siegbahn, G. Lindström Ark. Fysik **1**, 193 (1949) Nature, Lond. **163**, 211 (1949)	Brücke	a) LiNO₃, w.L. b) H₂O	$\mu_{\text{Li}^7}/\mu_{\text{H}^1}$ $= 1,165827$ $\pm 0,000060$		
			J. R. Zimmerman et al. Phys. Rev. **76**, 163 (1949) Phys. Rev. **76**, 350 (1949)	Pendelrückk.	a) Li(C₂H₃O₂) + CuCl₂, w.L. b) H₂O	$\nu_{\text{Li}^7}/\nu_{\text{H}^1}$ $= 0,38862$ $\pm 0,00008$		
			G. Lindström Ark. Fysik **4**, 1 (1951) Physica, Haag **17**, 412 (1951)	Brücke	LiNO₃, ges. w.L.	$\nu_{\text{Li}^7}/\nu_{\text{H}^1}$ $= 0,3886341$ $\pm 0,0000010$		3,25633 $\pm 0,00009$
			G. D. Watkins, R. V. Pound Phys. Rev. **82**, 343 (1951)	Empfindlicher Oszillator	LiCl, w.L.	$g_{\text{Li}^7}/g_{\text{Li}^6}$ $= 2,64091$ $\pm 0,00001$		
			T. Kanda et al. Phys. Rev. **85**, 938 (1952)	Empfindlicher Oszillator	LiCl + 0,1 m × MnSO₄, w.L.	$\nu_{\text{Li}^7}/\nu_{\text{H}^1}$ $= 0,388637$ $\pm 0,000010$		
			H. E. Walchli ORNL-1775 (1954)			$\nu_{\text{Li}^7}/\nu_{\text{Na}^{23}}$ $= 1,469225$ $\pm 0,000003$	+ 3,256003	
Be_4^9	100	$\frac{3}{2}$	P. Kusch et al. Phys. Rev. **55**, 666 (1939)	Atomstrahl	NaF · BeF₂ u. KF · BeF₂	$g = 0,783 \pm 0,003$	− 1,175	
			W. H. Chambers et al. Phys. Rev. **76**, 638 (1949)	Pendelrückk.	BeCl₂, w.L.	$\nu_{\text{Be}^9}/\nu_{\text{H}^1}$ $= 0,14034$ $\pm 0,00007$, Korr.		
			W. G. Dickinson et al. Phys. Rev. **75**, 1769 (1949)	Brücke	BeF₂, w.L.	$\nu_{\text{Be}^9}/\nu_{\text{H}^1}$ $= 0,1405187$ $\pm 0,000002$		
			J. R. Zimmerman et al. Phys. Rev. **75**, 699 (1949)	Pendelrückk.	BeCl₂, w.L.	$g = 0,783$		
			F. Alder, F. G. Yu Phys. Rev. **82**, 105 (1951)	Gekreuzte Spulen	Be(NO₃)₂, w.L.		< 0	
			H. S. Gutowsky et al. Phys. Rev. **81**, 635 (1951)	Brücke	K₂BeF₄- und BeF₂-Kristalle	$I = \frac{3}{2}$		
			J. Hatton et al. Phys. Rev. **83**, 672 (1951)		Be₃Al₂Si₆O₁₈- Einkristall	$I = \frac{3}{2}$		
			N. A. Schuster, G. E. Pake Phys. Rev. **81**, 886 (1951)		BeAl₂O₄-Ein- kristall	$I = \frac{3}{2}$		
			R. E. Sheriff, D. Williams Phys. Rev. **82**, 651 (1951)	Pendelrückk.	a) BeCl₂, w.L. b) D₂O + NiCl₂	$\nu_{\text{Be}^9}/\nu_{\text{D}^2}$ $= 0,915475$ $\pm 0,00007$		1,17746 $\pm 0,00009$

Kern	Vork. %	Spin	Zitat	Methode	Substanz	Ergebnisse	μ_{unkorr}	μ_{korr}
			L. C. BROWN, D. WILLIAMS Phys. Rev. 95, 1110 (1954) J. Chem. Phys. 24, 751 (1956)	Pendelrückk.	$Be_3Al_2Si_6O_{18}$- Einkristall	I	$\|\mu\| = 1,1774$ $\pm 0,0002$	
B_5^{10}	18,83	3	S. MILLMAN et al. Phys. Rev. 56, 165 (1939) Phys. Rev. 56, 213 (1939)	Atomstrahl	$Li_2B_4O_7$, $Na_2B_4O_7$, $K_2B_4O_7$, $NaBO_2$, KBO_2	$g = +0,597$ $\pm 0,003$	$+0,597$ $\pm 0,003 \ (I=1)$	
			F. BITTER Phys. Rev. 75, 1326 (1949)	Brücke		ν_{B10}/ν_{B11} $= 0,33488$ $\pm 0,00010$		
			Y. TING, D. WILLIAMS Phys. Rev. 82, 131 (1951)	Pendelrückk.	$Na_2B_2O_4$, w. L.		1,8006	
			Y. TING, D. WILLIAMS Phys. Rev. 89, 595 (1953)	Pendelrückk.	a) $Na_2B_2O_4$, w. L. b) $NiCl_2$ in D_2O	ν_{B10}/ν_{D2} $= 0,700065$ $\pm 0,00007$		1,80066 $\pm 0,00018$
					a) $Na_2B_2O_4$, w. L. b) RbCl, w. L.	ν_{B10}/ν_{Rb85} $= 1,11282$ $\pm 0,00005$		1,80066 $\pm 0,00015$
B_5^{11}	81,17	$\frac{3}{2}$	S. MILLMAN et al. Phys. Rev. 56, 165 (1939) P. KUSCH, S. MILLMAN Phys. Rev. 56, 213 (1939)	Atomstrahl	$Li_2B_4O_7$, $Na_2B_4O_7$, $K_2B_4O_7$, $NaBO_2$, KBO_2	$g = 1,788$ $\pm 0,005$	$+2,682$ $\pm 0,008$	
			S. MILLMAN, P. KUSCH Phys. Rev. 90, 91 (1941)	Berechnung aus früheren Daten				$-2,698$
			J. K. ZIMMERMAN et al. Phys. Rev. 74, 1885 (1948)	Pendelrückk.	$Na_2B_2O_4$, w. L.			2,700 $\pm 0,008$
			N. I. ADAMS et al. Report (October 1949), S. 24					
			D. A. ANDERSON Phys. Rev. 76, 434 (1949)	Brücke	Kaliummeta- borat, ges. w. L.	ν_{B11}/ν_{H1} $= 0,320827$ $\pm 0,000004$		2,68939 $\pm 0,00083$
			F. BITTER Phys. Rev. 75, 1326 (1949)	Brücke		ν_{B11}/ν_{H1} $= 0,32085$ $\pm 0,00006$		
			J. R. ZIMMERMAN et al. Phys. Rev. 76, 163 (1949) Phys. Rev. 76, 350 (1949)	Pendelrückk.	a) $Na_2B_2O_4$, w. L. b) H_2O	ν_{B4}/ν_{H1} $= 0,32076$ $\pm 0,00010$		
			G. LINDSTRÖM Ark. Fysik 4, 1 (1951) Physica, Haag 17, 412 (1951)	Brücke	$Na_2B_2O_4$, $K_2B_2O_4$, ges. w. L.	ν_{B11}/ν_{H1} $= 0,3208381$ $\pm 0,0000020$		2,68853 $\pm 0,00007$
			R. E. SHERIFF, D. WILLIAMS Phys. Rev. 82, 651 (1951)	Pendelrückk.	a) $Na_2B_2O_4$ b) $LiC_2H_3O_2 + Mn$ $(C_2H_3O_2)_2$, w. L.	ν_{B11}/ν_{Li7} $= 0,825615$ $\pm 0,00004$		2,68840 $\pm 0,00018$
C_6^{12}	98,9	0	H. A. BETHE, R. F. BACHER Rev. Mod. Phys. 8, 82 (1936)					
C_6^{13}	1,1	$\frac{1}{2}$	R. H. HAY Phys. Rev. 58, 180 (1940) Phys. Rev. 60, 75 (1941)	Atomstrahl	KCN, NaCN	$g = 1,401$ $\pm 0,004$	$+0,700$ $\pm 0,002$	
			H. L. POSS Phys. Rev. 75, 600 (1949)	Brücke	Methyljodid, 56,7% C^{13}	ν_{C13}/ν_{H1} $= 0,25143$ $\pm 0,00005$		0,7016 $\pm 0,0004$
			V. ROYDEN Phys. Rev. 96, 543 (1954)	Gekreuzte Spulen	CH_3J mit 51% C^{13}	ν_{C13}/ν_{H1} $= 0,2514431$ $\pm 0,0000005$		
C_6^{14}	—	0						
N_7^{14}	99,62	1	R. F. BACHER Phys. Rev. 43, 1001 (1933)	Hfs.			$\leq 0,2$	
			S. MILLMAN et al. Phys. Rev. 54, 968 (1938) P. KUSCH, S. MILLMAN et al. Phys. Rev. 55, 1176 (1939)	Atomstrahl	NaCN, KCN, RbCN		$+0,402$ $\pm 0,002$	$+0,402$ $\pm 0,002$
			W. G. PROCTOR, F. C. YU Phys. Rev. 77, 716 (1950) Phys. Rev. 81, 20 (1951)	Gekreuzte Spulen	a) HNO_3, konz. w. L. b) 0,6 m $MnSO_4$ in 85% D_2O	ν_{N14}/ν_{D2} $= 0,47070$ $\pm 0,00005$	$+0,40369$ $\pm 0,00006$	
			Y. TING, D. WILLIAMS Phys. Rev. 89, 595 (1953)	Pendelrückk.	a) HNO_3, w. L. b) RbCl, w. L.	ν_{N14}/ν_{Rb85} $= 0,74837$ $\pm 0,00004$		0,40369 $\pm 0,00003$

Kern	Vork. %	Spin	Zitat	Methode	Substanz	Ergebnisse	μ_{unkorr}	μ_{korr}
N_7^{15}	0,38	$\frac{1}{2}$	J. R. ZACHARIAS et al. Phys. Rev. **57**, 570 (1940)	Atomstrahl	N_2-Gas mit 50% N^{15}, 50at	$g = 0,560$ $\pm 0,006$	0,280 $\pm 0,003$	
			W. G. PROCTOR, F. G. YU Phys. Rev. **77**, 716 (1950)	Gekreuzte Spulen	a) NH_3, flüssig, 7,5% N^{15} $+ NH_4(NO_3)$ $+ Cr(NO_3)_3$ b) 1,8 m $MnSO_4$ in 25% D_2O	$\nu_{N^{15}}/\nu_{D^2}$ $= 0,66004$ $\pm 0,00006$	$-0,28302$ $\pm 0,00003$	
			W. G. PROCTOR, F. G. YU Phys. Rev. **81**, 20 (1951)	Gekreuzte Spulen	NH_3, flüssig, 7,5% N^{15} $+ NH_4NO_3$ $+ Cr(NO_3)_3$	$\nu_{N^{15}}/\nu_{N^{14}}$ $= 1,4026$ $\pm 0,0001$	$-0,28312$ $\pm 0,00004$	
					6,2 m $NaNO_3$ mit 31,4% N^{15} $+ 1,0m MnSO_4$ w. L.	$\nu_{N^{15}}/\nu_{N^{14}}$ $= 1,4027$ $\pm 0,0001$	$-0,28312$ $\pm 0,00004$	
O_8^{16}	99,7575	0	H.A.BETHE, R.F.BACHER Rev. Mod. Phys. **8**, 82 (1936)					
O_8^{17}	0,0392	$\frac{5}{2}$	F. ALDER, F. G. YU Phys. Rev. **81**, 1067 (1951)	Gekreuzte Spulen	D_2O, H_2O, CH_3OH, C_2H_5OH, CH_3COOH	$\nu_{O^{17}}/\nu_{D^2}$ $= 0,88313$ $\pm 0,00004$, I	$-1,8928$ $\pm 0,00019$	
O_8^{18}	0,2033	0						
F_9^{19}	100	$\frac{1}{2}$	J. S. CAMPBELL Z. Physik **84**, 393 (1953)	Hfs.		$g_{F^{19}}/g_{Br} = 3,2$		
			F. W. BROWN et al. Phys. Rev. **45**, 527 (1934)	Berechnung aus früheren Daten			1,9 bis 3,8 je nach Term	
			S. MILLMAN Phys. Rev. **55**, 628 (1939)	Atomstrahl		$\mu > 0$	> 0	
			I. I. RABI et al. Phys. Rev. **55**, 526 (1939)	Atomstrahl		$g = 5,243$ $\pm 0,0025$	$+2,622$ $\pm 0,001$	
			S. MILLMAN, P. KUSCH Phys. Rev. **60**, 91 (1941)	Atomstrahl	NaF	$g = 5,2496$ $\pm 0,0050$		
			S. MILLMAN, P. KUSCH Phys. Rev. **60**, 91 (1941)	Berechnung aus früheren Daten			$+2,625$	
			J. R. ZIMMERMAN et al. Phys. Rev. **73**, 94 (1948)	Pendelrückk.			2,625 $+0,017$	
			H. L. Poss Phys. Rev. **75**, 600 (1949)	Brücke	$HF + MnCl_2$ u. $FeCl_3$, w.L.	$\nu_{F^{19}}/\nu_{H^1}$ $= 0,94077$ $+0,0001$		2,626 $\pm 0,001$
			K. SIEGBAHN, G. LINDSTRÖM Ark. Fysik **1**, 193 (1949) Nature, Lond. **163**, 211(1949)	Brücke	a) $C_2F_3Cl_3$, flüssig b) H_2O	$\mu_{F^{19}}/\mu_{H^1}$ $= 0,940334$ $\pm 0,000015$		
			J. R. ZIMMERMAN et al. Phys. Rev. **76**, 163 (1949) Phys. Rev. **76**, 350 (1949)	Pendelrückk.	a) SbF_3, w.L. b) H_2O	$\nu_{F^{19}}/\nu_{H^1}$ $= 0,94086$ $\pm 0,00017$		
			E. W. GUPTILL et al. Canad. J. Res. A **28**, 359 (1950)	Empfindlicher Oszillator	HF, w.L.	$\nu_{H^1}/\nu_{F^{19}}$ $= 1,062917$ $\pm 0,00001$		
			G. J. BÉNÉ Helv. phys. Acta **24**, 367 (1951) und G. J. BÉNÉ et al. Helv. phys. Acta **24**, 304 (1951)	Gekreuzte Spulen	BeF_2, ges. w.L.	$\gamma_{F^{19}}/\gamma_{H^1}$ $= 0,940760$ $\pm 0,000050$		
			G. LINDSTRÖM Ark. Fysik **4**, 1 (1951)	Brücke	$CFCl_3$, flüssig	$\nu_{F^{19}}/\nu_{H^1}$ $= 0,9409330$ $\pm 0,0000030$		2,62895 $\pm 0,00010$
			T. KANDA et al. Phys. Rev. **83**, 1066 (1951) T. KANDA J. Phys. Soc. Japan **7**, 296 (1952)	Empfindlicher Verstärker	HF, 22% w.L.	$\mu_{F^{19}}/\mu_{H^1}$ $= 0,940814$ $\pm 0,000003$		
			R.E. SHERIFF, D.WILLIAMS Phys. Rev. **82**, 651 (1951)	Pendelrückk.	HF, w.L.	$\nu_{F^{19}}/\nu_{H^1}$ $= 0,94086$ $\pm 0,00005$		2,62842 $\pm 0,00014$
			P. S. FARAGÓ et al. Nuovo Cim. **2**, 10, 1110(1955) P. S. FARAGÓ Suppl. Nuovo Cim. **1**, 249 (1955)	Brücke	HF, konz.w.L.	$\nu_{F^{19}}/\nu_{H^1}$ $= 0,940875$ $\pm 0,000070$	2,62756 $\pm 0,0003$	2,6287 $\pm 0,0007$

Kern	Vork. %	Spin	Zitat	Methode	Substanz	Ergebnisse	μ_{unkorr}	μ_{korr}
* F_9^{19}	—	$\frac{5}{2}$	P. Lehmann et al. C. R. Acad. Sci., Paris 241, 700 (1955)	Winkelvert. der γ-Strahlung			$+4,0$ $\pm 0,9$	
			P. B. Treacy Nature, Lond. 176, 923 (1955)	Winkelvert. der γ-Strahlung		$g = +1,35$ $\pm 0,35$		
			P. Lehmann et al. J. Phys. Radium 17, 560 (1956) Phys. Rev. 104, 411 (1956)	Winkelvert. der γ-Strahlung			$+3,70$ $\pm 0,45$	
			W. R. Phillips et al. Phil. Mag. (8) 1, 576 (1956)	Winkelvert. der γ-Strahlung			$+3,50$ $\pm 0,24$	
			K. Sugimoto et al. Phys. Rev. 103, 739 (1956)	Winkelvert. der γ-Strahlung			$+3,0$ $\pm 0,7$	
Ne_{10}^{20}	90,51	0	G. Weinreich et al. Phys. Rev. 87, 229 (1952)	Atomstrahl	Ne		$< 0,0002$	
Ne_{10}^{21}	0,28	$\frac{3}{2}$	J. Koch et al. Phys. Rev. 76, 1417 (1949)	Hfs.			< 0	
Ne_{10}^{22}	9,21	0	H. Hausen Naturwiss. 15, 163 (1927)	Hfs.			≈ 0	
Na_{11}^{22}	$<3\cdot10^{-6}$	3	L. Davis jr. Phys. Rev. 74, 1193 (1948) Electronics Tech. Rep. No. 88 (1948). NP-631 L. Davis jr. et al. Phys. Rev. 76, 1068 (1949)	Atomstrahl	$300\mu C$ Na^{22} in NaN_3 $(1:10^4)$		$+1,746$ $\pm 0,003$	
Na_{11}^{23}	100	$\frac{3}{2}$	E. Fermi, E. Segrè Z. Physik, 82, 729 (1933)	Berechnung aus früheren Daten			μ_B/μ 880	
			L. P. Granath et al. Phys. Rev. 44, 935 (1933)	Hfs.			2,7	
			A. Ellet et al. Phys. Rev. 46, 583 (1934)	Hfs.			$1,9-2,6$ je nach Term	
			H. Schüler Z. Physik 88, 323 (1934)	Berechnung aus früheren Daten			$+2,14$ $\pm 0,2$	
			M. Fox, I. I. Rabi Phys. Rev. 48, 746 (1935)	Atomstrahl			2,08	
			K. W. Meissner, K. F. Luft Ann. Phys. 28, 667 (1937)	Hfs.			$1,96-2,08$ je nach Term	
			P. Kusch et al. Phys. Rev. 55, 1176 (1939)	Atomstrahl	Na_2, NaF, NaCN		2,214 $\pm 0,010$	2,216 $\pm 0,010$
			S. Millman, P. Kusch Phys. Rev. 58, 438 (1940)	Atomstrahl		$g = -1,477$		
			S. Millman, P. Kusch Phys. Rev. 60, 91 (1941)	Atomstrahl	Na_2	$g = 1,4765$ $\pm 0,0015$		
			A. Bolle, G. Zanotelli Ric. Sci. 18, 847 (1948)	Gekreuzte Spulen	NaOH, konz. w. L.	$\gamma H^1/\gamma Na^{23}$ $= 3,82$ $\pm 0,02$		2,19 $\pm 0,01$
			F. Bitter Phys. Rev. 75, 1326 (1949)	Brücke		$\nu Na^{23}/\nu H^1$ $= 0,264\,50$ $\pm 0,00003$		
			J. R. Zimmerman et al. Phys. Rev. 76, 163 (1949) Phys. Rev. 76, 350 (1949)	Pendelrückk.	a) NaJ, NaBr, $NaAsO_2$, $Na_2B_2O_4$, w. L. b) H_2O	$\nu Na^{23}/\nu H^1$ $= 0,264\,54$ $\pm 0,00007$		
			E. W. Guptill et al. Canad. J. Res. A 28, 359 (1950)	Empfindlicher Oszillator	Natrium- aluminat	$\nu Na/\nu Al$ $= 1,015\,081$ $\pm 0,00001$		
			G. Lindström Ark. Fysik 4, 1 (1951) Physica, Haag 17, 412 (1951)	Brücke	NaBr, $Na_2B_2O_4$, ges. w. L.	$\nu Na^{23}/\nu H^1$ $= 0,264\,518\,2$ $\pm 0,000001\,5$		2,217 54 $\pm 0,00010$
			R. A. Logan, P. Kusch Phys. Rev. 81, 280 (1951)	Hfs.		$g Na^{23}/g H^1$ $= 0,264\,51$ $\pm 0,00002$		
			R. E. Sheriff, D. Williams Phys. Rev. 82, 651 (1951)	Pendelrückk.	$Na_2B_2O_4 +$ $ScCl_3$, w. L.	$\nu Na^{23}/\nu Sc^{45}$ $= 1,088\,83$ $\pm 0,00005$		2,217 14 $\pm 0,00012$
					$NaBr + ScCl_3$, w. L.	$\nu Na^{23}/\nu Sc^{45}$ $= 1,088\,72$ $\pm 0,00006$		

Kern	Vork. %	Spin	Zitat	Methode	Substanz	Ergebnisse	μunkorr	μkorr
			T. Kanda et al. Phys. Rev. 85, 938 (1952)	Empfindlicher Oszillator	NaJ + 0,1 n MnSO$_4$, w. L.	νNa^{23}/νH^1 = 0,264 514 ± 0,000 009		
			H. E. Walchli ORNL-1775 (1954)			νNa^{23}/νD^2 = 1,723 167 ± 0,000 034	+ 2,216 124	
			P. S. Faragó et al. Nuovo Cim. 2, 10, 1110 (1955) P. S. Faragó Suppl. Nuovo Cim. 1, 249 (1955)	Brücke	Na$_2$S$_2$O$_3$ Konz. w. L., NaCl konz. w. L.	νNa^{23}/νH^1 = 0,264 553 ± 0,000 050	2,216 49 ± 0,000 25	2,217 70 ± 0,000 25
Na$^{24}_{11}$	<0,002	4	E. H. Bellamy et al. Phil. Mag. (7) 44, 33 (1953)	Atomstrahl	Na		+ 1,688 ± 0,005	
Mg$^{24}_{12}$	78,60	0						
Mg$^{25}_{12}$	10,11	$\frac{5}{2}$	M. F. Crawford et al. Phys. Rev. 76, 1527 (1949) F. M. Kelly et al. Phys. Rev. 77, 745 (1950)	Hfs.				− 0,96 ± 0,07
			F. Alder, F. C. Yu Phys. Rev. 82, 105 (1951)	Gekreuzte Spulen	4,6 m MgCl$_2$, w. L.	νMg^{25}/νN^{14} = 0,847 14 ± 0,000 08	− 0,854 96 ± 0,000 15	
Mg$^{26}_{12}$	11,29	0						
Al$^{27}_{13}$	100	$\frac{5}{2}$	S. Goudsmit Phys. Rev. 43, 636 (1933)	Berechnung aus früheren Daten		$g = 4,2$		
			H. Schüler Z. Physik 88, 323 (1934)	Berechnung aus früheren Daten			+ 1,93 ± 0,2 ($I = \frac{1}{2}$)	
			H. A. Bethe, R. F. Bacher Rev. Mod. Phys. 8, 82 (1936)	Berechnung aus früheren Daten			+ 2,2	
			D. A. Jackson, H. Kuhn Nature, Lond. 140, 110 (1937)	Hfs.			3,6 bis 4,1 je nach Term. ($I = \frac{9}{2}$)	
			M. Heyden, R. Ritschl Z. Physik 108, 739 (1938)	Hfs.			3,7 ± 0,3	
			S. Millman, P. Kusch Phys. Rev. 56, 214 (1939) Phys. Rev. 56, 303 (1939)	Atomstrahl	NaCl · AlCl$_3$, KCl · AlCl$_3$	$g = 1,451$ ± 0,004	+ 3,628 ± 0,010	
			F. Bitter Phys. Rev. 75, 1326 (1949)	Brücke		νAl^{27}/νH^1 = 0,260 56 ± 0,000 03		
			J. R. Zimmerman et al. Phys. Rev. 76, 163 (1949) Phys. Rev. 76, 350 (1949)	Pendelrückk.	a) AlCl$_3$ w. L. b) H$_2$O	νAl^{27}/νH^1 = 0,260 62 ± 0,000 10		
			G. Lindström Ark. Fysik 4, 1 (1951) Physica, Haag 17, 412 (1951)	Brücke	a) AlCl$_3$, ges.w.L. b) AlCl$_3$, ges.w.L. und H$_2$O dest.	νAl^{27}/νH^1 = 0,260 5694 ± 0,000 001 0		3,641 33 ± 0,000 15
			R. E. Sheriff, D. Williams Phys. Rev. 82, 651 (1951)	Pendelrückk.	a) AlCl$_3$, w. L. b) ScCl$_3$, w. L.	νAl^{27}/νSc^{45} = 1,072 61 ± 0,000 05		3,641 01 ± 0,000 17
			T. Kanda et al. Phys. Rev. 85, 938 (1952)	Empfindlicher Oszillator	AlCl$_3$, w. L.	νAl^{27}/νH^1 = 0,260 579 ± 0,000 008		
			H. E. Retch et al. Canad. J. Phys. 31, 837 (1953)					
			L. C. Brown, D. Williams Phys. Rev. 95, 1110 (1954)	Pendelrückk.	Be$_3$Al$_2$Si$_6$O$_{18}$- Einkristall C-Achse in Feldrichtung		3,6401 ± 0,0003	
					AlCl$_3$, w. L.		3,6408	
			H. E. Walchli ORNL-1775 (1954)			νAl^{27}/νNa^{23} = 0,985 055 ± 0,000 012	+ 3,638 360	
			L. C. Brown, D. Williams J. Chem. Phys. 24, 757 (1956)	Pendelrückk.	Be$_3$Al$_2$Si$_6$O$_{18}$- Einkristall		3,6385 ± 0,0003	
Si$^{28}_{14}$	92,28	0	R. A. Ogg jr., J. D. Ray J. Chem. Phys. 22, 147 (1954)	Gekreuzte Spulen	SiH$_4$, flüssig	$I = 0$		
			G. A. Williams et al. Phys. Rev. 93, 1428 (1954)	Brücke	SiF$_4$, angereichert	$g < 0,05$		

Kern	Vork. %	Spin	Zitat	Methode	Substanz	Ergebnisse	μ_{unkorr}	μ_{korr}
Si_{14}^{29}	4,67	$\frac{1}{2}$	S. S. DHARMATTI et al. Phys. Rev. **84**, 843 (1951)	Gekreuzte Spulen	a) Si-Pulver, SiO_2 in NaOH, Glas u.a. b) D_2O	$\nu_{Si^{29}}/\nu_{D^2}$ = 1,294 10 ± 0,000 07	− 0,554 92 ± 0,000 04	
			J. HATTON et al. Phys. Rev. **83**, 672 (1951)		$Be_3Al_2Si_6O_{18}$- Einkristall und Glas		0,55	
			R. H. SANDS, G. E. PAKE Bull. Amer. Phys. Soc. **27**, 11 (1952)	Empfindlicher Oszillator	SiO_2	$I = 4,3$ ± 0,5		
			R. A. OGG, jr., J. D. RAY J. Chem. Phys. **22**, 147 (1954)	Gekreuzte Spulen	SiH_4, flüssig	$I = \frac{1}{2}$		
Si_{14}^{30}	3,05	0	R. A. OGG, jr., J. D. RAY J. Chem. Phys. **22**, 147 (1954)	Gekreuzte Spulen	SiH_4, flüssig	$I = 0$		
P_{15}^{31}	100	$\frac{1}{2}$	R. V. POUND Phys. Rev. **73**, 1112 (1948)	Empfindlicher Oszillator	fester roter Phosphor und P_2O_5, konz. w.L.	$\nu_{P^{31}}/\nu_{Na^{23}}$ = 1,5310 ± 0,0003	1,1314 ± 0,0013	
			F. BITTER Phys. Rev. **75**, 1326 (1949)	Brücke		$\nu_{P^{31}}/\nu_{H^1}$ = 0,404 81 ± 0,000 04		
			W. H. CHAMBERS et al. Phys. Rev. **76**, 461 (1949) Phys. Rev. **76**, 638 (1949)	Pendelrückk.	H_3PO_4, w.L.	$\nu_{P^{31}}/\nu_{H^1}$ = 0,404 98 ± 0,000 11 Korr.		
			M. F. CRAWFORD et al. Canad. J. Res. A **27**, 156 (1949)	Hfs.			+ 1,15 ± 0,05 + 1,11 ± 0,01 je nach Term	
			R. E. SHERIFF, D.WILLIAMS Phys. Rev. **82**, 651 (1951)	Pendelrückk.	a) $H_3PO_4 + CuCl_2$, w.L. b) $LiC_2H_3O_2 + Mn(C_2H_3O_2)_2$, w.L.	$\nu_{P^{31}}/\nu_{Li^7}$ = 1,041 82 ± 0,000 05		1,131 65 ± 0,000 07
			T. KANDA et al. Phys. Rev. **85**, 938 (1952)	Empfindlicher Oszillator	$P_2O_5 + 0,1$ n $Fe(NO_3)_2$, w.L.	$\nu_{P^{31}}/\nu_{H^1}$ = 0,404 804 ± 0,000 010		
			H. E. WALCHLI ORNL-1775 (1954)			$\nu_{P^{31}}/\nu_{Na^{23}}$ = 1,530 366 ± 0,000 040	+ 1,130 500	
			P. S. FARAGO et al. Nuovo Cim. **2**, 10, 1110 (1955) P. S. FARAGO Suppl. Nuovo Cim. **1**, 249 (1955)	Brücke	H_3PO_4, konz. w.L.	$\nu_{P^{31}}/\nu_{H^1}$ = 0,404 868 ± 0,000 040	1,130 665 ± 0,000 12	1,131 695 ± 0,000 12
S_{16}^{32}	95,060	0	H. A. BETHE, R. F. BACHER Rev. Mod. Phys. **8**, 82 (1936)					
S_{16}^{33}	0,742	$\frac{3}{2}$	G. K. JEN Phys. Rev. **78**, 339 (1950)	Mikrowellen	OCS		$|\mu| = 0,9$	
			S. S. DHARMATTI et al. Phys. Rev. **83**, 845 (1951) H. E. WEAVER jr. Phys. Rev. **89**, 923 (1953)	Gekreuzte Spulen	a) CS_2 b) 3,2 m · HNO_3, w.L.	$\nu_{S^{33}}/\nu_{N^{14}}$ = 1,061 74 ± 0,000 13	+ 0,642 92 ± 0,000 14	
			G. K. JEN Physica, Haag **17**, 378 (1951)	Mikrowellen	OCS		$|\mu| = 0,63$ ± 0,02	
			J. R. ESHBACH et al. Phys. Rev. **85**, 532 (1952) NP-4368,Tech.Rep. No.224 U-24932 (1951)	Mikrowellen	OCS, angereichert		+ 0,633 ± 0,010	
S_{16}^{34}	4,182	0	G. H. TOWNES Phys. Rev. **71**, 909 (1947)				≈ 0	
S_{16}^{35}	< 0,002	$\frac{3}{2}$	B. F. BURKE et al. Phys. Rev. **93**, 193 (1954) BNL-1561 (1952)	Mikrowellen	OCS		1,00 ± 0,04 − 1,07 ± 0,04	
S_{16}^{36}	0,016	0						

Kern	Vork. %	Spin	Zitat	Methode	Substanz	Ergebnisse	μ_{unkorr}	μ_{korr}
Cl^{35}_{17}	75,4	$\frac{3}{2}$	T. Schmidt Z. Physik 108, 408 (1938)	Berechnung aus früheren Daten			< 0,3	
			P. Kusch, S. Millman Phys. Rev. 56, 527 (1939) Phys. Rev. 58, 925 (1940)	Atomstrahl	LiCl, RbCl	$g = +0,546$ $\pm 0,002$		
			F. Bitter Phys. Rev. 75, 1326 (1949)	Brücke		$\nu Cl^{35}/\nu Na^{23}$ $= 0,370\,51$ $\pm 0,000\,11$		
			W. H. Chambers et al. Phys. Rev. 76, 638 (1949) Phys. Rev. 76, 461 (1949)	Pendelrückk.	LiCl, w. L.	$\nu Cl^{35}/\nu H^1$, korr. $= 0,097\,99$ $\pm 0,000\,07$		
			L. Davis, jr. et al. Phys. Rev. 76, 1076 (1949)	Atomstrahl	Cl	$\mu > 0$	> 0	
			J. G. King, V. Jaccarino Phys. Rev. 84, 852 (1951)	Atomstrahl		gCl^{35}/gCl^{37} $= 1,201\,357$ $\pm 0,000\,013$		
			V. Jaccarino, J. G. King Phys. Rev. 83, 471 (1951)	Atomstrahl	Cl	gCl^{35}/gCl^{37} $= 1,201\,36$ $\pm 0,000\,05$		
			W. G. Proctor, F. C. Yu Phys. Rev. 81, 20 (1951)	Gekreuzte Spulen	a) HCl konz.w.L. b) 0,6 m MnSO₄ in 85% D_2O	$\nu Cl^{35}/\nu D^2$ $= 0,638\,27$ $\pm 0,0006$ $\nu Cl^{35}/\nu Cl^{37}$ $= 1,2014$ $\pm 0,0001$	0,8211 $\pm 0,0001$	
			G. D. Watkins et al. Phys. Rev. 82, 343 (1951)	Empfindlicher Oszillator	LiCl, w. L.	gCl^{35}/gCl^{37} $= 1,2013$ $\pm 0,0001$		
			H. E. Walchli et al. Phys. Rev. 85, 922 (1952)	Gekreuzte Spulen	a) VOCl₃ b) RbCl, ges. L. in 15% D_2O	$\nu Cl^{35}/\nu H^1$ $= 0,097\,985$ $\pm 0,000\,01$		
			Y. Ting et al. Phys. Rev. 92, 1581 (1953) Phys. Rev. 96, 408 (1954)	Quadrupol- Resonanz	NaClO₃- Einkristall		0,8215 $\pm 0,0001$	
			Y. Ting, D. Williams Phys. Rev. 89, 595 (1953)	Pendelrückk.	a) LiCl, w.L. b) RbCl, w.L. LiCl, w.L.	$\nu Cl^{35}/\nu Rb^{85}$ $= 1,014\,81$ $\pm 0,000\,05$ $\nu Cl^{35}/\nu Cl^{37}$ $= 1,201\,28$ $\pm 0,000\,06$	0,821\,80 $\pm 0,000\,05$	0,821\,80 $\pm 0,000\,05$
			H. E. Walchli ORNL-1775 (1954)			$\nu Cl^{35}/\nu D^2$ $= 0,638\,302$ $\pm 0,000\,008$		$+0,820\,905$
Cl^{36}_{17}	—	2	L. C. Aamodt et al. Phys. Rev. 98, 1317 (1955) Phys. Rev. 99, 613 (1955)	Quadrupol- Resonanz	CH_3Cl^{36}		$+1,32$ $\pm 0,08$	
			P. B. Sogo, C. D. Jeffries Phys. Rev. 98, 1316 (1955) Phys. Rev. 99, 613 (1955)	Gekreuzte Spulen	a) 4n HCl, w. L. b) 1 m Mn⁺⁺ in D_2O	$\nu Cl^{36}/\nu D^2$ $= 0,748\,73$ $\pm 0,0002$	$+1,2838$ $\pm 0,0002$	
Cl^{37}_{17}	24,6	$\frac{3}{2}$	T. Schmidt Z. Physik 108, 408 (1938)	Berechnung aus früheren Daten			< 0,3	
			P. Kusch, S. Millman Phys. Rev. 56, 527 (1939) E. F. Shrader et al. Phys. Rev. 58, 925 (1940)	Atomstrahl	LiCl, RbCl, HCl	$g = 0,454$ $\pm 0,002$		
			W. G. Proctor, F. C. Yu Phys. Rev. 77, 716 (1950) Phys. Rev. 81, 20 (1951)	Gekreuzte Spulen	HCl, konz. w. L.	$\nu Cl^{35}/\nu Cl^{37}$ $= 1,2014$ $\pm 0,0001$	$+0,6835$ $\pm 0,0001$	
			Y. Ting, D. Williams Phys. Rev. 89, 595 (1953)	Pendelrückk.	a) LiCl, w.L. b) RbCl, w.L.	$\nu Cl^{37}/\nu Rb^{85}$ $= 0,844\,77$ $\pm 0,000\,05$		0,684\,10 $\pm 0,000\,05$
Ar^{36}_{18}	0,35	0	H. Kopfermann et al. Z. Physik 105, 389 (1937)	Hfs.			≈ 0	
Ar^{38}_{18}	0,08	0						
Ar^{40}_{18}	99,57	0	H. Kopfermann et al. Z. Physik 105, 389 (1937)	Hfs.			≈ 0	

Kern	Vork. %	Spin	Zitat	Methode	Substanz	Ergebnisse	μ_{unkorr}	μ_{korr}
K_{19}^{39}	93,260	$\frac{3}{2}$	D. A. Jackson, H. Kuhn Nature, Lond. 134, 25 (1934) Nature, Lond. 137, 108 (1936)	Hfs.	Atomstrahl		$-0,39$	
			S. Millman, M. Fox et al. Phys. Rev. 46, 320 (1934)	Atomstrahl	K		0,38	
			H. Schüler Z. Physik 88, 323 (1934)	Berechnung aus früheren Daten			$<0,613$	
			M. Fox, I. I. Rabi Phys. Rev. 48, 746 (1935)	Atomstrahl	K		0,397	
			J. J. Gibbons jr. et al. Phys. Rev. 47, 692 (1935)	Berechnung aus früheren Daten			1,2	
			S. Millman Phys. Rev. 47, 739 (1935)	Atomstrahl	K		0,39	
			P. Kusch et al. Phys. Rev. 55, 1176 (1936) Phys. Rev. 57, 765 (1940)	Atomstrahl	KCN, K	$g = 0,261$ $\pm 0,001$	$+0,390$ $\pm 0,002$	$+0,391$ $\pm 0,002$
			K. Meissner, K. Luft Z. Physik 106, 362 (1937)	Hfs.			$+0,30$ bis $+0,40$ je nach Term	
			H. C. Torrey Phys. Rev. 51, 501 (1937)	Atomstrahl	K		$\mu > 0$	
			T. L. Collins Phys. Rev. 80, 103 (1950)	Empfindlicher Oszillator	a) KNO_2, ges. w.L. b) HNO_3, konz. w.L.	$\nu_{K^{39}}/\nu_{N^{14}}$ $= 0,645\,80$ $\pm 0,000\,06$		
			S. A. Ochs et al. Phys. Rev. 78, 184 (1950)	Atomstrahl		$g_{K^{39}}/g_{K^{41}}$ $= 1,8218$ $\pm 0,0002$		
			J. T. Eisinger et al. Phys. Rev. 86, 73 (1952)	Atomstrahl	K	$g_{K^{40}}/g_{K^{39}}$ $= -1,24346$ $\pm 0,00024$	0,39097 $\pm 0,00015$	
			E. Brun, J. Oeser et al. Phys. Rev. 93, 172 (1954)	Gekreuzte Spulen	a) 15 m KCO_2H, w.L. b) 0,1 m $MnSO_4$, w.L.	$\nu_{K^{39}}/\nu_{H^1}$	$+0,390873$ $\pm 0,000013$	
			L. C. Brown, D. Williams Phys. Rev. 98, 1537 (1955)		a) KF, ges.w.L. b) HNO_3	$\nu_{K^{39}}/\nu_{N^{14}}$ $= 0,645\,88$		
K_{19}^{40}	0,011	4	J. R. Zacharias Phys. Rev. 60, 168 (1941) Phys. Rev. 61, 270 (1942)	Atomstrahl	K		$-1,290$	
			L. Davis jr. et al. Phys. Rev. 76, 1068 (1949)	Atomstrahl	K		$-1,290$ $\pm 0,005$	
			J. T. Eisinger et al. Phys. Rev. 85, 716 (1952)	Atomstrahl	K	$g_{K^{40}}/g_{K^{39}}$ $= -1,2434$ $\pm 0,0003$	$-1,2965$ $\pm 0,0004$	
			J. T. Eisinger et al. Phys. Rev. 86, 73 (1952) Np-4123 Tech. Rep. No.212 U 24409 (1952)	Atomstrahl	K	$g_{K^{40}}/g_{K^{39}}$ $= -1,24346$ $\pm 0,00024$	$-1,2964$ $\pm 0,0004$	$-1,2982$
K_{19}^{41}	6,729	$\frac{3}{2}$	S. Millman Phys. Rev. 47, 739 (1935)	Atomstrahl	K	$\mu_{K^{41}}/\mu_{K^{39}}$ $= 0,42$ bis 0,88 je nach Spin von K^{41}		
			J. H. Manley Phys. Rev. 49, 921 (1936)	Atomstrahl	K		0,22	
			P. Kusch, S. Millman et al. Phys. Rev. 57, 765 (1940)	Atomstrahl	K	$\mu_{K^{41}}/\mu_{K^{39}}$ $= 0,55012$ $\pm 0,00006$		
			S. Millman, P. Kusch Phys. Rev. 60, 91 (1941)	Berechnung aus früheren Daten			$+0,215$	
			S. A. Ochs et al. Phys. Rev. 78, 184 (1950)	Atomstrahl		$\nu_{K^{41}}/\nu_{K^{39}}$ $= 0,54891$ $\pm 0,00005$		
			E. Brun, J. Oeser et al. Phys. Rev. 93, 172 (1954)	Gekreuzte Spulen	a) 15 m $KHCO_2$ b) 0,1 m $MnSO_4$, w.L.	$\nu_{K^{41}}/\nu_{K^{39}}$ $= 0,54886$ $\pm 0,00008$	0,214 53 $\pm 0,00003$	
K_{19}^{42}	$<4\cdot10^{-6}$	2	E. H. Bellamy et al. Phil. Mag. (7) 44, 33 (1953)	Atomstrahl	K		$-1,137$ $\pm 0,005$	
Ca_{20}^{40}	96,92	0	S. Frisch Z. Physik 68, 758 (1931)	Hfs.			≈ 0	
Ca_{20}^{42}	0,64	0						

Kern	Vork. %	Spin	Zitat	Methode	Substanz	Ergebnisse	μ_{unkorr}	μ_{korr}
Ca_{20}^{43}	0,129	$\frac{7}{2}$	C. D. JEFFRIES Phys. Rev. 90, 1130 (1953)	Gekreuzte Spulen	0,7 m $CaBr_2$ mit 68% Ca^{43} w.L. + 4,5 m $MnCl_2$ w.L. mit etwas D_2O, gemischt	$\nu Ca^{43}/\nu D^2$ = 0,438 32 ± 0,00004, I	− 1,3152 ± 0,0002	
Ca_{20}^{44}	2,13	0						
Ca_{20}^{46}	0,003	0						
Ca_{20}^{48}	0,178	0						
Sc_{21}^{45}	100	$\frac{7}{2}$	H. KOPFERMANN et al. Z. Physik 92, 82 (1934)	Hfs.			≈ 3,6	
			H. KOPFERMANN et al. Z. Physik 105, 16 (1937)	Hfs.			4,8	
			D. M. HUNTEN Phys. Rev. 78, 806 (1950)	Brücke	a) 0,1 m $ScCl_3$, ges. saure w.L. b) H_2O	$\nu Sc^{45}/\nu H^1$ = 0,242939 ± 0,000003		+ 4,7560 ± 0,0002
			W. G. PROCTOR, F. C. YU Phys. Rev. 78, 471 (1950) Phys. Rev. 81, 20 (1951)	Gekreuzte Spulen	a) 0,65 m $Sc(NO_3)_3$, w.L. b) NaCl, ges.w.L. + 0,5 m $MnSO_4$	$\nu Sc^{45}/\nu Na^{23}$ = 0,9183 ± 0,0001	+ 4,7497 ± 0,0008	
			R. E. SHERIFF, D. WILLIAMS Phys. Rev. 79, 175 (1950) N. F. RAMSEY Phys. Rev. 79, 1010 (1950)	Pendelrückk.	a) $ScCl_3$ ges.w.L. b) NaBr ges.w.L.	$\nu Sc^{45}/\nu Br^{79}$ = 0,969 54 ± 0,00006		4,7564 ± 0,0010
			D. M. HUNTEN Canad. J. Phys. 29, 463 (1951)	Brücke	$Sc(NO_3)_3$ + HNO_3, w.L.	$\nu Sc^{45}/\nu H^1$ = 0,242939 ± 0,000003	4,749 16 ± 0,00012	4,756 ± 0,007
Ti_{22}^{46}	7,95	0						
Ti_{22}^{47}	7,75	$\frac{5}{2}$	C. D. JEFFRIES Phys. Rev. 92, 1096 (1953)	Gekreuzte Spulen	$TiCl_4$, flüssig	$\nu Ti^{47}/\nu Cl^{35}$ = 0,574 93 ± 0,00006, I	− 0,7866	
			C. D. JEFFRIES Phys. Rev. 92, 1262 (1953)	Gekreuzte Spulen	a) $TiCl_4$, flüssig b) 1 m $MnSO_4$ in D_2O	$\nu Ti^{47}/\nu D^2$ = 0,367 21, ± 0,00006, I	− 0,787 06 ± 0,0001	
Ti_{22}^{48}	73,45	0	C. F. COLEMAN Phys. Rev. 103, 1647 (1956)	Winkelverteilung der γ-Strahlung		$I = 0$	≈ 0	
Ti_{22}^{49}	5,51	$\frac{7}{2}$	C. D. JEFFRIES et al. Helv. phys. Acta 24, 643 (1951)	Gekreuzte Spulen	$TiCl_4$, $TiBr_4$, H_2TiF_6		− 1,101	
			C. D. JEFFRIES et al. Phys. Rev. 85, 478 (1952)	Gekreuzte Spulen	$TiCl_4$, H_2TiF_6 w.L.	$\nu Ti^{49}/\nu H^1$ = 0,056 38 ± 0,00001, I	− 1,1022 ± 0,0003	
			C. D. JEFFRIES Phys. Rev. 12, 1096 (1953)	Gekreuzte Spulen	$TiCl_4$	$\nu Ti^{49}/\nu Cl^{35}$ = 0,575 08 ± 0,00006	− 1,102	
			C. D. JEFFRIES Phys. Rev. 92, 1262 (1953)	Gekreuzte Spulen	a) $TiCl_4$ b) 1 m Mn^{++} in D_2O TiCl$_4$	$\nu Ti^{49}/\nu D^2$ = 0,367 31 ± 0,00006, I $\nu Ti^{49}/\nu Ti^{47}$ = 1,000 26 ± 0,00002	− 1,1022 ± 0,0002	
Ti_{22}^{50}	5,34	0						
V_{23}^{49}	<5·10⁻⁵	$\frac{7}{2}$	M. M. WEISS et al. Bull. Amer. Phys. Soc. II 2, 31 (1957)	Paramagnetische Resonanz	Vanadium-(IV)-Kupferion-Chelat in Xylol gelöst	$\mu V^{49}/\mu V^{51}$ = 0,867 ± 0,01		
V_{23}^{50}	0,25	6	H. E. WALCHLI et al. Phys. Rev. 85, 922 (1952) Phys. Rev. 86, 618 (1952)	Gekreuzte Spulen	$VOCl_3$ + RbCl, ges. w.L. mit 15% D_2O	$\nu V^{50}/\nu Cl^{35}$ = 1,017 58 ± 0,0001 $\nu V^{50}/\nu Rb^{85}$ = 1,03262 ± 0,0001 $\nu V^{50}/\nu D^2$ = 0,649 530 ± 0,00007		

Kern	Vork. %	Spin	Zitat	Methode	Substanz	Ergebnisse	μ_{unkorr}	μ_{korr}
			H. E. Walchli, et al. Phys. Rev. 87, 541 (1952)	Gekreuzte Spulen	$NaVO_3$, ges. w. L.	$\nu V^{50}/\nu D^2$ = 0,649203 ± 0,000012	+ 3,898 mit $I = \frac{7}{2}$	+ 3,905
			G. Kikuchi et al. Phys. Rev. 92, 109 (1953)	Paramagnetische Resonanz	Vanadium-Ammonium-Doppelsalz (23% V^{50})	gV^{50}/gV^{51} = 0,3792 ± 0,0008		
			H. E. Walchli ORNL-1775 (1954)			$\nu V^{50}/\nu D^2$ = 0,649518 ± 0,000008	+ 3,34128	
V^{51}_{23}	100	$\frac{7}{2}$	W. D. Knight, V. W. Cohen Phys. Rev. 76, 1421 (1949)	Empfindlicher Oszillator	$Pb(VO_3)_2$-Pulver, V_2O_5-Pulver $Pb(VO_3)_2$ und V_2O_5 saure w. L.	gV^{51}/gNa^{23} = 0,99394 ± 0,00003	5,150	
			W. G. Proctor, F. C. Yu Phys. Rev. 81, 20 (1951)	Gekreuzte Spulen	$NaVO_3$, w. L.	$\mu > 0$	> 0	
			R. E. Sheriff, D. Williams Phys. Rev. 82, 651 (1951)	Pendelrückk.	a) V_2O_5-Pulver b) $ScCl_3$, w.L.	$\nu V^{51}/\nu Sc^{45}$ = 1,08156 ± 0,00005		5,14503 ± 0,00023
					$NaVO_3$, ges. w. L.	$\nu V^{51}/\nu Na^{23}$ = 0,993855 ± 0,000025		
			H. E. Walchli et al. Phys. Rev. 87, 541 (1952)		Na_3VO_4, NH_4VO_3, $VO(NO_3)_2$, $VOSO_4$, $K_2V_4O_9$,$VOCl_2$, $VOCl_2$; ges. w. L. V_2O_5-Pulver	$\nu V^{51}/\nu Na^{23}$ = 0,993855 ± 0,000035	$\mu > 0$	
					$VOCl_3$, flüssig	$\nu V^{51}/\nu Na^{23}$ = 0,994358 ± 0,000026		
					V-Metall-Pulver	$\nu V^{51}/\nu Na^{23}$ = 0,999960 ± 0,00001		
Cr^{50}_{24}	4,31	0						
Cr^{52}_{24}	83,76	0						
Cr^{53}_{24}	9,55	$\frac{3}{2}$	B. Bleaney et al. Proc. Phys. Soc. Lond. A 64 1135 (1951) Proc. Phys. Soc. Lond. A 65, 860 (1952)	Paramagnetische Resonanz	KCr-Selenat-Alaun in KAl-Alaun-Kristall KCr^{III}-Cyanid-KCo^{III}-Cyanid-Kristall		$|\mu| = 0,45$ ± 0,1	
			F. Alder, K. Halbach Helv. phys. Acta 26, 426 426 (1953)	Gekreuzte Spulen	a) Na_2CrO_4 b) HNO_3	$\nu Cr^{53}/\nu N^{14}$ = 0,78226 ± 0,00005	− 0,47351 ± 0,00006	
			C. D. Jeffries, P. B. Sogo Phys. Rev. 91, 1286 (1953)	Gekreuzte Spulen	a) 1,1 m Na_2CrO_4 w. L. und Na_2CrO_4, ges. w.L. b) 1 m $MnCl_2$ in D_2O	$\nu Cr^{53}/\nu D^2$ = 0,36820 ± 0,00003	− 0,47351 ± 0,00007	
			K. Halbach Helv. phys. Acta 27, 259 (1954)	Gekreuzte Spulen	a) 5,18 m Na_2CrO_4 w. L. b) 1,08 m $MnSO_4$ in 24,6% D_2O	$I = \frac{3}{2}$		
Cr^{54}_{24}	2,38	0						
Mn^{53}_{25}	<4·10⁻⁴	$\frac{7}{2}$	W. Dobrowolski et al. Phys. Rev. 104, 1378 (1956)	Paramagnetische Resonanz	$SrCl_2$-Pulver mit Mn^{++}	$|\mu Mn^{53}|/|\mu Mn^{55}|$ = 1,455 ± 0,002		$|\mu| = 5,050$ ± 0,007
Mn^{55}_{25}	100	$\frac{5}{2}$	R. A. Fisher, E. R. Peck Phys. Rev. 55, 270 (1939)	Hfs.			3,0	
			W. H. Chambers et al. Phys. Rev. 78, 640 (1950)	Pendelrückk.	$Ba(MnO_4)_2$, $Ca(MnO_4)_2$, $KMnO_4$ jeweils w. L.	$\nu Mn^{55}/\nu H^1$ = 0,24786 ± 0,00012		

Kern	Vork. %	Spin	Zitat	Methode	Substanz	Ergebnisse	μ_{unkorr}	μ_{korr}
			W. G. Proctor, F. C. Yu Phys. Rev. 77, 716 (1950) Phys. Rev. 81, 20 (1951)	Gekreuzte Spulen	a) 2 m LiMnO$_4$ w. L. b) 0,25 m NaCl in 1,0 m MnSO w. L.	$\nu Mn^{55}/\nu Na^{23}$ = 0,9372 $\pm$ 0,0001, I	+ 3,4624 $\pm$ 0,0006	
			R. E. Sheriff, D. Williams Phys. Rev. 82, 651 (1951)	Pendelrückk.	a) Ca(MnO$_4$)$_2$·w.L. b) ScCl$_3$ w. L.	$\nu Mn^{55}/\nu Sc^{45}$ = 1,02028 $\pm$ 0,00005		3,467 53 $\pm$ 0,000 17
Fe^{54}_{26}	5,81	0						
Fe^{56}_{26}	91,64	0						
Fe^{57}_{26}	2,21	$(\frac{3}{2}; \frac{5}{2})$	M. Gurevitch et al. Phys. Rev. 76, 151 (1949)	Hfs.			≈ 0	
			R. S. Trenam Proc. Phys. Soc. Lond. A 66, 414 (1953) B. Bleaney, R. S. Trenam Proc. Roy. Soc. Lond., Ser. A 223, 1 (1954)	Paramagnetische Resonanz	KAl-Selenat-Alaun-Kristall mit 0,1% Fe RbAl-Alaun mit Fe	$I = \frac{3}{2}$ oder $\frac{5}{2}$	< 0,05	
Fe^{58}_{26}	0,34	0						
Co^{56}_{27}		(4; 5)	L. Gallaher et al. Physica, Haag 21, 117 (1955)	Winkelverteilung der γ-Strahlung	(Co, Cu, Zn) (NH$_4$)$_2$(SO$_4$)$_2$× 6H$_2$O Einkristall		2,8 ($I = 4$) $\pm$ 0,9	
			J. M. Baker et al. Proc. Phys. Soc. Lond. A 69, 353 (1956)	Paramagnetische Resonanz	ZnK$_2$(SO$_4$)$_2$× 6D$_2$O Kristall mit Co-Gehalt	$\mu Co^{56}/\mu Co^{59}$ = 0,828 $\pm$ 0,003	3,848 ($I=4$) $\pm$ 0,015	
			R. V. Jones et al. Phys. Rev. 102, 738 (1956)	Paramagnetische Resonanz	(Co, Zn) K$_2$(SO$_4$)$_2$·6D$_2$O Einkristall	$\Delta \nu Co^{59}/\Delta \nu Co^{56}$ = 1,205 $\pm$ 0,002	$|\mu| = 3,855$ ($I=4$) $\pm$ 0,007	
			R. C. Sapp et al. Bull. Amer. Phys. Soc. II 1, 91 (1956)	Winkelverteilung der γ-Strahlung	CeMg-Nitrat Einkristall mit Co-Gehalt		3,30 ($I=5$) $\pm$ 0,20	
Co^{57}_{27}	<0,003	$\frac{7}{2}$	J. M. Baker et al. Proc. Phys. Soc. Lond. A 66, 305 (1953)	Paramagnetische Resonanz	Co^{57} aus Fe^{56} (d, n) Co^{57} Reaktion		4,6 $\pm$ 0,2	
			J. M. Baker et al. Proc. Phys. Soc. Lond. A 69, 353 (1956)	Paramagnetische Resonanz	ZnK$_2$(SO$_4$)$_2$× 6D$_2$O Kristall mit Co-Gehalt		4,65 $\pm$ 0,05	
Co^{58}_{27}		2	J. M. Daniels et al. Phil.Mag.(7) 43,1297(1952) M. A. Grace, H. Halban Physica, Haag 18, 1227 (1952)	Winkelverteilung der γ-Strahlung			3,5 $\pm$ 0,3	
Co^{59}_{27}	100	$\frac{7}{2}$	K. R. More Phys. Rev. 46, 470 (1934)	Hfs.			2 − 3	
			K. R. More Phys. Rev. 47, 256 (1935)	Hfs.			2,7	
			W. G. Proctor, F. C. Yu Phys. Rev. 77, 716 (1950) Phys. Rev. 81, 20 (1951)	Gekreuzte Spulen	a) 1 m K$_3$Co(CN)$_6$ w. L. b) 0,25 m NaCl+ 1,0 m MnSO$_4$ w. L.	$\nu Co^{59}/\nu Na^{23}$ = 0,897 09 $\pm$ 0,00009, I	+ 4,6399 $\pm$ 0,0009	
Co^{60}_{27}		5	B. Bleaney et al. Phys. Rev. 85, 688 (1952) Phil. Mag. (7) 43, 1297 (1952) Proc. Roy. Soc. Lond., Ser. A 221, 170 (1954) M. A. Grace, H. Halban Physica, Haag 18, 1227 (1952)	Winkelverteilung der γ-Strahlung	(1% Co, 12% Cu, 87% Zn)SO$_4$× Rb$_2$SO$_4$·6H$_2$O Einkristall		$\mu = 3,5$ $\pm$ 0,5	
			O. J. Poppema et al. Physica, Haag 21, 233 (1955)	Winkelverteilung der γ-Strahlung	(Co, Cu, Zn) (NH$_4$)$_2$(SO$_4$)$_2$× 6H$_2$O Ein-kristall und Vielkristall		4,3 $\pm$ 0,2	
			J. C. Wheatley et al. Physica, Haag 21, 841 (1955)	Zirkulare Polarisation der γ-Strahlung	2Ce(NO$_3$)$_3$× 3Mg(NO$_3$)$_2$× 24H$_2$O Ein-kristall mit 110 μC Co^{60}		> 0	

Kern	Vork. %	Spin	Zitat	Methode	Substanz	Ergebnisse	μ_{unkorr}	μ_{korr}
			W. Dobrowolski et al. Phys. Rev. 101, 1001 (1956)	Paramagnetische Resonanz	(Co, Zn) (NH$_4$)$_2$ × (SO$_4$)$_2$ · 6D$_2$O Einkristall	$\mu_{Co^{60}}/\mu_{Co^{59}}$ = 0,8191 ± 0,0016	$\|\mu\|$ = 3,800 ± 0,007	
Ni_{28}^{58}	69,76	0						
Ni_{28}^{60}	26,16	0						
Ni_{28}^{61}	1,25		K. G. Kessler Phys. Rev. 79, 167 (1950)	Hfs.			≤0,25	
Ni_{28}^{62}	3,66	0						
Ni_{28}^{64}	1,16	0						
Cu_{29}^{63}	68,94	$\frac{3}{2}$	E. Fermi, E. Segrè Z. Physik 82, 729 (1933)	Berechnung aus früheren Daten			μ_B/μ = 780	
			S. Goudsmit Phys. Rev. 43, 636 (1933)	Berechnung aus früheren Daten			2,5	
			H. Schüler Z. Physik 88, 323 (1934)	Berechnung aus früheren Daten			+ 2,74	
			H. Schüler, T. Schmidt Z. Physik 100, 113 (1936)	Hfs.			2,5	
			R. V. Pound Phys. Rev. 73, 523 (1948)	Empfindlicher Oszillator	a) Cu-(I)-Cl Pulver gepreßt b) NaBr-Pulver gepreßt	$\nu_{Cu^{63}}/\nu_{Na^{23}}$ = 1,0022 ± 0,0002		2,2265 ± 0,0025
			F. Bitter Phys. Rev. 75, 1326 (1949)	Brücke		$\nu_{Cu^{63}}/\nu_{H^1}$ = 0,265056 ± 0,00005		
			J. R. Zimmermann et al. Phys. Rev. 76, 163 (1949) Phys. Rev. 76, 350 (1949)	Pendelrückk.	a) Cu$_2$Cl$_2$+CuCl$_2$ w.L. b) H$_2$O	$\nu_{Cu^{63}}/\nu_{H^1}$ = 0,26515 ± 0,00005		
			R. E. Sheriff, D. Williams Phys. Rev. 82, 651 (1951)	Pendelrückk.	a) CuCl$_2$ Pulver b) ScCl$_3$ w. L.	$\nu_{Cu^{63}}/\nu_{Sc^{45}}$ = 1,09125 ± 0,00006		2,22628 ± 0,00013
			H. E. Walchli ORNL-1775 (1954)			$\nu_{Cu^{63}}/\nu_{Na^{23}}$ = 1,002008 ± 0,000016	+ 2,220586	
			H. L. Cox jr., D. Williams Bull. Amer. Phys. Soc. II 2, 30 (1957)	Quadrupol-Resonanz	Kuprit-Einkristall		2,226 ± 0,007	
Cu_{29}^{64}		1	A. Lemonick et al. Phys. Rev. 95, 1356 (1954)	Berechnung aus früheren Daten			$\|\mu\|$ = 0,40 ± 0,05	
Cu_{29}^{65}		$\frac{3}{2}$	E. Fermi, E. Segrè Z. Physik 82, 729 (1933)	Berechnung aus früheren Daten			μ_B/μ = 780	
			S. Goudsmit Phys. Rev. 43, 636 (1933)	Berechnung aus früheren Daten			2,5	
			H. Schüler Z. Physik 88, 323 (1934)	Berechnung aus früheren Daten			+ 2,74	
			H. Schüler, T. Schmidt Z. Physik 100, 113 (1936)	Hfs.			2,6	
			R. V. Pound Phys. Rev. 73, 523 (1948)	Empfindlicher Oszillator	Kupfer-I-Chlorid Pulver gepreßt	$\nu_{Cu^{65}}/\nu_{Cu^{63}}$ = 1,0711 ± 0,0002		2,3847 ± 0,0030
			F. Bitter Phys. Rev. 75, 1326 (1949)	Brücke		$\nu_{Cu^{65}}/\nu_{H^1}$ = 0,28391 ± 0,00006		
			J. R. Zimmerman et al. Phys. Rev. 76, 163 (1949) Phys. Rev. 76, 350 (1949)	Pendelrückk.	a) Cu$_2$Cl$_2$+CuCl$_2$ w.L. b) H$_2$O	$\nu_{Cu^{65}}/\nu_{H^1}$ = 0,28404 ± 0,00009		
			R. E. Sheriff, D. Williams Phys. Rev. 82, 651 (1951)	Pendelrückk.	a) Cu$_2$Cl$_2$ Pulver b) ScCl$_3$ Pulver	$\nu_{Cu^{65}}/\nu_{Sc^{45}}$ = 1,16951 ± 0,00006 $\nu_{Cu^{65}}/\nu_{Cu^{63}}$ = 1,07178 ± 0,00005		2,38594 ± 0,00013
			H. E. Walchli ORNL-1775 (1954)			$\nu_{Cu^{65}}/\nu_{H^1}$ = 0,283954 ± 0,000004	+ 2,378967	
			H. L. Cox jr., D. Williams Bull. Amer. Phys. Soc. II 2, 30 (1957)	Quadrupol-Resonanz	Kuprit-Einkristall		2,376 ± 0,007	

Kern	Vork. %	Spin	Zitat	Methode	Substanz	Ergebnisse	μ_{unkorr}	μ_{korr}
Zn_{30}^{64}	48,89	0	K. Murakawa Z. Physik 72, 793 (1931)	Hfs.			≈ 0	
Zn_{30}^{66}	27,81	0	K. Murakawa Z. Physik 72, 793 (1931)	Hfs.			≈ 0	
Zn_{30}^{67}	4,07	$\frac{5}{2}$	H. A. Bethe, R. F. Bacher Rev. Mod. Phys. 8, 82 (1936)	Berechnung aus früheren Daten			$-1,7$	
			J. M. Lyshede et al. Z. Physik 104, 434 (1937)	Hfs.			$+0,9$	
			S. S. Dharmatti et al. Phys. Rev. 85, 927 (1952) H. E. Weaver jr. Phys. Rev. 89, 923 (1953)	Gekreuzte Spulen	$2\,m\,Zn(NH_3)_4^{++}$, $ZnCl_2$, $ZnSO_4$, $Zn(NO_3)_2$ alles w. L.	$\nu Zn^{67}/\nu N^{14}$ $=0,86580$ $\pm0,00001$, I	$+0,87378$ $\pm0,00013$	
Zn_{30}^{68}	18,61	0	K. Murakawa Z. Physik 72, 793 (1931)	Hfs.			≈ 0	
Zn_{30}^{70}	0,62	0						
Ga_{31}^{69}	60,16	$\frac{3}{2}$	E. Fermi, E. Segrè Z. Physik 82, 729 (1933)	Berechnung aus früheren Daten			$\mu_B/\mu = 860$	
			S. Goudsmit Phys. Rev. 43, 636 (1933)	Berechnung aus früheren Daten			2,01	
			H. Schüler Z. Physik 88, 323 (1934)	Berechnung aus früheren Daten			$+2,14$	
			H. A. Bethe, R. F. Bacher Rev. Mod. Phys. 8, 82 (1936)	Berechnung aus früheren Daten			$+2,1$	
			H. Schüler, H. Korsching Z. Physik 103, 434 (1936)	Hfs.			$+2,0$	
			N. A. Renzetti Phys. Rev. 57, 753 (1940)	Atomstrahl	Ga		2,11	
			G. E. Becker, P. Kusch Phys. Rev. 73, 584 (1948)]	Atomstrahl	Ga	$\Delta\nu Ga^{71}/\Delta\nu Ga^{69}$	1,994 $\pm0,005$	
			R. V. Pound Phys. Rev. 73, 1112 (1948) Phys. Rev. 74, 228 (1948)	Empfindlicher Oszillator	$GaCl_3$ w. L.	$\nu Ga^{71}/\nu Ga^{69}$ $=1,2701$ $\pm0,0004$	2,0165 $\pm0,0035$	
			R. T. Daly jr., et al. Phys. Rev. 96, 539 (1954)	Atomstrahl		$\mu_3 = (0,107$ $\pm0,02) \times$ $10^{-24}\,\mu_K\,cm^2$		
			H. E. Walchli ORNL-1775 (1954)			$\nu Ga^{69}/\nu Na^{23}$ $=0,907349$ $\pm0,000020$	$+2,010809$	
			A. Lurio, A. G. Prodell Phys. Rev. 101, 79 (1956)	Atomstrahl	Ga	$\Delta\nu Ga^{69}/\Delta\nu Ga^{71}$ $-0,7870196$ $\pm0,0000006$		
Ga_{31}^{71}	39,84	$\frac{3}{2}$	J. S. Campbell Nature, Lond. 131, 204 (1933)	Hfs.		$g Ga^{71}/g Ga^{69} = 1,27$		
			E. Fermi, E. Segrè Z. Physik 82, 729 (1933)	Berechnung aus früheren Daten			$\mu_B/\mu = 670$	
			S. Goudsmit Phys. Rev. 43, 636 (1933)	Berechnung aus früheren Daten			2,55	
			H. Schüler Z. Physik 88, 323 (1934)	Berechnung aus früheren Daten			$+2,74$	
			H. A. Bethe, R. F. Bacher Rev. Mod. Phys. 8, 82 (1936)	Berechnung aus früheren Daten			$+2,7$	
			H. Schüler, H. Korsching Z. Physik 103, 434 (1936)	Hfs.		$\mu Ga^{71}/\mu Ga^{69}$ $=1,269$	2,5	
			N. A. Renzetti Phys. Rev. 57, 753 (1940)	Atomstrahl	Ga	$\mu Ga^{71}/\mu Ga^{69}$ $=1,270$ $\pm0,006$	2,69	
			G. E. Becker, P. Kusch Phys. Rev. 73, 584 (1948)	Atomstrahl	Ga	$\Delta\nu Ga^{71}/\Delta\nu Ga^{69}$ $=1,27059$ $\pm0,00008$	2,540 $\pm0,005$	
			R. V. Pound Phys. Rev. 73, 1112 (1948) Phys. Rev. 74, 228 (1948)	Empfindlicher Oszillator	$GaCl_3$ w. L.	$\nu Ga^{61}/\nu Na^{23}$ $=1,1529$ $\pm0,0004$	2,5611 $\pm0,0030$	
			R. T. Daly jr., et al. Phys. Rev. 96, 539 (1954)	Atomstrahl		$\mu_3 = (0,146$ $\pm0,02) \times$ $10^{-24}\mu_K\,cm^2$		
			H. E. Walchli ORNL-1775 (1954)			$\nu Ga^{71}/\nu Na^{23}$ $=1,152872$ $\pm0,000008$	$+2,554922$	
			M. Rice, R. V. Pound Phys. Rev. 99, 1036 (1955)	Empfindlicher Oszillator	$GaCl_3$ in 6n HCl	$\mu Ga^{71}/\mu Ga^{69}$ $=1,2706242$ $\pm0,0000020$		

Kern	Vork. %	Spin	Zitat	Methode	Substanz	Ergebnisse	μ_{unkorr}	μ_{korr}
Ge^{70}_{32}	21,2	0	C. H. Townes et al. Phys. Rev. 76, 700 (1949)	Hfs.			≈ 0	
Ge^{72}_{32}	27,3	0	C. H. Townes et al. Phys. Rev. 76, 700 (1949)	Hfs.			≈ 0	
Ge^{73}_{32}	7,9	$\frac{9}{2}$	C. D. Jeffries Phys. Rev. 92, 1262 (1953)	Gekreuzte Spulen	a) GeCl$_4$ rein b) TiCl$_4$	$\nu_{Ge^{73}}/\nu_{Cl^{35}}$ = 0,35572 ± 0,00004	−0,87675 ± 0,00012
			S. I. Aksenow et al. Dokl. Akad. Nauk SSSR. 96, 37 (1954)	Empfindlicher Oszillator	a) GeCl$_4$ flüssig b) 1m MnCl$_2$ in D$_2$O	$\nu_{Ge^{73}}/\nu_{D^2}$ = 0,22724 ± 0,00002	$\|\mu\|$ = 0,87677 ± 0,00009
Ge^{74}_{32}	37,1	0	C. H. Townes et al. Phys. Rev. 76, 700 (1949)	Hfs.			≈ 0	
Ge^{76}_{32}	6,5	0	C. H. Townes et al. Phys. Rev. 76, 700 (1949)	Hfs.			≈ 0	
As^{75}_{33}	100	$\frac{3}{2}$	S. Goudsmit Phys. Rev. 43, 636 (1933)	Berechnung aus früheren Daten			0,9	
			M. F. Crawford et al. Canad. J. Res. 10, 693 (1934)	Berechnung aus früheren Daten		$g = 1,1$		
			H. A. Bethe, R. F. Bacher Rev. Mod. Phys. 8, 82 (1936)	Berechnung aus früheren Daten			+1,5	
			H. Schüler, M. Marketu Z. Physik 102, 703 (1936)	Hfs.			+1,5 ±0,3	
			H. Schüler, T. Schmidt Z. Physik 98, 430 (1936)	Hfs.			∼0,8	
			S. S. Dharmatti et al. Phys. Rev. 84, 367 (1951) H. E. Weaver jr. Phys. Rev. 89, 923 (1953)	Gekreuzte Spulen	a) 2 m Na$_2$HAsO$_4$ in 3 m NaOH b) NaCl + 0,5 m MnCl$_2$ w.L.	$\nu_{As^{75}}/\nu_{Na^{23}}$ = 0,64745 ± 0,00015	+1,4347 ±0,0003	
			C. D. Jeffries et al. Helv. phys. Acta 24, 643 (1951) Phys Rev. 85, 478 (1952)	Gekreuzte Spulen	1,2 m Na$_3$AsS$_4$ w L., Na$_3$AsO$_4$ w.L.	$\nu_{As^{75}}/\nu_{H^1}$ = 0,17129 ± 0,00003, I	+1,4350 ±0,0003	
			H. Krüger et al. Z. Physik 132, 221 (1952)	Quadrupol-Resonanz	As$_2$O$_3$-Polykristall As$_2$O$_4$-Einkristall		1,44 ±0,03	
			K. Murakawa et al. Rep.-Inst. Sci. Technol. Univ. Tokyo 6, 209 (1952)				+1,45 ±0,15	
			Y. Ting, D. Williams Phys. Rev. 89, 595 (1953)	Pendelrückk.	a) Na$_2$HAsO$_4$ + NaOH b) D$_2$O + NiCl$_2$	$\nu_{As^{75}}/\nu_{D^2}$ = 1,11569 ± 0,00005		1,43893 ±0,00008
Se^{74}_{34}	0,87	0	M. W. P. Strandberg et al. Phys. Rev. 75, 827 (1949)	Mikrowellen	OCSe		≈ 0	
Se^{75}_{34}		$\frac{5}{2}$	L. C. Aamodt et al. Phys Rev. 98, 1224 (1955)	Mikrowellen	OCSe	$I = \frac{5}{2}$		
Se^{76}_{34}	9,02	0	M. W. P. Strandberg et al. Phys. Rev. 75, 827 (1949)	Mikrowellen	OCSe		≈ 0	
Se^{77}_{34}	7,58	$\frac{1}{2}$	S. S. Dharmatti et al. Phys. Rev. 86, 259 (1952) H. E. Weaver jr. Phys. Rev. 89. 923 (1953)	Gekreuzte Spulen	a) 12 m H$_2$SeO$_3$ w.L. b) NaCl	$\nu_{Se^{77}}/\nu_{Na^{23}}$ = 0,72193 ± 0,00002	+0,53326 ±0,00005	
			H. E. Walchli Phys. Rev. 90, 331 (1953)		a) H$_2$Se b) 0,1 m MnSO$_4$ in D$_2$O	$\nu_{Se^{77}}/\nu_{D^2}$ = 1,24211 ± 0,00010		
			H. E. Walchli ORNL-1775 (1954)			$\nu_{Se^{77}}/\nu_{D^2}$ = 1,242100 ± 0,000019		+0,5324786
Se^{78}_{34}	23,52	0	S. P. Davis Phys. Rev. 93, 159 (1954)	Hfs.			≈ 0	
Se^{79}_{34}		$\frac{7}{2}$	W. A. Hardy et al. Phys. Rev. 92, 1532 (1953)	Mikrowellen	OCSe		−1,015 ±0,015	
Se^{80}_{34}	49,82	0	S. P. Davis Phys. Rev. 93, 159 (1954)	Hfs.			≈ 0	
Se^{82}_{34}	9,19	0	M. W. P. Strandberg et al. Phys. Rev. 75, 827 (1949)	Mikrowellen	OCSe		≈ 0	

Kern	Vork. %	Spin	Zitat	Methode	Substanz	Ergebnisse	μunkorr	μkorr
Br_{35}^{79}	50,53	$\frac{3}{2}$	T. Schmidt Z. Physik **108**, 408 (1938)	Berechnung aus früheren Daten			2,6	
			S. B. Brody et al. Phys. Rev. **72**, 258 (1947)	Atomstrahl	CsBr, LiBr		2,110 $\pm$ 0,021	
			R. V. Pound Phys. Rev. **72**, 1273 (1947)	Empfindlicher Oszillator	LiBr, NaBr w.L.	$\nu Br^{81}/\nu Br^{79}$ $= 1,0778$ $\pm 0,0003$		2,1066 $\pm$ 0,003
			J. R. Zimmerman et al. Phys. Rev. **76**, 163 (1949) Phys. Rev. **76**, 350 (1949)	Pendelrückk.	a) NaBr w.L. b) H_2O	$\nu Br^{79}/\nu H^1$ $= 0,250\,59$ $\pm 0,00005$		
			J. D. Ranade Phil. Mag. (7) **42**, 279 (1951)	Hfs.			1,8	
			R. E. Sheriff et al. Phys. Rev. **82**, 651 (1951)	Pendelrückk.	a) NaBr w.L. b) $ScCl_3$ w.L.	$\nu Br^{79}/\nu Sc^{45}$ $= 1,031\,45$ $\pm 0,00005$		2,105\,74 $\pm$ 0,000\,10
			H. E. Walchli ORNL-1775 (1954)			$\nu Br^{79}/\nu Na^{23}$ $= 0,947\,140$ $\pm 0,000009$	$+ 2,098\,991$	
Br_{35}^{81}	49,47	$\frac{3}{2}$	T. Schmidt Z. Physik **108**, 408 (1938)	Berechnung aus früheren Daten			2,6	
			S. B. Brody et al. Phys. Rev. **72**, 258 (1947)	Atomstrahl	CsBr, LiBr		2,271 $\pm$ 0,023	
			R. V. Pound Phys. Rev. **72**, 1273 (1947)	Empfindlicher Oszillator	LiBr, NaBr, w.L.	$\nu Br^{81}/\nu Na^{23}$ $= 1,0209$ $\pm 0,0003$		2,2706 0,003
			F. Bitter Phys. Rev. **75**, 1326 (1949)	Brücke		$\nu Br^{81}/\nu H^1$ $= 0,270\,03$ $\pm 0,00008$		
			J. R. Zimmerman et al. Phys. Rev. **76**, 163 (1949) Phys. Rev. **76**, 350 (1949)	Pendelrückk.	a) NaBr w.L. b) H_2O	$\nu Br^{81}/\nu H^1$ $= 0,270\,14$ $\pm 0,00006$		
			J. D. Ranade Phil. Mag. (7) **42**, 279 (1951)	Hfs.			1,8	
			R. E. Sheriff et al. Phys. Rev. **82**, 651 (1951)	Pendelrückk.	a) NaBr w.L. b) $ScCl_3$ w.L.	$\nu Br^{81}/\nu Sc^{45}$ $= 1,111\,65$ $\pm 0,00006$ $\nu Br^{81}/\nu Br^{79}$ $= 1,077\,75$ $\pm 0,00005$		2,269\,47 $\pm$ 0,000\,13
			H. E. Walchli ORNL-1775 (1954)			$\nu Br^{81}/\nu Na^{23}$ $= 1,020\,965$ $\pm 0,000014$	$\pm 2,262\,597$	
Kr_{36}^{78}	0,342	0	H. Kopfermann et al. Z. Physik **85**, 353 (1933)	Hfs.			≈ 0	
Kr_{36}^{80}	2,228	0	H. Kopfermann et al. Z. Physik **85**, 353 (1933)	Hfs.			≈ 0	
Kr_{36}^{82}	11,50	0	H. Kopfermann et al. Z. Physik **85**, 353 (1933)	Hfs.			≈ 0	
Kr_{36}^{83}	11,48	$\frac{9}{2}$	H. Kopfermann et al. Z. Physik **85**, 353 (1933)	Hfs.			< 0	
			H. A. Bethe, R. F. Bacher Rev. Mod. Phys. **8**, 82 (1936)	Berechnung aus früheren Daten			$- 1$	
			Retherford, Kellogg, unveröffentlicht; nach: I. B. M. Kellog, S. Millman Rev. Mod. Phys. **18**, 323 (1946)				$- 0,967$	
			E. Brun et al. Helv. phys. Acta **27**, 173 (1954)	Gekreuzte Spulen			$+ 0,967\,06$ $\pm 0,00004$	
Kr_{36}^{84}	57,02	0	H. Kopfermann et al. Z. Physik **85**, 353 (1933)	Hfs.			≈ 0	
Kr_{36}^{85}		$\frac{9}{2}$	E. Rasmussen et al. Z. Physik **141**, 160 (1955)	Hfs.		$\mu Kr^{85}/\mu Kr^{83}$ $= 1,035$ $\pm 0,002$		
Kr_{36}^{86}	17,43	0	H. Kopfermann et al. Z. Physik **85**, 353 (1933)	Hfs.			≈ 0	

23*

Kern	Vork. %	Spin	Zitat	Methode	Substanz	Ergebnisse	μ_{unkorr}	μ_{korr}
Rb^{81}_{37}		$\frac{3}{2}$	J. P. HOBSON et al. Phys. Rev. 96, 1450 (1954) J. C. HUBBS et al. Phys. Rev. 99, 612 (1955)	Atomstrahl	RbBr		2,00 ± 0,06	
Rb^{82}_{37}		5	J. P. HOBSON et al. Phys. Rev. 104, 101 (1956)	Atomstrahl	Rb	$I = 5$		
Rb^{83}_{37}		$\frac{5}{2}$	J. P. HOBSON et al. Phys. Rev. 104, 101 (1956)	Atomstrahl	Rb	$I = \frac{5}{2}$		
Rb^{84}_{37}		2	J. P. HOBSON et al. Phys. Rev. 104, 101 (1956)	Atomstrahl	Rb	$I = 2$		
Rb^{85}_{37}	72,8	$\frac{5}{2}$	E. FERMI, E. SEGRÈ Z. Physik 82, 729 (1933)	Berechnung aus früheren Daten			$\mu_B/\mu = 1350$	
			S. GOUDSMIT Phys. Rev. 43, 636 (1933)	Berechnung aus früheren Daten			1,3	
			H. KOPFERMANN Z. Physik 83, 417 (1933) H. KOPFERMANN, H. KRÜGER Z. Physik 103, 485 (1936)	Hfs.			1,4	
			H. SCHÜLER Z. Physik 88, 323 (1934)	Berechnung aus früheren Daten			+ 1,49	
			S. MILLMAN, M. FOX Phys. Rev. 50, 220 (1936)	Atomstrahl	Rb		1,44	
			A. V. HOLLENBERG Phys. Rev. 52, 139 (1937)	Hfs.			1,4 − 1,7	
			S. MILLMAN et al. Phys. Rev. 51, 1049 (1937)	Atomstrahl			> 0	
			P. KUSCH, S. MILLMAN Phys. Rev. 56, 527 (1939)	Atomstrahl	Rb_2		1,340 ± 0,005	1,345 ± 0,005
			S. MILLMAN, P. KUSCH Phys. Rev. 58, 438 (1940)	Atomstrahl		$g = -0,539$		
			F. BITTER Phys. Rev. 75, 1326 (1949)	Brücke		$\nu Rb^{85}/\nu Na^{23}$ = 0,36512 ± 0,00011		
			W. H. CHAMBERS et al. Phys. Rev. 76, 638 (1949)	Pendelrückk.	Rb_2CO_3 w. L.	$\nu Rb^{85}/\nu H^1$ = 0,09661 ± 0,00004		
			J. R. ZIMMERMAN et al. Phys. Rev. 75, 699 (1949)	Pendelrückk.	Rb-Salz w. L.		1,34	
			E. YASAITIS, B. SMALLER Phys. Rev. 82, 750 (1951)	Brücke	a) RbCl ges. w.L. b) −	$\nu H^1/\nu Rb^{85}$ = 10,357105 ± 0,000030		
			H. E. WALCHLI et al. Phys. Rev. 85, 922 (1952)	Gekreuzte Spulen	RbCl in 15% D_2O ges. L.	$\nu Rb^{85}/\nu Cl^{35}$ = 0,98541 ± 0,00015		
			H. E. WALCHLI ORNL-1775 (1954)			$\nu Rb^{85}/\nu D^2$ = 0,628985 ± 0,000005	+ 1,348217	
Rb^{86}_{37}		2	E. H. BELLAMY Nature, Lond. 168, 556 (1951)	Atomstrahl			− 1,68 ± 0,01	
			E. H. BELLAMY et al. Phil. Mag. (7) 44, 33 (1953)	Atomstrahl			− 1,69 ± 0,01	
Rb^{87}_{37}	27,2	$\frac{3}{2}$	E. FERMI, E. SEGRÈ Z. Physik 82, 729 (1933)	Berechnung aus früheren Daten			$\mu_B/\mu = 660$	
			S. GOUDSMIT Phys. Rev. 43, 636 (1933)	Berechnung aus früheren Daten			2,7	
			H. KOPFERMANN Naturwiss. 21, 24 (1933)	Hfs.		$\mu Rb^{87}/\mu Rb^{85}$ = 2,3		
			H. KOPFERMANN Z. Physik 83, 417 (1933) H. KOPFERMANN et al. Z. Physik 103, 485 (1936)	Hfs.		$\mu Rb^{87}/\mu Rb^{85}$ = 2,03	2,8	
			H. SCHÜLER Z. Physik 88, 323 (1934)	Berechnung aus früheren Daten			+ 3,06	
			S. MILLMAN, M. FOX Phys. Rev. 50, 220 (1936)	Atomstrahl	Rb	$\mu Rb^{87}/\mu Rb^{85}$ = 2,026 ± 0,004	2,92	
			S. MILLMAN et al. Phys. Rev. 51, 1049 (1937)	Atomstrahl			> 0	

Kern	Vork. %	Spin	Zitat	Methode	Substanz	Ergebnisse	μ_{unkorr}	μ_{korr}
			P. Kusch, S. Millman Phys. Rev. **56**, 527 (1939)	Atomstrahl	Rb$_2$	$\mu_{\text{Rb}^{87}}/\mu_{\text{Rb}^{85}}$ = 2,038 ± 0,01	2,730 ± 0,009	2,741 ± 0,009
			S. Millman, P. Kusch Phys. Rev. **58**, 438 (1940)	Atomstrahl		$\mu_{\text{Rb}^{87}}/\mu_{\text{Rb}^{85}}$ = 2,0261 ± 0,0003		
			S. Millman, P. Kusch Phys. Rev. **60**, 91 (1941)	Berechnung aus früheren Daten			2,733	
			F. Bitter Phys. Rev. **75**, 1326 (1949)	Brücke		$\nu_{\text{Rb}^{87}}/\nu_{\text{H}^1}$ = 0,327 18 ± 0,00006		
			J. R. Zimmerman et al. Phys. Rev. **75**, 699 (1949)	Pendelrückk.	Rb-Salz w. L.		2,74	
			J. R. Zimmerman et al. Phys. Rev. **76**, 163 (1949) Phys. Rev. **76**, 350 (1949)	Pendelrückk.	a) Rb$_2$CO$_3$ w. L. b) H$_2$O	$\nu_{\text{Rb}^{87}}/\nu_{\text{H}^1}$ = 0,327 18 ± 0,00016		
			F. Bitter et al. Res. Lab. of Electronics, MIT, Progr. Rep. Dec. 1, 1950, NP-1947 N. I. Adams, III et al. Phys. Rev. **82**, 343 (1951)			$\mu_{\text{Rb}^{87}}/\mu_{\text{Rb}^{85}}$ = 2,033 380 ± 0,000028		
			W. H. Chambers et al. Phys. Rev. **78**, 482 (1950) D. Williams Phys. Rev. **83**, 858 (1951)	Pendelrückk.	a) Pr(NO$_3$)$_3$, ges. w. L.verun- reinigt mit Rb b) Natriumborat	$\nu_{\text{Rb}^{87}}/\nu_{\text{H}^1}$ = 0,326 98 ± 0,000 16		
			R. E. Sheriff et al. Phys. Rev. **82**, 651 (1951)	Pendelrückk.	a) Rb$_2$CO$_3$ + CuCl$_2$, w. L. b) AlCl$_3$, w. L.	$\nu_{\text{Rb}^{87}}/\nu_{\text{Al}^{127}}$ = 1,255 29 ± 0,00006		2,749 37 ± 0,000 19
			E. Yasaitis, B. Smaller Phys. Rev. **82**, 750 (1951)	Brücke	a) RbCl ges w. L. b) —	$\nu_{\text{H}^1}/\nu_{\text{Rb}^{87}}$ = 3,056 109 7 ± 0,000 005 5		
			H. E. Walchli ORNL-1775 (1954)			$\nu_{\text{Rb}^{87}}/\nu_{\text{Na}^{23}}$ = 1,237 041 ± 0,000008	+ 2,741 451	
Sr^{84}_{38}	0,55	0						
Sr^{86}_{38}	9,75	0						
Sr^{87}_{38}	6,96	$\frac{9}{2}$	H. Schüler et al. Naturwiss. **21**, 561 (1933)	Hfs.		$\mu_B/\mu = 1570$		
			H. Schüler Z. Physik **88**, 323 (1934)	Berechnung aus früheren Daten		$\approx -0,86$		
			H. Westmeyer Z. Physik **94**, 590 (1935)	Hfs.		0,86 $(I = \frac{3}{2})$		
			M. Heyden et al. Z. techn. Phys. **18**, 534 (1937) Z. Physik **108**, 232 (1938)	Hfs.		$-1,1$		
			C. D. Jeffries, P. B. Sogo Phys. Rev. **91**, 1286 (1953)	Gekreuzte Spulen	a) 3,1 m SrBr$_2$ w. L. mit 60% Sr^{87} b) 1 m MnCl$_2$ in D$_2$O	$\nu_{\text{Sr}^{87}}/\nu_{\text{D}^2}$ = 0,282 32 ± 0,00003	$-1,089 2$ ± 0,000 15	
Sr^{88}_{38}	82,74	0						
Y^{89}_{39}	100	$\frac{1}{2}$	H. Wittke Z. Physik **166**, 547 (1940)	Hfs.		$\leq 0,1$		
			M. F. Crawford et al. Phys. Rev. **76**, 1528 (1949)	Hfs.		$-0,14$		
			H. Kuhn et al. Proc. Phys. Soc. Lond. A **63**, 830 (1950)	Hfs.		< 0		
			E. Brun et al. Phys. Rev. **93**, 172 (1954)	Gekreuzte Spulen	a) 3,3 m Y(NO$_3$)$_3$ w. L. b) 0,1 m MnSO$_4$ w. L.	$\nu_{\text{Y}^{89}}/\nu_{\text{H}^1}$ = 0,048 994 ± 0,000001	$-0,136 825$ ± 0,000 004	
Zr^{89}_{40}	< 0,02	$\frac{9}{2}$	P. Axel et al. Phys. Rev. **97**, 975 (1955)	Zerfallsreihe		$I = \frac{9}{2}$		
Zr^{90}_{40}	51,46	0						

Kern	Vork. %	Spin	Zitat	Methode	Substanz	Ergebnisse	μunkorr	μkorr
Zr_{40}^{91}	11,23	$\frac{5}{2}$	S. Suwa Phys. Rev. 86, 247 (1952)	Hfs.			$-1,1$ $\pm 0,3$	
			S. Suwa J. Phys. Soc. Japan 8, 734 (1953)	Hfs.			$-1,3$ $\pm 0,3$	
			K. Murakawa Phys. Rev. 100, 1369 (1955)	Hfs.			$-1,9$ $\pm 0,2$	
Zr_{40}^{92}	17,11	0						
Zr_{40}^{94}	17,40	0						
Zr_{40}^{96}	2,80	0						
Nb_{41}^{93}	100	$\frac{9}{2}$	S. S. Ballard Phys. Rev. 46, 806 (1934)	Hfs.			3,7	
			W. W. Meeks, R. A. Fisher Phys. Rev. 72, 169 (1947) Phys. Rev. 72, 451 (1947)	Hfs.			5,3	
			R. E. Sheriff et al. Phys. Rev. 78, 476 (1950)	Pendelrückk.	a) Nb_2O_5 in HF b) B^{11}	$\nu Nb^{93}/\nu B^{11}$ $= 0,76187$ $\pm 0,00040$		6,1659 $\pm 0,0032$
			R. E. Sheriff et al. Phys. Rev. 82, 651 (1951)	Pendelrückk.	a) Nb_2O_5 in HF b) $ScCl_3$, w. L.	$\nu Nb^{93}/\nu Sc^{45}$ $= 1,00613$ $\pm 0,00005$		6,16670 $\pm 0,00030$
Mo_{42}^{91}	<0,002	$\frac{9}{2}$	P. Axel et al. Phys. Rev. 97, 975 (1955)	Zerfallsreihe		$I = \frac{9}{2}$		
Mo_{42}^{92}	15,84	0	O. H. Arroe, unveröffent- licht; nach: N. F. Ramsey, Nuclear Moments. NewYork: Wiley & Sons 1953				≈ 0	
Mo_{42}^{94}	9,04	0	O. H. Arroe, unveröffent- licht; nach: N. F. Ramsey, Nuclear Moments. NewYork: Wiley & Sons 1953				≈ 0	
Mo_{42}^{95}	15,72	$\frac{5}{2}$	N. S. Grace, K. R. More Phys. Rev. 45, 166 (1934)	Hfs.			sehr klein	
			W. G. Proctor, F. C. Yu Phys. Rev. 81, 20 (1951)	Gekreuzte Spulen	K_2MoO_4 ges. w. L.	$\nu Mo^{97}/\nu Mo^{95}$ $= 1,02110$ $\pm 0,0001$	$-0,9098$ $\pm 0,0002$	
			K. Murakawa Phys. Rev. 100, 1369 (1955)	Hfs.		$I = \frac{9}{2}$	$-1,3$ $\pm 0,2\ (I = \frac{9}{2})$	
Mo_{42}^{96}	16,53	0	O. H. Arroe, unveröffent- licht; nach: N. F. Ramsey, Nuclear Moments. NewYork: Wiley & Sons 1953				≈ 0	
Mo_{42}^{97}	9,46	$\frac{5}{2}$	N. S. Grace, K. R. More Phys. Rev. 45, 166 (1934)	Hfs.			sehr klein	
			W. Proctor, F. C. Yu Phys. Rev. 81, 20 (1951)	Gekreuzte Spulen	a) K_2MoO_4 ges. w. L. b) HNO_3	$\nu Mo^{97}/\nu N^{14}$ $= 0,9208$ $\pm 0,0001$	$-0,9289$ $\pm 0,0002$	
			E. C. Woodward, jr. Phys. Rev. 93, 954 (1954)	Hfs.		$\mu Mo^{97}/\mu Mo^{95}$ $= 1,022$		
			K. Murakawa Phys. Rev. 100, 1369 (1955)	Hfs.		$I = \frac{9}{2}$	$-1,3$ $\pm 0,2\ (I = \frac{9}{2})$	
Mo_{42}^{98}	23,78	0	O. H. Arroe, unveröffent- licht; nach: N. F. Ramsey, Nuclear Moments. NewYork: Wiley & Sons 1953				≈ 0	
Mo_{42}^{100}	9,63	0	O. H. Arroe, unveröffent- licht; nach: N. F. Ramsey, Nuclear Moments. NewYork: Wiley & Sons 1953				≈ 0	
Tc_{43}^{99}		$\frac{9}{2}$	K. G. Kessler et al. Meet. Phys. Soc. New York 1, 3 (1950) Phys. Rev. 82, 341 (1951)	Hfs.			5,2 $\pm 0,5$	

Kern	Vork. %	Spin	Zitat	Methode	Substanz	Ergebnisse	μ_{unkorr}	μ_{korr}
			H. Walchli et al. Phys. Rev. 85, 479 (1952)	Gekreuzte Spulen	NH_4TcO_4 w.L.+D_2O	$\nu Tc^{99}/\nu D2$ = 1,46628 ± 0,0001		+ 5,6805 ± 0,0004
			K. G. Kessler, R. E. Trees Phys. Rev. 92, 303 (1953)	Hfs.			5,5 ± 0,3	
Ru_{44}^{96}	5,68	0						
Ru_{44}^{98}	2,22	0						
Ru_{44}^{99}	12,81	$\frac{5}{2}$	K. Murakawa J. Phys. Soc. Japan 10, 919 (1955)	Hfs.			− 0,63 ± 0,15	
Ru_{44}^{100}	12,70	0						
Ru_{44}^{101}	16,98	$\frac{5}{2}$	J. H. E. Griffiths et al. Proc. Phys. Soc. Lond A 65 951 (1952)	Paramagnetische Resonanz	$Ru(NH_3)_6Cl_3$ in $Co(NH_3)_6Cl_3$ Kristall	$\mu Ru^{101}/\mu Ru^{99}$ = 1,09 ± 0,03		
			K. Murakawa J. Phys. Soc. Japan 10, 919 (1955)	Hfs.			− 0,69 ± 0,15	
Ru_{44}^{102}	31,34	0						
Ru_{44}^{104}	18,27	0						
Rh_{45}^{103}	100	$\frac{1}{2}$	L. Sibaiya Proc. Ind. Acad. Sci. A 6, 229 (1937)	Hfs.			>0 klein	
			H. Kuhn, G. K. Woodgate Nature, Lond. 166, 906(1950) Proc. Phys. Soc. Lond. A 64, 1090 (1950)	Hfs.			− 0,10 ± 0,025	
			P. B. Sogo, C. D. Jeffries Phys. Rev. 98, 265 (1955) Phys. Rev. 98, 1316 (1955)		a) Rh-Metall- pulver b) 1 m $MnCl_2$ in D_2O	$\nu Rh^{103}/\nu D2$ = 0,205574 ± 0,000007		− 0,08790 ± 0,00007
Pd_{46}^{102}	0,8	0						
Pd_{46}^{104}	9,3	0						
Pd_{46}^{105}	22,6	$\frac{5}{2}$	P. Brix, A. Steudel Naturwiss. 38, 431 (1951)	Hfs.			− 0,6	
			A. Steudel Z. Physik 132, 429 (1952)	Hfs.			− 0,57 ± 0,05	
			J. Blaise, H. Chantrel J. Phys. Radium 14, 135 (1953)	Hfs.			− 0,57	
Pd_{46}^{106}	27,1	0						
Pd_{46}^{108}	26,7	0						
Pd_{46}^{110}	13,5	0						
Ag_{47}^{105}		$\frac{1}{2}$	H. B. Silsbee et al. Bull. Amer. Phys. Soc. II 1, 389 (1956)	Atomstrahl		$I = \frac{1}{2}$		
Ag_{47}^{107}	51,92	$\frac{1}{2}$	S. Frisch, V. Matvejev C. R. Acad. Sci. URSS. 1, 462 (1934)	Hfs.			< 1/5000 μ_B	
			H. Schüler Z. Physik 88, 323 (1934)	Berechnung aus früheren Daten			< 0,24	
			H. A. Bethe, R. F. Bacher Rev. Mod. Phys. 8, 82 (1936)	Berechnung aus früheren Daten			+ 0,2	
			M. F. Crawford et al. Phys. Rev. 75, 1112 (1949) M. F. Crawford et al. Canad. J. Res. A 28, 558 (1950)	Hfs.	Atomstrahl		− 0,084	− 0,087

Kern	Vork. %	Spin	Zitat	Methode	Substanz	Egebnisse	μ_{unkorr}	μ_{korr}
			P. BRIX et al. Naturwiss. **38**, 68 (1951) Z. Physik **30**, 88 (1951)	Hfs.			$-0,111$ $\pm 0,008$	
			G. WESSEL, H. LEW Phys. Rev. **92**, 641 (1953)	Atomstrahl	Ag		$-0,111$	
			E. BRUN et al. Phys. Rev. **93**, 172 (1954)	Gekreuzte Spulen	a) 7 m AgNO$_3$+ 1m Mn(NO$_3$)$_2$ w.L. b) 0,1 m MnSO$_4$ w.L.	$\nu_{Ag^{107}}/\nu_{H^1}$ $= 0,040468$ $\pm 0,000001$	$-0,113014$ $\pm 0,000004$	
			P. B. SOGO, C. D. JEFFRIES Phys. Rev. **93**, 174 (1954)	Gekreuzte Spulen	6 m AgNO$_3$+ 2 m Mn(NO$_3$)$_3$ w.L.	$\nu_{Ag^{107}}/\nu_{Ag^{109}}$ $= 0,86985$ $\pm 0,00001$	$-0,113042$ $\pm 0,000013$	
$\mathbf{Ag^{109}_{47}}$	48,08	$\frac{1}{2}$	S. FRISCH, V. MATVEJEV C. R. Acad. Sci. URSS. **1**, 462 (1934)	Hfs.			$< 1/5000\mu_B$	
			H. SCHÜLER Z. Physik **88**, 323 (1934)	Berechnung aus früheren Daten			$< 0,24$	
			H. A. BETHE, R. F. BACHER Rev. Mod. Phys. **8**,82 (1936)	Berechnung aus früheren Daten			$+ 0,2$	
			M. F. CRAWFORD et al. Phys. Rev. **75**, 1112 (1949) M. F. CRAWFORD et al. Canad. J. Res. A **28**, 558 (1950)	Hfs.	Atomstrahl		$-0,155$	$-0,160$
			P. BRIX et al. Naturwiss. **38**, 68 (1951) Z. Physik **130**, 88 (1951)	Hfs.		$\mu_{Ag^{109}}/\mu_{Ag^{107}}$ $= 1,16$ $\pm 0,03$	$-0,129$ $\pm 0,008$	
			G. WESSEL, H. LEW Phys. Rev. **92**, 641 (1953)	Atomstrahl	Ag		$-0,129$	
			E. BRUN et al. Phys. Rev. **93**, 172 (1954)	Gekreuzte Spulen	a) 7 m AgNO$_3$+ 1 m Mn(NO$_3$)$_2$ w.L. b) 0,1 m MnSO$_4$ w.L.	$\nu_{Ag^{109}}/\nu_{H^1}$ $= 0,046523$ $\pm 0,000001$	$-0,129924$ $\pm 0,000004$	
			P. B. SOGO, C. D. JEFFRIES Phys. Rev. **93**, 174 (1954)	Gekreuzte Spulen	a) 6 m AgNO$_3$+ 2 m Mn(NO$_3$)$_2$ w.L. b) 1 m MnCl$_2$ in D$_2$O	$\nu_{Ag^{109}}/\nu_{D^2}$ $= 0,30316$ $\pm 0,00003$	$-0,129955$ $\pm 0,000013$	
$\mathbf{Ag^{111}_{47}}$		$\frac{1}{2}$	A. LEMONICK, F. M. PIPKIN Phys. Rev. **95**, 1356 (1954)	Atomstrahl			$|\mu| = 0,144$ $\pm 0,007$	
			G. K. WOODGATE et al. Nature, Lond. **167**, 395 (1955) Proc. Phys. Soc. Lond. A **69**, 581 (1956)	Atomstrahl	Ag		$-0,145$ $\pm 0,001$	
$\mathbf{Cd^{106}_{48}}$	1,215	0						
$\mathbf{Cd^{108}_{48}}$	0,875	0						
$\mathbf{Cd^{110}_{48}}$	12,39	0	H. SCHÜLER, H. BRUCK Z. Physik **56**, 291 (1929)	Hfs.				
$\mathbf{Cd^{111}_{48}}$	12,75	$\frac{1}{2}$	E. FERMI, E. SEGRÈ Z. Physik **82**, 729 (1933)	Berechnung aus früheren Daten			$\mu_B/\mu = -3500$	
			S. GOUDSMIT Phys. Rev. **43**, 636 (1933)	Berechnung aus früheren Daten			$-0,67$	
			E. G. JONES Proc. Phys. Soc. Lond. **45**, 625 (1933)	Hfs.			$-0,62$	
			H. A. BETHE, R. F. BACHER Rev. Mod. Phys. **8**, 82 (1936)	Berechnung aus früheren Daten			$-0,65$	
			W. G. PROCTOR, F. C. YU Phys. Rev. **76**, 1728 (1949) W. G. PROCTOR Phys. Rev. **79**, 35 (1950)	Gekreuzte Spulen	a) 0,3 m MnSO$_4$+ CdCl$_2$ ges. w.L. b) 0,1 m NaCl+ 0,2 m MnSO$_4$ w.L.	$\nu_{Cd^{111}}/\nu_{Na^{23}}$ $= 0,8016$ $\pm 0,0001$, I	$-0,5922$ $\pm 0,0002$	
$\mathbf{*Cd^{111}_{48}}$		$\frac{5}{2}$	H. AEPPLI et al. Phys. Rev. **84**, 370 (1951)	Winkelverteilung der γ-Strahlung			$-0,85$ $\pm 0,22$	
			H. AEPPLI et al. Helv. phys. Acta **25**, 339 (1952)	Winkelverteilung der γ-Strahlung	In^{111} in polykristalliner Silberfolie		$-0,70$ $\pm 0,12$	

Kern	Vork. %	Spin	Zitat	Methode	Substanz	Ergebnisse	μ_{unkorr}	μ_{korr}
			H. Albers-Schönberg et al. Helv. phys. Acta 27, 547 (1954) E. Heer J. Phys. Radium 16, 600 (1955)	Winkelverteilung der γ-Strahlung			$-0,725$ $\pm 0,047$	
			R. M. Steffen, W. Zobel Phys. Rev. 97, 1188 (1955) W. Zobel, R. M. Steffen Phys. Rev. 98, 1186 (1955) R. M. Steffen, W. Zobel Phys. Rev. 103, 126 (1956)	Winkelverteilung der γ-Strahlung	$In^{111}Cl_3$ w. L.		$-0,783$ $\pm 0,028$	
Cd_{48}^{112}	24,07	0	H. Schüler, H. Bruck Z. Physik 56, 291 (1929)	Hfs.			≈ 0	
Cd_{48}^{113}	12,26	$\frac{1}{2}$	E. Fermi, E. Segrè Z. Physik 82, 729 (1933)	Berechnung aus früheren Daten			$\mu_B/\mu = -3500$	
			S. Goudsmit Phys. Rev. 43, 636 (1933)	Berechnung aus früheren Daten			$-0,67$	
			E. G. Jones Proc. Phys. Soc. Lond. 45, 625 (1933)	Hfs.			$-0,62$	
			H. Schüler Z. Physik 88, 323 (1934)	Berechnung aus früheren Daten			$-0,63$	
			H. A. Bethe, R. F. Bacher Rev. Mod. Phys. 8, 82 (1936)	Berechnung aus früheren Daten			$-0,65$	
			W. G. Proctor, F. C. Yu Phys. Rev. 76, 1728 (1949) W. G. Proctor Phys. Rev. 79, 35 (1950)	Gekreuzte Spulen	a) $CdCl_2$ ges. w.L. $+0,3$ m $MnSO_4$ w. L. b) 0,1 m $NaCl+$ 0,2 m $MnSO_4$ w. L.	$\nu Cd^{113}/\nu Na^{23}$ $=0,8386$ $\pm 0,0001$ $\nu Cd^{113}/\nu Cd^{111}$ $=1,0461$ $\pm 0,0001, I$	$-0,6196$ $\pm 0,0001$	
			E. C. Woodward et al. Phys. Rev. 96, 529 (1954)	Hfs.		$\mu Cd^{113}/\mu Cd^{111}$ $=1,046$ $\pm 0,027$		
Cd_{48}^{114}	28,86	0	H. Schüler, H. Bruck Z. Physik 56, 291 (1929)	Hfs.			≈ 0	
Cd_{48}^{116}	7,58	0	H. Schüler, H. Bruck Z. Physik 56, 291 (1929)	Hfs.			≈ 0	
In_{49}^{113}	1,23	$\frac{9}{2}$	T. C. Hardy Phys. Rev. 60, 167 (1941) T. C. Hardy, S. Millman Phys. Rev. 61, 459 (1942)	Atomstrahl	In	$\mu In^{113}/\mu In^{115}$ $=0,998$	$+5,48$ $\pm 0,04$	
			W. G. Proctor, F. C. Yu Phys. Rev. 81, 20 (1951)	Gekreuzte Spulen	a) 0,38 m $In(NO_3)_2$ w.L. b) 0,25 m $NaCl$ $+0,5$ m $MnSO_4$ w.L.	$\nu In^{113}/\nu Na^{23}$ $=0,82667$ $\pm 0,00008$	$5,4972$ $\pm 0,0010$	
			Y. Ting, D. Williams Phys. Rev. 89, 595 (1953)	Pendelrückk.	$In(NO_3)_3$ w.L.	$\nu In^{115}/\nu In^{113}$ $=1,00213$ $\pm 0,00004$		$5,5222$ $\pm 0,0005$
$*In_{49}^{113}$		$\frac{1}{2}$	W. J. Childs et al. Bull. Amer. Phys. Soc. II 1, 342 (1956)	Atomstrahl			$|\mu| = 0,217$ $\pm 0,002$	
In_{49}^{114}	$<2\cdot10^{-5}$	5	L. S. Goodman, S. Wexler Phys. Rev. 100, 1245 (1955)	Atomstrahl			$+47$	
In_{49}^{115}	95,77	$\frac{9}{2}$	E. Fermi, E. Segrè Z. Physik 82, 729 (1933)	Berechnung aus früheren Daten			$\mu_B/\mu = 350$	
			S. Goudsmit Phys. Rev. 43, 636 (1933)	Berechnung aus früheren Daten			$5,4$	
			H. Schüler Z. Physik 88, 323 (1934)	Berechnung aus früheren Daten			$+5,25$	
			F. Paschen Sitzgsber. preuß. Akad. Wiss. Phys.-Math. Kl. 1935, 430	Hfs.			$6,05$	
			H. A. Bethe, R. F. Bacher Rev. Mod. Phys. 8, 82 (1936)	Berechnung aus früheren Daten			$+5,7$	
			H. Schüler, T. Schmidt Z. Physik 104, 468 (1937)	Hfs.			$+5,3$ $\pm 0,5$	
			S. Millman et al. Phys. Rev. 53, 384 (1938)	Atomstrahl			$6,65$	$6,40$ $\pm 0,20$

Kern	Vork. %	Spin	Zitat	Methode	Substanz	Ergebnisse	μunkorr	μkorr
			T. C. HARDY Phys. Rev. **59**, 686 (1941) Phys. Rev. **60**, 167 (1941)	Atomstrahl			+ 5,43 ± 0,03	
			T. C. HARDY, S. MILLMAN Phys. Rev. **61**, 459 (1942)	Atomstrahl	In		+ 5,49 ± 0,04	
			W. G. PROCTOR, F. C. YU Phys. Rev. **81**, 20 (1951)	Gekreuzte Spulen	a) 0,38 m In(NO$_3$)$_3$ w.L. b) 0,25 m NaCl + 0,5 m MnSO$_4$ w.L.	νIn^{115}/νNa^{23} = 0,828 41 ± 0,00008 νIn^{115}/νIn^{113} = 1,0021 ± 0,0001	5,5088 ± 0,0010	
			Y. TING et al. Phys. Rev. **86**, 618 (1952)	Kernabsorption	a) In(NO$_3$)$_3$ in 30% HNO$_3$ + Mn(NO$_3$)$_2$ in 30% HNO$_3$, b) ScCl$_3$	νIn^{115}/νSc^{45} = 0,902 292 ± 0,000010	5,50945 ± 0,00011	
			Y. TING, D. WILLIAMS Phys. Rev. **89**, 595 (1953)	Pendelrückk.	a) In(NO$_3$)$_3$ + Mn(NO$_3$)$_2$ w.L. b) ScCl$_3$ w.L.	νIn^{115}/νSc^{45} = 0,901 877 ± 0,00005		5,5339 ± 0,0004
			C. SCHWARTZ Phys. Rev. **97**, 380 (1955)			$\mu_3 = (0,31 \pm 0,01)$ $\times 10^{-24} \mu_K$ cm^2		
In_{49}^{116}	< 10^{-5}	5	L. S. GOODMAN, S. WEXLER Phys. Rev. **100**, 1796 (1955)	Atomstrahl			+ 4,4	
			P. B. NUTTER Phil. Mag. (8) **1**, 587 (1956)	Atomstrahl			4,21 ± 0,08	
Sn_{50}^{112}	0,94	0						
Sn_{50}^{114}	0,65	0						
Sn_{50}^{115}	0,33	$\frac{1}{2}$	M. GUREVITCH Phys. Rev. **75**, 767 (1949)	Hfs.			− 0,86	
			W. G. PROCTOR, F. C. YU Phys. Rev. **76**. 1728 (1949) W. G. PROCTOR Phys. Rev. **79**, 35 (1950)	Gekreuzte Spulen	a) 12 m SnCl$_2$ + 0,7 m MnSO$_4$ w.L. b) 0,2 m NaCl + 0,2 m MnSO$_4$ w.L.	νSn^{115}/νNa^{23} = 1,2362 ± 0,0001, I	− 0,9134 ± 0,0002	
Sn_{50}^{116}	14,36	0	K. MURAKAWA Z. Physik **72**, 793 (1931)	Hfs.			≈ 0	
Sn_{50}^{117}	7,51	$\frac{1}{2}$	H. SCHÜLER et al. Naturwiss, **21**, 660 (1933)	Hfs.			$\mu_B/\mu = 2030$	
			S. TOLANSKY Nature, Lond. **132**, 318 (1933) Proc. Roy. Soc. Lond., Ser. A **144**, 574 (1934)	Hfs.			− 0,89	
			W. G. PROCTOR Phys. Rev. **76**, 684 (1949) Phys. Rev. **79**, 35 (1950)	Gekreuzte Spulen	a) 5,3 m SnCl$_2$ + 1 m MnCl$_2$ w.L. b) 0,69 m NaCl + 1 m MnCl$_2$ w.L.	νSn^{117}/νNa^{23} = 1,3468 ± 0,0001, I	− 0,9951 ± 0,0002	− 1,000 ± 0,001
Sn_{50}^{118}	24,21	0	K. MURAKAWA Z. Physik **72**, 793 (1931)	Hfs.			≈ 0	
Sn_{50}^{119}	8,45	$\frac{1}{2}$	H. SCHÜLER et al. Naturwiss. **21**, 660 (1933)	Hfs.			$\mu_B/\mu = 2030$	
			S. TOLANSKY Nature, Lond. **132**, 318 (1933) Proc. Roy. Soc. Lond., Ser. A **144**, 574 (1934)	Hfs.			− 0,89	
			H. SCHÜLER Z. Physik **88**, 323 (1934)	Berechnung aus früheren Daten			− 0,95	
			W. G. PROCTOR Phys. Rev. **76**, 684 (1949) Phys. Rev. **79**, 35 (1950)	Gekreuzte Spulen	a) 5,3 m SnCl$_2$ + 1,0 MnCl$_2$ w.L. b) 0,69 m NaCl + 1,0 m MnCl$_2$ w.L.	νSn^{119}/νNa^{23} = 1,4090 ± 0,0001 νSn^{119}/νSn^{117} = 1,0465 ± 0,0001, I	− 1,0411 ± 0,0002	− 1,047 ± 0,001
Sn_{50}^{120}	33,11	0	K. MURAKAWA Z. Physik **72**, 793 (1931)	Hfs.			≈ 0	

Kern	Vork. %	Spin	Zitat	Methode	Substanz	Ergebnisse	μ_{unkorr}	μ_{korr}
Sn_{50}^{122}	4,61	0						
Sn_{50}^{124}	5,83	0						
Sb_{51}^{121}	57,25	$\frac{5}{2}$	S. Goudsmit Phys. Rev. **43**, 636 (1933)	Berechnung aus früheren Daten			2,7	
			M. F. Crawford et al. Canad. J. Res. **10**, 693 (1934)	Hfs.		g_{Sb121}/g_{Sb123} = 1,84 ± 0,02	4,0	
			H. A. Bethe, R. F. Bacher Rev. Mod. Phys. **8**, 82 (1936)	Berechnung aus früheren Daten			3,7	
			D. H. Tomboulian et al. Phys. Rev. **58**, 52 (1940)	Hfs.		μ_{Sb121}/μ_{Sb123} = 1,316		
			V. W. Cohen et al. AECU-902 (1950) Phys. Rev. **79**, 191 (1950)	Empfindlicher Oszillator	a) HSbCl$_6$ in HCl b) NaCl fest	ν_{Sb121}/ν_{Na23} = 0,90469 ± 0,00004		3,3595 ± 0,0004
			T. L. Collins Phys. Rev. **79**, 226 (1950)	Empfindlicher Oszillator	SbCl$_3$ konz. + NaCl w.L.	ν_{Sb121}/ν_{Na23} = 1,0041 ± 0,0003		3,730 ± 0,002
			W. G. Proctor, F. C. Yu Phys. Rev. **78**, 471 (1950) Phys. Rev. **81**, 20 (1951)	Gekreuzte Spulen	a) NaSbF$_6$ in HF w.L. b) 0,25 m NaCl + 0,5 mMnSO$_4$	ν_{Sb121}/ν_{Na23} = 0,90480 ± 0,00009	3,3427 ± 0,0006	
Sb_{51}^{123}	42,75	$\frac{7}{2}$	S. Goudsmit Phys. Rev. **43**, 636 (1933)	Berechnung aus früheren Daten			2,1	
			M. F. Crawford et al. Canad. J. Res. **10**, 693 (1934)	Hfs.			3,1	
			H. A. Bethe, R. F. Bacher Rev. Mod. Phys. **8**, 82 (1936)	Berechnung aus früheren Daten			2,8	
			V. W. Cohen et al. AECU-902 (1950) Phys. Rev. **79**, 191 (1950)	Empfindlicher Oszillator	a) HSbCl$_6$ in HCl b) D$_2$O	ν_{Sb123}/ν_{D2} = 0,8442 ± 0,0001		2,5470 ± 0,0003
			W. G. Proctor, F. C. Yu Phys. Rev. **78**, 471 (1950) Phys. Rev. **81**, 20 (1951)	Gekreuzte Spulen	a) NaSbF$_6$ in HF w.L. b) 1,8m MnSO$_4$ in 25% D$_2$O	ν_{Sb123}/ν_{D2} = 0,84423 ± 0,00008	2,5341 ± 0,0004	
Te_{52}^{120}	0,09	0						
Te_{52}^{122}	2,43	0						
Te_{52}^{123}	0,85	$\frac{1}{2}$	S. S. Dharmatti et al. Phys. Rev. **84**, 843 (1951) H. E. Weaver jr. Phys. Rev. **89**, 923 (1953)	Gekreuzte Spulen	a) 3,1 m TeO$_2$ in HCl, Te Metall in Königs-wasser b) NaCl	ν_{Te123}/ν_{Na23} = 0,99085 ± 0,00003	− 0,73188 ± 0,00004	
			J. S. Ross, K. Murakawa Phys. Rev. **83**, 229 (1951) Phys. Rev. **85**, 559 (1952)	Hfs.			− 0,6 ± 0,2	
Te_{52}^{124}	4,59	0						
Te_{52}^{125}	6,98	$\frac{1}{2}$	J. E. Mack, O. H. Arroe Phys. Rev. **76**, 1002 (1949)	Hfs.		μ_{Te125}/μ_{Te123} = 1,208 ± 0,06		
			S. S. Dharmatti et al. Phys. Rev. **84**, 843 (1951) H. E. Weaver jr. Phys. Rev. **89**, 923 (1953)	Gekreuzte Spulen	a) 3,1 m TeO$_2$ in HCl, Te Metall in Königs-wasser b) NaCl	ν_{Te125}/ν_{Na23} = 1,19457 ± 0,00004	− 0,88235 ± 0,00004	
			J. S. Ross, K. Murakawa Phys. Rev. **83**, 229 (1951) Phys. Rev. **85**, 559 (1952)	Hfs.		μ_{Te125}/μ_{Te123} = 1,186 ± 0,007	− 0,7 ± 0,2	
Te_{52}^{126}	18,70	0	S. Rafalowski Acta phys. polon. **2**, 119 (1933)	Hfs.			≈ 0	
Te_{52}^{128}	31,85	0	S. Rafalowski Acta phys. polon. **2**, 119 (1933)	Hfs.			≈ 0	

Kern	Vork. %	Spin	Zitat	Methode	Substanz	Ergebnisse	μ_{unkorr}	μ_{korr}
Te^{130}_{52}	34,51	0	S. Rafalowski Acta phys. polon. 2, 119 (1933)	Hfs.			≈ 0	
J^{125}_{53}		$\frac{5}{2}$	P. Fletcher, E. Amble Bull. Amer. Phys. Soc. II 2, 30 (1957)	Mikrowellen	CH_3J^{125}	$I = \frac{5}{2}$		
J^{127}_{53}	100	$\frac{5}{2}$	T. Schmidt Z. Physik 108, 408 (1938)	Berechnung aus früheren Daten			≈ 3	
			T. Schmidt Z. Physik 112, 199 (1939)	Hfs.			2,8	
			R. V. Pound Phys. Rev. 73, 1112 (1948)	Empfindlicher Oszillator	NaJ fest	$\nu J^{127}/\nu Na^{23}$ $= 0,75664$ $\pm 0,0002$		2,8122 $\pm 0,0030$
			W. Gordy et al. Phys. Rev. 76, 443 (1949)	Mikrowellen	CH_3J^{127}		2,792 $\pm 0,062$	2,810 $\pm 0,062$
			J. R. Zimmerman et al. Phys. Rev. 76, 163 (1949) Phys. Rev. 76, 350 (1949)	Pendelrückk.	a) KJ w.L. b) H_2O	$\nu J^{127}/\nu H^1$ $= 0,20003$ $\pm 0,00007$		
			R. E Sheriff, D. Williams Phys. Rev. 82, 651 (1951)	Pendelrückk.	a) KJ w.L. b) $D_2O + NiCl_2$	$\nu J^{127}/\nu D^2$ $= 1,30317$ $\pm 0,00006$		2,80838 $\pm 0,00013$
			H. Walchli et al. Phys. Rev. 82, 97 (1951)	Gekreuzte Spulen	J^- in Hydrazin $+ D_2O$	$\nu J^{127}/\nu D^2$ $= 1,30337$ $\pm 0,0002$		2,8090 $\pm 0,0004$
			E. Yasaitis, B. Smaller Phys. Rev. 82, 750 (1951)	Brücke	a) KJ ges. w.L. b) —	$\nu H^1/\nu J^{127}$ $= 4,99763$ $\pm 0,00015$		
			V. Jaccarino et al. Phys. Rev. 94, 1798 (1954)	Atomstrahl		$\mu_3 = +0,3 \times$ $10^{-24}\mu_K cm^2$		
			C. Schwartz Phys. Rev. 97, 380 (1955)	Berechnung aus früheren Daten		$\mu_3 = (0,17 \pm 0,03)$ $\times 10^{-24} \mu_K cm^2$		
J^{129}_{53}	<0,0025	$\frac{7}{2}$	W. Gordy et al. Phys. Rev. 76, 443 (1949)	Mikrowellen	CH_3J^{129}		2,74 $\pm 0,14$	
			H. Walchli et al. Phys. Rev. 82, 97 (1951)	Gekreuzte Spulen	J^- in Hydrazin $+ D_2O$ L.	$\nu J^{129}/\nu D^2$ $= 0,86744$ $\pm 0,0001$		2,6173 $\pm 0,0003$
J^{131}_{53}	<0,0004	$\frac{7}{2}$	R. Livingston et al. Phys. Rev. 92, 1271 (1953)	Mikrowellen	CH_3J^{127}	$I = \frac{7}{2}$		
Xe^{124}_{54}	0,095	0	H. Kopfermann Naturwiss. 21, 704 (1933)	Hfs.			≈ 0	
Xe^{126}_{54}	0,088	0	H. Kopfermann Naturwiss. 21, 704 (1933)	Hfs.			≈ 0	
Xe^{128}_{54}	1,916	0	H. Kopfermann et al. Z. Physik 87, 460 (1934)	Hfs.			≈ 0	
Xe^{129}_{54}	$26,23_5$	$\frac{1}{2}$	H. Kopfermann Naturwiss. 21, 704 (1933) H. Kopfermann et al. Z. Physik 87, 460 (1934)	Hfs.		$\mu Xe^{129}/\mu Xe^{131}$ $= -1,1$	<0	
			H. A. Bethe, R. F. Bacher Rev. Mod. Phys. 8, 82 (1936)	Berechnung aus früheren Daten			$-0,9$	
			W. G. Proctor, F. C. Yu Phys. Rev. 78, 471 (1950) Phys. Rev. 81, 20 (1951)	Gekreuzte Spulen	a) Xe Gas 12 at $+ Fe_2O_3$Pulver b) 0,25 m NaCl $+0,5$ m $MnSO_4$ w.L.	$\nu Xe^{129}/\nu Na^{23}$ $= 1,0457$ $\pm 0,0001$		$-0,7726$ $\pm 0,0001$
			E. Brun et al. Phys. Rev. 93, 904 (1954)	Gekreuzte Spulen	a) Xe Gas 50 at b) 0,1 m $MnSO_4$ w.L.	$\nu Xe^{129}/\nu H^1$ $= 0,276633$ $\pm 0,000005$		$-0,77255$ $\pm 0,00002$
Xe^{130}_{54}	4,051	0	H. Kopfermann Naturwiss. 21, 704 (1933)	Hfs.				
Xe^{131}_{54}	21,24	$\frac{3}{2}$	H. Kopfermann Naturwiss. 21, 704 (1933) H. Kopfermann et al. Z. Physik 87, 460 (1934)	Hfs.			>0	
			E. G. Jones Proc. Roy. Soc. Lond., Ser. A 144, 587 (1934)	Hfs.		$\mu Xe^{131}/\mu Xe^{129} < 0$		
			H. A. Bethe, R. F. Bacher Rev. Mod. Phys. 8, 82 (1936)	Berechnung aus früheren Daten			$+0,8$	

Kern	Vork. %	Spin	Zitat	Methode	Substanz	Ergebnisse	μ_{unkorr}	μ_{korr}
			A. Bohr et al. Ark. Fysik 4, 455 (1952)	Hfs.		$\nu Xe^{129}/\nu Xe^{131}$ $= -1{,}131$ $\pm 0{,}005$	$+0{,}683$ $\pm 0{,}003$	
			E. Brun et al. Phys. Rev. 93, 904 (1954)	Gekreuzte Spulen	a) Xe Gas 50 at b) 0,1 m MnSO$_4$ w.L.	$\mu Xe^{131}/\mu H^1$ $= 0{,}081976$ $\pm 0{,}000001$	$+0{,}68680$ $\pm 0{,}00002$	
Xe^{132}_{54}	26,92$_5$	0	H. Kopfermann Naturwiss. 21, 704 (1933)	Hfs.			≈ 0	
Xe^{134}_{54}	10,52	0	H. Kopfermann Naturwiss. 21, 704 (1933)	Hfs.			≈ 0	
Xe^{136}_{54}	8,93	0	H. Kopfermann Naturwiss. 21, 704 (1933)	Hfs.			≈ 0	
Cs^{127}_{55}		$\frac{1}{2}$	W. A. Nierenberg et al. Bull. Amer. Phys. Soc. II 1, 343 (1956) Bull. Amer. Phys. Soc. II 2, 30 (1957)	Atomstrahl			$+1{,}41$ $\pm 0{,}04$	
Cs^{129}_{55}		$\frac{1}{2}$	W. A. Nierenberg et al. Bull. Amer. Phys. Soc. II 1, 343 (1956) Bull. Amer. Phys. Soc. II 2, 30, (1957)	Atomstrahl			$+1{,}47$ $\pm 0{,}04$	
Cs^{130}_{55}	<2·10^{-6}	1	W. A. Nierenberg et al. Bull. Amer. Phys. Soc. II 1, 343 (1956)	Atomstrahl		$I = 1$		
Cs^{131}_{55}	<10·10^{-6}	$\frac{5}{2}$	E. H. Bellamy et al. Phil. Mag. (7) 44, 33 (1953)	Atomstrahl	CsCl; (Cs^{131}/Cs^{133} $\approx 3 \cdot 10^{-4}$)		$+3{,}48$ $\pm 0{,}04$	
Cs^{132}_{55}	<30·10^{6}	2	W. A. Nierenberg et al. Bull. Amer. Phys. Soc. II 1, 343 (1956)	Atomstrahl		$I = 2$		
Cs^{133}_{55}	100	$\frac{7}{2}$	E. Fermi, E. Segrè Z. Physik 82, 729 (1933)	Berechnung aus früheren Daten			$\mu_B/\mu = 700$	
			L. P. Granath et al. Phys. Rev. 46, 317 (1934) Phys. Rev. 48, 725 (1935)	Hfs.			$2{,}66 - 3{,}01$ je nach Term	
			N. P. Heydenburg Phys. Rev. 46, 802 (1934)	Hfs.			$2{,}40 - 2{,}50$ je nach Term	
			D. A. Jackson Proc. Roy. Soc. Lond., Ser. A 147, 500 (1934)	Hfs.			$2{,}75$ $\pm 0{,}15$	
			H. Schüler Z. Physik 88, 323 (1934)	Berechnung aus früheren Daten			$+2{,}10$	
			S. Millman et al. Phys. Rev. 51, 1049 (1937)	Atomstrahl			>0	
			P. Kusch et al. Phys. Rev. 55, 1176 (1939)	Atomstrahl	CsF, CsCl, Cs$_2$		$+2{,}555$ $\pm 0{,}013$	$+2{,}572$ $\pm 0{,}013$
			S. Millman, P. Kusch Phys. Rev. 58, 438 (1940)	Atomstrahl			$-2{,}569$	
			S. Millman, P. Kusch Phys. Rev. 60, 91 (1941)	Berechnung aus früheren Daten			$2{,}558$	
			F. Bitter Phys. Rev. 75, 1326 (1949)	Brücke		$\nu Cs^{133}/\nu Li^7$ $= 0{,}33743$ $\pm 0{,}00010$		
			W. H. Chambers et al. Phys. Rev. 76, 461 (1949) Phys. Rev. 76, 638 (1949)	Pendelrückk.	a) CsCl, w.L. b) —	$\nu Cs^{133}/\nu H^1$ $= 0{,}13093$ (korr.) $\pm 0{,}00014$		
			J. R. Zimmerman et al. Phys. Rev. 75, 699 (1949)	Pendelrückk.	Cs-Salz, w.L.		$+2{,}57$	
			R. E. Sheriff et al. Phys. Rev. 79, 175 (1950) Phys. Rev. 82, 651 (1951)	Pendelrückk.	a) CsCl + 0,03 m CuCl$_2$ w.L. b) 0,03 m NiCl$_2$ in D$_2$O	$\nu Cs^{133}/\nu D^2$ $= 0{,}85449$ $\pm 0{,}00004$		$2{,}57877$ $\pm 0{,}00012$
			H. E. Walchli ORNL-1775 (1954)			$\nu Cs^{133}/\nu D^2$ $= 0{,}854496$ $\pm 0{,}000018$	$+2{,}56421$	

Kern	Vork. %	Spin	Zitat	Methode	Substanz	Ergebnisse	μ_{unkorr}	μ_{korr}
Cs_{55}^{134}	$<5\cdot10^{-5}$	4	V. Jaccarino et al. Phys. Rev. 87, 676 (1952)	Atomstrahl	$(Cs^{134})_2 CO_3$, $(Cs^{134}/Cs^{133}$ $\sim 2\cdot 10^{-4}$		$+2,96$ $\pm0,01$	
			E. H. Bellamy et al. Phil. Mag. (7) 44, 33 (1953)	Atomstrahl	CsCl; $(Cs^{134}/Cs^{133}$ $\approx 2\cdot10^{-5})$		$+2,95$ $\pm0,01$	
$*Cs_{55}^{134}$		8	V. W. Cohen et al. Phys. Rev. 95, 569 (1954)	Atomstrahl	CsCl		$1,10$ $\pm0,01$	
			D. A. Gilbert et al. Phys. Rev. 97, 243 (1955) Phys. Rev. 98, 1194 (1955)	Atomstrahl			$+1,10$ $\pm0,01$	
			L. S. Goodmann et al. Phys. Rev. 95, 570 (1954) Phys. Rev. 97, 242 (1955) Phys. Rev. 99, 192 (1955)	Atomstrahl	Cs		$+1,10$ $\pm0,01$	
Cs_{55}^{135}	$<3\cdot10^{-5}$	$\frac{7}{2}$	L. Davis jr. et al. Phys. Rev. 76, 1068 (1949)	Atomstrahl	CsCl		$2,724$ $\pm0,010$	
			D. E. Nagle Phys. Rev. 76, 847 (1949)	Atomstrahl			$2,721$ $\pm0,010$	
Cs_{55}^{137}		$\frac{7}{2}$	L. Davis jr. Phys. Rev. 76, 435 (1949)	Atomstrahl	$CsCl + NaN_3$		$+2,837$ $\pm0,010$	
			L. Davis jr. et al. Phys. Rev. 76, 1068 (1949)	Atomstrahl	$CsCl + NaN_3$		$2,837$ $\pm0,010$	
			D. E. Nagle Phys. Rev. 76, 847 (1949)	Atomstrahl			$2,833$ $\pm0,010$	
Ba_{56}^{130}	0,102	0						
Ba_{56}^{132}	0,098	0						
Ba_{56}^{134}	2,42	0	O. H. Arroe Phys. Rev. 79, 836 (1950)	Hfs.			≈ 0	
Ba_{56}^{135}	6,59	$\frac{3}{2}$	H. A. Bethe, R. F. Bacher Rev. Mod. Phys. 8, 82 (1936)	Berechnung aus früheren Daten			$+1,0$	
			R. H. Hay Phys. Rev. 59, 686 (1941)	Atomstrahl				$+0,837$ $\pm0,003$
			R. H. Hay Phys. Rev. 60, 75 (1941)	Atomstrahl	Ba			$+0,8363$ $\pm0,0026$
			H. E. Walchli et al. Phys. Rev. 102, 1334 (1956)	Gekreuzte Spulen	$BaCl_2$ ges. w.L. (58,5% Ba^{135})	$\nu Ba^{135}/\nu Cl^{35}$ $=1,01387$ $\pm0,00002$	$0,832293$ $\pm0,000025$	
Ba_{56}^{136}	7,81	0	O. H. Arroe Phys. Rev. 79, 836 (1950)	Hfs.			≈ 0	
Ba_{56}^{137}	11,32	$\frac{3}{2}$	E. Fermi, E. Segrè Z. Physik 82, 729 (1933)	Berechnung aus früheren Daten			$\mu_B/\mu = 1750$	
			H. Schüler Z. Physik 88, 323 (1934)	Berechnung aus früheren Daten			$+0,94$	
			H. A. Bethe et al. Rev. Mod. Phys. 8, 82 (1936)	Berechnung aus früheren Daten			$+1,0$	
			R. H. Hay Phys. Rev. 59, 686 (1941)	Atomstrahl				$+0,936$ $\pm0,003$
			R. H. Hay Phys. Rev. 60, 75 (1941)	Atomstrahl	Ba			$+0,9354$ $\pm0,0029$
			H. E. Walchli et al. Phys. Rev. 102, 1334 (1956)	Gekreuzte Spulen	$BaCl_2$ ges. w.L. (43,5% Ba^{137})	$\nu Ba^{137}/\nu Cl^{35}$ $=1,13420$ $\pm0,00005$	$0,931074$ $\pm0,000055$	
Ba_{56}^{138}	71,66	0	O. H. Arroe Phys. Rev. 79, 836 (1950)	Hfs.			≈ 0	
La_{57}^{138}	0,089	5	P. B. Sogo, C. D. Jeffries Phys. Rev. 99, 613 (1955)	Gekreuzte Spulen		$\nu La^{138}/\nu La^{139}$ $=0,93407$ $\pm0,00003, I$	$+3,6844$ $\pm0,0004$	
La_{57}^{139}	99,911	$\frac{7}{2}$	O. E. Anderson Phys. Rev. 46, 473 (1934)	Hfs.			$|\mu| = 2,5$	
			M. F. Crawford et al. Phys. Rev. 47, 536 (1935)	Hfs.			$2,8$	
			M. F. Crawford Phys. Rev. 47, 768 (1935)	Berechnung aus früheren Daten			$2,5$	

Kern	Vork. %	Spin	Zitat	Methode	Substanz	Ergebnisse	μ_{unkorr}	μ_{korr}
			H. Wittke Z. Physik **116**, 547 (1940)	Hfs.			2,76	
			W. H. Chambers et al. Phys. Rev. **76**, 461 (1949) Phys. Rev. **76**, 638 (1949)	Pendelrückk.	a) $LaCl_3$ w. L. b) —	$\nu_{La^{139}}/\nu_{H^1}$ $= 0,14116$ (korr.) $\pm 0,00014$		
			W. C. Dickinson Phys. Rev. **76**, 1414 (1949)	Brücke	$LaCl_3$ w. L.	$\nu_{La^{139}}/\nu_{H^1}$ $= 0,141251$ $\pm 0,000014$		
			R. E. Sheriff et al. Phys. Rev. **82**, 651 (1951)	Pendelrückk.	a) $LaCl_3 + 0,03$ m $CuCl_2$ w. L. b) 0,03 m $NiCl_2$ in D_2O	$\nu_{La^{139}}/\nu_{D^2}$ $= 0,92025$ $\pm 0,00006$		2,77802 $\pm 0,00018$
Ce_{58}^{136}	0,19	0						
Ce_{58}^{138}	0,25	0						
Ce_{58}^{140}	88,49	0						
Ce_{58}^{141}	< 0,02	$\frac{7}{2}$	E. Ambler et al. Phys. Rev. **97**, 1212 (1955)	Winkelverteilung der γ-Strahlung	Ce-Mg-Nitrat-Einkristall		0,16 $\pm 0,06$	
			C. F. M. Cacho et al. Phil. Mag. (7) **46**, 1287 (1955)	Winkelverteilung der γ-Strahlung	Nd-Äthyl-Sulfat-Kristall mit 1% Ce-Äthyl-Sulfat		0,75 0,66 $\pm 0,20$ $\pm 0,16$ für für $\Delta I = 1$ $\Delta I = 0$	
Ce_{58}^{142}	11,07	0						
Pr_{59}^{141}	100	$\frac{5}{2}$	P. Brix Phys. Rev. **89**, 1245 (1953)	Hfs.			$+ 3,9$ $\pm 0,3$	
			H. Lew Phys. Rev. **89**, 530 (1953) Phys. Rev. **91**, 619 (1953)	Atomstrahl	a) $Pr + ThO_2$ b) $CsCl + Na$		$+ 3,8$ $\pm 0,4$	
			K. Murakawa J. Phys. Soc. Japan **9**, 93 (1954)	Hfs.			$+ 4,0$ $\pm 0,1$	
			J. M. Baker, B. Bleaney Proc. Phys. Soc. Lond. A **68**, 936 (1955)	Paramagnetische Resonanz	Y-Äthyl-Sulfat-Kristall mit Pr-Äthyl-Sulfat		3,92 $\pm 0,2$	
Nd_{60}^{142}	26,80	0						
Nd_{60}^{143}	12,12	$\frac{7}{2}$	B. Bleaney et al Proc. Phys. Soc. Lond. A **63**, 1369 (1950) Proc. Roy. Soc. Lond., Ser. A **223**, 15 (1954)	Paramagnetische Resonanz	La-Äthyl-Sulfat-Kristall mit Nd-Äthyl-Sulfat	$\mu_{Nd^{143}}/\mu_{Nd^{145}}$ $= 1,6083$ $\pm 0,0012$		
			K. Murakawa, J. S. Ross Phys. Rev. **82**, 967 (1951)	Hfs.		$\mu_{Nd^{143}}/\mu_{Nd^{145}}$ $= 1,60$ $\pm 0,06$	$- 1,0$ $\pm 0,2$	
			R. J. Elliott et al. Proc. Roy. Soc. Lond., Ser. A **219**, 387 (1953)	Berechnung aus früheren Daten			$|\mu| = 1,0$ $\pm 0,25$	
			K. Murakawa Phys. Rev. **96**, 1543 (1954)	Hfs.		$\mu_{Nd^{143}}/\mu_{Nd^{145}}$ $= 1,60$ $\pm 0,06$	$- 1,1$ $\pm 0,1$	
			B. Bleaney Proc. Phys. Soc. Lond. A **68**, 937 (1955)	Berechnung aus früheren Daten			1,03	
Nd_{60}^{144}	23,91	0						
Nd_{60}^{145}	8,35	$\frac{7}{2}$	K. Murakawa, J. S. Ross Phys. Rev. **82**, 967 (1951)	Hfs.			$- 0,62$ $\pm 0,09$	
			R. J. Elliott et al. Proc. Roy. Soc. Lond., Ser. A **219**, 387 (1953)	Berechnung aus früheren Daten			$|\mu| = 0,62$ $\pm 0,16$	
			K. Murakawa Phys. Rev. **96**, 1543 (1954)	Hfs.			$- 0,69$ $\pm 0,10$	
			B. Bleaney Proc. Phys. Soc. Lond. A **68**, 937 (1955)	Berechnung aus früheren Daten			0,64	

Kern	Vork. %	Spin	Zitat	Methode	Substanz	Ergebnisse	μ_{unkorr}	μ_{korr}
Nd_{60}^{146}	17,35	0						
Nd_{60}^{147}	<0,0002	$\frac{9}{2}$	E. AMBLER, R.P. HUDSON Phys. Rev. 97, 1212 (1955)	Winkelverteilung der γ-Strahlung	Nd-Mg-Nitrat-Einkristall		0,22 ±0,05	
Nd_{60}^{148}	5,78	0						
Nd_{60}^{150}	5,69	0						
Pm_{61}								
Sm_{62}^{144}	2,95	0						
Sm_{62}^{147}	14,62	$\frac{7}{2}$	K. MURAKAWA, J. S. ROSS Phys. Rev. 82, 967 (1951)	Hfs.		$\mu_{Sm^{147}}/\mu_{Sm^{149}}$ =1,177 ±0,015		−0,30 ±0,05 $(I=\frac{5}{2})$
			G. S. BOGLE, H. E. D SCOVIL Proc. Phys. Soc. Lond. A 65, 368 (1952)	Paramagnetische Resonanz	Sm-Äthyl-Sulfat-Kristall	$\mu_{Sm^{147}}/\mu_{Sm^{149}}$ =1,222 ±0,008		
			R. J. ELLIOTT et al. Proc. Roy. Soc. Lond., Ser. A 219, 387 (1953)	Berechnung aus früheren Daten				−0,83 ±0,15
			K. MURAKAWA Phys. Rev. 93, 1232 (1954)	Hfs.	Sm^{147}, 81,63% anger.	$\mu_{Sm^{147}}/\mu_{Sm^{149}}$ =1,198 ±0,015		−0,76 ±0,08
Sm_{62}^{148}	10,97	0						
Sm_{62}^{149}	13,56	$\frac{7}{2}$	K. MURAKAWA, J. S. ROSS Phys. Rev. 82, 967 (1951)	Hfs.				−0,25 ±0,04 $(I=\frac{5}{2})$
			R. J. ELLIOTT et al. Proc. Roy. Soc. Lond., Ser. A 219, 387 (1953)	Berechnung aus früheren Daten				−0,68 ±0,10
			K. MURAKAWA Phys. Rev. 93, 1232 (1954)	Hfs.	Sm^{149}, 71,53% anger.			−0,64 ±0,06
Sm_{62}^{150}	7,27	0						
Sm_{62}^{152}	27,34	0						
Sm_{62}^{154}	23,29	0						
Eu_{63}^{151}	47,77	$\frac{5}{2}$	H. SCHÜLER, T. SCHMIDT Z. Physik 94, 457 (1935)	Hfs.		$\mu_{Eu^{151}}/\mu_{Eu^{153}}$ =2,2		
			T. SCHMIDT Z. Physik 108, 408 (1938)	Hfs.				3,4
			B. BLEANEY, W. LOW Proc. Phys. Soc. Lond. A 68, 55 (1955)	Paramagnetische Resonanz	SrS-Pulver mit Eu, Sm in Spuren	$\mu_{Eu^{151}}/\mu_{Eu^{153}}$ =2,23_5 ±0,03		
Eu_{63}^{153}	52,23	$\frac{5}{2}$	T. SCHMIDT Z. Physik 108, 408 (1938)	Hfs.				1,5
Eu_{63}^{154}	<0,0002	3	R. W. KEDZIC et al. Bull. Amer. Phys. Soc. II 1, 390 (1956)	Paramagnetische Resonanz	KCl-Pulver mit Eu^{2+}-Ionen	$\mu_{Eu^{154}}/\mu_{Eu^{153}}$ =1,308 ±0,004	$|\mu|=2,0$	
Gd_{64}^{152}	0,20	0						
Gd_{64}^{154}	2,15	0						
Gd_{64}^{155}	14,78	$\frac{3}{2}$	S. SUWA Phys. Rev. 86, 247 (1952)	Hfs.			$|\mu|=0,25$ ±0,15	
			K. MURAKAWA Phys. Rev. 96, 1543 (1954)	Hfs.				−0,19 $(I=\frac{7}{2})$ ±0,05
			F.A. JENKINS, D.R. SPECK Phys. Rev. 100, 973 (1955)	Hfs.				−0,31
			W. LOW Phys. Rev. 103, 1309 (1956)	Paramagnetische Resonanz	$(Bi_2Mg_3)(NO_3)_{12}\times$ $24 H_2O + 0,02\%$ Gd Einkristall (u. a.)	$\mu_{Gd^{155}}/\mu_{Gd^{157}}$ =0,75 ±0,07		0,24

Kern	Vork. %	Spin	Zitat	Methode	Substanz	Ergebnisse	μ_{unkorr}	μ_{korr}
			D. R. Speck Phys. Rev. 101, 1725 (1956)	Hfs.	Gd^{155}, 72,3% anger.	μGd^{155}/μGd^{157} = 0,80 ± 0,02	− 0,30	
Gd$_{64}^{156}$	20,59	0						
Gd$_{64}^{157}$	15,71	$\frac{3}{2}$	S. Suwa Phys. Rev. 86, 247 (1952)	Hfs.			$|\mu| = 0,3$ ± 0,2	
			K. Murakawa Phys. Rev. 96, 1543 (1954)	Hfs.			− 0,33 ($I = \frac{7}{2}$) ± 0,06	
			F. A. Jenkins, D. R. Speck Phys. Rev. 100, 973 (1955)	Hfs.			− 0,38	
			W. Low Phys. Rev. 103, 1309 (1956)	Paramagnetische Resonanz	(Bi$_2$Mg$_3$)(NO$_3$)$_{12}$× 24 H$_2$O + 0,02% Gd Einkristall (u. a.)	μEu^{151}/μGd^{157} = 11,3	0,32	
			D. R. Speck Phys. Rev. 101, 1725 (1956)	Hfs.	Gd^{157} 69,7% anger.		− 0,37 ± 0,04	
Gd$_{64}^{158}$	24,78	0						
Gd$_{64}^{160}$	21,79	0						
Tb$_{65}^{159}$	100	$\frac{3}{2}$	J. M. Baker, B. Bleaney Proc. Phys. Soc. Lond. A 68, 257 (1955)	Paramagnetische Resonanz	Y-Äthyl-Sulfat-Kristall mit 0,1% Tb-Äthyl-Sulfat		$|\mu| = 1,5$ ± 0,4	
			B. Bleaney Proc. Phys. Soc. Lond. A 68, 937 (1955)	Berechnung aus früheren Daten			1,52	
Dy$_{66}^{156}$	0,05	0						
Dy$_{66}^{158}$	0,05	0						
Dy$_{66}^{160}$	0,1	0						
Dy$_{66}^{161}$	21,1	$(\frac{5}{2}, \frac{3}{2})$	K. Murakawa et al. Phys. Rev. 92, 325 (1953)	Hfs.		μDy^{161}/μDy^{163} ≈ 1		
			A. H. Cooke, J. G. Park Proc. Phys. Soc. Lond. A 69, 282 (1956)	Paramagnetische Resonanz	Y-Azetat-Kristall mit 1,5% Dy-Azetat		$|\mu| = 0,38$ ± 0,05	
			K. Murakawa J. Phys. Soc. Japan 11, 804 (1956)	Hfs.				
Dy$_{66}^{162}$	26,6	0						
Dy$_{66}^{163}$	24,8	$(\frac{5}{2}, \frac{3}{2})$	K. Murakawa, T. Kamei Phys. Rev. 92, 325 (1953)	Hfs.		μDy^{163}/μDy^{161} ≈ 1		
			A. H. Cooke, J. G. Park Proc. Phys. Soc. Lond. A 69, 282 (1956)	Paramagnetische Resonanz	Y-Azetat-Kristall mit 1,5% Dy-Azetat	μDy^{163}/μDy^{161} = 1,41 ± 0,02	$|\mu| = 0,53$ ± 0,05	
			K. Murakawa J. Phys. Soc. Japan 11, 804 (1956)	Hfs.		μDy^{163}/μDy^{161} = 1,07 ± 0,05		
D$_{66}^{164}$	27,3	0						
Ho$_{67}^{165}$	100	$\frac{7}{2}$	J. M. Baker, B. Bleaney Proc. Phys. Soc. Lond. A 68, 1090 (1955)	Paramagnetische Resonanz	Y-Äthyl-Sulfat-Kristall mit 1% Äthyl-Sulfat		3,29 ± 0,17	
			B. Bleaney Proc. Phys. Soc. Lond. A 68, 937 (1955)	Berechnung aus früheren Daten			3,31	
Er$_{68}^{162}$	0,1	0						
Er$_{68}^{164}$	1,5	0						
Er$_{68}^{166}$	32,9	0						

Kern	Vork. %	Spin	Zitat	Methode	Substanz	Ergebnisse	μ_{unkorr}	μ_{korr}
Er_{68}^{167}	24,4	$\frac{7}{2}$	R. J. ELLIOTT et al. Proc. Roy. Soc. Lond., Ser. A **219**, 387 (1953)	Berechnung aus früheren Daten			$\|\mu\| = 0,50$ $\pm 0,12$	
			B. BLEANEY Proc. Phys. Soc. Lond. A **68**, 937 (1955)	Berechnung aus früheren Daten			0,48	
Er_{68}^{168}	26,9	0						
Er_{68}^{170}	14,2	0						
Tm_{69}^{169}	100	$\frac{1}{2}$	H. SCHÜLER, T. SCHMIDT Naturwiss. **22**, 838 (1934)	Hfs.			sehr klein	
			K. H. LINDENBERGER et al. Naturwiss. **42**, 41 (1955)	Hfs.			$-0,20$ $\pm 0,03$	
			K. H. LINDENBERGER Z. Physik **141**, 476 (1955)	Hfs.			$-0,20_5$ $\pm 0,02$	
Yb_{70}^{168}	0,06	0						
Yb_{70}^{170}	4,21	0						
Yb_{70}^{171}	14,26	$\frac{1}{2}$	H. SCHÜLER, H. KORSCHING Z. Physik **111**, 386 (1938)	Hfs.			$+0,45$	
			B. BLEANEY Proc. Phys. Soc. Lond. A **68**, 937 (1955)	Berechnung aus früheren Daten			0,49	
			K. KREBS, H. NELKOWSKI Z. Physik **141**, 254 (1955)	Hfs.			$+0,49$ $\pm 0,06$	
			A. H. COOKE, J. G. PARK Proc. Phys. Soc. Lond. A **69**, 282 (1956)	Paramagnetische Resonanz	Y-Azetat-Kristall mit 1% Yb-Azetat		$\|\mu\| = 0,43$ $\pm 0,05$	
Yb_{70}^{172}	21,49	0						
Yb_{70}^{173}	17,02	$\frac{5}{2}$	H. SCHÜLER, H. KORSCHING Z. Physik **111**, 386 (1938)	Hfs.		μ_{Yb173}/μ_{Yb171} $= 1,4$	$-0,65$	
			B. BLEANEY Proc. Phys. Soc. Lond. A **68**, 937 (1955)	Berechnung aus früheren Daten			0,68	
			K. KREBS, H. NELKOWSKI Z. Physik **141**, 254 (1955)	Hfs.			$-0,67$ $\pm 0,01$	
			A. H. COOKE, J. G. PARK Proc. Phys. Soc. Lond. A **69**, 282 (1956)	Paramagnetische Resonanz	Y-Azetat-Kristall mit 1% Yb-Azetat	μ_{Yb173}/μ_{Yb171} $= 1,39$ $\pm 0,01$	$\|\mu\| = 0,60$ $\pm 0,05$	
Yb_{70}^{174}	29,58	0						
Yb_{70}^{176}	13,38	0						
Cp_{71}^{175}	97,5	$\frac{7}{2}$	H. SCHÜLER, T. SCHMIDT Z. Physik **95**, 265 (1935) Z. Physik **98**, 430 (1935)	Hfs.			1,7	
			H. GOLLNOW Z. Physik **103**, 443 (1936)	Hfs.			$\approx 2,6$ $\pm 0,5$	
Cp_{71}^{176}	2,5	(≥ 7)	H. SCHÜLER, H. GOLLNOW Z. Physik **113**, 1 (1939)	Hfs.			$\approx 3,8$ $\pm 0,7$	
Hf_{72}^{174}	0,18	0						
Hf_{72}^{176}	5,30	0						
Hf_{72}^{177}	18,47	$\frac{7}{2}$	D. R. SPECK, F. A. JENKINS Phys. Rev. **101**, 1831 (1956) D. R. SPECK Bull. Amer. Phys. Soc. II **1**, 282 (1956)	Hfs.		μ_{Hf177}/μ_{Hf179} $= 1,276$ $\pm 0,008$	$+0,61$ $\pm 0,03$	
Hf_{72}^{178}	27,10	0	E. RASMUSSEN Naturwiss. **23**, 69 (1935)	Hfs.			≈ 0	

Kern	Vork. %	Spin	Zitat	Methode	Substanz	Ergebnisse	μ_{unkorr}	μ_{korr}
Hf_{72}^{179}	13,84	$\frac{9}{2}$	D. R. SPECK Bull. Amer. Phys. Soc. II 1, 282 (1956)	Hfs.			$-0,47$ $\pm 0,03$	
Hf_{72}^{180}	35,11	0	E. RASMUSSEN Naturwiss. 23, 69 (1935)	Hfs.			≈ 0	
Ta_{73}^{181}	100	$\frac{7}{2}$	J. H. GISOLF Diss. Amsterdam 1935					
			B. M. BROWN et al. Phys. Rev. 88, 1158 (1952)	Hfs.			1,9	2,1
$*Ta_{73}^{181}$		$\frac{5}{2}$	S. RABOY, V. E. KROHN Phys. Rev. 95, 1689 (1954)	Winkelverteilung der γ-Strahlung	Bestrahltes Hf in HF		3,00 $\pm 0,30$	
			E. HEER et al. Helv. phys. Acta 28, 336 (1955)	Winkelverteilung der γ-Strahlung	Bestrahltes HfF_4 in HF		3,1	
			E. HEER et al. Z. Naturforsch. 10a, 834 (1955)	Winkelverteilung der γ-Strahlung	Bestrahltes HfF_4 in HF		3,25 $\pm 0,17$	
W_{74}^{180}	0,16	0						
W_{74}^{182}	26,35	0	N. S. GRACE, K. R. MORE Phys. Rev. 45, 166 (1934)	Hfs.				
W_{74}^{183}	14,32	$\frac{1}{2}$	J. A. VREELAND et al. Phys. Rev. 83, 229 (1951) Bull. Amer. Phys. Soc. 26, 38 (1951)	Hfs.			$+0,080$ $\pm 0,02$	
			P. B. SOGO, C. D. JEFFRIES Phys. Rev. 98, 265 (1955) Phys. Rev. 98, 1316 (1955)	Gekreuzte Spulen	a) W^{183} in Pulverform b) 1m $MnCl_2$ in D_2O	$\nu W^{183}/\nu D^2$ $=0,27395$ $\pm 0,00003$	$+0,115$ $\pm 0,001$	
W_{74}^{184}	30,68	0	N. S. GRACE, K. R. MORE Phys. Rev. 45, 166 (1934)	Hfs.				
W_{74}^{186}	28,49	0	N. S. GRACE, K. R. MORE Phys. Rev. 45, 166 (1934)	Hfs.				
Re_{75}^{185}	37,07	$\frac{5}{2}$	T. SCHMIDT Z. Physik 108, 408 (1938)	Berechnung aus früheren Daten			3,3	
			F. ALDER, F. C. YU Phys. Rev. 82, 105 (1951)	Gekreuzte Spulen	a) $NaReO_4$ w.L. b) 0,25m NaCl $+1$m $MnSO_4$ w.L.	$\nu Re^{185}/\nu Na^{23}$ $=0,85114$ $\pm 0,00009$	$+3,1433$ $\pm 0,0006$	
Re_{75}^{187}	62,93	$\frac{3}{2}$	H. SCHÜLER, H. KORSCHING Z. Physik 105, 168 (1937)	Hfs.		$\mu Re^{187}/\mu Re^{185}$ $=1,01069$ $\pm 0,00043$ $(\lambda=4889)$ $\mu Re^{187}/\mu Re^{185}$ $=1,01140$ $\pm 0,00077$ $(\lambda=5275)$		
			T. SCHMIDT Z. Physik 108, 408 (1938)	Berechnung aus früheren Daten			3,3	
			F. ALDER, F. C. YU Phys. Rev. 82, 105 (1951)	Gekreuzte Spulen	a) $NaReO_4$, w.L. b) 0,25m NaCl $+1$m $MnSO_4$ w.L.	$\nu Re^{187}/\nu Re^{185}$ $=1,01026$ $\pm 0,00008$ $\nu Re^{187}/\nu Na^{23}$ $=0,85987$ $\pm 0,00009$	$+3,1755$ $\pm 0,0006$	
Os_{76}^{184}	0,018	0						
Os_{76}^{186}	1,582	0						
Os_{76}^{187}	1,64	$\frac{1}{2}$	K. MURAKAWA Phys. Rev. 98, 1285 (1955)	Hfs.			$+0,12+0,04$ $-0,03$	
Os_{76}^{188}	13,27	0						
Os_{76}^{189}	16,14	$\frac{3}{2}$	S. SUWA Phys. Rev. 83, 1258 (1951)	Hfs.			$+0,6$ $\pm 0,1$ $(I=\frac{1}{2})$	
			K. MURAKAWA, S. SUWA Phys. Rev. 87, 1048 (1952)	Hfs.			$+0,70$ $\pm 0,09$	

24*

Kern	Vork. %	Spin	Zitat	Methode	Substanz	Ergebnisse	μ_{unkorr}	μ_{korr}
			H. R. Löliger, L R. Sarles Phys. Rev. 95, 291 (1954)	Gekreuzte Spulen	a) OsO$_4$-Schmelze b) TiCl$_4$	$\nu Os^{189}/\nu Cl^{35}$ = 0,791 896 ± 0,000 093, I	+ 0,650 655 ± 0,000 081	
Os_{76}^{190}	26,38	0						
Os_{76}^{192}	40,97	0						
Ir_{77}^{191}	38,5	$\frac{3}{2}$	B. Venkatesacher et al. Nature, Lond. 136, 437 (1935) Proc. Ind. Acad. Sci. A 2, 203 (1935)			$\mu_{Ir^{191}}/\mu_{Ir^{193}}$ ≈ −1,0		
			L. Sibaiya Phys. Rev. 56, 768 (1939)	Hfs.		$\mu_{Ir^{191}}/\mu_{Ir^{193}}$ = −0,92 ($I_{Ir^{191}} = \frac{1}{2}$)		
			P. Brix et al. Naturwiss. 37, 397 (1950)	Hfs.			>0 klein	
Ir_{77}^{193}	61,5	$\frac{3}{2}$	P. Brix et al. Naturwiss. 37, 397 (1950)	Hfs.			>0 klein	
			K. Murakawa, S. Suwa Phys. Rev. 87, 1048 (1952)	Hfs.		$\mu_{Ir^{193}}/\mu_{Ir^{191}}$ = 1,04 ± 0,04	0,17 ± 0,03	
			W. v. Siemens Ann. Phys. 13, 136 (1953)	Hfs.		$\mu_{Ir^{193}}/\mu_{Ir^{191}}$ = 1,0 ± 0,1	+ 0,2 ± 0,1	
Pt_{78}^{192}	0,8	0						
Pt_{78}^{194}	30,2	0	B. Fuchs, H. Kopfermann Naturwiss. 23, 372 (1935)	Hfs.			≈ 0	
Pt_{78}^{195}	35,2	$\frac{1}{2}$	T. Schmidt Z. Physik 101, 486 (1936)	Berechnung aus früheren Daten			+ 0,6	
			W. G. Proctor, F. C. Yu Phys. Rev. 76, 1728 (1949) Phys. Rev. 81, 20 (1951)	Gekreuzte Spulen	a) 1 m H$_2$PtCl$_6$ + 0,5 m MnCl$_2$ w. L. b) 0,2 m NaCl + 0,2 m MnSO$_4$ w. L.	$\nu Pt^{195}/\nu Na^{23}$ = 0,812 73 ± 0,000 08, I	+ 0,6005 ± 0,0001	
Pt_{78}^{196}	26,6	0	B. Fuchs, H. Kopfermann Naturwiss. 23, 372 (1935)	Hfs.			≈ 0	
Pt_{78}^{198}	7,2	0						
Au_{79}^{197}	100	$\frac{3}{2}$	E. Fermi, E. Segrè Z. Physik 82, 729 (1933)	Berechnung aus früheren Daten			μ_B/μ = 1010	
			H. Schüler Z. Physik 88, 323 (1934)	Berechnung aus früheren Daten			+ 0,23	
			H. A. Bethe, R. F. Bacher Rev. Mod. Phys. 8, 82 (1936)	Berechnung aus früheren Daten			+ 0,3	
			R. M. Elliott, J. Wulff Phys. Rev. 55, 170 (1939)	Hfs.			0,195 ± 0,004	
			F. M. Kelly Proc. Phys. Soc. Lond. A 65, 250 (1952)	Hfs.			0,134 ± 0,008	0,136 ± 0,008
			W. v. Siemens Ann. Phys. 13, 159 (1953)	Hfs.			0,14 ± 0,02	
			G. Wessel, H. Lew Phys. Rev. 92, 641 (1953)	Atomstrahl	Au + ThO$_2$		0,13 ± 0,01	
Au_{79}^{198}		2	R. L. Christensen et al. Phys. Rev. 101, 1389 (1956)	Atomstrahl	Au^{198}		$|\mu|$ = 0,50 ± 0,04	
Au_{79}^{199}	< 0,01	$\frac{3}{2}$	R. L. Christensen et al. Phys. Rev. 101, 1389 (1956)	Atomstrahl	Au^{199}		$|\mu|$ = 0,24 ± 0,02	
Hg_{80}^{196}	0,15	0	S. Tolansky Proc. Roy. Soc. Lond., Ser. A 130, 558 (1931)	Hfs.			≈ 0	
Hg_{80}^{197}	<0,0037	$\frac{1}{2}$	F. Bitter et al. Phys. Rev. 96, 1531 (1954)	Hfs.		$\mu_{Hg^{197}}/\mu_{Hg^{199}}$ = 1,033 ± 0,016	0,52	

Kern	Vork. %	Spin	Zitat	Methode	Substanz	Ergebnisse	μ_{unkorr}	μ_{korr}
Hg_{80}^{198}	10,12	0	S. TOLANSKY Proc. Roy. Soc. Lond., Ser. A **130**, 558 (1931)	Hfs.			≈ 0	
Hg_{80}^{199}	17,04	$\frac{1}{2}$	E. FERMI, E. SEGRÈ Z. Physik **82**, 729 (1933)	Berechnung aus früheren Daten			$\mu_B/\mu = 4000$	
			S. GOUDSMIT Phys. Rev. **43**, 636 (1933)	Berechnung aus früheren Daten			0,55	
			H. A. BETHE, R. F. BACHER Rev. Mod. Phys. **8**, 82 (1936)	Berechnung aus früheren Daten			$+0,5$	
			H. SCHÜLER, T. SCHMIDT Z. Physik **98**, 239 (1936)	Hfs.		$\mu Hg^{199}/\mu Hg^{201}$ $= -0,9018$		
			S. MROZOWSKI Phys. Rev. **57**, 207 (1940)	Hfs.			0,547 $\pm 0,002$	
			W. G. PROCTOR, F. C. YU Phys. Rev. **76**, 1728 (1949) Phys. Rev. **81**, 20 (1951)	Gekreuzte Spulen	a) 3,5 m $Hg_2(NO_3)_2$ $+0,2$ m $Mn(NO_3)_2$ in verd. HNO_3 b) 1,8m $MnSO_4$ in 25% D_2O	$\nu Hg^{199}/\nu D2$ $= 1,1647$ $\pm 0,0001$, I	$+0,4994$ $\pm 0,0001$	
Hg_{80}^{200}	23,25	0	S. TOLANSKY Proc. Roy. Soc. Lond., Ser. A **130**, 558 (1931)	Hfs.			≈ 0	
Hg_{80}^{201}	13,18	$\frac{3}{2}$	E. FERMI, E. SEGRÈ Z. Physik **82**, 729 (1933)	Berechnung aus früheren Daten			$\mu_B/\mu = -3600$	
			S. GOUDSMIT Phys. Rev. **43**, 636 (1933)	Berechnung aus früheren Daten			$-0,62$	
			H. A. BETHE, R. F. BACHER Rev. Mod. Phys. **8**, 82 (1936)	Berechnung aus früheren Daten			$-0,6$	
			S. MROZOWSKI Phys. Rev. **57**, 207 (1940)	Hfs.			$-0,607$ $\pm 0,003$	
Hg_{80}^{202}	29,45	0	S. TOLANSKY Proc. Roy. Soc. Lond., Ser. A **130**, 558 (1931)	Hfs.			≈ 0	
Hg_{80}^{204}	6,72	0	S. TOLANSKY Proc. Roy. Soc. Lond., Ser. A **130**, 558 (1931)	Hfs.			≈ 0	
Tl_{81}^{198}		7	G. O. BRINK et al. Bull. Amer. Phys. Soc. II **1**, 343 (1956)	Atomstrahl				
Tl_{81}^{199}		$\frac{1}{2}$	G. O. BRINK et al. Bull. Amer. Phys. Soc. II **1**, 343 (1956)	Atomstrahl				
Tl_{81}^{203}	29,46	$\frac{1}{2}$	J. C. MCLENNAN et al. Nature, Lond. **128**, 301 (1931)	Hfs.		$gTl/gPb^{207} \sim 4$		
			J. C. MCLENNAN et al. Proc. Roy Soc. Lond., Ser. A **133**, 652 (1931)	Hfs.		$gTl/gBi \approx 3-4$ $gTl/gPb \approx 3,7-5$		
			E. FERMI, E. SEGRÈ Z. Physik **82**, 729 (1933)	Berechnung aus früheren Daten			$\mu_B/\mu = 1300$	
			S. GOUDSMIT Phys. Rev. **43**, 636 (1933)	Berechnung aus früheren Daten			1,8	
			H. SCHÜLER Z. Physik **88**, 323 (1934)	Berechnung aus früheren Daten			$+1,47$	
			L. A. WILLS Phys. Rev. **45**, 883 (1934)	Berechnung aus früheren Daten			$1,35-1,45$	
			H. A. BETHE et al. Rev. Mod. Phys. **8**, 82 (1936)	Berechnung aus früheren Daten			$+1,4$	
			H. SCHÜLER, T. SCHMIDT Z. Physik **104**, 468 (1937)	Berechnung aus früheren Daten			$+1,45$ $\pm 0,1$	
			H. L. POSS Phys. Rev. **75**, 600 (1949)	Brücke	a) $Tl(C_2H_3O_2)$ ges. w.L. b) —	$\nu Tl^{203}/\nu H^1$ $= 0,571499$ $\pm 0,00005$		1,612
			W. G. PROCTOR, F. C. YU Phys. Rev. **75**, 522 (1949) Phys. Rev. **79**, 35 (1950)	Gekreuzte Spulen	2,6 m Tl $(C_2H_3O_2)$ $+0,03$ m $MnSO_4$ w.L.	$\nu Tl^{203}/\nu H^1$ $= 0,5714$ $\pm 0,0001$ $\nu Tl^{203}/\nu Tl^{205}$ $= 0,9903$ $\pm 0,0002$	$+1,5962$ $\pm 0,0003$	$+1,614$ $\pm 0,003$

Kern	Vork. %	Spin	Zitat	Methode	Substanz	Ergebnisse	μ_{unkorr}	μ_{korr}
			R. E. SHERIFF et al. Phys. Rev. 82, 651 (1951)	Pendelrückk.	$Tl(C_2H_3O_2)$ $+ Mn(C_2H_3O_2)_2$ w. L.	$\nu Tl205/\nu Tl203$ $= 1{,}00983$ $\pm 0{,}00005$		$1{,}61136$ $\pm 0{,}00011$
			A. LURIO, A. G. PRODELL Phys. Rev. 101, 79 (1956)	Atomstrahl	Tl	$\Delta\nu Tl203/\Delta\nu Tl205$ $= 0{,}9903622$ $\pm 0{,}0000005$		
Tl_{81}^{204}	<0,003	2	G. O. BRINK et al. Bull. Amer. Phys. Soc. II 1, 343 (1956)	Atomstrahl				
Tl_{81}^{205}	70,54	$\frac{1}{2}$	E. FERMI, E. SEGRÈ Z. Physik 82, 729 (1933)	Berechnung aus früheren Daten			$\mu_B/\mu = 1300$	
			S. GOUDSMIT Phys. Rev. 43, 636 (1933)	Berechnung aus früheren Daten			1,8	
			H. SCHÜLER Z. Physik 88, 323 (1934)	Berechnung aus früheren Daten			$+ 1{,}47$	
			L. A. WILLS Phys. Rev. 45, 883 (1934)				$1{,}35 - 1{,}45$	
			H. A. BETHE et al. Rev. Mod. Phys. 8, 82 (1936)	Berechnung aus früheren Daten			$+ 1{,}4$	
			H. SCHÜLER et al. Z. Physik 104, 468 (1937)	Berechnung aus früheren Daten			$1{,}45$ $\pm 0{,}1$	
			H. SCHÜLER et al. Z. Physik 105, 168 (1937)	Hfs.		$\mu Tl205/\mu Tl203$ $= 1{,}00966$ $\pm 0{,}00046$		
			H. L. POSS Phys. Rev. 72, 637 (1947) Phys. Rev. 75, 600 (1949)	Brücke	a) $Tl(C_2H_3O_2)$ ges. w. L. b) —	$\nu Tl205/\nu H1$ $= 0{,}577135$ $\pm 0{,}00005$		1,628
			W. G. PROCTOR Phys. Rev. 75, 522 (1949) Phys. Rev. 79, 35 (1950)	Gekreuzte Spulen	2,6 m $Tl(C_2H_3O_2)$ $+ 0{,}03$ m $MnSO_4$ w. L.	$\nu Tl205/\nu H1$ $= 0{,}5770$ $\pm 0{,}0001$ $\nu Tl205/\nu Tl203$ $= 1{,}0098\ I$ $\pm 0{,}0002$	$+ 1{,}6118$ $\pm 0{,}0003$	$+ 1{,}629$ $\pm 0{,}003$
			R. E. SHERIFF et al. Phys. Rev. 82, 651 (1951)	Pendelrückk.	a) $Tl(C_2H_3O_2)$ $+ Mn(C_2H_3O_2)_2$ w. L. b) $FeCl_2$ w. L.	$\nu Tl205/\nu H1$ $= 0{,}57702$ $\pm 0{,}00003$		$+ 1{,}62733$ $\pm 0{,}00008$
			H. S. GUTOWSKY et al. Phys. Rev. 91, 81 (1953)	Brücke	a) 1,4 m $Tl(C_2H_3O_2)$ w. L. u. a. b) Tl_2O_3 in Königswasser	$\mu Tl205/\mu Tl203$ $= 1{,}009838$ $\pm 0{,}000001$		
			H. E. WALCHLI ORNL-1775 (1954)			$\nu Tl205/\nu Tl203$ $= 1{,}009816$ $\pm 0{,}000022$		
Pb_{72}^{204}	1,54	0	F. E. GEIGER, unveröffent-licht; nach: N. F. RAMSEY, Nuclear Moments. NewYork: Wiley & Sons 1953				≈ 0	
$**Pb_{82}^{204}$		(4,2)	H. FRAUENFELDER et al. Phys. Rev. 93, 1126 (1954)	Winkelverteilung der γ-Strahlung	Pb^{204} in Tl-Metall		$\|\mu\| = 0{,}14 + 0{,}12$ $- 0{,}06$ $(I = 2)$	
			V. E. KROHN, S. RABOY Phys. Rev. 95, 608 (1954) Phys. Rev. 97, 1017 (1955)	Winkelverteilung der γ-Strahlung	D_2-bestrahlte Tl-Folie		$0{,}22$ $\pm 0{,}02 (I = 4)$	
Pb_{82}^{206}	22,62	0	K. MURAKAWA Z. Physik 72, 793 (1931)	Hfs.			≈ 0	
Pb_{82}^{207}	22,62	$\frac{1}{2}$	G. BREIT, L. A. WILLS Phys. Rev. 44, 470 (1933)	Berechnung aus früheren Daten			$0{,}67 - 0{,}75$	
			E. FERMI, E. SEGRÈ Z. Physik 82, 729 (1933)	Berechnung aus früheren Daten			$\mu_B/\mu = 3500$	
			S. GOUDSMIT Phys. Rev. 43, 636 (1933)	Berechnung aus früheren Daten			0,60	
			H. SCHÜLER Z. Physik 88, 323 (1934)	Berechnung aus früheren Daten			$+ 0{,}60$	
			H. A. BETHE et al. Rev. Mod. Phys. 8, 82 (1936)	Berechnung aus früheren Daten			$+ 0{,}6$	
			A. M. CROOKER Canad. J. Res. 14, 115 (1936)	Hfs.			0,60	

Kern	Vork. %	Spin	Zitat	Methode	Substanz	Ergebnisse	μ_{unkorr}	μ_{korr}
			W. G. PROCTOR Phys. Rev. 76, 684 (1949) Phys. Rev. 79, 35 (1950)	Gekreuzte Spulen	a) 1,0 m Pb(C₂H₃O₂)₂+ 0,8 m Mn(C₂H₃O₂)₂ w. L. b) 0,69 m NaCl+ 1,0 m MnCl₂ w. L.	$\nu_{Pb^{207}}/\nu_{Na^{23}}$ = 0,7901 ± 0,0001, I	+ 0,5837 ± 0,0001	
Pb^{208}_{82}	53,22	0	K. MURAKAWA Z. Physik 72, 793 (1931)	Hfs.			≈ 0	
Bi^{209}_{83}	100	$\frac{9}{2}$	E. FERMI, E. SEGRÈ Z. Physik 82, 729 (1933)	Berechnung aus früheren Daten			μ_B/μ = 520	
			S. GOUDSMIT Phys. Rev. 43, 636 (1933)	Berechnung aus früheren Daten			4,0	
			H. SCHÜLER Z. Physik 88, 323 (1934)	Berechnung aus früheren Daten			+ 3,60	
			H. A. BETHE, R. F. BACHER Rev. Mod. Phys. 8, 82 (1936)	Berechnung aus früheren Daten			+ 4,0	
			H. WITTKE Z. Physik 116, 557 (1940)	Hfs.			3,45 – 4,2 je nach Term	
			F. M. KELLY et al. Phys. Rev. 80, 295 (1950)	Hfs.			4,10 ± 0,08	
			W. G. PROCTOR, F. C. YU Phys. Rev. 78, 471 (1950) Phys. Rev. 81, 20 (1951)	Gekreuzte Spulen	a) 0,69 m Bi(NO₃)₃ w. L. b) 1,8 m MnSO₄+ 25% D₂O w.L.	$\nu_{Bi^{209}}/\nu_{D^2}$ = 1,0468 ± 0,0001	+ 4,0400 ± 0,0007	
			Y. TING, D. WILLIAMS Phys. Rev. 89, 595 (1953)	Pendelrückk.	a) Bi(NO₃)₂ w.L. b) D₂O	$\nu_{Bi^{209}}/\nu_{D^2}$ = 1,04684 ± 0,00005		4,0810 ± 0,0004
Bi^{210}_{83}	<0,002	(1 od.0)	R. W. KING Phys. Rev. 94, 795 (1954)					
Po^{209}_{84}		$\frac{1}{2}$	K. L. VAN DER SLUIS et al. J. Opt. Soc. Amer. 45, 1087 (1955)	Hfs.				
At_{85}								
Rn_{86}								
Fr_{87}								
Ra_{88}								
Ac^{227}_{89}		$\frac{3}{2}$	M. FRED, F. S. TOMKINS Phys. Rev. 98, 1514 (1955)	Hfs.			+ 1,1 ± 0,1	
Th^{232}_{90}	100	0						
Pa^{231}_{91}	100	$\frac{3}{2}$	H. SCHÜLER, H. GOLLNOW Naturwiss. 22, 511 (1934)	Hfs.				
U^{233}_{92}	<8·10⁻⁵	$\frac{5}{2}$	L. A. KAROSTYLEVA et al. J. exp. theor. Phys. USSR. 28, 471 (1955)	Hfs.	U₃O₈ (reines U²³³)	$\frac{\mu_{U^{233}}}{\mu_{U^{235}}}$ ≈ 1,5	> 0	
U^{234}_{92}	0,006	0						
U^{235}_{92}	0,720	$\frac{7}{2}$	C. A. HUTCHINSON et al. Phys. Rev. 102, 292 (1956)	Paramagnetische Resonanz	UCl₃+2% LaCl₃, wasserfreier Einkristall		0,31 – 0,38	
U^{237}_{92}	<10⁻⁴	$\frac{1}{2}$	S. A. BARNOV et al. Sessions of the Academy of Sciences USSR, on the peaceful application of atomic energy, OFMR, p. 251, June 1955					
U^{238}_{92}	99,274	0						

Kern	Vork. %	Spin	Zitat	Methode	Substanz	Ergebnisse	μunkorr	μkorr
Np_{93}^{237}		$\frac{5}{2}$	B. Bleaney et al. Phil. Mag. **45**, 992 (1954)	Paramagnetische Resonanz	$UO_2Rb(NO_3)_3$- Einkristall mit 400 µg Np^{237} als $(NpO_2) Rb(NO_3)_3$		$|\mu| = 6$ $\pm 2,5$	
			V. E. Krohn et al. Phys. Rev. **98**, 1187 (1955)	Winkelverteilung der γ-Strahlung	$Am^{241} \alpha$-γ-Zerfall		$2,0$ $\pm 0,5$	
Np_{93}^{239}		$\frac{1}{2}$	J. G. Conway et al. Phys. Rev. **96**, 541 (1954)	Hfs.				
Pu_{94}^{239}		$\frac{1}{2}$	M. v. d. Berg et al. Physica, Haag **20**, 37 (1954)	Hfs.			$\mu < 0$	
			B. Bleaney et al. Phil. Mag. (7) **45**, 773 (1954) Phil. Mag. (7) **45**, 991 (1954)	Paramagnetische Resonanz	$UO_2Rb(NO_3)_3$- Einkristall mit 40 µg PuO_2 $Rb(NO_3)_3$		$|\mu| = 0,4$ $\pm 0,2$	
			L. A. Korostyleva et al. J. exp. theor. Phys. USSR. **28**, 471 (1955)					
Pu_{94}^{241}		$\frac{5}{2}$	B. Bleaney et al. Phil. Mag. (7) **45**, 991 (1954)	Paramagnetische Resonanz	$UO_2Rb(NO_3)_3$- Einkristall mit 40 µg Pu^{241}	$\mu Pu^{241}/\mu Pu^{239}$ $= 5,53$ $\pm 0,02$	$|\mu| = 1,4$ $\pm 0,6$	
Am_{95}^{241}		$\frac{5}{2}$	T. E. Manning et al. Phys. Rev. **102**, 1108 (1956)	Hfs.			$+1,4$	
Am_{95}^{243}		$\frac{5}{2}$	T. E. Manning et al. Phys. Rev. **102**, 1108 (1956)	Hfs.		$\mu Am^{243}/\mu Am^{241}$ $= 1,00$ $\pm 0,1$		
Cm_{96}								
Bk_{97}								
Cf_{98}								
E_{99}								
Fm_{100}								
Mv_{101}								

Zusammenfassende Berichte über das Gebiet der kernmagnetischen Hochfrequenz-Spektroskopie.

Bücher.

Kopfermann, H.: Kernmomente. Leipzig: Akademische Verlagsgesellschaft 1940.
Ramsey, N. F.: Nuclear Moments. New York: John Wiley & Sons, Inc. 1953.
Andrew, E. R.: Nuclear Magnetic Resonance. Cambridge: Cambridge University Press 1955.
Grivet, P.: La Résonance Paramagnétique Nucléaire. Moments Dipolaires et Quadripolaires. Paris 1955.
Kopfermann, H.: Kernmomente 2. Aufl. Frankfurt a. M.: AkademischeVerlagsgesellschaft 1956.
Lösche, A.: Kerninduktion. Berlin: VEB Deutscher Verlag der Wissenschaften 1957.

Zeitschriftenartikel.

Roberts, A.: Nucleonics **1** (1947).
Purcell, E. M.: Science, Lancaster, Pa. **107**, 433 (1948).
Rollin, B. V.: Rep. Progr. Phys. **12**, 22 (1949).
Soutif, M.: J. Phys. Radium **10**, 61 D (1949).
Braunbek, W.: Phys. Bl. **6**, 5 (1950).
Pake, G. E.: Amer. J. Phys. **18**, 438, 473 (1950).
Purcell, E. M.: Physica, Haag **17**, 282 (1951).
Pound, R. V.: Progr. Nucl. Phys. **2**, 21 (1952).
Darrow, K. K.: Bell. Syst. Techn. J. **32**, 74 (1953).
Giulotto, L.: R. C. Semin. mat. fis. Milano **24** (1953).
Purcell, E. M.: Science, Lancaster, Pa. **118**, 431 (1953). — Phys. Bl. **9**, 453 (1953).
Smith, J. A. S.: Quart. Rev. Chem. Soc. **7**, 279 (1953).
Gutowsky, H. S.: Ann. Rev. Phys. Chem. **5**, 333 (1954).
Purcell, E. M.: Amer. J. Phys. **22**, 1 (1954).
Hausser, K. H.: Angew. Chem. **68**, 729 (1956).

Determination of Nuclear Quadrupole Moments.

By

CHARLES H. TOWNES.

With 15 Figures.

1. Introduction. The nuclear electric quadrupole moment gives a measure of the deviation of the nucleus, or more precisely of its charge distribution, from a spherical shape and hence is an important source of information about nuclear structure. Effects due to the non-spherical distribution of charge in a nucleus were first discovered only about twenty years ago, and for the next decade nuclear quadrupole moments were detected only rather rarely and by specialized efforts. However, during the past ten years observation and recognition of the effects of quadrupole moments have become a normal and commonplace occurrence, and large amounts of data on quadrupole moments are now available, especially from work in radiofrequency and microwave spectroscopy. In fact, effects of quadrupole moments are now so familiar and amenable to experiments that they have become very important and much used tools in the study of other physical phenomena.

Availability of extensive data and developments in understanding complex nuclei have recently led also to a broad theoretical understanding of nuclear quadrupole moments. In most cases the sign and approximate magnitude of the quadrupole moment of a given nucleus can now be predicted from theoretical and semi-empirical information. Measured nuclear quadrupole moments have in turn furnished many tests of nuclear models and theory.

Although observation of effects due to nuclear quadrupole moments are now commonplace and abundant, and some of them have been made with great precision, this does not mean that the quadrupole moments themselves are yet very completely or precisely determined. Unfortunately the theoretical connection between measured effects and the magnitudes of quadrupole moments usually involves uncertainties. It often occurs also that the largest uncertainties of this type occur for the effects which are susceptible to most accurate measurement. As a result, errors of the order of 20% in actual determinations of quadrupole moments are typical. On the other hand, such large errors do not generally occur in measurement of the ratios of quadrupole moments of two isotopes, where an accuracy better than $\frac{1}{10}$% can usually be achieved. Theoretical understanding of quadrupole moments is still somewhat coarse and unprecise, and in the past the errors in experimental determination of these moments have not limited seriously their usefulness. However, nuclear theory seems to be reaching the point where values of nuclear quadrupole moments are needed with a precision beyond that given by present methods of measurement, and it is increasingly important to appreciably refine these methods or to find new ones.

Determinations of quadrupole moments may be divided into two general classes. The first class, which is the older and has so far yielded the great majority of results, involves measurement of the energy required for reorientation of a nucleus in an atomic or molecular system. This energy is almost always measured by spectroscopic techniques, and it was the effect of this energy on the hyperfine structure of atomic spectra which allowed the first determination of a nuclear quadrupole moment by SCHÜLER and SCHMIDT [1] in 1935. Optical spectroscopy is still useful for this type of measurement, but more recently the higher resolution of radiofrequency and microwave spectroscopy has made them more fruitful sources of data on this "hyperfine" energy, particularly for the lighter nuclei where the energy involved is often too small to be readily detected by classical spectroscopic techniques. Effects of this first class involve interaction of the nuclear charge distribution, and in particular the non-spherical part of the charge distribution, with what might be considered static charge distributions of electrons or of other nuclei in atomic or molecular systems.

The second broad class of phenomena which yield data on quadrupole moments falls in the domain which is more generally recognized as nuclear physics—that is, the interactions between nuclei and fast particles, either charged particles or high-energy photons. Measurements of nuclear cross-section and lifetime have only recently been applied to determining nuclear quadrupole moments and the results are still somewhat limited. However, these methods will very probably be of increasing importance, and certainly will provide a valuable supplement to what can be learned from effects of the first class.

Since the two classes of phenomena involve quite different techniques and different theoretical approaches, they will be considered somewhat separately. Basic theory for interaction of nuclear quadrupole moments in atomic and molecular systems will first be presented, followed by a theoretical analysis of the many types of experiments which allow measurement of these interactions, and discussion of methods for evaluating quadrupole moments themselves from the measured coupling constant (Part A). The general theory of interactions between nuclear quadrupole moments and fast particles will then be treated, and evaluation of quadrupole moments from measurement of such interactions will be discussed (Part B). Actual numerical results of all determinations of quadrupole moments are reserved for the final sections, where some of the determinations will be discussed, and values of quadrupole moments will be summarized and compared.

A. Determination of nuclear quadrupole moments from atoms or molecules.

2. Basic theory of nuclear quadrupole interaction in an atomic or molecular system. If the finite size of the nucleus is taken into account, electrostatic interactions between the nuclear charge distribution and other charges are not precisely described by a Coulomb field surrounding the center of the nucleus, but may be written

$$W = \iint \frac{\varrho_N(\mathbf{r}_N)\,\varrho_e(\mathbf{r}_e)}{r}\,dv_N\,dv_e \tag{2.1}$$

where ϱ_N is the nuclear charge density, which is positive since it is due to protons, $\mathbf{r}_N$ the position vector for the volume dv_N or the element of charge $\varrho_N\,dv_N$, ϱ_e, $\mathbf{r}_e$, and dv_e are corresponding quantities for other charges in the atomic or molecular system being considered, and r is the distance between volume elements dv_N and dv_e. These quantities are illustrated in Fig. 1. It is convenient and natural to take the origin as the center of mass of the nucleus. Since it is clear that effects due to quadrupole moments are very much smaller than the normal separation between electron energy levels or *a fortiori* between nuclear energy levels, the charge densities in (2.1) may be taken as averaged over motions of the nuclear particles and of the surrounding electrons.

If only charges outside the nucleus, or outside a sphere of radius R surrounding the nucleus, give important interactions, then $1/r$ may be expanded as a power series in r_N/r_e.

$$
\begin{aligned}
\frac{1}{r} &= \frac{1}{\sqrt{r_e^2 + r_N^2 - 2\,r_e\,r_N\cos}} \\
&= \frac{P_0(\cos\vartheta)}{r_N} + \frac{r_N}{r_e^2}P_1(\cos\vartheta) + \frac{r_N^2}{r_e^3}P_2(\cos\vartheta) + \cdots + \frac{r_N^l}{r_e^{l+1}}P_l(\cos\vartheta) + \cdots
\end{aligned}
\quad (2.2)
$$

where P_l represents a Legendre polynomial, and in particular

$$
\begin{aligned}
P_0(\cos\vartheta) &= 1, \\
P_1(\cos\vartheta) &= \cos\vartheta, \\
P_2(\cos\vartheta) &= \tfrac{1}{2}(3\cos^2\vartheta - 1).
\end{aligned}
\quad (2.3)
$$

Using the expansion (2.2), the energy (2.1) may be expressed as a series of terms, each involving a particular P_l, which is described as due to a multipole moment of order 2^l. Thus the first is a monopole term,

$$
\iint \frac{\varrho_N\,\varrho_e\,dv_e\,dv_N}{r_e}
$$
$$
= Z\,e\int \frac{\varrho_e\,dv_e}{r_e},
$$

which is the ordinary Coulomb energy due to the nuclear charge $Z\,e$. The second term is the dipole term, and thought to be identically zero

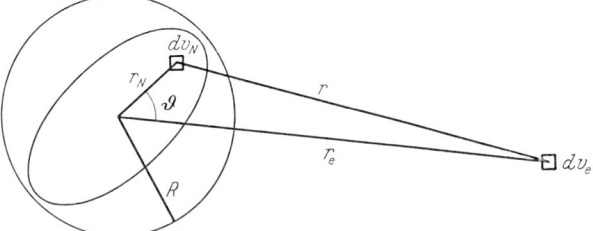

Fig. 1. Coordinates of elementary nuclear volume dv_n and volume dv_e of charge producing electrostatic interaction with nucleus. Nuclear boundary is represented by an ellipse.

(see below). The quadrupole term involves $P_2(\cos\vartheta)$. As may be seen from the expansion (2.2), it is smaller than the monopole or Coulomb term by the order of magnitude $\left(\frac{r_N}{r_e}\right)^2 \approx \left(\frac{10^{-12}}{10^{-9}}\right)^2 \sim 10^{-6}$. Since the Coulomb energy for a valence electron is near 20000 cm^{-1}, the quadrupole energy is of the order 0.02 cm^{-1} or 600 Mc/sec, expressed in frequency units.

Although it is usually appropriate to neglect the charge density ϱ_e which may exist inside the nucleus, or inside a small sphere of radius R immediately surrounding the nucleus, this charge density gives rise to some interesting effects, which need examination. For this purpose the expansion (2.2) is not valid. However, the interaction (2.1) may be expanded in Cartesian coordinates as

$$
\begin{aligned}
W = \int \varrho_N(x, y, z)&\left[V + x\frac{\partial V}{\partial x} + y\frac{\partial V}{\partial y} + z\frac{\partial V}{\partial z} + \frac{1}{2}x^2\frac{\partial^2 V}{\partial x^2} + \frac{1}{2}y^2\frac{\partial^2 V}{\partial y^2} + \right. \\
&+ \frac{1}{2}z^2\frac{\partial^2 V}{\partial z^2} + x\,y\frac{\partial^2 V}{\partial x\,\partial y} + y\,z\frac{\partial^2 V}{\partial y\,\partial z} + z\,x\frac{\partial^2 V}{\partial z\,\partial x} + \cdots \\
&\left. \cdots \frac{x^n\,y^m\,z^p}{u!\,m!\,p!}\frac{\partial^{n+m+p}V}{\partial x^n\,\partial y^m\,\partial z^p} + \cdots\right]dv,
\end{aligned}
\quad (2.4)
$$

where V is the electrostatic potential due to all non-nuclear charges which, with all its derivatives, is to be evaluated at the center of nuclear charge. x, y, and z are Cartesian coordinates of the nuclear charge ϱ_N. The first term of this expansion is again the monopole or normal Coulomb interaction of a point nucleus. The second three terms are due to the nuclear dipole moment. The average electric field ΔV is zero at the nucleus if the average nuclear acceleration is zero, and in addition, no known nuclear forces can produce an electric dipole moment if the nucleus is not in a degenerate state (see, for example [2], pp. 8, 23). Hence the dipole terms, linearly dependent on the coordinates, can be assumed to be zero. Terms dependent on the second power of the nuclear coordinates may be written in two parts $W_i + W_Q$, where

$$W_i = \frac{1}{6} \int \varrho_N r^2 \left(\frac{\partial^2 V}{\partial x^2} + \frac{\partial^2 V}{\partial y^2} + \frac{\partial^2 V}{\partial z^2} \right) dv = \frac{\nabla^2 V}{6} \int \varrho_N r^2 \, dv \tag{2.5}$$

and

$$W_Q = \frac{1}{6} \int \varrho_N \left[(3x^2 - r^2) \frac{\partial^2 V}{\partial x^2} + (3y^2 - r^2) \frac{\partial^2 V}{\partial y^2} + (3z^2 - r^2) \frac{\partial^2 V}{\partial z^2} + \right. \\ \left. + 6xy \frac{\partial^2 V}{\partial x \partial y} + 6yz \frac{\partial^2 V}{\partial y \partial z} + 6zx \frac{\partial^2 V}{\partial z \partial x} \right] dv = -\frac{1}{6} \mathbf{Q} : \nabla \mathbf{E} \tag{2.6}$$

where $\mathbf{Q} : \nabla \mathbf{E}$ represents the inner product of the dyadics $\mathbf{Q} = \int (3\mathbf{rr} - r^2 \mathbf{1}) \varrho_N \, dv$ and $\nabla \mathbf{E}$. $\mathbf{1}$ is the unit dyadic, and $\mathbf{E}$ the electric fieldstrength.

W_i is associated with the "isotope shift" of atomic spectra, and is an energy term due to the finite size of the nucleus, but independent of the nuclear orientation. In an atom, s electrons make by far the largest contribution to W_i because they have a large charge density ϱ_e at the nucleus and hence a large value of $\nabla^2 V$, which by LAPLACE's equation is $-4\pi \varrho_e$. On the other hand, because of their spherical symmetry, and as will be demonstrated more formally below, the s electrons make no contribution to W_Q, which is energy associated with the nuclear quadrupole moment and dependent on orientation of the nucleus with respect to the surrounding charge distribution.

Consider now the energy W_Q from a classical point of view. Since $\mathbf{Q}$ is symmetric, it may be diagonalized by a suitable choice of axes. Since also the rapid precession of nuclear particles about the spin axis makes it an axis of cylindrical symmetry, the direction of the nuclear spin must be one of the principal axes of $\mathbf{Q}$, which we shall designate the z_N axis, and the two perpendicular directions must be equivalent to each other. That is,

$$\int \varrho_N (3x_N^2 - r^2) \, dv = \int \varrho_N (3y_N^2 - r^2) \, dv = -\tfrac{1}{2} \int \varrho_N (3z_N^2 - r^2) \, dv. \tag{2.7}$$

Hence the entire dyadic $\mathbf{Q}$ can be specified by one constant, called "The" nuclear quadrupole moment

$$Q = \frac{1}{e} \int \varrho (3z_N^2 - r^2) \, dv \tag{2.8}$$

where e is the charge of one proton, and Q as defined by (2.8) has the dimensions of an area. From expression (2.8) it can be seen that a nucleus whose charge distribution is spherical has zero quadrupole moment, and that Q is positive for a nucleus elongated along the z or spin axis, negative for one flattened along this axis. From (2.6), the quadrupole energy becomes

$$W = \frac{e}{6} Q \left[\frac{\partial^2 V}{\partial z_N^2} - \frac{1}{2} \left(\frac{\partial^2 V}{\partial x_N^2} + \frac{\partial^2 V}{\partial y_N^2} \right) \right]. \tag{2.9}$$

If the potential V is due to a spherical distribution of charge, such as an s electron, $\frac{\partial^2 V}{\partial x_N^2} = \frac{\partial^2 V}{\partial y_N^2} = \frac{\partial^2 V}{\partial z_N^2}$ and hence $W_Q = 0$. Since s electrons contribute nothing to (2.9), and since p electrons or other angular momentum states have only a very small density at the nucleus, it is reasonable to consider in place of B a potential V' which is due only to charges outside the nucleus. Then $\nabla^2 V' = 0$ from LAPLACE's equation, and

$$W_Q = \frac{e}{4} Q \left(\frac{\partial^2 V'}{\partial z^2} \right). \tag{2.10}$$

A p electron, or a combined s and d wavefunction, could give a non-spherical charge density in the nucleus of the order of $e \frac{r_N^2}{r_e^5}$ which contributes to W_Q. Since from POISSON's equation $\frac{\partial^2 V}{\partial x_N^2}$ or $\frac{\partial^2 V}{\partial y_N^2}$ due to this charge density is also of the order $e \frac{r_N^2}{r_e^5}$, its total contribution to W_Q would be near $e Q \frac{e r_N^2}{r_e^5} \sim \frac{e^2 r_N^4}{r_e^5}$. This is only $(r_N/r_e)^2$ or about 10^{-6} of the main quadrupole energy, and in fact is of the order of higher terms in the expansion (2.4) which have already been neglected. This justifies the expansion (2.2), or the neglect of all charges within a small sphere immediately surrounding the nucleus.

A precise evaluation of the quadrupole energy must of course come from a quantum mechanical treatment, which was first given by CASIMIR[3]. The Hamiltonian may be taken from (2.6) as $H = -\frac{1}{6}Q : VE$, and in most cases Q and VE can be replaced by operators having known values for the usual nuclear and electron wave functions by the following method. Consider the operator $\frac{3}{2}(II + \widetilde{II} - I^2 1)$, where I is the nuclear spin and $\widetilde{II}$ represents the transpose of II. This operator is symmetric and traceless as is Q, and hence has the same angular dependence with respect to nuclear orientation; that is, its matrix elements between states of various orientations must be identical with those of Q except for a proportionality constant[1]:

$$(I\, m_I | Q | I\, m_{I'}) = \text{const} \, (I\, m_I | \tfrac{3}{2}[II + \widetilde{II}] - I^2 1 | I\, m_{I'}) \qquad (2.11)$$

where I, m_I are quantum numbers indicating the nuclear spin and its projection on a direction fixed in space. To evaluate the proportionality constant, consider the zz component of these operators for the state $m_I = I$:

$$(I I | Q_{zz} | I I) = \text{const} \, [3 I^2 - I(I + 1)] \qquad (2.12)$$

the quantity $(I I | Q_{zz} | I I)$ corresponds approximately to the classical quadrupole moment $e Q$, and hence is taken as the definition of $e Q$. That is, $e Q = \text{const} \, I(2I - 1)$, or

$$(Q)_{\text{op}} = \frac{e Q}{I(2I - 1)} \left[\frac{3}{2}(II + \widetilde{II}) - I^2 1 \right]. \qquad (2.13)$$

Similarly, since VE is also symmetric, and is traceless if one excludes from VE effects of charge density within the nucleus, it can be shown that

$$(VE)_{\text{op}} = \frac{q_J}{J(2J - 1)} \left[\frac{3}{2}(JJ + \widetilde{JJ}) - J^2 1 \right] \qquad (2.14)$$

where J is the angular momentum of the system of charge external to the nucleus and

$$q_J = \left(J J \left| \frac{\partial^2 V}{\partial z^2} \right| J J \right). \qquad (2.15)$$

It should be noted, however, that (2.14) is correct only for matrix elements between states of the same J. If J is not a "good" quantum number, the more basic form of $(VE)_{\text{op}}$ is necessary.

The quadrupole interaction when J is a "good" quantum number may be expressed now in the convenient form

$$W_Q = \frac{e q_J Q}{2 I(2I - 1) J(2J - 1)} \left[3(I \cdot J)^2 + \frac{3}{2}(I \cdot J) - I^2 J^2 \right] \qquad (2.16)$$

---

[1] A detailed demonstration of this can be found in Ref. [2], p. 16.

by using (2.13), (2.14) and the commutation rules for components of angular momentum. Very commonly, the nuclear spin I is coupled to the angular momentum J of electrons and other charges to form the total angular momentum F of the system. That is, $\boldsymbol{I} + \boldsymbol{J} = \boldsymbol{F}$. In this case I^2, J^2, and $\boldsymbol{I} \cdot \boldsymbol{J}$ are all diagonal matrices with values $I(I+1)$, $J(1+1)$, and $\boldsymbol{I} \cdot \boldsymbol{J} = \frac{1}{2}[F(F+1) - I(I+1) - J(J+1)]$ respectively. Hence, letting $C = F(F+1) - I(I+1) - J(J+1)$,

$$W_Q = \frac{e\, q_J\, Q}{2I(2I-1)\, J(2J-1)} \left[\frac{3}{4} C(C+1) - I(I+1)\, J(J+1) \right]. \qquad (2.17)$$

Since the electrostatic potential at the nucleus due to charge e is $\dfrac{e}{r} = \dfrac{e}{(x^2+y^2+z^2)^{\frac{1}{2}}}$, where r is the distance from nucleus to charge,

$$\frac{\partial^2 V}{\partial z^2} = e\, \frac{3\cos^2 \vartheta - 1}{r^3}, \qquad \text{where } \vartheta = \text{arc cos}\, \frac{z}{r}$$

is the angle between the z axis fixed in space and the radius vector $\boldsymbol{r}$. Therefore for a charge density distribution $\varrho(x, y, z)$,

$$q_J = \int \varrho_{JJ}\, \frac{3\cos^2 \vartheta - 1}{r^3}\, dv \qquad (2.18)$$

where ϱ_{JJ} is the average charge density for the state $m_J = J$. In the case of a single electron near a nucleus,

$$q_J = -e \int \psi_{JJ}^*\, \frac{3\cos^2 \vartheta - 1}{r^3}\, \psi_{JJ}\, dv \qquad (2.19)$$

where e is the magnitude of the electronic charge and ψ_{JJ} its wavefunction. In the case of an atom, expressions (2.16) and (2.17) or (2.18) are more or less immediately applicable. However for a molecule, further manipulation provides a more convenient expression of the energy associated with nuclear quadrupole moments.

3. Nuclear quadrupole interactions in molecules and the effects of molecular rotation. The quantity q_J for a molecule, depending on the charge distribution with respect to the angular momentum J, varies with the state of rotation of the molecule. However, the charge distribution with respect to molecule-fixed axes is, to a good approximation, identical in the various rotational states. This is quite different from the atomic case, where a change of electronic angular momentum involves always an appreciable change in the electronic charge distribution. It also suggests that q_J should be expressed in terms of the essentially invariable charge distribution in the molecule, and the dependence of q_J on rotational state made evident.

Consider first a nucleus on the axis of a symmetric top molecule—that is, a molecule with a three- or more-fold axis of symmetry. This includes, of course, diatomic and linear molecules, which have an ∞-fold axis of symmetry. Assume also that this is the only nucleus in the molecule which produces appreciable hyperfine structure. If now the dyadic $\boldsymbol{VE}$ is referred to coordinates fixed in the molecule, the symmetry axis must be one of its principal axes, which may be assumed to be the coordinate direction z_m. Since $\boldsymbol{VE}$ remains unchanged by a rotation about this axis less than $180°$ because of the molecular symmetry, the x and y components of $\boldsymbol{VE}$ must be equal, or, using this fact and LAPLACE's equation,

$$\frac{\partial^2 V}{\partial x_m^2} = \frac{\partial^2 V}{\partial y_m^2} = -\frac{1}{2}\, \frac{\partial^2 V}{\partial z_m^2}.$$

From (2.15), the quantity q_J is $\dfrac{\partial^2 V}{\partial z^2}$ referred to a space-fixed axis z, or $\left(JJ \left| \dfrac{\partial^2 V}{\partial z^2} \right| JJ \right)$ and is related by a simple transformation of coordinates to $\dfrac{\partial^2 V}{\partial z_m^2}$.

$$q_J = \frac{\partial^2 V}{\partial z_m^2} \left(\frac{3\cos^2 \vartheta - 1}{2} \right)_{av} \tag{3.1}$$

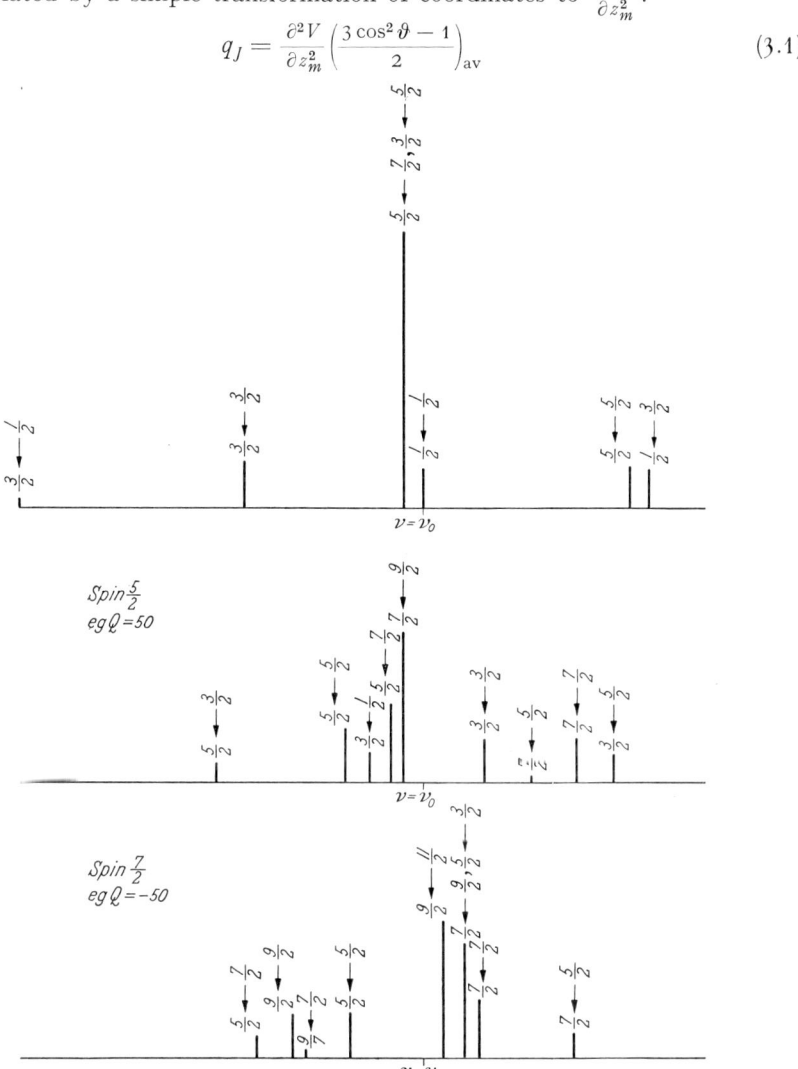

Fig. 2. Spectra due to quadrupole hyperfine structure in the rotational transition $J = 2 \leftarrow 1$ of a linear molecule, assuming various spins and coupling constants. Values of the total angular momentum F involved in each transition are indicated. ν_0 would be the frequency for the transition if there were no hyperfine structure.

where ϑ is the angle between the molecular axis and the angular momentum J. The quantity $\left(\dfrac{3\cos^2 \vartheta - 1}{2} \right)_{av}$ may be evaluated from the angular dependence of the wave functions of a symmetric top, which gives

$$q_J = \left[\frac{3 K^2}{J(J+1)} - 1 \right] \frac{J}{2J+3} \frac{\partial^2 V}{\partial z_m^2}. \tag{3.2}$$

Here K is the quantum number for the component of J along the molecular symmetry axis. For a linear molecule (in a $^1\Sigma$ state and with no vibrational angular momentum), K is necessarily zero, so that

$$q_J = -\frac{J}{2J+3} \frac{\partial^2 V}{\partial z_m^2}. \tag{3.3}$$

The quadrupole "hyperfine" energy for a symmetric top can be written, from (2.17) and (3.2),

$$W_Q = \frac{eqQ\left[3\dfrac{K^2}{J(J+1)}-1\right]}{2I(2I-1)\,(2J-1)\,(2J+3)\}}\left[\frac{3}{4}C(C+1)-I(I+1)\,J(J+1)\right] \tag{3.4}$$

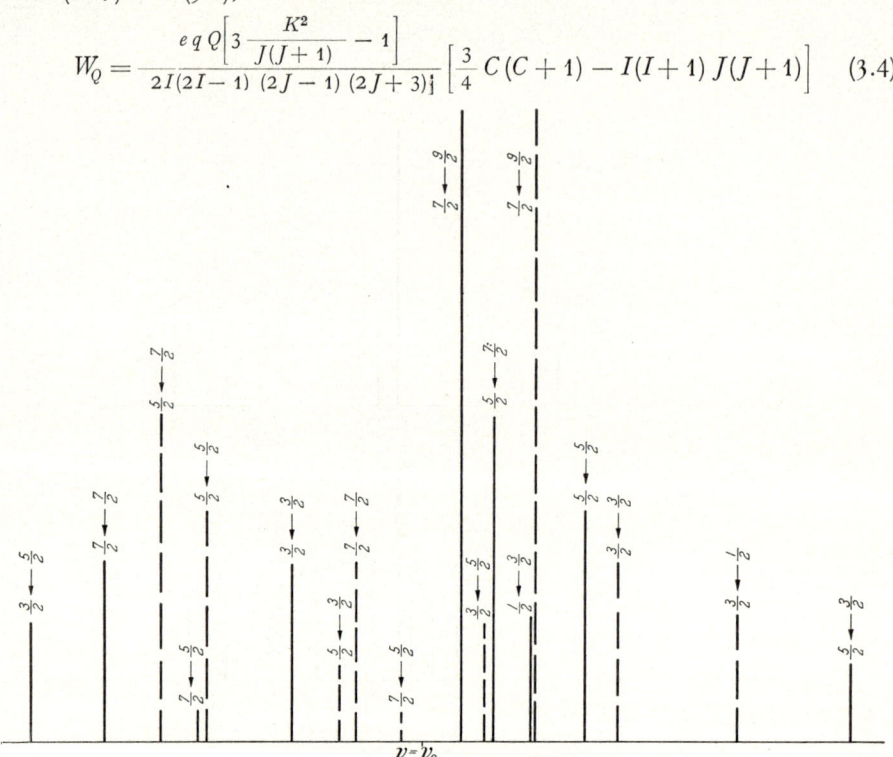

Fig. 3. Quadrupolar hyperfine structure in the $J=2\leftarrow1$ transition of symmetric top molecule. Solid lines are transitions with $K=0$, and dashed lines those with $K=1$.

where

$$C = F(F+1) - I(I+1) - J(J+1),$$

e is the charge of one proton, and

$$eqQ = e\frac{\partial^2 V}{\partial z_m^2}Q$$

is usually called the quadrupole coupling constant. It is a measure of the energy of reorientation of the nuclear quadrupole moment or spin in the molecule. The quantity q is normally used for $\dfrac{\partial^2 V}{\partial z_m^2}$, the second derivative of the electrostatic potential due to extranuclear charges with respect to the direction of the molecular axis. However, Kellogg, Rabi, Ramsey, and Zacharias [4] have used q for what is called above q_J, and Nordsieck [5] and Ramsey [6] have sometimes used $q'=\dfrac{1}{2e}\dfrac{\partial^2 V}{\partial z_m^2}$. The function multiplying $eqQ\left[\dfrac{3K^2}{J(J+1)}-1\right]$, that is

$$f(I,J,F) = \frac{\frac{3}{4}C(C+1)-I(I+1)\,J(J+1)}{2I(2I-1)\,(2J-1)\,(2J+3)} \tag{3.5}$$

has been called CASIMIR's function, and is so often used that it has been tabulated for the most usual values of I and J. TOWNES and SCHAWLOW [7] give this function for $I=1$ to $\frac{11}{2}$ and $J=0$ to 10. GORDY, SMITH, and TRAMBARULO [8] list it for $I=1$ to $\frac{9}{2}$ and $J=0$ to 20. With these tables and others giving relative intensities of transitions [7], [8], simple quadrupole hyperfine structure observed in symmetric tops can be quickly solved and the coupling constant eqQ evaluated. Fig. 2 shows the appearance of hyperfine structure in the $J=2\leftarrow1$ transition of a linear molecule (or when $K=0$) for several values of spin; Fig. 3 indicates the pattern obtained for the $J=2\leftarrow1$ transition of a symmetric top when the nucleus producing quadrupolar hyperfine structure has a spin $\frac{5}{2}$.

Expression (3.4) is accurate when the quadrupole hyperfine structure is small compared with the frequency of rotational transition so that J and K are "good" quantum numbers. This is often the case, since the coupling constant eqQ ranges from a small fraction of one Mc/sec, to several thousands of Mc/sec, while the rotational frequencies are typically many thousands of Mc/sec. However, there are frequent cases where (3.4) is not sufficiently accurate. Better approximations may be obtained from the matrix elements non-diagonal in J, which have been given by BARDEEN and TOWNES [9].

$$(I,J,K,F,M_F|H_Q|I,J+1,K,F,M_F)=3\,eq\,Q\,K\left[\frac{F(F+1)-I(I+1)-J(J+2)}{8\,I(2\,I-1)\,J(J+2)}\right]\times \tag{3.6}$$
$$\times\left\{\left[1-\frac{K^2}{(J+1)^2}\right]\frac{(I+J+F+2)(J+F-I+1)(I+F-J)(I+J-F+1)}{(2J+1)(2J+3)}\right\}^{\frac{1}{2}}$$

$$(I,J,K,F,M_F|H_Q|I,J+2,K,F,M_F)=\frac{3eq\,Q}{16\,I(2\,I-1)(2\,J+3)}\times \tag{3.7}$$
$$\times\left\{\left[1-\frac{K^2}{(J+1)^2}\right]\left[1-\frac{K^2}{(J+2)^2}\right](I+J+F+3)(I+J+F+2)(I+J-F+2)\times\right.$$
$$\times(I+J-F+1)(J+F-I+2)\frac{(J+F-I+1)(I+F-J)(I+F-J-1)}{(2J+1)(2J+5)}\right\}^{\frac{1}{2}}.$$

To second order in eqQ, the energy may be calculated by the usual perturbation expression

$$W_Q-(IJKFM_F|H_Q|IJKFM_F)+\sum_{J'K'}\frac{|(IJKFM_F|H_Q|IJ'K'FM_F)|^2}{W_{JK}-W_{J'K'}} \tag{3.8}$$

where the first term is the first order perturbation, and is given by expression (3.4). $W_{JK}-W_{J'K'}$ is the energy difference between the states characterized by JK and $J'K'$. H_Q is $-\frac{1}{6}\,\mathbf{Q}:\mathbf{V}\mathbf{E}$ as indicated above. For a nucleus on the axis of a true symmetric top, there are no matrix elements off-diagonal in K, so that the elements (3.6) and (3.7) suffice for use in (3.8). Since microwave techniques allow rather precise measurement of microwave hyperfine structure, the second order terms of (3.8) may be quite appreciable even when $\dfrac{eq\,Q}{W_{JK}-W_{J'K'}}$ is as small as $\frac{1}{20}$.

Quadrupole hyperfine structure in asymmetric rotors can be much more complicated than that of the symmetric rotor discussed above. So also can the case of a quadrupolar nucleus in a symmetric top, but not on the symmetry axis. However, in this latter case the symmetry requires that there be at least two other similar nuclei with appreciable hyperfine structure, and discussion of hyperfine structure due to more than one nucleus in a single molecule will be postponed until later. In the general case of an asymmetric rotor, each of the three elements $\dfrac{\partial^2 V}{\partial x_m^2}$, $\dfrac{\partial^2 V}{\partial y_m^2}$, and $\dfrac{\partial^2 V}{\partial z_m^2}$ of $\mathbf{V}\mathbf{E}$ referred to principal axes fixed in the molecule may be different. A single quadrupole coupling constant no longer can describe the hyperfine structure, but since these three quantities are related by LAPLACE's equation, only two independent quantities are needed, which are often taken to be $q = \dfrac{\partial^2 V}{\partial z_m^2}$ and an asymmetry constant

$$\eta = \frac{\dfrac{\partial^2 V}{\partial x_m^2} - \dfrac{\partial^2 V}{\partial y_m^2}}{\dfrac{\partial^2 V}{\partial z_m^2}}. \tag{3.9}$$

Here $\dfrac{\partial^2 V}{\partial z_m^2}$ is the larger of the three elements of $\boldsymbol{VE}$. An actual calculation of q_J is made by expressing this quantity in terms of components of $\boldsymbol{VE}$ referred to the principal axes of inertia of the molecule. These in turn can be related to the principal values of the components of $\boldsymbol{VE}$ mentioned above. Thus

$$q_J = \frac{\partial^2 V}{\partial z_J^2} = \left[\alpha_{z_J a}^2 \frac{\partial^2 V}{\partial a^2} + \alpha_{z_J b}^2 \frac{\partial^2 V}{\partial b^2} + \alpha_{z_J c}^2 \frac{\partial^2 V}{\partial c^2} + 2\alpha_{z_J a} \alpha_{z_J b} \frac{\partial^2 V}{\partial a\, \partial b} + \right. $$
$$\left. + 2\alpha_{z_J a} \alpha_{z_J c} \frac{\partial^2 V}{\partial a\, \partial c} + 2\alpha_{z b} \alpha_{z c} \frac{\partial^2 V}{\partial b\, \partial c} \right]_{\mathrm{av}} \tag{3.10}$$

where a, b, and c represent Cartesian coordinates along the directions of the principal moments of inertia of the molecule. z_J is a direction fixed in space. $\alpha_{z_J a}$, etc. represent cosines of the angles between z_J and the molecule-fixed axes.

Since symmetry of the Hamiltonian for rotational energy ensures equality of the probability for two orientations of the molecule which differ by a 180° rotation about one of the principal axes of inertia, while the α's in (3.10) may change sign with such a rotation, one can show that only the first three terms of (3.10) are non-zero. That is

$$q_J = (\alpha_{z_J a}^2)_{\mathrm{av}} \frac{\partial^2 V^2}{\partial a^2} + (\alpha_{z_J b}^2)_{\mathrm{av}} \frac{\partial^2 V}{\partial b^2} + (\alpha_{z c}^2)_{\mathrm{av}} \frac{\partial^2 V}{\partial^2 c}. \tag{3.11}$$

The averages are taken over molecular orientations.

The quantities $(\alpha_{z_J a}^2)_{\mathrm{av}}$, etc. cannot be easily evaluated in the general case, but q_J may be expressed in several forms which are helpful in its evaluation. BRAGG [10] has shown that (3.11) can be written

$$q_J = \left[\frac{2J}{(2J+1)\,(2J+3)} \right] \sum_{K'_{-1} K'_1} \left[\left(\frac{\partial^2 V}{\partial a^2} \right)^a S_{J K_{-1} K_1\, J K'_{-1} K'_1} + \right.$$
$$\left. + \left(\frac{\partial^2 V}{\partial b^2} \right)^b S_{J K_{-1} K_1\, J K'_{-1} K'_1} + \left(\frac{\partial^2 V}{\partial c^2} \right)^c S_{J K_{-1} K_1\, J K'_{-1} K'_1} \right] \tag{3.12}$$

where $^a S_{J K_{-1} K_1\, J K'_{-1} K'_1}$ is a quantity called the "line strength", which is proportional to the intensity of transition between two asymmetric rotor levels designated by quantum numbers J, K_{-1}, K_1 and J, K'_{-1}, K'_1, and due to a unit dipole moment parallel to the direction of the a axis. These line strengths are tabulated for J between 0 and 12, and for five different values of the molecular asymmetry parameter $\varkappa$ by CROSS, HAINER, and KING [11], and by TOWNES and SCHAWLOW [7]. The asymmetry parameter $\varkappa$ is not that associated with quadrupole effects, but is the ratio

$$\varkappa = \frac{\dfrac{2}{I_b} - \dfrac{1}{I_a} - \dfrac{1}{I_c}}{\dfrac{1}{I_a} - \dfrac{1}{I_c}} \tag{3.13}$$

where I_a, I_b, and I_c are the smallest, intermediate, and largest principal moments of inertia respectively of the molecule. $1/I_a$, $1/I_b$, and $1/I_c$ are proportional to the molecular rotational constants A, B and C respectively. The value of $\varkappa$ can lie only between -1 and $+1$. Although the line strengths are tabulated only for certain values of $\varkappa$, they may be at least roughly evaluated for other cases by interpolation.

BRAGG and GOLDEN [12] have expressed q_J also in the form

$$
\begin{aligned}
q_J = \frac{1}{(J+1)(2J+3)} \frac{\partial^2 V}{\partial a^2} \left[J(J+1) - E(\varkappa) - (\varkappa+1) \frac{\partial E(\varkappa)}{\partial \varkappa} \right] + \\
+ \frac{2}{(J+1)(2J+3)} \frac{\partial^2 V}{\partial b^2} \frac{\partial E(\varkappa)}{\partial \varkappa} + \frac{1}{(J+1)(2J+2)} \frac{\partial^2 V}{\partial c^2} \left[J(J+1) - E(\varkappa) + (\varkappa-1) \frac{\partial E(\varkappa)}{\partial \varkappa} \right]
\end{aligned}
\tag{3.14}
$$

where $E(\varkappa)$ is a tabulated quantity related to the rotational energy of the particular state $J_{K_{-1}K_1}$ in question and dependent on the asymmetry parameter $\varkappa$. This function is given for J from 0 to 12, and for one hundred values of $\varkappa$ (steps of 0.01) by TOWNES and SCHAWLOW [7]. ERLANDSON [13] has listed $E(\varkappa)$ for values of J as high as 40, but with only ten values of $\varkappa$ (steps of 0.1). Each of these tables is effectively doubled in size by the relation

$$
E_{mn}(\varkappa) = - E_{mn}(-\varkappa) \tag{3.15}
$$

where the first subscript is the value of the quantum number K_{-1} and the second that of K_1. These quantum numbers specify the state $J_{K_{-1}K_1}$ for a given value of J.

The wave function for an asymmetric rotor may be expanded in terms of those for a symmetric rotor as follows

$$
\psi_{J K_{-1} K_1} = \sum_{K = K_{-1} \pm 2n} a_{JK} \psi_{JK} \tag{3.16}
$$

where ψ_{JK} is a symmetric top wavefunction specified by the total angular momentum J and its projection K on the symmetry axis; J, K_{-1}, and K_1 are the quantum numbers for the state of the asymmetric rotor, and n is an integer. From this expansion and matrix elements of H_Q for a symmetric rotor, q_J may be written [10]

$$
q_J = \frac{q}{(J+1)(2J+3)} \sum_K \{ a_{JK}^2 [3K^2 - J(J+1)] - 2 a_{JK} a_{J,K+2} [f(J,K+1)]^{\frac{1}{2}} \eta \} \tag{3.17}
$$

where

$$
f(J,n) = \tfrac{1}{4}(J^2 - n^2)[(J+1)^2 - n^2],
$$

$$
q = \frac{\partial^2 V}{z_m^2} \quad \text{as above,}
$$

$$
\eta = \frac{\dfrac{\partial^2 V}{\partial x_m^2} - \dfrac{\partial^2 V}{\partial y_m^2}}{q} \quad \text{as defined by (3.9).}
$$

Although expression (3.17) can give q_J exactly, it is not convenient, since evaluation of the coefficients a_{JK} in the wavefunction (3.16) is troublesome. However, for molecules which are not very asymmetric, (3.17) leads to convenient approximations which involve an expansion in terms of an asymmetry parameter b. This asymmetry parameter is

$$
b = \frac{\dfrac{1}{I_C} - \dfrac{1}{I_B}}{\dfrac{2}{I_A} - \dfrac{1}{I_B} - \dfrac{1}{I_C}} = \frac{C-B}{2A-B-C} = \frac{\varkappa+1}{\varkappa+3} \tag{3.18a}
$$

25*

when the molecule is a slightly asymmetric prolate top, that is, $I_C \approx I_B$. When it is an asymmetric oblate top, $(I_A \approx I_B)$, then b is defined as

$$b = \frac{\dfrac{1}{I_A} - \dfrac{1}{I_B}}{\dfrac{2}{I_C} - \dfrac{1}{I_B} - \dfrac{1}{I_A}} = \frac{A - B}{2C - B - A} = \frac{\varkappa - 1}{\varkappa + 3}. \tag{3.18b}$$

Expressions for q_J in these cases, neglecting terms of order b^3 or higher, are as follows:

For $K = 0$

$$q_J = \frac{q}{(J+1)(2J+3)}\left[-J(J+1) + \left(\frac{3}{2}b^2 - b\eta\right)f(J,1)\right]. \tag{3.19}$$

For $K = 1$

$$q_J = \frac{q}{(J+1)(2J+3)}\left[3 - J(J+1) \mp \frac{\eta}{2}J(J+1) + \right. \\ \left. + \left(\frac{3}{2}b^2 - b\eta\right)\frac{f(J,2)}{4} \pm \frac{3}{128}b^2\eta J(J+1)f(J,2)\right]. \tag{3.20}$$

For $K = 2$

$$q_J = \frac{q}{(J+1)(2J+3)} \times \\ \times \left\{12 - J(J+1) + \left[\frac{3}{2}b^2 - b\eta\right]\left[\frac{f(J,3)}{6} - \frac{f(J,1)}{2} \mp \frac{f(J,1)}{2}\right]\right\}. \tag{3.21}$$

For $K = 3$

$$q_J = \frac{q}{(J+1)(2J+3)}\left\{27 - J(J+1) + \right. \\ \left. + \left[\frac{3}{2}b^2 - b\eta\right]\left[\frac{f(J,4)}{8} \mp \frac{f(J,2)}{4}\right] \mp \frac{3}{128}f(J,2)J(J+1)b^2\eta\right\}. \tag{3.22}$$

For $K > 3$

$$q_J = \frac{q}{(J+1)(2J+3)} \times \\ \times \left\{3K^2 - J(J+1) + \left[\frac{3}{2}b^2 - b\eta\right]\left[\frac{f(J,K+1)}{2(K+1)} - \frac{f(J,K-1)}{2(K-1)}\right]\right\}. \tag{3.23}$$

Here q, η, and $f(J, n)$ have the same meaning as in Eq. (3.17). Each energy level of a symmetric top with a quantum number K which is not zero is doubly degenerate, and is split into two levels when asymmetry occurs $(b \neq 0)$. The upper signs of $\mp$ occurring in Eqs. (3.20) to (3.23) apply to the upper energy level of these K-type doublets if the molecule is prolate and to the lower of the doublets if it is oblate. These expressions usually give adequate approximations to q_J when the asymmetry parameter b is less than a few tenths.

It is important to notice that even when the asymmetry parameter b is zero, there remains a term proportional to η in (3.20) which is different from the case of a symmetric top discussed above. This is because the nucleus was previously considered to be on the axis of a perfectly symmetric top, so that not only $b = 0$, but also $\eta = 0$. It is, however, possible to have an "accidental" symmetric top for which two principal moments of inertia are nearly the same so that b is very small, while there is no particular symmetry to the charge distribution so that η may be far from zero. An interesting case where the term in η must be taken into account even though the asymmetry b has a negligible effect on the hyperfine structure is that of a polyatomic linear molecule in the first excited state of a bending model of vibration. In this case there is an angular momentum around the molecular axis due to vibration, but which corresponds to $K=1$, and there is a small asymmetry η in the molecular electric field because the molecule is bent.

The quadrupole energy so far discussed for asymmetric rotors has been that given by first-order perturbation theory, i.e., first order in the quadrupole coupling constants. As in symmetric rotors, the quadrupole energy is sometimes large enough compared with the separation between rotational energy levels that second or higher order perturbations are important. To second-order approximation, the quadrupole energy may be written, as in (3.8),

$$W_Q = (I, J_{K_{-1}K_1}, F|H_Q|I, J_{K_{-1}K_1}, F) + \sum_{J' K'_{-1} K'_1} \frac{|(I, J_{K_{-1}K_1}, F|H|I, J'_{K_{-1}K_1}, F)|^2}{W_{J_{K_{-1}K_1}} - W_{J_{K'_{-1}K'_1}}} \quad (3.24)$$

where $W_{J_{K_{-1}K_1}}$ is the rotational energy, and the summation is made over all states excepting the unperturbed state $J_{K_{-1}K_1}$. The first term of (3.24) is the first-order energy already discussed.

The off-diagonal matrix elements indicated in (3.24) can be obtained by use of the wavefunction expansion (3.16), but an actual calculation may be quite tedious. It is possible to show that all matrix elements involved in the sum of (3.24) are zero except those for which $J'=J\pm1$ or $J'=J\pm2$. These matrix elements have been discussed by BRAGG [10], but have not yet been expressed in any very convenient form. In cases of near degeneracy, where two asymmetric top levels of appropriate symmetry lie close together, second-order quadrupole effects will, however, be quite important.

4. Hyperfine structure due to two or more nuclei in the same molecule. A molecule may contain more than one nucleus which produces an observable hyperfine structure in its spectrum. This usually occurs when it includes more than one nucleus of spin greater than $\frac{1}{2}$, and hence with quadrupole moments which can interact with molecular electrostatic fields. In such cases the quadrupole energies can no longer be computed by assuming $\boldsymbol{J}+\boldsymbol{I}=\boldsymbol{F}$, since two spins are involved which both interact with J and form the total angular momentum F.

Consider first the case of only two nuclei which produce hyperfine structure. If nucleus 1 is coupled to the molecule much more strongly than is nucleus 2, the entire system can be fairly well described by the vector model. The spin $\boldsymbol{I}_1$ of the first nucleus then adds vectorially to the molecular angular momentum $\boldsymbol{J}$ to form a resultant $\boldsymbol{F}_1$, and the spin $\boldsymbol{I}_2$, more weakly coupled to the molecule, adds vectorially to $\boldsymbol{F}_1$ to form the total angular momentum $\boldsymbol{F}$. The interaction energy due to the quadrupole moment of nucleus 1 is in this case given by the previous discussion and is independent of nucleus 2. If the interaction between $\boldsymbol{I}_2$ and $\boldsymbol{J}$ is proportional to the cosine of the angle between them, then a very simple calculation can be made of its average value. However, if it depends on the square of the cosine of this angle, as in the case of a quadrupole coupling, a more complicated quantum-mechanical calculation is necessary. Furthermore, if the couplings of nucleus 1 and nucleus 2 to the molecule are not widely different, then the vector model is a poor approximation in any case and a more complete treatment is needed. That outlined below follows the method of BARDEEN and TOWNES [14].

Let the Hamiltonian for the two nuclear interactions be

$$H = H_1(I_1, J) + H_2(I_2, J) \quad (4.1)$$

and first consider wavefunctions appropriate to the case where H_1 is much larger than H_2, as discussed above. Such wavefunctions may be indicated by $\varPsi_1(F_1, F)$

and are eigenfunctions of H_1, with eigenvalues which are those obtained by neglecting completely the spin of nucleus 2. There will be several functions $\Psi_1(F_1, F)$ having the same total angular momentum F, but different values of F_1, and of H_1, since the spin I_2 adds to F_1 to form F, and if I_2 is not zero the same value of F can result from more than one F_1. Consider next a set of wavefunctions appropriate to the case where H_2 is so much larger than H_1 that $\boldsymbol{I_2}$ and $\boldsymbol{J}$ couple to form the angular momentum $\boldsymbol{F_2}$. These wavefunctions may be designated $\Psi_2(F_2, F)$ and are similarly eigenfunctions of H_2. The number of different functions with the same value of F will be the same as before, and the two sets of wavefunctions are linearly related. That is,

$$\Psi_1(F_1, F) = \sum_{F_2} c(F_1, F_2)\, \Psi_2(F_2, F) \tag{4.2}$$

where the matrix $c(F_1, F_2)$ is unitary, and the phases of the wave functions may be chosen so that these coefficients are real. Hence the reverse transformation is

$$\Psi_2(F_2, F) = \sum_{F_1} c(F_1, F_2)\, \Psi_1(F_1, F). \tag{4.3}$$

Let the correct wavefunction for the system be

$$\Psi(F) = \sum_{F_1} a(F_1)\, \Psi_1(F_1, F). \tag{4.4}$$

Use of the Hamiltonian H_Q, the expressions (4.2), (4.3) and (4.4), and some manipulation yields the group of homogeneous equations

$$[A(F_1, F) + W_1(F_1) - W]\, a(F_1) + \sum_{F_1' \neq F_1} A(F_1, F_1')\, a(F_1') = 0 \tag{4.5}$$

for each value of F_1 which, when added to I_2, gives some particular value F of the total angular momentum. Here

$$A(F_1, F_1') = \sum_{F_2} c(F_1, F_2)\, c(F_1', F_2)\, W_2(F_2);$$

$W_1(F_1)$ is the eigenvalue of H_1 for the wavefunction $\Psi_1(F_1, F)$;
$W_2(F_2)$ is the eigenvalue of H_2 for $\Psi_2(F_2, F)$;
W is the total hyperfine energy due to H_1 and H_2.

Solutions of the secular determinant from Eqs. (4.5) give the possible values of the energy W. If H_2 is much smaller than H_1, and there is no degeneracy in W_1, the energy values are given to first order in H_2 by

$$W = W_1(F_1) + A(F_1, F_1) = W_1(F_1) + \sum_{F_2} [c(F_1, F_2)]^2\, W_2(F_2). \tag{4.6}$$

This is the case where the eigenfunction $\Psi_1(F_1, F)$ is essentially correct, and the interaction of nucleus 2 is just the sum of the various possible energies $W_2(F_2)$ weighted by their probability of occurrence $|c(F_1, F_2)|^2$.

The coefficients $c(F_1, F_2)$ are related as follows to certain W functions defined by RACAH [15]

$$c(F_1, F_2) = (-1)^{F+J-I_1-I_2}\, [(2F_1 + 1)\,(2F_2 + 1)]^{\frac{1}{2}}\, W(F_1 F J F_2; I_2 I_1). \tag{4.7}$$

The W functions are tabulated for most values of the variables which are of interest [16], [17]. However, the coefficients $c(F_1 F_2)$ can be obtained much more simply and directly from tabulation of these coefficients themselves given by TOWNES and SCHAWLOW [7] for I_1 arbitrary and $I_2 = \frac{1}{2}$, 1, or $\frac{3}{2}$. Of course the roles of $I_1 I_2$ may be reversed, so that these latter tables suffice if either one of the two spins is less than 2.

An example of the behavior of energy levels when two nuclear quadrupole moments interact with the molecular rotation is given in Fig. 4, which shows the variation of hyperfine energy as a function of the relative size of the two coupling constants. Fig. 5 illustrates a spectrum in which hyperfine structure of two nuclei is evident, but where the two coupling constants differ in size by about a factor of 20, so that the hyperfine pattern due to the nucleus with largest coupling is not distorted beyond recognition.

If H_1 and H_2 differ in magnitude by a factor of more than four or five, relative intensities of transitions can rather simply be obtained with the accuracy needed

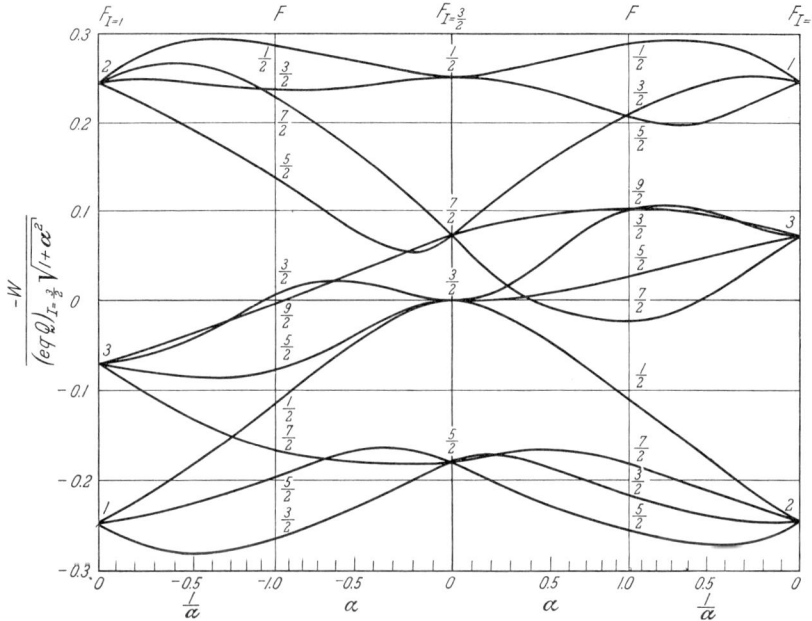

Fig. 4. Energies W resulting from quadrupole coupling of two nuclei of spin 1 and $\frac{3}{2}$, when $J = 2$. The parameter α is $(eqQ)_{I=1}/(eqQ)_{I=\frac{3}{2}}$ (from BARDEEN and TOWNES [14]).

for identifying the hyperfine structure components. Thus if H_1 is much larger than H_2, the relative intensity of a group of transitions from states characterized by J, F_1 to those with J', F_1' can be obtained from tables which apply when $I_2 = 0$. Relative intensities of the more closely spaced components of these groups, involving transitions between states J, F_1, F and J', F_1', F', can likewise be found from the same tables by replacing J with F_1, and I with I_2. When H_1 and H_2 are not very different in magnitude, evaluation of relative intensities is much more difficult. The coefficients $a(F_1)$ of the wavefunctions must be evaluated from Eqs. (4.5) for the various values of W, and from these wavefunctions, the intensities of transitions may be calculated.

The case of three nuclei coupled by quadrupole effects to a molecule is not very important, except for symmetric tops involving three halogens, such as $AsCl_3$, $HCBr_3$, etc. The resulting hyperfine structure is so complex that there are as yet no clear examples where this type of hyperfine structure has been measured and solved. Hence these cases are probably of little importance for determination of nuclear quadrupole moments. However, they have been discussed theoretically by MIZUSHIMA and ITO [18], and by BERSOHN [19].

5. Nuclear quadrupole coupling in solids. In principle, the interaction of a nuclear quadrupole moment with the surrounding charges in a solid may be calculated by treating the solid as a molecule with large angular momenta, and for which the rotational states are spaced much more closely than are the levels due to hyperfine energy. However, calculation of the interaction by this approach is awkward in some cases, and is preferably replaced by a fresh beginning which

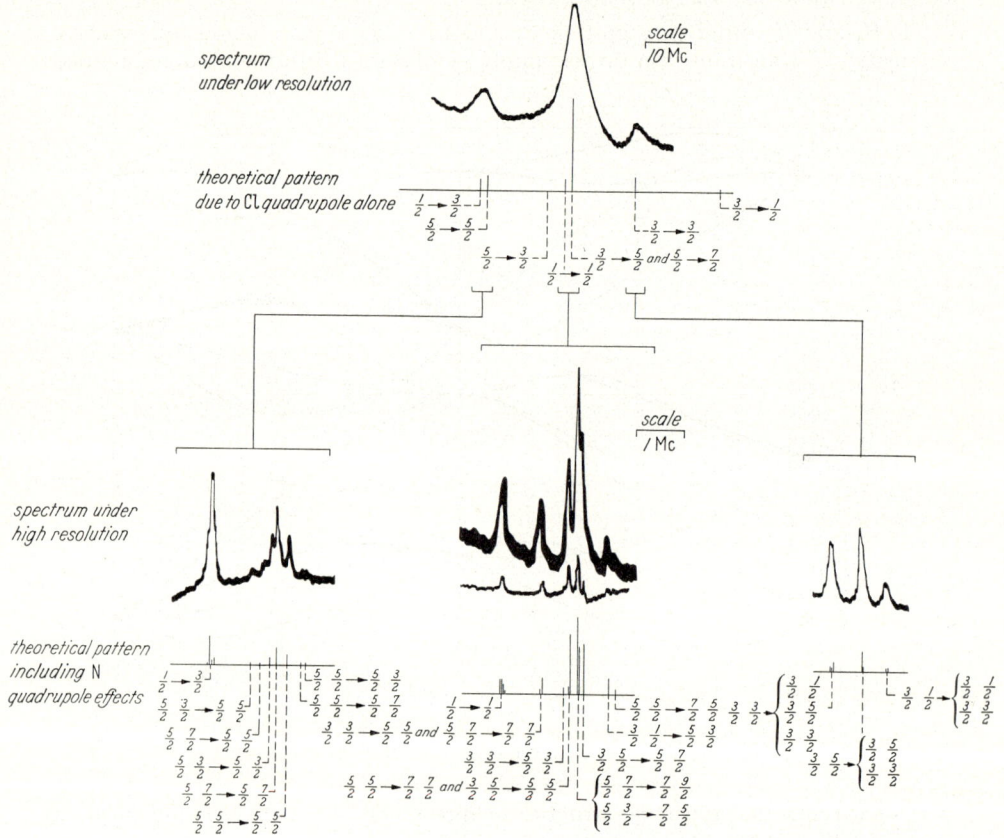

Fig. 5. Hyperfine structure involving two quadrupole interactions in the $J = 2 \leftarrow 1$ transition of Cl³⁵ CN¹⁴ (ground vibrational state). Experimental observations under low and high dispersion are compared with the theoretical pattern. Spins for Cl³⁵ and N¹⁴ are 3/2 and 1, quadrupole coupling constants -83.3 Mc and -3.63 Mc respectively. (From TOWNES, HOLDEN, and MERRITT [17 a].)

considers the solid as fixed in orientation *ab initio*. From the Hamiltonian $H_{\mathrm{op}} = -\frac{1}{6}\mathbf{Q}:\mathbf{\nabla E}$ and the expression (2.13) of $\mathbf{Q}$ in terms of I, the following form can be readily derived.

$$H = \frac{eQ}{2I(2I-1)}\left[I_x^2 \frac{\partial^2 V}{\partial x^2} + I_y^2 \frac{\partial^2 V}{\partial y^2} + I_z^2 \frac{\partial^2 V}{\partial z^2}\right] \tag{5.1}$$

where the coordinates x, y, and z are the principal axes of $\mathbf{\nabla E}$, which are fixed in space. Since all quantities in (5.1) are constants except $I_x^2 + I_y^2 + I_z^2$, this Hamiltonian is essentially the same as that for rotation of a rigid molecule, for which the rotational kinetic energy is $H = \frac{J_x^2}{2I_a} + \frac{J_y^2}{2I_b} + \frac{J_z^2}{2I_c}$, where I_a, I_b, and I_c are moments of inertia about the principal axes of inertia and J_x, J_y, J_z are the corresponding angular momenta. Because of the similarity of these two

Hamiltonians, theoretical discussion, formulae and tables of energies [7] for molecular rotational levels apply to the quadrupole coupling of a nucleus in a solid. However, the nuclear case is somewhat different from the usual molecular case in that values of angular momentum which are half-integral may be involved as well as those of integral values. In addition, the nuclear spin I is usually not larger than $\frac{9}{2}$, so that determination of eigenvalues of (5.1) is somewhat simpler than the case of a rotating molecule, where J may be much larger. Consequently, it is more convenient in most cases to calculate quadrupole energies directly without use of the considerable literature (see, for example, Chap. 4 of [7]) available for the eigenvalues of molecular rotation energy. Tables for quadrupole energies when $I = \frac{5}{2}, \frac{7}{2}$, or $\frac{9}{2}$ have also been given [20] for asymmetries η of 0.1, 0.2, 0.3, ... 1.0.

The Hamiltonian (5.1) may be put in the following form for convenient evaluation of matrix elements.

$$H_{op} = \frac{e\,q\,Q}{4\,I(2I-1)} \left[3\,I_z^2 - I^2 + \frac{1}{2}\,\eta(I_+^2 + I_-^2) \right]_{op} \tag{5.2}$$

where $I_\pm = I_x \pm i I_y$. Letting M be the projection of I on the z direction, the matrix elements for operators involved in (5.2) are the following

$$(M\,|I_z|\,M) = M, \tag{5.3}$$

$$(M\,|I^2|\,M) = I(I+1), \tag{5.4}$$

$$(M\,|I_\pm^2|\,M \mp 2) = [(I \pm M)(I \pm M - 1)(I \mp M - 1)(I \mp M + 2)]^{\frac{1}{2}}. \tag{5.5}$$

When the asymmetry η of the electric field gradient is zero, the eigenvalues of H_{op} are easily obtained as

$$W = \frac{e\,q\,Q}{4\,I(2I-1)} \left[3\,M^2 - I(I+1) \right]. \tag{5.6}$$

When η is not zero, the energy values can be written in closed form for $I = 1$, $\frac{3}{2}$, 2, and 3. For other values of I, they may be expanded in powers of the asymmetry η. In these expansions, which are given below, $W_{|M|}$ will be used to denote the quadrupole energies of the two states which, when η is zero, have projection $\pm M$ of I on the z axis. Except when $M = 0$, these levels occur in pairs, which are degenerate when $\eta = 0$. Because of this degeneracy, the projection of I on the z axis is not a good quantum number even for small non-zero values of η, although $|M|$ is not much changed by the assymmetry.

For $I = 1$
$$\left. \begin{aligned} W_0 &= \frac{-e\,q\,Q}{2}, \\ W_1 &= \frac{e\,q\,Q(1 \pm \eta)}{4}. \end{aligned} \right\} \tag{5.7}$$

For $I = \frac{3}{2}$
$$\left. \begin{aligned} W_{\frac{1}{2}} &= -\frac{1}{4}\,e\,q\,Q\left(1 + \frac{1}{3}\,\eta^2\right)^{\frac{1}{2}}, \\ W_{\frac{3}{2}} &= \frac{1}{4}\,e\,q\,Q\left(1 + \frac{1}{3}\,\eta^2\right)^{\frac{1}{2}}. \end{aligned} \right\} \tag{5.8}$$

For $I = 2$
$$\left. \begin{aligned} W_0 &= -\frac{e\,q\,Q}{4}\sqrt{1 + \frac{1}{3}\,\eta^2}, \\ W_1 &= -\frac{e\,q\,Q}{8}\,(1 \pm \eta), \\ W_2 &= \frac{e\,q\,Q}{4}; \quad \frac{e\,q\,Q}{4}\sqrt{1 + \frac{1}{3}\,\eta^2}. \end{aligned} \right\} \tag{5.9}$$

For $I = \frac{5}{2}$

$$W_{\frac{1}{2}} = \frac{-eqQ}{5}\left(1 + \frac{4}{9}\eta^2 - \frac{172}{329}\eta^4 + \frac{15332}{59044}\eta^6 - \frac{1717580}{4782969}\eta^8 + \cdots\right),$$

$$W_{\frac{3}{2}} = \frac{-eqQ}{20}\left(1 - \frac{3}{2}\eta^2 - \frac{23}{24}\eta^4 - \frac{449}{432}\eta^6 + \frac{44675}{31104}\eta^8 + \cdots\right), \qquad (5.10)$$

$$W_{\frac{5}{2}} = \frac{eqQ}{4}\left(1 + \frac{1}{18}\eta^2 + \frac{17}{5832}\eta^4 - \frac{143}{944784}\eta^6 - \frac{12587}{612220032}\eta^8 + \cdots\right).$$

For $I = 3$

$$W_0 = \frac{-eqQ}{10}\left(1 + \sqrt{1 + \frac{5}{3}\eta^2}\right),$$

$$W_1 = \frac{eqQ}{20}\left(1 \pm \eta - 2\sqrt{4 \mp 2\eta + \frac{2}{3}\eta^2}\right), \qquad (5.11)$$

$$W_2 = 0; \qquad \frac{-eqQ}{10}\left(1 - \sqrt{1 + \frac{5}{3}\eta^2}\right),$$

$$W_3 = \frac{eqQ}{20}\left(1 \pm \eta + 2\sqrt{4 \mp 2\eta + \frac{2}{3}\eta^2}\right).$$

For $I = \frac{7}{2}$

$$W_{\frac{1}{2}} = -\frac{5}{28}eqQ\left(1 + \frac{5}{6}\eta^2 - \frac{311}{216}\eta^4 + \frac{20557}{3888}\eta^6 - \frac{762019}{31104}\eta^8 + \cdots\right),$$

$$W_{\frac{3}{2}} = -\frac{3}{28}eqQ\left(1 - \frac{31}{30}\eta^2 + \frac{21967}{9000}\eta^4 - \frac{3975973}{450000}\eta^6 + \frac{734970487}{18000000}\eta^8 + \cdots\right),$$

$$W_{\frac{5}{2}} = \frac{eqQ}{28}\left(1 + \frac{5}{6}\eta^2 + \frac{25}{216}\eta^4 - \frac{275}{3888}\eta^6 - \frac{25}{93312}\eta^8 + \cdots\right), \qquad (5.12)$$

$$W_{\frac{7}{2}} = \frac{7}{28}eqQ\left(1 + \frac{1}{30}\eta^2 + \frac{29}{27000}\eta^4 + \frac{1241}{12150000}\eta^6 + \frac{2263}{1458000000}\eta^8 + \cdots\right).$$

For $I = \frac{9}{2}$

$$W_{\frac{1}{2}} = \frac{-eqQ}{6}\left(1 + \frac{4}{3}\eta^2 - \frac{764}{135}\eta^4 + \frac{12580}{243}\eta^6 - \frac{18185884}{30375}\eta^8 + \cdots\right),$$

$$W_{\frac{3}{2}} = \frac{-eqQ}{8}\left(1 - \frac{37}{30}\eta^2 + \frac{69161}{9000}\eta^4 - \frac{280555399}{4050000}\eta^6 + \frac{387985068547}{486000000}\eta^8 + \cdots\right),$$

$$W_{\frac{5}{2}} = \frac{-eqQ}{24}\left(1 - \frac{43}{42}\eta^2 - \frac{28979}{74088}\eta^4 + \frac{48841813}{65345616}\eta^6 - \frac{9454170085}{76846444416}\eta^8 + \cdots\right), \qquad (5.13)$$

$$W_{\frac{7}{2}} = \frac{eqQ}{12}\left(1 + \frac{7}{30}\eta^2 + \frac{287}{27000}\eta^4 + \frac{39767}{12150000}\eta^6 - \frac{1761851}{1458000000}\eta^8 + \cdots\right),$$

$$W_{\frac{9}{2}} = \frac{eqQ}{4}\left(1 + \frac{1}{42}\eta^2 + \frac{29}{41160}\eta^4 + \frac{971}{21781872}\eta^6 + \frac{14490601}{3201935184000}\eta^8 + \cdots\right).$$

It may be noted that when I is half integral, the energy levels are always degenerate in pairs, but this degeneracy is removed for integral values of I by the presence of asymmetry. In the latter case, the resulting splitting of the $|M|$ $=1$ levels is always greatest, being proportional to the first power of η, while splitting of other levels depends on higher powers of the asymmetry.

In most cases, the asymmetry η is small enough that the only strong transitions between quadrupole states in a crystal are those which are allowed when $\eta = 0$, that is, $\Delta M = \pm 1$. Hence the observed spectrum is usually quite simple and easily fitted to the theoretical energy levels. In some cases, however, the asymmetry may be sufficiently large that other transitions are observed, and the spectrum becomes much more complex.

Except in the case of $I = \frac{3}{2}$, measurement of transitions between the energy levels indicated in Eqs. (5.7) to (5.13) gives independently values of the magnitudes of eqQ and η. However, the sign of eqQ and η cannot thus be determined. In fact, symmetry considerations show [21] that the sign of eqQ in a solid cannot be determined by any combination of electromagnetic fields unless the temperature is so low that kT is comparable with the quadrupole energy. In this case, which requires a temperature so low as to be very inconvenient, the difference in population between the various hyperfine levels may be large enough to indicate which state is of lowest energy.

The sign of η may be determined in solids by application of a fixed magnetic field which produces a Zeeman effect [22], [23]. The Zeeman effect allows also determination of the magnitudes of both eqQ and η when $I = \frac{3}{2}$ [22], [23]. BERSOHN [22] has given a general expression for the first-order Zeeman effect for half-integral spins when the asymmetry η is sufficiently small so that terms in powers of η greater than the second may be neglected. The energy due to magnetic field and quadrupole effects is

$$
\begin{aligned}
W_M = W_M^0 &+ M\,\frac{\mu_I}{I}\,H_z + \eta^2\,\frac{\mu_I}{I}\,\frac{H_z(M+2)\,f(I,\,M+1)}{144\,(M+1)^2} + \\
&+ \eta^2\,\frac{\mu_I}{I}\,\frac{H_z(M-2)\,f(I,\,M-1)}{144\,(M-1)^2} \qquad \text{for } |M| > \frac{5}{2}\,,
\end{aligned}
$$

$$
\begin{aligned}
W_{\frac{3}{2}} = W_{\frac{3}{2}}^0 &\pm \frac{\mu_I}{I}\left\{ \frac{9}{4}\,H_z^2 + \frac{7}{600}\,\eta^2 f\left(I,\,\frac{5}{2}\right)H_z^2 - \frac{1}{24}\,\eta^2 H_z^2 f\left(I,\,\frac{1}{2}\right) + \right. \\
&\left. + \frac{1}{36}\,\eta^2\,(H_x^2 + H_y^2)\left(I - \frac{1}{2}\right)\left(I + \frac{3}{2}\right)f\left(I,\,\frac{1}{2}\right)\right\}^{\frac{1}{2}},
\end{aligned}
$$

$$
\begin{aligned}
W_{\frac{1}{2}} = W_{\frac{1}{2}}^0 &\pm \frac{\mu_I}{I}\left\{ \frac{1}{4}\,H_z^2 - \frac{1}{6}\,\eta\,(H_x^2 - H_y^2)\left(I + \frac{1}{2}\right)\left(I + \frac{3}{2}\right)^{\frac{1}{2}}\left(I - \frac{1}{2}\right)^{\frac{1}{2}}f^{\frac{1}{2}}\left(I,\,\frac{1}{2}\right) + \right. \\
&+ \frac{1}{4}\left(I + \frac{1}{2}\right)^2(H_x^2 + H_y^2) + \frac{5}{648}\,\eta^2 f\left(I,\,\frac{3}{2}\right)H_z^2 - \frac{1}{24}\,\eta^2 f\left(I,\,\frac{1}{2}\right)H_z^2 + \\
&+ \frac{\eta^2}{36}\,(H_x^2 + H_y^2)\left(I + \frac{3}{2}\right)\left(I - \frac{1}{2}\right)f\left(I,\,\frac{1}{2}\right) + \\
&\left. + \frac{\eta^2}{108}\,(H_x^2 + H_y^2)\left(I + \frac{1}{2}\right)\left(I + \frac{5}{2}\right)^{\frac{1}{2}}\left(I - \frac{3}{2}\right)^{\frac{1}{2}}f^{\frac{1}{2}}\left(I,\,\frac{1}{2}\right)f^{\frac{1}{2}}\left(I,\,\frac{3}{2}\right)\right\}^{\frac{1}{2}}
\end{aligned}
$$

$$(5.14)$$

where

W_M^0 is the quadrupole energy for zero magnetic field,

μ_I is the magnetic moment of the nucleus,

H_x, H_y, and H_z, are the components of the magnetic field along the principal axes of the field gradient dyadic $\boldsymbol{\nabla E}$,

$$f(I,\,M) = \tfrac{1}{4}\,(I^2 - M^2)\,[(I+1)^2 - M^2].$$

Since the energy levels depend on relative orientations of the magnetic field $\boldsymbol{H}$ and the axes of the dyadic $\boldsymbol{\nabla E}$ when both ZEEMAN and quadrupole effects are present, it is necessary to use a single crystal for accurate measurements in this case. Otherwise, in polycrystalline material, the lines are broadened and usually not detectable. Further expressions for the Zeeman effect in certain cases have been given by KOJIMA et al. [23], and numerical tables by COHEN [20] and by PARKER [24].

When the quadrupole coupling is rather small, direct transitions between quadrupole energy levels are not easy to observe, and such quadrupole interactions may be measured as a small perturbation of the precessional frequency of the nucleus in a large magnetic field [25]. When the interaction between the nuclear magnetic moment and magnetic field is very much larger than the quadrupole coupling, the spin is quantized with respect to $\boldsymbol{H}$ regardless of quadrupole effects, and the energy is

$$W = M \frac{\mu_I}{I} H + \frac{eQ}{8I(2I-1)} [3M^2 - I(I+1)][3\alpha_{z'z}^2 - 1 + (\alpha_{z'x}^2 - \alpha_{z'y}^2)\eta] \frac{\partial^2 V}{\partial z^2} \quad (5.15)$$

where the x, y, and z directions are the principal axes of the dyadic $\boldsymbol{V}\boldsymbol{E}$ and z' is the direction of the magnetic field $\boldsymbol{H}$. $\alpha_{z'x}$, $\alpha_{z'y}$, and $\alpha_{z'z}$ are cosines of the angles between z' and the axes x, y, and z respectively. BERSOHN [22] has given an expression for the energy when $eqQ < \mu_I H$ which includes terms proportional to the cube of eqQ, and hence should be used when eqQ is so large that (5.15) is not sufficiently accurate.

In case the quadrupole coupling is very small or rapidly varying, its presence may be evident only as a broadening of the resonant frequency of nuclear precession in a magnetic field. This may occur, for example, for an atom or ion which would have zero quadrupole coupling when isolated, but which experiences a smal fluctuating quadrupole energy in a liquid or solid. Such fluctuations induce transitions between Zeeman levels, and hence broaden them [25], [26]. Although rather rough measurement of quadrupole coupling constants, or of their ratios for two isotopes, can be made from the breadths of transitions between Zeeman levels, only in very unusual cases can such measurements be expected to yield new information on quadrupole moments themselves.

6. Quadrupole hyperfine structure in electronic paramagnetic resonance. Paramagnetic materials often show microwave resonances when placed in a magnetic field, which correspond roughly to transitions between the Zeeman levels of an isolated atom. It is of course only paramagnetic substances which produce resonances of this type, since only they show large Zeeman effects, because of the presence of electronic magnetic moments. The electron or electrons which are reoriented in such a transition also interact with the atomic nuclei, producing a hyperfine energy which may change and split or shift the observed resonant frequency.

Although Zeeman levels in a paramagnetic material are quite analogous to those in an isolated atom, there are pronounced differences between the two because electron states of a paramagnetic ion or atom in a solid may be very much affected by the presence of neighbouring atoms. For example, the electronic orbital angular momentum is "quenched" by interactions with neighbouring atoms so that its average value is zero. The Zeeman effect is then primarily due to unpaired electron spins. If there is axial symmetry, the Hamiltonian for a paramagnetic system with hyperfine structure and in a magnetic field is not excessively complicated and has been given by ABRAGAM and PRYCE [27] as

$$\left. \begin{aligned} H = \mu_0 \{g_\parallel H_z S_z + g_\perp (H_x S_x + H_y S_y)\} + D\left\{S_z^2 - \frac{1}{3}S(S+1)\right\} + AS_z I_z + \\ + B(S_x I_x + S_y I_y) + \frac{3eQ}{4I(2I-1)} \frac{\partial^2 V}{\partial z^2} \left\{I_z^2 - \frac{1}{3}I(I+1)\right\} - \frac{\mu_I}{I} \boldsymbol{H} \cdot \boldsymbol{I}. \end{aligned} \right\} \quad (6.1)$$

Here z is the axis of symmetry, μ_0 is the Bohr magneton, μ_I the nuclear magnetic moment, e the electron charge, and H_x, H_y, H_z, S_x, S_y, S_z, I_x, I_y, I_z vector components of the magnetic field H, the electron spin operator S, and the nuclear spin operator I respectively. The other quantities are constants which depend on the particular substance involved, and which may largely be determined from experimental measurements. $g_\parallel$ and $g_\perp$ are dimensionless g-factors giving the Zeeman effect on the electrons for a field parallel or perpendicular respectively to the z axis. The nuclear magnetic moment μ_I may in most cases be accurately measured by other types of experiments. A and B give the magnetic hyperfine interaction between S and I for orientations parallel or perpendicular to the axis respectively, and $eQ\,\dfrac{\partial^2 V}{\partial^2 z}$ is the same quadrupole coupling constant as used above[1]. Possibilities of measuring nuclear quadrupole moments from interactions given by (6.1) have been discussed in some detail by BLEANEY [28].

In the typical case of a strong magnetic field, the most intense Zeeman transitions involve a change of one unit in M_s, the projection of $\boldsymbol{S}$ on $\boldsymbol{H}$, and no change in M_I, the projection of the nuclear spin on the effective magnetic field which it experiences. For these transitions, the frequencies ν may be written [28] when $M_s \to M_s - 1$:

$$
\left.
\begin{aligned}
h\nu = h\nu_0(M_s) + K\,M_I &+ \frac{B^2}{4g\mu_0 H}\left(\frac{A^2+K^2}{K^2}\right)\{I(I+1) - M_I^2\} + \\
&+ \frac{B^2}{2g\mu_0 H}\frac{A}{K}M_I(2M_s-1) + \frac{1}{2g\mu_0 H}\left(\frac{A^2-B^2}{K}\right)^2\left(\frac{g_\parallel g_\perp}{g^2}\right)^2\sin^2\vartheta\cos^2\vartheta\,M_I^2+ \\
&+ X^2\frac{\cos^2\vartheta\sin^2\vartheta}{2KM_s(M_s-1)}\left(\frac{AB g_\parallel g_\perp}{K^2 g^2}\right)^2 M_I\{4I(I+1) - 8M_I^2 - 1\} - \\
&- X^2\frac{\sin^4\vartheta}{8KM_s(M_s-1)}\left(\frac{Bg_\perp}{Kg}\right)^4 M_I\{2I(I+1) - 2M_I^2 - 1\}
\end{aligned}
\right\} \quad (6.2)
$$

where

$$
X = \frac{3eQ}{4I(2I-1)}\frac{\partial^2 V}{\partial z^2},
$$

$$
K^2 g^2 = A^2 g_\parallel^2 \cos^2\vartheta + B^2 g_\perp^2 \sin^2\vartheta,
$$

$$
g^2 = g_\parallel^2 \cos^2\vartheta + g_\perp^2 \sin^2\vartheta.
$$

ϑ is the angle between the external magnetic field $\boldsymbol{H}$ and the z axis. Hence K is proportional to the magnetic field produced by the electrons at the nucleus, and with respect to which the nuclear spin is approximately quantized.

$\nu_0(M_s)$ is the frequency associated with the electronic Zeeman effect, which would be observed if the hyperfine structure were negligibly small. Direct interaction between the field $\boldsymbol{H}$ and the nuclear magnetic moment is usually quite small, and has been consequently neglected in (6.2).

Although the quadrupole coupling contributes to the energy of each hyperfine level in first order, this contribution depends only on the value of M_I, and hence does not change for transitions involving $\Delta M_I = 0$. However, there are second order terms, as can be seen from (6.2), which are proportional to Q^2. These allow a determination of the magnitude but not the sign of the quadrupole coupling constant when their effects are large enough to be detected experimentally. When $eqQ \ll K$, there are no first-order effects due to the quadrupole interaction and when $\boldsymbol{H}$ is parallel to the symmetry axis, no effects due to quadrupole interaction are detected at all, since this interaction is then diagonal and dependent only on M_s.

If $\boldsymbol{H}$ is at an angle with respect to the z axis, the quadrupole interaction breaks down the ordinary selection rule when eqQ is comparable in size with K, and the additional transitions $\Delta M_I = \pm 1, \pm 2$ are allowed. If eqQ is not too

---

[1] This differs from the original notation of ABRAGAM and PRYCE [27] who let Q be the entire energy coefficient multiplying $I_z^2 - \frac{1}{3}I(I+1)$ rather than the nuclear quadrupole moment itself.

large, the intensity of these transitions is of the order $\left(\frac{eqQ}{K}\right)^2$. For the transitions M_s, $M_I \to M_s - 1$, $M_I \pm 1$, or M_s, $k \pm \frac{1}{2} \to M_s - 1$, $k \mp \frac{1}{2}$ where k has the values $I - \frac{1}{2}$, $I - \frac{3}{2} \cdots - (I - \frac{1}{2})$, the frequencies are given by [28]

$$h\nu = h\nu_0(M_s) + Kk \pm \{K(M_s - \tfrac{1}{2}) + X'k - \gamma'\}. \tag{6.3}$$

For M_s, $M_I \pm 1 \to M_s - 1$, $M_I \mp 1$

$$h\nu = h\nu_0(M_s) + KM_I \pm \{K(2M_s - 1) + 2X'M_I - 2\gamma'\} \tag{6.4}$$

where

$$X' = \frac{3eqQ}{4I(2I-1)}\left(\frac{3A^2 g_{\parallel}^2}{K^2 g^2}\cos^2\vartheta - 1\right),$$

$$\gamma' = \frac{\mu_I H}{KIh}(A g_{\parallel}\cos^2\vartheta + B g_{\perp}\sin^2\vartheta).$$

The small terms dependent on powers of eqQ higher than the first have been omitted from (6.3) and (6.4). Intensities of the $\Delta M_I = \pm 1$ transitions are proportional to $\sin^2 2\vartheta$, and those for $\Delta M = \pm 2$ to $\sin^4\vartheta$.

If the electron spin is $\frac{1}{2}$, so that $M = \frac{1}{2}$ only, and if the term γ', which is proportional to $\mu_I H$, is too small to be observed, the frequencies given by (6.3) and (6.4) are $h\nu = h\nu_0 + (K \pm X')k$ and $h\nu = h\nu_0 + (K \pm 2X')M_I$ respectively. Since both k and M_I take on positive and negative values, these expressions show that the resulting spectra are completely symmetric and independent of the signs of K and X', or of their relative signs. Under these conditions it is thus impossible to determine experimentally the sign of X' or hence of eqQ.

An illustration of the effects of quadrupole interactions for the particular case of $S = \frac{1}{2}$ and $I = \frac{3}{2}$ is given in Fig. 6. There the four lines corresponding to $\Delta M_I = \pm 1$ may be seen to appear when ϑ is different from zero, to shift in position and reach maximum intensity when $\vartheta = 45°$, and to disappear again when $\vartheta = 90°$. The four $\Delta M_I = \pm 2$ transitions are much weaker, and have their maximum intensity near $\vartheta = 90°$. As indicated in the discussion above, the entire hyperfine pattern is symmetric. In a few cases, however, this procedure may be expected to fail since the magnetic field H may perturb the electronic state of the paramagnetic atom, which in turn changes the hyperfine interaction in a way which would vary linearly with magnetic field [28a]. This effect is then superficially similar to a direct interation between the nuclear magnetic moment and the magnetic field. However, it may have either sign and in some special cases may be large enough to give a misleading direction to the asymmetry of the observed spectral pattern.

In magnetic fields as large as 10000 oersteds, the direct effect of the external field on the magnetic moment of the nucleus is no longer negligible, and this produces an asymmetry in the spectra given by (6.3) and (6.4). From the asymmetry, the relative signs of $K\mu_I$ and X' can be determined. The sign of K depends on the sign of the magnetic field produced by the electrons at the nucleus and on the sign of μ_I, the nuclear magnetic moment. Hence the sign of $K\mu_I$ depends

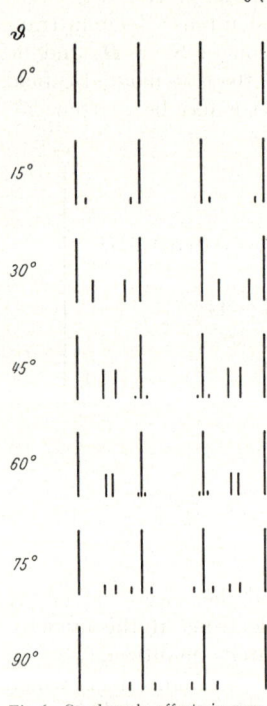

Fig. 6. Quadrupole effects in paramagnetic resonance for the case $S = \frac{1}{2}$, $I = \frac{3}{2}$, with no anisotropy and $A = B = 5 \times \dfrac{3eQ}{4I(2I-1)}\dfrac{\partial^2 V}{\partial z^2}$. (From Bleaney [28].)

only on the sign of the internal field, and if this is known from a sufficient theoretical understanding of the electronic wavefunctions involved, the sign of X' or of the quadrupole coupling eqQ is then determinable.

Another interesting and readily soluble case occurs when the magnetic field is zero or along the z-axis. If this is true and S is $\frac{1}{2}$, the energy levels are given by [28]

$$W = -\frac{A}{A} + X\left\{k^2 + \frac{1}{4} - \frac{1}{3}I(I+1)\right\} \pm$$
$$\pm \frac{1}{2}\left\{[k(A-2X) + g_\parallel \mu_0 H]^2 + B^2\left[\left(I+\frac{1}{2}\right)^2 - k^2\right]\right\}^{\frac{1}{2}} \tag{6.5}$$

where $k = I+\frac{1}{2}, I-\frac{1}{2}, \ldots I+\frac{1}{2}$ and only the positive sign in front of the radical must be taken for the particular case when $k = I+\frac{1}{2}$, only the negative sign for $k = -(I+\frac{1}{2})$. Allowed transitions are rather complicated when H is small and $B \neq 0$. However, if such transitions for very low fields are observed, they allow determination of the quadrupole coupling constant.

7. Effects of quadrupole coupling on angular correlation of successive particles from nuclear decay[1]. In some cases it is possible to determine the quadrupole coupling constant of an unstable nucleus by studying the angular correlation between radiation emitted by a radioactive nucleus which is its parent, and that emitted shortly thereafter by the nucleus itself. This type of determination is practical only when the unstable nucleus is the product of some radioactive parent of lifetime more than a few seconds, and when the reciprocal of its own lifetime τ falls in the range $\frac{1}{eqQ} < \tau < T_1$. Here T_1 is the relaxation time for reorientation of the nuclear spin and eqQ is the usual quadrupole coupling constant. Hence τ must usually be in the range 10^{-5} to 10^{-9} sec.

Suppose that a particular nucleus in a state indicated by a radiates a particle (or γ-ray) in a particular direction Ω, as a result of the perturbing Hamiltonian $H_1(\Omega_1)$, which leaves the nucleus in a state indicated by b. Suppose further that the nucleus in state b is not subject to any appreciable torques from external fields, so that it remains in state b until it radiates a second particle (or γ ray) in a direction Ω_2 as a result of the perturbation $H_2(\Omega_2)$, leaving the nucleus in a final state c. Then the probability of these two successive radiations occurring in the particular directions indicated is proportional to

$$p = \sum_{a,c,b,b'} (a|H_1(\Omega_1)|b)(b|H_2(\Omega_2)|c)(c|H_2(\Omega_2)b')(b'|H_1(\Omega_1)|a). \tag{7.1}$$

This relation can also be expressed in the form [29]

$$p = \sum_k A_k P_k(\cos\vartheta) \tag{7.2}$$

when A_k is a constant, P_k a Legendre polynomial of order k where k is even, and ϑ the angle between directions Ω_1 and Ω_2. From (7.2), it is evident that the relative directions of the two successive radiations are not necessarily random, but show a correlation. However, this correlation may be changed or destroyed if hyperfine effects act on the nucleus and reorient it during the intermediate state b, and from such changes the hyperfine interactions can at times be evaluated.

If the states b, b' of the nucleus are subjected to an external field, such as that due to a hyperfine interaction, these states may conveniently be taken as

---

[1] For the general theory of angular correlations cf. the article of S. DEVONS and L. J. B. GOLDFARB in Vol. XLII of this Encyclopedia.

eigenstates of the Hamiltonian describing the hyperfine interaction. Then the relative phases of states b and b' will vary with time as $e^{i(W_b - W_{b'}t/\hbar)}$, where E_b and $E_{b'}$ are the energies of these two eigenstates due to the perturbing field. Hence at some particular time t after the first radiation has taken place, the angular correlation between it and the second radiation will be given by

$$p(t) = \sum_{a,c,b,b'} e^{i(W_b - W_{b'}t/\hbar)} (a|H_1(\Omega_1)|b)(b|H_2(\Omega_2)|c) \times \\ \times (c|H_2(\Omega_2)|b')(b'|H_1(\Omega_1)a). \qquad (7.3)$$

If the perturbation is due to a fixed magnetic field, then expression (7.3) can be shown [31], [32] to represent a steady precession of the angular correlation pattern about the direction of the magnetic field with a frequency $\nu = \dfrac{\mu_I H}{I \hbar}$, where μ_I is the nuclear magnetic moment, I the spin, and H the magnetic field.

If the perturbation of states b, b' is due to a nuclear quadrupole interaction, there is no simple frequency of precession, and variation in the angular correlation pattern may be rather complex, although it shows some fairly simple features when the electric field gradient has an axis of symmetry ($\eta = 0$). In this case, the energies are given, from (5.6), by $E = \dfrac{eqQ}{4I(2I-1)}[3M^2 - I(I+1)]$ so that every energy difference $E_b - E_{b'}$ is a multiple of the smallest difference, namely $W = \dfrac{3eqQ}{2I(2I-1)}$ when the nuclear spin I is half-integral and $\dfrac{3eqQ}{4I(2I-1)}$ where the spin is integral. Hence from (7.3) $p(t)$ is periodic with a frequency

$$\nu = \frac{3eqQ}{2I(2I-1)} \quad \text{or} \quad \frac{3eqQ}{4I(2I-1)} \qquad (7.4)$$

for I half-integral or integral respectively. The quadrupole coupling constant in (7.4) is of course expressed in frequency units, whereas above it was in terms of energy.

In any practical experiment, the angular correlation is measured for particles emitted during a certain interval of time, e.g., between t_1 and t_2. During this time the rate of emission of particles decreases as $e^{-t/\tau}$, where τ is the lifetime of the radioactive nucleus. Hence the angular correlation for all particles emitted between t_1 and t_2 is given by a simple integration of (7.3)

$$P = \sum_{a,b,b',c} \frac{1}{\tau} \int_{t_1}^{t_2} e^{-\frac{t}{\tau} + \frac{i}{\hbar}(W_b - W_{b'})t} (a|H_1|b)(b|H_2|c)(c|H|b')(b'|H|a)\,dt. \qquad (7.5)$$

If the time interval $t_1 - t_2$ for counting particles is much smaller than $\hbar/(W_b - W_{b'})$, this expression is not very different from (7.3). If particles emitted at all times are counted, then the integral is to be taken from zero to infinity, and (7.5) becomes

$$P = \sum_{a,b,b',c} \frac{(a|H_1|b)(b|H_2|c)(c|H_2|b')(b'|H_1|a)}{1 - \frac{i}{\hbar}(W_b - W_{b'})\tau}. \qquad (7.6)$$

Expression (7.6) gives less correlation than occurs at $t=0$ or during the periodic repetitions of the initial correlation. However, this average over all time still shows angular correlation which can be used to evaluate quadrupole hyperfine structure if $(W_b - W_{b'}/\hbar)\tau$ is comparable with unity, or hence if eqQ and $1/\tau$ are comparable in magnitude. If $eqQ \gg 1/\tau$, the angular correlation does not become zero, but reaches what is called the "hard core" limit, from which nothing can be deduced about the coupling constant, except that it is much larger than $1/\tau$.

Angular correlation between successive radiations given by (7.3) may also be expressed in the form

$$p(t) = \sum_{k_1 k_2, \mu_1 \mu_2} \text{I}(k_1)\, \text{II}(k_2)\, \text{III}(k_1 k_2, \mu_1 \mu_2, t)\, Y_{k_1}^{\mu_1}(\Omega_1)\, Y_{k_2}^{\mu_2}(\Omega_2). \tag{7.7}$$

Here Y_k^μ are ordinary spherical harmonics and Ω_1, Ω_2 specify the directions of emission of the first and second radiations respectively. The indices k_1 and k_2 are even integers which have maximum values determined by the nuclear spin and the multipole orders of the emitted radiation. The functions $\text{I}(k_1)$ and $\text{II}(k_2)$ are given by ALDER [32], who also gives the rather lengthy derivation of a somewhat simplified form of (7.7). The function III, given by ABRAGAM and POUND[33] is

$$\left. \begin{aligned} \text{III}(k_1 k_2, \mu_1 \mu_2, t) = \sum_{m\,m'm''m'''\,b\,b'} (I\,k_1\,m'\mu_1 | I\,k_1\,I_m)\,(I\,k_2\,m'''\mu_2 | I k_2\,I\,m'') \times \\ \times (m\,|\,b)\,(b\,|\,m'')\,(m'''\,|\,b')\,(b'\,|\,m')\,e^{i(W_b - W_{b'})t/\hbar}. \end{aligned} \right\} \tag{7.8}$$

Here $(I\,k_1\,m'\mu_1 | I\,k_1\,I\,m)$ is a Clebsch-Gordon coefficient, and $(m\,|\,b)$ represents the projection of the eigenstate b of some Hamiltonian perturbing the nucleus in question on the state having a value m of I along any arbitrary axis. Expression (7.8) appears formidable partly because it is not restricted to the case where there is axial symmetry of the perturbing Hamiltonian, for which the eigenstates b and m may be taken as identical. This expression is hence also more general than (7.3) in not being restricted to axial symmetry.

Many special cases of expression (7.8) have been discussed. ALBERS-SCHÖNBERG, HEER, NOVEY, and SCHERRER [34] have studied angular correlations of radiation from a single crystal involving quadrupole hyperfine structure by counting all particles emitted, i.e., using the integral $\int_0^\infty \varrho(t)\,e^{-t/\tau}\,dt$. They also made measurements for the case of an additional applied magnetic field.

Perhaps one of the most interesting and useful cases occurs when there is a perturbation due to pure quadrupole coupling in polycrystalline material. If an average is made over all orientations of the crystalline electric field, (7.7) becomes independent of any crystalline axis, and depends only on the angle ϑ between the two successive radiations. It then has the form [33]

$$p(t) = \sum_k G_k(t)\, A_k\, P_k(\cos \vartheta) \tag{7.9}$$

where the $G_k(t)$ are "attenuation coefficients" of the angular correlation which are unity when $t = 0$, and represent the randomizing effects of the perturbing hyperfine interaction. A_k are constants which give the initial angular correlation. For axially symmetric quadrupole interactions in polycrystalline material ABRAGAM and POUND [33] have given the following expressions for the G_k which can occur with spins I between 1 and 3.

$$I = 1: \quad G_2(t) = \frac{1}{5}(3 + 2\cos\omega_0 t). \tag{7.10}$$

$$I = \frac{3}{2}: \quad G_2(t) = \frac{1}{5}(1 + 4\cos\omega_0 t). \tag{7.11}$$

$$\left. \begin{aligned} I = 2: \quad G_2(t) &= \frac{1}{5}\left(\frac{13}{7} + \frac{2}{7}\cos\omega_0 t + \frac{12}{7}\cos 3\omega_0 t + \frac{8}{7}\cos 4\omega_0 t\right), \\ G_4(t) &= \frac{1}{9}\left(\frac{29}{7} + \frac{12}{7}\cos\omega_0 t + \frac{16}{7}\cos 3\omega_0 t + \frac{6}{7}\cos 4\omega_0 t\right). \end{aligned} \right\} \tag{7.12}$$

$$I = \frac{5}{2}: \quad G_2(t) = \frac{1}{5}\left(1 + \frac{13}{7}\cos\omega_0 t + \frac{10}{7}\cos 2\omega_0 t + \frac{5}{7}\cos 3\omega_0 t\right),$$

$$G_4(t) = \frac{1}{9}\left(1 + \frac{15}{7}\cos\omega_0 t + \frac{18}{7}\cos 2\omega_0 t + \frac{23}{7}\cos 3\omega_0 t\right). \tag{7.13}$$

$$I = 3: \quad G_2(t) = \frac{1}{5}\left(\frac{11}{7} + \frac{2}{21}\cos\omega_0 t + \frac{5}{7}\cos 3\omega_0 t + \frac{20}{21}\cos 4\omega_0 t + \right.$$

$$\left. + \frac{25}{21}\cos 5\omega_0 t + \frac{10}{21}\cos 8\omega_0 t\right),$$

$$G_4(t) = \frac{1}{9}\left(\frac{17}{7} + \frac{30}{77}\cos\omega_0 t + \frac{92}{77}\cos 3\omega_0 t + \frac{6}{77}\cos 4\omega_0 t + \right.$$

$$\left. + \frac{60}{77}\cos 5\omega_0 t + \frac{192}{77}\cos 8\omega_0 t + \frac{126}{77}\cos 9\omega_0 t\right), \tag{7.14}$$

$$G_6(t) = \frac{1}{13}\left(5 + \frac{50}{33}\cos\omega_0 t + \frac{23}{11}\cos 3\omega_0 t + \frac{32}{33}\cos 4\omega_0 t + \right.$$

$$\left. + \frac{67}{33}\cos 5\omega_0 t + \frac{34}{33}\cos 8\omega_0 t + \frac{4}{11}\cos 9\omega_0 t\right).$$

In these expressions ω_0 is 2π times the frequency of repetition ν of the correlation given already by (7.5). The first (constant) term of each G_k represents the hard core correlation which is the limiting value when the coupling constant eqQ or ω_0 is much larger than $1/\tau$. Fig. 7 shows the variation of $G_2(t)$ with time for a nucleus of spin $\frac{5}{2}$ and comparison with actual measurements of angular correlation as a function of time. Periodic repetition of the angular correlation as shown by expression (7.5) is also demonstrated in this figure.

8. False quadrupole effects. There are some other effects which contribute to the variation in energy with nuclear orientation in much the same way as does a nuclear quadrupole moment. These are not easily distinguished from true quadrupole effects, but fortunately they are very small in most cases. They include the polarization of nuclei by electric fields and certain second-order effects due to nuclear magnetic hyperfine interactions. In molecules, the latter are known as "pseudo-quadrupole" effects, and are usual very small. However in atoms, where these effects are occasionally quite large and troublesome, they do not bear this name, but are usually discussed as a breakdown of the Landé interval rule. Nuclear polarization probably contributes to the apparent quadrupole moment of all nuclei, but its contribution should generally not be larger than about 1%, and has not yet been isolated.

If the polarizability of the nucleus in an electric field $\boldsymbol{E} = i E_x + j E_y + k E_z$ is described by a polarizability tensor α_{lm} or dyadic $\boldsymbol{\alpha}$, then the energy associated with polarization is

$$W_p = -\tfrac{1}{2}\boldsymbol{\alpha} : \boldsymbol{E}\boldsymbol{E} \tag{8.1}$$

or

$$W_p = -\tfrac{1}{2}(\boldsymbol{\alpha} - \alpha\,\mathbf{1}) : (\boldsymbol{E}\boldsymbol{E} - \tfrac{1}{3}E^2\,\mathbf{1}) - \tfrac{1}{2}\alpha\,E^2 \tag{8.2}$$

where

$\alpha = \tfrac{1}{3}(\alpha_{xx} + \alpha_{yy} + \alpha_{zz})$ is the average polarizability,

$\mathbf{1}$ is the unit dyadic,

$E^2 = E_x^2 + E_y^2 + E_z^2.$

The last term of (8.2), $-\tfrac{1}{2}\alpha E^2$, represents the average polarization for all orientations of the nucleus, and is isotropic. The first part may be seen to have the same form as the quadrupole energy given by (2.6) since $\boldsymbol{\alpha} - \alpha 1$ and $\boldsymbol{EE} - \frac{1}{3}E^2 1$ are symmetric traceless dyadics corresponding to $\boldsymbol{Q}$ and $\boldsymbol{VE}$.

From (8.2) and the previous discussion of quadrupole energy it is not difficult to obtain the expression for variation in polarization energy given by GUNTHER-MOHR, GESCHWIND, and TOWNES [36]:

$$W_P = \frac{e}{3} \frac{(\alpha_z - \alpha_x) p_J}{I(2I-1) J(2J-1)} \left[\frac{3}{4} C (C+1) - I(I+1) J(J+1) \right], \tag{8.3}$$

where again I and J are respectively the nuclear spin and angular momentum of the system producing the field $\boldsymbol{E}$. $\boldsymbol{I}$ and $\boldsymbol{J}$ add vectorially to produce $\boldsymbol{F}$, and $C = F(F+1) - I(I+1) - J(J+1)$. The coupling constant $eq_J(\alpha_z - \alpha_x)$ corresponds to $\frac{3}{2}eqQ$ in the case of quadrupole coupling and p_J, corresponding to q_J, is defined as

$$\left. \begin{aligned} p_J &= \int \varrho_{JJ} \frac{3\cos^2\vartheta - 1}{r^4}\,dv \\ &= \frac{2}{e}(E_z^2 - E_x^2)_{\mathrm{av}}. \end{aligned} \right\} \tag{8.4}$$

$\alpha_z - \alpha_x$, classically the difference in polarizability along the spin axis and perpendicular to it, is defined by

$$\left. \begin{aligned} \alpha_z - \alpha_x &= \frac{2I(I+1)}{2I-1} \times \\ &\times \sum_n \frac{|\mu_{0n}|^2_{M=I} - |\mu_{0n}|^2_{M=I-1}}{W_n - W_0} \end{aligned} \right\} \tag{8.5}$$

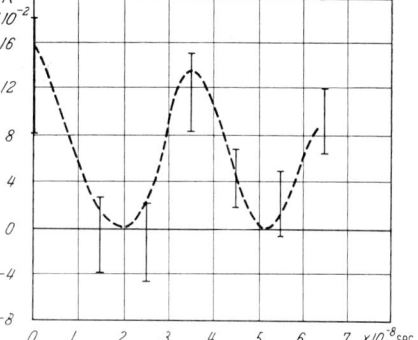

Fig. 7. Variation of angular correlation with time. Comparison of theoretical expectations and experimental points for Cd111, of spin $\frac{5}{2}$, in polycrystalline Cd SO$_4$. $A = A_2 G_2(t)$ where A_2 is the correlation at $t = 0$. (From LEHMANN and MILLER [35].)

where $|\mu_{0n}|_{M=I}$ is the dipole matrix element between the nuclear ground excited states for an orientation of the nuclear spin with respect to the electric field such that its projection M on the field direction equals I. $|\mu_{0n}|_{M=I-1}$ is the corresponding quantity with projection $M = I - 1$, and $W_0 - W_n$ is the difference in energy of the two states.

If the nuclear matrix elements and energy levels were known, the polarizabilities could be readily calculated. However, usually only very rough magnitudes are known for the matrix elements. A crude estimate of $\alpha_z - \alpha_x$ is given by equating it to the nuclear volume, or alternatively $|\mu_{0n}|$ may be taken as approximately equal to one proton charge times the nuclear radius, and $W_n - W_0$ as 1 Mev. From such estimates and evaluation of p_J from atomic wavefunctions, it can be shown that anisotropic polarization effects are usually about 1% of the nuclear quadrupole effects, and hence less than other experimental uncertainties in determination of quadrupole moments. However, future improvement in the accuracy of measurement of quadrupole moments may make correction due to polarizability of some importance. Since quadrupole coupling depends on $(1/r^3)_{\mathrm{av}}$, the average inverse cube of the distance between nucleus and external charges, while polarization effects depend on $(1/r^4)_{\mathrm{av}}$, it may be possible to discriminate between these two types of effects by studying the same nucleus or pairs of isotopes in different environments [37], [38].

The second-order effects of interactions between electrons and the nuclear magnetic moment can also have a result similar to that of quadrupole coupling. This may be understood qualitatively from the fact that in first order, interactions between a field and a dipole moment depend on the cosine of the angle between the two, and that second order energy terms depend on the square of the inter-

26*

action, which may be expected to involve the square of the cosine of this angle just as does the quadrupole energy. The nuclear polarization discussed above may be regarded as a second order interaction of this type between the electric field of the electrons and the off-diagonal matrix elements of the nuclear electric dipole moment. Although the nucleus has no permanent electric dipole moment, as indicated above, it does have a permanent magnetic dipole moment which can interact with off-diagonal elements of the magnetic field produced by electrons and hence give additional second-order effects of the type with which we are concerned.

Interaction between the nuclear magnetic moment μ_I and the magnetic field due to surrounding charges may be written [3], [39], [40]

$$H_m = -\frac{\mu_I}{IJ}(H_z)_{M=J}\,\boldsymbol{I}\cdot\boldsymbol{J} \qquad (8.6)$$

where J is the total angular momentum of the system, excepting the nuclear spin I, and $(H_z)_{M=J}$ is the z component of the magnetic field produced at the nucleus by the surrounding charges when the projection M of J on the z-axis has its maximum value J. The characteristic values of $\boldsymbol{I}\cdot\boldsymbol{J}$ are, as usual,

$$\boldsymbol{I}\cdot\boldsymbol{J} = \tfrac{1}{2}[F(F+1) - I(I+1) - J(J+1)] \qquad (8.7)$$

where F is the total angular momentum, or $\boldsymbol{F}=\boldsymbol{I}+\boldsymbol{J}$. Energy due to second-order effects of perturbing interaction (8.6) may be written

$$\Delta W = \sum_{n'} \frac{|(n\,|H_m|\,n')|^2}{W_n - W_{n'}} \qquad (8.8)$$

where n represents quantum numbers of the perturbed state and n' those of other states. W_n and $W_{n'}$ are the corresponding unperturbed energies.

It may be seen from expression (8.8) that these perturbations are of the order of the square of the magnetic hyperfine structure divided by the separation between energy levels of the system, and hence their size may vary a great deal, depending on the particular system involved. Most molecules whose hyperfine structure has been studied are in $^1\Sigma$ electronic states, and the matrix elements of H_m diagonal in such an electronic state are all zero. Hence the rather close spacing of rotational or vibrational levels is of no importance in contributing to the second-order energy (8.8) for these cases. Off-diagonal matrix elements of H_m exist between certain electronic states, and it is the difference in energy between these electronic states which is important in determining the magnitude of ΔW from (8.8). The ground state of most molecules, in which hyperfine structure is typically measured, is usually well separated from any excited electronic state and these perturbations tend to be small for molecular cases of interest. However, they may be much larger in certain cases of atomic spectra.

For the atomic case, (8.8) may be written

$$\Delta W = \sum_{n',\,J'\neq J} \frac{|(n,\,J,\,F\,|H_m|\,n',\,J',\,F)|^2}{W(n,\,J,\,F) - W(n',\,J',\,F)}\,. \qquad (8.9)$$

Here n is the principal quantum number and J represents the electronic angular momentum. No sum is indicated over the total angular momentum F nor over the magnetic quantum numbers M_F, because both F and M_F are constants of the motion and matrix elements off-diagonal in these quantities are zero. Since the interaction H_m has even parity (with respect to inversion of spatial coordinates), its matrix elements are also zero unless the two states n, J and n', J' have the same parity. Finally, as is typical of interactions of dipolar symmetry, the matrix elements of (8.9) are non-zero only where $J'=J$ or $J'=J\pm1$.

Although the values of the matrix elements in question depend on the particular case, their variation with F can be given in general by the following expressions:

$$(n, J, F |H_m| n', J, F)$$
$$= - (n, J, J |H_m| n', J, J) \frac{F(F + 1) - I(I + 1) - J(J + 1)}{I(I + 1)} , \qquad (8.10)$$

$$(n, J, F |H_m| n', J + 1, F) = (n, J, J |H_m| n', J + 1, J) \times$$
$$\times \frac{[(F + J + I + 2) (F + J - I + 1) (J + I - F + 1) (F - J + 1)]^{\frac{1}{2}}}{[(2 J + I + 2) (2 J - I + 1) (I) (I + 1)]^{\frac{1}{2}}} . \qquad (8.11)$$

Here the matrix element in parentheses on the right is for the particular case when $F=J$.

A typical case where second-order magnetic hyperfine effects are important in atomic spectra is for an sd electronic configuration, where a large magnetic interaction is provided by the s electron, but the fine structure of the d orbit is not large, and hence two fine structure components may have a rather small energy separation. Such is the case, for example, for the hyperfine structure of the 6^3D_1 and 6^1D_2 levels of Hg^{201}, where these effects were first detected experimentally by SCHÜLER and JONES [41]. The observed spacing of these levels and their hyperfine structure is indicated in Fig. 8. The pairs of levels with F values of $\frac{1}{2}$, $\frac{3}{2}$, and $\frac{5}{2}$ repel each other, producing shifts as large as about 10% of the hyperfine structure. The $F=\frac{7}{2}$ level is not perturbed, since there is no nearby level with the same total angular momentum.

From the above discussion, some helpful general rules about second-order magnetic interactions in atomic spectra may be stated.

(1) Only those states perturb each other which
    a) have the same total angular momentum F;
    b) have the same parity;
    c) have electronic angular momentum J which are the same, or which differ by one unit.

(2) When two states perturb each other there is a mutual repulsion, each state being shifted by the same amount, for which the order of magnitude is given by the square of the magnetic hyperfine structure divided by the energy separation.

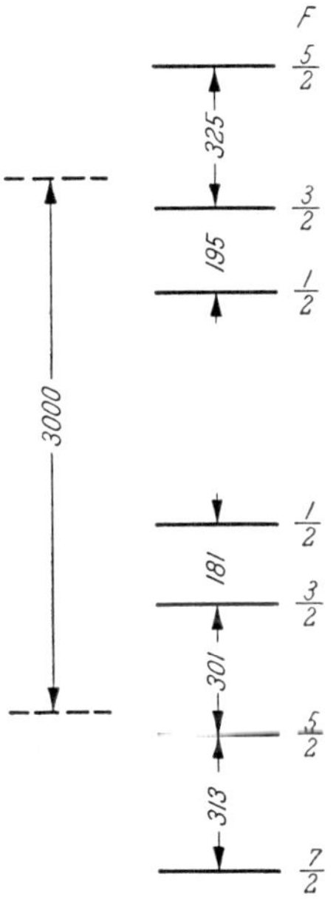

Fig. 8. Energy level diagram of 6^1D_1 and 6^3D_2 levels of Hg^{201}, which show second-order magnetic hyperfine perturbations. Energy separations are given in 10^{-3} cm^{-1}. (From measurements by SCHÜLER and JONES [41].)

(3) Ratios of the repulsions of pairs of levels with the same n, J and n', J' but different F can be obtained from expressions (8.10) and (8.11).

For evaluation of quadrupole moments from atomic hyperfine structure, perturbations of this type must either be shown to be negligible for the particular states involved, or they must be calculated with sufficient accuracy to allow appropriate corrections.

In molecular states which have electronic spin momentum, i.e. in states which are not singlet, second order perturbations between fine structure components may occur in a way rather similar to the case of atomic spectra. An example is afforded by the hyperfine structure of O^{17} in the molecule O_2 [42] which is in a $^3\Sigma$ state. The usual molecular case involves a $^1\Sigma$ state, which, if strictly pure, is spherically symmetric and can show no variation of hyperfine interaction with nuclear orientation. However, any molecular rotation produces a preferred axis (that of the angular momentum J) for the molecule, a non-spherical distribution of electron charge, and a second-order magnetic interaction between the nucleus and the electron orbital momentum which varies with nuclear orientation in a way similar to a quadrupole interaction [43]. The energy of this "pseudo-quadrupole" interaction has the form

$$\Delta W = \sum_n \frac{|(0\,|\,a\,\boldsymbol{I}\cdot\boldsymbol{L}\,|\,n)|^2}{W_0 - W_n} \tag{8.12}$$

where L is the electronic orbital momentum, and W_0 and W_n are the energies of ground and excited states respectively. The quantity a is

$$a = \frac{2\mu_I\,\mu_0}{I}\left(\frac{1}{r^3}\right),$$

in which μ_I is the nuclear magnetic moment, μ_0 the Bohr magneton, and r the distance from electron to nucleus.

In a linear molecule, $\boldsymbol{L}$ may be thought of as precessing around the molecular axis with a projection of zero on the axis for the $^1\Sigma$ ground state. $\boldsymbol{I}\cdot\boldsymbol{L}$ is then proportional to the cosine of the angle between the spin and the axis, or $|(0\,|\,a\,\boldsymbol{I}\cdot\boldsymbol{L}\,|\,n)|^2$ to the square of this cosine. Hence experimentally the energy term (8.12) cannot be distinguished from a quadrupole effect. Fortunately, however, this pseudoquadrupole interaction is almost always negligibly small. As in the atomic case, it is proportional to the square of the magnetic hyperfine interaction divided by the separation between electronic levels, which makes it usually less than a few cycles per second, or of the order of 10^{-6} as large as normal quadrupole interactions. It is less important than in the atomic case because the rather large hyperfine interaction between the nucleus and an s electron is not involved, and also because excited molecular electronic levels are usually well separated from the ground state.

9. Determination of $(\partial^2 V/\partial z^2)_{\mathrm{av}}$ in atoms. Determination of the effective field gradients at a nucleus, or $(\partial^2 V/\partial z^2)_{\mathrm{av}}$, is essential in order to evaluate nuclear quadrupole moments from experimental measurement of quadrupole coupling constants. As indicated in (2.18), $(\partial^2 V/\partial z^2)_{\mathrm{av}}$ or q involves an average over the charge distributions outside a small sphere around the nucleus, which for a single electron may be written

$$q_j = -\,e\int \psi_{jj}^* \frac{3\cos^2\vartheta - 1}{r^3}\,\psi_{jj}\,dv. \tag{9.1}$$

Only in the case of a hydrogen-like atom can a very precise value of q_j be obtained from (9.1). For many-electron atoms, it cannot be very accurately evaluated because of lack of knowledge of the wavefunctions, and furthermore the total q_j cannot be attributed to a single electron alone.

For the hydrogenic case, Eq. (9.1) can be evaluated as

$$q_j = e\,\frac{2j-1}{2j+2}\left(\frac{1}{r^3}\right)R_r. \tag{9.2}$$

Here j is the sum of the orbital angular momentum l and electron spin, and

$$R_r = \frac{l(l+1)(2l+1)}{\varrho(\varrho^2-1)(4\varrho^2-1)}\left[3k(k+1)-\varrho^2+1\right] \tag{9.3}$$

is a relativistic correction which approaches unity when Z, the nuclear charge, is small. This quantity is tabulated as a function of Z in Table 1. The quantity ϱ is

$$\varrho = \sqrt{k^2 - Z^2 \alpha^2} \tag{9.4}$$

where α is the fine structure constant and $k = l+1$ for levels with $j = l + \frac{1}{2}$, $k = -1$ for cases when $j = l - \frac{1}{2}$. $\overline{(1/r^3)}$ is the average value of $1/r^3$ for the non-relativistic hydrogenic case, i.e.,

$$\overline{\left(\frac{1}{r^3}\right)} = \frac{Z^3}{n^3 l (l + \frac{1}{2}) (l+1) a_0^3} \tag{9.5}$$

where a_0 is $\dfrac{\hbar^2}{m e^2}$, the "Bohr radius" of the hydrogen atom. Using (9.5), Eq. (9.2) becomes

$$q_j = e \, \frac{2j - 1}{2j + 2} \, \frac{Z^3 \, R_r}{n^3 l (l + \frac{1}{2}) (l+1) a_0^3} . \tag{9.6}$$

Consider now a many-electron atom such as a neutral alkali, with a single electron outside a closed shell of screening electrons. Its angular distribution is the same for the hydrogenic case, but its radial wavefunction and the average of $1/r^3$ are considerably modified. The value of q might be approximated by choosing a Z_i and Z_0, the effective values of Z for the interior and exterior parts respectively of the penetrating orbit, and using the approximation

$$\overline{\left(\frac{1}{r^3}\right)} = \frac{Z_i Z_0^2}{n^{*3} l (l + \frac{1}{2}) (l+1) a_0^3} \tag{9.7}$$

where the effective principal quantum number n^* may be determined from the electronic energy. However, better accuracy can be obtained by use of one of the following two methods which determine $\overline{(1/r^3)}$ more directly from experimental measurements.

Magnetic hyperfine structure also depends on $\overline{(1/r^3)}$, and of course on the nuclear magnetic moment μ_I. If μ_I is known from other measurements, then an experimental determination of the magnetic hyperfine structure allows a fairly direct evaluation of $\overline{(1/r^3)}$. For a hydrogenic atom, the magnetic hyperfine energy due to an electron with orbital momentum l and total angular momentum j is $W = a\,\boldsymbol{I} \cdot \boldsymbol{j}$ with

$$a = \frac{2 \mu_I \mu_0}{I} \overline{\left(\frac{1}{r^3}\right)} \frac{l(l+1)}{j(j+1)} F_r \tag{9.8}$$

when μ_0 is the Bohr magnetron and F_r is a relativistic factor which approaches unity for small Z.

$$F_r = \frac{4j (j + \frac{1}{2}) (j+1)}{\varrho (4 \varrho^2 - 1)} . \tag{9.9}$$

ϱ is as given by Eq. (9.4). Using (9.8) and (9.2),

$$q_j = \frac{e \, a \, I \, j (2j - 1)}{4 l (l+1) \mu_I \mu_0} \frac{R_r}{F_r} . \tag{9.10}$$

Z_i, the effective value of Z for the electron during the inner part of its orbit, should probably best be used for the relativistic correction. This correction R_r/F_r, is tabulated in Table 2.

Still another quantity which can provide a type of experimental determination of $\overline{(1/r^3)}$ is the fine structure. The separation between two fine structure levels

Table 1. *Relativistic correction factor for quadrupole hyperfine structure (after* KOPFERMANN [39])

$$R_r = \frac{l(l+1)(2l+1)}{\varrho(\varrho^2-1)(4\varrho^2-1)} [3k(k+1) - \varrho^2 + 1] .$$

Z	R_r			Z	R_r		
	$j=\frac{3}{2},\, l=1,\, k=2$	$j=\frac{3}{2},\, l=2,\, k=-2$	$j=\frac{5}{2},\, l=2,\, k=3$		$j=\frac{3}{2},\, l=1,\, k=2$	$j=\frac{3}{2},\, l=2,\, k=-2$	$j=\frac{5}{2},\, l=2,\, k=3$
1	1.0000₄	1.0000₇	1.0000₂	47	1.0992	1.1334	1.0399
2	1.0001₆	1.0003	1.0000₇	48	1.1037	1.1395	1.0416
3	1.0004	1.0005	1.0001₆	49	1.1084	1.1459	1.0434
4	1.0006	1.0009	1.0003	50	1.1133	1.1524	1.0453
5	1.0010	1.0014	1.0004	51	1.1182	1.1591	1.0472
6	1.0015	1.0021	1.0006	52	1.1233	1.1660	1.0492
7	1.0020	1.0027	1.0009	53	1.1284	1.1730	1.0511
8	1.0027	1.0036	1.0011	54	1.1337	1.1801	1.0531
9	1.0034	1.0046	1.0014	55	1.1392	1.1876	1.0552
10	1.0042	1.0057	1.0018	56	1.1448	1.1952	1.0572
11	1.0051	1.0068	1.0021	57	1.1505	1.2030	1.0594
12	1.0061	1.0081	1.0025	58	1.1564	1.2110	1.0616
13	1.0071	1.0095	1.0030	59	1.1624	1.2192	1.0638
14	1.0083	1.0111	1.0035	60	1.1686	1.2275	1.0661
15	1.0095	1.0128	1.0039	61	1.1748	1.2361	1.0684
16	1.0109	1.0146	1.0045	62	1.1813	1.2449	1.0708
17	1.0123	1.0165	1.0051	63	1.1879	1.2539	1.0732
18	1.0138	1.0184	1.0057	64	1.1946	1.2631	1.0756
19	1.0155	1.0206	1.0064	65	1.2016	1.2726	1.0782
20	1.0171	1.0228	1.0070	66	1.2086	1.2822	1.0807
21	1.0189	1.0252	1.0078	67	1.2159	1.2922	1.0833
22	1.0207	1.0277	1.0086	68	1.2233	1.3024	1.0859
23	1.0226	1.0303	1.0094	69	1.2309	1.3128	1.0887
24	1.0246	1.0330	1.0102	70	1.2387	1.3234	1.0913
25	1.0268	1.0359	1.0110	71	1.2467	1.3344	1.0941
26	1.0291	1.0389	1.0119	72	1.2547	1.3454	1.0970
27	1.0314	1.0420	1.0129	73	1.2631	1.3569	1.0999
28	1.0338	1.0453	1.0139	74	1.2716	1.3686	1.1027
29	1.0363	1.0486	1.0150	75	1.2803	1.3805	1.1058
30	1.0389	1.0522	1.0160	76	1.2892	1.3928	1.1088
31	1.0417	1.0558	1.0171	77	1.2983	1.4053	1.1119
32	1.0444	1.0595	1.0182	78	1.3077	1.4182	1.1151
33	1.0473	1.0634	1.0194	79	1.3172	1.4314	1.1182
34	1.0504	1.0675	1.0206	80	1.3268	1.4447	1.1215
35	1.0535	1.0717	1.0219	81	1.3369	1.4586	1.1248
36	1.0566	1.0760	1.0231	82	1.3470	1.4726	1.1281
37	1.0600	1.0805	1.0245	83	1.3574	1.4870	1.1315
38	1.0634	1.0851	1.0258	84	1.3681	1.5018	1.1350
39	1.0670	1.0899	1.0272	85	1.3790	1.5169	1.1385
40	1.0706	1.0948	1.0286	86	1.3902	1.5324	1.1421
41	1.0743	1.0998	1.0302	87	1.4015	1.5482	1.1458
42	1.0782	1.1050	1.0317	88	1.4139	1.5645	1.1495
43	1.0821	1.1104	1.0333	89	1.4252	1.5811	1.1532
44	1.0863	1.1159	1.0349	90	1.4373	1.5980	1.1570
45	1.0905	1.1216	1.0365	91	1.4498	1.6154	1.1609
46	1.0947	1.1274	1.0381	92	1.4626	1.6332	1.1647

is given by

$$\Delta\nu = Z_i(l + \tfrac{1}{2})\left(\frac{1}{r^3}\right) R\,\alpha^2 H\,\frac{dn^*}{dn} \tag{9.11}$$

where R in the Rydberg constant and H is a relativistic correction approaching unity for small Z. H is given by

$$H = [\varrho(l+1) - \varrho(-l) - 1]\,\frac{Z^2\alpha^2}{2l(l+1)},$$

Table 2. *Relativistic correction to q_j evaluated by use of magnetic hyperfine structure* [see Eq.(9.10)]

$$\frac{R_r}{F_r} = \frac{l(l+1)(2l+1)}{4j(j+\frac{1}{2})(j+1)(\varrho^2-1)} [3k(k+1)-\varrho^2+1].$$

Z	R_r/F_r			Z	R_r/F_r		
	$j=\frac{3}{2},l=1,k=2$	$j=\frac{3}{2},l=2,k=-2$	$j=\frac{5}{2},l=2,k=3$		$j=\frac{3}{2},l=1,k=2$	$j=\frac{3}{2},l=2,k=-2$	$j=\frac{5}{2},l=2,k=3$
1	1.0000	1.0000	1.0000	47	1.0490	1.0816	1.0192
2	1.0000	1.0002	1.0000	48	1.0511	1.0852	1.0200
3	1.0002	1.0003	1.0000	49	1.0534	1.0891	1.0208
4	1.0003	1.0006	1.0002	50	1.0558	1.0928	1.0218
5	1.0005	1.0009	1.0002	51	1.0581	1.0968	1.0227
6	1.0007	1.0013	1.0003	52	1.0605	1.1008	1.0236
7	1.0010	1.0017	1.0005	53	1.0629	1.1049	1.0245
8	1.0014	1.0023	1.0005	54	1.0655	1.1091	1.0254
9	1.0017	1.0029	1.0007	55	1.0681	1.1134	1.0265
10	1.0021	1.0036	1.0009	56	1.0707	1.1178	1.0274
11	1.0026	1.0043	1.0010	57	1.0734	1.1224	1.0284
12	1.0031	1.0051	1.0012	58	1.0762	1.1270	1.0295
13	1.0036	1.0060	1.0015	59	1.0790	1.1317	1.0305
14	1.0042	1.0070	1.0017	60	1.0819	1.1365	1.0315
15	1.0048	1.0081	1.0019	61	1.0849	1.1415	1.0327
16	1.0055	1.0092	1.0022	62	1.0879	1.1464	1.0338
17	1.0062	1.0103	1.0025	63	1.0909	1.1515	1.0349
18	1.0070	1.0115	1.0028	64	1.0941	1.1568	1.0360
19	1.0078	1.0129	1.0031	65	1.0973	1.1621	1.0372
20	1.0086	1.0143	1.0034	66	1.1005	1.1675	1.0384
21	1.0095	1.0158	1.0038	67	1.1039	1.1731	1.0396
22	1.0104	1.0173	1.0042	68	1.1073	1.1789	1.0408
23	1.0114	1.0190	1.0046	69	1.1107	1.1846	1.0421
24	1.0124	1.0207	1.0050	70	1.1143	1.1905	1.0433
25	1.0134	1.0224	1.0053	71	1.1179	1.1966	1.0446
26	1.0146	1.0243	1.0058	72	1.1216	1.2026	1.0459
27	1.0158	1.0262	1.0063	73	1.1254	1.2089	1.0473
28	1.0169	1.0282	1.0068	74	1.1292	1.2153	1.0486
29	1.0182	1.0303	1.0072	75	1.1331	1.2218	1.0500
30	1.0194	1.0325	1.0077	76	1.1371	1.2284	1.0514
31	1.0209	1.0347	1.0082	77	1.1412	1.2352	1.0528
32	1.0222	1.0370	1.0088	78	1.1453	1.2421	1.0543
33	1.0237	1.0394	1.0094	79	1.1495	1.2491	1.0557
34	1.0252	1.0419	1.0100	80	1.1537	1.2563	1.0572
35	1.0267	1.0444	1.0106	81	1.1582	1.2636	1.0587
36	1.0282	1.0471	1.0112	82	1.1626	1.2710	1.0602
37	1.0299	1.0498	1.0119	83	1.1672	1.2786	1.0617
38	1.0315	1.0526	1.0124	84	1.1718	1.2863	1.0633
39	1.0333	1.0555	1.0131	85	1.1765	1.2942	1.0649
40	1.0351	1.0585	1.0138	86	1.1813	1.3022	1.0666
41	1.0369	1.0615	1.0146	87	1.1862	1.3104	1.0682
42	1.0388	1.0646	1.0153	88	1.1918	1.3187	1.0699
43	1.0407	1.0679	1.0160	89	1.1963	1.3272	1.0715
44	1.0427	1.0711	1.0168	90	1.2014	1.3358	1.0733
45	1.0447	1.0745	1.0176	91	1.2068	1.3446	1.0750
46	1.0468	1.0780	1.0183	92	2.2122	1.3536	1.0767

where the two quantities ϱ are those given by (9.4) for $k=l+1$ or $k=-l$ as indicated. The quantity dn^*/dn is the change of n^* with principal quantum number n, and is usually quite near unity. Combining (9.11) and (9.2) one obtains

$$q_j = \frac{e\,\Delta v\,(2j-1)}{(j+1)(2l+1)Z_i R\alpha^2} \frac{R_r}{H}. \tag{9.12}$$

Here dn^*/dn has been assumed to be unity.

Table 3. *Relativistic correction for q_j evaluated by use of fine structure* [see Eq. (9.12)]

$$\frac{R_r}{H} = \frac{\alpha^2 Z^2 (2l+1)}{\varrho (\varrho' - \varrho'' - 1)(\varrho^2 - 1)(4\varrho^2 - 1)} [3k(k+1) - \varrho^2 + 1].$$

Z	R_r/H			Z	R_r/H		
	$j=\frac{3}{2},l=1,k=2$	$j=\frac{3}{2},l=2,k=-2$	$j=\frac{5}{2},l=2,k=3$		$j=\frac{3}{2},l=1,k=2$	$j=\frac{3}{2},l=2,k=-2$	$j=\frac{5}{2},l=2,k=3$
1	1.0000	1.0000	1.0000	47	1.0418	1.1158	1.0237
2	1.0000	1.0003	1.0000	48	1.0436	1.1210	1.0247
3	1.0002	1.0005	1.0001	49	1.0454	1.1265	1.0258
4	1.0002	1.0008	1.0002	50	1.0474	1.1320	1.0268
5	1.0004	1.0012	1.0002	51	1.0493	1.1379	1.0281
6	1.0007	1.0018	1.0003	52	1.0513	1.1438	1.0292
7	1.0008	1.0023	1.0005	53	1.0532	1.1498	1.0303
8	1.0012	1.0031	1.0006	54	1.0552	1.1158	1.0314
9	1.0015	1.0040	1.0008	55	1.0573	1.1623	1.0327
10	1.0019	1.0050	1.0011	56	1.0594	1.1688	1.0338
11	1.0023	1.0059	1.0012	57	1.0615	1.1755	1.0352
12	1.0027	1.0071	1.0015	58	1.0637	1.1822	1.0363
13	1.0032	1.0083	1.0018	59	1.0658	1.1892	1.0377
14	1.0037	1.0097	1.0021	60	1.0682	1.1963	1.0390
15	1.0042	1.0112	1.0023	61	1.0703	1.2036	1.0403
16	1.0049	1.0128	1.0027	62	1.0726	1.2111	1.0417
17	1.0055	1.0145	1.0031	63	1.0750	1.2188	1.0432
18	1.0062	1.0161	1.0034	64	1.0774	1.2265	1.0445
19	1.0069	1.0180	1.0038	65	1.0798	1.2346	1.0460
20	1.0076	1.0199	1.0042	66	1.0822	1.2427	1.0474
21	1.0084	1.0220	1.0047	67	1.0847	1.2512	1.0489
22	1.0092	1.0241	1.0051	68	1.0872	1.2598	1.0504
23	1.0100	1.0265	1.0057	69	1.0897	1.2687	1.0521
24	1.0109	1.0288	1.0061	70	1.0923	1.2775	1.0535
25	1.0118	1.0314	1.0066	71	1.0949	1.2869	1.0552
26	1.0128	1.0339	1.0071	72	1.0974	1.2961	1.0568
27	1.0138	1.0367	1.0078	73	1.1002	1.3058	1.0585
28	1.0148	1.0396	1.0084	74	1.1029	1.3156	1.0600
29	1.0159	1.0423	1.0089	75	1.1055	1.3255	1.0617
30	1.0169	1.0455	1.0095	76	1.1083	1.3359	1.0635
31	1.0182	1.0487	1.0102	77	1.1110	1.3463	1.0652
32	1.0193	1.0518	1.0108	78	1.1139	1.3571	1.0671
33	1.0206	1.0553	1.0116	79	1.1166	1.3682	1.0688
34	1.0219	1.0588	1.0123	80	1.1194	1.3792	1.0706
35	1.0232	1.0625	1.0131	81	1.1223	1.3909	1.0726
36	1.0244	1.0662	1.0138	82	1.1250	1.4025	1.0744
37	1.0259	1.0701	1.0147	83	1.1279	1.4144	1.0763
38	1.0273	1.0740	1.0153	84	1.1308	1.4268	1.0783
39	1.0288	1.0783	1.0162	85	1.1337	1.4392	1.0802
40	1.0303	1.0825	1.0170	86	1.1366	1.4521	1.0823
41	1.0318	1.0868	1.0180	87	1.1394	1.4651	1.0843
42	1.0334	1.0913	1.0189	88	1.1429	1.4786	1.0864
43	1.0349	1.0959	1.0192	89	1.1454	1.4922	1.0883
44	1.0366	1.1007	1.0208	90	1.1482	1.5061	1.0905
45	1.0384	1.1056	1.0217	91	1.1511	1.5204	1.0926
46	1.0400	1.1106	1.0227	92	1.1540	1.5351	1.0947

A tabulation of the relativistic correction R_r/H can be found in Table 3. Use of the fine structure and Eq. (9.12) does not usually yield as accurate a value of q_j as does the magnetic hyperfine structure and expression (9.10), since the quantity Z_i cannot be very precisely determined. However, accurate experimental values are more often available for $\Delta \nu$ than for a/μ_I, and in addition we shall see below that the screening electrons produce other uncertainties which have not yet been discussed, but which may frequently be larger than uncertainties in Z_i.

For many atoms with more than one valence electron, fine structure can also be simply used to evaluate $\overline{(1/r^3)}$. Thus if the valence electron shell is missing one electron instead of having a single electron, the fine structure is just inverted and (9.12) still applies where Δv is the fine structure splitting. If there are two valence electrons in the same subshell, each with angular momentum l, the splitting $\Delta v'$ between the two extreme components of the triplet level may be used. Expression (9.12), without the refinement of relativistic corrections R_r and H, is still applicable if

$$\Delta v = \frac{\Delta v'(l + \frac{1}{2})}{2(L + \frac{1}{2})}$$

is used. Here L is the total orbital angular momentum and L-S coupling has been assumed. A subshell which lacks two electrons of being complete corresponds to the same case, but with inverted fine structure.

Appropriate values of Z_i for expression (9.12) may be obtained from examination of atomic spectra. For heavy atoms Z_i is not far from $Z-4$ for a p electron, and $Z-10$ for a d electron. Errors involved in the difference between Z and Z_i for heavy atoms are less serious than those for light atoms, since Z itself is so large. BARNES and SMITH [44] have examined values of Z_i for electrons in p states and suggest as the best rough approximation the value $Z_i = Z - n$ for neutral atoms, where n is the principal quantum number. They give the values of Z_i listed in Table 4 for specific states.

When an atom has more than one electron outside a closed shell, evaluation of q_j is of course somewhat more complex. Let the total angular momentum of such a system be J. For evaluation of q_j, the state with projection of the total angular momentum J on the z axis equal to J itself must be considered. If for such a case A_{Jnlm} is the probability for an electron to be in a state characterized

Table 4. *Values of Z_i for p electrons of various elements and degrees of ionization* (from BARNES and SMITH [44]).

Atom or Ion	Configuration	Z_i	Atom or Ion	Configuration	Z_i	Atom or Ion	Configuration	Z_i
Li I	$2p$	0.94	Si III	$3s\ 4p$	11.70	As IV	$4s\ 5p$	30.0
Be I	$2s\ 2p$	1.77	Si IV	$6p$	12.30	As V	$5p$	30.5
Be II	$2p$	2.06	P III	$4p$	12.20	Se IV	$5p$	30.8
B I	$2p$	3.40	P IV	$3s\ 3p$	13.05	Se VI	$4p$	31.8
B II	$2s\ 2p$	2.91	P V	$5p$	13.45	Rb I	$7p$	31.3
B III	$3p$	3.17	S VI	$5p$	14.14	Sr I	$5s\ 7p$	33.2
C II	$3p$	4.11	Cl V	$3p$	14.80	Sr II	$6p$	34.5
C III	$2s\ 3p$	3.92	Cl VII	$4p$	15.10	Ag I	$8p$	42.2
C IV	$4p$	4.21	A VI	$3p$	15.70	Cd I	$5s\ 6p$	42.7
N III	$3p$	5.06	A VIII	$5p$	15.50	Cd II	$6p$	45.0
N IV	$2s\ 3p$	4.98	K I	$7p$	15.10	In I	$8p$	45.2
N V	$3p$	5.14	Ca I	$4s\ 6p$	16.20	In II	$5s\ 7p$	45.0
O IV	$3p$	6.30	Ca II	$6p$	17.00	Sn II	$8p$	46.6
O VI	$4p$	6.19	Ca I	$4p$	23.4	Sn III	$5s\ 5p$	48.0
F V	$3p$	7.12	Zn I	$4s\ 4p$	24.0	Sb III	$6p$	49.4
F VII	$3p$	7.20	Zn II	$5p$	26.0	Sb IV	$5s\ 5p$	48.6
Na I	$6p$	7.62	Ga I	$7p$	27.4	Sb V	$5p$	49.4
Mg I	$3s\ 3p$	9.80	Ga II	$4s\ 5p$	27.2	Te IV	$6p$	48.3
Mg II	$6p$	9.85	Ga III	$5p$	27.6	Te V	$5s\ 5p$	48.3
Al I	$6p$	10.05	Ge II	$5p$	28.2	Te VI	$5p$	49.8
Al II	$3s\ 6p$	10.63	Ge III	$4s\ 5p$	28.4	Cs I	$7p$	49.2
Al III	$6p$	11.12	Ge IV	$5p$	29.0	Ba I	$6s\ 8p$	50.6
Si II	$4p$	11.40	As III	$5p$	30.2	Ba II	$8p$	53.6

by principal quantum number n, orbital angular momentum l, and projection m of l on the z-axis, then

$$q_j = \sum_{nlm} A_{jnlm} q_{nlm} \qquad (9.13)$$

where q_{nlm} is the value of q_j for the state nlm, or

$$q_{nlm} = - e \int \psi_{nlm}^* \frac{(3\cos^2\vartheta - 1)}{r^3} \psi_{nlm} dv . \qquad (9.14)$$

Evaluation of the angular part of the integral in (9.14) gives

$$q_{nlm} = -\frac{2e\left[3m^2 - l(l+1)\right]}{(2l-1)(2l+3)} \overline{\left(\frac{1}{r^3}\right)} = \left[1 - \frac{3m^2}{l(l+1)}\right] q_{nl0} . \qquad (9.15)$$

A large fraction of the important cases of quadrupole coupling in many-electron spectra occur when a p or d electron is coupled to an s electron, and where Russell-Saunders coupling holds approximately. In these cases, q_j may be written $-e(3\cos^2\vartheta - 1)_{j,j}\overline{(1/r^3)}$, where $(3\cos^2\vartheta - 1)_{j,j}$ is given by Table 5 (cf. [3]). The quantity β, indicating departure from Russell-Saunders coupling, is Δ/d, where d is the separation between the singlet and triplet terms, and Δ is the deviation of the central triplet level ($^3D_2$ or $^3P_1$) from the position it should have according to the Landé interval rule.

Table 5. *Values* $- q_j \Big/ e \overline{\left(\dfrac{1}{r^3}\right)}$ *or* $(3\cos^2\vartheta - 1)_{jj}$ *for sp and sd configurations.* β *is zero for pure L-S coupling (see text).*

Term	$(3\cos^2\vartheta-1)_{jj}$	Term	$(3\cos^2\vartheta-1)_{jj}$
$^3D_3$	$-\frac{4}{7}$	$^3P_2$	$-\frac{2}{5}$
$^3D_2$	$-\frac{2}{7}(1+\beta^2)$	$^3P_1$	$\frac{1}{5}(1-3\beta^2)$
$^3D_1$	$-\frac{1}{5}$	$^3P_0$	0
$^1D_2$	$-\frac{4}{7}(1-\beta^2/2)$	$^1P_1$	$-\frac{2}{5}(1-\frac{3}{2}\beta^2)$

For quadrupolar effects, great precision in evaluation of the angular part of the wavefunction is not justified because of uncertainties in $\overline{(1/r^3)}$ and in the shielding factor to be discussed below. Furthermore, electrons with orbital angular momentum higher than 2 cannot usually give a meaningful contribution to the coupling because the value of $\overline{(1/r^3)}$ for such electrons is quite small, and shielding factors are rather large and uncertain.

10. Shielding effects due to closed shells of electrons. The discussion so far has attributed atomic hyperfine structure to interactions between nuclei and their valence electrons, allowing the closed shells of electrons the sole function of changing the effective values of Z and thereby modifying the valence electron wavefunctions. However, there is another type of shielding effect due to the closed shells which can be described as a distortion of their spherical symmetry by the valence electron which then gives hyperfine interactions between the nucleus and the closed shells themselves. These shielding effects can to some extent be calculated and appropriate corrections allowed. However, precise calculations of this type are very difficult, and the effects remain one of the most troublesome sources of uncertainty in determination of nuclear quadrupole moments.

Consider a single valence electron with a p-type wavefunction $\psi_{n,1,1}$ interacting with a spherical distribution of electrons which might be simply represented by one electron in an s-orbit, or by the wavefunction $\psi_{n',0,0}$. To a first approximation the nuclear quadrupole moment interacts only with the valence electron and not with the spherical shell. However, as a result of Coulomb repulsion the s electron will be distorted by the addition of a small amount of d wavefunction,

$\varepsilon \psi_{n'',2,2}$. Now if the repulsion between the two electrons is most important at some distance from the nucleus comparable with the atomic radius r_0, then certainly at this radius the s electron density will be slightly decreased in the regions where the density of the p electron is high, i.e., in a plane through the nucleus and perpendicular to the z-axis. This corresponds to a negative value of ε if $\psi_{n'',2,2}$ and $\psi_{n',0,0}$ have the same sign in this region, since the perturbed wavefunction is $\psi_{n',0,0} + \varepsilon \psi_{n'',2,2}$. The decrease in density of the s electron then compensates to some extent for the presence of the p electron, and the net value of q produced at the nucleus by electron charges near r_0 would be decreased. This effect is then very naturally called *shielding*. It is a second-order effect which may alternately be described as a perturbation of the closed shells of electrons by the nuclear quadrupole moment which results in an induced quadrupole moment of the closed shells. This induced quadrupole moment then interacts with the valence electron, contributing to the quadrupole hyperfine structure.

Consider now the perturbed s electron in a region much nearer the nucleus than r_0. Its distribution is still non-spherical, but its density in the plane perpendicular to the z-axis may have been either increased or decreased because of the perturbation, depending on the relative signs of $\psi_{n',0,0}$ and $\psi_{n'',2,2}$ at the particular radius in question. Thus if $\psi_{n',0,0}$ has changed sign in going from the radius r_0 to the smaller radius and $\psi_{n'',2,2}$ has not, the charge density of the s-electron will then be larger in just those regions where the density of the p-electron is large. This magnifies the effect of the valence electron on q_j, corresponding to *antishielding*. It should be noted again that charge rather near the nucleus gives much larger contributions to q_j than the same charge further away from the nucleus, because q_j depends on $\overline{(1/r^3)}$. Hence it is quite possible for antishielding effects near the nucleus to dominate over the more normally expected shielding effects if the latter occur only in regions far from the nucleus. Whether shielding or antishielding actually occurs when contributions to q_j from all regions are considered, the magnitude of such effects will of course depend on detailed properties of the wavefunctions.

A number of theoretical calculations have been made by STERNHEIMER and others [45] to [51], [56], [57] on the modification of hyperfine interactions by the shielding or antishielding effects indicated above. For quadrupole hyperfine structure these consist of two types: (a) Calculations of modifications of hyperfine structure due to certain valence electrons in selected atoms and (b) calculation of antishielding of an external electric field of quadrupolar symmetry applied to various ions. Calculations of either type are rather involved and difficult to carry out with completeness and precision. Hence different computations of type (a) have not always been in agreement even as to sign, although they do show consistently that these effects are of importance. Table 6 lists the results of the most recent calculations of type (a) for various cases. Corrections for shielding or antishielding may be made by multiplying the value of the quadrupole moment Q obtained without corrections by the factor $1/(1-R)$ listed in Table 6. The uncertainty in this factor may be judged to some extent by the differences between values given for states labelled p_a and those labelled p_b for the alkalis. These correspond simply to two different sets of wavefunctions for the valence electrons which, so far as can be definitely established, each represent the true wavefunctions with comparable accuracy. However, STERNHEIMER [49] offers some reasons for preferring the wavefunction represented by p_a in the case of potassium.

Table 6. *Shielding factors for various atomic states* (from STERNHEIMER [46], [48], [49]).

R represents the negative ratio of the contribution to q_0 of the "spherical" core of electrons to that of the valence electron, being positive for shielding and negative for antishielding. For the alkalis, a state indicated by p_a represents calculations with one type of wavefunction, and that by p_b corresponds to a different, but perhaps equally valid wavefunction.

Element	State	R	$\dfrac{1}{1-R}$	Element	State	R	$\dfrac{1}{1-R}$
B	$2p$	0.143	1.17	Ag	$4d^9 5s^2$	0.115	1.13
F	$2p^5$	0.110	1.12	I	$5p^5$	0.028	1.03
Na	$3p$	-0.243	0.805	Cs	$6p_a$	-0.204	0.831
Cl	$3p^5$	0.112	1.13	Cs	$6p_b$	-0.332	0.751
K	$4p_a$	-0.188	0.842	Cs	$7p_a$	-0.178	0.849
K	$4p_b$	-0.009	0.991	Cs	$7p_b$	-0.275	0.784
K	$5p_a$	-0.142	0.876	La	$5d$	0.164	1.18
K	$5p_b$	-0.003	0.997	La	$4f$	0.229	1.30
Cl	$3d^9 4s^2$	0.253	1.34	W	$5d^4$	-0.510	0.663
Br	$4p^5$	0.038	1.04	Pt	$5d^9$	0.085	1.09
Rb	$5p_a$	-0.141	0.876	Tl	$6p$	0.019	1.03
Rb	$5p_b$	-0.271	0.787	At	$6p$	0.019	1.02
Rb	$6p_a$	-0.120	0.893	U	$5f$	0.225	1.29
Rb	$6p_b$	-0.332	0.751				

Table 7 gives values for the factors R and $1/(1-R)$ for certain ions in an external quadrupolar field. For the larger ions, very large antishielding seems to occur, so that the quadrupole hyperfine structure produced by a charge located outside the ion would be multiplied by factors as large as 50 or 100 due to the antishielding. This factor of increase, $1-R$, is often designated as $\varkappa$.

Table 7. *Antishielding factors for an external quadrupolar field acting on ions* (after STERNHEIMER and FOLEY [56], and DAS and BERSOHN [57]).

R is as defined in Table 6.

Ion	R	$\dfrac{1}{1-R}$
Li$^+$	0.256	1.33
Be^{++}	0.185	1.23
B^{+++}	0.146	1.17
Na$^+$	$-$ 4.1	0.176
Cl$^-$	$-$ 56.6	0.0173
Cu$^+$	$-$ 15.0	0.0625
Br$^-$	$-$ 99.0	0.01
Rb$^+$	$-$ 70.7	0.0139
I$^-$	-179	0.00556
Cs$^+$	-143.5	0.00691

11. Experimental tests of shielding and antishielding. Experimental tests of the effects of shielding on quadrupole hyperfine structure are not yet very precise, but some pertinent data are available. Perhaps the only experimental determination of shielding effects to a really satisfactory precision comes from a comparison of the magnetic hyperfine structure of Cl with a theoretical calculation of the rather similar shielding effects of the "closed" electron shells on magnetic hyperfine structure. In this case, calculations indicate [47] that shielding effects should increase the expected ratio of hyperfine structure of the $3p^5\,^2P_{\frac{3}{2}}$ state of Cl to that of the $3p^5\,^2P_{\frac{1}{2}}$ state by 8%. Experimental measurements show an actual decrease of one percent [47]. This rather disappointing discrepancy presumably comes from the very delicate dependence of the theoretical calculations on the nature of the atomic wavefunctions used [47].

For quadrupole hyperfine structure in atoms, a coarse experimental test of shielding effects is allowed by comparison of quadrupole moments determined from hyperfine structure and those measured by Coulomb excitation. This latter method of measurement will be described in a later section. Although subject to various uncertainties, the values obtained from Coulomb excitation are not affected by screening. Table 8 gives a comparison between the two sets of values for all nuclei whose quadrupole moments have so far been measured by both techniques. In this table, no corrections have been made for shielding so that, in spite of large experimental uncertainties, a rough magnitude and sign of shielding effects may be obtained by comparison of the average difference. According to the table, antishielding seems to occur in all of these cases, and quadrupole values from hyperfine structure are on the average about 18% larger than those from Coulomb excitation experiments. This is not an unreasonable magnitude in view of the calculated shielding or antishielding corrections for p and d electrons in atoms near the weight of those in Table 8. However, it must be remembered that this difference may possibly came from some other systematic errors not yet clearly recognized in the Coulomb excitation measurements.

Table 8. *Comparison between quadrupole moments obtained from hyperfine structure with those from Coulomb excitation.*
Ho^{165} is omitted from the average because Q_s for it is much less accurately determined than are values for other nuclei.

Nucleus	Q_c (Value from Coulomb excitation)	Q_s (Value from h.f.s.)	$\dfrac{Q_s - Q_c}{Q_s}$
Eu^{153}	2.4×10^{-24} cm^2	2.5×10^{-24} cm^2	0.04
Ho^{165}	3.6	~2	
Lu^{175}	3.6	5.7	0.37
Ta^{181}	3.1	4.3	0.28
Re^{185}	2.6	2.8	0.08
Re^{187}	2.3	2.6	0.11
			average: 0.18

Table 8 at least indicates that errors due to shielding are not much greater than 18%. Some additional evidence on shielding effects can be obtained by comparison of quadrupole couplings which have now been measured [52], [53] for the Rb^{85} nucleus in both the $5p$ and $6p$ atomic states. These measurements give a ratio 0.97 ± 0.10 for the quadrupole moment, determined without shielding corrections from the $5p$ state to that from the $6p$ state. According to Table 6 this ratio should be 1.00 ± 0.03. The experimental results show only that there is no large difference in shielding between the two states. A somewhat similar situation occurs for the case of K^{39}, for which the quadrupole coupling has been measured in the $4p$ and $5p$ states [54a], [54b]. Without shielding corrections, Q yielded by the $5p$ state is 1.5 times larger than that from the $4p$ state, although this difference is not clearly outside of experimental error for the two measurements. The ratio of shielding corrections for the two states has not yet been calculated.

Antishielding effects for charges external to an atom or ion can be tested experimentally by examination of quadrupole coupling in molecules. For example, one may consider, as an approximation, the quadrupole coupling in some of the alkali halides as due to the effect of a nearby ionic charge with antishielding. This is because the alkali halides may for many purposes be considered to consist simply of two nearby ions of opposite charge. Each ion, if undistorted by its partner, has a spherical distribution of charge and hence contributes nothing to the quadrupole coupling constant of its own nucleus. Of course it may be somewhat distorted by various effects in addition to the electrostatic field of its neighboring ion, and hence an experimentally measured coupling constant can never give simply the effect on an external charge and its antishielding. On the other hand, the situations where one is most tempted to use the results of antishielding calculations are just such molecular cases. Hence it is important to examine, as far as possible, the extent of agreement between the calculations and experimental observation in these molecules. Table 9 gives pertinent figures derived from all known coupling constants for the halogens in alkali halides. Differences in electronegativity between the alkali and halogen for each molecule are also listed. In each case, the measured value of $q = \partial^2 V/\partial z^2$ is determined from the quadrupole coupling constant and the known quadrupole moment. The ratio of q to the value calculated for $\partial^2 V/\partial z^2$ due to one (positive) electron charge at the position of the alkali nucleus is the value of $\varkappa$, the antishielding factor.

Table 9. *Comparison between theoretical and "experimental" antishielding factors for halogens in the alkali halides.*

The "experimental" factors $\varkappa$ represent the ratio of observed quadrupole coupling constants to those which would be present if the alkali and halogen were undistorted ions separated by the observed internuclear distance. Theoretical factors $\varkappa$ are those calculated by allowing for distortion of the halogen ion by the electric field of the alkali.

Molecule	Nucleus	eqQ (Mc/sec)	Electro-negativity difference	$\varkappa$ (experimental)	$\varkappa$ (theoretical)
CsCl	Cl^{35}	$\lvert eqQ \rvert < 3$	2.4	<16	58
RbCl	Cl^{35}	-0.77	2.3	3	58
KCl	Cl^{35}	$\lvert eqQ \rvert < 0.04$	2.3	<1	58
NaCl	Cl^{35}	-5.4	2.2	13	58
KBr	Br^{79}	10.2	2.1	10	100
NaBr	Br^{79}	58	2.0	39	100
LiBr	Br^{79}	37.2	1.9	16	100
KI	I^{127}	-60	1.8	42	180
NaI	I^{127}	-260	1.7	126	180
LiI	I^{127}	-198	1.6	66	180

Table 9 shows that for molecules with electronegativity differences greater than 2.0, $\varkappa$ is actually considerably (a factor of four or more) less than the calculated values. For electronegativity differences smaller than 2.0, there is good evidence, some of which will be given below in the discussion of molecular quadrupole coupling constants, that the alkali halides can no longer be considered ionic, but are appreciably covalent. This is reflected in Table 9 by the fact that for such molecules the quadrupole coupling is very considerably larger than for the more ionic cases with electronegativity differences greater than 2.0. Hence any apparent agreement between calculation and measurements for NaBr and for the alkali iodides must be discounted.

There are, of course, a number of additional molecular distortions which have not been allowed for, and some of these might be supposed to almost cancel the value of q contributed by antishielding. Any covalent character which is present would tend to increase the observed $\varkappa$ rather than make it too small.

Table 10 gives corresponding data for the alkali nuclei in alkali halides. In this case, since the valence electrons are s electrons, a small amount of covalent character is unimportant in influencing the quadrupole coupling constant (see below). It is evident that much better agreement is obtained between the measured and calculated values of the antishielding factor for the alkalis than for the halogen cases. Furthermore, the experimental $\varkappa$ factors are less erratic for these ions.

Still further removed from the direct calculation of $\varkappa$ for ions are results of experiments on antishielding in ionic crystals. For example, one may ask how large a value of q is generated by displacement of a Na^+ ion in the NaCl crystal with respect to some adjacent Cl ion. Here of course the calculations of $\varkappa$ made for atoms cannot be highly reliable, but one may still expect antishielding factors of the same order of magnitude as those found for atoms.

A number of experiments have been used to determine $\varkappa$ in crystals, of which three will be mentioned. The first is a measurement of relaxation times for the Cl nuclear magnetic resonance in NaCl. This relaxation can be interpreted as due to thermal vibrations of the lattice which induce quadrupole fields at the Cl nucleus. VAN KRANENDONK [54] has calculated the resulting relaxation and found a value of $\varkappa$ which is greater than 100.

A much more direct experiment due to PROCTOR and ROBINSON [55] is to excite elastic vibrations in the crystal of frequency near that of resonance and

Table 10. *Comparison between theoretical and "experimental" antishielding factors for alkalis in the alkali halides.*

The "experimental" factors $\varkappa$ represent the ratio of observed quadrupole coupling constants to those which would be present if the alkali and halogen were undistorted ions separated by the observed internuclear distance. Theoretical factors $\varkappa$ are those calculated by allowing for distortion of the alkali ion by the electric field of the halogen.

Molecule	Nucleus	$\varkappa$ (experimental)	$\varkappa$ (calculated)	Molecule	Nucleus	$\varkappa$ (experimental)	$\varkappa$ (calculated)
NaF	Na^{23}	7	5.6	KCl	K^{39}	10	—
NaCl	Na^{23}	10	5.6	KBr	K^{39}	10	—
NaBr	Na^{23}	10	5.6	RbF	Rb^{85}	40	50
NaI	Na^{23}	11	5.6	RbCl	Rb^{85}	50	50
KF	K^{39}	8	—	CsF	Cs^{133}	90	90

to observe the resulting enhanced relaxation. An experiment of this type yields a value near unity for $\varkappa$ [55]. This latter experiment seems to allow a much more direct and simple evaluation of $\varkappa$ than does the interpretation of relaxation times, and hence the actual value of $\varkappa$ for Cl in NaCl may be near unity, rather than very large as might be inferred from calculations.

The third experiment involves distortion of the cubically symmetric electric field about a Na^+ ion in crystalline NaCl when some of the Cl is replaced by Br. These distortions decrease the observed intensity of the pure quadrupole resonance line of Na^{23}, and this decrease in intensity has been used by KAWAMURA, OTSUKA, and ISHIWATARI [58] to obtain the approximate value of 10 for $\varkappa$. This value is in resonable agreement with the calculated value from Table 10.

The above experimental evidence is not as complete or direct as is desirable. However, these results coupled with known theoretical difficulties indicate enough uncertainty in the calculated correction factors due to shielding that their present general use for evaluating nuclear quadrupole moments does not seem justified. Hence no corrections of this type will be applied in the present work for evaluation of quadrupole moments from interactions between nuclei and their valence electrons. However, one may expect quadrupole moments determined without such corrections to be frequently in error as much as 15%. Use of antishielding factors calculated for isolated atoms or ions to evaluate quadrupole moments from hyperfine structure in molecules or crystals seems at present very hazardous.

12. Evaluation of nuclear quadrupole moments from their coupling in molecules. Determination of a nuclear quadrupole moment from its hyperfine interaction in an atom involves some imprecision, as has been seen above, because of incomplete knowledge of the radial electronic wavefunctions and of shielding or antishielding effects. Evaluation of a quadrupole moment from hyperfine inter-actions in molecules is susceptible to these same errors plus usually more serious uncertainties due to incomplete knowledge of the molecular structure. However, when suitable atomic hyperfine structure is not known, determination of nuclear quadrupole moments can frequently be made with useful accuracy from molecular hyperfine structure. In fact, the wealth of information available on hyperfine effects in molecules and solids due to nuclear quadrupole moments puts a high premium on the development of methods for extracting values of the moments themselves from the measured coupling constants.

In principle, a direct calculation of the electronic wavefunction for a molecule will allow a complete determination of the distribution of charges, and hence of $\partial^2 V/\partial z^2$ or $\boldsymbol{V}\boldsymbol{E}$, which is the quantity needed to evaluate quadrupole moments from coupling constants such as $e \dfrac{\partial^2 V}{\partial z^2} Q$. However, in most cases a direct and complete calculation is too complicated to be practical. Only in the case of the hydrogen molecule has such a method been used with much success. A number of calculations on the electronic wavefunction of HD establish the value of $q \equiv \partial^2 V/\partial z^2$ at the deuterium nucleus and hence of Q by use of the experimental determination of eqQ, to an accuracy of about one percent [5], [59]. Calculations beginning from basic principles have also been made to evaluate q from the molecule Li_2, but they yield rather uncertain and varying results [60] to [62].

Although a direct and complete calculation of the field gradients in a molecule is not usually practical, in many cases empirical knowledge of atomic and mole-cular structure may be combined with some theoretical considerations to yield evaluations of q with useful accuracy. This type of approach has been emphasized by TOWNES and DAILEY [63], [7], who pointed out in particular that in many cases the value of q can be obtained from a few rather simple parameters of the wavefunction for the valence electrons, such as the probability that they occur in p orbitals of the lowest atomic state. Evaluation of the field gradients at a nucleus in a molecule from known molecular and atomic parameters may usually be conveniently approached by analyzing separately contributions from the following three sources:

1. Valence electrons of the nucleus or atom in question.
2. Distortion of the closed shells of electrons around the nucleus.
3. Charge distributions associated with adjacent atoms or ions, i.e., charges essentially outside the atomic radius.

It will be shown that in many cases, contributions of type 1 can be evaluated rather easily with useful accuracy, and that the other two types of contributions are small enough to be neglected.

In an atom, the state of a valence electron may usually be well described by quantum numbers n, l, m, and the atomic wavefunction Ψ_{nlm}. In a molecule, the wavefunction of such an electron will be modified—in some cases radically changed—but it may be expressed in terms of atomic wavefunctions of a parti-cular nucleus by the expansion

$$\Psi = \sum_{nlm} a_{nlm} \Psi_{nlm}. \tag{12.1}$$

The advantage of such an expansion is that frequently its larger terms are fairly well known from molecular structure, and further that only the first terms, cor-responding to the lowest atomic states, affect q appreciably.

An expression for q, or $\partial^2 V/\partial z^2$, can be obtained from Ψ as

$$\left. \begin{aligned} q &= e \int \Psi^* \left(\frac{3\cos^2\vartheta - 1}{r^3} \right) \Psi\, dv \\ &= \sum_{nlm} |a_{nlm}|^2\, q_{nlm,nlm} + \sum_{nlm,\,n'l'm'} a_{nlm}\, a^*_{n'l'm'}\, q_{nlm,\,n'l'm'} \end{aligned} \right\} \tag{12.2}$$

where

$$q_{nlm,\,n'l'm'} = e \int \Psi^*_{n'l'm'} \left(\frac{3\cos^2\vartheta - 1}{r^3} \right) \Psi_{nlm}\, dv. \tag{12.3}$$

The second sum in (12.2) is over all cases where $n'l'm'$ is not identical with $n\,l\,m$; those where $n=n'$, $l=l'$, and $m=m'$ are written in the first sum. Since the

Table 11. *Relative values of $q_{nl0} = \partial^2 V/\partial z^2$ for various atomic states, assuming hydrogenic wavefunctions, and actual values of q_{nl0} (after TOWNES and SCHAWLOW [7]).*

Electronic state	Atom from which experimental value of $\Delta\nu$ is taken	Values of q_{nl0} from fine structure, in e.s.u.	Relative values of q_{nl0} assuming hydrogen wavefunctions and no screening
			Relative to $5p$
$5p$	I	$-45 \quad \times 10^{15}$	1.00
$5d$	Cs	$- 0.31 \times 10^{15}$	0.14
$5f$	Cs		0.048
$6p$	Cs	$- 3.4 \quad \times 10^{15}$	0.58
$6d$	Cs	$- 0.16 \times 10^{15}$	0.08
$6f$	Cs		0.028
$7p$	Cs	$- 1.1 \quad \times 10^{15}$	0.36
$7d$	Cs	$- 0.09 \times 10^{15}$	0.05
			Relative to $2p$
$2p$	F	$-21 \quad \times 10^{15}$	1.00
$3p$	Na	$- 0.7 \quad \times 10^{15}$	0.30
$4p$	Na	$- 0.2 \quad \times 10^{15}$	0.12

angular variations of the wavefunctions Ψ_{nlm} are given by spherical harmonics, $q_{nlm,n'l'm'}$ is zero unless $m=m'$ and either $l=l'\neq 0$ or $l=l'\pm 2$. This eliminates a large number of terms from expression (12.2).

Consider now the relative sizes of the various quantities $q_{nlm,nlm}$ (which will be shortened to q_{nlm}) and $q_{nlm,n'l'm'}$. As an example, q_{nl0} is given, from (9.2) and (9.15) by

$$q_{nl0} = \frac{2l(l+1)\, e}{(2l-1)\,(2l+3)} \left(\frac{1}{r^3} \right)_{av} \tag{12.4}$$

which for hydrogenic wavefunctions is

$$q_{nl0} = \frac{4\, Z^3\, e}{n^3\, a_0^3\, (2l-1)\,(2l+1)\,(2l+3)}. \tag{12.5}$$

Clearly q_{nl0} decreases rapidly with increasing n or l, because it depends on the average inverse cube of distance from the nucleus to the electron. For non-hydrogenic atoms, where screening is involved, q_{nl0} decreases even more rapidly with increasing n or l than is indicated by (12.5). This behavior is illustrated in Table 11, which gives relative values for some hydrogenic cases, and actual values for corresponding cases of real atoms with screening. For the screened atoms, $(1/r^3)_{av}$ is evaluated by use of experimental values of $\Delta\nu$, as indicated by (9.11). It is evident from this table that q_{nl0} for the lowest p level of the halogens is much larger than q_{nl0} for higher states, which are usually less by more than a factor of ten. Hence if the coefficient $|a_{nl0}|^2$ in (12.2) for the lowest p state is fairly

27*

large, all other terms in the first sum which represent higher states can be neglected without involving an error of more than 5 or 10%.

The non-zero terms in the second summation of (12.2), that is, those for which $m'=m$ and $l'=l\neq0$ or $l'=l\pm2$, are usually also small compared with the contribution of the lowest p wavefunction. This is partly because $q_{nlm,nl\pm2\,m}$ is rather small, and partly because the coefficient $a_{nlm}a^*_{nl\pm2,m}$ can be expected to be small. Consider, for example, the case where a low-lying p state is the largest component of the wavefunction ψ of (12.1). Cross-product terms which involve its amplitude can be expected to be the largest parts of the second sum in (12.2). The only terms of this type which can exist involve mixing this p function with higher p states and with f states. The corresponding values of $q_{nlm,n'l'm}$ can be expected to be somewhat smaller than the geometric mean of q_{nlm} and $q_{n'l'm}$. Furthermore, it is well known that the most prevalent type of bond hybridization involves mixtures of states of different parity, such as an s state with a p state, since this transfers charge to the bond from the far side of the nucleus. Hybridization of a p bond with other p or f orbitals should be small, so that $a_{nlm}a^*_{n'l'm}$ is also small, and terms of this type can probably also be neglected. In some unusual cases, however, such cross-terms may contribute an important part of q.

Direct contributions to q from charges on neighboring atoms or ions are quite small and may be neglected whenever contributions from the valence electrons are of normal magnitude and do not cancel each other to a large extent. For example, a neighboring ion with an average charge one-half that of an electron and at a distance 2.0 Å from the nucleus being considered produces a value of q of only 3×10^{13} e.s.u. It may be seen from Table 11 that this is less than one percent of the value due to a p electron in I or F. Indirectly, a neighboring ion can have a somewhat more important effect by distorting the valence shell or closed shells of electrons surrounding the nucleus.

Closed shells of electrons are subject to distortion either by valence electrons of the same atom, or by charges on neighboring atoms and ions. Distortion by valence electrons is essentially the same as that which occurs in atoms, giving the shielding or antishielding effects discussed above. The effect of a neighboring ion can be judged either from the calculations of FOLEY, STERNHEIMER, and TYCKO [51] for antishielding of an external charge, or from a certain amount of empirical evidence. These have been discussed above, where it was found that antishielding effects of the order of 10 are typical. Hence the ion with one-half of an electron charge and at a distance of 2.0 Å may be expected to contribute about 3×10^{14} e.s.u. to q. This is still less than a few percent as large as the contribution due to a low-lying p state, and hence may often be neglected.

It may be seen from the above that rather frequently one or a few large terms due to the lowest lying states dominate not only all other terms in expression (12.2), but also any other contributions to q. If in such favorable cases the probabilities $|a_{nlm}|^2$ for these states alone can be determined, then uncertainties in evaluation of nuclear quadrupole moments from quadrupole coupling constants are not appreciably greater than for the case of atoms, both being limited by the poorly-known shielding effects to perhaps 10 or 20% accuracy. On the other hand, in some molecules this approximation for obtaining q will have little success. For example, in hydrogen the largest term in the expansion (12.1) is the $1S$ wavefunction, which gives $q_{100}=0$, and any contributions to q must come from higher terms in the expansion which are normally discarded in the approximation discussed. This approach leads then only to the conclusion that q should be small, but not to an evaluation of q. In general, the type of approximation outlined above is at its best when the valence electrons fill a p shell incompletely and when these p orbitals are rather pentrating, as in all but the very lightest elements. In heavier elements d orbitals can also at times give couplings which are large enough to make the approximation favorable.

Consider now two atoms A and B, each containing spherical shells of electrons plus one p electron. If the two atoms are allowed to approach and form a chemical bond, the spatial part of the wavefunction of each valence electron might as a first approximation be taken in the molecular orbital form

$$\psi = \frac{2}{\sqrt{1}} \frac{(\psi_A + \psi_B)}{\sqrt{1+S}} \tag{12.6}$$

where ψ_A and ψ_B are the p wavefunctions about atoms A and B respectively with a projection $m=0$ of orbital angular momentum on the interatomic axis. For simplicity these wavefunctions will be taken as real, and S may be written

$$S = \int \psi_A \psi_B \, dv. \tag{12.7}$$

If ψ is expanded as indicated by expression (12.1), $a_{n l 0} = \sqrt{\frac{1}{2}(1+S)}$ and hence in the approximation which neglects all but the lowest valence p state, the contribution of each valence electron to q is $\dfrac{1+S}{2} q_{n l 0}$, so that one would be led to the value

$$q = (1 + S) q_{n l 0}. \tag{12.8}$$

Choice of a HEITLER-LONDON rather than a molecular orbital form for the wavefunction ψ leads to somewhat different, but comparable results.

Whether or not the overlap integral S actually contributes as much to q as is indicated by (12.8) has been discussed by several authors [63] to [66]. It is easy to see that, while ψ_A approximates rather well a solution of SCHRÖDINGER's equation near the nucleus of atom A, ψ_B cannot be expected to be a good approximation in this region. Hence, although the wavefunction (12.6) may be adequate for approximate calculations of binding energy or other properties which depend on the electronic distribution far from the nucleus, the part ψ_B cannot be expected to give correct hyperfine structure for the nucleus of atom A, since this depends on the electronic distribution very near this nucleus. ITOH [67] has calculated q from a form for ψ similar to (12.6), but after modifying ψ_B to make it orthogonal, as it must be, to all electrons of atom A in nonbonding states. This necessitates a particularly large modification of ψ_B near the nucleus of A, and specific calculations for wavefunctions of this type in the case of Cl indicate that the contribution of the overlap to q is reduced to a few percent of $q_{n l 0}$. Furthermore, experimental evidence indicates that there is no large contribution to q from the overlap, so that q may be obtained from a bonding orbital of the form (12.6) by neglecting overlap completely. This gives, instead of (12.8)

$$q = q_{n l 0}. \tag{12.9}$$

It should perhaps be noted that in obtaining the coefficient $a_{n l 0}$ of the electronic wavefunction, it is expanded in terms of true atomic wavefunctions as indicated in (12.1), and not in terms of the various types of functions which are sometimes chosen to calculate molecular energies or other molecular properties. Hence $a_{n l 0}$ may not be very different, but is not necessarily identical with what is often indicated as the p component of the bonding orbital in calculations of molecular energies or dipole moments, since the p wavefunction chosen for such calculations is usually only an approximation to the true atomic wavefunction.

Consider now as an example a chlorine atom bonded in a molecule by a single bond. When Cl is negatively ionic, its valence electrons fill a closed shell and hence in the present approximation contribute nothing to q. There is, of course, distortion of this closed shell by neighboring atoms or ions, but the resulting contribution to q is small, as seen above, and will be neglected. When the Cl is covalently bonded, the wavefunction of the bonding electrons may be represented as in (12.6) where ψ_A can be expected to contain a large component of the atomic $3p$ state with $m=0$, smaller components of $3s$ and $3d$ states, and very small components of a large number of higher atomic states. Thus

$$\psi_A = a_p \psi_{310} + a_s \psi_{300} + a_d \psi_{320} + \cdots, \tag{12.10}$$

and the resulting contribution of each bonding electron to q is approximately $\frac{1}{2} a_p^2 q_{310}$, where $a_p^2 = 1 - a_S^2 - a_d^2$. Here a_p^2 is evaluated by normalizing ψ_A and neglecting small contributions from higher states other than those written down in (12.10). The non-bonding valence electrons also contribute. There are four in p orbits which contribute $2q_{311}$ and $2q_{31-1}$ or $4q_{311}$ since q_{311} and q_{31-1} are equal. The two electrons in $3S$ orbits also affect q appreciably. Their wavefunctions must be orthogonal to ψ_A, and hence should have the form $\psi = \frac{1}{\sqrt{a_p^2 + a_s^2}} [a_p \psi_{300} - a_s \psi_{310}]$ so that each non-bonding "S" electron contributes $\frac{a_s^2}{a_p^2 + a_s^2} q_{310}$ to q. Adding up the effects of the entire valence shell, one obtains

$$q = 4q_{311} + \left(\frac{1 + a_s^2 - 2a_d^2}{1 - a_d^2} \right) q_{310} \tag{12.11}$$

which, from Eq. (9.15) gives

$$q = - \frac{(1 - a_s^2)}{1 - a_d^2} q_{310},$$

or, since a_d^2 is usually small compared to unity,

$$q = - (1 - a_s^2 + a_d^2) q_{310}. \tag{12.12}$$

When Cl is positively ionic, with no electrons occupying the ψ_{310} orbit,

$$q = 4 \overset{+}{q}_{311} = - 2 \overset{+}{q}_{310}. \tag{12.13}$$

Here the plus sign over q_{310} indicates that, for the positive ion, q_{310} may be somewhat different from its value for the atom. This difference is not large, and will be discussed below in more detail.

An actual Cl atom, singly bonded in a molecule, is normally considered to be in some combination of the three states indicated above. If the probability of the negatively ionic state, the covalently bonded state, and the positively ionic state are A, B, and C, respectively, then

$$q = - B (1 - a_s^2 + a_d^2) q_{310} - 2C \overset{+}{q}_{310}, \tag{12.14}$$

where $A + B + C = 1$. This result may be compared with typical cases given in Table 12. The quadrupole coupling constant listed there is eqQ in Mc/sec for Cl^{35}, and $eq_{310}Q$ is known to be 109.7 Mc/sec [68], from measurements on the ground state of atomic Cl. It may be seen that for cases such as NaCl, where Cl is expected to be nearly like a negative ion ($B = C = 0$), eqQ is very small. For FCl, where Cl is expected to be somewhat positively ionic (C not small), the magnitude of eqQ is appreciably larger than for the atomic case, and for those examples where Cl is expected to be essentially covalent ($B = 1$, $C = 0$), its magnitude is smaller than q_{310}, as if $a_s^2 - a_d^2$ were small but not negligible. A systematic study of many cases for which quadrupole coupling constants have been measured, combined with consideration of other types of evidence on the structure of the bonds involved, indicates that most covalent Cl bonds are hybridized to the extent that $a_s^2 - a_d^2 \approx 0.15$, and that probably a_d^2 is rather small.

As indicated above, the value of q_{310} will depend on whether the p orbit involved is in an ion or a neutral atom. A comparison of fine structure in atoms with that in corresponding ions indicates that, for each electron removed from an atom, q_{310} increases approximately by the factor $\frac{5}{4}$, and for an added electron, it decreases by approximately the factor $\frac{4}{5}$. Thus one may write $\overset{+}{q}_{310} = (1 + \varepsilon)\, q_{310}$ and $\overset{-}{q}_{310} = \dfrac{1}{1+\varepsilon}\, q_{310}$, where $\varepsilon \approx 0.25$. The value of ε varies somewhat with atomic number, and more accurate values are given by Table 13.

In most cases, both positive and negative ionic structures do not need to be considered in the same bond. For example, an equal probability of positive and negative ionic character for the case of Cl discussed above gives a quadrupole coupling essentially the same as neutral Cl, and neutral Cl may be assumed

Table 12. *Values of eqQ for Cl^{35} in a few typical symmetric molecules.*

For the atomic state,
$-eq_{310}Q = -109.6$ Mc/sec.

Molecule	eqQ (Mc/sec)
KCl	$|eqQ| < 0.04$
CsCl	$|eqQ| < 3$
GaCl	
TlCl	-15.8
CH$_3$Cl	-74.7
ClCN	-83.3
BrCl	-103.6
ICl	-82.5
C$_2$HCl	-79.7
FCl	-145.99

to describe such a case. If Cl has a probability i of being negatively ionic, which is the usual case, and otherwise is covalently bonded, then (12.14) may be written

$$q = -(1 - i)(1 - s + d)\, q_{310} \tag{12.15}$$

where $s \equiv a_s^2$ is the amount of s hybridization of the covalent bond and $d \equiv a_d^2$ is the amount of d hybridization. The factor multiplying $-q_{310}$ in (12.15) is

Table 13. *Values of ε for a number of elements* (from TOWNES and SCHAWLOW [7]).
The coupling constant due to a valence p electron is modified by a factor approximately $1 + \varepsilon$ for each stage of ionization, being larger for positive ionization.

Be	0.90	B	0.50	C	0.45	N	0.30	O	0.25	F	0.20
Mg	0.70	Al	0.35	Si	0.30	P	0.20	S	0.20	Cl	0.15
Ca	0.60	Sc	0.30	Ge	0.25	As	0.15	Se	0.20	Br	0.15
Sr	0.60	Ga	0.20			Sb	0.15	Te	0.20	I	0.15

designated U_p, the "number" of unbalanced electrons. If U_p electrons were added in the ψ_{310} orbit of Cl, the Cl would be spherically surrounded by valence electrons and hence q would be zero. Thus the "number" of unbalanced p electrons is proportional to q and a characteristic of the type of bonding for a particular column of the periodic table. U_p is tabulated for a number of common cases in Table 14. These include multiple as well as single bonds. In the case of multiple bonds between two atoms, the p wavefunctions indicated by $\psi_{31\pm1}$ or p_π may form covalent bonds as well as those designated ψ_{310} or p_σ.

Some bonds are best described as a combination of single and multiple bonds, each perhaps with certain amounts of hybridization and ionic character. U_p and hence q for most of these cases can be obtained from combinations of values given in Table 14 if the probability of each bond structure is estimated from other known properties of the bond. There are a number of helpful, but not usually

Table 14. *The number of unbalanced p electrons, U_p, for various bond structures.*
The function of s or d hybridization is indicated by the symbols s or d, respectively. It is assumed that the hybridizing s or d wave functions have the same principal quantum number as the p function. U_p is with respect to the axis of the bond or bonds unless otherwise stated. The quantity ε which appears here is given in Table 13 (from TOWNES and SCHAWLOW [7]).

Electron configuration of atom	Type of bond	Hybridization	U_p
s^2p^5 (like Cl)	Single covalent	s and d	$1 - s + d$
s^2p^6 (like Cl$^-$)	Single ionic	—	0
s^2p^4 (like Cl$^+$)	Single ionic	—	$2(1 + \varepsilon)$
s^2p^4 (like O)	Double covalent	$\begin{cases} p_\sigma s \text{ and } d \\ p_\pi \text{ none} \end{cases}$	$\frac{1}{2} - s + d$
s^2p^3 (like N)	Triple covalent	$\begin{cases} p_\sigma s \text{ and } d \\ p_\pi \text{ none} \end{cases}$	$-s + d$
s^2p^4 (like O)	Two single covalent, each of ionic character i, with O positive	s	$\left(\frac{1}{2} + \frac{i}{2} + 2s\right)(1 + 3i\varepsilon)$ (along direction bisecting bond angle) $(s - 1 - i)(1 + 2i\varepsilon)$ (along direction perpendicular to plane of bonds)
	With O negative	s	$\dfrac{(\frac{1}{2} - 2s)(1 - i)}{1 + 2i\varepsilon}$ (along direction bisecting bond angle) $\dfrac{(s - i)(1 - i)}{1 + 2i\varepsilon}$ (along direction perpendicular to plane of bonds)
s^2p^3 (like N)	Three single bonds, each of ionic character i, with N positive	s	$-3s(1 + i)(1 + 3i\varepsilon)$
	With N negative	s	$-\dfrac{3s(1 - i)}{(1 + 3i\varepsilon)}$
s^2p^2 (like C)	Four covalent bonds	Any $s - p$ hybridization	0
s^2p (like B)	Three bonds in a plane	s	1

very accurate, ways of determining the structure of bonds and hence estimating q. Some of these will be indicated below.

The ionic character of a covalent bond in a diatomic molecule may be judged from Fig. 9, where ionic character is plotted against electronegativity of the two bonded atoms. Electronegativity differences may in turn be found in Table 15. A measure of the uncertainty involved in relating electronegativity differences to ionic character is indicated in Fig. 9 by the differences between the curve and points for various molecules. For polyatomic molecules, the ionic character of a bond between two atoms is given approximately by the curve of Fig. 9, but this relation is not expected to be as accurate as for diatomic molecules.

In some cases, electric dipole moments of molecules or of bonds may be used to deduce the ionic character. Thus it is often assumed that $i = \left|\dfrac{\mu}{eR}\right|$, where μ is the dipole moment associated with two bonded atoms, e the electron charge, and R the internuclear distance. Dipole moments are actually rather strongly affected by hybridization, overlap, and depolarization effects and hence values of ionic character deduced from them cannot be expected to be very reliable. However, in many cases, there does appear to be a useful approximate correlation between ionic character and dipole moment, which, allowing for depolarizing effects [65] is closer to

$$i = 1.5 \left| \frac{\mu}{eR} \right| \qquad (12.16)$$

than to the grosser approximation indicated above.

The amount of hybridization is difficult to evaluate and varies so much that no very general rules are applicable. For Cl, Br and I, 15% s hybridization can usually be assumed when these atoms are more electronegative than the atom to which they are bonded by 0.25 units, and otherwise the hybridization is appreciably smaller [65]. As indicated above, d hybridization for these cases appears to be negligible so far as q is concerned. Approximately 15% s hybridization seems to be fairly common in other elements, but there is evidence that oxygen bonds have as much as 40% s character [71] and that s hybridization is somewhat less than 15% for many of the heavier elements. If an atom is singly bonded to two or more other atoms, the angle between the bonds often gives an estimate of the amount of hybridization. If only s and p orbitals are involved in two or more equivalent bonds, the s hybridization can be expected to be

$$s = \frac{\cos \vartheta}{\cos \vartheta - 1} \qquad (12.17)$$

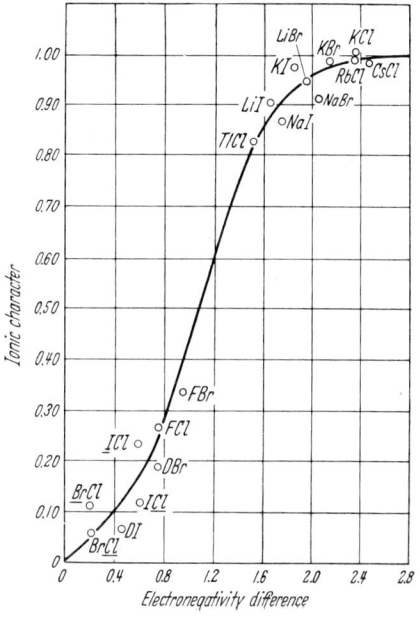

Fig. 9. Ionic character of bonds in diatomic molecules as a function of difference in electronegativity between the two bonded atoms. (After DAILEY and TOWNES [65]).

where ϑ is the angle between two of the equivalent bonds. Expression (12.17) appears to give reasonably accurate values of s hybridization except for the hydrides, where the hybridization is usually appreciably larger than that indicated by Eq. (12.17) [72], [71].

Whether an atom is bonded singly or multiply, or in some combination of the two can usually be determined from its bond length. Single, double, and triple-bond radii of many atoms are listed in Table 15. The actual internuclear distance R between two bonded atoms is given approximately by [69]

$$R = \frac{x_1 R_1 + 3 x_2 R_2 + 6 x_3 R_3}{x_1 + 3 x_2 + 6 x_3} \qquad (12.18)$$

where R_1, R_2, R_3 are the sums of single, double, and triple bond radii respectively for both atoms, and x_1, x_2, x_3 are the fractional importances of single, double, and triple bond states respectively.

The above rules allow calculations of q for many molecules to an accuracy of about 10%. In other cases, particularly where there are multiple bonds or considerable cancellation between several contributions to q, the fractional accuracy may be considerably worse. A number of typical cases are illustrated in

Table 15. *Radii of covalent bonds and electronegativities of atoms.*

Bond radii are those of PAULING [69] and are given in angstrom units. Values of electronegativities are from HUGGINS [70] and PAULING [69], and are given in arbitrary units. Differences in electronegativities may be considered approximately as electron volts.

	H						
Electronegativity	2.2						
Single-bond radius . . .	0.30						
	Li	Be	B	C	N	O	F
Electronegativity	1.0	1.5	2.0	2.6	3.0	3.5	3.9
Single-bond radius . . .			0.88	0.771	0.70	0.66	0.72
Double-bond radius . . .			0.76	0.665	0.60	0.55	0.62
Triple-bond radius . . .			0.68	0.602	0.547	0.50	
	Na	Mg	Al	Si	P	S	Cl
Electronegativity	0.9	1.2	1.5	1.9	2.1	2.6	3.1
Single-bond radius . . .				1.17	1.10	1.04	0.99
Double-bond radius . . .				1.07	1.00	0.94	0.89
Triple-bond radius . . .				1.00	0.93	0.87	
	K	Ca	Sc	Ge	Aa	Se	Br
Electronegativity	0.8	1.0	1.3	1.9	2.1	2.5	2.9
Single-bond radius . . .				1.22	1.21	1.17	1.14
Double-bond radius . . .				1.12	1.11	1.07	1.04
	Rb	Sr	Y	Sn	Sb	Te	I
Electronegativity	0.8	1.0	1.3	1.9	2.0	2.3	2.6
Single-bond radius . . .				1.40	1.41	1.37	1.33
Double-bond radius . . .				1.30	1.31	1.27	1.23
	Cs	Ba					
Electronegativity	0.7	0.9					

Table 16. For cases in which the molecule is symmetric about a bond, U_p referred to the internuclear axis is given, and is sufficient to specify completely the quadrupole coupling tensor, since then $q_{xx}=q_{yy}=-\frac{1}{2}q_{zz}$. For non-symmetric cases, U_p is specified for three directions, and $q_{xx}=-U_p q_{n10}$, $q_{yy}=-U_{py} q_{n10}$, $q_{zz}=-U_{pz} q_{n10}$.

In some cases, U_p or q are dependent on only one or two parameters of the electronic structure of the molecule, and rather independent of others. Thus in AsH$_3$ or AsCl$_3$, q is primarily dependent on the s hybridization of each bond, being zero if this hybridization is zero. Thus also the present approximation gives $q=0$ for any quadruply bonded nitrogen, as in the central atom of NNO, as long as only s and p orbitals are used.

Since the electronic structures of molecules are not usually very precisely known, it is very useful to compare the quadrupole coupling of a given isotope in a wide variety of molecules in order to afford a check on estimates of the bond structures involved, and to minimize the possibility of errors in determining q, or hence in evaluating quadrupole moments from the coupling constants. Illustrations of the usefulness of such comparisons are the cases of oxygen [71] and of sulphur [72] bonds and the quadrupole moments of O^{17} and S^{33}. For these nuclei, certain molecules allow only a very crude evaluation of q. Others are more favorable and a comparison of a number of cases probably affords reasonable accuracy.

Table 16. *Calculation of U_p, the "number" of unbalanced valence electrons, for various molecules.*

Quadrupole coupling constants may be obtained by multiplying U_p by $-eq_{n10}Q$, the quadrupole coupling per p electron.

Nucleus	Molecule	Partial structures	Comments	U_p for each structure	% importance	Net U_p
Cl	FCl	F—Cl	No hybridization	1.00	75	1.37
		$\bar{\text{F}}$ $\overset{+}{\text{Cl}}$		2.50	25	
	ICl	I—Cl	15% s hybridization	0.85	85	0.72
		$\overset{+}{\text{I}}$ $\bar{\text{Cl}}$		0	15	
	TlCl	$\overset{+}{\text{Tl}}$—$\bar{\text{Cl}}$	15% s hybridization	0.85	18	0.15
		$\overset{+}{\text{Tl}}$ $\bar{\text{Cl}}$		0	82	
	SiH₃Cl	H_3Si—Cl	15% hybridization	0.85	30	0.38
		$\overset{+}{\text{H}_3\text{Si}}\,\bar{\text{Cl}}$		0	40	
		$\overset{-}{\text{H}_3\text{Si}}=\overset{+}{\text{Cl}}$	p_σ bond with 15% s hybridization	0.40	30	
	C₂H₃Cl	(structure: H/H–C=C–Cl/H)	U_{pz} refers to z axis along C—Cl bond. y axis is in Cl—C—H plane and x axis is perpendicular. 15% s hybridization	$U_{pz}=0.85$ $U_{py}=-0.42$ $U_{px}=-0.42$	75	$U_{pz}=0.66$ $U_{py}=-0.38$ $U_{px}=-0.28$
		(structure: H/H–$\bar{\text{C}}$–C=$\overset{+}{\text{Cl}}$/H)	p_π bond assumed perpendicular to Cl—C—H plane	$U_{pz}=0.55$ $U_{py}=-1.16$ $U_{px}=0.72$	5	
		(structure: H/H–C=C–$\bar{\text{Cl}}/\text{H}$, $\overset{+}{\text{C}}$)		0	20	
S	OCS	$\bar{\text{O}}$—C≡$\overset{+}{\text{S}}$	No hybridization	0	14	0.51
		O=C=S		0.5	58	
		$\overset{+}{\text{O}}$≡C—$\bar{\text{S}}$		0.8	28	
		or $\bar{\text{O}}$—C≡$\overset{+}{\text{S}}$	25% s hybridization or p_σ bond	-0.31	14	0.27
		O=C=S		0.25	58	
		$\overset{+}{\text{O}}$≡C—$\bar{\text{S}}$		0.60	28	
N	NH₃	$\bar{\text{N}}$—H (with $\overset{+}{\text{H}}$, H)	25% hybridization	-0.40	100	-0.40
	N₂O	$\bar{\text{N}}$=$\overset{+}{\text{N}}$=O	45% hybridization of p_σ bondy only	0.05	55	-0.17
		$\overset{+}{\text{N}}$≡N—$\overset{-}{\text{O}}$	End nitrogen	-0.45	45	
		$\bar{\text{N}}$=$\overset{+}{\text{N}}$=O	Central nitrogen	0	55	0
		N≡$\overset{+}{\text{N}}$—$\bar{\text{O}}$		0	45	
As	AsH₃	As—H (with H, H)	10% hybridization	-0.30	100	-0.30
	AsCl₃	$\overset{+}{\text{As}}$—$\bar{\text{Cl}}$ (with Cl, Cl)	10% hydribization	-0.25	50	-0.28
		As—Cl (with Cl, Cl)		-0.30	50	

13. Determination of ΔV in solids. ΔV or VE at a nucleus in a solid may be treated by the same general methods discussed for a nucleus in an isolated or gaseous molecule since the solid may itself be considered a large and complex molecule. However, it is usually convenient to discuss a solid as made of molecular components, and to consider the modifications in ΔV which occur when such a molecular component is condensed from a gaseous state of relative isolation into a crystalline lattice. These modifications are often rather small — particularly for molecular crystals which do not have strong intermolecular forces. In any cases, they may be conveniently divided into the following three types [73]:

1. Direct contributions to ΔV by ions or other charges in neighboring molecules. As was found above, the value of ΔV produced by an ion at a distance of two atomic radii or more is usually considerably smaller than that produced by a valence electron. However, in cases where the effects of valence electrons almost cancel, and ΔV is small, nearby ions may play a dominant role in determining the quadrupole coupling.

2. Modification of the electronic structure or a molecular unit as a result of interactions with neighboring molecules. The most characteristic modifications of this type are increases in ionic character of molecular bonds when the molecule is condensed into a crystal lattice. In many cases, electrostatic interactions between the dipole moments of adjacent molecules play an important role in determining the crystal structure of solids. Hence these structures tend to allow interactions between dipole moments which lower the energy of the crystal. This implies further stabilization of ionic bond structures and some increase in ionic character of the molecular bonds. Examples will be discussed below.

Van der Waals interactions between molecules may also cause some modification of electronic structures. When such modifications become large, they may usually be described under the third category which follows.

3. Additional bonds may occur between molecules in solids which imply in some cases a very large change in the molecular electronic structure. These may be either covalent electronic bonds or hydrogen bonds. Quadrupole coupling is in fact a very sensitive indicator of intermolecular bonds. However, there are often other useful indications, such as geometrical arrangement and internuclear distances of the atoms involved, or crystal binding energy.

The normal behavior of quadrupole coupling in molecular crystals may be judged from Table 17. With the exception of ICN and the homonuclear molecules Cl_2, Br_2, and I_2 which have no dipole moments, each molecule listed becomes more ionic in passing from the gaseous to the solid phase. In almost all cases listed, this results in a decrease of a few percent in the quadrupole coupling.

Such increases in ionic character are quite typical of effects of type 2. They are to be expected for molecules with dipole moments, since interactions between these moments will frequently affect the crystalline form in such a way that the electrostatic interactions tend to lower the crystalline energy. If the crystalline energy is lowered as a result of dipole-dipole interactions, the ionic bonds are more stable than for the case of isolated molecules, and the ionic character must therefore increase somewhat.

Effects of type 3 are evident [73] in I_2, Br_2, and ICN—in the latter case they are strong enough to overcome the tendency towards increased ionic character noted above. The large asymmetry parameters in I_2 and Br_2 show a contribution to the quadrupole coupling which cannot come from the covalent bond of a diatomic molecule because of symmetry, but which is comparable in magnitude to the contribution of a valence p electron. Examination of their

Table 17. *Quadrupole coupling constants of halogen-containing molecules which have been measured in both the gaseous and the solid state (see [75] for source of data).*

Molecule	Nucleus	Quadrupole atomic state	Coupling constant gaseous molecule	Solid[1] $eQq_{zz}\,\eta$	
Cl_2	Cl^{35}	-109.74		-108.95	
CH_3Cl	Cl^{35}	-109.74	-74.74	-68.4	
CF_3Cl	Cl^{35}	-109.74	-78.05	-77.58	
ICl	Cl^{35}	-109.74	-82.5	-74.4	
Br_2	Br^{79}	769.76		765.86	-0.20
CH_3Br	Br^{79}	769.76	577.0	528.9	
CF_3Br	Br^{79}	769.76	619	604	
I_2	I^{127}	-2292.71		-2156	-0.16
CH_3I	I^{127}	-2292.71	-1929	-1766	
CF_3I	I^{127}	-2292.71	-2143.8	-2069	
ICl	I^{127}	-2292.71	-2944	-3037	
ICN	I^{127}	-2292.71	-2420	-2549	

crystal structure shows further evidence of strong interaction between adjacent
molecules. For example, twice the van der Waals radius of iodine is 4.35 Å, but
each iodine atom of the I_2 group has, in the crystal, two additional neighbors
at a distance of only 3.54 Å. This distance is close enough to indicate partial
covalent bonding between these neighboring atoms. The structure of these
bonds—that is, the orbitals used—is not completely clear, but their existence
and some information about the orbitals seem well established from their effects
on quadrupole coupling [73], [74]. Similar intermolecular bonding occurs in Br_2
and ICN mentioned above. It has also been found prominently in compounds
of Sb and Bi [75], [76].

Contributions of type 1 to the quadrupole coupling in solids can be considered
in much the same way as in isolated molecules. If there is a large coupling constant due to a valence p electron, the direct contribution of nearby ions in solids
(even with antishielding effects) is usually small by comparison. However, if
the coupling due to valence electrons is not large, the effect of nearby ions may
dominate. Such would be the case, for example, for Na in $NaClO_3$, where
contributions of valence electrons are very small, and those of nearby ClO_3 ions
are quite significant.

Effects of rotational motion of a gaseous molecule on quadrupole interactions
are abstracted, as indicated earlier, so that the coupling constant is obtained for
an idealized stationary molecule. In solids, the rotational motions of molecules
are replaced by torsional vibration, or sometimes by hindered rotation, and their
effects on hyperfine structure are sufficiently complicated that they cannot
easily be abstracted from the experimental results. These motions always decrease the magnitude of the quadrupole coupling constant [77]—usually by a
small amount, but sometimes, if large torsional motions or hindered rotation
occurs, by a major amount. In ordinary cases of small torsional vibration, this
decrease is not of much importance in determination of nuclear quadrupole
moments, but it may be of importance in determining the ratio of quadrupole
moments of two isotopes [75]. Since the different masses of the two isotopes
result in slightly different amounts of torsional vibration, the effective value
of q is not the same for the two cases, and some additional uncertainty is introduced into the ratio of quadrupole moments even when the ratio of coupling
constants can be very accurately measured.

---

[1] The sign of eQq_{zz} in the solid state is not experimentally measured, but inferred from
that of the gaseous state.

Calculation of ΔV in a solid is often not much more difficult, nor usable approximations more uncertain than for the case of an isolated molecule. The approximation proposed by TOWNES and DAILEY [63], [73], and discussed above for molecules, may be used with some success for molecular crystals when penetrating valence electrons give the largest part of ΔV. Appropriate modifications of the case for isolated molecules may often be made to allow for intermolecular bonding, or some increase in ionic character due to condensation. For ionic crystals, a rather similar treatment is useful when valence electrons are predominant, as in the case of Cu in $CuK_2(SO_4)_2 \cdot 6H_2O$ [78].

When nearby ions rather than valence electrons give the largest contributions to ΔV, evaluation of ΔV takes on a different character. If the positions and charges of nearby ions in the crystal are known, their direct contribution to ΔV at the nucleus in question can easily be calculated. There remains to be evaluated or estimated the antishielding factor $\varkappa$, by which this value of ΔV must be multiplied. The uncertainties in evaluation of $\varkappa$ have already been emphasized in a detailed discussion above. Nevertheless, some calculations of ΔV and $\varkappa$ in ionic crystals seem to have met with success [79], and it is possible that for ions in a crystalline lattice, the calculated value of $\varkappa$ fits the observed coupling constants better than in diatomic molecules such as the alkali halides, because in a solid the ions are more symmetrically surrounded and hence less distorted.

B. Determination of nuclear quadrupole moments from nuclear transitions.

14. General discussion. The techniques so far discussed have measured the expectation values of nuclear quadrupole moments in an energy characteristic state (normally the ground state). They utilize the rather gentle interactions of nuclei with surrounding atomic electrons, which perturb the nuclear state being measured by an amount which is usually quite negligible. Another class of effects involves stronger interactions with the nucleus, such as those due to nuclear bombardment, and allows determinations of off-diagonal matrix elements of quadrupole moments, or other properties closely related to the quadrupole moment. In many cases the diagonal matrix element of the quadrupole moment, or hence "the" quadrupole moment of an energy characteristic state, may be derived from these related quantities by the assumption of some particular nuclear model. The most useful model of this type is the "collective" model of a rotating deformed nucleus[1] which has been discussed and exploited particularly by BOHR, MOTTLESON and their associates [80] to [82]. The model is not always applicable, but is often sufficiently accurate to give useful determinations of quadrupole moments. For a nucleus in a state having spin 0 or $\frac{1}{2}$, the diagonal matrix element of the quadrupole moment is zero, but the off-diagonal matrix elements are not usually zero. For such cases, and when the nucleus can be considered with some accuracy to behave like a rigid rotor, the off-diagonal matrix elements allow a determination of an "intrinsic" quadrupole moment or shape parameter which is directly given by the diagonal elements of the quadrupole moment in any state of spin higher than $\frac{1}{2}$.

The class of quadrupole moment determinations involving nuclear transitions and the techniques which are typical of nuclear physics has been rapidly growing in importance since about 1952, and has already yielded a considerable

---

[1] Details on this model will be found in S. A. MOSZKOWSKI's contribution to Vol. XXXIX of this Encyclopedia.

quantity of information. The effects utilized include γ-decay probabilities, photonuclear reactions, Coulomb excitation, neutron excitation, and energies of nuclear states.

15. Nuclear rotational energy levels. If individual particles in the nucleus are sufficiently tightly bound, and the nucleus is non-spherical, the lowest energy levels are approximately given by those of a rigid symmetric rotator. The angular momentum Ω about the symmetry axis may be assumed constant for these levels, since the nucleus does not normally act like a rigid rotator when Ω is changed. Hence the energy for these lowest levels is, except for a constant [81],

$$W = \frac{\hbar^2}{2\mathfrak{J}} I(I+1) \tag{15.1}$$

where I is the nuclear spin and $\mathfrak{J}$ the effective moment of inertia. Energy levels of this type are rather closely approximated for many medium or heavy nuclei [81], [82] with large deformations from spherical shape, or with resulting large quadrupole moments. This indicates the validity of treating the nucleus as a rotator in such cases.

The most important deformation of the nucleus is of the quadrupolar type, and hence the effective moment of inertia is related, though somewhat indirectly, to the electric quadrupole moment. Assuming a spheroidal nucleus of constant mass and charge density, a first approximation to this relation is

$$\mathfrak{J} = \frac{5}{8} \frac{mA\,Q_0^2}{R^2 Z^2} \tag{15.2}$$

where Q_0 is the "intrinsic" quadrupole moment of the nucleus, defined by

$$Q_0 = \frac{1}{e} \int \varrho\,(3z'^2 - r'^2)\,dv, \tag{15.3}$$

where ϱ is the nuclear charge density and z' is the Cartesian coordinate along the symmetry axis of the deformation of the nucleus (not the spin axis). Thus Q_0 may be regarded as the quadrupole moment of the nucleus if it were held stationary rather than allowed to rotate freely. mA is the total nuclear mass, and R the nuclear radius.

If the nuclear spin is greater than $\frac{1}{2}$, Q_0 is directly related to Q, the quadrupole moment as usually defined, by the expression [80], [82]

$$Q = \frac{I(2I-1)}{(I+1)(2I+3)} Q_0. \tag{15.4}$$

This relation illustrates the fact that the quadrupole moment Q is always zero if the nuclear spin I is 0 or $\frac{1}{2}$. However, the nucleus may still have an intrinsic quadrupole moment Q_0, or hence an intrinsic non-spherical shape which, because of precession of its direction in space, gives on the average a spherical shape. This intrinsic quadrupole moment was first pointed out by BRIX and KOPFERMANN [83]. It should be noted, however, that Q_0 has this straight forward meaning only when the nucleus is rather tightly bound and its rotational energy levels closely spaced so that it may be considered to perform an end-over-end rotation as does a molecule. Otherwise, excitation of individual particles in the nucleus may occur, the expression (15.1) for energy levels will not apply, and such a simple intrinsic shape cannot be attributed to the nucleus.

From (15.1) and (15.2) above, the square of the intrinsic quadrupole moment may be obtained from the energy ΔW of a transition $I \rightarrow I+2$ as

$$Q_0^2 = \frac{8}{5} \frac{\hbar^2 [2I+3] R_0^2 Z^2}{mA\,\Delta W}. \tag{15.5}$$

However, comparison of values of $|Q_0|$ obtained from (15.5) with those obtained by other methods shows that they are characteristically too large by about a factor of two [84], [82]. The effective moment of inertia given by the approximate relation (15.2) is evidently too small by a factor of approximately four for most nuclei, and any values of $|Q_0|$ obtained from (15.5) should be reduced by a factor of about 2. Because of this correction, which must at present be empirical, expression (15.5) cannot be used with much confidence for evaluation of $|Q_0|$ [or hence $|Q|$ from (15.4)], but in some cases where better determinations are not available, it may be of value.

16. Probability of γ-ray transitions. The probability per second for a γ-decay may be written [81], [82]

$$P(E\,2) = \frac{4\pi}{75}\,\frac{1}{\hbar}\left(\frac{\Delta W}{\hbar c}\right)^5 B\,(E\,2) \tag{16.1}$$

where ΔW is the γ-ray energy, and $B\,(E\,2)$ is a reduced transition probability which is closely related to the quadrupole moment matrix elements. For an even-even nucleus which exhibits collective rotation, having energy levels of the type indicated by expression (15.1), $B\,(E\,2)$ may be evaluated in terms of the square of the intrinsic quadrupole moment Q_0 [81], [82], and for the transition $I+2 \rightarrow I$,

$$B\,(E\,2) = \frac{15}{32\pi}\,e^2\,Q_0^2\,\frac{(I+1)\,(I+2)}{(2I+3)\,(2I+5)} \tag{16.2}$$

where e is the electron charge and I the nuclear spin of the final state.

For a nucleus with an odd number of neutrons or protons, the difference in spin between successive energy levels is usually unity, and an excited state may decay by a combination of magnetic dipole and electric quadrupole radiation. The probability per second for a magnetic dipole decay $I+1 \rightarrow I$ is given by [81], [82]

$$P(M\,1) = \frac{16\pi}{9\hbar}\left(\frac{\Delta W}{\hbar c}\right)^3 B\,(M\,1) \tag{16.3}$$

where, for nuclei which exhibit collective rotation except for the addition of an odd particle [81], [82]

$$B\,(M\,1) = \frac{3}{4\pi}\left(\frac{e\,\hbar}{2m\,c}\right)(g_\Omega - g_R)^2\,\frac{\Omega^2\,(I+1-\Omega)\,(I+1+\Omega)}{(I+1)\,(2I+3)}. \tag{16.4}$$

Here Ω is the quantum number representing the component of angular momentum of the odd particle about the symmetry axis of the remainder of the deformed nucleus, and Ω must be greater than $\frac{1}{2}$ for expression (16.4) to apply. g_Ω is the g-factor for the odd particle, and g_R that for the remaining part of the nucleus, which rotates in a collective motion. The decay probability for the transition $I+1 \rightarrow I$ due to an electric quadrupole is given [81], [82] by (16.1) with

$$B\,(E\,2) = \frac{15}{16\pi}\,e^2\,Q_0^2\,\frac{\Omega^2\,(I+1-\Omega)\,(I+1+\Omega)}{I(I+1)\,(2I+3)\,(I+2)}. \tag{16.5}$$

If the nuclear quadrupole moment is due to the orbit of a single particle rather than to deformation of the entire nucleus, the magnetic dipole transitions are several orders of magnitude more rapid than those due to quadrupole moments. However, for strongly deformed nuclei, such as those which show rotational states, $|Q_0|$ is sufficiently large that it contributes appreciably to transitions of the type $I+1 \rightarrow I$, and at the same time may produce cross-over transitions of the type $I+2 \rightarrow I$, which are due solely to an intrinsic quadrupole moment.

For a nucleus with an odd number of neutrons or protons, the electric quadrupole transition $I+2\to I$ is given as above by expression (16.1), with [82]

$$B(E2) = \frac{15}{32\pi} e^2 Q_0^2 \frac{(I+1-\Omega)(I+1+\Omega)(I+2-\Omega)(I+2+\Omega)}{(I+1)(2I+3)(I+2)(2I+5)}. \quad (16.6)$$

This expression is valid even when $\Omega = \frac{1}{2}$ or 0. In the latter case it reduces to (16.2).

Table 18. *Comparison of values of $B(E2)$ from determinations of γ-decay lifetimes and from measurements of yields by Coulomb excitation (from ALDER et al. [82]).*

Nucleus	Half-life sec	$B(E2)$ From lifetime	[units of $e^2\times 10^{-48}\,\text{cm}^2$] From Coulomb excitation	Nucleus	Half-life sec	$B(E2)$ From lifetime	[units of $e^2\times 10^{-48}\,\text{cm}^2$] From Coulomb excitation
$_9\text{F}^{19}$	6×10^{-8}	0.01	0.003	$_{72}\text{Hf}^{180}$	1.4×10^{-9}	4.9	5.0
$_{23}\text{V}^{51}$	1×10^{-10}	0.008	0.006	$_{74}\text{W}^{182}$	1.27×10^{-9}	4.3	5.6
$_{32}\text{Ge}^{72}$	3.2×10^{-12}	0.19	0.26	$_{80}\text{Hg}^{198}$	2.1×10^{-11}	1.1	0.8
$_{32}\text{Ge}^{74}$	1.3×10^{-11}	0.28	0.30	$_{80}\text{Hg}^{199}$	2.4×10^{-9}	0.35	0.26
$_{62}\text{Sm}^{152}$	1.4×10^{-9}	3.3	3.1	$_{80}\text{Hg}^{202}$	2.2×10^{-11}	0.8	0.5
$_{64}\text{Gd}^{154}$	1.2×10^{-9}	3.6	4.5	$_{81}\text{Tl}^{203}$	1.4×10^{-10}	0.28	0.12
$_{68}\text{Er}^{166}$	1.7×10^{-9}	5.7	6.8	$_{82}\text{Pb}^{207}$	9×10^{-11}	0.031	0.028
$_{72}\text{Hf}^{176}$	1.35×10^{-9}	5.3	6.0				

The decay rate through γ-emission has not yet been very much used for evaluation of $|Q_0|$ or $|Q|$ primarily because of experimental difficulties in obtaining accurate determinations of the very short lifetimes which are involved. However, available measurements yield values of $B(E2)$ and hence of quadrupole moments, when (15.4) applies, which agree well with other determinations [85], [82]. Table 18 indicates magnitudes of the lifetimes involved, and the extent to which half-life and Coulomb excitation measurements agree. Furthermore, recent improvements in techniques of measuring very short decay times appear to provide for an increasing importance of direct measurements of γ-ray transition probabilities.

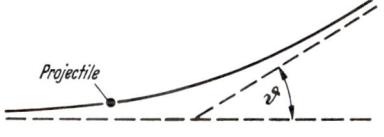

Fig. 10. Classical orbit of a charged projectile in the Coulomb field of a nucleus having a quadrupole moment.

17. Coulomb excitation[1]. An experimentally convenient way of measuring the transition probabilities between nuclear levels due to quadrupole effects is by Coulomb excitation, the excitation of a nucleus by Coulomb interactions with a fast particle. γ-rays resulting from Coulomb excitation were first seen in 1952 by DAY and HUUS [86], and by McCLELLAND [87]. Since that time, the technique of Coulomb excitation has been extensively exploited and has added considerably to knowledge about nuclear quadrupole moments of the medium-weight and heavy nuclei (see, for example, Ref. [82]).

The classical behavior of a charged projectile which passes close by a non-spherical nucleus is indicated in Fig. 10. The projectile is scattered at an angle ϑ, given approximately by the Rutherford scattering formula, and at the same time, the nucleus may be excited to some rotational energy level as the result of electrostatic interactions between the charged projectile and the nuclear quadrupole

---

[1] There is a special article on Coulomb excitation of nuclei by G. BREIT and R. L. GLUCKSTERN in Vol. XLI of this Encyclopedia which may be consulted for more information.

moment. This latter exchange of energy will modify the path of the projectile somewhat, but usually by a minor amount. If the projectile approaches so close to the nucleus that it is within range of the nuclear forces, then they may also produce nuclear transitions. However, the Coulomb forces are usually sufficiently strong and long-range when the nuclear deformation is large, and even more so when the projectile has more than one proton charge, that the excitation cross section due to electromagnetic forces can be much larger than that due to specific nuclear interactions. In order to minimize nuclear excitation due to specific nuclear forces, the energy of the projectile, which is positively charged, may be made sufficiently small that it does not approach close enough to the nucleus for these forces to be important. A simple calculation shows that the kinetic energy W, of the projectile should then be limited by the relation

$$W_1 < \frac{Z_1 Z_2}{A_2^{\frac{1}{3}}} \text{ Mev} \tag{17.1}$$

where Z_1 and Z_2 are the atomic numbers of the projectile and nucleus respectively, and A_2 the mass number of the nucleus. In this relation, the radius of the projectile is assumed small compared with that of the nucleus.

The total cross section for Coulomb excitation from a level of spin I to one of spin $I+2$ is of course closely related to the γ-ray transition probability already discussed, and may be written [88] to [90], [82]

$$\sigma = C_{E2}(W_1 - \Delta W_1) \, B(E2) \, f_{E2}(\eta, \xi) \tag{17.2}$$

where

$W_1 =$ the initial energy of the projectile in Mev,

$$\Delta W_1 = \left(1 + \frac{A_1}{A_2}\right) \Delta W,$$

$\Delta W =$ the energy difference in Mev between the two nuclear levels,

$A_1, A_2 =$ mass number of the projectile and nucleus, respectively,

$$C_{E2} = \frac{4.819}{\left(1 + \frac{A_1}{A_2}\right)^2} \frac{A_1}{Z_2^2} \times 10^{-24} \text{ cm}^2,$$

$Z_1, Z_2 =$ the atomic number of the projectile and nucleus respectively,

$$\eta = \frac{Z_1 Z_2}{2} \left(\frac{A_1}{10.008 \, W_1}\right)^{\frac{1}{2}},$$

$$\xi = \frac{Z_1 Z_2}{2} \left(\frac{A_1}{10.008 \, W_1}\right)^{\frac{1}{2}} \left[\left(1 - \frac{\Delta W_1}{W_1}\right)^{-\frac{1}{2}} - 1\right]$$

$$= \frac{Z_1 Z_2 A_1^{\frac{1}{2}} \Delta W_1}{12.65 \left(W_1 - \frac{\Delta W_1}{2}\right)^{\frac{3}{2}}} \left[1 + \frac{5}{32}\left(\frac{\Delta W_1}{W_1}\right)^2 + \cdots\right].$$

$B(E2)$ has already been defined in Eqs. (16.2), (16.5), or (16.6) for transitions of the type $I+2 \to I$ or $I+1 \to I$. For transitions involving an increase ΔI of I by 1 or 2 rather than a decrease, $B(E2)$ may be obtained by the relation

$$B(E2)_{I \to I'} = \frac{2I' + 1}{2I + 1} (E2)_{I' \to I}. \tag{17.3}$$

The function $f(\eta, \xi)$ is listed in Table 19 for a number of values of η and ξ. Frequently the value of η is large enough that $\eta = \infty$ is a reasonable approximation.

Table 19. $f(\eta, \xi)$ for Coulomb excitation by quadrupolar interaction [see Eq. (17.2)]. (After ALDER and WINTHER [91], and BIEDENHARN et al. [92].)

η	ξ												
	0	0.1	0.2	0.3	0.4	0.5	0.6	0.8	1.0	1.2	1.4	1.6	2.0
0.5	0.321	0.344	0.307	0.243									
1	0.620	0.614	0.528	0.409	0.295	0.203	0.1350	0.0553					
1.5	0.754	0.732	0.624	0.480	0.346	0.237	0.1570	0.0640	0.0244	0.00887	0.00312		
2	0.812	0.784	0.666	0.512	0.368	0.253	0.1672	0.0680	0.0259	0.00939	0.00330	0.001130	
2.5	0.842	0.810	0.688	0.529	0.380	0.261	0.1726	0.0702	0.0267	0.00968	0.00340	0.001162	0.0001287
3	0.858	0.825	0.700	0.538	0.387	0.266	0.1759	0.0715	0.0272	0.00986	0.00345	0.001181	0.0001306
3.5	0.869	0.834	0.708	0.545	0.392	0.269	0.1779	0.0724	0.0275	0.00996	0.00349	0.001194	0.0001319
4	0.875	0.840	0.713	0.548	0.395	0.271	0.1793	0.0730	0.0277	0.01005	0.00352	0.001203	0.0001328
5	0.881	0.847	0.719	0.553	0.398	0.273	0.1810	0.0737	0.0280	0.01015	0.00356	0.001214	0.0001339
6	0.886	0.851	0.722	0.556	0.400	0.275	0.1819	0.0741	0.0282	0.01021	0.00358	0.001221	0.0001345
7	0.888	0.854	0.724	0.558	0.401	0.276	0.1825	0.0743	0.0283	0.01024	0.00359	0.001225	0.0001350
8	0.890	0.855	0.726	0.559	0.402	0.276	0.1829	0.0745	0.0283	0.01027	0.00360	0.001228	0.0001353
	0.895	0.859	0.729	0.561	0.405	0.278	0.1844	0.0751	0.0286	0.01035	0.00363	0.001238	0.0001363

The derivation of expression (17.2) involves a number of approximations, none of which are ordinarily large enough to be of importance at present in the determination of quadrupole moments. These include the omission of relativistic effects in the velocity of the projectile, screening of the nucleus by atomic electrons, and various higher order effects in the interaction between projectile and nucleus [82]. Of considerably more importance so far as determination of quadrupole moments are concerned are inaccuracies in the theoretical connection between $B(E2)$ and $|Q|$ as given by (16.6) and (15.4).

Nuclei which are normally considered to furnish excellent examples of collective rotation have energy levels which agree with expression (15.1) to an accuracy of about 1%. This indicates that wavefunctions which describe the energy characteristic states of these nuclei correspond largely to a particular rotational state on the collective model, but with admixtures of somewhere near 10% of the neighboring rotational states. Such a 10% admixture of other rotational states in the amplitude of the wavefunction would make a contribution of only about 1% to the energy, but a contribution of the order of 10% to the matrix elements for an electric quadrupole transition, since the off-diagonal matrix elements of the quadrupole moment connect adjacent rotational states. Thus even in the cases which best represent the approximation of collective rotation, the value of $|Q|$ obtained from (15.4) and (16.6) cannot be expected to be more accurate than about 10%. In general, if the energy levels fit a scheme indicated by (15.1) to a fractional error ε, the fractional error in the diagonal element Q of the quadrupole moment obtained from Coulomb excitation can be expected to be approximately $\sqrt{\varepsilon}$. On the other hand, determination of $B(E2)$, and hence of the off-diagonal matrix elements of the quadrupole moment is usually limited primarily by experimental errors.

Experimental errors in determination of cross-sections for Coulomb excitation vary considerably from case to case, depending on details of the nuclear systems and the experimental techniques used. The common types of measurements involve either a measurement of de-excitation γ-rays emitted while a target is being bombarded, or of conversion electrons. Uncertainties in conversion coefficients, background radiation, thick target effects, and other normal hazards of nuclear spectroscopy give errors of varying amounts, and normally limit the results of the best experimental work to an accuracy of about 20% in the Coulomb excitation cross section. This corresponds to an error of 10% in the quadrupole moments.

A measure of the accuracy of experimental determinations of cross sections for Coulomb excitation, and of expression (16.6) for $B(E2)$ is afforded by Table 20. This table gives the measured ratios of $B(E2)$ for the transitions $I \to I + 2$ and $I \to I + 1$ of odd-A nuclei, and compares them with the ratios from expression (16.6), obtained on the basis of a collective model. This comparison may give, however, a somewhat overoptimistic view of the errors involved, since the absolute values of both theoretical and experimental errors are expected to be somewhat larger than their relative values. It is primarily relative values which are involved in Table 20.

Table 20. *A comparison between experimental ratios of $B(E2)$ from Coulomb excitation, and those expected on the basis of a collective model, and expression (16.6). Values come from references [82] and [93].*

Nucleus	Spin of ground state	$\dfrac{B(E2)_{I \to I+2}}{B(E2)_{I \to I+1}}$	
		Experimental values	Calculated values
$_{64}\mathrm{Gd}^{159}$	$\frac{3}{2}$	0.38	0.56
$_{65}\mathrm{Tb}^{159}$	$\frac{3}{2}$	0.56	0.56
$_{63}\mathrm{Eu}^{153}$	$\frac{5}{2}$	0.28	0.35
$_{75}\mathrm{Re}^{185}$	$\frac{5}{2}$	0.35	0.35
$_{75}\mathrm{Re}^{187}$	$\frac{5}{2}$	0.28	0.35
$_{92}\mathrm{U}^{233}$	$\frac{5}{2}$	0.18	0.35
$_{93}\mathrm{Np}^{237}$	$\frac{5}{2}$	0.44	0.35
$_{67}\mathrm{Ho}^{165}$	$\frac{7}{2}$	0.20	0.26
$_{71}\mathrm{Lu}^{175}$	$\frac{7}{2}$	0.23	0.26
$_{72}\mathrm{Hf}^{177}$	$\frac{7}{2}$	0.26	0.26
$_{73}\mathrm{Ta}^{181}$	$\frac{7}{2}$	0.29	0.26
$_{92}\mathrm{U}^{235}$	$\frac{7}{2}$	0.16	0.26
$_{72}\mathrm{Hf}^{179}$	$\frac{9}{2}$	0.22	0.20

The collective rotation model of the nucleus, which allows an easy correlation between $|Q|$ and Coulomb excitation cross sections, applies best for the heavier nuclei with large deformations. It can hence be expected to be particularly useful between mass numbers about 140 and 200, for uranium and transuranium elements, and to a lesser extent for some of the intermediate-weight nuclei which are well away from closed shells.

The applicability of Coulomb excitation techniques is superficially somewhat similar to that of hyperfine structure measurements by optical spectroscopy. Both are most usable for medium-weight or heavy nuclei, particularly those with large quadrupole moments. In both cases the experimental accuracy is not great, although present experimental techniques are more or less adequate for the still rather primitive theories of quadrupole moments which are available. Both techniques are also subject to theoretical uncertainties of the order of 10 to 20% which are very difficult to eliminate—for optical spectroscopy these uncertainties come from electronic wavefunctions and shielding factors; for Coulomb excitation from the nuclear wavefunctions.

Coulomb excitation has the disadvantage over optical spectroscopy of not being very useful even for some of the heavier nuclei which do not approximate good collective rotation, and of not allowing a determination of the sign of Q. This last handicap is, however, not a serious restriction since enough is now known about quadrupole moments that their signs may be predicted as positive in almost all cases to which Coulomb excitation techniques apply. These techniques have the very great advantage of being applicable to deformed nuclei of spin 0 and $\frac{1}{2}$, and of somewhat easier application than the methods of hyperfine structure to every atomic number within a certain range.

18. Scattering of negatively charged and neutral projectiles. Positively charged projectiles are effective in exciting rotational modes of a nucleus by Coulomb interaction without penetrating the nucleus itself. Negatively charged or neutral projectiles, such as electrons or neutrons, penetrate the nucleus and are scattered in a way which may depend on details of the nuclear shape. Effects of nuclear shapes on scattering are particularly prominent when the wavelength of the projectile inside the nucleus is comparable with the nuclear diameter. For electrons, this involves energies of the order of hundreds of Mev; for neutrons, it involves energies less than a few Mev; and for negative mesons, appropriate energies are in the range of tens of Mev.

Electrons with energies of a few hundred Mev are scattered in a diffraction pattern by the Coulomb potential of a spherical nucleus. The resulting fluctuations in the differential cross section with scattering angle may be seen in the curves for Pb, Bi, or Au of Fig. 11. For a deformed nucleus with a large quadrupole moment this pattern is appreciably changed, and depends on the orientation of the nuclear axis with respect to the path of the approaching electron. At normal temperatures, the various possible orientations of the nuclear spin are equally populated, so that electron scattering is an average of the scattering for all nuclear spin orientations, and fluctuations in the differential scattering cross section are largely averaged out. This is illustrated by the curves for W, Ta, or U in Fig. 11.

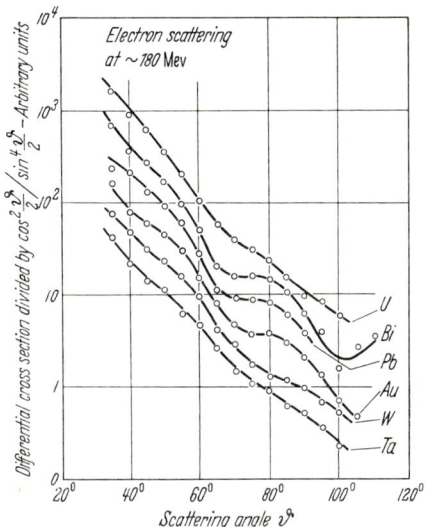

Fig. 11. Effect of large quadrupole moments on high-energy electron scattering. U, W, and Ta have rather large Q_0; Bi, Pb, and Au small Q_0. (From Downs, Ravenhall, and Yennie [95]).

For scattering of electrons of very high energies, off-diagonal as well as diagonal matrix elements of the quadrupole moment are important. SCHIFF [94] and others [95] have shown that in the Bohr-Mottelson approximation of collective motion, which has been discussed above, the sum of effects associated with both diagonal and off-diagonal matrix elements on fast electrons can be represented by scattering from a classical deformed charge distribution, regardless of the particular nature of the nuclear spin. This may be expected when the electron collision with the nucleus is short compared with the time of rotation of the nucleus.

Details of the differential scattering cross section from a deformed nucleus depend on the variation of charge distribution with distance r from the center of the nucleus[1] as well as on the quadrupole moment. DOWNS, RAVENHALL and YENNIE [95] have taken a radial charge density of the form

$$\varrho(r) = \varrho(0)\left[1 + e^{\frac{r-c}{0.228\,t}}\right]^{-1}. \tag{18.1}$$

They evaluate the constants c and t essentially from experiments on undeformed nuclei, and then choose a value of eccentricity for the deformed nucleus which will fit the observed scattering cross sections. In the case of Ta^{181}, they obtain a value for the intrinsic quadrupole moment Q_0 of 10×10^{-24} cm^2, which is in satisfactory agreement with other determinations.

It appears possible, but not easy, to obtain considerable information about nuclear quadrupole moments from refined electron scattering experiments. One desirable refinement would involve scattering from aligned nuclei, since then an averaging over the various possible orientations of quadrupole moment is not necessary, and a much more direct separation between quadrupole and monopole effects is possible.

---

[1] For charge distribution see D. L. HILL in Vol. XXXIX of this Encyclopedia.

Scattering of μ mesons by nonspherical nuclei also offers attractive possibilities for determination of nuclear deformations. In this case the particle wavelength tends to be shorter than that of an electron because of the larger mass, and effects of nuclear deformation on scattering become evident at a much lower energy. Use of μ meson scattering for determination of quadrupole moments is less advanced, however, than application of energetic electrons because of experimental difficulties. In addition, theoretical analysis of μ meson scattering from deformed nuclei at moderate energies is somewhat more complex than the high-energy electron case.

Neutrons interact with the nucleus primarily through specific nuclear forces, and hence do not detect very directly the nuclear charge distribution. However, on the reasonable approximation of constant charge to density ratio throughout the nucleus, scattering of neutrons by deformed nuclei may also be used to obtain some information about quadrupole moments. Low-energy neutron scattering by a spheroidal complex potential representing a deformed nucleus has been studied by MARGOLIS and TROUBETSKOY [96] and by others [97]. They show rather clearly the effects of nuclear quadrupole moments on the total scattering cross section, and are able to obtain a rather rough evaluation of the amount of deformation in the region where quadrupole moments are large. However, details of nuclear levels and neutron-nuclear interactions are important to such scattering, and are still insufficiently known to allow very useful determinations of quadrupole moments of individual nuclei.

19. Influence of quadrupolar deformations on the giant γ-ray resonances of nuclei. The very intense γ-ray resonances of nuclei which lie in the region 10 to 30 Mev are thought to represent collective oscillations of the nuclear protons with respect to the center of mass of the nucleus [98], [99]. The mean energy of these oscillations follows fairly closely the relation [100]

$$E_m = 42\, A^{-\frac{1}{3}}\,\text{Mev}. \tag{19.1}$$

This form fits one rough theory [98] of the collective oscillations. A hydrodynamic model [101], [102] of the collective nuclear oscillations gives E_m proportional to $A^{-\frac{1}{6}}$. This also does not fit badly the experimental data, which can be also expressed approximately by [103]

$$E_m = 80\, A^{-\frac{1}{6}}\,\text{Mev}. \tag{19.2}$$

Widths of the (γ, n) reaction cross sections at these very intense resonances are a few Mev, but broaden appreciably in the region where nuclei are known to have large quadrupolar deformations. DANOS [104] and OKAMOTO [105] have pointed out that this apparent broadening is due to an actual splitting of the otherwise degenerate collective vibrations when the nucleus becomes nonspherical.

In a general spheroidal nucleus, three different frequencies of collective motion of protons with respect to neutrons may be expected, corresponding to motion along the three principal axes. In an actual nucleus, there is rotational symmetry about one principal axis, so that two of the vibrational modes coincide. If the nucleus has a positive quadrupole moment, as is normal in the strongly distorted heavy nuclei, the higher frequency vibrations, corresponding to motion along the two short principal axes, are the ones that coincide. For a negative quadrupole moment, the two low frequency vibrations would coincide. The mean energy corresponding to these giant resonances is given by

$$E_m = \frac{E_a + 2E_b}{3}, \tag{19.3}$$

where E_a corresponds to vibrations along the unique (normally the longer) axis, and E_b to those along the two equal axes.

DANOS [104] has shown that, on the basis of a hydrodynamic model,

$$\frac{E_b}{E_a} = 0.911\, x + 0.089 \quad (19.4)$$

where $x = a/b$, the ratio of the two axes. The intrinsic quadrupole moment Q_0 is given by

$$\left.\begin{aligned} Q_0 &= \frac{2}{5} Z(a^2 - b^2) \\ &= \frac{2}{5} ZR^2 \frac{(x^2 - 1)}{x^{\frac{2}{3}}} \end{aligned}\right\} \quad (19.5)$$

where R is the nuclear radius, which is approximately $1.2 \times 10^{-12} A^{\frac{1}{3}}$ cm, with A the mass number.

Variation with energy E for the total cross section of the reaction (γ, n) may be fitted to an expression of the form

$$\sigma = \frac{\sigma_m}{1 + \left(\dfrac{E^2 - E_m^2}{E\,\Gamma}\right)^2} \quad (19.6)$$

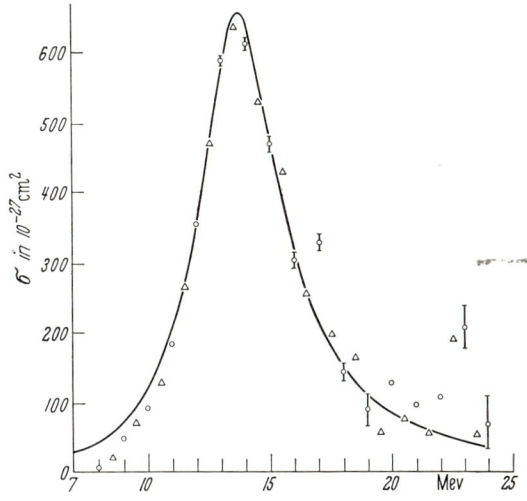

Fig. 12. Giant γ-ray resonance in gold, an approximately spherical nucleus. Theoretical curve is expression (19.6). (From FULLER and WEISS [106].)

where Γ represents the resonance width, and σ_m, E_m, and Γ are all constants to be evaluated by fitting the experimental data. A fit of this type is illustrated

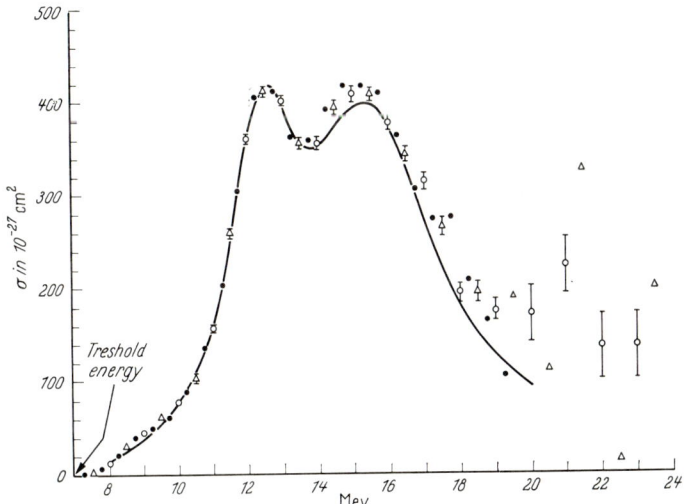

Fig. 13. Giant γ-ray resonance of Ta^{181}, a nucleus with large positive quadrupole moment. (From FULLER and WEISS) [106].)

by Fig. 12, for the nearly spherical nucleus of Au. When the nucleus is non-spherical, expression (19.6) must be replaced by the sum of two resonances, corresponding to $E_m = E_a$ and $E_m = E_b$. For the component with $E_m = E_b$, the product $\sigma_m \Gamma$ must be twice as large as that for the component $E_m = E_a$, since it represents two resonances rather than one. The case of this type which has been most carefully studied is Ta^{181}, illustrated in Fig. 13.

Constants used for the curves of Figs. 12 and 13 are given in Table 21, along with the vibrational energies of Tb^{159}. From these, relations (19.4) and (19.5), and using $R = 1.2 \times 10^{-12} A^{\frac{1}{3}}$, the value $Q_0 = 5.7 \times 10^{-24}$ cm² is obtained for Ta^{181} and $Q_0 = 5.6 \times 10^{-24}$ cm² for Tb^{159}. The experimental error of such a measurement of Q_0 can, in favorable cases, be quite small. For Tb^{159} it is about 25%, and for Ta^{181} only 4%. Theoretical uncertainties have not yet been examined very closely. They include questions about what nuclear radius should be used, as well as adequacy of a hydrodynamic model.

Although measurement of (γ, n) cross sections with sufficient accuracy for determinations of nuclear quadrupole moments is rather difficult, it appears to

Table 21. *Constants for giant (γ, n) resonances of several nuclei.*
For definition of constants, see Eqs. (19.3) and (19.6). (After FULLER and WEISS [*106*].)

Au^{197}:	$E_m = 13.75$ Mev $\Gamma = 4.1$ Mev $\sigma_m = 650 \times 10^{-27}$ cm²	Ta^{181}:	$E_b = 15.45$ Mev $\Gamma_b = 4.4$ Mev $\sigma_{ma} = 348 \times 10^{-27}$ cm²
Ta^{181}:	$E_a = 12.45$ Mev $\Gamma_a = 2.3$ Mev $\sigma_{ma} 308 = \times 10^{-27}$ cm²	Tb^{159}:	$E_a = 12.5$ Mev $E_b = 16.3$ Mev

be a very useful method for nuclei with particularly large distortions. It has the advantage over Coulomb excitation measurements of determining the sign as well as the magnitude of Q_0.

C. Mesonic quadrupole effects.

20. Effects of quadrupole moments on the hyperfine structure of μ mesonium. A negative μ meson may come to rest and be captured by an atom. It radiates rather quickly in falling through the various quantum states about the atomic nucleus, and finally rests a short time in the $1s$ state before decaying or being captured by the nucleus. The last few transitions downward toward the $1s$ state produce γ-radiation which can be detected, and for which there is reasonable hope of detecting the fine and hyperfine structure.

Hyperfine structure of the μ meson in a $1s$, $2s$, or $2p$ state about a nucleus (μ mesonium) may be enormously large because the μ meson orbit is, at least for heavy nuclei, not much larger than the dimensions of the nucleus. This close interaction and considerable penetration of the nucleus allows the spectrum of μ mesonium to yield considerable detailed information about the nucleus, as pointed out by WHEELER [*107*].

The effect of quadrupole moments on the hyperfine structure of μ mesonium has been discussed in some detail by WHEELER [*107*], JACOBSON [*108*], and WILETS [*109*]. In contrast to the case of an electron, the quadrupole hyperfine energy is much larger than that due to interaction between the nuclear and mesonic magnetic moments, since the magnetic moment of the μ meson is less by a factor of about three hundred than that of the electron. The hyperfine structure is, in fact, of the same order of magnitude as the fine structure of an excited state, e.g., the $2p$ state, of mesonium. Furthermore, the quadrupolar hyperfine interaction is, for heavy nuclei, as large as several hundred kilovolts, and hence comparable to the separation between nuclear energy levels.

Under these conditions, the hyperfine eigenstates involve appreciable mixtures of nuclear energy states and of fine structure states, so that their calculation is a little complex. No interactions between μ mesons and nucleons of non-electromagnetic origin have yet been detected, and it is generally assumed that any such unknown interactions would be negligible. In this case, everything which enters into such a calculation of μ mesonium is known except certain properties of the nucleus, so that an experimental determination of the hyperfine structure can give very valuable information about characteristics of the nucleus.

WHEELER [107] has given the fine and hyperfine energy for the $2p$ state of mesonium involving a nucleus somewhat like that of tantalum in the approximation that the nuclear energy levels are widely separated by comparison with the hyperfine structure. The result is

$$W = a\left(\frac{Z}{101}\right)^4 f_s + b\,\frac{Q}{10^{-24}}\left(\frac{Z}{237}\right)^3 f_q\,\frac{[\tfrac{3}{4}C(C+1)-I(I+1)\,J(J+1)]}{2I(2I-1)\,J(2J-1)} \tag{20.1}$$

where

$a = \tfrac{1}{3}$ Mev when $J = \tfrac{3}{2}$
and $-\tfrac{2}{3}$ Mev when $J = \tfrac{1}{2}$,

$b = 4$ Mev when $J = \tfrac{3}{2}$
and 0 when $J = \tfrac{1}{2}$,

$f_s \approx \dfrac{1}{(1+0.2\,x^2)}$,

$f_q \approx \dfrac{1}{(1+0.1\,x^2)^2}$,

$x = RZ\mu\,e^2/\hbar^2$,

$R =$ the nuclear radius,

$\mu =$ the meson mass,

$C = F(F+1) -$
$\quad - I(I+1) - J(J+1)$
as in Eq. (2.17),

$Q =$ the nuclear quadrupole moment.

Table 22. *The factor of $(I_1 I_2; I_0)$ of Eq. (20.2).*

I_0, the spin of the ground state of the nucleus, is zero for an even-even nucleus.

I_2	$f(I_1 I_2; I_0)$
I_1	$\left[\dfrac{2I_1+1}{5I_1(2I_1-1)(I_1+1)(2I_1+3)}\right][3I_0^2 - I_1(I_1+1)]$
I_1+1	$\left\{\dfrac{3I_0^2[(I_1+1)^2 - I_0^2]}{5I_1(I_1+1)(I_1+2)}\right\}^{\frac{1}{2}}$
I_1+2	$\left\{\dfrac{3[(I_1+1)^2 - I_0^2][(I_1+2)^2 - I_0^2]}{10(I_1+1)(I_1+2)(2I_1+3)}\right\}^{\frac{1}{2}}$

The first term of expression (20.1) represents the fine structure and the second term has the familiar form of the quadrupole interaction.

To obtain this expression, an approximate model for the nuclear charge distribution must be assumed, since the effect of a nuclear quadrupole moment for a particle which has a high probability of being inside the nucleus depends on the particular distribution of the quadrupole-producing charges.

If the nuclear levels are not widely spaced by comparison with the hyperfine interaction, as is frequently the case for nuclei in the region where large quadrupole moments occur, off-diagonal matrix elements of the quadrupole moment contribute to the hyperfine structure. On the basis of the Bohr-Mottelson model for deformed nuclei, these matrix elements for the $2p$ state of the meson may be written [108]

$$\left. \begin{aligned} (I_1 J_1 F |H_Q| I_2 J_2 F) &= e^2 (-1)^{I_1 + \frac{3}{2} - F}\, W(I_1 J_1 I_2 J_2; F\,2) \times \\ &\quad \times f\,(I_1 I_2; I_0)\, Q_0 \langle P(r)/r^3 \rangle_{2p}. \end{aligned} \right\} \tag{20.2}$$

Here I_1 and I_2 are spins of the two nuclear states involved, and I_0 is the spin of the ground state of the same nucleus. J_1 and J_2 are the angular momenta of the two meson states. $W(I_1 J_1 I_2 J_2; F\,2)$ is a function given by RACAH [110]. $f\,(I_1 I_2; I_0)$ may be obtained from Table 22. $\langle P(r)/r^3 \rangle_{2p}$ is an expectation value for the $2p$ state of the inverse cube of the distance from the nuclear center of mass multiplied by a form factor $P(r)$ which depends on the distribution of quadrupole-producing charge in the nucleus. If a uniform charge distribution is assumed for a spheroidal nucleus, $P(r) = (r/R)^5$ inside the nucleus $(r < R)$, and $P(r) = 1$ outside the nucleus. Q_0 is again related to the quadrupole moment Q by expression (15.4), assuming collective rotation as described by BOHR and MOTTELSON [80]. In the more general case, it represents an off-diagonal matrix element of the quadrupole moment.

The off-diagonal matrix elements allow large effects of the nuclear quadrupole moment on the mesonium fine structure for nuclei with spin zero or $\frac{1}{2}$, as

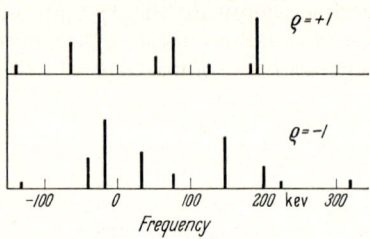

Fig. 14. Fine and hyperfine structure of the $2p-1s$ transition for an even isotope of tungsten, assuming $Q_0 = -7{,}0\varrho \times 10^{-24}$ cm². ϱ is $+1$ or -1 as indicated. (From JACOBSON [108].)

well as for nuclei with larger spins. The energy contributions due to such effects depend on both the magnitude and the sign of Q_0, as indicated in Fig. 14. In this respect the fine structure of μ mesonium gives information which is completely unavailable (i.e., Q_0 for nuclei of spins 0 or $\frac{1}{2}$) from normal electronic hyperfine structure. In contrast to Coulomb excitation, it also allows determination of the sign of Q_0.

D. Results.

21. Experimentally determined values of nuclear quadrupole moments. Values of nuclear quadrupole moments which have been measured are summarized in Table 23. Both "the" quadrupole moment, that is, the diagonal matrix element Q, and the "intrinsic" quadrupole moment, Q_0, are listed for cases where they have been measured. Where only one has been measured, the other may be obtained in the approximation of collective nuclear rotation by the relation

$$Q_0 = \frac{(I + 1)(2I + 3)}{I(2I - 1)} Q.$$

In some cases, more accurate values of Q may possibly be obtained by use of this relation and measured values of Q_0; or vice versa, experimental values of Q may be used to yield improved values of Q_0. However, since some uncertainties are involved in such a procedure, it has not been used in the table.

For many nuclei, a variety of values of Q have been published. References are given to each different author or set of authors who have evaluated Q, but in these cases the actual value listed represents the writer's opinion of the most probable value. Corrections for shielding and antishielding of valence electrons by inner shells have not been made since it appears that most of these corrections, though important, have not yet been determined with enough surety to justify their use. Such corrections will be found in Sect. 10 and can be applied at the discretion of the reader. Many values of quadrupole moments have been given to considerable accuracy by authors who have not adequately recognized the errors due to uncertainties in shielding. It is for this reason that the errors listed here will sometimes be appreciably larger than those given in the original references.

References for data on Q, Q_0, or ratios are listed together, and are indicated first by the year, and then in arbitrary order by following letters. Some additional earlier references for the same authors or laboratories quoted may be found in the compilations of MACK [111], or of KYULTS, KUNTS, and KHARTMAN [112]. Q_0 has in each case been measured by Coulomb excitation. In some cases an additional determination has been made from decay times, and for Ta¹⁸¹ and Tb¹⁵⁹ Q_0 has been obtained from measurement of giant γ-ray resonances. All values of Q_0

Table 23. *Summary of measured values of quadrupole moments.* See text for discussion of this table.

Nucleus	Spin	Q $(10^{-24}\,cm^2)$	Q_0 $(10^{-24}\,cm^2)$	Ratios of quadrupole moments	Technique used in measurement of Q	References
$_1H^2$	1	0.002738 ± 0.000014		$\dfrac{Li^6}{Li^7} = 1.9 \pm 0.1 \times 10^{-2}$	Mol.	52a, 50a, 40a
$_3Li^6$	1	$(-)\,0.0005$			Mol., Solid	53a, 53b, 53c, 51a
$_3Li^7$	3/2	$(-)\,0.02$			Mol., Solid	53c, 53d, 53e, 49a
$_4Be^9$	3/2	$(+)\,0.02$			Solid	53f, 51b
$_5B^{10}$	3	0.074		$\dfrac{B^{10}}{B^{11}} = 2.084 \pm 0.002$	Atom, Mol., Solid	53g, 53h, 52b, 50b
$_5B^{11}$	3/2	0.036			Atom, Mol., Solid	53g, 53h, 53i, 50b
$_7N^{14}$	1	0.02			Mol.	57a, 55a, 54a, 50c, 49b
$_8O^{17}$	5/2	-0.0265 ± 0.003			Atom, Mol.	57b, 57c, 52c
$_{11}Na^{23}$	3/2	0.11 ± 0.01			Atom	55b, 54b
$_{13}Al^{27}$	5/2	0.15 ± 0.01			Atom	53j, 52d
$_{16}S^{33}$	3/2	-0.064 ± 0.01		$\dfrac{S^{33}}{S^{35}} = -1.330 \pm 0.002$	Mol., Solid	54c, 53k, 53l
$_{16}S^{35}$	3/2	0.054 ± 0.01			Mol.	54c, 54d, 54e, 53k, 51c
$_{17}Cl^{35}$	3/2	-0.080 ± 0.002		$\dfrac{Cl^{35}}{Cl^{37}} = 1.26878 \pm 0.00015$	Atom, Mol., Solid	53m, 53n, 52d, 52e, 52f, 52g, 52h, 51d, 51e, 51f, 51g, 50d, 49b, 49c
$_{17}Cl^{36}$	2	-0.017 ± 0.001		$\dfrac{Cl^{36}}{Cl^{35}} = 0.211 \pm 0.001$	Mol.	52i, 51h, 49d
$_{17}Cl^{37}$	3/2	-0.063 ± 0.002			Atom, Mol., Solid	53m, 53n, 52d, 52e, 52f, 52g, 52h, 51d, 51e, 51f, 51g, 50d, 49b, 49c
$_{19}K^{39}$	3/2	0.09 ± 0.02		$\dfrac{K^{39}}{K^{41}} = 0.820 \pm 0.002$	Atom, Mol.	57d, 57q, 52g
$_{19}K^{41}$	3/2	0.11 ± 0.02			Atom, Mol.	57d, 57q, 52g
$_{23}V^{51}$	7/2	0.2 ± 0.15	1.1		Atom	56b
$_{25}Mn^{55}$	5/2	0.4 ± 0.2	0.9		Atom, Mol.	56l, 54q, 53o, 53p
$_{26}Fe^{57}$						56l
$_{27}Co^{59}$	7/2	0.5 ± 0.2			Atom	53o
$_{29}Cu^{63}$	3/2	0.16 ± 0.03		$\dfrac{Cu^{63}}{Cu^{65}} = 1.0806 \pm 0.0003$	Atom, Solid	55c, 53r, 52j, 51i, 51j, 49e, 36a
$_{29}Cu^{65}$	3/2	0.15 ± 0.03			Atom, Solid	55c, 53r, 52j, 51i, 51j, 49e, 36a
$_{30}Zn^{67}$	5/2	0.18		$\dfrac{Ga^{67}}{Ga^{71}} = 1.82 \pm 0.1$	Atom	57e
$_{31}Ga^{67}$	3/2	0.219 ± 0.012			Atom	57f

Table 23. (Continued.)

Nucleus	Spin	Q $(10^{-24}\,\mathrm{cm^2})$	Q_0 $(10^{-24}\,\mathrm{cm^2})$	Ratios of quadrupole moments	Technique used in measurement of Q	References
$_{31}\mathrm{Ga}^{69}$	$3/2$	0.190 ± 0.01		$\dfrac{\mathrm{Ga}^{69}}{\mathrm{Ga}^{71}} = 1.58690 \pm 0.00001$	Atom, Solid	56c, 54f, 53s, 52d, 49c, 48b, 36b
$_{31}\mathrm{Ga}^{71}$	$3/2$	0.120 ± 0.01			Atom, Solid	52d, 48b, 36b
$_{32}\mathrm{Ge}^{73}$	$9/2$	0.2 ± 0.1			Mol.	51k
$_{33}\mathrm{As}^{75}$	$3/2$	0.3 ± 0.2			Atom, Mol.	49b, 36c
$_{34}\mathrm{Se}^{75}$	$5/2$	1.1 ± 0.2		$\dfrac{\mathrm{Se}^{75}}{\mathrm{Se}^{79}} = 1.25783 \pm 0.0006$	Mol.	55d
$_{34}\mathrm{Se}^{79}$	$7/2$	0.9 ± 0.2			Mol.	54c, 53t
$_{35}\mathrm{Br}^{79}$	$3/2$	0.33 ± 0.02		$\dfrac{\mathrm{Br}^{79}}{\mathrm{Br}^{81}} = 1.19707 \pm 0.00003$	Atom, Mol., Solid	54g, 54h, 54i, 54j, 54k, 53u, 53v, 53w, 51l, 51m, 49b
$_{35}\mathrm{Br}^{81}$	$3/2$	0.28 ± 0.02			Atom, Mol., Solid	54g, 54h, 54i, 54j, 54k, 53u, 53v, 53w, 51l, 51m, 49b
$_{35}\mathrm{Br}^{82}$	5	$|Q| = 0.7 \pm 0.1$			Atom	57g
$_{36}\mathrm{Kr}^{83}$	$9/2$	0.22 ± 0.02			Atom	55e, 38a
$_{37}\mathrm{Ru}^{85}$	$5/2$	0.28 ± 0.02		$\dfrac{\mathrm{Ru}^{85}}{\mathrm{Ru}^{87}} = 2.07 \pm 0.01$	Atom, Mol.	56d, 55f, 50e
$_{37}\mathrm{Ru}^{87}$	$3/2$	0.136 ± 0.01			Atom, Mol.	56d, 56e, 55f
$_{41}\mathrm{Nb}^{93}$	$9/2$	-0.2 ± 0.1			Atom	55g
$_{43}\mathrm{Tc}^{99}$	$9/2$	0.34 ± 0.17			Atom	53x
$_{47}\mathrm{Ag}^{107}$	$1/2$		1.9			56l
$_{47}\mathrm{Ag}^{109}$	$1/2$		1.9			56l
$_{49}\mathrm{In}^{113}$	$9/2$	1.14 ± 0.05		$\dfrac{\mathrm{In}^{113}}{\mathrm{In}^{115}} = 0.9863649 \pm 0.0000013$	Atom	57h, 52d, 50f
$_{49}\mathrm{In}^{115}$	$9/2$	1.16 ± 0.05			Atom	52d, 50f
$_{51}\mathrm{Sb}^{121}$	$5/2$	-0.53 ± 0.10		$\dfrac{\mathrm{Sb}^{121}}{\mathrm{Sb}^{123}} = 0.78447 \pm 0.00005$	Atom, Mol., Solid	55h, 55i, 53y, 51g, 51n
$_{51}\mathrm{Sb}^{123}$	$7/2$	-0.68 ± 0.10			Atom, Mol., Solid	55h, 55i, 53y, 51g, 51n
$_{53}\mathrm{I}^{125}$	$5/2$	-0.88 ± 0.05		$\dfrac{\mathrm{I}^{125}}{\mathrm{I}^{127}} = 1.1270 \pm 0.0005$	Mol.	57i
$_{53}\mathrm{I}^{127}$	$5/2$	-0.78 ± 0.05		$\dfrac{\mathrm{I}^{129}}{\mathrm{I}^{127}} = 0.701213 \pm 0.000015$	Atom, Mol., Solid	55h, 54l, 52k, 51o, 49b, 39a
$_{53}\mathrm{I}^{129}$	$7/2$	-0.55 ± 0.04			Atom, Mol., Solid	53z, 51o
$_{53}\mathrm{I}^{131}$	$7/2$	-0.39 ± 0.04		$\dfrac{\mathrm{I}^{131}}{\mathrm{I}^{127}} = 0.5036 \pm 0.0005$	Mol.	57j, 53aa
$_{54}\mathrm{Xe}^{131}$	$3/2$	-0.12			Atom	52l, 38a
$_{55}\mathrm{Cs}^{133}$	$7/2$	-0.003 ± 0.002			Atom	56a, 55j, 40b

Isotope	I	$|Q|$	Ratio	State	References
$_{57}\text{La}^{138}$	5	$|Q| \sim 1.5$	$\dfrac{\text{La}^{138}}{\text{La}^{139}} \sim 3$	Solid	551
$_{57}\text{La}^{139}$	7/2	0.5 ± 0.2		Atom	57k, 55k
$_{59}\text{Pr}^{141}$	5/2	-0.054		Atom	53ab
$_{60}\text{Nd}^{143}$	7/2	$|Q| \sim 1$		Solid	54n, 51p
$_{60}\text{Nd}^{145}$	7/2	$|Q| \sim 1$		Solid	54n, 51p,
$_{60}\text{Nd}^{150}$	0				56k, 561
$_{62}\text{Sm}^{147}$	7/2	$|Q| < 0.7$	$\dfrac{\text{Sm}^{147}}{\text{Sm}^{149}} < 1$	Atom, Solid	54o, 54p, 52m
$_{62}\text{Sm}^{149}$	7/2	$|Q| < 0.7$		Atom, Solid	54p, 52m
$_{62}\text{Sm}^{152}$	0				56k, 561
$_{62}\text{Sm}^{154}$	0				56k, 561
$_{63}\text{Eu}^{151}$	5/2	1.2		Atom	35a
$_{63}\text{Eu}^{153}$	5/2	2.5		Atom	56k, 561, 56m, 35a
$_{64}\text{Gd}^{154}$	0				561
$_{64}\text{Gd}^{155}$	3/2	1.1		Atom	56f, 56k, 56m
$_{64}\text{Gd}^{156}$	0				56k, 561
$_{64}\text{Gd}^{157}$	3/2	1.0		Atom	56f, 56k, 56m
$_{64}\text{Gd}^{158}$	0				56k, 561
$_{65}\text{Tb}^{159}$	3/2	$|Q| \sim 2$			57q, 56k, 561, 56m
$_{66}\text{Dy}^{160}$	0				56k
$_{66}\text{Dy}^{162}$	0				56k, 561
$_{66}\text{Dy}^{164}$	0			Solid	56k, 561
$_{67}\text{Ho}^{165}$	7/2	$|Q| = 10 \pm 2$			56k, 561, 56m, 55m
$_{68}\text{Er}^{164}$	0				56k
$_{68}\text{Er}^{166}$	0				56k
$_{68}\text{Er}^{167}$	7/2			Solid	53ac
$_{68}\text{Er}^{168}$	0				56k
$_{68}\text{Er}^{170}$	0				56k
$_{69}\text{Tm}^{169}$	1/2				56k, 561
$_{70}\text{Yb}^{170}$	0				56k
$_{70}\text{Yb}^{172}$	0				56k
$_{70}\text{Yb}^{173}$	5/2	3.9 ± 0.4		Atom	38a
$_{70}\text{Yb}^{174}$	0				56k
$_{70}\text{Yb}^{176}$	0				56k
$_{71}\text{Lu}^{175}$	7/2	5.1 ± 0.3		Atom	57k, 56k, 561, 56m, 36d
$_{71}\text{Lu}^{176}$	$\geqq 7$	7 ± 1		Atom	39b
$_{72}\text{Hf}^{176}$	0				56k
$_{72}\text{Hf}^{177}$	7/2		$\dfrac{\text{Hf}^{177}}{\text{Hf}^{179}} = 0.99 \pm 0.02$	Atom	56g, 56k, 561, 56m

Table 23. (Continued.)

Nucleus	Spin	Q (10^{-24} cm^2)	Q_0 (10^{-24} cm^2)	Ratios of quadrupole moments	Technique used in measurement of Q	References
$_{72}\text{Hf}^{178}$	0		7.5		Atom	56k, 56l
$_{72}\text{Hf}^{179}$	$\frac{9}{2}$		7.0			56g, 56k, 56l, 56m
$_{72}\text{Hf}^{180}$	0		7.0			56k, 56l
$_{73}\text{Ta}^{181}$	$\frac{7}{2}$	3.9 ± 0.4	6.8		Atom	57k, 57q, 56k, 56l, 56m
$_{74}\text{W}^{182}$	0		7.1			56k, 56l
$_{74}\text{W}^{183}$	$\frac{1}{2}$		8.4			56l
$_{74}\text{W}^{184}$	0		6.5			56k, 56l
$_{74}\text{W}^{186}$	0		6.4	$\frac{\text{Re}^{185}}{\text{Re}^{187}} = 1.056 \pm 0.005$		56k, 56l
$_{75}\text{Re}^{185}$	$\frac{5}{2}$	2.8	5.4		Atom, Mol., Solid	57l, 56k, 56l, 54q, 37a
$_{75}\text{Re}^{187}$	$\frac{5}{2}$	2.6	5.0		Atom, Mol., Solid	56k, 56l, 37a
$_{76}\text{Os}^{186}$	0		5.5			56k
$_{76}\text{Os}^{188}$	0		5.1			56k
$_{76}\text{Os}^{189}$	$\frac{3}{2}$	0.8 ± 0.2	3.3	$\frac{\text{Ir}^{191}}{\text{Ir}^{193}} = 1.0 \pm 0.3$	Atom	57k
$_{77}\text{Ir}^{191}$	$\frac{3}{2}$	1.5 ± 0.5	2.7		Atom	56k, 53q, 52n, 50g
$_{77}\text{Ir}^{193}$	$\frac{3}{2}$	1.5 ± 0.1	3.4		Atom	56k, 53q, 53ae, 52n
$_{78}\text{Pt}^{195}$	$\frac{1}{2}$		2.2			56l
$_{79}\text{Au}^{197}$	$\frac{3}{2}$	0.56 ± 0.10				56l, 53ad
$_{80}\text{Hg}^{201}$	$\frac{3}{2}$	0.50 ± 0.05			Atom, Solid	57m, 56n, 55g, 54r, 35b
$_{83}\text{Bi}^{209}$	$\frac{9}{2}$	-0.4			Atom	36e
$_{89}\text{Ac}^{227}$	$\frac{3}{2}$	-1.7 ± 0.2			Atom	55n
$_{90}\text{Th}^{232}$	0		7.0			56k
$_{92}\text{U}^{233}$	$\frac{5}{2}$	$|Q| = 2.7$	14	$\frac{\text{U}^{233}}{\text{U}^{235}} = 0.70$	Atom, Solid	57n, 57o, 56i, 56k, 56l
$_{92}\text{U}^{235}$	$\frac{7}{2}$	$|Q| = 3.8$	10.1		Atom, Solid	57p, 56h, 56k, 56l, 54s
$_{92}\text{U}^{238}$	0		10.3			57p, 56k
$_{93}\text{Np}^{237}$	$\frac{5}{2}$		9			56k, 56l
$_{94}\text{Pu}^{239}$	$\frac{1}{2}$		9.2	$\frac{\text{Am}^{241}}{\text{Am}^{243}} = 1.00 \pm 0.01$		57p, 56k, 56l
$_{95}\text{Am}^{241}$	$\frac{5}{2}$	4.9			Atom	56j
$_{95}\text{Am}^{243}$	$\frac{5}{2}$	4.9			Atom	56j

References to Table 23.

The figure gives the year of publication, i.e., 35 means "published in 1935", the letters a, b, ... refer to the papers published in that year.

35a. H. Schüler and T. Schmidt: Z. Physik **98**, 430.
35b. H. Schüler and T. Schmidt: Z. Physik **98**, 239.
36a. H. Schüler and T. Schmidt: Z. Physik **100**, 113.
36b. H. Schüler and H. Korsching: Z. Physik **103**, 434.
36c. H. Schüler and M. Marketu: Z. Physik **102**, 703.
36d. H. Gollnow: Z. Physik **103**, 443.
36e. H. Schüler and T. Schmidt: Z. Physik **99**, 717.
37a. H. Schüler and H. Korsching: Z. Physik **105**, 168.
38a. Schüler, Roig and Korsching: Z. Physik **111**, 165.
39a. T. Schmidt: Z. Physik **112**, 199.
39b. H. Schüler and H. Gollnow: Z. Physik **113**, 1.
40a. Kellog, Rabi and Ramsey: Phys. Rev. **57**, 677.
40b. T. Schmidt: Naturwiss. **28**, 565.
48a. Smith, Ring, Smith and Gordy: Phys. Rev. **74**, 370.
48b. G. E. Becker and P. Kusch: Phys. Rev. **73**, 584.
49a. P. Kusch: Phys. Rev. **76**, 138.
49b. C. H. Townes and B. P. Dailey: J. Chem. Phys. **17**, 782.
49c. Davis, Feld, Zabel and Zacharias: Phys. Rev. **76**, 1076.
49d. C. H. Townes and L. C. Aamodt: Phys. Rev. **76**, 691.
49e. P. Brix: Z. Physik **126**, 725.
50a. G. F. Newell: Phys. Rev. **78**, 711.
50b. Gordy, Ring and Burg: Phys. Rev. **78**, 512.
50c. J. Sheridan and W. Gordy: Phys. Rev. **79**, 513.
50d. Smith, Tidwell and Williams: Phys. Rev. **79**, 1007.
50e. V. Hughes and L. Grabner: Phys. Rev. **79**, 314.
50f. A. K. Mann and P. Kusch: Phys. Rev. **77**, 427.
50g. Brix, Kopfermann and Siemens: Naturwiss. **37**, 397.
51a. N. A. Schuster and G. E. Pake: Phys. Rev. **81**, 157.
51b. Hatton, Rollin and Seymour: Phys. Rev. **83**, 672.
51c. Wentink, Koski and Cohen: Phys. Rev. **81**, 948.
51d. V. Jaccarino and J. G. King: Phys. Rev. **84**, 852.
51e. Geschwind, Gunther-Mohr and Townes: Phys. Rev. **82**, 343.
51f. R. Livingston: Phys. Rev. **82**, 289.
51g. H. G. Dehmelt and H. Krüger: Z. Physik **130**, 385.
51h. Johnson, Gordy and Livingston: Phys. Rev. **83**, 1249.
51i. G. Becker: Z. Physik **130**, 415.
51j. A. Abragam and M. H. L. Pryce: Proc. Roy. Soc. Lond. **206**, 164.
51k. J. M. Mays and C. H. Townes: Phys. Rev. **81**, 940.
51l. W. Gordy: J. Chem. Phys. **19**, 792.
51m. H. G. Dehmelt and H. Krüger: Z. Physik **130**, 480.
51n. C. C. Loomis and M. W. P. Strandberg: Phys. Rev. **81**, 798.
51o. W. Gordy: J. Chem. Phys. **19**, 792.
51p. R. J. Elliott and K. W. H. Stevens: Proc. Phys. Soc. Lond. A **64**, 205.
52a. Kolsky, Phipps, Ramsey and Silsbee: Phys. Rev. **87**, 395.
52b. H. G. Dehmelt: Z. Physik **133**, 528.
52c. Geschwind, Gunther-Mohr and Silvey: Phys. Rev. **85**, 474.
52d. C. F. Koster: Phys. Rev. **86**, 148.
52e. Wang, Townes, Schawlow and Holden: Phys. Rev. **86**, 809.
52f. C. Dean and R. V. Pound: J. Chem. Phys. **20**, 195.
52g. Carlson, Lee and Fabricand: Phys. Rev. **85**, 784.
52h. H. Zeiger and D. Bolef: Phys. Rev. **85**, 788.
52i. D. A. Gilbert: Phys. Rev. **85**, 716.
52j. H. Krüger and U. Meyer-Berkhoot: Z. Physik **132**, 171.
52k. T. Kamei: J. Phys. Soc. Japan **7**, 649.
52l. Bohr, Koch and Rasmussen: Ark. Fysik **4**, 455.
52m. R. J. Elliott and K. W. H. Stevens: Proc. Phys. Soc. Lond. A **65**, 370.
52n. K. Murakawa and S. Suwa: Phys. Rev. **87**, 1048.
53a. P. Kusch: Phys. Rev. **92**, 268.
53b. N. G. Cranna: Canad. J. Phys. **31**, 1185.
53c. D. R. Inglis: Rev. Mod. Phys. **25**, 390.

53d. E. G. HARRIS and M. A. MELKANOFF: Phys. Rev. **90**, 585.
53e. R. M. STERNHEIMER and H. M. FOLEY: Phys. Rev. **92**, 1460.
53f. W. D. KNIGHT: Phys. Rev. **92**, 539.
53g. G. WESSEL: Phys. Rev. **92**, 1581.
53h. H. G. DEHMELT: Z. Physik **134**, 642.
53i. A. BASSOMPIERRE: C. R. Acad. Sci., Paris **237**, 1224.
53j. H. LEW and G. WESSEL: Phys. Rev. **90**, 1.
53k. H. G. DEHMELT: Phys. Rev. **91**, 313.
53l. C. A. BURRUS and W. GORDY: Phys. Rev. **92**, 274.
53m. H.C. MEAL and H. C. ALLEN: Phys. Rev. **90**, 348.
53n. M. BUGLE-BODIN and A. MONFILS: C. R. Acad. Sci., Paris **236**, 1157.
53o. K. MURAKAWA and T. KAMEI: Phys. Rev. **92**, 325.
53p. R. E. TREES: Phys. Rev. **92**, 308.
53q. W. V. SIEMENS: Ann. d. Phys. **13**, 136.
53r. KOPFERMANN, STEUDEL, WAGNER and WALCHER: Nachr. Akad. Wiss. Göttingen, math.-phys. Kl. IIa, No. 1, 1.
53s. H. G. DEHMELT: Phys. Rev. **92**, 1240.
53t. HARDY, SILVEY, TOWNES, BURKE, STRANDBERG, PARKER and COHEN: Phys. Rev. **92**, 1532.
53u. KOJIMA, TSUKADA, OGAWA and SHIMAUCHI: J. Chem. Phys. **21**, 1415.
53v. MANRING, BROWN and WILLIAMS: Phys. Rev. **90**, 348.
53w. FABRICAND, CARLSON, LEE and RABI: Phys. Rev. **91**, 1403.
53x. K. G. KESSLER and R. E. TREES: Phys. Rev. **92**, 303.
53y. G. SPRAGUE and D. H. TOMBOULIAN: Phys. Rev. **92**, 105.
53z. R. LIVINGSTON and H. ZELDES: Phys. Rev. **90**, 609.
53aa. LIVINGSTON, BENJAMIN, COX and GORDY: Phys. Rev. **92**, 1271.
53ab. H. LEW: Phys. Rev. **91**, 619.
53ac. R. J. ELLIOTT and K. W. H. STEVENS: Proc. Roy. Soc. Lond. A **219**, 387.
53ad. W. V. SIEMENS: Ann. d. Phys. **13**, 158.
53ae. W. V. Siemens, Ann. d. Phys. **13**, 136.
54a. P. KISLIUK: J. Chem. Phys. **22**, 86.
54b. P. L. SAGALYN: Phys. Rev. **94**, 885.
54c. G. BIRD and C. H. TOWNES: Phys. Rev. **94**, 1203.
54d. BURKE, STRANDBERG, COHEN and KOSKI: Phys. Rev. **93**, 193.
54e. R. L. WHITE: Rev. Mod. Phys. **27**, 276.
54f. R. T. DALY jr., and J. H. HOLLOWAY: Phys. Rev. **96**, 539.
54g. J. G. KING and V. JACCARINO: Phys. Rev. **94**, 1610.
54h. H. BENOIT and M. BUGLE-BODIN: C. R. Acad. Sci., Paris **238**, 671.
54i. DEHMELT, ROBINSON and GORDY: J. Chem. Phys. **22**, 511.
54j. D. DAUTREPPE and A. BLAISE: C. R. Acad. Sci., Paris **239**, 493.
54k. A. L. SCHAWLOW: J. Chem. Phys. **22**, 1211.
54l. JACCARINO, KING, SATTEN and STROKE: Phys. Rev. **94**, 1798.
54m. K. MURAKAWA and S. SUWA: J. Phys. Soc. Japan **9**, 93.
54n. BLEANEY, SCOVIL and TRENAM: Proc. Roy. Soc. Lond. A **223**, 15.
54o. K. MURAKAWA: Phys. Rev. **93**, 1232.
54p. A. R. BODMER: Proc. Phys. Soc. Lond. A **67**, 622.
54q. A. JAVAN and A. ENGELBRECHT: Phys. Rev. **96**, 649.
54r. DEHMELT, ROBINSON and GORDY: Phys. Rev. **93**, 920.
54s. K. L. VANDERSLUIS and J. R. McNALLY: J. Opt. Soc. Amer. **44**, 87.
55a. A. BASSOMPIERRE: C. R. Acad. Sci., Paris **240**, 285.
55b. PERL, RABI and SENITSKY: Phys. Rev. **97**, 838.
55c. BLEANEY, BOWERS and PRYCE: Proc. Roy. Soc. Lond., Ser. A **228**, 116.
55d. L. C. AAMODT and P. C. FLETCHER: Phys. Rev. **98**, 1224.
55e. E. RASMUSSEN and V. MIDDELBOE: Kgl. danske Vidensk. Selsk. **30**, No. 13.
55f. U. MEYER-BERKHOUT: Z. Physik **141**, 185.
55g. K. MURAKAWA: Phys. Rev. **98**, 1285.
55h. K. MURAKAWA: Phys. Rev. **100**, 1369.
55i. T. C. WANG: Phys. Rev. **99**, 566.
55j. K. ALTHOFF: Z. Physik **141**, 33.
55k. G. LUHRS: Z. Physik **141**, 486.
55l. P. B. SOGO and C. D. JEFFRIES: Phys. Rev. **99**, 613.
55m. J. M. BAKER and B. BLEANEY: Proc. Phys. Soc. Lond. A **68**, 1090.
55n. FRED, TOMKINS and MEGGERS: Phys. Rev. **98**, 1514.
56a. BUCK, RABI and SENITSKY: Phys. Rev. **104**, 553.
56b. K. MURAKAWA: J. Phys. Soc. Japan **11**, 422.

56c. KNIGHT, HEWITT and POMERANTZ: Phys. Rev. **104**, 271.
56d. B. SENITSKY and I. I. RABI: Phys. Rev. **103**, 315.
56e. KOPFERMANN, STEUDEL and TRIER: Z. Physik **144**, 1.
56f. D. R. SPECK: Phys. Rev. **101**, 1725.
56g. D. R. SPECK and F. A. JENKINS: Phys. Rev. **101**, 1831.
56h. BLEANEY, HUTCHINSON, LLEWELLYN and POPE: Proc. Phys. Soc. Lond. B **69**, 1167.
56i. N. I. KALITEEVSKII and N. P. CHAIKA: Optika i Spektroska **1**, 809.
56j. MANNING, FRED and TOMKINS: Phys. Rev. **102**, 1108.
56k. ALDER, BOHR, HUUS, MOTTELSON and WINTHER: Rev. Mod. Phys. **28**, 432.
56l. HUUS, BJERREGAARD and ELBEK: Kgl. danske Vidensk. Selsk., math.-fys. Medd. **30**, no. 17.
56m. N. P. HEYDENBERG and G. M. TEMMER: Phys. Rev. **104**, 981.
56n. R. V. POUND and G. K. WERTHEIM: Phys. Rev. **102**, 396.
57a. M. MIZUSHIMA: Phys. Rev. **105**, 1262.
57b. KAMPER, LEA and LUSTIG: Proc. Phys. Soc. Lond. **70**, 897.
57c. M. J. STEVENSON and C. H. TOWNES: Phys. Rev. **107**, 635.
57d. G. J. RITTER and G. W. SERIES: Proc. Roy. Soc. Lond., Ser. A **238**, 367.
57e. BÖCKMANN, KRÜGER and RECKNAGEL: Naturwiss. **44**, 7.
57f. WORCESTER, HUBBS and NIERENBERG: Bull. Amer. Phys. Soc. **2**, 316.
57g. GREEN, GARVIN, LIPWORTH and NIERENBERG: Bull. Amer. Phys. Soc. **2**, 383.
57h. T. G. ECK and P. KUSCH: Phys. Rev. **106**, 958.
57i. P. C. FLETCHER and E. AMBLE: Bull. Amer. Phys. Soc. **2**, 30.
57j. P. C. FLETCHER: Thesis, Columbia University.
57k. K. MURAKAWA and T. KAMEI: Phys. Rev. **105**, 671.
57l. S. L. SIEGEL and R. G. BARNES: Phys. Rev. **107**, 638.
57m. J. BLAISE and H. CHANTREL: J. Phys. Radium **18**, 193.
57n. BLAISE, GERSTENKORN and LOUVEGNIES: J. Phys. Radium **18**, 318.
57o. DORAIN, HUTCHINSON and WANG: Phys. Rev. **105**, 1307.
57p. J. O. NEWTON: Nuclear Phys. **3**, 345.
57q. P. BUCK and I. I. RABI: Phys. Rev. **107**, 1291.

listed will be found in Ref. 56k, 56l, 56m, or 57p. A very wide variety of techniques have been used to determine the values of Q listed. However, they all involve measurement of hyperfine structure in an atom, a molecule, or a crystalline solid, and subsequent evaluation of q for the environment involved. Which of these three types of environments was used is indicated in the column designated "Technique". In a few cases, the sign of Q as given in the literature is rather uncertain. This is indicated by a parenthesis around the sign given for Q.

22. Systematic behavior of nuclear quadrupole moments[1]. Nuclear quadrupole moments vary in a somewhat regular and understandable way with variation in the number of protons or of neutrons [*113*], [*114*] as indicated by Fig. 15. In this figure, the value of Q_0/R^2 is plotted as a function of the number of odd neutrons or of odd protons. For even-even nuclei, Q_0/R^2 is plotted against both the number of protons and the number of neutrons. Odd-odd nuclei are omitted from this Fig. 1. For the lighter nuclei, Q_0 has no simple physical meaning as it does for heavy deformed nuclei, but in any case it is defined as

$$Q_0 = \frac{(I+1)(2I+3)}{I(2I-1)} Q.$$

R is taken as $1.2 \times 10^{-13} A^{\frac{1}{3}}$. A value near unity for Q_0/R^2 is to be expected if the quadrupole moment is due to the equivalent of a single charged particle. Some values near unity are seen to occur for light nuclei, while for heavier nuclei Q_0/R^2 increases considerably above this value, indicating large contributions to Q_0 from collective distortion of the nucleus. It may also be noted that the majority of quadrupole moments are positive, and that the positive quadrupole moments often have appreciably greater magnitudes than those of the typical

---

[1] For more detailed information see Vol. XXXIX of this Encyclopedia.

negative quadrupole moment. This tendency towards positive quadrupole moments is in part associated with Coulomb energy [115]. An elongated nuclear form, corresponding to a positive Q_0, has lower electrostatic energy than does a flattened form, with negative Q_0. There are also other, somewhat less obvious, reasons why deformations corresponding to positive quadrupole moments tend to be lower in energy than those for negative quadrupole moments [115].

Another striking feature of Fig. 15 is the more or less cyclic pattern, with quadrupole moments reversing in sign where the number of particles coincides with the closing of one of the major shells on the shell model. The positions of

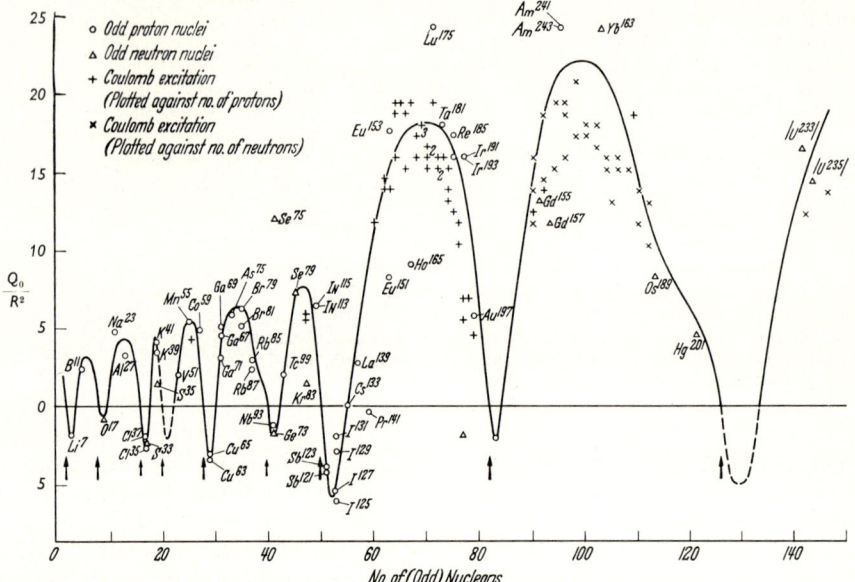

Fig. 15. Systematic behavior of nuclear quadrupole moments. $Q_0 = \dfrac{(I+1)(2I+3)}{I(2I-1)} Q$ and $R = 1.2 \times 10^{-13} A^{\frac{1}{3}}$. Arrows mark closed shells, the longer arrows indicating shells of major importance.

these closed shells are indicated by arrows; those for less prominent shells by shorter arrows. If the last shell of particles is assumed to move in a spherical potential, or one which would be spherical except for some small polarization of the closed shells of nucleons, the behavior of quadrupole moments as the shell fills up is easily predicted. The first particle introduced into such a shell should produce a negative quadrupole moment (unless it has spin $\frac{1}{2}$ and hence zero moment). The magnitude of such a moment is that produced by the single particle itself plus that due to polarization or distortion of the core by the single particle [114], [116]. The quadrupole moment due to distortion of the core will be of the same sign as that of the single particle, and at times much larger in size. In case the single particle is a neutron, it is this moment due to distortion which alone may be expected. As the shell is filled on this simple model, and when all particles are paired except the last odd particle, the quadrupole moment should increase linearly, becoming zero for a half-filled shell, and reaching a maximum positive value when the shell is filled except for one particle.

The qualitative features of the variation of quadrupole moments with number of particles can be fairly well understood in terms of the tendency for cyclic variations with each shell indicated above, and that for positive quadrupole moments. For the lighter nuclei, the effects of shells tend to dominate. For the heavier nuclei, the differences in energy between single particle levels become less and the lowering of energy due to distortion of the nucleus becomes greater. Hence, the quadrupole moment is not negative except immediately after the closing of major shells, and nuclear ground states tend to be those with large positive quadrupole moments rather than those expected with a regular filling of each subshell.

Fig. 15 indicates primarily the variation in Q_0 with number of odd particles. However, values of Q_0 for even-even nuclei are also plotted for comparison. Since they cannot be included unambiguously in such a plot against either number of protons or of neutrons, they are entered in both positions. Their values in fact fall reasonably well on the curve of systematic behavior established by the odd-even and even-odd nuclei. This indicates the tendency for distortion of the nucleus, especially in the heavier nuclei, more or less regardless of details of the last few particles added.

Odd-odd nuclei have been omitted from the figure. They are rather few in number and their quadrupole moments, particularly in the lighter nuclei, vary widely in accordance with details of coupling of the two odd particles.

It is possible to make an estimate with useful accuracy of the value and sign of the quadrupole moment for cases where the value is not known simply by use of Fig. 15. It may be seen that there are rather few cases which fall distinctly off of the general curve indicated. Most of such exceptions appear to have reasonable bases in terms of shell structure phenomena. Two cases have been omitted entirely from Fig. 15, since their experimentally determined values are unusual enough that some error may be indicated. One is $_{89}Ac^{227}$. It is reported to have a negative Q_0/R^2 of -0.23, which is several times larger than any other negative value, and occurs at a position where positive moments are expected. The other is $_{68}Er^{167}$, which has a value 0.7 for $|Q_0/R^2|$, twice as large as the largest other known value.

More accurate and more complete knowledge of quadrupole moments will probably not improve very much their fit on any plot such as Fig. 15. Precise values depend on more variables and more details than can be thus indicated. Nor will more complete experimental knowledge necessarily improve immediately our understanding of nuclear quadrupole moments. Their values are so intimately connected with the interesting but complicated structure of the entire nucleus that we must, for the predictable future, be satisfied with slow progress towards understanding, with theories which are rather imprecise if they are comprehensive, or ones which, if more nearly precise, have only limited applicability.

Acknowledgment.

It is a pleasure to acknowledge the considerable help of Mr. PATRICK THADDEUS in collecting and examining information on nuclear quadrupole monents which is summarized in Table 23 and in Sects. 21 and 22.

References.

[1] SCHÜLER, H., and T. SCHMIDT: Z. Physik **94**, 457 (1935).
[2] RAMSEY, N. F.: Nuclear Moments. New York: Wiley 1953.
[3] CASIMIR, H. B. G.: On the Interaction between Atomic Nuclei and Electrons. Harlem: Teyler's Tweede Genootschap, E. F. Bohn 1936.
[4] KELLOGG, RABI, RAMSEY and ZACHARIAS: Phys. Rev. **56**, 728 (1939); **57**, 677 (1940).
[5] NORDSIECK, A.: Phys. Rev. **58**, 310 (1940).
[6] RAMSEY, N. F.: Phys. Rev. **74**, 286 (1948).
[7] TOWNES, C. H., and A. L. SCHAWLOW: Microwave Spectroscopy. New York-Toronto-London: McGraw-Hill 1955.

[8] GORDY, SMITH and TRAMBARULO: Microwave Spectroscopy. New York: Wiley 1953.
[9] BARDEEN, J., and C. H. TOWNES: Phys. Rev. 73, 627; Errata 73, 1204 (1948).
[10] BRAGG, J. K.: Phys. Rev. 74, 533 (1948).
[11] CROSS, HAINER and KING: J. Chem. Phys. 12, 210 (1944).
[12] BRAGG, J. K., and S. GOLDEN: Phys. Rev. 75, 735 (1949).
[13] ERLANDSON, G.: Ark. Fysik (to be published).
[14] BARDEEN, J., and C. H. TOWNES: Phys. Rev. 73, 97 (1948).
[15] RACAH, G.: Phys. Rev. 62, 438 (1942).
[16] BIEDENHARN, BLATT and ROSE: Rev. Mod. Phys. 24, 249 (1952).
[17] OBI, ISHIDZU, YANAGAWA, TANABE and SATO: Ann. Tokyo Astron. Observ. 3, 89 (1953).
[17a] TOWNES, HOLDEN and MERRITT: Phys. Rev. 74, 1113 (1948).
[18] MIZUSHIMA, M., and T. ITO: J. Chem. Phys. 19, 739 (1951).
[19] BERSOHN, R.: J. Chem. Phys. 18, 1124 L (1950).
[20] COHEN, M. H.: Phys. Rev. 96, 1278 (1954).
[21] ABRAGAM, A.: Private Communication.
[22] BERSOHN, R.: J. Chem. Phys. 20, 1505 (1952).
[23] KOJIMA, TSUKADA, SHIMAUCHI and HINAGA: J. Phys. Soc. Japan 9, 795 (1954).
[24] PARKER, P. M.: J. Chem. Phys. 24, 1096 (1956).
[25] POUND, R. V.: Phys. Rev. 79, 685 (1950).
[26] POUND, R. V.: Phys. Rev. 73, 1247 (1948).
[27] ABRAGAM, A., and M. H. L. PRYCE: Proc. Roy. Soc. Lond., Ser. A 205, 135 (1951).
[28] BLEANEY, B.: Phil. Mag. 42, 441 (1951).
[28a] FOLEY, H. M.: Phys. Rev. 80, 288 (1950). — LOW, W.: Private communication.
[29] GARDNER, J. W.: Proc. Phys. Soc. Lond. A 62, 763 (1949).
[30] GOERTZEL, G.: Phys. Rev. 70, 897 (1946).
[31] LLOYD, S. P.: Phys. Rev. 82, 277 (1951).
[32] ALDER, K.: Helv. phys. Acta 25, 235 (1952).
[33] ABRAGAM, A., and R. V. POUND: Phys. Rev. 92, 943 (1953).
[34] ALBERS-SCHÖNBERG, HEEV, NOVEY and SCHERRER: Helv. phys. Acta 27, 547 (1954).
[35] LEHMANN, P., and J. MILLER: C. R. Acad. Sci., Paris 240, 298 (1955).
[36] GUNTHER-MOHR, GESCHWIND and TOWNES: Phys. Rev. 81, 289 (1951).
[37] GESCHWIND, GUNTHER-MOHR and TOWNES: Phys. Rev. 81, 288 (1951).
[38] WANG, TOWNES, SCHAWLOW and HOLDEN: Phys. Rev. 86, 809 (1953).
[39] KOPFERMANN, H.: Kernmomente. Akademische Verlagsgesellschaft 1956.
[40] WHITE, H. E.: Introduction to Atomic Spectra. New York-Toronto-London: McGraw-Hill 1934.
[41] SCHÜLER, H., u. E. JONES: Z. Physik 77, 802 (1932).
[42] MILLER, S. L., and C. H. TOWNES: Phys. Rev. 90, 537 (1953).
[43] FOLEY, H. M.: Phys. Rev. 72, 504 (1947).
[44] BARNES, R. G., and W. V. SMITH: Phys. Rev. 93, 95 (1954).
[45] STERNHEIMER, R.: Phys. Rev. 80, 102 (1950).
[46] STERNHEIMER, R.: Phys. Rev. 84, 244 (1951).
[47] STERNHEIMER, R.: Phys. Rev. 86, 316 (1953).
[48] STERNHEIMER, R.: Phys. Rev. 95, 736 (1954).
[49] STERNHEIMER, R.: Bull. Amer. Phys. Soc. 1, 193 (1956) and private communication.
[50] GOLDMAN, I. I.: Dokl. Akad. Nauk SSSR. 88, 241 (1953).
[51] FOLEY, STERNHEIMER and TYCKO: Phys. Rev. 93, 734 (1954).
[52] MEYER-BERKHOUT, V.: Z. Physik 141, 185 (1955).
[53] SENITSKY, B., and I. I. RABI: Phys. Rev. 103, 315 (1956).
[54] KRANENDONK, J. VAN: Physica, Haag 20, 781 (1954).
[54a] BUCK, P., and I. I. RABI: Phys. Rev. 107, 1291 (1951).
[54b] RITTER, G. J., and G. W. SERIES: Proc. Roy. Soc. Lond., Ser. A 238, 473 (1957).
[55] PROCTOR, W. G., and W. A. ROBINSON: Phys. Rev. 102, 1183 (1956) and private communication.
[56] STERNHEIMER, R. M., and H. M. FOLEY: Phys. Rev. 102, 731 (1956).
[57] DAS, T. P., and R. BERSOHN: Phys. Rev. 102, 733 (1956). — DAS, T. P.: Private communication.
[58] KAWAMURA, OTSUKA and ISHIWATARI: J. Phys. Soc. Japan 11, 1064 (1956).
[59] NEWELL, G. F.: Phys. Rev. 78, 711 (1950). — ISHIGURO, E.: J. Phys. Soc. Japan 3, 129, 133 (1948).
[60] HARRIS, E. G., and M. A. MELKANOFF: Phys. Rev. 90, 585 (1953).
[61] STERNHEIMER, R. M., and H. M. FOLEY: Phys. Rev. 92, 1460 (1953).
[62] MANNARI, I., and T. ARAI: To be published.
[63] TOWNES, C. H., and B. P. DAILEY: J. Chem. Phys. 17, 782 (1949).
[64] SCHATZ, P. N.: J. Chem. Phys. 22, 695, 755 (1954).

[65] DAILEY, B. P., and C. H. TOWNES: J. Chem. Phys. 23, 118 (1955).
[66] GORDY, W.: Disc. Faraday Soc. 19, 14 (1955).
[67] ITOH, T.: Private communication.
[68] JACCARINO, V., and J. G. KING: Phys. Rev. 83, 471 (1951).
[69] PAULING, L.: Nature of the Chemical Bond. Ithaca, New York: Cornell University Press 1939.
[70] HUGGINS, M. L.: J. Amer. Chem. Soc. 75, 4123 (1953).
[71] STEVENSON, M. J., and C. H. TOWNES: Phys. Rev. 107, 635 (1957).
[72] BIRD, G. R., and C. H. TOWNES: Phys. Rev. 94, 1203 (1954).
[73] TOWNES, C. H., and B. P. DAILEY: Phys. Rev. 20, 35 (1952).
[74] DAS, T. P., and E. L. HAHN: Nuclear Quadrupole Resonance Spectroscopy. Academic Press 1958.
[75] WANG, T. C.: Phys. Rev. 99, 566 (1955).
[76] ROBINSON, H. G.: Phys. Rev. 100, 1731 (1955).
[77] BAYER, H.: Z. Physik 130, 227 (1951).
[78] BLEANEY, BOWERS and PRYCE: Proc. Roy. Soc. Lond., Ser. A 228, 166 (1955).
[79] DAS, T. P., and R. BERSOHN: Phys. Rev. 100, 1792 (1955). — BERSOHN, R.: J. Chem. Phys. (to be published).
[80] BOHR, A., and B. R. MOTTELSON: Kgl. danske Vidensk. Selsk., mat.-fys. Medd. 27, No. 16 (1953).
[81] BOHR, A.: Rotational States of Nuclei. Copenhagen: Ejnar Munksgaards Forlag 1954.
[82] ALDER, BOHR, HUUS, MOTTELSON and WINTHER: Rev. Mod. Phys. 28, 432 (1956).
[83] BRIX, P., and H. KOPFERMANN: Z. Physik 126, 344 (1949).
[84] BOHR, A., and B. MOTTELSON: Kgl. danske Vidensk. Selsk., mat.-fys. Medd. 30, No. 1 (1955).
[85] BOHR, A., and B. MOTTELSON: Phys. Rev. 89, 316 (1953).
[86] DAY, R. B., and T. HUUS: Phys. Rev. 85, 761 (1952).
[87] McCLELLAND, C. L.: S. M. thesis, Mass. Inst. of Technology 1952.
[88] MULLIN, C. J., and E. GUTH: Phys. Rev. 82, 141 (1951).
[89] HUBY, R., and H. C. NEWNS: Proc. Phys. Soc. Lond. A 64, 619 (1951).
[90] TER-MARTIROSYAN, K. A.: J. exp. theor. Phys. USSR. 22, 284 (1952).
[91] ALDER, K., and A. WINTHER: CERN report T/KA-AW-4, 1955.
[92] BIEDENHARN, GOLDSTEIN, McHALE and THALER: Phys. Rev. 101, 662 (1956).
[93] GOLDRING, G., and G. T. PAULISSEN: Phys. Rev. 103, 1314 (1956).
[94] SCHIFF, L. I.: Phys. Rev. 96, 765 (1954).
[95] DOWNS, RAVENHALL and YENNIE: Phys. Rev. 106, 1285 (1957).
[96] MARGOLIS, B., and E. S. TROUBETZKOY: Phys. Rev. 106, 105 (1957).
[97] CHASE, WILETS and EDMONDS: To be published.
[98] GOLDHABER, M., and E. TELLER: Phys. Rev. 74, 1046 (1948).
[99] WILKINSON, D. H.: Physica, Haag 22, 1039 (1956).
[100] NATHANE, R., and J. HALPERN: Phys. Rev. 92, 207 (1952).
[101] STEINWEDEL, H., and J. H. JENSEN: Z. Naturforsch. 5a, 413 (1950).
[102] DANOS, M.: Ann. Phys. 10, 265 (1952).
[103] FULLER, PETREE and WEISS: National Bureau of Standards Report 315 1, 1957.
[104] DANOS, M.: Bull. Amer. Phys. Soc. (II) 1, 135 (1956). — Nuclear Phys. 5, 23 (1958).
[105] OKAMOTO, K.: Prog. Theor. Phys. 15, 75 (1956).
[106] FULLER, E. G., and M. S. WEISS: Phys. Rev. to be published.
[107] WHEELER, J. A.: Phys. Rev. 92, 812 (1953).
[108] JACOSON, B. A.: Phys. Rev. 96, 1637 (1954).
[109] WILBETS, L.: Dan. Mat.-Fys. Medd. 29, no. 3 (1954).
[110] RACAH, G.: Phys. Rev. 62, 438 (1942).
[111] MACK, J. E.: Rev. Mod. Phys. 21, 64 (1950).
[112] KYULTS, KUNTS and KHARTMAN: Uspekhi Fiz. Nauk 55, 537 (1955).
[113] GORDY, W.: Phys. Rev. 76, 139 (1949).
[114] TOWNES, FOLEY and LOW: Phys. Rev. 76, 1415 (1949).
[115] MOSZKOWSKI, S. A., and C. H. TOWNES: Phys. Rev. 93, 306 (1954).
[116] RAINWATER, J.: Phys. Rev. 79, 432 (1950).

Sachverzeichnis.

(Deutsch-Englisch.)

Bei gleicher Schreibweise in beiden Sprachen sind die Stichwörter nur einmal aufgeführt.

30*

Subject-Index.

(English-German.)

Where English and German spelling of a word is identical the German version is omitted.

30*